Radioactive Air Sampling Methods

Edited by
Mark L. Maiello and
Mark D. Hoover

CRC Press
Taylor & Francis Group
Boca Raton London New York

CRC Press is an imprint of the
Taylor & Francis Group, an **informa** business

CRC Press
Taylor & Francis Group
6000 Broken Sound Parkway NW, Suite 300
Boca Raton, FL 33487-2742

© 2011 by Taylor and Francis Group, LLC
CRC Press is an imprint of Taylor & Francis Group, an Informa business

No claim to original U.S. Government works

Printed in the United States of America on acid-free paper
10 9 8 7 6 5 4 3 2 1

International Standard Book Number: 978-0-8493-9717-2 (Hardback)

This book contains information obtained from authentic and highly regarded sources. Reasonable efforts have been made to publish reliable data and information, but the author and publisher cannot assume responsibility for the validity of all materials or the consequences of their use. The authors and publishers have attempted to trace the copyright holders of all material reproduced in this publication and apologize to copyright holders if permission to publish in this form has not been obtained. If any copyright material has not been acknowledged please write and let us know so we may rectify in any future reprint.

Except as permitted under U.S. Copyright Law, no part of this book may be reprinted, reproduced, transmitted, or utilized in any form by any electronic, mechanical, or other means, now known or hereafter invented, including photocopying, microfilming, and recording, or in any information storage or retrieval system, without written permission from the publishers.

For permission to photocopy or use material electronically from this work, please access www.copyright.com (http://www.copyright.com/) or contact the Copyright Clearance Center, Inc. (CCC), 222 Rosewood Drive, Danvers, MA 01923, 978-750-8400. CCC is a not-for-profit organization that provides licenses and registration for a variety of users. For organizations that have been granted a photocopy license by the CCC, a separate system of payment has been arranged.

Trademark Notice: Product or corporate names may be trademarks or registered trademarks, and are used only for identification and explanation without intent to infringe.

Library of Congress Cataloging-in-Publication Data

Radioactive air sampling methods / editors, Mark L. Maiello, Mark D. Hoover.
    p. cm.
    "A CRC title."
    Includes bibliographical references and index.
    ISBN 978-0-8493-9717-2 (hardcover : alk. paper)
    1. Radioactive pollution of the atmosphere--Measurement. 2. Atmospheric radioactivity--Measurement. 3. Radioactive substances--Detection. 4. Nuclear counters. I. Maiello, Mark L. II. Hoover, Mark D. III. Title.

TD887.R3R27 2011
628.5'350287--dc22
                                                      2010028255

Visit the Taylor & Francis Web site at
http://www.taylorandfrancis.com

and the CRC Press Web site at
http://www.crcpress.com

Contents

Preface vii
Acknowledgments ix
Editors xi
Contributors xiii
Method Authors, Reviewers, and Editors xv
Technical Reviewers xvii

PART I Objectives, Safety Issues, Standards, and a Life-Cycle Approach for Sampling Airborne Radioactivity

Chapter 1 Objectives for Sampling Airborne Radioactivity 3

Mark D. Hoover, George J. Newton, and Mark L. Maiello

Chapter 2 Radiation Safety Issues for Air Sampling 11

Mark L. Maiello and Mark D. Hoover

Chapter 3 Standards, Guidelines, Regulations, and Recommendations for Measuring Airborne Radioactivity 21

Mark D. Hoover, Morgan Cox, Cynthia G. Jones, Liliane Grivaud, Michelle L. Johnson, Mark L. Maiello, and George J. Newton

Chapter 4 A Life-Cycle Approach to Development and Application of Air Sampling Methods and Instrumentation 43

Mark D. Hoover and Morgan Cox

PART II Fundamentals of Radioactivity and Radioactive Aerosols

Chapter 5 Review of Radioactivity, Detection, and Measurement 55

Mark L. Maiello

Chapter 6 The Physics of Aerosols 77

Erno Sajo

Chapter 7 Behavior of Radioactive Aerosols and Gases 135
Mark D. Hoover

Chapter 8 Filtration 157
Mark D. Hoover

Chapter 9 Behavior of Radon and Its Decay Products 181
Phillip Jenkins

Chapter 10 Internal Dosimetry of Inhaled Radioactive Aerosols 209
Charles A. Potter

PART III Fundamentals of Sampling System Design and Operation for Airborne Radioactivity

Chapter 11 Basic Air Sampling Equipment 221
Mark L. Maiello

Chapter 12 Calibration of Air Samplers and Monitors 245
James T. Voss and Jeffrey J. Whicker

Appendix A: Generic Calibration Procedure 260
James T. Voss and Jeffrey J. Whicker

Appendix B: A Multiple-Frame-of-Reference Method for Rotameter Correction Factors 264
Mark D. Hoover

Chapter 13 Principles of Air Sampler Placement in the Workplace 271
Jeffrey J. Whicker

Chapter 14 The Practice of Continuous Air Monitoring for Alpha-Emitting Radionuclides 285
John C. Rodgers

Chapter 15 Principles of Sampling Airborne Radioactivity from Stacks 315
John Glissmeyer

Chapter 16 Methods for Comprehensive Characterization of Radioactive Aerosols: A Graded Approach 341
Mark D. Hoover

Contents

PART IV Nonroutine Radioactive Air Sampling

Chapter 17 Emergency Situation Air Sampling 357
Robert B. Hayes

Appendix: First Responder Radiological Monitoring 362
Thomas F. O'Connell and Stephen P. Clendenin

Chapter 18 Monitoring Nuclear Fallout 369
Harold L. Beck

Part V Example Air Sampling Methods for Airborne Radioactivity

Introduction to the Methods 389
Mark L. Maiello and Mark D. Hoover

Method 1 Determination of the Gross Alpha-Radioactivity Content of the Atmosphere 391

Method 2 Determination of the Gross Beta-Radioactivity Content of the Atmosphere 399

Method 3 Determination of the Tritiated Water Vapor Content of the Atmosphere 407

Method 4 Determination of the Elemental Tritium Content of the Atmosphere 419

Method 5 Determination of Carbon-14 in Air 425

Method 6 Determination of the Iodine-131 Content of the Atmosphere 435

Method 7 Sampling Air for Argon-41, Krypton-85, and Other Gamma-Emitting Radioactive Gases Using Gamma-Spectroscopy 447

Method 8 Determination of the ^{222}Rn Content of the Atmosphere 457

Method 9 A Procedure for Continuous Air Monitoring of Plutonium 481

Method 10 Personal Air Sampling for Particulate Radioactivity 497

Method 11 Real-Time Breathing Zone Monitoring for Personal Respiratory Protection 505

Appendix: Radionuclide Characteristics 511
Mark L. Maiello

Glossary 543
Mark L. Maiello and Morgan Cox

Index 563

Preface

This book takes its origin from the fourth revision of the book *Methods of Air Sampling and Analysis*. One of us (Dr. Mark Maiello), with the blessing of Andrew Hull of Brookhaven National Laboratory, worked with the editor, James P. Lodge, Jr., to update the radioactive air sampling section of that large compilation. Unfortunately, Dr. Lodge passed away in 2001 and the publication of the fourth edition was never accomplished. By that time, new radioactive sampling methods had been drafted as had an introductory chapter on the general principles of radioactivity. With this foundation laid, it was decided to appeal to a potential publisher about producing a much-expanded book solely on the subject of air sampling for radioactive materials. A contributor to the radioactive section of the Lodge effort, Dr. Mark Hoover, offered the expertise of the Air Monitoring User Group (AMUG), a group of radioactive air sampling experts largely, but not exclusively drawn from the ranks of the U.S. Department of Energy and the private industries that support air sampling technology. This very active professional association, in part founded by Dr. Hoover, has been meeting regularly for over 20 years helping to write national and international standards, uncovering and solving measurement problems, and working with commercial suppliers to improve radioactive air sampling techniques. Thus armed with experience, expertise, and a sense of enterprise, what started out as a section within a much larger reference, became a book in its own right. We think such a book is long overdue.

Air sampling in general, apart from radionuclides, is fraught with potential errors, especially for the novice. To its advantage, radioactive air sampling has been maturing for many years primarily driven by human dosimetry, environmental, and regulatory concerns. But, there has been no introduction to the subject matter written by experts in the field that could give the novice a stepping-stone across its treacherous waters. Conference proceedings and book chapters do exist, but a stand-alone introduction to the material has been lacking. It is hoped that this book will find use as a guide for those recently charged with the difficult task of measuring the varied forms of airborne radioactivity that result from industrial, research, and nuclear power operations—and that are also present naturally in the environment. It may also find use by administrators of radioactive air sampling programs, providing the background information they need to conduct appropriate oversight.

In a manner compatible with the *Guidelines for Assessment and Instruction in Statistics Education (GAISE)* developed by the American Statistical Association (ASA) and available at http://www.amstat.org/education/gaise/, we hope that all users of our book will learn from, apply, teach, and build on its contents to

1. Emphasize literacy and develop critical thinking about air sampling
2. Develop and use real-life data examples
3. Stress conceptual understanding rather than mere knowledge of procedures
4. Foster life-long learning and active discussions
5. Use technology for developing conceptual understanding and for analyzing and sharing data (e.g., databases, simulation and modeling, etc.)
6. Use assessments to improve and evaluate the efficacy and impact of these activities on occupational and environmental safety and health

The book is divided into several parts to reflect the varied information needed to successfully employ airborne radioactivity sampling techniques. Within these topical areas are chapters covering the contemporary thinking of experts in the field. As implied above, radioactive air sampling is a mature field with a deep literature base. With the results of applied research it continues to advance and transform. This evolution also originates with improvements in measurement equipment, refinements in human dose modeling of inhaled radioactivity, and the consequent and inevitable changes to national and local radiation-safety regulations. We have attempted to present the most recent information about this subject matter.

This book is largely about fundamental air sampling technique and practical knowledge. Comparatively little is written about specific instrumentation, for example, tritium or iodine real-time monitors, although some information about the basic components of air samplers such as pumps and rotameters is provided. Introductory material such as a description of the myriad national and international standards that apply to air sampling is included to assist the novice in the discovery and application of the standards that are appropriate to his or her measurement concerns. Other introductory material covers the process of air filtration and the behavior of aerosols. Supporting information about radioactivity, detectors, the ever-present concern about radon, and the dosimetry of inhaled radionuclides (the ultimate reason we do air sampling) rounds out this foundation. Then, practical information is supplied about sampler calibration and placement for accurate measurements, the proper techniques for stack sampling, the new concerns about air sampling under emergency situations, and sampling under extreme environmental conditions. Our one venture into instrumentation, continuous air sampling (CAM), is justified by its preeminent use in projects involving the control and processing of plutonium. It also is an excellent technique to illustrate the interfering effects of natural airborne radioactivity, for example, radon, on radioactive air sampling measurements. Finally, we have supplied air sampling methods for various radionuclides both as means to illustrate the techniques involved and as actual didactic instructions for conducting the measurements. They were intended for the 4th edition of *Methods of Air Sampling and Analysis* but have found a new home here instead.

This work could not have been accomplished without the help of many individuals who worked on it in their spare time. The editors thank the membership of AMUG for their voluntary assistance in the production of this book (the book was a lively agenda topic at several meetings of AMUG both at Health Physics Society annual conventions and at stand-alone AMUG meetings). Our gratitude is also extended to the other experts who devoted much time to the writing and review of material for this book. Special thanks for reviewer help go to Robert M. Castellan, Hueng-Cheng Chiou, Lawrence T. Dauer, Andrew Hull, Adam R. Hutter, Paul Linsalata, Donald M. Mayer, Robert P. Miltenberger, Dennis M. Quinn, Theodore E. Rahon, and Derek Lane-Smith. The views, opinions, and content in the book are those of the contributing authors and editors and do not necessarily represent the views, opinions, or policies of their respective employers or organizations. Mention of company names or products does not constitute endorsement. Of course, the editors assume responsibility for any errors or omissions in this work. We welcome feedback at any level on how to improve this work.

Mark L. Maiello
Mark D. Hoover

Acknowledgments

The editors sincerely express their gratitude to the many technical experts in the radioactive air monitoring community who encouraged us to pursue this project and who earnestly and constructively engaged with us in developing and refining the technical content of this book. We are indebted to them for their unselfish investment of many long days, long nights, weekends, and holidays in this effort. It was a group effort in which we can all take great pride.

Mark L. Maiello is particularly grateful for this opportunity to express his deep gratitude to his wife Jenny for her patience during the production of this book. He also is grateful for the fortitude and grace of his mother Ida Maiello, because she too could have benefited from the family time otherwise taken up by this project. Her quiet backing of this and other professional efforts of her sons, and her strength of will during a particularly difficult time in her life and the life of her family, will remain an inspiration. Sincere love and affection are felt for his brother Michael J. Maiello and late father, Michael E. Maiello both of whom taught him the value of perseverance. Mark also gratefully thanks his teachers and professors at Archbishop Stepinac High School (White Plains, New York), Manhattan College and New York University, who provided him with the inspiration and tools to make a scientific career. He also acknowledges the late Andrew P. Hull of Brookhaven National Laboratory and the late James P. Lodge both of whom involved him in the editing of the 4th edition of *Methods of Air Sampling and Analysis* in the late 1990s, thus, laying the ground work for the present book. Lastly, Dr. Maiello acknowledges the technical and professional expertise of his coeditor Dr. Mark Hoover. Without him you would not be reading this book.

Mark D. Hoover is particularly appreciative of this opportunity to acknowledge his deep gratitude to his wife Martha for her unfailing love, patience, and wisdom at all times during the more than 40 years that she has graced his life. To the extent that "life is what happens when you're busy making other plans" (particularly when you are perpetually mentally and incessantly physically off in science-land), he knows the "real" parts of his life have been with her, their children Alison and Ryan, and now most recently with their wonderful grandson Riley. And he thanks his late parents Merton and Lucille for encouraging him always and in all ways. Dr. Hoover also gratefully acknowledges the many mentors, sponsors, colleagues, and friends who gave him the encouragement, skills, insights, and professional opportunities that enabled an exciting and rewarding career in the field of aerosol science to protect workers and the public. Mentors like Ronald Knief, Morgan Cox, George Newton, Dick Cuddihy, Bruce Boecker, Roger McClellan, and Werner Stoeber are unique in the world and deeply appreciated. The completion of this technical book is a tribute to them, and to so many of the other "giants on whose shoulders I have stood." Lastly, Dr. Hoover acknowledges the technical and professional expertise and focus of his coeditor Dr. Mark Maiello. Without him you would not be reading this book.

Editors

Mark L. Maiello holds a bachelor's degree in physics from Manhattan College in the Bronx, New York. He started his environmental health science career at New York University with a master's degree on the investigation of low-energy x-ray emission from cathode ray tubes and a doctoral dissertation on the development of a thermoluminescent dosimeter-based radon gas detector that was awarded a patent shared with his thesis advisor, Dr. Naomi Harley. In 1986, he joined the U.S. Department of Energy (DOE) at the Environmental Measurements Laboratory in New York City performing natural background radiation measurements and quality assessment testing of thermoluminescent dosimeters under the guidance of Dr. Gail de Planque. He was also part of a DOE team that conducted radon gas measurements atop Mauna Loa volcano in Hawaii in support of a larger meteorological study of the island. Dr. Maiello later joined CoPhysics Corporation in Monroe, New York, a radiological consulting firm specializing in decommissioning and decontamination of various academic and industrial facilities. He has published both peer-reviewed and nontechnical papers on various topics concerning health physics and nuclear weapons. His work has appeared in *Radiation Protection Dosimetry*, the *Health Physics* journal, *Nuclear News*, the online journal *Weapons of Mass Destruction Insights*, and in *Health Physics News* where he is a contributing editor. Dr. Maiello is a member of the New York Academy of Sciences, Sigma Xi, the Union of Concerned Scientists, and both the American Nuclear and Health Physics Societies. He has been a guest lecturer in radiation safety at New York University and the New Jersey Institute of Technology. Presently employed as a radiation safety officer at the Pearl River, New York facility of Wyeth, a subsidiary of Pfizer, a major pharmaceutical research and development company, Dr. Maiello conducts, among many other functions, radioactive air sampling for regulatory compliance purposes.

Mark D. Hoover earned a bachelor of science (BS) degree in mathematics and English from Carnegie-Mellon University in 1970, and master of science (MS) and doctor of philosophy (PhD) degrees in nuclear engineering from the University of New Mexico in 1975 and 1980. He is certified in the comprehensive practice of health physics (CHP) and certified in the comprehensive practice of industrial hygienist (CIH). From 1975 to 2000, Mark was an aerosol scientist at the Inhalation Toxicology Research Institute (currently Lovelace Respiratory Research Institute) in Albuquerque, New Mexico, where he focused on health and safety research in support of U.S. Department of Energy programs. In 2001, he assumed his current position as a senior research scientist in the Division of Respiratory Disease Studies at the National Institute for Occupational Safety and Health in Morgantown, West Virginia. NIOSH is part of the Centers for Disease Control and Prevention in the U.S. Department of Health and Human Services.

Mark's career goals have been to develop, apply, and improve the professional practice of protecting workers and the public from respiratory disease by establishing a technical basis for anticipating, measuring, modeling, and mitigating toxic aerosols in the workplace and environment. Special concerns have included aerosols of beryllium, toxic chemicals, biological agents, fibers, plutonium, uranium, and nanomaterials. He has developed improved methods and instrumentation for exposure assessment, aerosol characterization, and generation of representative aerosols for industrial hygiene, health physics, drug delivery, bioaerosol, and inhalation toxicology studies. He has served as an aerosol scientist and/or principal investigator for studies on respiratory health and safety and the inhalation hazards of routine operations or accidents involving radioactive, biological, and chemically toxic materials and dispersal of aerosols during operation, waste management, and environmental restoration of chemical, nuclear, biological, and nanotechnology facilities. Mark is also an active contributor to professional society activities and to the development of national and international standards. He is coeditor of a 10-volume book series on space nuclear power and author or coauthor of more than 150 open literature publications.

Contributors

Harold L. Beck
Nuclear Fallout Consultant, formerly of the
 U.S. Department of Energy, Environmental
 Measurements Laboratory
New York, New York

Stephen P. Clendenin
Massachusetts Department of Fire Services
Stow, Massachusetts

Morgan Cox
Radiological, Aerosol Measurements and
 International Standards Consultant
Moreland Hills, Ohio

John Glissmeyer
Pacific Northwest National Laboratory
Richland, Washington

Liliane Grivaud
Aerosol Measurements Consultant, formerly
 of the Institut de Radioprotection et Sûreté
 Nucléaire
Gif-Sur-Yvette Cedex, France

Robert B. Hayes
Remote Sensing Laboratory
U.S. Department of Energy
Las Vegas, Nevada

Mark D. Hoover
Division of Respiratory Disease Studies
National Institute for Occupational Safety
 and Health
Morgantown, West Virginia

Phillip Jenkins
Bowser-Morner Inc.
Dayton, Ohio

Michelle L. Johnson
Pacific Northwest National Laboratory
Richland, Washington

Cynthia G. Jones
U.S. Nuclear Regulatory Commission
Washington, DC

Mark L. Maiello
Pfizer Inc.
Pearl River, New York

George J. Newton
Aerosol Measurements Consultant, formerly
 of the Lovelace Respiratory Research
 Institute
Albuquerque, New Mexico

Thomas F. O'Connell
Massachusetts Department of Public Health
 and Massachusetts Department of Fire
 Services
Stow, Massachusetts

Charles A. Potter
Sandia National Laboratories
Albuquerque, New Mexico

John C. Rodgers
Health Physicist Emeritus
Santa Fe, New Mexico

Erno Sajo
Nuclear Science Center
Louisiana State University
Baton Rouge, Louisiana

James T. Voss
Los Alamos National Laboratory
Los Alamos, New Mexico

Jeffrey J. Whicker
Los Alamos National Laboratory
Los Alamos, New Mexico

Method Authors, Reviewers, and Editors

David Baltz
Bladewerx LLC
Albuquerque, New Mexico

Lawrence T. Dauer
Department of Medical Physics
Memorial Sloan Kettering Cancer Center
New York, New York

Mark D. Hoover
National Institute for Occupational Safety
 and Health
Morgantown, West Virginia

Andrew Hull
Brookhaven National Laboratory
Upton, New York

Adam R. Hutter
U.S. Department of Homeland Security
National Urban Security Technology
 Laboratory
New York, New York

Phillip Jenkins
Bowser-Morner Inc.
Dayton, Ohio

Derek Lane-Smith
Durridge Company Inc.
Bedford, Massachusetts

Paul Linsalata
Pfizer Inc.
Pearl River, New York

Mark L. Maiello
Pfizer Inc.
Pearl River, New York

Donald M. Mayer
Entergy Nuclear Indian Point Energy Center
Buchanan, New York

Robert P. Miltenberger
Sandia National Laboratories
Albuquerque, New Mexico

James W. Neton
National Institute of Occupational Safety
 and Health
Cincinnati, Ohio

George J. Newton
Aerosol Measurements Consultant
Albuquerque, New Mexico

Dennis M. Quinn
DAQ Inc.
Hopewell Junction, New York

Theodore E. Rahon
CoPhysics Corporation
Monroe, New York

Technical Reviewers

Dennis N. Brown
Shaw Environmental and Infrastructure Inc.
Centennial, Colorado

Hung-Cheng Chiou
Waste Isolation Pilot Plant
U.S. Department of Energy
Carlsbad, New Mexico

Part I

Objectives, Safety Issues, Standards, and a Life-Cycle Approach for Sampling Airborne Radioactivity

1 Objectives for Sampling Airborne Radioactivity

Mark D. Hoover, George J. Newton, and Mark L. Maiello

CONTENTS

Introduction 3
Specific Sampling Objectives 3
    Basic Aerosol Characterization 5
    Worker Health Protection 5
    Environmental Monitoring 6
    Process Quality Assurance and Control 6
    Emergency Preparedness and Response 6
    Demonstration of Compliance 7
Research 8
Conclusion 8
References 8

INTRODUCTION

Understanding the objectives of air sampling and how different objectives relate to and dictate the critical details of specific sampling activities is essential to the design and implementation of any successful sampling program. Before decisions can be made about *how* to sample, an understanding must be reached about *why* the sample will be taken and *under what conditions* it will be taken. As noted in the ANSI/ASTM E 1370-96 (2008) *Standard Guide for Air Sampling Strategies for Worker and Workplace Protection*, sampling may be done for single or multiple purposes, and conflicts arise when a single air sampling strategy is expected to satisfy multiple purposes. Similar views are expressed in the international standard ISO 16000-1 (2004) *Indoor Air—Part 1: General Aspects of Sampling Strategy*. The importance of beginning with clear objectives applies in any setting to all sampling programs to assess any type of airborne material, both radioactive and nonradioactive. In the following discussion of objectives, the term "sampling" is used in a general context to include collection of airborne material followed by off-line (e.g., laboratory) analysis, as well as collection of airborne material with real-time detection. The term "monitoring" is used in a more limited context to denote either collection with real-time detection or sequential sampling.

SPECIFIC SAMPLING OBJECTIVES

As illustrated in Figure 1.1, objectives for sampling airborne radioactivity can be viewed as a collection of building blocks (Hoover and Newton, 1993, 2001). Selection of the appropriate objective or objectives can help ensure that airborne materials are properly and successfully assessed and that measurement results are meaningful. Although interrelated, the categories of objectives presented here and detailed in the sections below are sufficiently distinct to warrant separate attention.

```
┌─────────────────────────────────────────────────────────────────────────────┐
│                        BASIC AEROSOL CHARACTERIZATION                        │
│    Understanding relevant physicochemical and biological properties of       │
│                          the aerosols of interest                            │
└─────────────────────────────────────────────────────────────────────────────┘

┌──────────────┬──────────────────┬──────────────┬──────────────┐
│   WORKER     │  ENVIRONMENTAL   │   PROCESS    │  EMERGENCY   │
│   HEALTH     │    MONITORING    │   QUALITY    │ PREPAREDNESS │
│  PROTECTION  │                  │  ASSURANCE   │     AND      │
│              │   Ensuring that  │  AND CONTROL │   RESPONSE   │
│ Ensuring that│ environmental    │              │              │
│ worker       │ releases of      │ Ensuring that│ Providing a  │
│ exposures are│ aerosols are     │ processes and│ basis for    │
│ within       │ within allowed   │ process      │ appropriate  │
│ allowed      │ limits and ALARA │ controls are │ actions when │
│ limits and As│ for environmental│ operating    │ things go    │
│ Low As       │ and public       │ properly     │ wrong        │
│ Reasonably   │ health concerns  │              │              │
│ Achievable   │                  │              │              │
│ (ALARA)      │                  │              │              │
└──────────────┴──────────────────┴──────────────┴──────────────┘

┌─────────────────────────────────────────────────────────────────────────────┐
│                      DEMONSTRATION OF COMPLIANCE                             │
│    Documenting that administrative and regulatory requirements are met       │
└─────────────────────────────────────────────────────────────────────────────┘

┌─────────────────────────────────────────────────────────────────────────────┐
│                                RESEARCH                                      │
│ Advancing a comprehensive understanding of the behavior, measurement, and    │
│                          control of aerosols                                 │
└─────────────────────────────────────────────────────────────────────────────┘
```

FIGURE 1.1 Illustration of aerosol sampling objectives as a collection of building blocks. Selection of the appropriate objectives can help ensure that airborne materials are properly and successfully assessed and that measurement results are meaningful. (Adapted from Hoover, M.D. and Newton, G.J., *Aerosol Measurement: Principles, Techniques, and Applications*, Willeke, K. and Baron, P.A., Eds., Van Nostrand Reinhold, New York, 1993; Hoover, M.D. and Newton, G.J., *Aerosol Measurement: Principles, Techniques, and Applications*, 2nd ed., Baron, P.A. and Willeke, K., Eds., John Wiley & Sons, New York, 2001.)

An initial and overarching sampling objective is *basic aerosol characterization* to understand the physicochemical and biological properties of aerosols that may be encountered. The understanding gained from this basic characterization step is then used to guide the need for design and execution of sampling initiatives for *worker health protection* to ensure that worker exposures are within allowed limits and As Low As Reasonably Achievable (ALARA); *environmental monitoring* to ensure that environmental releases of aerosols are within allowed limits and ALARA for environmental and public health concerns; *process quality assurance and control* to ensure that processes and process controls are working properly; and *emergency preparedness and response* to provide a basis for appropriate actions when things go wrong.

Because sampling that is conducted on a voluntary or exploratory basis for worker health protection, environmental protection, process quality assurance and control, or emergency planning and response may not necessarily meet formal administrative or regulatory requirements for demonstration of compliance, a distinct objective is identified for *demonstration of compliance* to document that administrative and regulatory requirements are met.

Finally, as a fundamental underpinning of all the sampling approaches, theories, methods, practice, and applications, a supporting and unifying objective is included for *research* to advance a comprehensive understanding of the behavior, measurement, and control of aerosols.

These objectives are not exhaustive, are not necessarily mutually exclusive, and other more specialized objectives may also exist. What is important to understand is that the selection of methods for measuring radioactive aerosols and gases should be determined by the underlying objective of the sampling effort. "What information is really needed?" "What conditions or interferences will be encountered when the measurements must be made?" "How frequently will the measurements need to be made?" "Who will be assigned to make the measurements?" "What level of training and experience will they have, or need to have?" "What data quality requirements must be met?" and "What other similar questions need to be answered?"

The concept of beginning with the objectives in mind seems trivial, but it must be emphasized so that needed data will not be missed, sampling errors will not be made, erroneous data will not be reported, and time and fiscal resources will not be wasted. Important concepts and introductory examples for each of the major sampling objectives are discussed in the following sections. References to additional details in other chapters of this book are also provided.

BASIC AEROSOL CHARACTERIZATION

Basic aerosol characterization, including related radiobiological and toxicological testing, is ideally done before the initiation of any industrial process, assessment, or activity involving radioactive materials. Unfortunately, historical experience has shown that basic aerosol characterization is often lacking. Toohey (2008) noted that "Perhaps the greatest challenge in a worker dose reconstruction program is obtaining adequate data to characterize site operations and all plausible sources of significant occupational exposure to radiation and radioactive materials."

The basic aerosol characterization process involves both collecting and characterizing relevant samples of selected materials and creating laboratory surrogate aerosols that have well-controlled and well-characterized properties. The full spectrum of aerosol sampling tools should be applied to measure parameters such as particle size, concentration, morphology, chemical composition, and solubility. This comprehensive approach provides a *defensible technical basis* for selecting appropriate requirements for designing and managing the entire process to keep exposures of people within limits and ALARA. This approach also provides a technical basis for balancing the relative risks of internal radiation exposures from inhalation or ingestion and external radiation exposures from working in a radiation area. Additional information on internal dosimetry issues can be found in Chapter 10.

Having the basic information needed to reduce and control risk also applies to the general public living in and around nuclear operations. As discussed further in Chapter 3, regulations such as the *National Emission Standards for Hazardous Air Pollutants* (40 CFR 61) from the U.S. Environmental Protection Agency that are proscribed in the U.S. Code of Federal Regulations and in many state codes specify allowable air concentrations and dose limits for the public from airborne radioactivity released to the environment.

WORKER HEALTH PROTECTION

Measurements done for the purpose of worker health protection can be taken in the breathing zone of workers or can be taken in the general work area. Results of breathing zone and area sampling can indicate the need for bioassay, the need for improved containment or work practices, and the need for air monitoring and personal protective equipment. Basic characterization and process control measurements provide the guidance for determining when, where, and for which radionuclides the health protection measurements should be made; processes involving potentially small source terms of low-toxicity materials do not require as extensive a measurement program as would be appropriate for potentially large source terms of high-toxicity materials. In all cases, the objective is to keep worker exposures within limits and ALARA. Concerns for determining particle solubility are generally substantial because of the influence that solubility has on the biological behavior of inhaled material and because knowledge of solubility is needed to correctly apply biokinetic models and interpret bioassay information from urine, fecal, or blood samples obtained from workers. Other measurement techniques such as radiation monitoring for hand and foot contamination or for contamination of workplace surfaces, are also part of a total health protection program because they often signal increased airborne concentrations before worker exposures become excessive. Additional information on radiation safety issues for air sampling can be found in Chapter 2.

Environmental Monitoring

Effluent monitoring or sampling for airborne radioactivity in outdoor areas surrounding a nuclear or radioactive-materials-handling facility is often routinely performed both on-site and off-site to identify when releases are occurring and to ensure that environmental releases are within regulatory limits and ALARA. Some measurements are designed to support operational and contamination-control goals established by facility management. Other measurements (as described further under the objective for demonstration of compliance) are designed to meet the formal requirements of regulatory compliance. Because concentration limits for environmental releases are generally lower than the allowable concentration limits for occupational exposures, greater sensitivity is usually required for environmental sampling. Greater sensitivity is typically achieved by using higher sampling flow rates, sampling for longer periods of time, or using more sensitive analytical techniques.

Environmental air monitoring presents its own set of challenges. The source term (the type and amount of radioactivity to be released and/or its release rate) may or may not be well characterized. The industrial process that produces the release may vary on a periodic basis. There are also "industrial meteorological" considerations that include the effects of buildings on the near-field meteorology (the airflow pattern around the release point and downwind). Other geospatial meteorological considerations include the prevailing wind patterns at the site of the source release and in the surrounding areas, the mean wind speed, and the locations of and distances to the receptors (the nearest occupationally exposed person and/or the nearest off-site person).

As was the case for sampling in the workplace to ensure worker health protection, methods for environmental monitoring also involve considerations beyond simple collection of the sample. Dose estimates of outdoor, occupationally exposed persons and off-site groups are routinely acquired by applying specialized software. Such software accounts for radionuclides incorporated into the food chain. Therefore, distances to and locations of local farm, grazing lands, and surface waters become important. In some special instances, resuspension of deposited radioactive materials on soils and plant life must also be factored into the estimate of ultimate dose to the receptor. Additional information about sampling in the environment for nuclear fallout is provided in Chapter 18.

Process Quality Assurance and Control

Routine sampling or monitoring for aerosols in and around processes such as nuclear fuel fabrication, reactor operations, or radioactive waste disposal assures that the processes are working properly, or provides an early warning that conditions are changing. Although process-related measurements can help to prevent exposures of workers and environment, process measurements are typically designed to identify and correct operational problems, not to quantify or document the adequacy of worker health protection or environmental protection. Process measurements are usually made where instantaneous readings are needed to trigger an immediate response (such as to correct a problem with the quality or quantity of process feedstocks or to repair or replace a leaking gasket) and generally focus on concentration of radioactivity or aerosol size measurements. Examples include measurements made inside radiopharmaceutical processing lines or nuclear reactor containment areas and stack monitoring for radiochemical operations. Other aerosol parameters may be measured periodically. The concept of a graded approach to total material and aerosol characterization is discussed further in Chapter 16.

Emergency Preparedness and Response

Good response for radiological emergencies involves graded levels of reaction, depending on the severity of any radioactive releases. This implies a greater reliance on real-time or near-real-time information than is necessary during normal facility or process operations.

For example, in typical occupational settings involving plutonium, continuous air monitors are installed in work areas to provide an alarm in the event of an airborne release of plutonium.

An emergency response would include a shutdown of manufacturing operations and an evacuation of the area. This would be followed by operations to contain and remediate the release. Aerosol samples collected during an accidental release can be analyzed to provide particle size, solubility, and composition information to estimate and provide appropriate medical response for exposed individuals (Cheng et al., 2004). To provide data to assess the magnitude of the release, air monitoring instruments need to remain operational throughout the emergency and be capable of providing useable results at aerosol concentrations much higher than are seen in routine process, health protection, or environmental monitoring. It is sometimes necessary to temporarily deploy portable or mobile instrumentation under such conditions. Samples taken during simulated accidents or emergencies are also important for developing appropriate emergency response plans (Parkhurst et al., 2005; Parkhurst and Guilmette, 2009).

Environmental air sampling instruments for emergency response may need more frequent cleaning, component replacement, or general attention due to outdoor exposure and duration of use. Air sampling instruments for emergency response in workplaces may need to operate under conditions of higher temperature, humidity, vibration, or electrical instability than instruments designed exclusively for use during routine operations. Additional information about sampling for airborne radioactivity in emergency situations is provided in Chapter 17.

DEMONSTRATION OF COMPLIANCE

Throughout all the measuring regimes, there will be times when it is necessary to demonstrate compliance. In a sense, demonstration of compliance underlies the preceding objectives for worker health protection and for environmental monitoring. However, much of the sampling that is typically conducted for the preceding objectives will not and does not need to meet formal administrative or regulatory requirements for demonstration of compliance. Therefore, a distinct objective is identified for compliance sampling. Such demonstration generally involves two issues:

- Demonstration that the air monitoring instruments are properly selected, positioned, calibrated, and operated in accordance with appropriate procedures
- Documentation of the actual releases of airborne radionuclides in comparison with statutory limits for release or exposure

Regarding procedures, compliance requires a monitoring effort to be performed according to procedures that have typically been agreed upon by the monitoring party and the regulating agency. In some instances, the facility emitting the effluent may also implement strict standard operating procedures to support the air sampling effort in order to ensure that compliance is achieved. Agreement on procedures is usually made when a permit-to-release radioactive effluent into the air is applied for with the regulator. In this application, monitoring techniques, frequency of monitoring, types of air samplers to be used, and other details are specified.

The documentation of sampling results is a key component to achieving regulatory compliance because the regulating body will not have a representative on site except in special circumstances, such as at some nuclear reactor power stations. Compliance is usually tested by a physical inspection conducted by the regulator at periodic intervals. Inspections commonly include both documentation reviews and visual inspections of air monitoring systems where feasible. Documentation inspections can include reviews of monitoring equipment logbooks, electronic data files, and data-processing spreadsheets. Inspections also include interviews with key personnel involved in the air monitoring effort and, in some instances, with those involved in the production of the effluent.

An important consideration for sustained regulatory compliance is the changing nature of regulations. Although regulations may remain unchanged for years, it would be unwise to consider them static. For a number of reasons that also include improved understanding of the health effects of radiation and radioactivity, regulations may be changed to reflect new scientific information. Compliance, therefore, demands that any changes in regulations be addressed promptly and appropriately. Demonstration of compliance with new requirements can be accomplished either by development and use of modified air sampling procedures or by other associated means that provide proof to the regulating agency that compliance is achieved. In keeping with this observation, any current air sampling methods should only be considered examples, and the larger focus should be on matching the science of aerosol measurement to the applicable requirements and circumstances.

RESEARCH

Air sampling research is performed by government, academia, instrument and safety equipment providers, and industry. Research can be conducted in the laboratory, in occupational and environmental settings, and on a theoretical basis to advance the state-of-understanding for both practical and purely academic reasons. As such, research is both related to and separate from the other objectives mentioned above. In fact, research supports the means to meet the objectives of basic aerosol characterization, worker health protection, environmental monitoring, process quality assurance and control, emergency preparedness, and regulatory compliance. Without research, the other sampling objectives could not have been met in the past, and would be difficult to meet in the future.

A major consideration for research is that the test materials, environmental conditions, and measurement principles and operating parameters should be fully documented so that research results can be understood in context and appropriately generalized. Measurement of radioactive and non-radioactive airborne materials in the indoor home environment is an example of recent research with direct bearing on human health assessments. Other examples include measurements of radon and radon-related aerosols in extreme environments such as those encountered in uranium mines or at high altitudes for meteorological reasons. Research can also include the characterization of new air sampling instruments and new techniques. Fundamental investigations may involve the investigation of associated phenomena such as aerosol electrostatic behavior or the airflow pattern in work areas and its impact on the proper placement of air sampling instrumentation. Such studies increase our overall knowledge of the behavior of airborne radioactivity and the means to measure it. As detailed by Phalen and Hoover (2006), additional research is also needed on how aerosol properties influence deposition, retention, and toxicity of aerosols in the respiratory tract.

CONCLUSION

Sampling for airborne radioactivity is an important method of occupational and environmental assessment and risk management for many types of organizations and facilities, including nuclear-powered electric generating stations, pharmaceutical research and manufacturing facilities, manufacturers of luminescent products containing tritium, government and university research laboratories, and environmental restoration companies involved with the removal and processing of radioactively contaminated soils and building materials. In addition, with the introduction, many years ago, of several simple at-home radon measurement techniques, monitoring for naturally occurring airborne radioactivity has become possible for the general public to perform. If the objectives and methods of sampling are properly matched, accurate assessment and management of the occupational and public health hazards of airborne radioactivity can be achieved. This chapter has introduced the major objectives for air sampling, and the following chapters describe how to achieve those objectives.

REFERENCES

ANSI/ASTM, *Standard Guide for Air Sampling Strategies for Worker and Workplace Protection*, E 1370-96, ASTM International, West Conshohocken, PA, 2008.

Cheng, Y.S., Guilmette, R.A., Zhou, Y., Gao, J., LaBone, T., Whicker, J.J., and Hoover, M.D., Characterization of plutonium aerosol collected during an accident, *Health Phys* 87(6):596–605, 2004.

Hoover, M.D. and Newton, G.J., Radioactive aerosols, in: *Aerosol Measurement: Principles, Techniques, and Applications*, Willeke, K. and Baron, P.A., Eds., Van Nostrand Reinhold, New York, 1993.

Hoover, M.D. and Newton, G.J., Radioactive aerosols, in: *Aerosol Measurement: Principles, Techniques, and Applications*, 2nd ed., Baron, P.A. and Willeke, K., Eds., John Wiley & Sons, New York, 2001.

International Organization for Standardization, Technical Committee 146, *Indoor Air—Part 1: General Aspects of Sampling Strategy*, ISO 16000-1, International Organization for Standardization, Geneva, 2004.

Parkhurst, M.A. and Guilmette, R.A., Overview of the Capstone depleted uranium study of aerosols from impact with armored vehicles: Test setup and aerosol generation, characterization, and application in assessing risk, *Health Phys* 96(3):207–220, 2009.

Parkhurst, M.A., Daxon, E.G., Lodde, G.M., Szrom, F., Guilmette, R.A., Roszell, L.E., Fallo, G.A., and McKee, C.B., *Depleted Uranium Aerosol Doses and Risks: Summary of U.S. Assessments*, Battelle Press, Richland, WA, 2005.

Phalen, R.F. and Hoover, M.D., Aerosol dosimetry research needs, *Inhal Toxicol* 18:841–843, 2006.

Toohey, R.E., Scientific issues in radiation dose reconstruction, *Health Phys* 95(1):26–35, 2008.

U.S. Environmental Protection Agency, *National Emission Standards for Hazardous Air Pollutants (NESHAPS)*, Title 40 Part 61, U.S. Code of Federal Regulations, 2008.

2 Radiation Safety Issues for Air Sampling

Mark L. Maiello and Mark D. Hoover

CONTENTS

Introduction 11
Fundamental Program Elements 11
    Regulatory Concerns 12
    Training 12
    Contamination Control 13
    Personal Protection 16
    Dosimetry 17
    Some Special Issues for Sampling and Handling Plutonium 17
Conclusion 18
References 19

INTRODUCTION

Radiation safety programs exist to maintain control of radioactive sources, to minimize human radiation doses, and to control or eliminate radioactive contamination in the workplace. Detailed records are generated to document these procedures because the use of radioactive materials is licensed and regulated by governmental authorities. For the individuals who collect air samples and for the analysts who operate the radioactivity counting laboratory, the above safety goals apply. In addition, due to the frequent requirement to measure low levels of radioactivity and to maintain low background count rates in measurement equipment, these safety considerations emphasize stringent control of contamination in the laboratory. The typical requirements of a radiation safety program are outlined below. Often, a professional such as an institutional health physicist will lead the program.

FUNDAMENTAL PROGRAM ELEMENTS

As described in the following sections, the fundamental elements for conducting an effective radiation safety program for air sampling include: *regulatory concerns*, including a written program that delineates the responsible individuals and the regulations and procedures to which they must adhere; *training* for the workers who collect the samples and the workers who analyze the samples; *contamination control* procedures and equipment such as ventilation hoods for the management of radioactive standards, sources, and samples; *personal protection* procedures and equipment to provide protection when engineering controls and administrative procedures alone are insufficient; *radiation dosimetry* procedures and methods; and attention to *special issues* such as those associated with the sampling and handling of plutonium (Pu). Proper attention to these elements helps to ensure that radioactivity from operational sources and laboratory calibration sources is controlled, that human radiation exposures are within statutory limits and as low as reasonably achievable (ALARA), and that inadvertent radioactive contamination of samples and contamination in the counting laboratory is controlled to levels that enable meaningful counting results at the required concentrations.

Regulatory Concerns

Regulatory bodies oversee radiation safety at facilities registered with them to use radioactive materials (i.e., the radioactive materials "licensees" under their jurisdiction). Most of the regulations to which radiation safety programs must adhere in the United States are promulgated by the U.S. Nuclear Regulatory Commission (NRC) and can be found in Title 10 of the Code of Federal Regulations (CFR), Part 20, *Standards for Protection Against Radiation* (CFR, 2006). (Additional information about other regulations and guidance, such as from the Occupational Safety and Health Administration, can be found in Chapter 3 of this book.) An NRC "Agreement State" is one that has signed an agreement with the NRC that authorizes the State to regulate radioactive materials within its borders. City authorities also exist in some locations such as New York City. States or cities that oversee licensees within their borders have regulations that match or, in some instances, exceed those of the NRC. Oversight is characterized by occasional interactions that are often carried out by official correspondence between the facility and the regulatory agency (e.g., the registration of sealed sources and radiation-generating devices, or requests to amend the radioactive materials license to accommodate changes in the use of radioactive materials). Every aspect of an institutional radiation safety program must be described in a written "Radiation Safety Manual," which is usually submitted with an application for a radioactive materials license.

The most crucial interaction is the periodic inspection of the facility by representatives of the regulating agency. These inspections are designed to assure that the radiation safety program and the institutional health physicist serving as the Radiation Safety Officer (RSO) are providing safeguards that protect employees, property, and the environment in a manner consistent with the regulations. Violations of the regulations may result in fines, public censure, employee dissatisfaction, and other consequences. Therefore, a radiation safety program must include at least the following basic components:

- A qualified RSO
- A structure—described in a written "Radiation Safety Manual"—that indicates how the program will be implemented and the conditions of the radioactive materials license will be met
- Inventory control of on-site radioactive sources
- Procedures to meet periodic regulatory deadlines such as biannual leak tests for sealed sources and annual training for authorized users of radioactivity and radiation
- Procedures to control contamination and human exposure (exposures are to be maintained ALARA)
- Procedures to measure radiation and radioactivity in the facility and, if necessary, in stacks or at outdoor locations such that doses to the public can be calculated
- Procedures to control and process radioactive waste
- A well-organized means of documenting all aspects of the program (i.e., survey measurements, air measurements, leak tests, bioassays on employees, etc.) that can be presented to an agency inspector to demonstrate compliance with regulations. (Note that although hardcopy documents are still maintained for this purpose, a hardcopy filing system is usually supplemented with an electronic database.)

Management commitment to safety is a significant predictor of the safety program success and it is the responsibility of facility management to support the RSO and the radiation safety program with the means to abide by license conditions, radiation safety manual protocols, and the city, state, or NRC regulations.

Training

Probably the most unheralded component of a safety program is the training of personnel who will be handling hazardous materials, in this case radioactive materials. Without training, the techniques

Radiation Safety Issues for Air Sampling

and equipment designed to mitigate personal and laboratory radiation contamination are useless. Initial radiation safety training and periodic refresher training are usually mandated by the applicable regulating agency. For example, both the City of New York and the NRC stipulate that annual refresher training is required. Both initial and refresher training must be documented, usually by employing a sign-in sheet and/or a means of electronic record keeping.

At minimum, radiation safety training should cover the following basic topics:

- Characteristics of radiation
- Biological effects of radiation
- Risks incurred from exposure
- Mitigation of exposure
- Measurement of radioactivity
- Contamination control
- Radioactive waste processing
- Radiation safety program requirements (i.e., required documentation such as inventory of radioactivity, survey data, training records, etc.)

CONTAMINATION CONTROL

The general radioactivity work areas and the counting laboratory and its equipment should be designed and operated to support the safe and orderly collection, handling, analysis, and disposal of radioactive sources and samples. All areas should be kept free of contamination to the extent possible. The principle of "starting clean and staying clean" mitigates against the general contamination of work and laboratory surfaces, the potential for cross-contamination of air samples, and the potential for human exposure.

All radioactive standards and sources used in the calibration or quality control of instruments should be secured in a locked facility when not in use. Standards, particularly radium standards or those of fragile construction, should be periodically checked for leakage to assure that damage to the source has not occurred. By regulation, this is done every six months. Radium standards, less commonly used these days than in the past, were a problem because the radioactive decay of radium gives rise to radon gas, which is also a radioactive α-particle (i.e., helium gas) emitter. The pressure of the accumulated gas has been known to alter the radium source container, creating openings for the radium to leak.

Standards and other sources should be handled carefully to preserve their physical integrity. Surface scratches or indentations on plated sources must be avoided. Damage or alteration of sources may result in the loss of source material or the change of the source geometry and the invalidation of the calibrated source intensity. Other sources, such as air-filter media spiked with a radioactive standard solution or cartridges containing spiked collection media, must also be handled carefully to prevent damage. In addition to taking precautions to prevent external physical damage to all radioactive media, additional care is required when handling high-specific-activity, α-emitting radionuclides, such as Pu-238, which can cause radiation damage and degradation over time to both the sample substrate media and its containment. The liquid sources in glass vials used in the calibration of liquid scintillation counters (LSCs) are usually of weak activity, but still require careful handling. All sources should be properly labeled with radioactive warning symbols, the radionuclide(s) name(s), and the date of source preparation.

Ventilation hoods are essential for contamination control during the unpacking of radioactive air samples or radioactively contaminated field sampling equipment, or during the manipulation of solutions used to make radioactive standards. A properly working and uncluttered fume hood provides a partially sealed-off work area and inhalation protection. Fume hood face velocities of approximately 100–125 ft min^{-1} (0.5–0.64 m s^{-1}) were advised in the past (Stewart, 1981), but in the light of considerations that airflow into and within the hood should not be excessively turbulent, a hood face velocity of 60–100 ft min^{-1} is now considered adequate (Saunders, 1993). Standards of performance for ventilation hoods are set forth by ANSI/AIHA Z9.5 (2003) and OSHA 29 CFR 1910.145 (2003). Face velocities should be checked periodically (monthly, quarterly, or semiannually) with a calibrated anemometer or velometer and recorded. Fume hoods can be fitted with low-flow alarms to warn the user of a potentially hazardous condition caused by a malfunctioning fan motor. Care should be taken to ensure that the amount of equipment and other items in a hood is minimized to the level essential to conduct the work at hand, and that the placement of equipment or other items does not interfere with proper airflow. Avoiding over-energetic hand and arm movements helps to prevent or minimize the spread of contamination when placing items into or removing waste or other items from hoods.

When work is done in glovebox enclosures, ultraviolet-resistant and chemically resistant glove materials such as Hypalon® and specially designed modular glovebox links may be used to minimize the spread of contamination and exposures of workers (Hoover et al., 1999). Temporary, special-use enclosures may be required for nonroutine activities such as maintenance or decommissioning (Newton et al., 1987). Portable aerosol sampling systems used to extract aerosols from special enclosures or radioactive work processes may include local ventilation and filtration around the sampling instruments to provide contamination control (Hoover et al., 1983).

The basic equipment employed to control the spread of contamination on laboratory surfaces, including within hoods, includes disposable spill trays and absorbent pads. These, along with the proper radioactive warning labels and signs, delineate the radioactive work area, warning employees about the potential for contamination of the workspace. The combination of good laboratory design and engineering controls, proper labels and signs, worker training, and adherence to good work practices can ensure that operations within the radioactive workspace will be carried out in a manner that results in effective contamination control.

An air sample (filter cassette, filter cartridge, bubbler, charcoal canister, etc.) brought in from the field can introduce contamination into sensitive counting instruments if the sample was obtained from a highly radioactive effluent stream. Depending on the sampler design, decontamination of cassettes, cartridges, or canisters may be necessary prior to counting. This can be accomplished using commercially available spray or liquid cleansers and disposable wipes. Some specialized liquid decontaminants for radioactivity exist, but any surfactant may be used so far as it is safe to the user and the equipment. All contaminated equipment that cannot be cleaned must be disposed of in designated radioactive waste containers according to governmental and/or institutional regulations. Decontamination must be verified by wipe tests (see below). Scanning potentially contaminated equipment with a hand-held Geiger counter or a sensitive α/β scintillation counter may be sufficient for some situations where a fast determination is needed (e.g., in the field), but collecting wipe samples and counting them in an LSC is a much more sensitive technique (see below).

Air sampling field equipment such as filter holders, sampling tubes, flow meters, pumps, ring stands, mobile carts, and other auxiliary equipment not normally thought of as coming in contact with radioactivity can often do so depending on the environment in which they are placed. The exterior surfaces of these components should be wipe tested on a periodic basis or anytime contamination is suspected. The contamination of the internal surfaces of air sampling components is important because of the problem of cross-contamination from sample to sample. Sampling lines and filter holders should be replaced periodically to prevent this contamination. Care should be taken to avoid activation or operation of a sampler without a suitable collection filter in place or without a proper seal between the filter media and the filter holder. To protect expensive equipment downstream in the sampling train (e.g., flow meters and pumps), it is prudent to install a low-pressure, cartridge-type, back-up filter to collect any particulate material that may penetrate or inadvertently bypass the sampler. A column containing a desiccant such as silica gel, activated charcoal, or Drierite™ may also be used to trap moisture, vapors, and destructive chemicals in addition to the radioactivity that may not be trapped by upstream devices such as bubblers.

Wipe testing for "removable" radioactivity may not be a very efficient means of measuring radioactive contamination, but it is a time-honored technique and is the only means to detect stray tritium (H-3) (L'Annunziata, 1998). Wipe testing supplements the detection of other weak β-emitters such as carbon-14 (C-14) and sulfur-35. One of the most common wipe-testing methods is to use a filter paper or a cotton swab moistened with a solvent such as methanol to wipe the surface of interest. The solvent is employed to more efficiently capture radioactivity on the wipe-sample media. Some RSOs prefer to take dry wipes, interpreting "fixed" radioactivity as radioactivity that remains on a surface without solubilization. The area tested should be no less than that proscribed by regulation [e.g., 100 cm^2, or 4 in × 4 in, is indicated in current U.S. regulations (CFR, 2006)]. The filter paper or swab is typically counted in an appropriate instrument, such as an LSC, to determine if significant net counts are present which would be indicative of contamination. The counter must be set to detect the radionuclide of interest. A media blank (a clean, unused wipe sample) should always be counted for background comparison purposes or to provide the LSC computer with a means to calculate net counts per minute (cpm). A field blank (a clean wipe sample, which is taken to the field and handled in the same way as the actual sample, but not wiped on a surface) should also be included to provide evidence of any inadvertent sample contamination in the field. Modern LSC equipment will also estimate the counting efficiency to convert the net cpm results to net disintegrations per minute (dpm). A manual conversion from cpm to dpm is needed if the wipe-test sample is counted with a device such as a Geiger counter that does not automatically provide the counting-efficiency correction. Determination of counting efficiency is accomplished by counting a standard source of similar area, radioactive emission, and energy to the nuclide in question. A cpm-to-dpm ratio is obtained and used to manually convert the cpm results into dpm. This conversion is required to compare the results to regulatory contamination limits that are written in units of dpm 100 cm^{-2} (refer to Table 2.1) (USAEC, 1974). All wipe testing should be performed in accordance with local governmental and/or institutional regulations and documented for inspection purposes.

Portable instrument scans are another means to control contamination in radioactive work areas and are often teamed with wipe testing. Similar to analyzing wipe-test samples with nonportable,

TABLE 2.1
Acceptable Surface Contamination Levels adopted by the U.S. NRC

Nuclide[a]	Average[b] (dpm 100 cm^{-2})	Maximum[c] (dpm 100 cm^{-2})	Removable[d] (dpm 100 cm^{-2})
U-nat, U-235, U-238, and associated decay products	5000 α	15,000 α	1000 α
Transuranics, Ra-226, Ra-228, Th-230, Th-228, Pa-231, Ac-227, I-125, and I-129	100	300	20
Th-nat, Th-232, Sr-90, Ra-223, Ra-224, U-232, I-126, I-131, and I-133	1000	3000	200
β/γ-emitters (nuclides with decay modes other than α-emission or spontaneous fission), except Sr-90 and others noted above	5000 βγ	15,000 βγ	1000 βγ

Source: Adapted from Atomic Energy Commission (USAEC), *Termination of Operating Licenses for Nuclear Reactors*, Regulatory Guide 1.86, Nuclear Regulatory Commission, June 1974.

[a] Where surface contamination by both α- and β–γ-emitting nuclide exists, the limits established for α- and β-emitting nuclide should apply independently.

[b] Measurements for determination of average contamination are restricted to areas less than 1 m^2. For objects of less surface area, the average should be derived for each such object.

[c] Measurements for determination of maximum contamination are restricted to areas not exceeding 100 cm^2.

[d] The removable limits should be reduced proportionately for objects less than 100 cm^2 and the entire object should be wipe tested.

laboratory-based counting systems, the concerns here are appropriate selection and proper use of the correct instrument or instruments for the radioisotopes to be detected. For example, the efficiency of C-14 detection for a typical pancake-style Geiger Mueller (GM) probe is between 3% and 6% with a 15-cm^2-area probe in "near contact" with the contamination. The GM meter is acceptable for detection of C-14 as long as it is realized that the weak β-particle emission from C-14 (49 keV mean energy) is not easily detected unless a "slow and low" (methodical, near-contact) scan of suspected surfaces is performed. Alternatively, a wide-area α/β scintillation probe can be used (efficiency about 5% for C-14) if one is willing to accept a higher background compared to a GM probe. A third alternative is to use a gas proportional counter (GPC) (efficiency for C-14 about 10% and available with 100 cm^2 and 400 cm^2 probes) to achieve both a higher detection efficiency and a larger detection area than obtainable with a GM meter. A disadvantage of the GPC method is that it requires more preparation time than the use of a GM meter because the GPC must be filled with P-10 counting gas (standard 90% argon/10% methane mixture by volume).

For detection of radioisotopes that emit γ-radiation and x-rays, a sodium iodide (NaI) scintillation detector is preferable. Some scintillation probes are dual functional β- and γ-detectors. Again, the probe must be capable of detecting the radionuclide(s) of concern and the efficiency of detection for each nuclide should be determined before use.

Survey meters are typically required by regulatory agencies to be recalibrated annually. Recalibration is sometimes accomplished using an in-house calibration facility, but very often is met through the services of a commercial vendor who will affix a dated label to the meter and provide a certificate detailing the response of the meter to the radionuclides to which it was calibrated. When requesting calibration, the owner of the meter must communicate a list of radionuclides with which the meter is to be used and the potential cpm or mrem h^{-1} ranges that may be encountered.

Prior to use, the operator should perform a functional check of any survey meter planned to be used. In this era of documentation and confirmation, results of the functional check should be recorded as part of the laboratory contamination report. The check can be carried out by exposing the probe at a controlled and repeatable distance to an appropriate long-lived radioactive "check" source. The response in cpm or mrem h^{-1} is compared to the results of previous functional checks to determine if a significant change has taken place. The acceptable range for the response check should be determined and specified by the RSO prior to use as part of the operating procedure for the survey meter. If the result of the response check falls outside the acceptable range, the meter should be returned for maintenance and recalibration; it should not be used for the survey at hand. Other aspects of the meter that should be verified prior to use are that sufficient battery power is available to complete the survey and that readings of the background radiation levels are not significantly different from previous uses of the meter.

PERSONAL PROTECTION

The first consideration of any safety program is occupational safety (i.e., the protection of employees or, in the case at hand, the protection of air sampling technicians and laboratory workers from the hazards associated with their work functions). The most effective approach to worker protection involves application of the traditional "hierarchy of control": *elimination* of the presence or magnitude of the hazard; *substitution* of a less hazardous material or procedure; use of *engineering controls* to prevent exposures; use of *administrative and work procedures* to prevent exposures; and finally, as the last barrier to exposure, use of *personal protective equipment*. Basic personnel protection equipment in the workplace where air samples are collected and in the laboratory using or admitting radioactive substances include eye protection, gloves, and laboratory coats. In most instances, both eye protection and laboratory coats need not be of any special design. This is true as long as chemical explosivity, risk of fire, or use of open flame is negligible or nonexistent. If not, face shields, goggles, and cotton laboratory coats rather than synthetics may be in order. In some cases, it may be possible to reduce the level of personal protective equipment needed for air sampling

Radiation Safety Issues for Air Sampling

personnel by using extractive sampling methods to transport the radioactive air to a filter or sampling location that is outside the radioactivity work area. In addition, proper facility and process design can increase the overall level of safety for air sampling personnel by ensuring that sampling and analytical work locations are easily accessible; free of slip, trip, and fall hazards; and free of other physical, chemical, thermal, and electrical hazards.

Gloves should be compatible with all chemicals encountered (if any) in the procedures for processing air samples. Chemical compatibility charts are available from some glove manufacturers (e.g., Best Manufacturing Company, 2006). Double gloving, though not always applicable to every laboratory situation, may be invoked to control radioactive contamination, especially when liquid radioactive sources are manipulated. Of course, this technique is useful only when coupled with use of a radioactivity monitoring device such as a Geiger counter. Gloves must be periodically scanned to prevent the spread of contamination.

Dosimetry

A personal dosimeter is usually assigned if an employee has a reasonable probability of exceeding a 50-mrem annual total effective dose equivalent or other dose limitation as stipulated in 10 CFR, Part 20.1502, whichever is more limiting (CFR, 2006). However, use of dosimeters is often less formally dictated. Other concerns, not the least of which is institutional liability, often contribute to a decision as to whether or not a dosimeter should be assigned. Dosimetry monitoring results are typically provided by commercial vendors on a periodic basis set by the customer. A monthly or quarterly deployment/return cycle for the dosimeters is typical and this dictates the frequency of the reporting schedule. A finding that a dose has exceeded a set point (e.g., 1250 mrem or 25% of the maximum 5000 mrem annual dose limitation in the United States) results in telephone delivery of an emergency message to the RSO so that an investigation may be promptly initiated. Dosimetry results must be reviewed by the RSO for accuracy and stored carefully for inspection by regulatory personnel.

Some Special Issues for Sampling and Handling Plutonium

For an air sampling laboratory, the measurement and contamination control of plutonium represents a challenge. This is due to the myriad emissions from plutonium isotopes. The many α, γ, and neutron emission energies associated with plutonium decay affect the measurement of these radiations when using energy-dependent probes.

The atomic number of plutonium is 94 and the important plutonium isotopes are Pu-238, 239, 240, 241, and 242 (see Table 2.2). With the exception of Pu-241 (which is a pure β-emitter), each of

TABLE 2.2
Some Characteristics of the Important Plutonium Isotopes

	Pu-238			Pu-239			Pu-240			Pu-241			Pu-242		
Half-life (y)	87.7			24,110			6537			14.4			376,000		
Decay modes	α	β	γ	α	β	γ	α^*	β	γ	α	β	γ	α	β	γ
Decay energy (MeV)	5.593	0.011	0.0118	5.244	0.0067	<	5.255	0.011	0.0017	<	0.021	<	4.983	0.0087	0.0014

Source: Adapted from Lederer, C.M. and Shirley, V.S., *Table of Isotopes*, 7th ed. John Wiley & Sons, New York, 1978.
Note: Atomic number of Pu is 94.
< Indicates less than 0.001 MeV.
α^* Indicates both α-decay and spontaneous fission, which means that neutron radiation is present. To some extent, all Pu isotopes undergo spontaneous fission.

these isotopes emits α-particles. All these isotopes emit γ-radiation (photons). There are, in fact, hundreds of plutonium photon energies emitted, but few of these are high-energy (i.e., highly penetrating) emissions. Most plutonium photon emissions consist of L x-rays that can be easily shielded and are therefore difficult to detect. Nonetheless, γ-doses associated with plutonium operations can be high, primarily due to the Americium-241 a progeny of Pu-241. Am-241 emits a 60-keV photon that is easily measured using NaI scintillation detectors.

Therefore, measurements to detect plutonium contamination in laboratories that handle trace amounts of plutonium such as air sampling laboratories include the following:

- α-measurements (for all plutonium isotopes except Pu-241)
- β-measurements (in the case of Pu-241)
- γ-measurements

Alpha counters may be of the fixed or portable types. Alpha-radiation emissions from a wipe-test medium (e.g., filter paper) can be detected using a light-sensitive photomultiplier tube covered with a permanent or removable layer of zinc sulfide (ZnS) scintillator material. Gas proportional detectors and scintillation-counting probes may also be employed. Lower background counting rates can be achieved if the units are not sensitive to β- and γ-radiation.

As a pure β-emitter (maximum energy of 0.022 MeV), Pu-241 is amenable to liquid scintillation detection (L'Annunziata, 1998) making wipe testing for this isotope feasible. Quantification of Pu-241 is possible assuming that the Pu-241 can be isolated from the other plutonium isotopes. Portable plastic scintillators are excellent β-detectors with little photon sensitivity.

Photon emissions from plutonium may be detected using the "FIDLER" (field instrument for the detection of low-energy radiation). This instrument is composed of a thin NaI crystal, 2 mm (0.08 in) thick, by about 127 mm (5 in) in diameter, which is portable enough for laboratory surveys. A scaler or ratemeter provides the readout with some energy discrimination (the 17-keV Pu-239 x-rays are the photons of interest). The 60-keV gamma photons from Am-241 can also be detected if the unit is calibrated for them. Other specialized detectors include the "phoswich," which is a NaI crystal coupled to a CsI scintillator that discriminates against background (Compton scattering of photons from charged particles) by means of pulse shape discrimination, yielding a lower background than can be achieved with NaI alone.

Problems with correlations between air sampling results and bioassay can be especially severe for Pu-238 and other high-specific-activity radionuclides because a small number of particles can be significant from a dose standpoint (Scott and Fencl, 1999; Scott et al., 1997). Stochastic concerns for whether or not an individual particle was inhaled do not exist for isotopes with lower specific activity where the aerosol cloud can be expected to be more homogenous at concentrations of concern.

The isotopes of plutonium with even numbers of protons and neutrons (Pu-238, -240, and -242) undergo fission more frequently, and therefore emit neutrons at a higher rate, than the other isotopes of plutonium. Pu-238 possesses the highest spontaneous fission rate (2.59×10^3 n s^{-1} g^{-1}), while Pu-239 has the lowest (2.18×10^{-2} n s^{-1} g^{-1}). As indicated, subgram amounts of Pu-238 can produce significant neutron doses. However, Pu-240 is the most abundant isotope and so is also an important source of spontaneous neutron emissions (1.02×10^3 n s^{-1} g^{-1}). Spontaneous neutron emission is supplemented (and exceeded) by alpha-neutron (α, n) reactions in the oxide and fluoride forms of plutonium. For example, the α, n yield for Pu-238 (per gram of nuclide, not compound) in oxides is 1.34×10^4 n s^{-1} g^{-1}, while that of Pu-240 is 1.41×10^2 n s^{-1} g^{-1} (DOE, 1998).

CONCLUSION

A commitment to safety by all levels of management and by workers, along with proper attention on a daily basis to all aspects of a well-developed and well-documented radiation safety program can ensure that (1) radioactivity from operational sources and laboratory calibration sources is controlled,

(2) human radiation exposures are within statutory limits and ALARA, and (3) inadvertent radioactive contamination of samples and contamination in the counting laboratory is controlled to levels that enable meaningful counting results at the required concentrations.

REFERENCES

ANSI/AIHA, *American National Standard for Laboratory Ventilation*, ANSI/AIHA Z9.5, American National Standards Institute, New York, 2003.

Atomic Energy Commission (USAEC), *Termination of Operating Licenses for Nuclear Reactors*, Regulatory Guide 1.86, Nuclear Regulatory Commission, June 1974.

Best Manufacturing Company, *Chemrest Chemical Guide*, www.bestglove.com, Menlo, GA, 2006.

Code of Federal Regulations, Title 10—Energy, Chapter I—Nuclear Regulatory Commission, Part 20, *Standards for Protection against Radiation*, Revised, January 1, 2006, www.access.gpo.gov, U.S. Government Printing Office, Washington, DC, 2006.

Code of Federal Regulations, Title 29—Labor, Chapter XVII—Occupational Safety and Health Administration, Department of Labor, Part 1910, *Occupational Safety and Health Standards*, Revised, 2003, www.access.gpo.gov, U.S. Government Printing Office, Washington, DC, 2003.

Department of Energy, *Guide of Good Practices for Occupational Radiological Protection in Plutonium Facilities*, DOE-STD-1128-98, DOE Standard, U.S. Department of Energy, Washington, DC, 1998.

Hoover, M.D., Mewhinney, C.J., and Newton, G.J., Modular glovebox connector and associated good practices for control of radioactive and chemically toxic materials, *Health Phys.* 76:66–72, 1999.

Hoover, M.D., Newton, G.J., Yeh, H.C., and Eidson, A.F., Characterization of aerosols from industrial fabrication of mixed-oxide nuclear reactor fuels, in *Aerosols in the Mining and Industrial Work Environments*, V. A. Marple and B. Y. H. Liu, Eds., Ann Arbor Science, Ann Arbor, MI, 1983.

L'Annunziata, M.F., *Handbook of Radioactivity Analysis*, Academic Press, San Diego, 1998.

Lederer, C.M. and Shirley, V.S., *Table of Isotopes*, 7th ed. John Wiley & Sons, New York, 1978.

Newton, G.J., Hoover, M.D., Barr, E.B., Wong, B.A., and Ritter, P.D., Collection and characterization of aerosols from metal cutting techniques typically used in decommissioning nuclear facilities, *Am. Ind. Hyg. Assoc. J.* 48:922–932, 1987.

Saunders, G.T., *Laboratory Fume Hoods*, John Wiley & Sons, Inc., New York, 1993.

Scott, B.R. and Fencl, A.F., Variability in PuO_2 intake by inhalation: Implications for worker protection at the US Department of Energy. *Radiat. Prot. Dosim.*, 83:221–232, 1999.

Scott, B.R., Hoover, M.D., and Newton, G.J., On evaluating respiratory tract intake of high-specific activity emitting particles for brief occupational exposure, *Radiat. Prot. Dosim.*, 69(1):43–50, 1997.

Stewart, D.C., *Handling Radioactivity*, John Wiley & Sons, New York, 1981.

3 Standards, Guidelines, Regulations, and Recommendations for Measuring Airborne Radioactivity

Mark D. Hoover, Morgan Cox, Cynthia G. Jones, Liliane Grivaud, Michelle L. Johnson, Mark L. Maiello, and George J. Newton

CONTENTS

Introduction	21
International Standards	23
The International Electrotechnical Commission	24
The International Organization for Standardization	32
Regional International Standards	32
National Standards	32
U.S. National Standards Organizations	33
Standards of Interest from Other ANSI-Accredited Organizations	34
The National Standards Organizations in France	34
Government Regulations and Guidance	35
U.S. Law—National Technology Transfer and Advancement Act (PL 104-113)	35
U.S. Radiation Protection Regulations and Guidance	35
Other Recommendations and Guidance	37
Perspectives on the Nature of Regulations	39
Conclusion	39
References	40

INTRODUCTION

Numerous standards at the international, regional, national, and local levels pertain to monitoring and measuring radioactive aerosols (Cox et al., 2003a, 2003b). The topics of these standards range from generic subjects such as sampling strategies and particle size determinations to specific issues such as how to sample for a particular radioactive particle or gas. As shown in Figure 3.1, there is a complex set of interrelationships and interactions among numerous private sector, national, and international interests in the development of industry-wide, voluntary consensus standards. As shown in Figure 3.2, there are also a myriad of regulations, recommendations, requirements, and

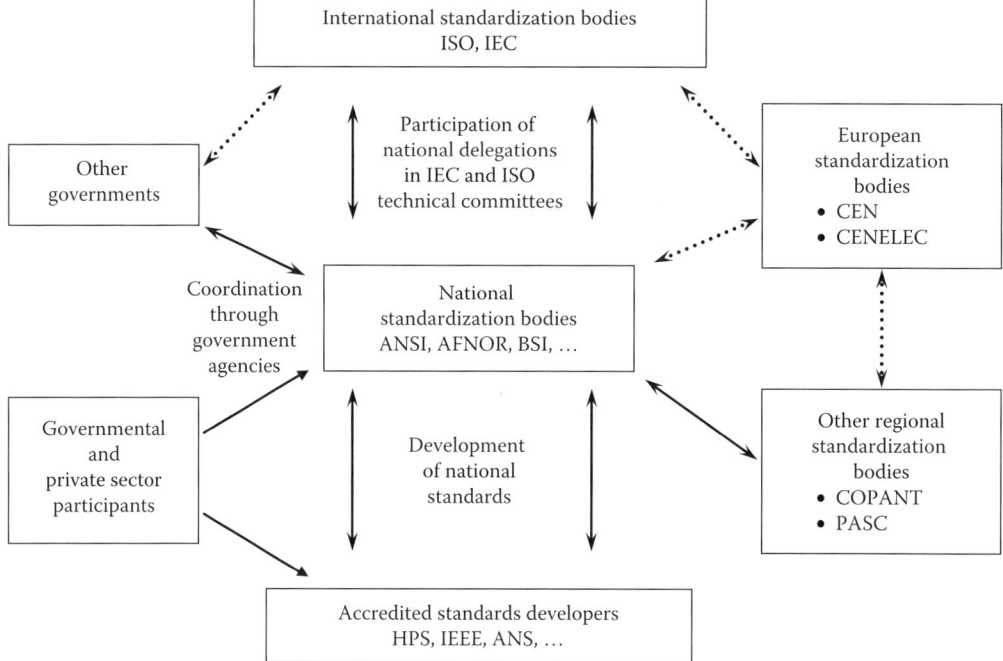

FIGURE 3.1 Interrelationships and interactions of private sector, national, and international interests in the development of industry-wide, voluntary consensus standards. (Adapted from Cox, M., et al., *An International Review of Currently Applicable Standards for Measuring Airborne Radioactivity*, Full Report, published online by the International Electrotechnical Commission, http://www.iec.ch/support/tcnews/2003/AirMonitoringStdsReview.htm, 2003a.)

FIGURE 3.2 Illustration of the relationships among standards, regulations, recommendations, requirements, and scientific resources for air monitoring equipment manufacturers and users. (Adapted from Cox, M., et al., *An International Review of Currently Applicable Standards for Measuring Airborne Radioactivity*, Full Report, published online by the International Electrotechnical Commission, http://www.iec.ch/support/tcnews/2003/AirMonitoringStdsReview.htm, 2003a.)

scientific resources for air monitoring equipment manufacturers and users. Recognizing the origins, content, and applicability of these requirements and resources can be daunting.

Regulatory authorities develop regulations and conduct inspections of radioactive materials licensees to assure compliance with radiological health and safety requirements. Inspections also occur routinely at industrial organizations, universities, commercial laboratories, and medical establishments with regulated air-effluent emission points. Regulators often require that published guidelines be met to assure that the air-sampling data are accurate. It is advisable to research the local and national radiological health regulations, standards, or recommendations. The following sections describe the standards organizations for airborne radioactivity and provide a snapshot of standards, guidelines, and regulations in common usage or development.

INTERNATIONAL STANDARDS

The overarching bodies for global standardization are the International Electrotechnical Commission (IEC) and the International Organization for Standardization (ISO). The national committees of the member nations who participate in the development of international standards work together to produce what are termed *industry-wide, voluntary consensus* standards. Taking into account the views of interests worldwide, consensus standards represent the negotiated agreements of the participants, but not necessarily the specific wishes of all participants.

Depending on the needs of the regulator, manufacturer, or user, standards can be categorized in several ways, including (1) committee of origin at the international, regional, or national level, (2) nature of the standard as either procedural or technical, (3) type or form of radioactive aerosol, (4) relevance to a particular instrument design or hardware, and (5) applicability for monitoring workplaces, effluent stacks, or the environment. Table 3.1 illustrates how the various national and international standards can be grouped according to the type of radioactive material and the circumstances under which the measurement is to be made.

TABLE 3.1
Matrix Showing Various Air Monitoring Standards According to the Type of Radioactive Material and the Circumstances under Which the Measurement is to be Made

	General	Particles	Noble Gases	Iodine	Tritium
General	ANSI N42.17B ANSI N323 ANSI N13.2	IEC 61578	IEC 62302		ANSI N42.30 IEC 62303
Effluent	IEC 60761-1 IEC 60951-4 ISO 2889 ANSI N13.1	IEC 60761-2	IEC 60761-3 IEC 62302	IEC 60761-4	IEC 60761-5 IEC 62303
Workplace	ANSI N42.18 ANSI N317	IEC 579 EN 481	IEC 62302		IEC 60710 IEC 62303
Environment	ANSI N13.9	IEC 1172	IEC 62302	IEC 1171	IEC 62303
Emergency	IEC 60951-1 ANSI N320	IEC 951-5	IEC 60951-2		

Source: Adapted from Cox, M., et al., *An International Review of Currently Applicable Standards for Measuring Airborne Radioactivity*, Full Report, published online by the International Electrotechnical Commission, http://www.iec.ch/support/tcnews/2003/AirMonitoringStdsReview.htm, 2003a.

THE INTERNATIONAL ELECTROTECHNICAL COMMISSION

The IEC (www.iec.ch) is the global organization that prepares and publishes international standards for electrical, electronic, and related technologies. It was founded in 1906 following passage of a resolution at the 1904 meeting of the International Electrical Congress in St. Louis, Missouri, USA. The IEC currently has 61 participating countries, including all the world's major trading nations and a growing number of industrializing countries. Table 3.2 presents a selected list of national standards organizations participating in IEC activities related to the measurement of airborne radioactivity.

As noted on the IEC Web site, the IEC charter embraces all electrotechnologies including electronics, magnetics and electromagnetics, electroacoustics, multimedia, telecommunication, and energy production and distribution, as well as associated general disciplines such as terminology and symbols, electromagnetic compatibility, measurement and performance, dependability, design and development, and safety and the environment.

TABLE 3.2
Selected List of National Standards Organizations Participating in IEC Activities Related to the Measurement of Airborne Radioactivity

Country	Symbol	Organization
Austria[a,c]	ON	Österreichisches Normungsinstitut
Belgium[a,c]	IBN	Institut Belge de Normalisation
Bulgaria[b]	BDS	State Agency for Standardization and Metrology
Canada[a,d]	SCC	Standards Council of Canada
China[a,d]	CSBTS	China State Bureau of Quality and Technical Supervision
Czech Republic[a]	CSNI	Czech Standards Institute
Denmark[b,c]	DS	Dansk Standardiseringsread
Egypt[a]	EOS	Egyptian Organization for Standardization and Quality Control
Finland[a,c]	SFS	Suomen Standardisoimisliitto
France[a,c]	AFNOR	Association Française de Normalisation;
	UTE	Union Technique de l'Electricité
Germany[a,c]	DIN	Deutsches Institut für Normung
Greece[c]	ELOT	Hellenic Organization for Standardization
Iceland[c]	STRI	Technological Institute of Iceland
India[b]	BIS	Bureau of Indian Standards
Ireland[b,c]	NSAI	National Standards Authority of Ireland
Israel[a]	SII	Standards Institution of Israel
Italy[a,c]	UNI	Ente Naziolale Italiano Unificazione
Japan[a,d]	JISC	Japanese Industrial Standards Committee
Korea (Republic of)[b,d]	KATS	Korean Agency for Technology and Standards
Luxembourg[c]	SEE	Service de l'Energie de l'Etat
Netherlands[b,c]	NNI	Nederlands Normalisatie-Instituut
New Zealand[b,d]	SNZ	Standards New Zealand
Norway[b,c]	NSF	Norges Standardiseringsforbund
Poland[b]	PKN	Ogloszenie Polskiego Komitetu Normalizacyjnego
Romania[a]	ASRO	Asociatia de Standardizare din România
Portugal[c]	IPQ	Instituto Portugues de Qualidade
Russian Federation[a,d]	GOST R	Gosstandart of Russia
Slovakia[b]	SUTN	Slovensky Ustav Technickej Normalizacie
South Africa[b,d]	SABS	South African Bureau of Standards
Spain[b,c]	AENOR	Associacion Español de Normalización y Certificacion

TABLE 3.2 (continued)
Selected List of National Standards Organizations Participating in IEC Activities Related to the Measurement of Airborne Radioactivity

Country	Symbol	Organization
Sweden[a,c]	SIS	Standardiseringskommissionen i Sverige
Switzerland[a,c]	SNV	Schweizerische Normen-Vereinigung
Ukraine[a]	DSTU	State Committee of Ukraine for Standardization, Metrology & Certification
United Kingdom[a,c]	BSI	British Standards Institution
United States of America[a,d]	ANSI	American National Standards Institute
Yugoslavia[b]	SZS	Savezni Zavod za Standardizaciju

Source: Adapted from Cox, M., et al., *An International Review of Currently Applicable Standards for Measuring Airborne Radioactivity*, Full Report, published online by the International Electrotechnical Commission, http://www.iec.ch/support/tcnews/2003/AirMonitoringStdsReview.htm, 2003a.

[a] Participating members of IEC SC45B.
[b] Observer status for IEC SC45B.
[c] Members of the CEN and the CENELEC.
[d] Members of the PASC.

In 1958, the IEC's Committee of Action proposed establishment of an IEC Technical Committee (TC) to develop international standards covering nuclear-related instrumentation. TC45 (Nuclear Instrumentation) first met in New Delhi, India, in 1960 and subsequently evolved to comprise two subcommittees (SCs): SC45A (Nuclear Reactor Instrumentation) and SC45B (Radiation Protection Instrumentation). Working Group (WG) 13 of SC45B has the primary responsibilities for Measurements of Airborne Radioactivity.

Table 3.3 contains an annotated list of relevant IEC standards for monitoring and measuring airborne radioactivity, including the scope for each standard as listed on the IEC Web site.

TABLE 3.3
Annotated List of Relevant Standards for Measuring Airborne Radioactivity

Applicable Standards from ANSI N13 (www.hps.org)

ANSI N13.1 (1969, Revised 1999) *Sampling and Monitoring Releases of Airborne Radioactive Substances from the Stacks and Ducts of Nuclear Facilities.* This standard was originally issued in 1969 and was substantially revised in 1999 to account for improved understanding of the physics and practice of extractive air sampling. ANSI N13.1 (1999) is a performance-based document rather than one based on prescriptive rules and specifications. It includes discussions of air moving methods and air flow and air flow rate measurement devices. It does not address instrumentation used to quantify or measure the radioactivity in air samples. The revision of this standard is contributing to the revision of ISO 2889-1975.

ANSI N13.2 (1969, Reaffirmed 1982) *Administrative Practices in Radiation Monitoring.* This standard is directed toward nuclear facility management, providing an administrative basis for programs to monitor and measure airborne radioactivity in the workplace and environment. This standard does not provide technical guidance for measuring or monitoring airborne radioactive materials.

ANSI N13.9 (In preparation) *Guide to Environmental Surveillance around Nuclear Facilities.* This standard is under development to address specific issues of sampling in the outdoor environment.

ANSI N13.39 (2001) *Design of Internal Dosimetry Programs.* This standard provides general policies and the framework for the design and implementation of an acceptable internal dosimetry program. Air sampling and monitoring results are critical components of such programs.

continued

TABLE 3.3 (continued)
Annotated List of Relevant Standards for Measuring Airborne Radioactivity

Applicable Standards from ANSI N42 (www.ieee.org)

ANSI N42.17B (1989) *Performance Specification for Health Physics Instrumentation—Occupational Airborne Radioactivity Monitoring Instrumentation.* The scope of this standard includes performance criteria and testing procedures for instruments and instrument systems designed to continuously sample and quantify concentrations of radioactivity in ambient air in the workplace. This standard does not specify which instruments or systems are required, nor does it address the specific locations or applications of such instruments.

ANSI N42.18 (1980, Reaffirmed 1991) *Specification and Performance of On-site Instrumentation for Continuously Monitoring Radioactivity in Effluents.* This standard provides recommendations for selecting instruments used to continuously monitor and measure radioactivity in effluents released to the environment. The scope includes all physical forms of radioactive materials such as gases, liquids, particulates, or dissolved solids singly or in combination, in effluent paths including liquid effluents. The standard does not provide specific requirements for instrument capabilities, emergency situations, or sample extractions and laboratory analyses. This standard is currently being reviewed and revised in light of a number of lessons learned and new technologies since development of the original standard.

ANSI N42.30 (2002) *Performance Specification for Tritium Monitors.* This standard provides the performance requirements and test procedures for tritium monitors used for monitoring airborne tritium radioactivity. The standard applies to tritium monitors used to measure tritium in the workplace. It does not specify which systems or instruments are required, nor does it address the specific locations or applications of such instruments. Work is underway to replace this standard by incorporating its content into a comprehensive revision of ANSI N42.18.

ANSI N42.50 (In preparation) Performance Specifications for Measurement Systems Designed to Measure Radon Progeny in Atmospheres. This standard is under development.

ANSI N42.51 (In preparation) *Performance Specifications for Systems Designed to Measure Radon Gas in Atmospheres.* This standard is being developed under the auspices of the American Association of Radon Scientists and Technologists.

ANSI N317 (1980) *Performance Criteria for Instrumentation Used for In-Plant Plutonium Monitoring.* The scope of this standard is limited to instruments used to monitor and measure airborne plutonium in handling and storage facilities. Reactors and irradiated fuel reprocessing facilities are specifically excluded. The standard includes requirements similar to those found in ANSI N42.17B, such as temperature response and alarm capabilities. Unlike N42.17B or N42.18, ANSI N317 includes a specific requirement for the sensitivity of instruments used to monitor or measure airborne radioactive materials. This standard is currently being updated in light of current technology.

ANSI N320 (1979, Reaffirmed 1985) *Performance Specifications for Reactor Emergency Radiological Monitoring Instrumentation.* This standard addresses performance parameters and general placement for monitors used to measure radionuclides released during an accident at a reactor facility. Included in the standard are general discussions of monitoring systems and specific requirements for lower and upper detection limits for emergency monitoring instruments. This standard is being revised.

ANSI N323C (2009) *Radiation Protection Instrumentation Test and Calibration—Air Monitoring Instruments.* The general requirements of ANSI Standard N323 (1983) were expanded in the new ANSI 323C to provide detailed guidance for calibration of air sampling and monitoring instruments used to meet the requirements and recommendations of governmental regulations and other ANSI standards that specify when, where, and how air monitoring instruments shall, should, or may be used to sample or monitor airborne radioactive substances. The standard applies to fixed, portable, and personal air samplers and monitors used for regulatory compliance, including continuous air monitors for stack, workplace, and environmental applications. Calibration of specialized air monitoring instruments such as working level monitors or process-monitoring instruments may involve issues that extend beyond the fundamental considerations presented in this standard. In such cases, designers and users should exercise professional judgment in the application of these requirements and should explicitly document the sampling objectives and the reasons for any exceptions to the requirements of this standard.

Applicable Standards from ASTM International (www.astm.org)

ANSI/ASTM E 741-93 *Standard Test Methods for Determining Air Change in a Single Zone by means of a Tracer Gas Dilution.* This standard provides guidance on assessment and understanding of air flow patterns that influences approaches to and results from air sampling.

ANSI/ASTM C 1086-95, *Standard Guide for Qualification of Measurement Methods by a Laboratory within the Nuclear Industry.* This standard includes measurement methods that can be used for retrospective assessment of air samples.

TABLE 3.3 (continued)
Annotated List of Relevant Standards for Measuring Airborne Radioactivity

ANSI/ASTM E 1893-97 Standard for Selection and Use of Portable Radiological Survey Instruments for Performing *In Situ* Radiological Assessments in Support of Decommissioning. Assessment of the radioactive content of surfaces and volumes contributes to a comprehensive radiation detection and protection program.

ANSI/ASTM D 1356-00 *Standard Terminology Relating to Sampling and Analysis of Atmospheres.* Ensuring the definition and use of common terminology is a continuing challenge.

ANSI/ASTM D 4532-97 *Standard Test Method for Respirable Dust in Workplace Atmospheres.* This standard is an example of a gravimetric approach to assessment of airborne dust. Limitations on the test method are a minimum weight of 0.2 mg of dust on the filter and a maximum loading of 0.3 mg cm^{-2} on the filter. The test method may be used at higher loadings if the flowrate can be maintained constant. Sampling of airborne radioactivity is not the focus of this standard, but if airborne radioactivity is present along with significant amounts of other dusts, the sampling limitations addressed in this standard may be applicable.

ANSI/ASTM D 6061-96 *Standard Practice for Evaluating the Performance of Respirable Aerosol Samplers.* This standard is another example of methods for the general assessment of airborne materials in different size ranges.

ANSI/ASTM D 6062M-96 *Standard Guide for Personal Samplers of Health-Related Aerosol Fractions.* This standard is a further example of methods for general assessment of airborne materials in size ranges that may be inhaled and deposited in the respiratory tract.

ANSI/ASTM E 1370-96 (2008) *Standard Guide for Air Sampling Strategies for Worker and Workplace Protection.* This recently updated guide describes standard approaches used to determine air sampling strategies before any actual air sampling occurs. The standard notes that air sampling in the workplace may be done for single or multiple purposes and wisely points out that conflicts arise when a single air sampling strategy is expected to satisfy multiple purposes.

Applicable Standards from the IEC (www.iec.ch)

IEC 60579 (1977-01) *Radioactive aerosol contamination meters and monitors.* This standard is of historical interest in the area of effluent monitoring, but has been superseded by the comprehensive IEC 60761 series. It was applicable only to assemblies equipped with filters. It was not applicable to meters or monitors designed for selective monitoring.

IEC 60710 (1981-01) *Radiation protection equipment for the measuring and monitoring of airborne tritium.* This IEC standard has served the nuclear industry well and is currently under revision by IEC SC45B. The standard provides requirements and gives examples of acceptable methods for measuring and monitoring equipment to enable the determination of the average value of the concentration of atmospheric tritium in working areas and its variation as a function of time, and to actuate an alarm system if necessary.

IEC 60761 (2002-01) *Equipment for continuously monitoring radioactivity in gaseous effluents.* This comprehensive set of standards covers five areas:
- IEC 60761-1 (2002-01) Part 1: General requirements.
- IEC 60761-2 (2002-01) Part 2: Specific requirements for radioactive aerosol monitors including transuranics.
- IEC 60761-3 (2002-01) Part 3: Specific requirements for radioactive noble gas monitors.
- IEC 60761-4 (2002-01) Part 4: Specific requirements for radioactive iodine monitors.
- IEC 60761-5 (2002-01) Part 5: Specific requirements for tritium monitors.

IEC 60951 (1988-08) *Radiation monitoring equipment for accident and post-accident conditions in nuclear power plants.* This set of standards was originally written in five parts in the aftermath of the Three Mile Island and Chernobyl nuclear reactor accidents. The scope and content are not intended to provide detailed direction or guidance for air monitoring under normal workplace conditions or in the environment. Parts 1–4 of this standard are being or have been recently updated. Part 5 has recently been eliminated, but its original content is listed here for historical completeness.

- **IEC 60951 (1988-08)** *Part 1: General requirements.* Part 1 provides guidance on the design principles and performance criteria for equipment to measure radiation and fluid (gases or liquids) radioactivity levels in reactor plants during and after an accident. Specifies general design and operating characteristics, general test procedures, radiation characteristics, electrical quantities, safety and environmental characteristics, as well as the documentation required.
- **IEC 60951 (1988-08)** *Part 2: Equipment for continuously monitoring radioactive noble gases in gaseous effluents.* Part 2 lays down specific requirements for monitoring equipment for noble gases and gaseous effluents, measuring in accident and post-accident conditions the volumetric activity of radioactive noble gases in effluents and the total discharge of noble gas activity over a given period. Part 2 does not deal with sampling procedures and laboratory analysis methods.

continued

TABLE 3.3 (continued)
Annotated List of Relevant Standards for Measuring Airborne Radioactivity

- **IEC 60951-3 (1989-12)** *Part 3: High range area γ-radiation dose rate monitoring equipment.* Part 3 lays down specific requirements, including technical characteristics and test conditions for area γ-radiation dose rate monitoring equipment for use in accident conditions. This may be used, for example, for detection of leaks in a system containing radioactivity and to give useful information for interpreting the accident conditions that are present, making, in some cases, an assessment of potential releases to the environment, and implementation of emergency procedures.
- **IEC 60951-4 (1991-12; R2007)** *Part 4: Process stream in light water nuclear power plants.* Part 4 applies to equipment for the monitoring of radioactive substances within plant process streams of stationary nuclear power plants using light water reactors during and after accident conditions. Provides criteria for the design, selection, functional location, testing, and calibration of stationary radiation monitoring equipment to be used for continuous monitoring of plant process streams in operation during and after accident conditions.
- **IEC 60951-5 (1994-02)** *Part 5: Radioactivity of air in light water nuclear power plants.* Part 5 has recently been eliminated, but it originally provided criteria for the design, selection, functional location, testing, and calibration of installed equipment for monitoring airborne radioactivity within nuclear power plants with light water reactors during and after accident conditions.

IEC 61171 (1992-09) *Radiation protection instrumentation—Monitoring equipment—Atmospheric radioactive iodines in the environment.* This standard is applicable to transportable or installed equipment used for monitoring, as a function of time, airborne radioactive iodines in the environments near nuclear facilities during normal operations, during anticipated operational occurrences, or during accident conditions.

IEC 61172 (1992-09) *Radiation protection instrumentation—Monitoring equipment—Radioactive aerosols in the environment.* This standard is applicable to transportable or installed equipment for continuous monitoring of radioactive aerosol in the environment for both normal and accident conditions. For purposes of this standard, monitoring includes continuous sample collection with, if desired, the capability to automatically initiate sampling.

IEC 61577 *Radiation protection instrumentation—Radon and radon decay product measuring instruments.* This set of four standards covers the important aspects of requirements for instrumentation to measure radon and radon decay products.

- **IEC 61577-1 (2006)** *Part 1: General principles.* Part 1 addresses only the instruments and associated methods for measuring isotopes 220 and 222 of radon and their subsequent short-lived decay products in gases. Its object is to help define type tests that must be conducted to qualify these instruments. Note that type tests for these instruments are described in IEC 61577-2-1, IEC 61577-2-2, IEC 61577-3-1, and IEC 61577-3-2. Part 1 also proposes a classification of the instruments measuring radon or radon decay products based on the duration of sampling.
- **IEC 61577-2 (2000)** *Part 2: Specific requirements for radon-measuring instruments.* Part 2 specifies the main performance characteristics of instruments intended for measurement of airborne radon volume activity, their specific method of testing, and documentation requirements. It is applicable to instruments used to measure radon in workplaces, dwellings, outdoor air, and soil. The method of measurement depends on the exact objective but the requirements are for general purpose instruments to be used for radiological protection or research applications. This standard applies to all types of radon-measuring instruments that are based on grab sampling, continuous sampling techniques, and integrated measurement methods. The activity can be measured continuously by pumping or by diffusing the air containing radon into the detector or at a particular moment by measuring the activity of an air sample (grab sampling).
- **IEC 61577-3 (2002)** *Part 3: Specific requirements for radon decay product measuring instruments.* Part 3 is applicable to instruments that are used to measure radon decay product activity concentration and/or volume potential α-energy in workplaces, dwellings, and outdoors. The standard applies practically to all types of instruments that are based on grab sampling, continuous sampling techniques, and integrating measurement methods that include electronic equipment. The measurement of activity retained by a sampling device, for example, a filtering device, can be made both during sampling or after the completion of a collection cycle. These requirements specify test procedures and identification certificates, as well as radiation, electrical, mechanical, safety, and environmental characteristics.
- **IEC 61577-4 (2007)** *Part 4: Equipment for the production of reference atmospheres containing radon isotopes and their decay products (STAR).* Part 4 concerns the System for Test Atmospheres with Radon (STAR) needed for testing, in a reference atmosphere, the instruments measuring radon and RnDP. The standard notes that it neither claims to solve all the problems involved in the production of equipment for setting up reference atmospheres for radon and its decay products, nor to describe all the methods for doing so. The standard further states that it is intended to be a guide enabling those faced with such problems to choose the best methods for adoption in full knowledge of the facts.

TABLE 3.3 (continued)
Annotated List of Relevant Standards for Measuring Airborne Radioactivity

IEC 61578 (1997-08) *Test methods for the calibration and verification of the effectiveness of radon compensation for α and/or β aerosol measuring instruments.* This standard applies to real-time air monitoring instruments that use compensation algorithms to remove interference from the alpha emissions of naturally occurring radon progeny.

IEC 62302 (2007) *Equipment for noble gas monitoring in the workplace, effluents, and the environment.* This standard is applicable to equipment used for sampling and continuous measurement of radioactive noble gases in the workplace, in gaseous effluents discharged into the environment, and in the environment itself.

IEC 62303 (2008) *Equipment for monitoring airborne tritium.* This standard is applicable to equipment used for sampling and continuous measurement of tritium in the workplace, in gaseous effluents discharged into the environment, and in the environment itself. It is applicable to installed, portable, and transportable equipment. This standard complements IEC 60761-5, which is only applicable to equipment for sampling and monitoring tritium in gaseous effluents. IEC 62303 expands coverage to include monitoring all possible locations where tritium could present a radiological hazard. The equipment is designed to be in operation during normal operation conditions, as well as under emergency conditions, both during and following an accident. Depending on the nature of the emergency conditions, it might be necessary to install specially designed equipment for normal operation conditions and other equipment for emergency conditions. The standard specifies the general characteristics; general testing procedures; mechanical, electrical, electronic, radiological, safety, and environmental characteristics of the equipment; and proper identification and certification of the equipment. The standard notes that if this equipment is part of a centralized system for continuous radiation monitoring in a nuclear facility, there may be additional requirements from other standards related to those systems.

Applicable Standards from the ISO (www.iso.ch)

ISO 2889 (1975) *General principles for sampling airborne radioactive materials.* This standard was originally issued in 1972, was revised in 1975, and is currently undergoing further revision. It comes under the purview of ISO TC85 *Nuclear Energy*. It is currently under revision to incorporate technological advances and lessons learned since the 1970s. Its scope is similar to that of ANSI N13.1-1999. Revisions being formulated in the 2009 time frame are expected to cover optimal sampling methodology, including sampling and monitoring techniques, and combining sampling, monitoring and control procedures. The revised 2889 is being expanded as follows: Part 1 (general requirements), Part 2 (stacks and ducts), Part 3 (workplace), and Part 4 (outdoors or in the environment).

ISO 7708 (1995) *Air quality—Particle size fraction definitions for health-related sampling.* This standard comes under the purview of ISO TC 146 *Air Quality*. It defines sampling conventions for particle size fractions for use in assessing possible health effects of airborne particles in the workplace and ambient environment. Definitions include the conventions for the inhalable, thoracic, and respirable fractions. The extrathoracic and tracheobronchial conventions may be calculated from the defined conventions. The standard notes that the conventions should not be used in association with limit values defined in other terms, for example, for limit values of fibers defined in terms of their length and diameter. Although exposure limits for radioactivity are not based on these conventions, considerations for penetration of particles to different portions of the respiratory tract are of interest for understanding particle behavior and health-related considerations, in general.

ISO 11665 (under development) *Measurement of radioactivity in the environment—Air.* This eight-part series (under development by ISO TC 85 SC 2 *Radiation Protection*) will cover the following topics:

- **ISO 11665-1** Measurement of radioactivity in the environment—Air—Part 1: Radon-222 and its short-lived decay products in the atmospheric environment: their origins and measuring methods.
- **ISO 11665-2** Measurement of radioactivity in the environment—Air—Part 2: Radon-222: Integrated measuring methods of the potential α energy concentration of short-lived radon decay products in the atmospheric environment.
- **ISO 11665-3** Measurement of radioactivity in the environment—Air—Part 3: Radon-222: Short-term measuring methods of the potential α energy concentration of short-lived radon decay products in the atmospheric environment.
- **ISO 11665-4** Measurement of radioactivity in the environment—Air—Part 4: Radon-222: Integrated measuring methods of the average radon activity concentration in the atmospheric environment using passive sampling and delayed analysis.
- **ISO 11665-5** Measurement of radioactivity in the environment—Air—Part 5: Radon-222: Continuous measuring methods of radon activity concentration in the atmospheric environment.
- **ISO 11665-6** Measurement of radioactivity in the environment—Air—Part 6: Radon-222: Methods of estimating the exhalation surface rate by accumulation in the environment.

continued

TABLE 3.3 (continued)
Annotated List of Relevant Standards for Measuring Airborne Radioactivity

- **ISO 11665-7** Measurement of radioactivity in the environment—Air—Part 7: Radon-222: Short-term measuring methods of radon activity concentration in the atmospheric environment.
- **ISO 11665-8** Measurement of radioactivity in the environment—Air—Part 8: Radon-222 in buildings: Methodologies for screening and additional investigations.

ISO 16000-1 (2004) *Indoor air—Part 1: General aspects of sampling strategy.* This standard is under the purview of ISO TC 146 SC 6 *Indoor Air*, and illustrates the content of an extensive ISO series related to assessment of nonradioactive particles and gases in the indoor environment. This part of the series is intended to aid the planning of indoor pollution monitoring. In a manner similar to the focus noted above for the ANSI/ASTM E 1370-96 (2008) *Standard Guide for Air Sampling Strategies for Worker and Workplace Protection*, ISO 16000-1 is another example of the generic considerations that underlie all types of air sampling initiatives. As noted in the standard, before a sampling strategy is devised for indoor air monitoring, it is necessary to clarify for what purposes, when, where, how often, and over what periods of time monitoring is to be performed. The answers to these questions depend, in particular, on a number of special characteristics of the indoor environments, on the objective of the measurement and, finally, on the environment to be measured. The standard deals with the significance of these factors and offers suggestions on how to develop a suitable sampling strategy. Other standards in the 16,000 series address issues for specific chemical compounds and materials.

Applicable European Standards (www.cenorm.be)

EN 481 (1993) *Workplace atmospheres—Size fraction definitions for measurement of airborne particles.* This standard defines sampling conventions for particle size fractions, which are used to assess the possible health effects from inhaling airborne particles in the workplace.

DIN 25423-1 (1999) *Sampling procedures for the monitoring of radioactivity in air—Part 1: General requirements.* This document from the DIN is applicable to the sampling of gas- or aerosol-bound radioactive materials in the air. This means the sampling of gas- or aerosol-bound radioactive materials in plants or installations where unsealed radioactive materials are handled in the sense of the German Radiation Protection Order. This includes the sampling of exhaust air out of rooms and elsewhere. The standard is also applicable to the sampling of other gases and gas mixtures.

DIN 25423-2 (2000) *Sampling procedures for the monitoring of radioactivity in air—Part 2*: Special requirements for sampling from air ducts and stacks. This document in the DIN series specifies special requirements relating to sampling procedures from air ducts and stacks for the monitoring of radioactivity in air for the purpose of controlling radioactive contamination. This involves in particular the sampling of airborne particles in plants handling radioactive materials and the sampling of exhaust air before this is released into the atmosphere.

DIN 25423-3 (1987) *Sampling procedures for the monitoring of radioactivity in air; sampling methods.* This standard in the DIN series specifies methods and auxiliary means relating to sampling procedures for the monitoring of radioactivity in air for the purpose of controlling radioactive contamination. This involves, in particular, the sampling of airborne particles in plants handling radioactive materials and the sampling of exhaust air before it is released into the atmosphere.

DIN 25441-1 (1983) *Monitoring of radioactivity in the inner atmosphere of nuclear power plants; safety requirements.* This standard which deals with the monitoring of radioactivity in the inner atmosphere of nuclear power plants and is used exclusively in stationary nuclear plants during normal operation.

Applicable French National Standards (www.afnor.fr)

The first four French standards listed below are generic and comparable to requirements of the U.S. Clean Air Act of 1990.

NF X 43-021 (December 1984) *Air Quality—Filter sampling of particulate material suspended in ambient air—Automatic sequential equipment.* This standard describes an automatic sequential apparatus for sampling with filters for particulate matter in environmental air. The concentration of various organic or minerals such as heavy metals, including lead, can thus be estimated.

NF X 43-022 (May 1985) Air Quality—Ambient air—Concepts relating to the sampling of particulate matter.

NF X 43-257 (August 1988) Air Quality—Air in workplaces—Individual sampling of inspirable fraction of particulate pollution.

NF X 44-051 (July 1978) Sampling of dust in a stream of gas (general case).

NF X 44-052 (May 2002) Stationary Source Emissions—Determination of high range mass concentration of dust—Manual gravimetric method.

TABLE 3.3 (continued)
Annotated List of Relevant Standards for Measuring Airborne Radioactivity

The following French standards are for specific applications for monitoring and measuring radioactive aerosols.

NF M 60-312 (October 1999) Nuclear energy—Measurement of environmental radioactivity—Air—Determination by liquid scintillation of the activity concentration of atmospheric tritium sampled by the sparging technique (air through water). This document describes a bubbler sampling method for determining the atmospheric concentration of tritium (vapor and gas). The detection limit of this method is about 1 Bq m^{-3}. It does not allow the measurement of the natural background. There are a number of passive or active sampling methods for determining the concentration of tritium in air. Most of them concern only tritium in water vapor form. These methods avoid the dilution of the atmospheric water vapor in the water of the bubblers and have a limit of detection of about 10 mBq m^{-3}.

NF M 60-760 (October 2001) Nuclear energy—Measurement of radioactivity in the environment—Air—Sampling of aerosols for measurement of radioactivity in the environment. This standard is concerned with sampling aerosols in the environment using filters to measure the specific activity (Bq m^{-3}) in the sampled air.

NF M 60-763 (March 1998) Nuclear energy—Measurement of radioactivity in the environment—Air—Radon and its short-lived decay products in the atmospheric environment: Origins and measuring methods. This document summarizes the available general knowledge regarding the origin and behavior of radon 222 in various atmospheric environments. The air sampling can be classified as occasional, continuous, and/or integrated measurements.

NF M 60-764 (December 1997) Nuclear energy—Measurement of radioactivity in the environment—Air—Radon 222: Integrated methods for measurement of α potential energy of short-life decay products in the atmospheric environment. This document describes a method for determining the α potential volumetric energy (EAP$_V$) of the short-half-life progeny of radon 222 in the atmosphere over a period of one month. This measurement allows for the evaluation of radiological exposure to man.

NF M 60-765 (December 1997) Nuclear energy—Measurement of radioactivity in the environment—Air—Radon 222: Methods for spot measurement of α potential energy of radon daughters in the atmospheric environment. This document describes a method for occasionally and quickly (in a few minutes) determining the α potential energy of the short-half-life daughter products of radon 222 in the atmosphere.

NF M 60-766 (December 1999) Nuclear energy—Measurement of radioactivity in the environment—Air—Radon 222: Integrated methods for measurement of the average volumic activity of radon in the atmospheric environment, with passive collection and a deferred analysis. This document gives some requirements concerning the measurement of the average volumic activity of radon 222 by passive sampling in the atmosphere and delayed analysis. Among the different existing methods, only the integrated methods of measurement of the average volumic activity from several weeks up to one year are covered by this document, which describes the mode of sampling, the mode of detection, and the conditions for use. This document is used jointly with NF M 60-763.

NF M 60-767 (August 1999) Nuclear energy—Measurement of environmental radioactivity—Air—Radon 222: Continuous measurement methods of the volumic (volumetric) activity of radon in the atmospheric environment. This document gives some requirements to measure the volumic (volumetric) activity of radon 222 in the atmosphere. Among the different methods used, only the methods of continuous measurement are covered in this document. This document describes the mode of sampling, the mode of detection, and the conditions for use. This document is used jointly with NF M 60-763.

NF M 60-768 (October 2002) Nuclear energy—Measurement of environmental radioactivity—Air—Radon 222: Methods for the estimation of the surface activity of exhalation by accumulation methods. This document gives some requirements for the estimation of the surface exhalation rate of radon at a given point and time at an interface with the atmosphere, from the measurement of the volumic (volumetric) activity of radon 222 inside a storage volume. It is used jointly with NF M 60-763 M 60-767 and M 60-769.

NF M 60-769 (November 2000) Nuclear energy—Measurement of environmental radioactivity—Air—Radon 222: Methods for spot measurement of the volumic activity of radon in the atmospheric environment. This document gives some requirements to measure the volumic (volumetric) activity of radon 222 in the atmosphere. Among the different existing methods, only the methods of spot measurement are covered by this document, which describes the mode of sampling, the mode of detection, and the conditions for use. It is used jointly with NF M 60-763.

NF M 60-770 (October 2000) Nuclear energy—Measurement of environmental radioactivity—Air—Determination of the activity concentration for atmospheric deposits on the soil. This document describes a passive device for collecting deposits of atmospheric radionuclides on soil to determine their identity and concentration. The sampling conditions and the periodicity are also given. The collection of the dried fallout is given specifically in NF X 43-007 (1973).

continued

TABLE 3.3 (continued)
Annotated List of Relevant Standards for Measuring Airborne Radioactivity

NF M 60-771 (July 2001) Nuclear energy—Measurement of environmental radioactivity—Air—Radon 222 in buildings—Methodologies for screening and complementary investigations. This document provides some requirements to screen for radon and to conduct complementary measurements necessary to identify the source and transfer of radon in buildings. It was prepared for use with standards NF M 60-763, M 60-764, M 60-765, M 60-766, M 60-767, and M 60-769.

THE INTERNATIONAL ORGANIZATION FOR STANDARDIZATION

The ISO (www.iso.ch) was formed in 1947 to fill the need for an international organization for standards outside of the electrical and electronic disciplines. ISO characterizes itself as "a network of national standards institutes from 140 countries working in partnership with international organizations, governments, industry, business and consumer representatives; a bridge between public and private sectors." As is the case with the IEC, the ISO has numerous TCs covering many disciplines.

ISO air monitoring standards for radioactive materials are developed by ISO TC85 (Nuclear Energy), SC2 (Radiation Protection), and WG14 (Air Control and Monitoring). There are common interests and collaborations between the IEC TC45/SC45B/WG13 and the ISO TC85/SC2/WG14 in terms of leadership and the personnel who have volunteered to develop and maintain related standards. Table 3.3 includes an annotated list of relevant ISO standards for monitoring and measuring airborne radioactivity, including the scope for each standard as listed on the ISO Web site.

REGIONAL INTERNATIONAL STANDARDS

The Comité Européen de Normalisation (CEN) (www.cenorm.be) was established in 1961 and is responsible for harmonization of national standards issued from the various countries of the European Union (EU). Standards in the electrotechnical sector are developed by the Comité Européen de Normalisation Electrotechnique (CENELEC) (www.cenelec.org), which was established in 1973 and comprises the electrotechnical committees of some 19 European countries. CENELEC works closely with the CEN, IEC, and other similar organizations. More than 80% of the European standards adopted by CEN and CENELEC are identical to or based on corresponding international standards. Table 3.3 contains an annotated list of relevant European air monitoring and measuring standards.

Although there are other regional organizations, such as the Pacific Area Standards Congress (PASC) (www.pascnet.org), they are not engaged in development of air monitoring standards.

NATIONAL STANDARDS

Most nations are represented in the international standards process by a national organization. These national organizations can (1) adopt and follow international standards, (2) adopt and follow standards provided by their regional standards bodies, and/or (3) prepare standards (or accredit qualified organizations to prepare them) as needed within their own country. Some organizations, such as the British Standards Institution (BSI) (www.bsi-global.com), have chosen not to develop their own set of standards for air monitoring and measuring, but instead contribute to and use the international standards of the ISO and IEC. In the member nations of the EU, the various national standards are gradually being replaced by European standards. For example, standards developed by CEN and CENELEC are automatically adopted as national standards by the member countries. Other nations, including the United States, develop their own national standards in addition to participating in the development of international standards.

U.S. National Standards Organizations

The American National Standards Institute (ANSI) (www.ansi.org) is the national standardizing body in the United States. Its staff serves as the Secretariat for the U.S. National Committee (USNC) for the ISO and the USNC for the IEC. The USNCs appoint delegates and WG experts, all volunteers, who represent U.S. interests in meetings of the IEC and ISO and participate as WG experts in the development of standards.

ANSI has accredited more than 250 professional societies or industrial firms to serve as the Secretariats for the committees that develop standards in specific topical areas and maintain them in contemporary form. ANSI then processes, adopts, and publishes national standards. In the areas of nuclear and health physics instrumentation and radiation safety, ANSI has accredited the Institute for Electrical and Electronic Engineers (IEEE) and the Health Physics Society (HPS). The specific committee responsibilities described below concerning development of standards for radiological protection instruments and instrumentation have been determined by memoranda of understanding among responsible committees. For example, see ANSI (1999) for the most recent agreement among ANSI N13, ANSI N42, and ANSI N43. For purposes of the agreements, the term "instrument" means any device used to detect or measure any characteristic of ionizing radiation. Note that although HPS has been accredited by ANSI to serve as the sponsor and secretariat for ANSI Committee N43 (Equipment for Non-Medical Radiation Applications), ANSI N43 is not responsible for standards for monitoring airborne radioactivity. However, ANSI N43 coordinates with ANSI N13 and ANSI N42 regarding its scope of activities. As delineated in the 1999 memorandum of understanding, ANSI N43 develops and maintains standards pertaining to radiation protection aspects of radiation-producing equipment used in industrial and nonmedical research and development activities (excluding nuclear reactors).

The IEEE (www.ieee.org) was formed in the late 1950s by a merger of the Institute of Electrical Engineers and the Institute of Radio Engineers. It is an international organization with most of its membership in the United States and a continually growing number of members from outside the United States. The IEEE is accredited by ANSI to serve as the sponsor and Secretariat of ANSI Committee N42 (Radiation Instrumentation), which is responsible for developing American National Standards in nuclear and health physics instrumentation including radiological safety instrumentation such as dosimeters, portable survey meters, and contamination monitors. As delineated in the 1999 memorandum of understanding, the standards concerning radiological protection instruments pertain to design and construction, design performance criteria, performance testing against design criteria, calibration, and field-response testing of instrumentation. Although the IEEE is an international organization, its standards in this area are generated to a great extent nationally and are processed as American National Standards through ANSI. Table 3.3 contains an annotated list of relevant N42 standards for monitoring and measuring airborne radioactivity, including the scope for each standard as listed on the ANSI Web site.

The HPS (www.hps.org) was founded in 1956 and is a society of occupational and environmental radiation safety professionals. ANSI has accredited HPS to serve as the sponsor and secretariat for ANSI Committee N13 (Radiation Protection). ANSI N13 is responsible for the development of standards concerned with radiation safety and health physics activities, such as air sampling, whole body counting, external and internal dosimetry, and bioassay. As delineated in the 1999 memorandum of understanding, ANSI N13 develops and maintains radiological protection standards pertaining to the selection, use, interpretation, application, and accreditation of radiological protection instruments. Because the radiological safety instrumentation standards developed by the IEEE-sponsored ANSI Committee N42 are of considerable utility to the HPS, close communication is maintained between N42 and the HPS-sponsored ANSI Committee N13. Most members of N42 are health physicists who are also members of the HPS. Table 3.3 contains an annotated list of relevant N13 standards for monitoring and measuring airborne radioactivity, including the scope for each standard as listed on the ANSI Web site.

Standards of Interest from Other ANSI-Accredited Organizations

The American Nuclear Society (ANS) (www.ans.org) is responsible for the development and maintenance of standards that address the design, analysis, and operation of components, systems, and facilities involved in or utilizing nuclear technology. For example, the standard on *Airborne Release Fractions at Non-Reactor Nuclear Facilities* (ANSI/ANS-5.10-1998; R2006) provides criteria for defining Airborne Release Fractions (ARFs) for radioactive materials under accident conditions (excluding nuclear criticalities) at nonreactor nuclear facilities. Criteria are also provided for gases and materials that can be converted into the form of vapor. ANS standards reference the ANSI/IEEE and ANSI/HPS standards for radiological safety instrumentation.

The American Industrial Hygiene Association (AIHA) (www.aiha.org) covers the nonradioisotope aspects of topics of interest to the users of ANSI/IEEE and ANSI/HPS standards. For example, the *American National Standard for Laboratory Ventilation* (ANSI/AIHA Z9.5-1992) provides guidance for controlling, monitoring, and measuring air contamination for laboratories or hoods other than those used for radioisotopes and the *American National Standard Practices for Respiratory Protection* (ANSI/AIHA Z88.2-1992) provides guidance for the selection and use of respirators.

The American Society of Heating, Refrigerating and Air-Conditioning Engineers (ASHRAE) (www.ashrae.org) has a method for testing air cleaning devices used in general ventilation for removing particulate matter (ANSI/ASHRAE 52-1968).

The ASTM International (www.astm.org) (formerly known as the American Society for Testing and Materials) develops standards on the characteristics and performance of materials, products, systems, and services. Table 3.3 contains an annotated list of relevant ASTM International standards for monitoring and measuring airborne radioactivity, including the scope for each standard as listed on the ASTM International Web site.

The National Institute of Standards and Technology (www.nist.gov) is a key technical contributor to the U.S. standards infrastructure.

The Standards Engineering Society (www.ses-standards.org) provides links to a number of standards organizations and information on the standards process.

The NSSN (www.nssn.org) (formerly known as the National Standards System Network) is a free, web-based information service administered by ANSI. The NSSN serves as "a national resource for global standards" and provides information on more that 250,000 approved standards from more than 600 national, foreign, regional, and international bodies. More than 400 of the listed standards are associated with the key words "air sampling."

The National Standards Organizations in France

The national standards organizations in France provide another example of a strong national standards effort that contributes significantly to international standards developments in both the procedural and electrotechnical areas. The Association Française de Normalisation (AFNOR) (www.afnor.fr) is the French member of the ISO and CEN and appoints experts to WGs of both. The AFNOR, organized in 1926 and controlled by the Ministry for Industry, manages and coordinates the entire system, which is comprised of some 30,000 experts from all walks of life, including companies, government, professional organizations, and trade unions. Some of these experts work at the European (CEN) and international (ISO) levels. The AFNOR has 31 Bureaux de Normalisation (BN), which prepare the AFNOR standards. One of them is the Bureau de Normalisation des Equipements Nucléaires (BNEN), which was established in 1990 to develop relevant nuclear standards. One of the three commissions in BNEN is M60.1, which is responsible for monitoring the work of ISO/TC85/SC2 on "Nuclear Energy-Health Physics."

France has generic standards in the *Air Quality* category with requirements comparable to those implemented in the United States under the authority of the U.S. Clean Air Act of 1990.

GOVERNMENT REGULATIONS AND GUIDANCE

U.S. LAW—NATIONAL TECHNOLOGY TRANSFER AND ADVANCEMENT ACT (PL 104-113)

In the mid-1990s, it was postulated that as the United States increased its participation in the global trading market, it would be faced with a series of problems resulting from its lack of an agreed-upon infrastructure for standards and conformity assessment in such areas as scientific investigations, engineering, manufacturing, commerce, and industry. In reality, problems began to arise because the systems in the United States for standards and conformity activities were decentralized and often competitive, representing a mixture of public and private participants. An approach comprising pluralistic and uncoordinated systems for various standards-related activities significantly hampered the U.S. trade. In contrast, the EU was very active in building an agreed-upon technical infrastructure among its members, although usually without considering input from outside the EU.

The costs of the various disjointed standards activities in the United States became increasingly burdensome as private industry and government entities contended with multiple, duplicate standards for similar activities. Because this type of system would continue to increase product cost, waste time and staff resources, and could be perceived by our international trading partners as a technical barrier to trade, the issue was raised in Congress. This legislative body provided a forum to discuss and resolve the need for the government and private sector to work together to create and maintain sound technical arrangements for the United States.

Consequently, in 1996, the National Technology Transfer and Advancement Act (NTTAA) [Public Law (P.L.) 104-113] became law. The NTTAA requires Federal agencies to use voluntary consensus standards to the extent practicable, to report development of agency-unique standards, and to participate in the development of voluntary consensus standards. To implement the NTTAA, the Office of Management and Budget (OMB) issued Circular A-119, "Federal Participation in the Development and Use of Voluntary Consensus Standards and in Conformity Assessment Activities." As a result, government agencies, in fulfilling their missions, reference a variety of technical standards in rulemaking, procurement, and in other activities. In the years since passage of NTTAA, NIST has reported significant progress by Federal agencies toward meeting the objectives of the Act (see, e.g., NIST 2008 and http://ts.nist.gov/Standards/Conformity/toolkit.cfm). In particular, the Act has successfully encouraged agencies to first look to voluntary consensus standards to meet their needs rather than to develop government-unique standards. Reported challenges are associated with the ability of Federal agencies to maintain their levels of participation in standards developing organizations due to competing organizational priorities, dwindling budget resources, and staff losses due to retirement and downsizing.

U.S. RADIATION PROTECTION REGULATIONS AND GUIDANCE

A number of U.S. federal agencies with responsibilities for radiation-related matters contribute chapters to the U.S. Code of Federal Regulations (CFR). The code is divided into "titles" that broadly define the subject matter. The regulations of the U.S. Department of Energy (DOE) and the U.S. Nuclear Regulatory Commission (NRC) fall under Title 10, *Energy*; those of the U.S. Environmental Protection Agency (EPA) come under Title 40, *Protection of Environment*; and those of the U.S. Occupational Safety and Health Administration (OSHA) come under Title 29, *Labor*. Jones (2005) provides a comprehensive review of the history of U.S. radiation protection regulations, recommendations, and standards. Important air-monitoring-related information from that review and history are summarized in the following paragraphs.

The U.S. Nuclear Regulatory Commission Regulatory (NRC) (www.nrc.gov) provided updated *Standards for Protection against Radiation* (10 CFR Part 20) in 1991, issued regulatory guidance on *Calibration and Error Limits of Air Sampling Instruments for Total Volume of Air Sampled* (Regulatory Guide 8.25) in 1992, published an associated document on *Air Sampling in the Workplace* (NUREG-1400) in 1993, and published *Consolidated Guidance: 10 CFR Part 20—Standards for Protection Against Radiation* (NUREG-1736) in 2001 (http://www.nrc.gov/reading-rm/doc-collections/nuregs/staff/sr1736/sr1736.pdf). In addition, NRC has issued numerous guidance documents into a NUREG-series of reports. More than 20 volumes of regulatory guidance have been issued in the NUREG-1556 series *Consolidated Guidance about Materials Licensees.* For a complete list of this series, see http://www.nrc.gov/reading-rm/doc-collections/nuregs/staff/sr1556/.

As an example, NUREG-1736 provides the following list of existing regulatory guidance associated with compliance for air sampling: *Air Sampling in the Workplace* (NUREG-1400), *Acceptable Concepts, Models, Equations, and Assumptions for a Bioassay Program* (Regulatory Guide 8.9), *Acceptable Programs for Respiratory Protection* (Regulatory Guide 8.15), *Air Sampling in the Workplace* (Regulatory Guide 8.25), and *Monitoring Criteria and Methods to Calculate Occupational Radiation Doses* (Regulatory Guide 8.34). Related guidance that involves the use of air sampling data to determine or control internal exposure limits include: *Interpretation of Bioassay Measurements* (NUREG/CR 4884), *Information for Establishing Bioassay Measurements and Evaluations of Tritium Exposure* (NUREG-0938), and *Radiation Dose to the Embryo/Fetus* (Regulatory Guide 8.36).

In addition to the licensing of nuclear power and fuel cycle facilities, approximately 22,000 licenses are issued for medical, academic, industrial and general uses of nuclear materials (see *U.S. Nuclear Regulatory Commission 2008–2009 Information Digest*, NUREG-1350, Vol. 20; at http://www.nrc.gov/reading-rm/doc-collections/nuregs/staff/sr1350/). Approximately 80% of these licenses are administered by the 35 States that participate in the NRC Agreement States Program. An NRC Agreement State is one that has signed an agreement with the NRC that authorizes the State to regulate radioactive materials within its borders. State or City authorities in the United States may have identical or more stringent requirements than those of the NRC. The reader is advised to investigate these requirements before beginning a new compliance-oriented air sampling program.

The U.S. Department of Energy (DOE) (www.doe.gov) published an *Operational Health Physics Training Manual* (ANL-88-26) in 1988; promulgated its regulation for *Nuclear Safety Management* (10 CFR 830) in 1991; promulgated *Standards for Protection against Radiation* (10 CFR 835) in 1993 and amended 10 CFR 835 in 2007 (Federal Register, Vol. 72, 31,904, 2007) using newer scientific information and recommendations of International Commission on Radiological Protection (ICRP Publication 60); summarized the characteristics of aerosols from a wide range of activities such as powder handling, spills, and fires in a 1994 handbook on *Airborne Release Fractions/Rates and Respirable Fractions for Nonreactor Nuclear Facilities* (DOE-HDBK-3010-94, Vols. 1 and 2); provided an updated *Implementation Guide for Air Monitoring for use with Title 10 Code of Federal Regulations Part 835 Occupational Radiation Protection* (DOE G 441.1-8) in 1999; and more recently issued *Radiation Protection Programs Guide for Use with Title 10 Code of Federal Regulations Part 835 Occupational Radiation Protection* (see Section 10: Air Monitoring) to reflect the recent amendments in 10 CFR 835 (DOE G 441.1-1C).

The U.S. Environmental Protection Agency (EPA) (www.epa.gov) issued *National Emission Standards for Hazardous Air Pollutants (NESHAPs)* (40 CFR 61) in 1991 to implement provisions of the Clean Air Act of 1990. EPA has also issued a number of Federal Guidance Reports that are relevant to airborne radioactivity. *Limiting Values of Radionuclide Intake and Air Concentration and Dose Conversion Factors for Inhalation, Submersion, and Ingestion* (Federal Guidance Report No. 11, EPA-520/1-88-020) was issued in 1988; *External Exposure to Radionuclides in Air, Water, and Soil* (Federal Guidance Report No. 12, EPA-402-R-93-081) was issued in 1993; and *Health Risks from Low-Level Environmental Exposure to Radionuclides* (Federal Guidance Report No. 13,

EPA-402-R-99-001) was issued in 1999. EPA further recommended in 2006 that CAP-88 PC Version 3 be used to demonstrate dose compliance under NESHAPs for the offsite maximally exposed individual. Version 3 implements dose conversion factors of Federal Guidance Report No. 13 using ICRP-60 weighting factors and the latest models (Federal Register, Vol. 71, 8854, 2006).

The *Occupational Safety and Health Administration (OSHA)* (www.osha.gov) addresses ionizing radiation in the areas of general industry, shipyard employment, and construction. Within the *OSHA General Industry Standards* (29 CFR 1910), the subtitle on *Personal Protective Equipment* includes requirements for *Respiratory Protection* (29 CFR 1910.134), and specific issues for ionizing radiation are covered in the portions on *Hazardous Waste Operations and Emergency Response* (29 CFR 1910.120); *Occupational Safety and Health Standard for Ionizing Radiation* (29 CFR 1910.1096); *Hazardous Materials* (29 CFR 1910 Subpart H); and *Toxic and Hazardous Substances* (29 CFR 1910 Subpart Z). Within the *Occupational Safety and Health Standards for Shipyard Employment* (29 CFR 1915), specific issues for ionizing radiation are covered in the portions on *Welding, Cutting, and Heating* (29 CFR 1915 Supart D) and *Uses of Fissionable Material in Ship Repairing and Shipbuilding* (29 CFR 1915.57). Within the *Safety and Health Regulations for Construction* (29 CFR 1926), specific issues for ionizing radiation are covered in the portions on *Occupational Health and Environmental Controls* (29 CFR 1926 Subpart D); *Ionizing Radiation (Construction)* (29 CFR 1926.53); and *Hazardous Waste Operations and Emergency Response* (29 CFR 1926.65).

There is an October 30, 1978, OSHA directive that delineates *OSHA Coverage of Ionizing Radiation Sources Not Covered by the Atomic Energy Act of 1954* (STD 01-04-001 [STD 1-4.1]). In addition, there is a December 22, 1989, *Memorandum of Understanding between the OSHA and the U.S. Nuclear Regulatory Commission* (CPL 02-00-086 [CPL 2.86], December 22, 1989) delineating the authorities, responsibilities, and other activities between OSHA and NRC for occupational health and safety at radiation sites. Section 18 of the *Occupational Safety and Health Act of 1970* (29 USC 667) encourages States to develop and operate their own job safety and health programs. OSHA approves and monitors State plans. Twenty-four states, Puerto Rico and the Virgin Islands, currently have OSHA-approved State plans and have adopted their own standards and enforcement policies. For most parts, these States adopt standards that are identical to Federal OSHA. However, some States have adopted different standards applicable to this topic or may have different enforcement policies. Additional information about OSHA and its standards in the area of ionizing radiation can be found at http://www.osha.gov/SLTC/radiationionizing/standards.html.

OTHER RECOMMENDATIONS AND GUIDANCE

The International Commission on Radiological Protection (ICRP) (www.icrp.org) provides recommendations that are internationally accepted as a coherent and consistent approach to radiation protection. Of particular interest from a historical perspective are *Principles of Environmental Monitoring related to the Handling of Radioactive Materials* (ICRP Publication 7) issued in 1965; *General Principles of Monitoring for Radiation Protection of Workers* (ICRP Publications 12 and 35) issued in 1968 and 1982; *Implications of Commission Recommendations that Doses be kept as Low As Reasonably Achievable* (ICRP Publication 22) issued in 1973; *Reference Man: Anatomical, Physiological and Metabolic Characteristics* (ICRP Publication 23) issued in 1975; *Recommendations of the International Commission on Radiological Protection* (ICRP Publication 26) issued in 1977; *Limits on Intakes of Radionuclides by Workers* (ICRP Publication 30 and addenda) issued beginning in 1979; *Principles of Monitoring for the Radiation Protection of the Population* (ICRP Publication 43) issued in 1985; several reports issued in 1994 on *1990 Recommendations of the International Commission on Radiation Protection* (Publication 60), *Human Respiratory Tract Model for Radiological Protection* (ICRP Publication 66), *Dose Coefficients for Intakes of Radionuclides by Workers* (ICRP Publication 68); and *Age-Dependent Doses to Members of the Public from Intake of Radionuclides, Part 5. Compilation of Ingestion and Inhalation Dose*

Coefficients (ICRP Publication 72) issued in 1996. In 2007, the ICRP issued an updated version of *The Recommendations of the International Commission on Radiological Protection* (ICRP Publication 103).

The recommendations of the ICRP have continually been informed and improved by the results of experiments, investigations, and research that have been published in the open literature. For example, Dorrian and Bailey (1995) summarized the particle size distribution of radioactive aerosols measured in a wide range of industrial operations. The typical particle size distribution had an activity median aerodynamic diameter (AMAD) of 5 µm with a geometric standard deviation of 2.5, although smaller size distributions were observed in operations involving high temperatures or fumes, and larger particle size distributions were observed in operations such as coarse powder handling. The typical size distribution values reported by Dorrian and Bailey are the accepted default values for use in ICRP Publication 66 on the new *Human Respiratory Tract Model for Radiological Protection*. The previous default assumption in ICRP Publication 30 for aerosol particle size in the workplace had been an AMAD of 1 µm with a geometric standard deviation of 2.5, which remains the default value for the particle size of radioactive aerosols in the environment.

The National Council on Radiation Protection and Measurements (NCRP) (www.ncrp.com) has produced several documents of importance to the air monitoring community. In 1978 NCRP produced a report on *Instrumentation and Monitoring Methods for Radiation Protection* (NCRP Report 57) and a *Handbook of Radiation Protection Measurements Procedures* (NCRP Report 58). In 1997, NCRP focused on fundamental considerations of human respiratory tract structure and function in deriving an alternate mathematical model to describe *Deposition, Retention and Dosimetry of Inhaled Radioactive Substances* (NCRP Report 125). In 2001, NCRP issued *Liver Cancer from Internally-Deposited Radionuclides* (NCRP Report 135) and *Management of Terrorist Events Involving Radioactive Material* (NCRP Report 138).

The International Commission on Radiation Units and Measurements (ICRU) (www.icru.org) issued a document on *Radiation Protection Instrumentation and Its Application* (ICRU Report 20) in 1971.

The International Atomic Energy Agency (IAEA) (www.iaea.org) produced a 1978 report on *Particle Size Analysis in Estimating the Significance of Airborne Contamination* (IAEA Technical Report Series No. 179) and a 1999 *Assessment of Occupational Exposure Due to Intakes of Radionuclides Safety Guide* (Safety Standards Series No. RS-G-1.2).

The American Conference of Governmental Industrial Hygienists (ACGIH) (www.acgih.org) periodically updates its book series on *Air Sampling Instruments for Evaluation of Atmospheric Contaminants*, which includes information on sampling radioactive aerosols (see, e.g., Cohen, 2001).

Other sources of information in the published literature are also available. Measurement of radioactive aerosols involves most of the standard tools of aerosol science and technology, as well as a number of specialized techniques that take advantage of the unique physical properties of radioactive materials. In addition to the guidance and standards noted above, instruments and techniques for characterizing radioactive aerosols have been described extensively in numerous books and reports (see, e.g., Price, 1965; Raabe, 1972; Dennis, 1976; Chamberlain, 1991; Ness, 1991; Turner, 1996; Schleien et al., 1998; Hoover and Newton, 2001; Schery, 2001; Ruzer, 2005; Ahmed, 2007; Papastefanou, 2007; Cember and Johnson, 2009). Voss (2007) has assembled a pocket-size handbook of useful radiation monitoring field information from a wide range of sources. Published software and computational programs, such as Anand et al. (1996) and Riehl et al. (1996), are available for calculating aerosol losses in transport lines. There are also a number of comprehensive monographs on the general properties, behavior, and measurement of airborne particles of all types (see, e.g., Hinds, 1999; Baron and Willeke, 2001; Ruzer and Harley, 2005; Vincent, 2007). Detailed review and application of this information, as well as the development of new techniques and applications, continue to occupy the careers of many aerosol scientists and health protection professionals.

PERSPECTIVES ON THE NATURE OF REGULATIONS

The regulatory situation is neither globally consistent nor static. Regulations vary internationally and during transition periods they may even vary between and among federal agencies. In addition, the manner in which regulations are promulgated and policed can be complicated. Using the United States as an example, no fewer than four federal agencies may be involved. These include the DOE, EPA, NRC, and OSHA. As noted above, each agency has regulations that affect human radiological dosimetry and different regulations apply in different situations. For example, in CFR Title 10, Chapter III, Parts 830 and 835 are relevant to DOE employees while Chapter I, Parts 1–171 (more specifically Parts 19, 20, 30–36, and 50) affect those working under the purview of the NRC. DOE and NRC issue regulations, and along with the EPA, also issue guidance documents and directives that outline methods that may be used to calculate and determine occupational and public dose. The EPA's Federal Guidance Recommendations (FGR) 11, 12, and 13 (EPA, 1988, 1993, 1999), numerous Nuclear Regulatory Guides, and NUREG-series documents relevant to dose (and air sampling) are examples. The primary sources of recommendations for occupational and public exposure limits for the regulations and guidance come from the NCRP and the ICRP.

For many years, the ICRP has summarized radionuclide dosimetric-related characteristics and produced metabolic models of ingestion and inhalation to be used with these data. As dosimetric knowledge is refined, regulations are changed over time to accommodate the new information. Those changes involve time. For example, a lag of more than a decade elapsed between the release of ICRP Publications 26 and 30 and incorporation in 1991 of those updated recommendations into 10 CFR 20. A further lag occurred until the recommendations were incorporated in 10 CFR 835 in 1993. ICRP Publications 26 and 30 were then used as the basis for many years for both NRC and DOE regulations.

In the 1994 ICRP Publication 66, an improved respiratory model was described that replaced the model described in ICRP Publication 30. In 1999, the EPA adopted the more recent data in ICRP Publications 60, 66, and 72 for their FGR Number 13 that tabulates cancer risk probabilities for intakes of many nuclides. The DOE updated Part 835 in 2007 to accommodate the new ICRP recommendations and implemented a three-year transition period for compliance at its facilities. The NRC has allowed it licensees, upon written request, to use the newer ICRP recommendations and guidance for calculating internal dose (e.g., use of ICRP Publication 66 and beyond).

In keeping with the continuous refinement of our knowledge, the ICRP has issued its latest recommendations in ICRP Publication 103, which formally replaces the recommendations in ICRP Publication 60. Those changes (although less extensive than in previous ICRP updates) may be expected to eventually be incorporated into regulatory updates, or into regulations, as they are reviewed by the various U.S. federal agencies.

These brief examples illustrate that, at times, specific regulations that need to be met are not immediately obvious. The reader must research the regulatory state of affairs at the moment, find the regulations to be met and then, through careful measurement and analysis, determine if compliance has been achieved. If it has not, appropriate reports to the relevant regulating body must be made that include corrective actions to be implemented to ultimately achieve compliance.

CONCLUSION

The depth and breadth of available standards in the area of airborne radioactivity sampling is substantial. The development and maintenance of credible technical standards is a dynamic process and the content, status, and applicability of the dozens of standards described above comprise a snapshot in time. Over time, new standards are developed or existing standards are revised to accommodate new industries, technologies, or expectations. This ongoing process presents opportunities for practitioners to engage in the standards development process to help make those advances.

Both manufacturers and users of instrumentation for measuring airborne radioactivity and manufacturers and users of all types of instrumentation work together and in concert with regulatory bodies and corporate entities to identify, interpret, and comply with the appropriate standards. The challenge to practitioners is to conduct appropriate due diligence and use caution to ensure that all reliance on standards is appropriate for each application.

REFERENCES

Ahmed, S.N., *Physics and Engineering of Radiation Detection*, Academic Press, New York, 2007.

Anand, N.K., McFarland, A.R., Dileep, V.R., and Riehl, J.D., *DEPOSITION: Software to Calculate Particle Penetration through Aerosol Transport Lines*, NUREG/GR-0006, U.S. Nuclear Regulatory Commission, Washington, DC, 1996.

ANSI, Memorandum of Understanding between ANSI-Accredited Committees N13, N42, and N43 Concerning Radiological Protection Instrument and Instrumentation Standards, American National Standards Institute, New York, October 21, 1999.

Baron, P.A. and Willeke, K., Eds., *Aerosol Measurement: Principles, Techniques, and Applications*, 2nd ed., John Wiley & Sons, New York, 2001.

Cember, H. and Johnson T.E., *Introduction to Health Physics*, 4th ed., McGraw-Hill, New York, 2009.

Chamberlain, A.C., *Radioactive Aerosols*, Cambridge University Press, Cambridge, 1991.

Cohen, B.S., Sampling airborne radioactivity, in: *Air Sampling Instruments for Evaluation of Atmospheric Contaminants*, 9th ed., Cohen, B.S. and McCammon, C.S., Eds., American Conference of Governmental Industrial Hygienists, Cincinnati, 2001.

Cox, M., Hoover, M.D., Grivaud, L., Johnson, M.L., and Newton, G.J., *An International Review of Currently Applicable Standards for Measuring Airborne Radioactivity*, Full Report, published online by the International Electrotechnical Commission, http://www.iec.ch/support/tcnews/2003/AirMonitoringStdsReview.htm, 2003a.

Cox, M., Hoover, M.D., Grivaud, L., Johnson, M.L., and Newton, G.J., Standards for measuring airborne radioactivity, *Health Phys* 85:236–241, 2003b.

Dennis, R., Ed., *Handbook on Aerosols*, Report TID-26608, U.S. Department of Energy, Washington, DC, 1976.

Dorrian, M.D. and Bailey, M.R., Particle size distributions of radioactive aerosols measured in workplaces, *Radiat Prot Dosim* 60:119–133, 1995.

Hinds, W.C., *Aerosol Technology—Properties, Behavior, and Measurement of Airborne Particles*, 2nd ed., John Wiley & Sons, New York, 1999.

Hoover, M.D. and Newton, G.J., Radioactive aerosols, in: *Aerosol Measurement: Principles, Techniques, and Applications*, 2nd ed., Baron, P.A. and Willeke, K., Eds., John Wiley & Sons, New York, 2001.

Jones, C.G., A review of the history of U.S. Radiation Protection Regulations, Recommendations, and Standards, *Health Phys* 88(2):105–124, 2005.

Ness, S.A., *Air Monitoring for Toxic Exposures—An Integrated Approach*, Van Nostrand Reinhold, New York, 1991.

NIST, *Eleventh Annual Report on Federal Agency Use of Voluntary Consensus Standards and Conformity Assessment*, National Institute of Standards and Technology, Gaithersburg, MD, June 2008; accessed October 10, 2009 at http://standards.gov/NTTAA/resources/nttaa_ar_2007.pdf

Papastefanou, C., *Radioactive Aerosols*, Elsevier, Amsterdam, 2007.

Price, W.S., *Nuclear Radiation Detection*, McGraw-Hill, New York, 1965.

Raabe, O.G., Instruments and methods for characterizing radioactive aerosols, *IEEE Trans Nucl Sci* NS-19(1):64–75, 1972.

Riehl, J.R., Dileep, V.R., Anand, N.K., and McFarland, A.R., *DEPOSITION 4.0: An Illustrated User's Guide*. Texas A&M University Department of Mechanical Engineering Aerosol Technology Laboratory Report 8838/7/96, College Station, TX, 1996.

Ruzer, L.S., Radioactive aerosols, in: *Aerosols Handbook—Measurement, Dosimetry, and Health Effects*, Ruzer, L.S. and Harley, N.H., Eds., CRC Press, Boca Raton, FL, 2005.

Ruzer, L.S. and Harley, N.H., Eds., *Aerosols Handbook—Measurement, Dosimetry, and Health Effects*, CRC Press, Boca Raton, FL, 2005.

Schery, S.D., *Understanding Radioactive Aerosols and Their Measurement*, Kluwer Academic Publishers, Boston, 2001.

Schleien, B.S., Slabeck, L.A. Jr., and Kent, B.K., Eds., *Handbook of Health Physics and Radiological Health*, 3rd ed., Williams & Wilkins, Baltimore, MD, 1998.
Turner, J.E., *Atoms, Radiation, and Radiation Protection*, 2nd ed., John Wiley & Sons, New York, 1996.
Vincent, J.H., *Aerosol Sampling—Science, Standards, Instrumentation and Applications*, John Wiley & Sons, New York, 2007.
Voss, J.T., *Handbook of Radiation Data*, 8th ed., A Risky Business LLC, Los Alamos, NM, 2007.

4 A Life-Cycle Approach to Development and Application of Air Sampling Methods and Instrumentation

Mark D. Hoover and Morgan Cox

CONTENTS

Introduction 43
Life-Cycle Steps 44
    Mission Evaluation 44
    Research and Development 45
    Prototype Testing 48
    Type Testing 48
    Production Control Testing 49
    Training 49
    Acceptance Testing 49
    Initial Calibration 49
    Functional Checks 50
    Operational Experience 50
    Maintenance and Recalibration 50
    Periodic Performance Testing 51
Benefits of Harmonization 51
    Harmonization across the Life-Cycle Process 51
    Harmonization among Instrument Types or Classes 51
    Harmonization of Life-Cycle Processes in a Broader Context 52
Conclusion 52
References 52

INTRODUCTION

It is useful to view the development and application of air sampling methods and instrumentation as a "life cycle" of logical and interrelated steps (Hoover and Cox, 2004). The cycle begins with evaluation of a mission or performance requirement (real or emerging); proceeds through research and development, prototype testing, type testing, production control testing, training, and acceptance of a method and the associated instrumentation to accomplish the mission; continues with initial calibration, functional checks, and accumulation and review of operational experience to conduct the mission in a scientifically defensible manner; proceeds through maintenance and recalibration and through periodic performance testing to ensure that the method is still working; and eventually ends with either the ultimate completion of the mission or with the replacement of the method by more efficient or more effective methods. Documentation and continuous improvement are essential at each step. This concept of a "life-cycle approach" provides a cradle-to-grave foundation for the development and application of methods and their associated instrumentation to accomplish any mission, including the meaningful sampling of airborne radioactivity.

The life-cycle approach presented in this chapter is derived from participation in and review of historical experience with health protection and industrial hygiene instrumentation for air monitoring, personal dosimetry, and detection of biological, chemical, and radioactive materials. This includes experience in development of consensus standards for radiation detection instruments intended for homeland security and a range of other applications. In all cases, developing documentation and national and international consensus standards adequate to guide, record, and demonstrate the scientific defensibility of methods and instrument use throughout the life-cycle process is considered essential to success.

Because it is likely that the reader will frequently employ off-the-shelf methods and instruments for many or perhaps all of his or her air sampling requirements, it is important to understand the level of maturity that air sampling systems may have in the cradle-to-grave regime of care that is required to attain accurate and meaningful data. Not all systems on the market have been fully validated and care must be taken to ensure that any system that is selected has the full complement of capabilities needed to accomplish the intended mission; "a chain is only as strong as its weakest link." It is equally important to ensure that proper attention is paid to complete implementation and documentation of each step in the life cycle. Chapter 3 of this book has described particular standards and test methods applicable to measuring airborne radioactivity. A number of those standards such as ANSI N42.33 (2003) and ANSI N323C (2009) include informative sections on the life-cycle approach described here.

As detailed below, the life-cycle approach provides a template for improved understanding and collaboration between all stakeholders involved in the process. This approach provides for the development of air sampling methods or instruments and their subsequent testing, training, calibration and maintenance, and periodic performance testing so that the user can be assured that the equipment provides accurate data over its lifetime. Stakeholders include risk assessment and risk management professionals, research and development scientists and engineers, instrument manufacturers and vendors, procurement specialists and their technical representatives in user organizations, instrument calibration and maintenance specialists, instrument users, and the workers and members of the public requiring health and safety protection. The approach is beneficial and relevant to new method and instrument concepts, as well as to refinement or modification and improvement of methods and commercially available instruments for specific purposes.

LIFE-CYCLE STEPS

General considerations of method and instrument life-cycle steps from conception to retirement are shown schematically in Figure 4.1, summarized in Table 4.1, and described in the following sections. Each step is described individually, but grouping and consideration of the steps as any logical sequence of "phases" such as development and testing, purchase and use, and so on can easily be done depending on the responsibilities of the developers, testers, purchasers, users, quality assurance managers, regulators, or others who may apply the steps. Although the approach is applicable to both methods and their associated instrumentation, for the sake of clarity the following discussion of life-cycle development focuses primarily on the development of instrumentation.

MISSION EVALUATION

The mission evaluation step identifies mission objectives and conditions, measurement objectives, and candidate technologies. It serves to clarify the need for the instrument and to define the ideal design endpoints. As noted in Chapter 1 of this book, understanding the objectives of air sampling and how different objectives relate to and dictate the critical details of specific sampling activities

A Life-Cycle Approach to Methods and Instrumentation

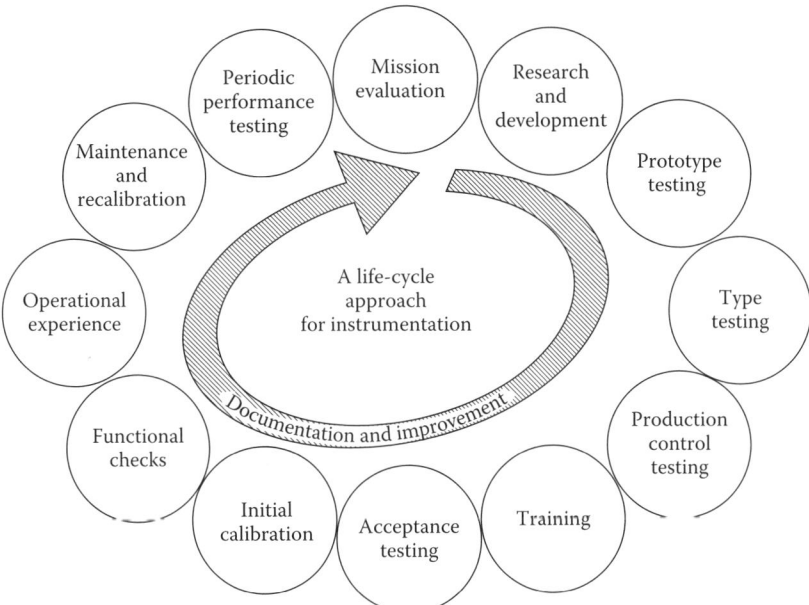

FIGURE 4.1 A life-cycle approach to development and application of air sampling methods and instrumentation. Understanding and following the comprehensive series of steps in this approach provides a cradle-to-grave foundation to ensure that airborne materials are properly and successfully assessed and that measurement results are meaningful. (Adapted from Hoover, M.D. and Cox, M., *Public Protection from Nuclear, Chemical, and Biological Terrorism*, Brodsky, A., Johnson, R.H. Jr., and Goans, R.E., Eds., Medical Physics Publishing, Madison, WI, 2004.)

is essential to the design and implementation of any successful sampling program. Mission evaluation includes identifying constraints on when, where, for what measurements, under what conditions, over what frequency and duration, and by whom the instrument is to be used. The evaluation and thorough understanding of mission objectives and the intended conditions of instrument use is the critical initial step of the instrument life-cycle process. In short, "What is the mission?" "What are the constraints and relevant regulations?" and "What candidate technologies might be used?"

A traditional feedback loop involving documentation and continuous improvement can be applied, in which technical advances and continued field use of early generation equipment illustrate and define the need for any refinements or design changes. These changing solution options, such as electronic advances and lessons learned from field use, can alter the vision of how the mission can be accomplished. For example, personal sampling pumps began as rather massive devices of limited flow rate and limited battery life. Newer generation pumps are both smaller in size and weight and provide greater capacities for flow rate and battery life. These improvements have enabled more extensive use of personal air sampling. In a similar fashion, advances and miniaturization of multichannel analyzer technologies have enabled isotope identification and alarming to be incorporated into personal air sampling methods. The new generations of real-time instrumentation can also be used with radiofrequency telemetry to provide remote continuous monitoring of personal dose and dose rate, including geospatial positioning.

RESEARCH AND DEVELOPMENT

Research and development is performed by the manufacturer or researcher to determine the likelihood that a specific instrument design or concept will meet the intended specifications. This includes

TABLE 4.1
Summary of a Life-Cycle Approach to Development and Application of Air Sampling Methods and Instrumentation

Life-Cycle Step	Purpose	Timing or Frequency	Units to be Tested	Specifications	Responsibility
Mission evaluation	To identify mission objectives and conditions, measurement objectives, performance requirements, and candidate technologies	Prior to initiation of the development process and continuously throughout the life cycle	None (this step is conceptual in nature)	Definition of specifications to be evaluated in subsequent life-cycle steps	Responsible officials, with input from research, development, manufacturing, and users
Research and development	To aid in development of a prototype likely to meet specifications	As needed	Individual components and assemblies	As selected by the manufacturer; or requested by the purchaser/user	Manufacturer
Prototype testing	To demonstrate that design of the instrument is likely to meet specifications	As needed prior to start of production	One or more prototype units	As selected by the manufacturer; or requested by the purchaser/user	Manufacturer (generally); purchaser/user (occasionally)
Type testing	To demonstrate that design of the instrument as manufactured meets specifications	A minimum of once prior to full production	Two or more initial production units	All specifications from the relevant standard, or as agreed upon between the manufacturer and the purchaser/user	Manufacturer (generally); purchaser/user (occasionally)
Production control testing	To control production, avoid defects, and confirm instrument compliance with specifications	Depending on acceptable failure rate agreed upon between manufacturer and purchaser/user	As determined by manufacturer or as agreed upon between the manufacturer and the purchaser/user	As selected by manufacturer, or requested by purchaser/user	Manufacturer
Training	To ensure proper execution of all tests and requirements	As appropriate to support proper execution of each step of the instrument life cycle	As needed to train individuals to properly carry out life-cycle steps	As appropriate for instrument and its use	All life-cycle participants

Acceptance testing	To demonstrate compliance with selected specifications	After the units are received and prior to their initial use	As agreed upon between manufacturer and purchaser/user	As selected by the purchaser/user	Purchaser/user
Initial calibration	To establish a traceable calibration relevant to expected conditions of use	Prior to initial use	Each unit	Selected instrument parameters and responses	Designated calibration staff of the users' organization (or selected vendor)
Functional checks	To provide indications that the instrument is operational	Before each use and periodically during use	Each unit	As appropriate for the instrument being used	User
Operational experience	To provide evidence of any operational deficiencies	During each operation and during periodic reviews of experience	A representative number of units	Selected instrument parameters and responses	User
Maintenance and recalibration	To provide preventive maintenance, make necessary repairs, and reestablish a traceable calibration	At a frequency, such as annually, based on design and reliability history of instrument	Each unit	As appropriate for the instrument	Designated maintenance staff of the user's organization (or selected vendor)
Periodic performance testing	To verify that instrument continues to meet specifications	As appropriate based on experience and anticipated modes of failure	A representative number of units	Selected specifications from the type test	As arranged by the purchaser

Source: Adapted from Hoover, M.D. and Cox, M., *Public Protection from Nuclear, Chemical, and Biological Terrorism*, Brodsky, A., Johnson, R.H. Jr., and Goans, R.E., Eds., Medical Physics Publishing, Madison, WI, 2004.

balancing trade-offs of user requirements for limits of detection and response time against instrument size, weight, reliability, ease of use, cost, and training requirements. Development tests are normally conducted under "breadboard" conditions, in which subsystem performance is evaluated without a complete instrument configuration. Understanding the mission objectives, performance requirements, and expected conditions of use allows the manufacturer to develop an instrument that is likely to meet mission objectives and performance requirements.

Selection of the polarity for the front face of a solid-state detector in an α-radiation continuous air monitor is an example of the value of having an early understanding of the expected conditions of use. Detectors with positive front surface and a grounded back surface are susceptible to radiofrequency interference; the configuration essentially acts as an antenna. A postdevelopment retrofit might require installation of a Faraday cage around the detector. Such a retrofit might interfere with air sampling efficiency or ease of use. An insightful development precaution would be to design the detector with a grounded front face, so that concerns for radiofrequency interference are "engineered out" from inception.

PROTOTYPE TESTING

Prototype testing demonstrates that the design of an instrument is likely to meet certain specifications, including ease of use. Although prototype testing is not formally required by national or international standards, it is industrial practice that decreases the possibility that production units will fail to meet requirements of applicable standards. This testing phase also provides opportunities to ensure functionality under any special operational controls that may be required for specific situations and field applications (e.g., provision of highly readable displays, audible alarms, and easy-to-operate dials for use with personal protective equipment that may restrict vision, hearing, or touch).

TYPE TESTING

Type testing is the first *formal* requirement of national and international standards (see Chapter 3 for additional details). To fully define the performance and limitations of an instrument, such standards require that the instrument be type tested in accordance with requirements detailed in the applicable standards. Type testing is typically performed on two or more production models of each instrument to fully characterize the performance of the instrument. The essential elements of instrument performance-type testing (in the most desirable sequence) are radiological (or chemical or biological), electrical/electronic, mechanical, and environmental. Instruments that fail at any point in the test sequence need not be additionally tested. The test sequence should be reinitiated if design modifications are made to correct the performance deficiency. In other words, the design of an instrument failing the radiological tests should be modified for improvement and retesting before electrical/electronic tests are initiated. The range of required tests addresses concerns that the accuracy of measurements during field use can be affected by variation in source or agent characteristics, instrument orientation with respect to the source, temperature, humidity, and instrument stability. Other tests, such as vibration and drop tests, address the ruggedness of the instrument. The instrument performance, as defined by results of the type test and the standard-specified interpretation of the test data, will document for the user how accurately the instrument can be expected to detect and quantify the radiological (or chemical or biological) agent of concern under a range of operational conditions.

For a number of instrument types, including those for noble gases, tritium, and particulate transuranium radionuclides, the type testing step includes establishing the initial calibration. This typically involves a single calibration with the radionuclide of concern, either as gas, vapor, or in particulate form. Concurrently with this initial calibration, solid photon sources and electroplated sources are also used so that they can serve in the future as secondary calibration standards.

Production Control Testing

Production control testing has not generally been a formal step in the standards process for individual classes of radiological (or chemical or biological) instrumentation. However, production quality management and quality assurance do fall under the manufacturer requirements delineated by the International Organization for Standardization (ISO) under the ISO 9000 standards (ISO, 2000). Compliance with ISO 9000 includes a Plan-Do-Check-Act (PDCA) approach (ISO, 2004) that is similar to the life-cycle approach described in this chapter. General attention to having and following a set of business procedures may not necessarily ensure that all critical factors for identifying and correcting instrument errors or defects are addressed. Production control testing represents good practice to ensure that instruments meet critical requirements. Higher performance and reliability are especially important for homeland security applications. Production control testing is performed by the manufacturer in accordance with documented procedures. In some cases, purchasers make agreements with the manufacturer to ensure that specific requirements are met.

Training

The most notable training issues for practitioners apply to individuals who will operate the instrumentation and interpret and act on its results. For that reason, the training element in the life-cycle model presented here is positioned between instrument manufacture and use. However, training is a cross-cutting activity that must be conducted in a task-appropriate manner during each element of the life cycle. As illustrated in the American National Standard for *Training Requirements for Homeland Security Purposes using Radiation Detection Instrumentation for Interdiction and Prevention* (ANSI N42.37, 2006), training requirements apply to a wide range of individuals, including not only those who will test, maintain, calibrate, and operate the equipment, but also to those who will conduct the training itself. If training needed to conduct any of the steps is inadequate, then the critical contributions of that step are likely to be inadequate.

Acceptance Testing

Acceptance testing should be performed by the purchaser or user on each new instrument before initial use. Acceptance testing should test each instrument against specific characteristics identified as critical or indicative of overall instrument performance. The purpose of acceptance testing is to demonstrate that the specifications of the end user, usually stated in a formal contractual agreement with the manufacturer, are met. An acceptance test generally consists of

- Physical inspection
- General operational tests
- Appropriate response tests

Some aspects of acceptance testing, such as response testing, may be completed under type testing (above) as agreed upon between the user and the manufacturer. An example would be type testing a radioactive gas detection system that need only be done one time with a calibrated radioactive gas.

Initial Calibration

The initial instrument calibration is frequently performed as part of the acceptance test. Initial calibration can also be performed at any time after general acceptance, but before initial use of the instrument. However, some instruments are factory calibrated and do not require additional calibration. For other instruments (as noted in the type testing section above), the "initial" calibration is performed as part of the type test using the radionuclide of concern. When initial calibration has been established during type testing or at the factory, calibration checks or verifications by the purchaser or user may be appropriate. When surrogate solid photon sources and/or electroplated sources have been used simultaneously during initial calibration, they can be used as secondary standard sources in subsequent calibrations. For instruments being held in reserve for future use, the initial calibration may be performed at a later date. Calibration of air sampling equipment is discussed further in Chapter 12 of this book.

FUNCTIONAL CHECKS

A functional check is often qualitative and determines that an instrument is operational and capable of performing its intended function. Functional checks may include battery check, zero setting, or source response check. Many modern instruments include automatic diagnostic and self-checking features. For radiation detection instrumentation, functional checks may include response to natural background radiation. For biological or chemical detection instrumentation, checks may include response to ambient materials, or may require the artificial introduction of test materials. All or a subset of these checks are typically performed at least daily, or prior to each intermittent use, whichever is less frequent. In many instances, functional checks must be documented to validate subsequently acquired data.

OPERATIONAL EXPERIENCE

The intelligent observation and review of operational experience can provide early evidence of instrument performance deficiencies. For example, deviations of instrument results during temperature or humidity excursions may indicate that the instrument seals or sensors are damaged or that the instrument itself does not have adequate stability features. Unexpectedly high readings in the presence of electric-arc welding operations may indicate that an instrument is susceptible to electromagnetic interference. The review and documentation of inconsistencies observed during operational experience is not necessarily a statutory or consensus standard requirement, but it is a recommended practice.

MAINTENANCE AND RECALIBRATION

Maintenance must be performed using components at least equivalent to those specified by the manufacturer. Replacement components must be manufacturer-specified or equivalent. Repairs made using unapproved instructions or components that may affect instrument performance constitute an instrument modification, and must render invalid any type tests made on the instrument model as applied to the specific instrument. Modified instruments must have their performance tested and documented prior to issuance for field use. If the user can document that the modification does not affect the instrument performance, additional testing is not required. For example, modifying the size or shape of a control knob to enable its use with protective gloves might be a valid modification that would not require additional testing. However, if modifications deal with the instrumental operating principle, then additional testing would be required.

Recalibration of an instrument or system may be performed by the user or by a contracted and qualified facility at a frequency either required by applicable regulation, suggested by recommendations for good practice, indicated as necessary and sufficient by operating experience, or as allowed by a combination of the instrument design and the conditions of use. The frequency of maintenance and recalibration is typically set at one year, but if performance history indicates that the instrument is sufficiently reliable (e.g., recalibration of as-found data is within tolerance >95% of the time), less frequent recalibrations may be performed, provided the recalibration frequency does not exceed three years. An example motivation for less frequent recalibrations is if the monitor is not routinely accessible (i.e., during periods of reactor operation).

PERIODIC PERFORMANCE TESTING

Periodic performance testing is needed to determine whether instruments continue to provide adequate performance under existing or altered conditions of use. Aging and degradation of critical instrument components are also concerns. Based on experience and anticipated modes of failure, the purchaser should test or arrange to test a representative number of units against selected specifications from the type test to verify that the instrument continues to meet relevant specifications. Individual units, models, or families of instruments should be modified or removed from service when they no longer meet operational needs. The life-cycle step of periodic performance testing should also be viewed as an opportune time to reevaluate mission needs and instrument options. It should also be viewed as an opportunity to adopt or develop improved tests or test-agent characteristics to better reflect instrument performance in the real world.

BENEFITS OF HARMONIZATION

Harmonization is the name given to international efforts by industry and other groups to develop uniform global standards and methods. Comprehensive adoption of a life-cycle approach to development and application of sampling methods and instrumentation can reduce the costs and improve the efficiency of conceiving, developing, verifying, and validating instruments, and conducting training and developing documentation for their use. To achieve the potential benefits, it is important that basic aspects of the life-cycle process be harmonized both between the steps of the cycle and among the various types and classes of instruments, as well as across the broader context of all systems that are created by humans. Note that some countries consider the instrument standards developed by the International Electrotechnical Commission, and perhaps even local standards, as the "law of the land," while other countries are less inclined to follow standards to the letter and are more inclined to adhere to the "spirit" of given standards, as guidance. Differences in these approaches reflect considerations of the ideal world versus the real world.

Harmonization across the Life-Cycle Process

Harmonization and integration of testing and evaluation involves employing common tests throughout the life cycle of the instrument to improve the likelihood that malfunctions will be identified, understood, and corrected. Harmonization and integration can save time and money by using the same test sources, reference materials, and test facilities. A radiological example is use of the same design and specifications for electroplated radioactivity samples at all test stages. This type of attention to the details of reference materials and reference methods is identical to prudent practice for evaluation of biological detection systems in which, for example, the same strains or forms of anthrax or anthrax surrogates would be used throughout the development, testing, and application of an instrument.

Harmonization among Instrument Types or Classes

Different detection methods or configurations of detection methods can give different results. To integrate detection results from many locations, under many conditions, with hand-held, portable, and fixed instrumentation, it is necessary that the life-cycle approach include common tests and calibration approaches to allow comparison and interpretation of all available data. When attempting to compare data from multiple instruments and locations, it is important to be aware of differences in instrument configuration or operating principle. For example, an instrument with a long averaging period may give a different response to temporal variation in radiation concentration than an instrument with a short averaging period. In addition, harmonization includes the recognition that several instrument types are frequently needed to cover different phases of incident response.

This may become apparent when responding to a mixed-radionuclide source undergoing temporal changes in the ratio of α-, β-, and γ-emitting components. Initial monitoring and protection strategies may rely on the measurement of one radiation type (which may dominate the initial radiological risk) as a metric of concern, but later decisions may need information about other radiation types that assume greater relative importance with time. Events involving multiple agents of biological, chemical, or radiological materials may require an evolving suite of monitoring instrumentation.

HARMONIZATION OF LIFE-CYCLE PROCESSES IN A BROADER CONTEXT

A useful reference for understanding life-cycle issues in a broader context is ISO/IEC 15288 (2008). As described in its abstract, this international consensus standard establishes a common framework, including a set of processes and associated terminology for describing the life cycle of a human-created system. The processes for managing a system's life cycle can be applied at any level in the hierarchy of a system's structure throughout the life cycle. This is accomplished through the involvement of all interested parties, with the ultimate goal of achieving customer satisfaction. A broad understanding and application of life-cycle assessment and management processes is likely to enhance the probability of success for life-cycle approaches to development and application of air sampling methods and instrumentation.

CONCLUSION

An integrated life-cycle approach expands the traditional views of standards and testing and will improve the ability to accomplish any type of mission that requires instrumentation and methods. Application of the life-cycle approach can guide the development of new instruments and improve the process for adapting or modifying existing instruments to current problems and new applications.

REFERENCES

ANSI, *American National Standard for Radiation Detection Instrumentation Test and Calibration—Air Monitoring Instruments*, ANSI N323C, American National Standards Institute, New York, 2009.

ANSI, *American National Standard for Portable Radiation Detection Instrumentation for Homeland Security*, ANSI N42.33, American National Standards Institute, New York, 2003.

ANSI, *American National Standard for Training Requirements for Homeland Security Purposes using Radiation Detection Instrumentation for Interdiction and Prevention*, ANSI N42.37, American National Standards Institute, New York, 2006.

Hoover, M.D. and Cox, M., A life-cycle approach for development and use of emergency response and health protection instrumentation, in *Public Protection from Nuclear, Chemical, and Biological Terrorism*, Brodsky, A., Johnson, R.H. Jr., and Goans, R.E., Eds., Medical Physics Publishing, Madison, WI, 2004.

ISO/IEC, *Systems and Software Engineering—System Life Cycle Processes*, ISO/IEC 15288, International Organization for Standardization and International Electrotechnical Commission, Geneva, 2008.

ISO, *Introduction and Support Package Guidance on the Concept and Use of the Process Approach for Management Systems*, ISO/TC 176/SC 2/N544R2(r), International Organization for Standardization, Geneva, 2004.

ISO, *9000 Series Standards*, International Organization for Standardization, Geneva, 2000.

Part II

Fundamentals of Radioactivity and Radioactive Aerosols

5 Review of Radioactivity, Detection, and Measurement

Mark L. Maiello

CONTENTS

Introduction 55
Radioactivity 55
    Radioactive Decay and Radiation 55
Radiation Detection 61
    Review of Radiation Detection Systems 61
Counting Statistics 69
    Significant Figures and Rounding, Negative and Zero Data 69
    Error Terms 69
    Detection Limits 71
    Quality Assurance 73
References 74

INTRODUCTION

This chapter provides a brief review of radioactivity, detection systems, and the statistics and error reporting associated with radioactivity. This knowledge is fundamental to performing and reporting the results obtained from radioactive air sampling.

RADIOACTIVITY

RADIOACTIVE DECAY AND RADIATION

Radioactive decay is the transformation of one atom into another by the emission of radiation. Radiation and the harm it could do are the reasons that airborne measurements are performed. The radiation may be an alpha (α) particle, a beta (β) particle, or a gamma (γ)-ray. In many cases, radioactive decay involves a combination of α and γ or β and γ emissions. These radiations are briefly described below:

- *Alpha particles:* A nucleus of the helium atom (helium is denoted as ^4_2He where subscript 2 indicates the number of positively charged atomic protons and superscript 4 indicates the number of atomic protons plus neutrons). A nucleus means that two protons and two neutrons are present but the two orbiting electrons that ordinarily are present and make the atom neutral overall are missing. Therefore, the α-particle carries a double positive charge (denoted $^4_2\text{He}^{2+}$). These particles, being relatively massive compared to other atomic particles and doubly charged, travel in linear paths through matter that are shortened by collisions and electrostatic interactions. Alpha particles usually possess energies between 4 and 10 million electron volts (MeV). This kinetic energy (energy of motion) is part of the total energy available for the transformation. The balance of that energy is divided among the kinetic energy of the recoiling nucleus and the energy of any γ-ray emitted. When an atom transforms by α-particle decay, it loses four nuclear particles (two protons and two neutrons). The loss of the protons transforms it into a different element. The transformation is symbolized as

$$^A_Z X \rightarrow {}^{A-4}_{Z-2} Y + {}^4_2 \alpha + \gamma,$$

where A indicates the mass number of element X and Z is the atomic number. As indicated, element Y has four less nucleons (protons + neutrons) than element X, the α-particle carries four nucleons, two of which are protons, and there is the possibility that a γ-ray also may be emitted.

- *Beta particles:* High-speed negatively charged electrons (e^-) and high-speed positively charged positrons (e^+). The emission of a negatively charged β-particle (also called a negatron) can be symbolized as

$$^A_Z X \rightarrow {}^A_{Z+1} Y + \beta^- + \nu + \gamma.$$

Here, the negative β-particle is denoted by β^-, the ν symbol indicates the emission of a neutrino, and the γ symbol stands for the γ-ray as before. The neutrino is of little consequence for radioactive measurement. This nuclear exchange involving the β-particle is equivalent to

$$^1_0 n \rightarrow {}^1_1 p + e^-,$$

where a neutron (n) is converted into a proton (p) with the ejection of an electron (the β⁻ particle). The formation of the proton is responsible for the transformation of the atom from one element to another. Elements are distinguished from each other by their unique atomic numbers.

Radioactive decay by emission of a positron (symbolized as either e^+ or β^+) is denoted by

$$^A_Z X \rightarrow {}^A_{Z-1} Y + \beta^+ + \nu + \gamma.$$

Here, the nuclear changes are as follows:

$$^1_1 p \rightarrow {}^1_0 n + \beta^+,$$

where the positron is emitted with the conversion of a proton into a neutron. The change of a positively charged proton into a neutral neutron will cause the release of an orbital electron to maintain the daughter atom at a neutral charge. Therefore, two electron masses are lost: that of the positron and that of the orbital electron. Each mass is equivalent to 0.511 MeV of energy; so 1.02 MeV in total is lost. This also means that the atomic mass of the daughter atom must be at least 1.02 MeV less than the parent, or positron decay will not be possible.

Negative and positive β-particles travel a twisted series of recoils through matter because they interact with the similarly massive orbiting electrons of atoms. Compared to an α-particle of equal energy, this interaction is much weaker primarily due to the high velocity of the β-particle that does not permit much time for atomic interactions to occur. Moreover, because the β-particle has only half the electrostatic charge of the α-particle, the rate of energy lost by the β-particle is less. Positron emission will always be accompanied by the production of two 0.511 MeV photons. This arises from the interaction of the released positron with an orbital electron encountered as the positron slows along its track. The two charges neutralize and the two electrons convert into two photons of equal energy directed in opposite trajectories by the law of conservation of momentum.

Note: Transformation by the process of electron capture (EC) is an alternative to positron emission, and for some radioactive elements, it is the only radioactive transformation energetically possible. Like positron emission, it occurs when nuclear stability is achieved by the transformation of a proton into a neutron. The transformation is realized when an orbital electron is captured and is equivalent to

$$^{1}_{1}p \rightarrow e^{-} + ^{1}_{0}n.$$

The result is

$$^{A}_{Z}X + e^{-} \rightarrow ^{A}_{Z-1}Y + \nu + \gamma.$$

The acquisition of the electron adds 0.511 MeV of energy to the nucleus. Since one of the electron orbitals is left vacant, it will be filled from an outer orbital electron. The subsequent vacancy must also be filled. Each time this occurs (until all orbitals are filled), x-rays are emitted, which is characteristic of the decay product. Another radiation emission originates with the capture of the orbital electron. When it is accelerated out of orbit, it will emit photons that range in energy from zero up to the energy available for the transition. This is called internal bremsstrahlung radiation. Since no charged particle and perhaps no γ-ray are emitted in EC, the characteristic x-rays may be the only means of detecting a nuclide that decays using this mode. Usually, the characteristic x-rays are only a few keV in energy which may make detection difficult.

- *Gamma rays:* Photons (or massless "packets") of electromagnetic energy. Gamma rays arise when a nucleus of an atom, residing at a higher than rest state energy level, returns to its stable state. The γ-ray carries off the excess energy. Radioactive decay often leaves the transformed atom in an excited state; therefore, the γ-ray is emitted from the "daughter" atom and not the originally radioactive "parent" atom. X-rays of the same energy are the same as γ-rays except that x-rays are formed by the de-excitation of electrons when transferring from one orbital position to another. X-rays may also be formed in x-ray tubes when high-speed electrons directed at a metal target suddenly convert their kinetic energy of velocity into heat and x-ray radiation (often referred to by the German word *bremsstrahlung*). Gamma rays and x-rays undergo collisions with orbital electrons mainly by interactions of the electromagnetic fields of the photons with electrostatic fields of the electrons. This interaction becomes weaker with photons of higher energy. Thus, higher energy penetrating photons pass through matter unattenuated. Those that do interact may give up some of their kinetic energy and be scattered or may be totally stopped with no scattering.

The basic decay data for the nuclides mentioned in the methods section of this book are shown in the Appendix of Radionuclide Characteristics. More complete decay schemes and data are given by Kocher (1981) and Firestone and Shirley (1996), but, those provided here should suffice for the purposes of illustrating radioactive decay. Other, descriptive explanations of radioactive decay may be found in many texts including those of Martin (2000), Turner (1995), Hallenbeck (1994), and Cember (1992).

Radioactive Decay and Air Sampling: It is important to understand the radioactive decay mode by which the radionuclide transforms in order to choose an analysis method for the air sample. The predominant type of radiation emitted by the nuclide is a consideration in the choice of a detector. A few examples may suffice to explain some of the choices involved. Radionuclides such as ^{3}H and

^{14}C emit β-particles of very low average energy (5.685 and 49.47 keV, respectively), and consequently of very short range. The detector must be in close contact with the collecting medium, for example, an air filter, in order to capture these weak β-energies for further signal processing. One such method in current use is liquid scintillation counting (LSC), whereby the collecting medium (an air filter or bubbler solution) is immersed or solubilized in a fluid that scintillates when the β-particles interact with it. To capture γ-rays, which do not deposit their energy as efficiently as particulate radiations, a more massive detector is needed to increase the chances of producing an interaction. A solid scintillation crystal such as sodium iodide may be used to achieve this. These crystals are suitable for counting air filters and charcoal canisters. When attempting the measurement of α-particles, one is presented with an even greater challenge, as these emissions have even shorter ranges than β-particles despite possessing greater energies. Moreover, the α-particles may be shielded from detection by the collecting medium itself, necessitating the use of special air filters. Again, LSC may be employed. An alternative is to place an extremely thin wafer of a solid scintillator such as zinc sulfide (ZnS) in intimate contact with the air filter and count the filter in a light-tight box. Similarly, if a grab air sample is to be collected, an evacuated flask with a coating of ZnS could be used. Small solid-state detectors made from silicon are also suitable for detecting α-particles. Larger solid-state devices made from germanium (Ge) are used for γ-ray detection.

Radiation Energy: Many analysis systems can totally or partially distinguish radiation energies because the electronic signal produced by the interaction with the detector is proportional to the deposited γ-ray or particle radiation energy. The proportionality allows the signal to be measured and recorded according to the amplitude of its voltage. Counts are recorded in a corresponding energy window or channel. An interfering radionuclide produces radiation with an energy similar to that of the radionuclide of interest so that a nondiscriminating analysis system would include counts from both radionuclides and produce an incorrect result. Consider the case of ^{3}H and ^{14}C. Beta particles are emitted over a continuum of energies from zero to some maximum value dependent on the isotope undergoing decay. Because the maximum energy of ^{3}H β-particles are weaker than the majority of the β-emissions of ^{14}C, ^{3}H, and ^{14}C β-particles would be counted as if they were emitted by one radionuclide if a nondiscriminating liquid scintillation counter were used. Fortunately, modern liquid scintillation counters make provision to discriminate at least partially between these radionuclides. Gamma ray detection systems employ computer-based multichannel analyzers (MCAs) to produce a spectrum of γ-energy for analysis by the operator. The γ-energy range of the detection system is projected on the *x*-axis while the number of counts in each energy range is shown on the *y*-axis of such a visual representation.

Half-Life: The half-life ($t_{1/2}$) of a radioactive substance is the amount of time required for 50% of the radioactivity (50% of the population of radioactive atoms) to have transformed into the daughter or decay product. The loss by radioactive decay can be expressed mathematically as

$$\Delta N = -\lambda N \Delta t, \qquad (5.1)$$

where ΔN is the number of atoms decaying, Δt is the time period (considered short enough that N can be considered constant), and λ is the decay constant (the fractional rate of decay over Δt).

Using calculus, the above formula can be transformed into

$$\ln \frac{N_0}{N} = \lambda t, \qquad (5.2)$$

where ln is the natural log, N_0 is the original number of atoms, and N is the number of atoms at time t.

To obtain a formula for half-life, we substitute $0.5N_0$ for N in the above equation to obtain

$$\ln \frac{N_0}{0.5N_0} = \ln 2 = \lambda t \qquad (5.3)$$

and, solving for time t,

$$t_{1/2} = \frac{\ln 2}{\lambda} = \frac{0.693}{\lambda}, \qquad (5.4)$$

where $t_{1/2}$ is the half-life.

The influence of half-life on the measurement process increases with decreasing $t_{1/2}$. If the half-life is short enough so that a significant fraction of the nuclide will decay while it is being counted, then a correction for this loss is necessary if the result is to be an accurate assessment of the air concentration during sampling. Even a correction for the duration of sampling may be necessary. These could be taken into consideration when counting the relatively short-lived ^{131}I. A short half-life must also be considered if the air sample cannot be counted immediately due to the time required for transportation or shipping to a laboratory or because the sample analysis is deliberately delayed to allow other short-lived radionuclides that may interfere with the analysis to decay. The latter is a consideration when ^{222}Rn-progeny may have been collected and the analysis system cannot distinguish their contribution to the sample activity.

Radioactive Decay Law: The mathematical relationship of radioactive decay is given by

$$A_t = A_0 e^{-\lambda t}, \qquad (5.5)$$

where A_t is the activity observed at time t, A_0 is the initial activity (or number of atoms) on hand at initial time t_0, λ is the radionuclide decay constant (units of inverse time; i.e., t^{-1}), and t is the amount of time (units of time t so that it cancels with λ) that has elapsed since A_0 was determined.

Note: the radionuclide decay constant is related to the half-life as follows: $\lambda = 0.693/t_{1/2}$.

This equation may be used to calculate the original activity of a sample by rearranging terms.

$$A_0 = \frac{A_t}{e^{-\lambda t}}. \qquad (5.6)$$

This equation is useful if the sample is analyzed at a date much later than the date of collection, and the half-life of the radionuclide is short enough to make a difference in the radioactivity over the period of storage. However, radioactive decay is an ongoing process. It occurs

- While the air sample is being acquired
- While it is stored or transited to a laboratory for analysis
- While it is being analyzed or counted

In order to obtain the air concentration at the time of sampling, the loss of the sample from decay during the above intervals must be accounted for. To do so is expedited by the knowledge of the half-life of the radioactive contaminant and the magnitude of the sampling interval. For example, the half-life may be

- Long compared to the sampling time
- Short compared to the sampling time
- Comparable to the sampling time

These variations in the half-life can be used to simplify the equations used to find the air concentration at the time of sampling.

Assume that we are collecting a particulate sample on an air filter. It will take a period of time for the contaminant to build up on the filter (Figure 5.1). This rate of collection (or deposition) depends on the sampling flow rate, the unknown air concentration, and the half-life of the radioactive contaminant. Helgeson (1963) developed an equation to calculate the radioactive air concentration by defining the rate of change (the buildup) of the radioisotope on the filter (or the collecting

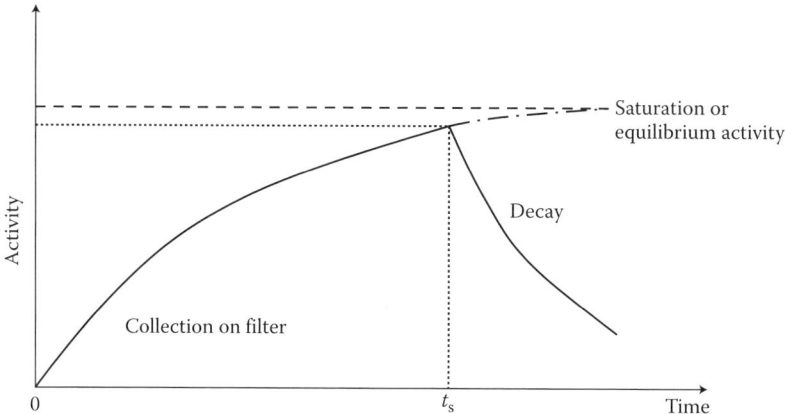

FIGURE 5.1 Activity collected on a filter during a sampling period t_s.

medium). This rate of change (dA/dt) is simply the amount of activity acquired by filtering the air minus the activity lost by radioactive decay:

$$\frac{dA}{dt} = \frac{CF}{E_f} - \lambda A, \tag{5.7}$$

where A is the activity of the radioactive contaminant (in units of activity, e.g., counts per minute)—A should be the *net* activity, that is, the activity of the sample minus the background radioactivity associated with naturally occurring radioactivity ($A = A_g - A_b$, where A_g is the "gross" radioactivity and A_b is the background radioactivity), F is the flow rate of the air sampler (in volume per unit time, e.g., cubic feet per minute, ft³ min⁻¹, or milliliters per minute, mL min⁻¹, such that the time unit is the same as for A), C is the concentration of radioactive contaminant (in activity per unit volume, e.g., μCi mL⁻¹ such that the volume unit is the same as in F), E_f is the collection efficiency of the filter; usually, E_f is very high (above 99%), so that this term may be ignored in many instances, λ is the decay constant of the radioisotope as defined above (in units of inverse time consistent with F and A, e.g., min⁻¹).

Assume that C does not vary with time and that at the beginning of sampling, $A = 0$ activity. Then, using calculus to solve for A, the equation for buildup of radionuclide activity on the filter is

$$A = \frac{CF}{E_f \lambda}(1 - e^{-\lambda t_s}), \tag{5.8}$$

where t_s is the time at the end of the sampling period.

Note: consistent units must be used for each variable to obtain the desired units for A.

The value of A can be considered the initial activity on the filter. During sampling, the radioactivity decays, but it is being replaced by radioactivity captured on the filter. Once sampling stops, replacement stops and loss of the sample continues by radioactive decay. To account for a period of decay between the end of sampling and the beginning of counting, substitute the expression for A (Equation 5.8) into A_0 in Equation 5.5.

$$A = \frac{CF}{E_f \lambda}(1 - e^{-\lambda t_s})e^{-\lambda t_d}, \tag{5.9}$$

where t_d is the delay time before counting.

Equation 5.9 can be rearranged as follows to solve for the concentration of radioactivity in air, we obtain

$$C = \frac{E_f \lambda A e^{\lambda t_d}}{F(1 - e^{-\lambda t_s})}. \tag{5.10}$$

This is a general equation for air sampling when the sample concentration and the flow rate are constant. It may be modified to account for effects associated with analysis of the filter (or other collection media). For example,

$$C = \frac{E_f \lambda A e^{\lambda t_d}}{E_c F k (1 - b)(1 - e^{-\lambda t_s})}, \tag{5.11}$$

where E_c is the counting efficiency of the radiation detector (the ratio of counts per minute recorded by the detection system to disintegrations per minute of the radioactivity on the sample, i.e., cpm/dpm), k is a factor to convert dpm into a unit of radioactivity such as micro-Curies (μCi) or Becquerels (Bq), b is the loss of radiation, and therefore, the loss of radioactive counts due to absorption of radiation, that is, α-particles, in the collection medium before it is detected (this loss can be included in the value for E_c if a proper calibration of the detector is performed).

Effect of Long Half-Life: Isotopes with long half-lives have very small decay constants (λ). This has an important effect when the half-life is long compared to t_s or t_d because (λt_s) or (λt_d) becomes small values. In fact, a very long half-life implies that radioactive decay is almost nonexistent. We have $e^{\lambda t_d} = e^0 = 1$. Therefore, the radioactive decay terms can be ignored in Equation 5.11 to give the following equation:

$$C = \frac{A}{E_c F k t_s}, \tag{5.12}$$

where we have assumed that $E_f \approx 1$, so that it may be ignored. In this form, the final result for C will be in units of activity (μCi or Bq) per unit volume of air.

Effect of Short Half-Life: If the half-life of the radioisotope is short compared with the sampling time, $e^{\lambda t_s} = e^{-\infty} = 0$ and Equation 5.11 simplifies to

$$C = \frac{0.693 A e^{\lambda t_d}}{E_c F k t_{1/2}} \tag{5.13}$$

RADIATION DETECTION

REVIEW OF RADIATION DETECTION SYSTEMS

Liquid Scintillation Counters: LSC can be used to detect α- and β-particle-emitting nuclides collected in bubbler solutions or on small air filters. LSC is often available at institutions with radiation safety programs, as it can be used for several purposes including assay of biological samples. Many texts contain overviews of LSC, including those of the NCRP (1985), L'Annunziata (1998), and Knoll (2000). A detailed monograph edited by Kessler (1989) has been available through one commercial vendor of LSC.

In oversimplified terms, LSC consists of placing the sample in a small volume of scintillation fluid (often referred to as a *cocktail*). α- or β-particle interactions with the cocktail cause the release of light photons (scintillations) that are detected by a photomultiplier tube (PMT) inside the counter. The PMT transforms the light pulses into an amplified electrical signal that undergoes further signal processing.

Cocktail scintillates because it consists of a combination of chemicals that work together to transform the kinetic energy of the radiation into a light pulse. An organic solvent makes up about 99.5% of the mixture. The molecules of the solvent are ionized and excited by the radiation. The energy of ionization and excitation is transferred to the fluor that largely makes up the remainder of the mixture. When the fluor molecules de-excite, they luminesce. A second fluor may be used to shift the wavelength of the luminescence (corresponding to about 3 eV) to one that can be detected by a PMT. The signal generated by the PMT is proportional to the radiation energy deposited in the scintillation fluid. Counting efficiencies for LSC can be nearly 100% depending on the radionuclide and the constituents of the sample.

Sample preparation is an extremely important step in the process of LSC. Although hundreds of milliliter of bubbler solution may be used to collect an air sample, perhaps 1–5 mL are actually counted due to the size of the counting vial (20 mL) and the typical volume of scintillation cocktail used (10–15 mL). The sample in the counting vial must not undergo phase separation, that is, it must be a stable, homogenous sample that maintains the cocktail and radioactive constituent in intimate contact. Phase separation will manifest itself as an unstable count rate (increasing or decreasing cpm) in repetitively counted samples. Homogeneity may be achieved by choosing a commercially available cocktail compatible with the chemistry of the sampling solution. Vigorous mixing of the radioactive sample solution with the cocktail or increasing the volume of cocktail may also help. LS counter manufacturers can often provide written and verbal guidance for these and other special counting situations.

LSC can be subject to other interfering effects. For example, increasing the opacity of the cocktail by the addition of a colored sample may cause absorption of the scintillation photons (color-quenching). This will reduce the counting efficiency. Chemical quenching will also result in reduced counting efficiency. It occurs when the β-energy is dissipated between the solvent and the fluor. It may be overcome by using more cocktail or by using a quench-resistant cocktail. LS counter manufacturers also provide electronic techniques and analysis software that can effectively compensate for quenching. Other sample chemicals may de-excite the cocktail so that photons are emitted without being caused by radiation interaction (chemiluminescence). Repetitively counted samples exhibiting chemiluminescence will show a slow count decay rate over a period of 0.5 to perhaps more than 24 h (Kessler, 1989). It can be eliminated using chemical and electronic means. Chemiluminescence photons are weak and will fall generally into the tritium energy window of the counter. All LS samples should be dark-adapted for 15 min before counting as a precaution against photoluminescence that can occur when the samples are exposed to ultraviolet light (L'Annunziata, 1998).

The application of computers to LSC, as with other nuclear counting systems, has greatly increased ease of use and reporting capabilities. Modern LS counters may be programmed with many analysis protocols to accommodate various samples and counting requirements. The counting efficiency, counting time, the option to subtract background counts from all the samples and the number of times a sample may be counted usually can all be preset and retained indefinitely in computer memory. Most importantly, a protocol to perform a dual count for both ^{3}H and ^{14}C is usually available for analyzing samples that may contain both radionuclides. Both gross counting results and net dpm may be reported. This capability can be extended to many LS counters for calculating sample concentrations (pCi/mL of sample solution) and downloading these results to other computers for further calculations.

The calibration of LS counters is described in the associated air sampling methods of this book. Standards are commercially available with known disintegration rates for the determination of counting efficiencies. Manufacturers also provide quality assurance techniques as part of the computer software of the LS counter that supplements the calibrations recommended here. LSC is somewhat dependent on scintillation fluid volume and the constituents of the sample (volume dependence can be compensated for using spectrum analysis techniques) (Kessler, 1989). Standards as similar as possible in both respects to a field sample permit an accurate counting efficiency value to be obtained. The procedure for calibrating one type of popular LSC for both α and β counting is described by Scarpitta and Fisenne (1996). This report also succinctly explains the quenching compensation procedures used by the LSC manufacturer.

Photon Detectors: Sodium Iodide. Detectors composed of thallium-activated sodium iodide [NaI(Tl)] crystals work by a scintillation process, which is described in detail by Mann et al. (1980), NCRP (1985), L'Annunziata (1998), and Knoll (2000). They are used for the detection of γ-rays and x-rays above about 3 keV (Mann et al., 1980). Thin crystals are used for low-energy photons. NaI (Tl) detectors can be used for the measurement of charcoal canisters that are employed for the atmospheric sampling of ^{131}I or ^{222}Rn (measurement methodologies for both are described in the methods section of this book).

Radiation detection using NaI(Tl) basically consists of the following steps:

1. Excitation and ionization of crystal atoms by incident γ-radiation
2. UV photon emission (peak wavelength about 430 nm) during de-excitation
3. Detection of the UV photons by a PMT
4. Production of photoelectrons by the PMT
5. Amplification of the current by the PMT and subsequent electronics

NaI(Tl) detectors typically consist of a NaI(Tl) crystal completely sealed in a metallic cylinder. A typical detector size is 76 mm (3 in.) diameter × 76 mm (3 in.) thick, but larger sizes such as 127 mm (5 in.) × 127 mm (5 in.) have been manufactured (PGT, 2002; Bicron, 2007). Other geometries are commercially available including well-type crystals for surrounding a sample with NaI(Tl) to achieve maximum radiation absorption, thin crystals for low-energy photon analyses, and hole-through shapes for the measurement of flowing media.

The metallic housing of the detector (aluminum, copper, or stainless steel) seals out air moisture that will destroy the hygroscopic NaI(Tl). A coating of Al_2O_3 packed powder on the inside surface of the can reflects the UV photons emitted by the crystal after radiation interaction back to where the PMT may detect them. The crystals are coupled to a PMT using an optical jelly and a UV-transparent window made of glass or quartz (3–76 mm thickness). NaI(Tl) detectors are usually housed in a lead cave (about 100-mm-thick lead is needed; Ortec, 1997) with a door or hatch available to introduce the sample and a separate pass-through for electronic cables. The lead cave reduces the background signal produced by natural terrestrial γ-ray emitters or anthropogenic radiation sources. However, care must be taken to obtain uncontaminated lead or lead of low background characteristics. The materials of construction for the crystal, the metallic housing, and the optical window are chosen by manufacturers to be low in natural emitters such as ^{40}K and isotopes of uranium and thorium.

The PMT is connected to a load resistor, the output of which is a negative voltage pulse. The magnitude of the voltage is proportional to the energy of the incident γ-ray. This pulse undergoes further signal processing, which is referred to as pulse-height analysis (PHA), by an MCA in order to produce an energy-discriminated spectrum that can be used to identify the radionuclide. The number of counts, that is, the spectral peak height, determines the amount of the radionuclide that is present in the sample (Mann et al., 1980; Knoll, 2000).

In brief, energy discrimination is accomplished by electronic circuitry that converts the analog output voltage of the PMT into a digital signal based essentially on its amplitude. The circuitry can resolve the voltage amplitude such that a tally of the amplitudes achieved can be kept. Channels are gated to accept only those PMT output voltages that are confined to a specified voltage range. A simple type of computer memory is assigned to each channel so that the memory will increment each time a pulse falls within the voltage range. Thus, the photopeak is assigned a channel based on its energy (voltage amplitude) and its height results from the tally of pulses for those channels. This is the basis of γ-ray spectroscopy.

Resolution of spectral peaks becomes important when the peak of the radionuclide of interest may be overlapped or fall very close to another. This may be of concern when sampling for airborne

^{131}I when other noble gases are present (Method 603). Resolution (peak broadening) is a function of several variables, including the uniform efficiency of the NaI(Tl) to produce light pulses throughout the body of the crystal and the efficiency of the PMT to produce and multiply photoelectrons. These variables are largely out of control of the operator, but resolution may be routinely observed as part of a quality assurance plan. The 662 keV peak of ^{137}Cs is usually employed for resolution measurements. The resolution (R) is the ratio of the full width of the peak (measured in energy) at half its maximum height [full-width *half-maximum* height (FWHM)] to the energy of the peak (Mann et al., 1980; L'Annunziata, 1998; Knoll, 2000).

$$R = \frac{(E_1 - E_2)}{E_p} \times 100\%, \tag{5.14}$$

where E_1 is the energy of the channel greater than E_p corresponding to half the peak height (keV), E_2 is the energy of the channel less than E_p corresponding to half the peak height (keV), and E_p is the peak channel energy (e.g., 662 keV).

Modern NaI(Tl) counting systems have a resolution of about 7% at 662 keV (Canberra, 2007).

Slight voltage changes in the PMT and other electronic components of the NaI(Tl) counting system can shift the spectrum across channels and render the calibration inaccurate. These drifts may be due to temperature changes in the detector and the electronics or due to high sample-count rates. Modern counting systems employ a constant source of light to produce an artificial peak which is used as a reference to automatically adjust the voltage to hold the spectrum in place. Other electronic means of spectrum stabilization are available (Knoll, 2000). Drift may be checked manually by taking position measurements of the ^{137}Cs photopeak using a 1000 cps reference source in order to compute the mean long-term drift in % per day:

$$\text{Drift} = \frac{\Sigma(P_{\text{avg}} - P_i)}{n} \times \frac{100}{P_{\text{avg}}}, \tag{5.15}$$

where P_i is one of n photopeak positions and P_{avg} is the mean position of the series of n positions.

The short-term drift due to encountering high count rates may be found using a 10,000 cps ^{137}Cs source immediately after performing the long-term shift measurement. Four position measurements are made at 10 min intervals and used in the following equation:

$$\text{Drift} = \frac{\Sigma(P_i - P_4)}{4} \times \frac{100}{P_4}, \tag{5.16}$$

where P_4 is the last photopeak position.

Computer-based γ-ray spectroscopy analysis systems work best when the laboratory power supply is free of voltage surges and changes in frequency. Voltage surge protectors and uninterruptible power supplies that can maintain operating voltages in the event of an outage are commercially available.

Calibration of a NaI(Tl) detection system is accomplished by using reference sources of several energies to produce photopeaks that may be locked into channel positions (computer memory) for use as energy reference locations. Thus, the channels are assigned energy values. Since γ-ray spectroscopy systems are geometry dependent, it is crucial to use the same container for the reference source(s) as is used for the field samples.

Modern computer-based γ-ray spectroscopy systems maintain many photopeak energies in memory to identify the nuclides in the sample spectrum. Determining the amount of a radionuclide in a sample by γ-ray spectroscopy is complicated by the presence of background counts generated by the radioactivity in the nearby environment and the background generated by Compton scattering of the incident radionuclide photons. Compton-scattered photons produce more counts in the lower energy region of the spectrum than in the higher energy region. Software to remove background counts accurately and compute the activity of the radionuclide(s) of interest is provided with computer-based spectroscopy systems.

Photon Detectors: Germanium. Germanium (Ge) detectors provide far superior photopeak resolution as compared to NaI detectors. However, they require a liquid nitrogen (LN_2) supply to cool the detector (to 77 K) and thus reduce the electronic noise of the system. Shielding, supporting electronics, and a computerized photopeak analysis system are also required. However, these detection systems lend themselves for use as portable field instruments, a true asset in emergency situations. Further advancements in detector cooling technology (electromechanical cooling) have eliminated the need for liquid nitrogen as long as electric power is available for battery recharge.

Radiation detection using Ge detectors consists of the following steps:

1. Excitation and ionization of Ge crystal atoms by incident γ-radiation
2. Elevation of current carriers (electrons and holes) from the valence energy level to the conduction energy level of the crystals
3. Collection of charge carriers under the influence of an applied electric field (the detector is placed under high voltage)
4. Integration of charge carriers by a preamplifier
5. Conversion of integrated charge into a voltage pulse with an amplitude proportional to the incident photon energy

Mann et al. (1980) and Knoll (2000) provide detailed explanations of the physics underlying the detection process. Catalogs containing detailed descriptions and performance characteristics of these detectors are produced by manufacturers such as Ortec (1997) and Canberra (1997). L'Annunziata (1998) provides a good general overview of semiconductor detectors and related sample analysis.

Ge detectors consist of the detector element and a field effect transistor mounted inside a vacuum enclosure called an endcap. These components are followed by a preamplifier and a high-voltage filter contained inside an electrical shield (the electronic shroud). Next is a tube-shaped vacuum enclosure (the cryostat) that provides a path to the dewar (another vacuum container) for LN_2 storage. The cryostat may be electronically cooled as an alternative to cooling by LN_2. Typical Ge detector diameters are about 2.3 in. (58 mm) to 2.8 in. (71 mm). A large detector of about 3.2 in. (81 mm) diameter is available (Ortec, 1998).

Various types of cryostats and dewars are available to provide a convenient counting geometry. There are vertical cryostats (straight throughput directly to the dewar) and horizontal cryostats with a 90° turn to accommodate a side-looking counting configuration. Thirty-liter dewars with centrally located, offset, and even side-looking throughputs for the cryostat are available. Fifty-liter volumes can also be purchased. One to five liter dewars usually accompany small, portable Ge detectors for field work. LN_2 loss rates for typical dewars are about 0.6 L day^{-1} (Canberra, 1997). As mentioned, an electrical cooling system that does not require LN_2 is available, but it is relatively expensive. Since a lead shield is required around a laboratory-based detector to reduce background radiation, it must be built to allow access to the detector and dewar for sample placement, electrical cable passthrough, and LN_2 supply.

Cooling systems are required to reduce noise levels and to achieve high photopeak resolution. The noise is created by thermally generated charge carriers (not γ or x-ray photon-induced charge carriers). Cooling to 77 K reduces this interference to an acceptable level. Both the field effect transistor and the detector must also be kept cool when high voltage is applied, otherwise both may be damaged. Automatic systems to provide voltage shutdown in the event of warm up due to LN_2 loss are commercially available.

Like NaI(Tl) gamma spectroscopy systems, Ge detection systems require high voltage supplies and analog-to-digital converters. They also rely on MCAs and computer software for system calibration and for sample radionuclide identification and measurement. Standards with multiple isotopes are available for energy calibration and counting efficiency determination. It should be noted that operation of Ge detectors in high ambient temperatures may cause a shift of the γ-ray spectrum so that the energy of sample photopeaks cannot be accurately determined. This could be especially an important phenomenon for portable Ge detectors used in the outdoor environment.

Superior resolution is achieved with Ge detectors because a relatively small amount of incident energy is required to produce a charge carrier. This output signal of a Ge detector is large compared to a NaI(Tl) detector for the same photon energy. At 1332 keV, a 76 mm × 76 mm NaI(Tl) detector has an FWHM of about 60 keV. At the same energy, the FWHM for a Ge detector is about 2 keV (Canberra, 1997). Guidelines for measuring the FWHM and other quality parameters of Ge detectors can be found in the literature (ANSI, 1999).

Ionization Chambers: Ionization chambers find use for continuous monitoring of certain radioactive gases (refer to the method for γ-emitting gases in Chapter 26 of this book). Airborne concentrations are instantly available, which is an advantage, particularly when an ongoing process is releasing a radioactive effluent that must be monitored, or when an integration of the total radioactivity released is required. The signal from ion chambers performing these tasks may be recorded using commercially available software that can calculate the amount of radioactivity released over time. This relieves the operator of the tedious work required to integrate the effluent signal recorded on a less sophisticated device such as a paper chart recorder.

For radon measurements, inexpensive electret-based direct current ionization chambers are used as integrating detectors capable of providing a short-term mean air concentration.

Direct current ionization chambers consist of two electrodes, one of which is the chamber itself (the cathode), and the other a thin wire which is located within the confines of the chamber (the anode). A source of electromotive force (a power supply), a ground, and a current measurement device complete the ionization chamber circuit. When the anode is energized, an electric field is created inside the chamber. The chamber gas, which may be air, will be ionized by either particulate or electromagnetic radiation. For α- and β-particles, x-rays and γ-rays about 34 eV are required to ionize air molecules. The electrons liberated in the ionization process and the remaining ions make their way by Coulomb attraction to the electrode of opposite sign. With a strong saturation voltage applied, virtually all the ions and electrons are collected, creating a current (I). This current is the result of a mean number of ion pairs (N) being produced per unit time. Thus, $I = Ne$, where e is the elementary electron charge. The current is a measure of the energy deposited (number of ion pairs formed) by the impinging γ-ray radiation or the radiation emanating from the airborne radioactivity introduced into the chamber. One means of measuring the current is to introduce a capacitor into the circuit, allow charge (Q) to build up on it, and then measure the change of the voltage (ΔV) on the capacitance (C). Thus, $\Delta V = \Delta Q/C$. Refer Mann et al. (1980) and Knoll (2000), for more details. This method will permit the measurement of an integrated ionization current over a finite time. It is the basis for small integrating radon detectors that use electrets. These dielectric materials, for example, Teflon, slowly lose their surface charge due to collection of ion pairs produced by radiation in air.

Pulse ionization chambers have an external circuit consisting of a capacitor and a resistor in parallel, along with a power supply and a ground as in the DC version. This arrangement makes the signal–voltage time dependent. By choosing values of the capacitance and resistance (R), the signal pulses can be made dependent only on the swift electron collection (the positive ions are essentially ignored). Thus, $V(t) = V_0 e^{-t/RC}$, where V_0 is the initial voltage imparted to the system by the radiation energy and t is any subsequent time. The values of RC are usually chosen so that $V(t)$ returns rapidly to zero. A voltage pulse is created that can be processed by counting electronics.

The response of ionization chambers depends on the incident radiation energy, operating voltage of the chamber, chamber dimensions, anode rod diameter, and chamber-wall material (Sauter et al., 1985). If a flow-through ionization chamber is calibrated for one radioactive gas, correction factors must be applied when other radio gases are monitored. An effective way to calibrate an ionization-type flow-through monitor is to pass a known radioactive concentration through the detector using a calibrated gas standard and observe the readings. A discussion of this is given by Wood et al. (1997). Those authors also show that ionization chambers are apparently not capable of accurate results when more than one radioactive gas is present.

Ionization chambers used for air monitoring will also respond to γ-ray radiation. Commercially available ^{3}H and ^{14}C monitors employ separate ionization chambers for air and γ-ray measurements. This allows the current due to γ-rays to be subtracted from the total current (airborne radioactivity and γ-ray signal) to yield the signal due only to airborne radioactivity (Wood et al., 1993).

Ionization chambers are affected by humidity and airborne particles. Humidity can reduce the resistance of the insulation around parts of the anode. This may be manifested by increased difficulty in obtaining the zero-rest position of the current measuring meter. The chambers should be kept as dry as possible. Drying cans (filled with Drierite or silica-gel) are sometimes provided by the manufacturers and are placed in close proximity to the ionization chambers, or other means are used to seal the chambers from atmospheric moisture. Both methodologies should be periodically inspected for efficacy. *Note*: If monitoring for ^{3}H, it is not advisable to connect a drying tube before the air inlet of the ionization chamber since some of the ^{3}H may be trapped and not detected.

If a typical, commercially available ionization chamber used for air monitoring becomes contaminated, it can be cleaned internally if great care is exercised. Recommended solutions include detergent diluted in warm water, ethanol, and methanol. No residue of any kind should be left on the internal surfaces. Great care must be exerted when cleaning the anode to prevent damaging it.

Commercially available ionization-type air monitors usually are equipped with an audible alarm to indicate that a preset concentration has been exceeded. Many have the option for a remote alarm, which may be placed in a room far from the monitor location for the purposes of warning response personnel.

At least one commercially available hardware/software package can accept an analog ion chamber output signal and, using an analog-to-digital converter, transform the signal for computerized data analysis (Scintco, 1996). The software is sophisticated enough to subtract the electronic noise background (baseline) from the radioactive effluent peak(s) and to integrate the peak area(s) to obtain the discharge activity concentration in units such as Ci-s m^{-3}. Using the airflow rate of the discharge stack (m^3 s^{-1}), one can have the computer determine the total radioactivity emitted (in this case, in Ci). The advantages of this system include accurate peak integration, substantial savings in time, and the ability to efficiently store and recall many hours of air monitoring data.

Gas Proportional Counters: Gas proportional counters work at voltages higher than ionization chambers (typically 500–5000 V). They are electronically similar to ionization chambers in that a resistance and capacitance exist in parallel across the gas-filled chamber. The chamber is filled to atmospheric pressure with an electropositive gas (typically 90% argon mixed with 10% methane) that yields positive ions and electrons rather than positive and negative ions after interaction with radiation. At these relatively higher voltages, the kinetic energy of the electrons is capable of ionizing the gas molecules. These secondary electrons also create ion pairs because they undergo acceleration under the influence of the electric field surrounding the anode. This process, known as gas multiplication, creates an avalanche of electrons. These avalanches are limited in physical size and they occur within 0.005 cm of the anode surface (if the anode diameter is 0.002 cm). The cloud of remaining positive ions surrounding the anode is small and does not greatly interfere with the collection of electrons. Certain values of R and C allow the detector to recover quickly after an ionizing event (in approximately 1×10^{-6} s) so that it can respond to the next radiation interaction. Thus, high rates of counting are possible. As in ionization chambers, the charge collected on the anode is proportional to the energy deposited in the gas by the incident radiation (Mann et al., 1980; NCRP, 1985; Knoll, 2000). The greater the energy deposited in the detector gas by an incident particle of radiation, the larger the number of primary ion pairs, the larger the number of avalanches, and the larger the pulse.

The total number of electrons (N) reaching the anode is given by

$$N = nm_{avg}(1 - m_{avg}P), \quad (5.17)$$

where n is the total number of ion pairs formed in the ionizing track of the radiation, m_{avg} is the average number of ion pairs formed in the multiplicative region of the anode, and P is the probability of the secondary ion pair in the avalanche forming a photoelectron.

Gas proportional detectors offer the advantage of windowless counting, that is, there is no barrier between the sample and the counting gas (Knoll, 2000). The detectors can be built to completely surround the sample. In fact, using two anode wires, one above and one below the sample (such as a thin air filter) allows for the so-called 4π counting since each detector subtends an angle of 2 steradians in such a geometry. Cylindrical, spherical, and pill-box configurations have all been used for 4π gas proportional counting. Both α- and β-particle counting is possible.

Alpha-Radiation Counters: Methods other than gas proportional counting may be used for α-detection. A straightforward method using a light-tight box, a PMT with a power supply, a scaler, and a fluorescing disc of ZnS in close contact with the sample (an air filter) has been in use for many years (Fisenne, 1981). ZnS is relatively inexpensive. Discs of varying diameters are commercially available so that a match, or a best approximation, can be made to the diameter of the air filter. The filter and the disc are placed in close contact to minimize loss of α-particles by interaction with air molecules. The discs are thin, permitting the transmission of fluorescence photons to a PMT with which the disc is also in close contact. The PMT–ZnS–filter sandwich is housed in a light-tight box so that the PMT may respond only to the photons emitted by the ZnS disc. A scaler-timer is employed to count the α-particle interactions in a preset counting interval.

Solid-state detectors (Ortec, 1997) have been developed that operate in vacuum to prevent α-particle loss from collisions with air molecules. Some models can function under atmospheric conditions. Solid state (charged-particle) α detectors are constructed of silicon. Some types are built with ultrathin windows to achieve energy resolutions of about 20 keV FWHM for 5.5 MeV α-particles (using modern spectroscopic electronics and an active detection area = 450 mm^2). This is important when the detector is used in α-spectroscopy for radionuclide identification. Background count rates of detectors of this size can be as little as 25 counts per day in the 3–8 MeV energy interval. Some models are ruggedized for nonvacuum operation in continuous air monitors (CAMs). Resolutions are poorer than for thin window types (35 keV at 5.5 MeV in vacuum) but should be of little concern for CAM operation. These surface barrier types are available in areas up to 3000 mm^2. Smaller area models produce better energy resolution than larger area types (Ortec, 1997).

Silicon surface barrier detectors have diameters ranging from 2 to 5 cm. A typical thickness is 1 cm. Therefore, they are suitable for small air filters only. Alpha spectroscopy requires mounting the detector in a vacuum chamber. These chambers are cylindrical with heights of about 20 cm and diameter of 15 cm. The detector is mounted to the inside of the top cover. The sample is mounted below on a support rod at a preselected distance from the detector. A vacuum pump is connected to the chamber through a single valve provided in the sidewall. More sophisticated chambers constructed using low-background materials with front-loading doors and racks accepting samples on trays are also available.

A counting efficiency must be determined for an α-counter. Disc-shaped α-emitting standards are available commercially for this purpose. The diameter of the standard and the sample air filters should be as similar as possible. Energies of the standard should be similar to the airborne radionuclide under investigation.

It is possible to backscatter α-particles 180°. This can occur if an α-standard is electroplated onto a backing such as a metallic disc. This can yield a greater efficiency than is attained when an unsupported air filter is counted. One must be careful to assure that the standard mimics the sample in size, geometry, composition and backing material as much as possible.

COUNTING STATISTICS

SIGNIFICANT FIGURES AND ROUNDING, NEGATIVE AND ZERO DATA

Significant figures in a number are those that are known with certainty plus the first doubtful or estimated digit. The rules for significant figures should be followed carefully when reporting airborne radioactivity (as well as any other scientific results). In brief, they include the following:

1. In addition and subtraction, the last digit retained in the result should correspond to the first doubtful decimal place of any of the added or subtracted numbers.
2. In multiplication or division, the result should contain no more figures than the least number of significant figures in any of the multiplied or divided numbers.
3. When rounding off in order to remove superfluous figures, increase the last digit retained by 1 if the following figure is 5 or more.
4. Intermediate calculations should use superfluous figures until the final result is achieved. Only the final result should be rounded to the appropriate number of significant figures to prevent the introduction of round-off error that occurs when rounding is performed on a series of intermediate results. Values calculated by computer often contain too many decimal places and should be rounded to the appropriate number when a final result is obtained.
5. When recording large values involving zeros, the number of significant figures should be indicated by using exponential notation.
6. Exact numbers that are not subject to the uncertainties of measurement, such as the number (n) of samples obtained, are not subject to the rules for significant figures. For example, if 10 measurements are obtained and a mean is required, the 10 is an exact number and is not considered in the determination of the significant figures for the mean.

Another method of choosing the correct number of significant figures is to determine the significant figures in the uncertainty and round the reported value to the same decimal place (Sanderson et al., 1980). The calculated standard deviation and rounded-off standard deviation must not differ by more than 20%. The number of decimal places required to keep this difference less than 20% is what is reported. The result and the uncertainty must have the last significant figure in the same decimal place.

When measuring low levels of airborne radioactivity, zero and negative net results obtained by subtracting a background count or count rate should never be discarded (EML, 1997). This is especially true if a mean air concentration is to be calculated from a series of measurements. Discarding negative and zero values will result in a biased average. Similarly, listing data as *below detection limits* results in a loss of information and could be misinterpreted unless thoroughly explained. It is better to report the actual result no matter whatever be the value (Sanderson et al., 1980). Databases with missing data can be analyzed with certain techniques to account for the lost information (Gilbert and Kinnison, 1981).

ERROR TERMS

Each air sampling result should be accompanied by an uncertainty (σ) (EML, 1997). Radioactivity measurements often include systematic errors and random errors. Systematic errors are repeated errors of the same magnitude that usually bias a sample in one direction. They are usually associated with the measurement system such as a calibration error. Random errors are temporally unpredictable and vary in a nonreproducible way around the mean. They are present in all radioactivity measurements because radioactivity is a random process. That is, the specific decay of a radioactive atom cannot be predicted. It has a probability that it will decay and a probability that it will not decay in any given time period. Because of this, radioactivity measurements are categorized as probability distributions. Statistics attempts to ascertain the probability of decay of a certain fraction of atoms in a population of radioactive atoms.

Radioactive decay is described by the complex formulae of the binomial distribution. The Poisson distribution is a suitable approximation of the binomial under virtually all counting circumstances. It applies to discrete, positive integers. At mean values as low as 10, it is approximately symmetric. The mean value, in fact, completely defines the distribution, for example, the square root of the mean yields the distribution standard deviation (see, e.g., Turner, 1995; Martin, 2000; Knoll, 2000 for more details).

The formula for the Poisson is cumbersome to use especially when the mean is >100. Fortunately, the normal distribution (the familiar bell or Gaussian curve) approximates the Poisson if the number of radioactive counts exceeds 30. This is advantageous since the normal distribution expressions for variance and standard deviation are the same as for the Poisson and binomial distributions. A description of a normal distribution requires that both the mean and the standard deviation be known. For a given mean value, there are infinite normal distributions, each defined by a standard deviation value.

For each parent population of data, there exists a true mean. Since we cannot obtain an unlimited number of data points, we obtain a sample to estimate the true mean. A typical measurement scenario would be the accumulation of 10, 1-min counts from which a mean and standard deviation would be computed. This process might be repeated five times. Each mean and standard deviation so computed will probably differ from each other. However, each is an estimate of the true mean. According to the Central Limit Theorem of statistics, a plot of these sample means (the sampling distribution) would be normally distributed even if the parent distribution is not a normal distribution. A plot of the standard deviations would follow the chi-square distribution. The sampling distribution of the means has its own standard deviation called the standard error. The standard error decreases inversely with the square root of the sample size. This implies that a significant increase in precision (decrease in the standard error) will occur with an initial increase in sample size, but further increases in precision are not as significant as the sample size is made larger.

The standard error (S_x) is defined as

$$S_x = \frac{S}{N^{1/2}}, \tag{5.18}$$

where the standard deviation (S) is defined as

$$S^2 = \frac{\Sigma(x_i - \bar{x})^2}{N - 1}. \tag{5.19}$$

Here S^2 is called the variance, x is the sample count, $\bar{x}$ is the sample mean, and N is the number of sample counts obtained.

A useful property of the Poisson distribution is that S is the square root of the mean.

$$S = (\bar{x})^{1/2}. \tag{5.20}$$

This estimate of S can even be used with a single measurement. Note that the results of Equations 5.19 and 5.20 will differ since as in Equation 5.19, the computation is based on the dispersion of data, while in Equation 5.20, it is based on the assumption that the data follow the Poisson distribution.

In other words, each sample count represents an estimate of the true mean of the sample activity. This estimate may be combined with a constant that indicates the confidence level. In a normal distribution, one standard deviation (1σ) encompasses 68.26% of the distribution, that is, 68.26% of the time; the actual value of the mean falls within 1σ of any estimate. Similarly, a range of 2σ encompasses a span in which the mean falls 95.46% of the time and at 3σ, 99.74% of the time. Hence, a value of 1σ may be written as

Review of Radioactivity, Detection, and Measurement

$$E_C = k(n)^{1/2}, \quad (5.21)$$

where n is the counts obtained in a measurement, k is the critical value for the chosen confidence level (e.g., 1.000 at the 68% confidence level), and E_C is the uncertainty associated with the count in the absence of systematic error.

At 2σ (95% confidence level), $k = 1.960$ and at 3σ (99% confidence level), $k = 2.576$. Similarly, count rate (counts per unit time) errors are calculated as

$$E_R = k\left(\frac{r}{t}\right)^{1/2}, \quad (5.22)$$

where r is the count rate, E_R is the uncertainty associated with the count rate in the absence of systematic error, t is the counting time (same time units as r), and k is the same as defined above depending on the chosen confidence level.

In the manipulation of counting data, the standard deviations must be propagated. When adding or subtracting counting data (e.g., when subtracting a background count from a gross count), the uncertainties are added together in the following way to obtain a total uncertainty E_t.

$$E_t = [k_1(E_1)^2 + k_2(E_2)^2 + k_3(E_3)^2 + \cdots + k_n(E_n)^2]^{1/2}, \quad (5.23)$$

where k_n is defined as above.

When multiplying or dividing, the following formula applies:

$$\%E_t = [k_1(\%E_1)^2 + k_2(\%E_2)^2 + k_3(\%E_3)^2 + \cdots + k_n(\%E_n)^2]^{1/2}, \quad (5.24)$$

where $\%E_t$ is the total uncertainty expressed as a fraction of the result and $\%E_n$ is the individual uncertainties expressed as fractions of the individual terms to be multiplied or divided.

To obtain an absolute value of E_t, multiply the counting result by the fractional uncertainty $\%E_t$.

The formulas for the uncertainty terms given in the radioactive air sampling methods consider only the uncertainties due to counting. Uncertainties usually exist in the airflow rates or total air volume sampled, in the timing of the sampling period, and in the calibration of the radiation counting device and should be worked into the total uncertainty calculation, especially if they are significant.

DETECTION LIMITS

The sensitivity of the air measurement should always be presented with a description or reference to the statistic describing the sensitivity. Discussions of the critical measurement level (L_C) and the lower limit of detection (LLD) statistics may be found in NCRP (1985) and Martin (2000). The reader is urged to consult other documents (Altshuler and Pasternack, 1963; Currie, 1968; Pasternack and Harley, 1971; NCRP, 1985; Martin, 2000) for detailed explanations of the critical level (L_C) and LLD concepts. Taken together, these limits are the basis for determining if the radioactive air effluent measurement is adequate to meet the investigation and/or regulatory air concentration limits. The minimum detectable activity (MDA) is the LLD with adjustments made for counting efficiency and other instrument factors (Martin, 2000). Similar names for these statistics are often encountered (refer NCRP, 1985). Also refer the equation for MDA used by the U.S. EPA (NUREG, 1992). The LLD can be calculated using slightly different formulae based on certain assumptions about the variation of background and gross counts [refer NCRP, 1988 for one definition and compare with those in NCRP (1985); also refer Martin, 2000]. The reader involved in compliance work is cautioned to use the appropriate formula and terminology needed to comply with the public health and environmental laws. For purely scientific publications, one should unambiguously define the formulae employed for sensitivity. Definitions of other terms used in data quality control are given in EML (1997). The formulae described for L_C and LLD below are applicable for the simple counting situation when radon and thoron progeny are not present. If they are present, more complicated expressions of L_C and LLD are required (Allen, 1997).

L_C (sometimes referred to as the *a posteriori decision limit*) is the *net* count rate, that is, gross count rate minus background count rate, for deciding if a measurement is statistically different from background. One concludes that there is no radioactivity in the sample if the actual net counts of the sample are $<L_C$ or that there is radioactivity present if the actual net counts are $>L_C$. A preselected risk for concluding falsely that activity is present (usually 5%) is incorporated into the formula for L_C (NCRP, 1985; Martin, 2000). For the situation where the background and sample count times are equal, we have

$$L_C = 2.33(S_0), \tag{5.25}$$

where L_C is the critical level (cpm), $S_0 = (R_B/t_B)^{1/2}$ is the standard deviation (1σ) of the background measurement (cpm), t_B is the background count time (min), and R_B is the background count rate (cpm).

Note: Any units of time for the count rates may be used as long as they are kept consistent in the equation.

The LLD (sometimes referred to as the "detection limit") (NCRP, 1985; Martin, 2000) is the minimum *a priori* net count rate necessary in a sample in order to detect its radioactivity with a specified degree of confidence. It accounts for the statistical variation of both the background and sample counts. The LLD incorporates a probability of incorrectly concluding that radioactivity is not present when in fact it is. Again, this probability is usually chosen to be small (5%) as is done in the following equation (Brodsky, 1992; NRC, 1997; Martin, 2000) where the counting time (*t*) for the background and the sample are equal:

$$\text{LLD} = \left(\frac{3}{t}\right) + 4.65(S_0). \tag{5.26}$$

EXAMPLE

If the investigation level or regulatory guideline specifies a concentration limit of 1×10^{-12} µCi mL^{-1}, then the LLD of the measurement system must yield an air concentration lower than this in order for the field measurement to show regulatory compliance. Assume that the volume of air sampled for the background measurement was 25 m^3, that a 1-min count was performed for both sample and background, that the background of the counter with the background filter in place is 75 cpm, and that the counting efficiency is 0.20 cpm dpm^{-1}. Then,

$$S_0 = \left(\frac{75}{1}\right)^{1/2} = 8.7\,\text{cpm},$$

$$L_C = 2.32(8.7) = 20\,\text{cpm},$$

$$\text{LLD} = \left(\frac{3}{1}\right) + 4.65(8.7) = 43\,\text{cpm}.$$

This LLD converts into an MDA of 3.9×10^{-12} µCi mL^{-1} using the given values of counting efficiency, and sample volume and conversion factors of 2.22×10^6 dpm µCi^{-1} and 1×10^6 mL m^{-3}, respectively.

A net count rate $<L_C$ (20 cpm) is not distinguishable from the background at the 95% confidence level. The LLD is the net count rate value a sample must have when counted for 1 min (in this case) that will be, 95 times out of 100, reported as activity detected. Comparing the LLD of 3.9×10^{-12} µCi mL^{-1} with the guideline value of 1×10^{-12} µCi mL^{-1} also indicates that the measurement system is not sensitive enough for the task at hand. To increase the sensitivity, one may

increase the air sampling time. However, the effects from loading the filter with excessive dust must be kept in mind for very long sampling periods. In this example, the original flow rate or the sampling time could be increased by *at least* a factor of 3.9 (3.9×10^{-12} µCi mL^{-1}/1×10^{-12} µCi mL^{-1}) = 3.9 to achieve an (barely) acceptable LLD. Alternatively, one may achieve similar results by decreasing the background count rate, increasing the count time, or increasing the counting efficiency independently or in combination so that the LLD is decreased by *at least* a factor of 3.9.

General versions of the formulas for L_C and LLD may be found in the literature to account for unequal background and counting times (Martin, 2000). A version in NCRP Report 97 (NCRP, 1988) takes into consideration that multiple background measurements may be used to decrease the LLD of the measurement system. The reader is urged to check that the formulas used are appropriate for the task at hand, especially if it involves regulatory compliance (see, e.g., NRC, 1992; NRC, 1997).

QUALITY ASSURANCE

Airborne radioactivity analyses should periodically be verified using blind methods such as intercomparisons with reference labs. The former EPA program for proficiency in Rn-222 measurements is an example (EPA, 1997). For many years, the U.S. Department of Homeland Security Environmental Measurements Laboratory* had run a Quality Assessment Program for government contractor labs and other participants that included air filter measurements of various radionuclides and quality assurance measurements for gamma-spectroscopy (EML, 1998).

Databases, especially large ones, are routinely maintained using computer spreadsheet software. These programs are useful for repetitive calculations. However, the details of the calculations are not usually explicit, making it necessary to validate the spreadsheet computations by inputting data with known results or by checking the results against manual calculations. It is perhaps best to have these validations conducted by an individual other than the one who inputs the sampling data. Spreadsheet data should be reviewed periodically during the course of data input with an end towards identifying outlying results that may be the result of computational error. Validations and reviews are necessary because a flawed spreadsheet, like any mis-programmed software, will repeat an error and cause continued miscalculations in subsequent results.

Operation of an air-monitoring program requires that certain quality controls be implemented daily. These include but are not limited to the following:

Operability checks of air-sampling and radiological counting equipment
Testing for air leaks in the components of a sampling system
Responding to equipment failures
Maintaining equipment operational logbooks
Implementing realistic concentration levels for air-monitoring alarms (to prevent false alarms)
Properly identifying, handling, and storing air samples
Properly maintaining sample records (ID numbers, results, and sample field conditions)

An implementation guide for use by the U.S. Department of Energy and its contractors (DOE, 1994) provides a brief overview of these considerations for occupational workplace monitoring that may be helpful in outlining a general quality-control program for air sampling.

The U.S. Nuclear Regulatory Commission has made recommendations for occupational air measurements that include such details as quality assurance of effluent data (NRC, 1979), air-sampling

---

* This laboratory was a Department of Energy facility until 2003 when it was incorporated into the Department of Homeland Security and its name was changed to the National Urban Security Technology Laboratory.

instrument calibration (NRC, 1980), and measurement location in the workplace (NRC, 1992). The American National Standards Institute has also issued performance guidelines for sampling airborne radioactive materials (ANSI, 1989, 1993).

REFERENCES

Allen, D.E., Determination of MDA for a two-count method for stripping short-lived activity out of an air sample, *Health Phys.* 73, 512–517, 1997.

Altshuler, B. and Pasternack, B., Statistical measures of the lower limit of detection of radioactivity counter, *Health Phys.* 9, 293–298, 1963.

ANSI, *Performance Specifications for Health Physics Instrumentation—Occupational Airborne Radioactivity Monitoring Instrumentation*, ANSI N42.17B, American National Standards Institute, Inc., New York, 1989.

ANSI, *Guide to Sampling Airborne Radioactive Materials in Nuclear Facilitates*, ANSI N13.1-1969 (R1993), American National Standards Institute, Inc., New York, 1993.

ANSI, *Calibration and Use of Germanium Spectrometers for Measurement of Gamma-Ray Emission Rates of Radionuclides*, ANSI N42.14, American National Standards Institute, Inc., New York, 1999.

Bicron, *On-line Product Catalog*, 2007 [available at www.bicron.com].

Brodsky, A., Exact calculation of probabilities of false positives and false negatives for low background counting, *Health Phys.* 63, 198–204, 1992.

Canberra, Meriden, CT, 2007 [www.canberra.com].

Cember, H., *Introduction to Health Physics*, 2nd ed., Pergamon Press, Elmsford, NY, 1992.

Currie, L.A., Limits for qualitative detection and quantitative determination. Applications to radiochemistry, *Anal. Chem.* 40, 586–593, 1968.

DOE, *Implementation Guide—Workplace Air Monitoring*, G-10 CFR 835/E2—Rev. 1, U.S. Department of Energy, Washington, DC, 1994.

EML, *EML Procedures Manual*, HASL-300, 28th ed. U.S. Department of Energy—Environmental Measurements Laboratory, New York, NY, 1997. [www.eml.st.dhs.gov].

EML, *Semi-Annual Report of the Department of Energy, Office of Environmental Management, Quality Assessment Program*, EML-594, U.S. Department of Energy—Environmental Measurements Laboratory, New York, NY, USA and National Technical Information Service, U.S. Department of Commerce, Springfield, VA, 1998 [www.eml.st.dhs.gov].

EPA, *National Radon Proficiency Program Guidance on Quality Assurance*, EPA 402-R-97-012, United States Environmental Protection Agency, Washington, DC, 1997 [www.epa.gov/iaq/radon/proficiency.htm].

Firestone, R.B. and Shirley, V.S., *Table of Isotopes*, 8th ed., and CD-ROM, John Wiley & Sons, New York, 1996.

Fisenne, I., A short history of ZnS, *Health Phys.* 40, 739–741, 1981.

Gilbert, R.O. and Kinnison, R.R., Statistical methods for estimating the mean and variance from radionuclide data sets containing negative, unreported or less-than values, *Health Phys.* 40, 377–390, 1981.

Hallenbeck, W.H., *Radiation Protection*, CRC Press LLC, Boca Raton, FL, 1994.

Helgeson, G.L., Determination of concentrations of airborne radioactivity, *Health Phys.* 9, 931–942, 1963.

Kessler, M.J., *Liquid Scintillation Analysis Science and Technology*, Packard Instrument Company (now Perkin Elmer), Meriden, CT, 1989.

Knoll, G.F., *Radiation Detection and Measurement*, 3rd ed., John Wiley & Sons, New York, 2000.

Kocher, D.C., *Radioactive Decay Data Tables*, DOE/TIC-11026, National Technical Information Service, U.S. Department of Commerce, Springfield, VA, 1981.

L'Annunziata, M.F., *Handbook of Radioactivity Analysis*, Academic Press, New York, 1998.

Mann, W.B, Ayres, R.L., and Garfinkel, S.B., *Radioactivity and Its Measurement*, 2nd ed., Pergamon Press, Oxford, 1980.

Martin, J.E., *Physics for Radiation Protection*, John Wiley & Sons, New York, 2000.

NCRP, *A Handbook of Radioactivity Measurements*, 2nd ed., National Council on Radiation Protection and Measurements Report No. 58, NCRP, Bethesda, MD, 1985.

NCRP, *Measurements of Radon and Radon Daughters In Air*, National Council on Radiation Protection and Measurements Report No. 97, NCRP, Bethesda, MD, 1988.

NRC, Quality Assurance For Radiological Monitoring Programs (Normal Operations)—Effluent Streams and the Environment, Regulatory Guide 4.15, U.S. Nuclear Regulatory Commission, Washington, DC, 1979.

NRC, *Calibration and Error Limits of Air Sampling Instruments for Total Volume of Air Sampled*, Regulatory Guide 8.25, U.S. Nuclear Regulatory Commission, Washington, DC, 1980.

NRC, *Air Sampling in the Workplace*, Regulatory Guide 8.25, U.S. Nuclear Regulatory Commission, Washington, DC, 1992.

NRC, *Multi-Agency Radiation Survey and Site Investigation Manual (MARSSIM)*, U.S. Nuclear Regulatory Commission, NUREG 1575, National Technical Information Service, Springfield, VA, 1997.

NUREG, *Manual for Conducting Radiological Surveys in Support of License Termination, NUREG/CR-5848 and ORAU-92/C57*, U.S. Nuclear Regulatory Commission, Division of Regulatory Applications, Office of Nuclear Regulatory Research, Washington, DC, 1992.

Ortec, *Modular Pulse-Processing Electronics and Semiconductor Radiation Detectors 97/98*, EG&G Ortec (now Ortec, part of Ametek, Inc.), Oak Ridge, TN, 1997 [www.ortec-online.com].

Pasternack, B.S. and Harley, N.H., Detection limits for radionuclides in the analysis of multi-component gamma-ray spectrometer data, *Nucl. Instrum. Methods* 91, 533–540, 1971.

PGT, *Princeton Gamma-Tech Nuclear Products Catalog*, Princeton Gamma-Tech Inc., Princeton, NJ 08542, 2002 [www.pgt.com].

Sanderson, C.G., Cohen, L.K., Goldin, A., Jarvis, A.N., Kanipe, L., Sill, C., Trautman, M., and Kahn, B., Quality Assurance for Environmental Monitoring Programs, U.S. Environmental Protection Agency Report EPA 520/1-80-012, U.S. Environmental Protection Agency, Washington, DC, 1980.

Sauter, R., Disalvo, R., Tice, G., and Pierzynski, E.J., *Calibration of Ion Chamber Air Monitor for Various Radioactive Gases*, Nucleonics Data AN-0012 5/17/85, Becton Dickinson Diagnostic Instrument Systems, Towson, MD, 1985.

Scarpitta, S.C and Fisenne, I.M., *Calibration of a Liquid Scintillation Counter for Alpha, Beta and Cerenkov Counting*, U.S. Department of Energy—Environmental Measurements Laboratory, Report EML-583, New York, NY and National Technical Information Service, Springfield, VA, 1996.

Scintco, Scintpack S.A. *Laboratory Data Acquisition, Evaluation & Reporting System Version 3.02*, Scintco, Augusta, NJ, 1996.

Turner, J. E., *Atoms, Radiation, and Radiation Protection*, 2nd ed., John Wiley & Sons, New York, 1995.

Wood, M.J., McElroy, R.G.C., Surette, R.A., and Brown, R.M., Tritium sampling and measurement, *Health Phys.* 65, 610–627, 1993.

Wood, M.J., Hong, A., Cross, W.G., Nunes, J.C., and Leon, J.W., Calibration of portable tritium-in-air monitor for various radioactive gases, *Health Phys.* 72, 423–430, 1997.

6 The Physics of Aerosols

Erno Sajo

CONTENTS

Introduction	77
Aerosol Size Distributions	78
Aerosol Characterization	79
Lognormal Distribution	81
The Junge Distribution	82
The Gamma Distribution	82
Multimodal Distributions	83
Particle Motion in Gas	84
Macroscopic and Microscopic Fluid Properties	84
The Mean Free Path	86
Diffusion of Particles and Molecules	88
Drag Force and Cunningham's Slip Correction	93
Nonspherical Particles	96
Aerosol Transport	102
Moments of the Particle Size Spectrum	102
Aerosol Phase Space	106
Particle Deposition and the General Dynamic Equation of Aerosol Transport	110
Coagulation	115
The Sectional Method of Solving the Coagulation Problem	117
The Method of Moments	120
Coagulation Kernels	124
Brownian Coagulation—Diffusion Regime	124
Brownian Coagulation—Slip Flow Regime	125
Brownian Coagulation—Free Molecular Flow Regime	126
Fuchs' Method for All Brownian Regimes	127
Gravitational Coagulation	127
Simultaneous Brownian and Gravitational Coagulation	129
Coagulation by Turbulent Diffusion	130
Simultaneous Coagulation Mechanisms	132
References	132

INTRODUCTION

Aerosols are solid or liquid particles suspended in gas. They belong to a group known as colloids—a mixture in which one substance is divided into very small particles and dispersed throughout a second substance. Colloidal particles are larger than molecules, and thus they do not form a solution. Colloids are classified according to the phase of the particles and that of the medium of dispersion. In this way, for example, a gas dispersed in liquid or solid forms a foam (consider whipped cream or Styrofoam), and a solid dispersed in a liquid forms a sol, for example, paint, sometimes referred to as liquisol or hydrosol as opposed to aerosol. In this chapter, the terms particle and aerosol will be used interchangeably.

Aerosols are ubiquitous in our environment. In 1 cm³ of clean air the number of particles is about 1000, and their size ranges from a few nanometers to tens of micrometers. These particles originate in both natural processes, such as condensation of water or generation of pollens, and anthropogenic sources, such as engineered combustion or grinding operations, to name only a few. Generally, they have complex shapes and may have complex chemical or radiochemical compositions.

The particles move through the air chiefly driven by extraneous forces, such as air currents and gravity, depending on their size. However, besides these transport phenomena, there are other forces that govern their movement and make them deposit, resuspend, and collide with each other. The size range of these particles often spans 15 orders of magnitude in volume, all simultaneously present in the same volume of air. Thus, apart from their macroscopic fluid dynamic behavior, it is important to understand their interaction with the surrounding air molecules, and with each other. The molecules of air, which are mainly composed of N_2 and O_2, are approximately 0.3 nm in diameter. The average separation distance of these molecules is about 10 times this. The number density of electrically charged air molecules or ions in 1 cm³ is about the same as that of the aerosol particles in clean air. Thus, it is obvious that molecular–ionic–particle interactions cannot be neglected for particles of small size, and that the size distribution of the particles is one of the most important determinants of the aerosol field.

AEROSOL SIZE DISTRIBUTIONS

Before embarking on the description of various size distributions that are most commonly encountered, the concept of size must be clarified. Most aerosols are not regularly shaped objects. Liquid particles at low velocity are almost spherical, but solid particles occur in many shapes and can agglomerate to form three-dimensional (3D) chains or amorphous aggregates (Figure 6.1). Thus, a measure of their size in terms of single distance, which is a 1D description, such as particle diameter, may not adequately describe the particle. The size of an irregular object depends on the length scale used to measure it. This problem was analyzed by Mandelbrot (1977) and others who proposed that the perimeter of an irregular object is proportional to the scale length raised to the

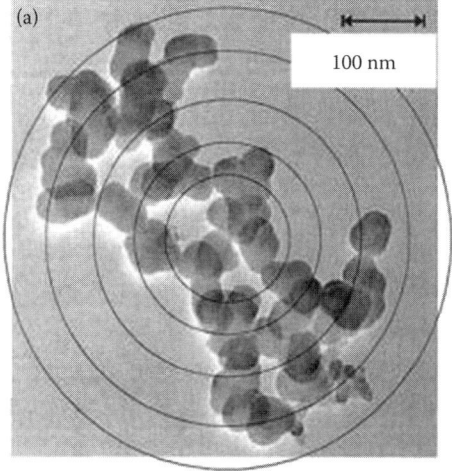
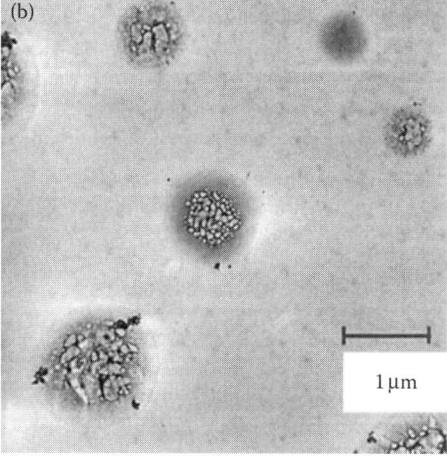

FIGURE 6.1 Electron micrographs of various aerosol particles. (a) shows a folded chain-like aggregate; (b) shows droplets evaporated in the electron microscope. (Adapted from Xiong, C. and Friedlander, S.K. *Proc. Natl. Acad. Sci.* 98:11851–11856, 2001.)

The Physics of Aerosols

power of $(1 - D)$, where D is a fractional dimension. A number of investigators have carried out measurements on the shape and dimensionality of aggregates as a function of the process that generates them. Most recently, Xiong and Friedlander (2001) measured the fractal dimensions of such particles. They found that for initial or primary particles of 0.1 µm, the fractal dimension, D, increased from nearly 1 to above 2, as the number of particles making up the aggregate ranged from 10 to just below 200.

The importance of particle morphology lies in the fact that their interaction with one another, the heat and mass transfer to them, and some of their chemical properties are functions of shape and size. Yet, because each particle is somewhat differently shaped, even in a monodisperse system with a single method of particle generation, it would be impractical to introduce a shape function into the most commonly used governing equations, although some of the formalism is capable of accepting it. A notable exception is the settling velocity that is usually generalized to include a dynamic shape factor, as will be discussed later in this chapter.

By considering an equivalent particle diameter, some of the important properties of the aerosol particle can be adequately described. An equivalent diameter may be defined as the diameter of a spherical particle that gives the same result as the nonspherical particle when measured the same way. The objective of the measurement may vary, thus the equivalent diameter will also vary depending on the property of interest. For example, the surface-equivalent diameter of a particle is the diameter of a spherical particle whose surface area is identical to the surface of the particle in question. Of course, there are many properties of interest that can be used to define an equivalent diameter.

AEROSOL CHARACTERIZATION

The aerosol is often characterized in terms of mass (µg cm^{-3}), volume (µm^3 cm^{-3}), or number concentration (cm^{-3}) with respect to unit volume of the carrier fluid. The number concentration is often referred to as number density. These are properties of the aerosol that may change in space and time, and they describe the aerosol "phase space." The aerosol phase space, as defined here, gives the detailed space–size–concentration–directional distribution of the aerosol field. A generalized aerosol property may be expressed in terms of the number concentration and the volume of the particles. Let $q(r, v, t)$ be a character or aspect of an aerosol, that is, an aerosol property. Then $q(r, v, t)$ may be written as

$$q(r,v,t) = \alpha v^\gamma n(r,v,t), \tag{6.1}$$

and the total or integral aerosol property may be written as

$$Q(r,t) = \int_0^\infty q(r,v,t)\,dv. \tag{6.2}$$

Here, r is the spatial variable (usually a vector), v is the volume of the particle, t is the time, and $n(r, v, t)$ is the space and time-dependent number density of particles having volume v. Often, the dependence on space and time is understood, and the above equations lack the variables r and t. By assigning appropriate values for the parameters α and γ, the aerosol property may represent a variety of distributions. For example, for $\alpha = 1$, $\gamma = 1$, the aerosol property, $q(r, v, t)$, represents the aerosol volume distribution, while $Q(r, t)$ expresses the total volume concentration, and for $\alpha = 1$, $\gamma = 0$, $q(r, v, t)$ represents the aerosol number distribution. Other distributions, such as mass and surface distributions, may also be defined in this way. Note that α and γ are not restricted to integers, but in the special case when $\alpha = 1$, $Q(r, t)$ represents the "moments" of the function

$n(r, v, t)$, which we may notate as $Q^{(\gamma)}(r, t)$. It is further possible to use a different independent variable than the particle volume and to include the direction of the moving particle, $\mathbf{\Omega}$; however, the function $q(r, v, t)$ is preferred because many experimental measurements are expressed in terms of particle volume and because it permits mathematical simplifications in the description of the phase space.

It is often convenient and practical to relate the aerosol size distribution to values that can be either easily observed or expressed by mathematical means. In this way the mean or average particle size, the median size, and the standard deviation corresponding to various distributions, for example, number, volume, surface, and mass, have gained widespread use. Some of these aerosol properties are related to the moments of the size spectrum, expressed in terms of particle diameter or volume. For example, the zeroth moment, $Q^{(0)}(r, t)$, is equivalent to the total particle number concentration, while the first moment, $Q^{(1)}(r, t)$, represents the total volume concentration, which is the volume fraction of the dispersed particles in the fluid. If the diameter (or volume) of the particle is denoted with ξ, and the total number of particles with N, the mean diameter (or volume) and the square of the standard deviation are written, respectively, as

$$N = \int_0^\infty n(\xi)\,d\xi,$$

$$\bar{\xi} = \frac{1}{N}\int_0^\infty \xi n(\xi)\,d\xi, \qquad (6.3)$$

$$\sigma^2 = \frac{1}{N}\int_0^\infty (\xi - \bar{\xi})^2 n(\xi)\,d\xi.$$

Note that the mean is given by the ratio of the two moments, $Q^{(1)}/Q^{(0)}$. The median diameter is simply the value of the particle diameter which halves the number distribution, that is, $N/2$ is above this size and $N/2$ is below the size. In a similar manner, particle diameter of average mass, ξ_m, may be expressed as

$$\xi_m = \left[\frac{1}{N}\int_0^\infty \xi^3 n(\xi)\,d\xi\right]^{1/3}, \qquad (6.4)$$

and the mean mass diameter, ξ_μ, is

$$\xi_\mu = \frac{1}{N\xi_m^3}\int_0^\infty \xi^4 n(\xi)\,d\xi. \qquad (6.5)$$

The above integrals are evaluated for all particles sizes present in the aerosol. The notation of ξ for the particle diameter is used here, lest the differential d is confused with the diameter. For this reason, this notation will be used whenever the meaning of the letter d is not self-explanatory.

Fractional moments also possess physical significance. For example, if the particles are assumed to be spherical and the moments are expressed in terms of particle volume, the 1/3rd moment, $Q^{(1/3)}$, is related to mean particle diameter, the 2/3rd moment, $Q^{(2/3)}$, is proportional to the total surface concentration and it is also related to the mean square diameter. When written in terms of diameters as opposed to volume, it is the second moment, which is proportional to the surface concentration, while the third moment is related to the volume concentration. Further details are presented in the *Aerosol Transport* section of this chapter.

The Physics of Aerosols

There is a wide range of methods for measuring particle sizes. However, because particle sizes cover many orders of magnitude, there is no single method or instrument that can provide reliable measurements of all particles. Further, since the number of particles per unit volume is large, having different physical and chemical properties, we are interested in the bulk behavior of aerosols in most applications, rather than in their individual description. In this way, we are interested in the distribution function of the aerosol property in terms of its size.

LOGNORMAL DISTRIBUTION

Most industrial and natural aerosols have been observed to have a geometric symmetry about the mean size of the particles. This is because intermediately sized particles remain suspended in the air for a longer time than large particles or small particles. Large particles are removed mainly by gravitational settling or filtering, while small particles are removed by coagulation, diffusion, and other phoretic effects. A geometrically normal distribution means that the particles are symmetrically distributed about the mean multiplicatively. Thus, whereas in a Gaussian or normal distribution, the same number of particles are observed equidistantly or arithmetically from the mean, $n(d_m + \delta) = n(d_m - \delta)$, in a lognormal distribution, the same number of particles are observed at the same fraction of the geometric mean: $n(d_g \delta) = n(d_g/\delta)$. In mathematical formalism, the aerosol number distribution may be written as

$$n(d,t) = \frac{N(t)}{\sqrt{2\pi}\ln(\sigma_g)d} \exp\left[-\frac{\{\ln(d)-\ln(d_m)\}^2}{2(\ln\sigma_g)^2}\right]. \tag{6.6}$$

Here, d represents the particle diameter, σ_g is the geometric standard deviation, d_m is the geometric mean diameter which is the mean of the logarithms of d, and $N(t)$ is the total volume concentration as a function of time. It is also possible to use a different independent variable than the particle diameter. For example, if the volume of the particle is used, then the equation for the number distribution becomes

$$n(v,t) = \frac{N(t)}{\sqrt{2\pi}\ln(\sigma_{g,v})v} \exp\left[-\frac{\{\ln(v)-\ln(v_m)\}^2}{2(\ln\sigma_{g,v})^2}\right],$$

$$\sigma_{g,v} = \sigma_g^3, \tag{6.7}$$

$$v_m = \frac{\pi d_m^3}{6}.$$

Here, v is the particle volume, $\sigma_{g,v}$ is the geometric standard deviation of the particle volume, and v_m is the geometric mean volume, equivalent to the mean of the logarithms of the volumes of the number distribution. Note that the above formalism uses the proper definition of the lognormal probability density function. In some literature, d or v is missing from the denominators of the above equations, effectively defining $n(d, t)d$ or $n(v, t)v$ as a particle distribution function, and plots of size distributions are sometimes given using decimal logarithm instead of natural logarithm.

As opposed to the Gaussian distribution where mean, median, and mode are identical quantities, the lognormal distribution is skewed with separate and distinct values for mean, median, and mode. Table 6.1 summarizes the parameters of the lognormal distribution when the number distribution is expressed in terms of particle diameter, and gives their relationships to each other (Aitchison and Brown, 1957).

TABLE 6.1
Relations of Mean, Median, Mode, and Standard Deviation in Lognormal Distributions

Symbol	Meaning	Relationship
d_m	Mean of natural logarithms of d	$\ln(d_\mu)$
σ_g	Geometric standard deviation	$\exp(\sigma)$
σ	Standard deviation of logarithms of d	$\ln(\sigma_g)$
d_μ	Median	$\exp(d_m)$
$\bar{d}$	Arithmetic mean = first moment of the distribution	$\exp(d_m + \sigma^2/2)$
$\hat{d}$	Mode = the diameter at which $n(d)$ is maximum	$\exp(d_m - \sigma^2)$

THE JUNGE DISTRIBUTION

The Junge distribution (Junge, 1963), or in its generalized form the inverse power distribution, describes an aerosol system where the mass is uniformly distributed in equal logarithmic intervals:

$$m \frac{\int_\xi^{2\xi} n(\xi) d\xi}{\Delta \ln(\xi)} = \text{constant}. \tag{6.8}$$

Here, m is the mass of the particles and ξ is the diameter of the particles. This means that the size range of $(\xi, 2\xi)$ contains the same mass of particles as the size range of $(2\xi, 4\xi)$, etc. The above equation, in its generalized form, yields an inverse power distribution as follows:

$$n(\xi) = A\xi^{-\alpha}.$$

The factor A is a constant, and α usually ranges from 3 to 5. The original Junge distribution used $\alpha = 4$. This distribution appears as a straight line when plotted in a log–log scale, and has been found useful in describing certain atmospheric aerosol data. However, in fitting a measured size distribution, due to inability of instrumentation to resolve the entire size spectrum, a selection of a minimum and a maximum particle size is needed. Because the choice of these extremis is dependent on the instrumentation, the computed mean and median will vary accordingly. A detailed examination of atmospheric aerosol size spectrum (Whitby et al., 1972) showed that simple functions such as the power law are inadequate to describe the complex and dynamic behaviors of atmospheric aerosols.

THE GAMMA DISTRIBUTION

Whereas on a linear scale both lognormal and gamma distributions appear skewed toward the smaller particles, on a logarithmic scale, the gamma distribution appears skewed to the higher particle sizes while the lognormal distribution is symmetric. Thus, the gamma distribution is a better fit to aerosol size distributions that have a more abrupt high-end tail in the logarithmic scale. This is often the case when the aerosol generation process involves an upper size limit, such as mechanical grinding. In mathematical formalism, the function is as follows:

$$n(d) = A \frac{d^c}{\Gamma(c+1)} \left(\frac{c}{\hat{d}}\right)^{c+1} \exp\left(-d \frac{c}{\hat{d}}\right), \tag{6.9}$$

where $\hat{d}$ is the mode of the distribution, A is a normalization constant, and c is a parameter of choice. Figure 6.2 illustrates the lognormal and gamma distributions. For comparison purposes, the parameters of the functions are selected such that the modes of the distributions coincide.

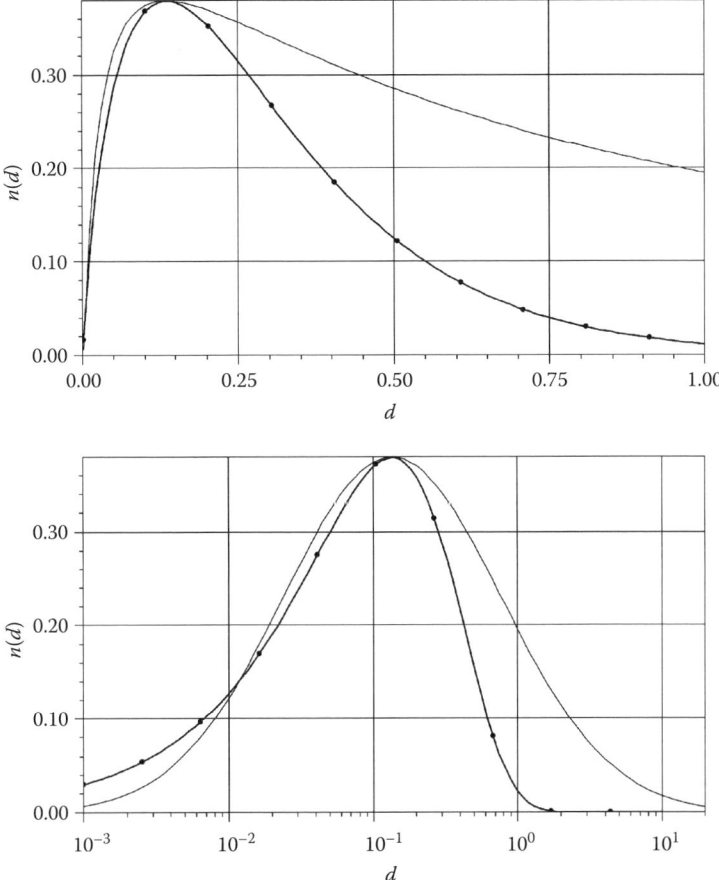

FIGURE 6.2 Lognormal and the gamma distributions in arbitrary units. The gamma distribution is shown with bullets. The mode of both distributions is at 0.1353. The mean and the variance (σ^2) of the lognormal distribution are 1 and 3, respectively. Parameters in the gamma distribution, Equation 6.9, are $A = 0.16$ and $c = 0.8$.

MULTIMODAL DISTRIBUTIONS

Measurements of various size distributions indicate a frequent occurrence of a multimodal distribution for the aerosol volume concentration. For example, common atmospheric aerosols exhibit a trimodal behavior (Figure 6.3). Invariably, the aerosol populations in the modes are formed by different mechanisms, and originate in different sources, thus they may have different chemical compositions. There is no universal mathematical form that would describe such a size distribution. The generally accepted practice is to superimpose the different modes of the same or different size distribution functions discussed above. In the case of atmospheric aerosols, the volume distribution, $V(d)$, which represents the aerosol volume concentration as a function of the particle diameter, can be written as the sum of three lognormal distributions:

$$V(d) = \frac{1}{d\sqrt{2\pi}} \sum_{i=1}^{3} \frac{V_i}{\ln \sigma_i} \exp\left[-\frac{\ln^2(d/d_i)}{2\ln^2 \sigma_i}\right], \tag{6.10}$$

where V_i is the total aerosol volume, d_i is the geometric mean aerosol diameter, and σ_i is the geometric standard deviation in mode i. The three modes here correspond to two fine modes, nucleation

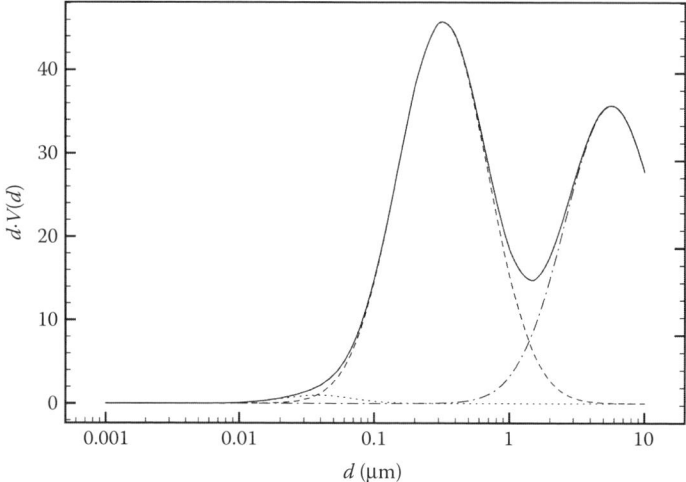

FIGURE 6.3 Trimodal size distribution of atmospheric aerosols in urban environment, based on Whitby's data (1978).

and accumulation, and a coarse mode. The coarse mode is thought to be affected by emissions, sedimentation, and deposition processes, while the fine modes are influenced by condensation and coagulation. Whitby (1978, 1981) measured these parameters in various atmospheric conditions and found that in an urban environment $V_i = (0.63, 38.4, 30.8)$ (µm)^3, $d_i = (0.038, 0.032, 5.7)$ µm, and $\sigma_i = (1.8, 2.16, 2.21)$ where the indices are $i = 1, 2, 3$ and they represent the nucleation, accumulation, and coarse modes, respectively.

The formula shown above, with a suitable upper limit of summation, may also be used in other cases where a simple superposition of lognormal distribution provides an adequate description of the size spectrum.

PARTICLE MOTION IN GAS

MACROSCOPIC AND MICROSCOPIC FLUID PROPERTIES

As aerosol particles are suspended in the gas, they experience various forces due to their interactions with the atoms and molecules of the medium, such as drag. In dilute suspensions, where the number density of particles is much smaller than that of the gas molecules, interparticle collisions are rare compared to interactions between particles and molecules. Therefore, it is important to understand the behavior of gas molecules. The nature of collisions between particles will be discussed later.

One of the fundamental concepts of classical statistical molecular theory was laid down by Ludwig Boltzmann (1844–1906) who formulated the distribution of the amount of energy shared among identical but distinguishable particles. The Boltzmann distribution function expresses the probability of a particle or molecule having a particular energy, E:

$$f(E) = A e^{-E/kT}. \tag{6.11}$$

Here, k is Boltzmann's constant (1.38065 × 10^{-23} J/K), T is the absolute temperature, and A is a normalization factor. The latter is needed because the probability that the particle or molecule has any energy is one. That is, the integral of the above equation for all energies equals 1, that is, normalized. Because a particle with low energy can be generated by many more ways than one with high energy, the distribution is skewed to the lower energies, which means that more particles can be found with lower energies and it is unlikely that a particle will attain far more energy than the average. If the distribution function is subjected to the constraint that the number of particles is constant and the energy is conserved, the normalization factor can be found as $A = 1/kT$. The integral average energy of the molecule, $\langle E \rangle$, when the energy is randomly distributed among the available energy states can be found as follows:

$$\langle E \rangle = \frac{\int_0^\infty E f(E) dE}{\int_0^\infty f(E) dE} = kT. \tag{6.12}$$

The Boltzmann probability distribution can be used to find the 1D velocity distribution function (e.g., in the x-direction) by substituting $E = 1/2\ mv_x^2$ in Equations 6.11 and 6.12, where m is the mass of the molecule, and performing normalization in the velocity domain of $v_x \in (-\infty, +\infty)$:

$$f(v_x) = \sqrt{\frac{m}{2\pi kT}}\, e^{-(mv_x^2/2kT)}. \tag{6.13}$$

This equation describes the motion of the gas molecules in one direction only. According to it, it is equally likely that the molecules move in positive versus negative directions, and thus the most likely speed of the molecules along this direction is zero. Obviously, a gas with nonzero pressure has nonzero molecular speed; however, in the absence of extraneous forces or convective fluid motion, such as airflow, it will have a zero net displacement. This apparent contradiction is resolved in 3D space. The mean square 1D velocity can be expressed as $\langle v_x^2 \rangle = kT/m$, yielding a mean 1D kinetic energy of $\langle E_k \rangle = kT/2$, and by conversion to three dimensions, the mean kinetic energy becomes

$$\langle E_k \rangle = \left\langle \tfrac{1}{2} mv^2 \right\rangle = \frac{3}{2} kT. \tag{6.14}$$

The 3D velocity distribution function can be similarly obtained, and it is known as the Maxwell–Boltzmann distribution, or Maxwellian for short:

$$f(v) = 4\pi \left(\frac{m}{2\pi kT}\right)^{3/2} v^2 e^{-(mv^2/2kT)}. \tag{6.15}$$

This distribution function forms the basis of the kinetic theory of gases. It is a probability density, which describes the speed distribution of particles in a dilute system where the dominant particle interaction process is collision, while radiative and quantum effects are negligible. The unit of the function is probability per speed, or s/m when expressed in the SI system. The 1D form of this equation was derived by James Clerk Maxwell (1831–1879) in 1866, independently of Boltzmann, which explains the naming of the formula.

Comparing $f(v_x)$ and $f(v)$, it is readily seen that while in the 1D case the probability of finding a molecule with zero velocity is the highest, in the 3D case it is the lowest. Thus in the 3D case the probability of finding a particle with zero velocity is zero, while in the 1D case the zero probability is associated with speeds $-\infty$ and $+\infty$. Integrating the Maxwellian over all velocities, the mean molecular speed becomes $\langle v \rangle = \sqrt{(8kT/\pi m)}$ and the mean square velocity is $\langle v^2 \rangle = 3kT/m$, which is three times the magnitude of the 1D mean square velocity.

A dynamic nonequilibrium property of the aerosol carrier fluid, which is closely related to kinetic theory of gases, is the viscosity. If a shear stress is applied to a fluid by an extraneous force, for example, by moving a layer in the x-direction with a constant force, a velocity gradient will develop along the perpendicular direction, y, with respect to the direction of motion. The fluid velocity will be the highest at the point where the shear stress is applied and it gradually decreases with increasing distance from this point. This is a lateral momentum transfer between layers of the fluid whose magnitude depends on the internal friction of the molecules. Small friction results in a large lateral velocity gradient, as the momentum transfer between the molecules is low. This internal friction is called viscosity, which may be expressed as the ratio of the applied shear stress per unit area to the velocity gradient. It is a macroscopic property, which is a function of the state of the fluid. Fluids in which the viscosity is independent of the rate of shear are called Newtonian fluids. There are two commonly used viscosities: dynamic viscosity (denoted μ) and kinematic viscosity (denoted ν). The unit of the dynamic viscosity follows its original definition, kg/m-s in SI and dyne-s/cm² in CGS. The latter unit is also called Poise, and 1 P = 0.1 kg/m-s. The kinematic viscosity is simply the ratio of the dynamic viscosity to the density of the fluid, and it carries a unit of m²/s.

Using kinetic theory to describe the net flux of momentum between two layers of the fluid, assuming the molecules interact as hard spheres, the viscosity can be derived from first principles (Reid et al. 1977):

$$\mu = \frac{5(\pi MRT)^{1/2}}{16(\pi d^2)\Omega} = 26.69 \frac{(MT)^{1/2}}{d^2 \Omega} \quad [\mu P] \tag{6.16}$$

In this formula, M is the molecular weight, T is the gas temperature in Kelvins, d is the molecular diameter in Ångstroms, and Ω is the collision integral. Ω is unity if there are no intermolecular forces, that is, the molecules do not attract or repel one another. Values for the collision integral are given by Reid et al. (1977).

THE MEAN FREE PATH

As it was seen in the discussion of particle size distributions, aerosol particles encompass a large size range, from nanometers to hundreds of micrometers, simultaneously present in the same compartment. When expressed in terms of volume rather than in diameter, which is common in computational developments, they can easily span 15 orders of magnitude or more. At the smallest extreme, the particles are not much larger than the air molecules surrounding them. Therefore, their interactions with the air molecules determine their motion. At the opposite extreme, however, the granular nature of air is no longer a determining factor, as large particles experience a continuum of carrier medium. For example, the effective size of molecules in air is roughly 3.6×10^{-10} m (3.6 Å). Small aerosol particles experience individual collisions with these molecules while large particles, in the absence of other forces, tend to drift with the continuum. The flow regime, as the varying behavior of particles with respect to the air molecules is called, can be conveniently described using a dimensionless number, the Knudsen number, which relates the particle size to the mean distance an air molecule travels between collisions, or mean free path, λ. The Knudsen number may be written as Kn = λ/a, where a is the aerosol particle radius.

The collision frequency in an equilibrium gas is related to the mean molecular speed, $\langle v \rangle$, which is obtained by integrating the Maxwellian for all speeds. Using first principles, it can be shown that if the molecules are approximated as rigid spheres, each having a radius r, the collision frequency, κ, between species i and j becomes

$$\kappa_{ij} = \pi(r_i + r_j)^2 n_j \sqrt{\langle v_i \rangle^2 + \langle v_j \rangle^2}. \tag{6.17}$$

Here, subscript i identifies the projectile molecule and subscript j signifies the other species with which molecule i collides, including its own kind. Further, n_j is the number of molecules per unit volume and $\langle v \rangle$ is the mean molecular speed. The term $\pi(r_i + r_j)^2$ is the total collision cross section of interacting spheres i and j, and $(r_i + r_j)$ is their closest approach, which is sometimes called the collision diameter. Note that the literature inconsistently uses the notation σ_{ij} for either the collision

TABLE 6.2
Molecular Parameters for Selected Elements and Compounds

Compound or Molecule (Element)	Molecular Weight	Molecular Collision Diameter (Å)	Atomic Volume, V (Diffusional Volume Increment, v_i) (cm³/mol)	Diffusion Coefficient in Air at 1 atm and at 273 K (cm²/s)	Temperature Correlation Exponent, m
H_2 (H)	2.016	2.915	7.07 (1.98)	0.610	1.75
O_2 (O)	32.00	3.433	16.6 (5.48)	0.178	1.75
N_2 (N)	28.02	3.681	17.9 (5.69)	0.175	1.90
Cl_2 (Cl)		4.40	37.7 (19.5)	0.103	1.75
NH_3	17.03	3.441	14.9	0.201	1.75
SO_2 (S)	64.07	4.29	41.1 (17.0)	0.108	1.75
CO_2	44.01	3.996	26.9	0.142	1.70
CO	28.01	3.59	18.9	0.190	1.75
H_2O	18.02	2.649	12.7	0.230	1.74
Air (dry)	28.97	3.617	20.1	—	—

cross section or for the collision diameter. In an N-component gas, the collision frequency of species i with all other species, including its own kind, can be obtained by summing over all other species, j: $\kappa_i = \sum_j \kappa_{ij}$. Table 6.2 lists the molecular collision diameters, $2r_i$, for selected substances.

The mean free path can be determined from the collision frequency in gas by using Boltzmann's kinetic theory. It may be defined as the length of the path of a moving molecule divided by the number of collisions while traveling the path. Using the collision frequency above, and substituting $\langle v \rangle = \sqrt{(8kT/\pi m)}$, the mean free path of species i of an N-component gas is

$$\lambda_i = \frac{1}{\sum_{j=1}^{N} \pi(r_i + r_j)^2 n_j \sqrt{1 + (m_i/m_j)}}. \qquad (6.18)$$

Here, m_j is the mass of a single target molecule of type j: $m_j = M_j/N_A$, where M_j is the molecular weight and N_A is Avogadro's number. In the case of a single-component gas, having a molecular diameter d, the above equation simplifies to

$$\lambda = \frac{RT}{\pi d^2 N_A P \sqrt{2}} = \frac{kT}{\pi d^2 P \sqrt{2}}. \qquad (6.19)$$

Here, the number of molecules per unit volume is expressed using the ideal gas law and Avogadro's number: $N_A = 6.0221 \times 10^{23}$/mol, $R =$ universal gas constant $= 8.3145$ J/mol K, and $k =$ Boltzmann constant $= R/N_A = 1.38066 \times 10^{-23}$ J/K $= 8.617385 \times 10^{-5}$ eV/K.

Jennings (1988) measured the mean free path in air under various temperature and humidity conditions at atmospheric pressure (101.325 kPa) and found that at 20°C and 50% relative humidity, the value is 0.06544 μm. This is about 20 times of the average molecular separation, which is 0.0033 μm and gives rise to a collision frequency of 0.7×10^{10} s^{-1}. The corresponding mean time between collisions is 1.4×10^{-10} s. As it is obvious, the mean free path depends on the pressure and temperature of the gas, and therefore on its density. Using the above formula, the mean free path at other temperatures and pressures can be determined from a known value, λ_0:

$$\lambda = \lambda_0 \frac{P_0}{P} \frac{T}{T_0}. \qquad (6.20)$$

This formalism, however, does not consider the effect of the changing mean molecular size in the air as a function of gas constituents, such as water vapor. A more accurate way is to consider the gas viscosity, μ, as well (Reist, 1993):

$$\lambda = \lambda_0 \frac{P_0}{P} \frac{\mu}{\mu_0} \sqrt{\frac{T}{T_0}}. \tag{6.21}$$

In air, however, under ambient conditions, by virtue of Equation 6.16, the ratio of viscosities may be well approximated with the square root of the ratio of temperatures, thus resulting in the same formula as Equation 6.20. To correct for the remaining discrepancies, which are on the order of 2% or less for ambient conditions, Willeke (1976) applies a correction factor incorporated in Equation 6.20:

$$\lambda = \lambda_0 \frac{P_0}{P} \frac{T}{T_0} \left(\frac{1 + 110.4/T_0}{1 + 110.4/T} \right). \tag{6.22}$$

The reference temperature, reference pressure, and reference mean free path suggested by the U.S. National Institute of Standards and Technology are $T_0 = 296.15$ K, $P_0 = 101.3$ kPa, and $\lambda_0 = 0.0673$ μm, respectively (Kim et al., 2005).

For particles ranging from 10^{-3} μm to 100 μm in equivalent radius, the Knudsen number varies between 65 and essentially zero. At high Knudsen numbers, the aerosol particles are in the free molecular flow regime and experience forces that depend on the Maxwell molecular speed distribution function, while at low Knudsen numbers, the particles are many mean free paths in size, and the gas act as a continuous fluid. This has important implications in the computation of aerosol transport: in high Kn flow regimes (Kn > 1), the Boltzmann transport equation must be solved, whereas in low Kn flow regimes, Boltzmann's equation can be applied at the hydrodynamic limit, which affords significant reductions in complexity.

DIFFUSION OF PARTICLES AND MOLECULES

The Maxwell–Boltzmann distribution gains significance when the ability of the gas molecule or an aerosol particle to move with respect to its surroundings is investigated. Mass transfer in general, and aerosol transport in particular, is the result of several simultaneous phenomena, which may be cast into three major categories: convection, diffusion, and movement due to extraneous forces, such as gravity or an electromagnetic field. Convection occurs on a macroscopic level, and it is the transfer of mass due to fluid flow in the system. For example, an aerosol moved by airflow is convective transport. If there is a concentration gradient of the molecular or aerosol species across the system, particles will move on a microscopic scale from the higher concentration region to the lower concentration region due to diffusion. This microscopic movement will continue until equilibrium is established and the concentration distribution is uniform.

Diffusion may be present independently of convection. However, convection-induced turbulence can give rise to enhanced diffusion owing to eddy mixing processes. A temperature gradient can also induce diffusion of particles, called thermophoresis or thermal diffusion. Because a concentration gradient can also result in a temperature gradient (known as the Dufour effect), these processes are coupled.

In 1855, Adolph Fick formulated the concept that the diffusion mass transfer of one particle or gas species through a second one, which may be itself, is proportional to the gradient in the concentration of that species. Fick's law may be written for the mass flux as follows:

$$\bar{j}_A = -D_{AB} \nabla C_A. \tag{6.23}$$

Here, $\bar{j}_A$ is the mass flux (kg/m²-s), C_A (kg/m³) is the mass concentration of species A, and D_{AB} (m²/s) is the diffusion coefficient of species A through species or medium B. ∇C_A represents the concentration gradient at the location of interest. The negative sign in the equation indicates that the diffusion occurs in the direction from high to low concentrations. It is seen that the diffusion coefficient is a proportionality constant, which scales the concentration gradient; therefore it is related to the particle's ability to move along the gradient. In general, it is a function of the molecular or atomic properties of the medium; therefore, in a nonhomogeneous system it is a spatially dependent function. Because the diffusion process takes place in two ways along the concentration gradient: gas A diffuses into gas B and vice versa and the net mass flux, which is $\bar{j}$ summed over all species at the point of interest, becomes zero. Therefore the diffusion coefficient obeys the reciprocity condition, $D_{AB} = D_{BA}$, and in a binary compound only a single diffusion coefficient is defined.

In a multicomponent gas mixture, Fick's law can be expressed in terms of the species velocity (V_A) and the mixture velocity (V) via the bulk density of the species (ρ_A): $\bar{j}_A = \rho_A(V_A - V)$. In this formalism, species A diffuses through a multicomponent mixture or media, which consists of N species, including A. This and the previous form of the law, coupled with Boltzmann's kinetic theory, can be used to approximate the diffusion coefficient based on first principles, as a function of the collision frequency, κ_A:

$$D_{AB} = \frac{\langle v_A \rangle \sqrt{\langle v_A^2 \rangle}}{\ell \kappa_A} = \frac{\lambda_A}{\ell}\sqrt{\frac{3kT}{m_A}}. \qquad (6.24)$$

In this equation, ℓ is a dimensional constant whose value is 3 in the case of 1D diffusion, and $64/3\pi\sqrt{2}$, when three dimensions are considered. Note that medium B is a multicomponent mixture and the terms relating to its components, j, are inside the sum that appears in the equation for κ_A or λ_A (Equations 6.16 and 6.17). As previously, the above equation is an approximation because in deriving it, it is assumed that the particles are nondeformable spheres. It is important to recognize that because the collision frequency is a function of the species concentration, the diffusion coefficient varies with concentration. In practice, however, the system is sufficiently dilute, or at least so assumed, to make the dependence on concentration negligibly small.

The diffusion coefficient of a single-species gas A in the multicomponent mixture B can also be expressed in terms of common binary diffusion coefficients D_{ij}, where i is the single component in gas A, and $j = 1 \ldots N$ are components in gas B. This is desirable because contrary to multispecies mixtures, much experimental data exists for binary mixtures. This type of diffusion is described by the Stefan–Maxwell equation, whose appearance although similar, is different from Fick's law:

$$\nabla f_i = \sum_{j=1}^{N} \frac{C_i C_j}{C^2 D_{ij}} \left(\frac{j_j}{C_j} - \frac{j_i}{C_i} \right), \qquad (6.25)$$

where f_i is the mole fraction of species i in gas A and ∇f_i is its gradient. j_i and j_j are the mass fluxes of species i and j, and C_i and C_j are their concentrations, respectively. C is the mixture concentration and D_{ij} is the binary diffusion coefficient of the ij system. Note that the mole fraction here is simply the ratio of the species concentration to the mixture concentration. In this way, $f_i = C_i/C$ and $f_j = C_j/C$, and the term $C_i C_j/C^2 = f_i f_j$. Often the approximation is made that if gas B is a *homogeneous* mixture, then the mass flux of its components are nearly zero: $j_j \approx 0$, which simplifies the above equation, while introducing a caveat of $j \neq i$. From Fick's law, we can define the mixture diffusion coefficient, $D_{iB} = -j_i/\nabla f_i$, which gives a relation known as Blanc's law:

$$\frac{1}{D_{iB}} = \sum_{\substack{j=1 \\ j \neq i}}^{N} \frac{f_j}{D_{ij}}. \qquad (6.26)$$

This formula provides a good approximation for ternary systems (Mathur and Saxena, 1966). Deviations from Blanc's law have been observed, however, in many systems by both measurements (Sandler and Mason, 1968; Takata, 1985) and, most recently, by Monte Carlo computations (Benhenni et al., 2006). With adjustments, however, its applicability can be extended to many gas mixtures (Jovanovic et al., 2004).

Many authors have proposed semiempirical equations and correlations to estimate the diffusion coefficient based on experimental data. One of the most widely used correlations, which provides a low average discrepancy from measured binary diffusion coefficients in dilute systems, was proposed by Fuller et al. (1966) based on the earlier work by Gilliland (1934), which follows the Chapman–Enskog theory for solving Boltzmann's transport equation:

$$D_{AB} = 10^{-3} \frac{T^m}{P\left(V_A^{1/3} + V_B^{1/3}\right)^2} \sqrt{\frac{1}{M_A} + \frac{1}{M_B}} \text{ (cm}^2\text{/s)}. \tag{6.27}$$

Here, P is the pressure of the system (atm), T is the temperature (K), M_j is the molecular weight of the components, and m is the temperature correlation exponent, whose value is invariably about 1.75 for most gas mixtures of interest. V is the molar volume of the compound at the normal boiling point and it is a function of phase and allotrope. It may be estimated as the sum of the atomic and diffusion volume increments, v_i, associated with each element in the compound, as $V = \Sigma v_i$. Table 6.2 lists the molar volumes, V, and the incremental molar volumes, v_i, for selected compounds and atoms along with the molecular collision diameters and experimental diffusion coefficients in air with the corresponding temperature correlation exponent, m. Although the diffusion correlation Equation 6.27 gives a very good agreement with observations, within 4.3% on average for small molecules in light gases, it should be used only when measurements for the compound of interest are not available. The performance of the Fuller–Schettler–Giddings correlation was found to be poor for polar gases, such as ammonia, and for large molecules, such as hydrocarbons.

As Equation 6.27 shows, at low pressures, the diffusion coefficient is inversely proportional to the gas pressure, while it is proportional to the mth power of the temperature. This provides a basis for temperature and pressure correction of experimental data:

$$D_{AB}(T, p) = D_{AB}(T_0, p_0) \left(\frac{T}{T_0}\right)^m \left(\frac{p_0}{p}\right)^n. \tag{6.28}$$

$D_{AB}(T_0, p_0)$ is the observed diffusion coefficient at T_0 and p_0 conditions, while $D_{AB}(T, p)$ is the diffusion coefficient at the desired temperature and pressure. The exponent m listed in Table 6.2 is seen to vary depending on the gas. It also varies when the temperature changes over wide ranges, first increasing then decreasing. Therefore the applicability of Equation 6.27 is restricted to temperatures below $0.1T_r$. For most monatomic gases, the exponent n is unity. Complex compounds, however, show a nontrivial discrepancy from this simple scaling, and n may be below or above unity. In addition, at pressures above the critical point of the gas, the self-diffusion is nonlinear with both pressure and temperature. Mathur and Thodos (1965) showed that for reduced densities less than unity, $D_{AA}\rho$ is nearly constant for constant temperature. Thus, the value of $D_{AA}\rho$ at high pressures may be estimated using data observed at low pressure. They proposed the following correlation:

$$D\rho_r = 10.7 \times 10^{-5} \frac{T_r T_c^{5/6}}{M^{1/2} P_c^{1/3}}. \tag{6.29}$$

Here, T_c and P_c are the critical temperature and critical pressure, respectively. T_r is the reduced temperature, $T_r = T/T_c$, and ρ_r is the reduced density, $\rho_r = \rho/\rho_c$, with T in units of Kelvin and P in atmospheres.

In Brownian motion the gas molecules move on an irregular path, with speed determined by their thermal energy as shown by the Maxwell–Boltzmann distribution. Aerosol particles, suspended in this fluid, are thus subjected to random impacts by the fluid molecules. The degree at which such impacts influence their motion depends on the relative size of the aerosol particles to the molecules. In the limiting case when the particles are large compared to the mean free path of the fluid atoms or molecules, the fluid acts as a continuum, which is a determining factor in the particles' trajectory. In this way, Boltzmann's transport equation may be approximated by the diffusion equation, which indicates that the time rate of change of the particle concentration at spatial location r is proportional to the movement of particles, as follows:

$$\frac{\partial C(r,t)}{\partial t} = D\nabla^2 C(r,t). \tag{6.30}$$

Here, ∇^2 is the Laplace operator, often written as $\nabla^2 \equiv \Delta = \partial^2/\partial x^2 + \partial^2/\partial y^2 + \partial^2/\partial z^2$. By multiplying Equation 6.30 by x^2, y^2, and z^2, and integrating it over all space, it can be shown that the diffusion coefficient of the aerosol particles is related to the mean square distance of travel:

$$\langle r^2 \rangle = 6Dt. \tag{6.31}$$

As above, the brackets $\langle ... \rangle$ indicate an integral average or mean quantity. The problem of Brownian motion was investigated by Einstein (1956), who showed that using Stokes' law for the drag force on a moving rigid sphere in the continuum regime gives the mean square distance as

$$\langle r^2 \rangle = \frac{kT}{\pi \mu a} t, \tag{6.32}$$

and that the diffusion coefficient of the particle is proportional to its mobility:

$$D = BkT. \tag{6.33}$$

Here, μ is the dynamic viscosity of the fluid, a is the radius of the aerosol particle, and B is its mobility. The mobility can be interpreted as the particle velocity per unit force acting upon it. For example, in the case when only gravity moves the particle, the mobility is expressed as $B = v_s/mg$, where v_s is the sedimentation velocity, m is the particle mass, and g is the gravitational acceleration. It follows that the diffusion coefficient of the particle is

$$D = \frac{kT}{6\pi \mu a}, \tag{6.34}$$

which is often called the Stokes–Einstein equation. For unit-density spherical aerosol particles the diffusion coefficient in air at standard temperature and pressure extends from about 5.1E–2–2.4E–9 cm²/s in the size range of 0.001 μm–100 μm in diameter, respectively. For a nonspherical particle, because the forces acting on it are nonuniformly distributed as a function of particle orientation with respect to its motion, the diffusion coefficients, derived in the foregoing, must be corrected. This will be shown later in this chapter.

The 1D form of Equation 6.31 can be also derived based on Fick's law, which will give us an insight into the diffusion process. In Brownian motion, a particle moves away from its original position with very low probability of returning to the same location. Thus, there is a net displacement. Note that when examined in large numbers under equilibrium, the sum of displacements for all particles is zero. The displacement of any single particle can be estimated by its root–mean-square displacement. Let δ notate the displacement of a particle in one dimension, where particles

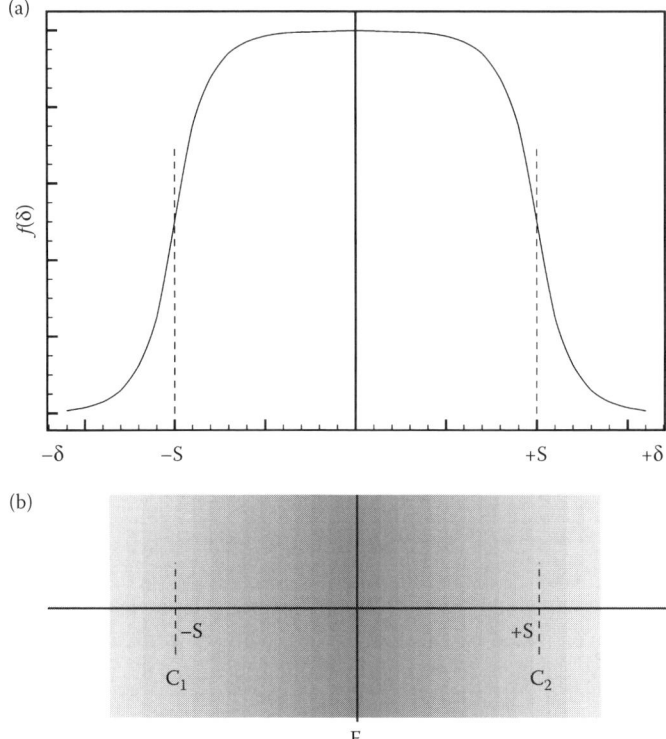

FIGURE 6.4 Particle diffusion in one dimension. (a) Shows the probability distribution of particle displacement (δ) in arbitrary units. (b) Shows the schematics of particle movement about the equilibrium point E.

are allowed to move only along the x-coordinate axis, back and forth. The root-mean-square displacement is then $s = \sqrt{\langle \delta^2 \rangle}$, which is the square root of the average of the squared displacements of all the particles. A schematic diagram of this is shown in Figure 6.4. This means that if the particles move on a line, on average, only particles within a distance of $\pm s$ may cross the equilibrium point E (Figure 6.4b). Assuming that the particle concentrations on the two sides of point E are C_1 (1/m) and C_2 (1/m), the average number of particles crossing point E is $1/2 C_1 s$ from the left and $-1/2 C_2 s$ from the right. The net flow across point E is

$$J(C,s) = \frac{1}{2}(C_1 - C_2)s. \qquad (6.35)$$

For small values of s, the concentration gradient may be written as

$$\frac{dC}{dx} = \lim_{\Delta x \to 0} \frac{\Delta C}{\Delta x} = \frac{C_2 - C_1}{s}. \qquad (6.36)$$

Thus, it follows that the net flow and net current are, respectively,

$$J(C,s) = -\frac{1}{2}s^2 \frac{dC}{dx}, \qquad (6.37)$$

$$J(C,s;t) = -\frac{1}{2}\frac{s^2}{t}\frac{dC}{dx}. \qquad (6.38)$$

But the latter equation is none other than Fick's law, Equation 6.22, written in one dimension and with $s^2/2t$ replacing the diffusion coefficient, D. In this way, the mean square displacement is written as

$$s^2 = 2Dt. \tag{6.39}$$

In three dimensions, it may be shown that $s^2 = 4Dt/\pi$. This suggests that particle motion by diffusion alone is a slow process. For example, the diffusion coefficient of a 1.0 μm spherical particle in air is about 2.74E–7 cm^2/s. For this particle, it would take almost 507 h to move a distance of just 1 cm if there were no other forces acting on it. In free space convection and gravitational settling dominate the aerosol transport compared to diffusion. However, in the absence of impaction, close to boundaries, such as a wall, diffusion is a more efficient process to deposit the particles than convection.

Note that by virtue of Equations 6.33 and 6.34, the diffusion coefficient does not depend on the particle mass. For particles having much larger mass than the surrounding fluid molecules, however, inertial effects may not be negligible. Therefore the Brownian motion of heavy particles follows a much less tortuous path than that of the fluid molecules, and their mean square displacement is different than that of the surrounding molecules or smaller particles. The inertia of the particle may be considered via the *relaxation time*, τ, which is the time the particle requires to attain a new velocity as it responds to external forces: $\tau = m/(6\pi\mu a) = mB$. Using this formalism, the diffusion coefficient, Equation 6.34, may be written as $D = \tau kT/m$. Fuchs (1989) showed that the mean square displacement could be adjusted for this inertial effect as $s^2 = 2D[t - \tau(1 - e^{-t/\tau})]$, as written for the 1D case, Equation 6.39. It is readily seen that this adjustment is significant only in the case when the observation time is comparable or smaller than the relaxation time. For example, the relaxation time of a 100 μm diameter unit-density sphere is approximately 0.03 s. Because the observation time is invariably much greater than the relaxation time for all, except for the heaviest particles, in most cases, the inertia effect is negligible.

DRAG FORCE AND CUNNINGHAM'S SLIP CORRECTION

As the fluid moves over the body of an object, it exerts a frictional force on it. Similarly, a particle moving through a fluid experiences a resisting force, called drag force. The drag varies according to the shape of the particle, the characteristics of the fluid, and the relative velocity. The most fundamental motion, of a rigid sphere in an infinite medium, was investigated by Stokes (1851). By solving the momentum balance equation, a.k.a. Navier–Stokes equation, for steady incompressible laminar flow, neglecting inertia, he found that for particles that are large compared to the mean free path in air, that is, Kn < 1, or in the continuum regime, the drag force is

$$F_d = 3\pi d\mu v, \tag{6.40}$$

which is known as Stokes' law. F_d is the drag force, d is the particle diameter, μ is the fluid dynamic viscosity, and v is the particle velocity relative to that of the fluid. Often, this formula is written in vector form with a negative sign on the right-hand side, indicating that the drag force points in the direction opposite the fluid velocity.

In this form, Stokes' law is valid for low Reynolds numbers, Re ≪ 1, only. At higher Re numbers, this equation underestimates the drag force, and there have been many authors who proposed amendments for Re ≥ 1 in the continuum regime. For example, by taking into account the inertial forces, Oseen (1927) obtained the expression

$$F_d = 3\pi d\mu v \left(1 + \frac{3}{16}\text{Re}\right), \tag{6.41}$$

which is valid for Re ≤ 5. This formula was asymptotically expanded by others (Goldstein, 1938; Proudman and Pearson, 1957) to include terms with higher orders of the Re number.

Although the Stokes' law is not applicable for Re > 5, the drag force, however, can be also written in terms of the drag coefficient, C_D, which is a dimensionless constant (Sutton, 1957):

$$F_D = C_D \frac{1}{2}\rho v^2 A = C_D \rho v^2 \frac{d^2 \pi}{8}. \quad (6.42)$$

Here, ρ is the density of the fluid and A is the projected area of the sphere. This equation is valid over a large range of Re numbers through the proper use of the drag coefficient, which depends on the Re number. For laminar flow (Re < 1), C_D is determined analytically, but for higher Re numbers, experimental correlations are normally used:

$$C_D = \frac{24}{Re} \quad \text{for Re} < 1, \quad (6.43)$$

$$C_D = \frac{24}{Re}(1 + 0.15 Re^{0.687}) \quad \text{for } 1 \le Re < 1000. \quad (6.44)$$

Further correlations and experimental measurements for a wide range or Re numbers are given, for example, by Perry and Green (2007).

Stokes' law can be written to obtain an expression for the terminal settling or sedimentation velocity of particles. When a particle, initially at rest, is released and allowed to fall under the sole influence of gravity and aerodynamic drag, it will accelerate until its speed reaches the terminal settling velocity. At this velocity, the aerodynamic drag balances the gravitational force, $F_d = mg$, and Equation 6.40 can be rearranged to express the terminal settling velocity:

$$v_s = \frac{mg}{3\pi d \mu}. \quad (6.45)$$

Here, m is the particle's mass and g is the gravitational acceleration. As seen in the previous section, the term $m/(3\pi d\mu)$ is distinguished as the relaxation time of the particle, τ, which carries a unit of seconds, and it is often separated in the above equation. Therefore, Equation 6.45 may be written as $v_s = \tau g$. Note that Oseen's formula Equation 6.41 could also be employed for cases when Re ≤ 5. Using Sutton's formalism, Equation 6.42, which is valid for higher Re numbers, the square of the settling velocity may be expressed as

$$v_s^2 = \frac{4d(\rho_0 - \rho)g}{3C_D \rho}. \quad (6.46)$$

Here, ρ_0 is the particle's density while ρ is the fluid density. Note that this is a nonlinear equation, as C_D is a function of the Re number which in turn depends on the particle velocity. Thus, in order to obtain the settling velocity one must know it. This problem may be solved iteratively, whereby one assumes an initial or guess value for v_s and using Equation 6.46 one obtains a new value for v_s. Using the new value as input, yet another value of v_s is obtained. This procedure is continued until two subsequent values of v_s are not substantially different. The convergence of this method depends on the quality of the initial guess. Another method is given by Reist (1993) where $C_D Re^2$ is computed without using v_s, and Re is obtained from a nomogram.

A common assumption in fluid mechanics is that the gas velocity at the surface of an object is zero. This is known as the no-slip condition. For small particles, when the mean free path is comparatively large and Kn > 1, this assumption may not hold: depending on their size, the particles can move among air molecules, sometimes without much interaction. This is known as slip, which introduces a deviation from Stokes' law as the particles settle faster than predicted by Equation 6.45. This was analyzed by Cunningham (1910), and was measured by Millikan in his famous oil drop experiment (1910). The degree of the deviation is expressed as the ratio of the actual terminal velocity of the particle to the settling velocity obtained from Stokes' law, Equation 6.45. This ratio is known as the Cunningham slip correction factor, $C = v_{s,m}/v_{s,Stokes}$. In this way, Stokes' equation is often rewritten as $F_d = 3\pi d\mu v/C$, and the terminal settling velocity Equation 6.45 is rearranged as $v_s = mgC/3\pi d\mu$. Note that via v_s, the diffusion coefficient and the particle mobility (Equation 6.33) are both adjusted for slip condition. The mathematical expression for the Cunningham factor is sought in the form of

$$C(Kn) = 1 + Kn(A_1 + A_2 e^{-A_3/Kn}), \qquad (6.47)$$

where A_1, A_2, and A_3 are empirical constants. Many investigators have measured these constants over the past century, including Millikan (1923), Allen and Raabe (1982, 1985), and Jennings (1988). A summary of the experimental methods and the different results, including the underlying theory, is given by Davies and Schweiger (2002). Recently, Hutchins et al. (1995) used light scattering measurements of solid microspheres of diameter ranging from 1.00 to 2.12 μm in dry air, and found $A_1 = 1.2310 \pm 0.0022$, $A_2 = 0.4695 \pm 0.0037$, and $A_3 = 1.1783 \pm 0.0910$. For nanoparticles, Kim et al. (2005) obtained results by differential electrical mobility analysis in the Kn number range of 0.5–83 using polystyrene latex spheres of 19.9 nm, 100.7 nm, and 269 nm in diameter. The constants in the Cunningham slip correction factor were evaluated with a relative uncertainty of 3% or less in the 95% confidence interval, and found $A_1 = 1.165$, $A_2 = 0.483$, and $A_3 = 0.997$. Table 6.3 summarizes some of the parameters reported by various authors. Note that in the range of particle diameters extending from 0.01 μm to 10 μm, the discrepancy in the slip correction factors using parameters from one versus another author, as given in Table 6.3, is generally within 2%.

Using this form of the Cunningham correction factor extends the validity of Stokes' equation up to Kn = 10. In the case of very small particles, when Kn ≫ 1, Williams and Loyalka (1991) showed that Stokes' law maybe modified as

$$F = F_d \left[Kn(A_1 + A_2) \right]^{-1}, \qquad (6.48)$$

where F_d is given in Equation 6.40. This agrees well with exact theoretical results, and the authors convincingly argue that by suitably adjusting the parameters A_i, a generalized formula may be developed, which is accurate over the entire range of Knudsen numbers.

The Cunningham factor is nonnegligible even at larger particle sizes. For example, at 7 μm, it is about 1.023. A value of 1 would mean that there is no slip. As a general rule of thumb, for particles smaller than 10 μm, Cunningham's factor should be included in the calculations.

TABLE 6.3
Parameters of Cunningham's Correction Factor as Reported by Various Authors

Parameter	Allen and Raabe (1982)	Allen and Raabe (1985)	Jennings (1988)	Hutchins et al. (1995)	Kim et al. (2005)
A_1	1.155	1.142	1.252	1.2310	1.165
A_2	0.471	0.558	0.399	1.4695	0.483
A_3	0.596	0.999	1.100	1.1783	0.997

Note: The slip correction factor multiplies the standard or Stokes' formula for settling velocity, Equation 6.45.

NONSPHERICAL PARTICLES

The assumption of spherical particles, which is predominant in the development leading to some of the formulas in the foregoing sections, does not always hold. Even liquid droplets, whose geometry can approach a theoretical sphere, tend to deform as a result of aerodynamic drag and other forces. In practice, almost none of the aerosol particles are spherical, but have complex morphologies. Depending on the nature of aerosol generation methods, particles can take any shape, and in most cases they form irregular 3D chains or amorphous aggregates, which may contain internal voids and can have multiple constituents. For such particles, the drag force depends on their geometry and on the orientation of the particles with respect to their velocity vector as they traverse the suspending medium; hence large deviations from Stokes' law may be observed. In addition, their mobility and diffusion coefficients may be appreciably different from their spherical counterparts.

Because most of the theoretical foundations of aerosol transport are based on spherical particle geometry, various properties of the irregular aerosol particle are often related to those of a regular spheroid or to those of an equivalently sized sphere. Depending on the aim and scope of the discussion, different equivalent particle diameters may be defined. In this way, the *volume-equivalent diameter*, d_{ve}, is defined as the diameter of a sphere having the same volume as the irregularly shaped particle of interest, including any void space within. The *mass-equivalent diameter*, d_{me}, has a similar definition, in that it is the diameter of a sphere whose mass is equivalent to the mass of the particle, and its material composition is the same as that of the nonvoid fraction of the irregular particle, assuming the density of the material filling the void is negligible compared to the rest of the particle. Note that this definition does not include internal voids, thus $d_{me} \leq d_{ve}$. By extension, the mass-equivalent volume, V_{me}, is that of a sphere whose diameter equals d_{me}. The particle density, ρ_p, is distinguished from the material density of the particle, ρ_m: the material density is the weighted average density of the particle over its constituent materials, excluding the void. $\rho_m = m/V_{me} = \Sigma \rho_i V_i / \Sigma V_i$, where m is the particle mass, ρ_i is the density, and V_i is the volume of the ith constituent. The particle density, on the other hand, includes the void, and it is $\rho_p = m/V = \Sigma \rho_i V_i / (V_{void} + \Sigma V_i)$, where V is the volume of the particle. This distinction is useful in defining the *internal void fraction* of the particle, $\delta = (\rho_m/\rho_p)^{1/3}$. Because $\rho_m \geq \rho_p$, $\delta \geq 1$. An in-depth review of the related concepts is given by DeCarlo et al. (2004).

In addition to the above definitions, the Stokes' diameter and the aerodynamic diameter are often used. The *Stokes' diameter*, d_s, is defined as the diameter of a sphere having the same density and settling velocity as the particle of interest. The *aerodynamic diameter*, d_a, is the diameter of a unit-density sphere with the same settling velocity as the particle of interest. Because the aerodynamic diameter does not depend on the particle density, which is difficult to measure, it has gained widespread use in experimental investigations. Note, however, that depending on the aerosol application and the dominant physical processes therein, not all equivalent diameters are appropriate measures of the particle size, and a new equivalent particle size may have to be defined to suit the needs of the analysis.

The irregular particle shape not only impacts the application of Stokes' law, but it has ramifications on the diffusion coefficient as well. Recall that the Stokes–Einstein equation for the diffusion coefficient Equation 6.34 entailed the assumption that the particle is a rigid sphere undergoing Brownian motion in the continuum regime. For the case of nonspherical particles, one must solve the Navier–Stokes equations for creeping flow (a.k.a. Stokes' flow) to obtain the drag force, and average the results in all orientations of the particle with respect to the direction of the flow.

In order to facilitate theoretical developments, and because complex shapes may make the governing equations mathematically intractable, irregular particles are often approximated by spheroids or ellipsoids. Consider a regular ellipsoid with semiaxes a and b, which are the major and minor axes parallel to the x- and y- axes, respectively, and with rotational symmetry around one of its axes. The axis of revolution is known as the polar axis, which is perpendicular to the equatorial plane. An oblate (thick or lentil-shaped) spheroid is formed by the rotation of an ellipse about its minor axis, b, and a prolate (slim or cigar-shaped) spheroid is formed by the rotation of the ellipse about its major axis, a. Introducing the slenderness factor, or axis ratio, q, which is the ratio of the polar axis to the equatorial axis, if $q > 1$, the ellipsoid is prolate, while if $q < 1$, the ellipsoid is oblate. Thus, for prolate ellipsoid, $q = a/b$, and for oblate ellipsoid, $q = b/a$. If we compare this ellipsoid to a volume-equivalent sphere whose radius is x_0, then using standard mensuration formulas for oblate ellipsoid, $x_0 = aq^{1/3} = bq^{-2/3}$, and for prolate ellipsoid, $x_0 = bq^{1/3} = aq^{-2/3}$. For motion of the particle or forces acting along its polar axis, the $\parallel$ subscript, and for motion or forces perpendicular to the polar axis, the $\perp$ subscript, is used in the following.

In general, the spheroid may be oriented at an arbitrary angle between its polar axis and the coordinate system, as outlined above. The net drag force acting on such a body may be written as the superposition of forces parallel and perpendicular to this axis. If the angle between the polar axis and the x-axis of the coordinate system is θ, the net drag force is $F = F_\parallel \cos(\theta) + F_\perp \sin(\theta)$, where the velocities in the evaluation of $\parallel$ and $\perp$ forces are replaced with $(v \cos \theta)$ and $(v \sin \theta)$, respectively. Using Equations 6.33, 6.34, 6.40, the particle velocity may be expressed in terms of the drag force and the mobility: $v = BF$, and the diffusion coefficient in terms of the drag force as $D = kTvF^{-1}$. Hence, the diffusion coefficient for an arbitrarily oriented spheroid is expressed as $D^{-1} = \cos(\theta)/D_\parallel + \sin(\theta)/D_\perp$.

In this way, for particles much greater than the mean free path in air, Perrin (1910) derived correction factors for the diffusion coefficient of ellipsoids of rotational symmetry undergoing Brownian motion in a fluid without shear. In his work, Perrin assumed that the particle is a spheroid obtained by rotating and ellipse around its axis a, and defined the axis ratio as $z = b/a$. The corrections factors are written as D/D_0, where D_0 is the diffusion coefficient of a volume-equivalent sphere. For this geometry, as seen above, if the radius of the volume-equivalent sphere is x_0, then $x_0 = az^{2/3}$. The correction factors are written separately for prolate ($z < 1$) and oblate ($z > 1$) ellipsoids as follows (Friedlander, 2000):

$$\frac{D}{D_0} = \frac{z^{2/3}}{(1-z^2)^{1/2}} \ln\left[\frac{1+(1-z^2)^{1/2}}{z}\right] \quad \text{for } z < 1$$
$$\frac{D}{D_0} = \frac{z^{2/3}}{(z^2-1)^{1/2}} \arctan\left[(z^2-1)^{1/2}\right] \quad \text{for } z > 1 \qquad (6.49)$$

Note that the separate equations are only necessary if one wants to avoid computations on the complex plain. Otherwise, they give identical results for all z. The diffusion coefficient of the ellipsoid is always smaller than that of a sphere. Figure 6.5 illustrates the behavior of the diffusion correction factor in the range $0.01 < z < 100$. The correction factor peaks at $z = 1$, having a value of 1, as this is the case when the ellipsoid is identical to the sphere. In the range $0.3 < z < 3$, D/D_0 is above 0.9 and in the range of $0.1 < z < 10$, D/D_0 is above 0.6. The results are not directly applicable to shear flow, because in that case the particle orientation is not random, there are preferred directions, and particles undergo characteristic rotational motions; thus Perrin's nonweighted averaging could provide erroneous results.

The settling velocity, which is related to the fluid drag on the particle, is also different for spherical versus nonspherical particles, and it is a direct function of particle density and size. Early experiments by Kunkel (1948) demonstrated that irregular particles always experience a greater drag than the equivalent sphere. Linear chains of particles tend to rotate such that their longitudinal axis is perpendicular to the particle motion, and plate-like particles settle with their flat surfaces oriented horizontally. In addition, it was observed that asymmetry causes sideways drift.

In practice, the increased drag on the nonspherical particle is taken into account by the *dynamic shape factor*, κ, which is a ratio of resistance force of the nonspherical particle to that of a sphere

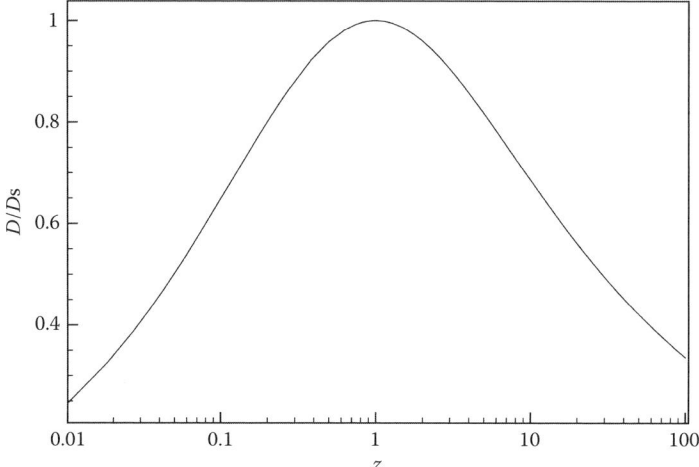

FIGURE 6.5 Correction factor in the diffusion coefficient for spheroid particles based on Perrin's (1910) approach, Equation 6.49.

having the same volume and moving with the same relative air velocity: $\kappa = F/F_d$. In this way, using Stokes' law, Equation 6.40, the drag force on an irregular particle may be written as

$$F = 3\pi d_e \mu v \kappa, \qquad (6.50)$$

where d_e is the volume-equivalent diameter. Using the above formalism with Equation 6.45, the terminal settling velocity of an irregular particle is written as

$$v_{si} = \frac{\rho d_e^2 g}{18 \mu \kappa}. \qquad (6.51)$$

For spherical particles, $\kappa = 1$, and d_e is replaced by the particle diameter.

Stöber (1972) showed that in the range of low Re numbers, the dynamic shape factor could be expressed as the ratio of the settling velocity of the volume-equivalent sphere to that of the irregular particle: $\kappa = v_{se}/v_{si}$. This flow regime is predominant for fine particles whose Knudsen number is in the range where slip correction becomes important:

$$\kappa = \frac{v_{se}}{v_{si}} = \frac{\rho d_e^2 C(d_e)}{\rho_0 d_a^2 C(d_a)} = \frac{d_e^2 C(d_e)}{d_s^2 C(d_s)}. \qquad (6.52)$$

Here, v_{si} is the settling velocity of the irregular particle, v_{se} is the settling velocity of the equivalent-volume particle, ρ_0 is the unit density, d_e is the volume-equivalent diameter, d_s is the Stokes' diameter, and d_a is the aerodynamic diameter. $C(d)$ is Cunningham's correction factor evaluated at the appropriate diameter. In this case, the slip correction must also be introduced into Equation 6.50 either implicitly via κ or explicitly including $C(d_e)$:

$$F = \frac{3\pi d_e \mu v \kappa}{C(d_e)} \qquad (6.53)$$

In using the above formula, the analyst must exercise care to avoid double correction: if a slip-corrected value for κ is used, then there is no need to use $C(d_e)$. Otherwise, if $C(d_e)$ is explicitly used,

The Physics of Aerosols

an uncorrected κ must be used. Equation 6.52 provides means to relate the aerodynamic diameter to the volume-equivalent diameter and to the Stokes' diameter:

$$d_a = \frac{d_e}{\sqrt{\kappa}} \left(\frac{\rho C(d_e)}{\rho_0 C(d_a)} \right)^{1/2},$$

$$d_a^2 = d_s^2 \frac{\rho C(d_s)}{\rho_0 C(d_a)}. \tag{6.54}$$

In most cases, the dynamic shape factor is written to include the particle size range where slip correction is not necessary:

$$\bar{\kappa} = \frac{\rho d_e^2}{\rho_0 d_a^2} = \frac{d_e^2}{d_s^2}. \tag{6.55}$$

Analytical evaluations of the dynamic shape factor are hindered by the large variety of different particle shapes, most of which are mathematically intractable. Most of the theoretical work in the literature is based on or utilizes Oberbeck's (1876) solution for the fluid drag on ellipsoids in uniform flow as surrogate for irregularly shaped particles. A review of the solutions is presented by Stöber (1972).

For oblate spheroids, $q < 1$ as defined above, moving parallel and perpendicular to the polar axis, the respective dynamic shape factors are

$$\bar{\kappa}_{\parallel,\text{oblate}} = \frac{4}{3q^{1/3}} \frac{1-q^2}{(1-2q^2)(1-q^2)^{-1/2} \cos^{-1}(q) + q} \quad (q<1), \tag{6.56}$$

$$\bar{\kappa}_{\perp,\text{oblate}} = \frac{8}{3q^{1/3}} \frac{1-q^2}{(3-2q^2)(1-q^2)^{-1/2} \cos^{-1}(q) - q} \quad (q<1). \tag{6.57}$$

For prolate spheroids, $q > 1$, moving parallel and perpendicular to the polar axis, the respective dynamic shape factors are

$$\bar{\kappa}_{\parallel,\text{prolate}} = \frac{4}{3q^{1/3}} \frac{q^2-1}{(2q^2-1)(q^2-1)^{-1/2} \ln\left(q+(q^2-1)^{1/2}\right) - q} \quad (q>1), \tag{6.58}$$

$$\bar{\kappa}_{\perp,\text{prolate}} = \frac{8}{3q^{1/3}} \frac{q^2-1}{(2q^2-3)(q^2-1)^{-1/2} \ln\left(q+(q^2-1)^{1/2}\right) + q} \quad (q>1). \tag{6.59}$$

The behavior of these functions is illustrated in Figure 6.6. Note that similar to Equation 6.49, the separate equations for oblate and prolate bodies Equations 6.56 and 6.58 for ∥, and Equations 6.57 and 6.59 for ⊥ geometries, are only necessary if one wants to avoid computations on the complex plain. Otherwise, these pairs of formulas give identical results for all q values.

The term $(1-q^2)^{1/2}$ is often called eccentricity, and the factor $2q^{-1/3}$ is related to the *sphericity*, ϕ. The latter, in turn, is connected to the *geometric shape factor, S*, which relates the volume of the particle to a projected area diameter and it equals the squared ratio of the volume-equivalent diameter to the surface-equivalent diameter: $S \equiv d_{ve}^2/d_{se}^2 = \phi\pi/2$. For prolate bodies with $q \gg 1$, $\phi\pi = 4q^{-1/3}$, and for oblate bodies with $q \gg 1$, $\phi\pi = 4q^{-4/3}$.

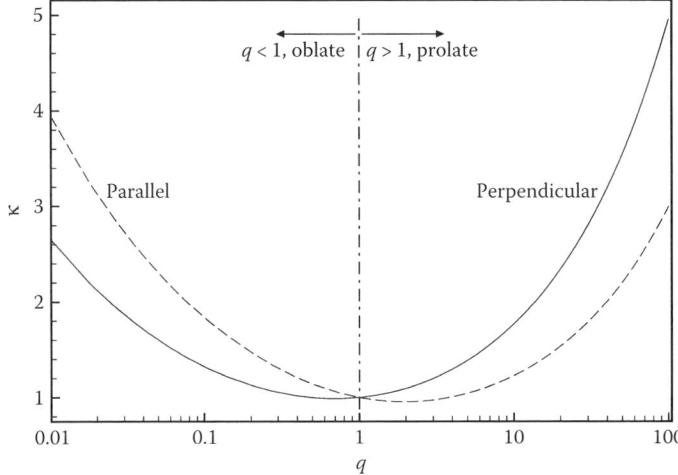

FIGURE 6.6 Dynamic shape factors for prolate and oblate spheroids moving in parallel (∥) and perpendicular (⊥) directions with respect to their polar axes, as shown in Equations 6.56 through 6.59.

Based on experimental investigations, Stöber (1972) found that for elongated bodies, the dynamic shape factor may be written universally as

$$\bar{\kappa}\left(\frac{d_{min}}{d_{max}}\right)^{1/3} \cong \text{const.} \tag{6.60}$$

where d_{min} and d_{max} are the minimum and maximum dimensions of the object. For oblate spheroids, in the limit of $q \ll 1$, Equations 6.55 and 6.56 yield

$$\kappa_{\parallel,oblate} = \frac{8}{3\pi}\frac{1}{q^{1/3}} \quad (q \ll 1),$$
$$\kappa_{\perp,oblate} = \frac{2}{3}\kappa_{\parallel,oblate} \quad (q \ll 1), \tag{6.61}$$

which agree well with his observations for chain aggregates and long fibers, which are prolate bodies:

$$\bar{\kappa}_{\parallel,prolate} \cong 0.82 q^{1/3} \quad (q \gg 1),$$
$$\bar{\kappa}_{\perp,prolate} \cong 0.57 q^{1/3} \quad (q \gg 1). \tag{6.62}$$

As it is seen, the dynamic shape factor not only depends on the actual shape of the particle, but also on its orientation with respect to the direction of its movement or settling. If the particle has a streamlined shape, it may align and assume a preferred orientation in the flow, and its dynamic shape factor becomes less than 1. This can be seen in Figure 6.6: for prolate ellipsoids moving parallel to their polar axis, the function minimum is $\kappa_{\parallel} = 0.955$, which occurs at about $q = 1.95$, and for oblate ellipsoids moving perpendicular to their polar axis, $\kappa_{\perp} = 0.988$ at $q = 0.702$. At small Re numbers, the flow does not tend to align the particles in any preferred direction. Small particles, however, often undergo random orientation due to Brownian motion, for which an average shape factor can be defined based on the orientation-averaged drag force which yields an unequal weighting of the parallel and perpendicular components (Stöber, 1972):

$$\bar{\kappa}_{prolate} = \frac{1}{3}\bar{\kappa}_{\parallel,prolate} + \frac{2}{3}\bar{\kappa}_{\perp,prolate},$$
$$\bar{\kappa}_{oblate} = \frac{1}{3}\bar{\kappa}_{\parallel,oblate} + \frac{2}{3}\bar{\kappa}_{\perp,oblate}.$$
(6.63)

A similar weighting was introduced by Leith (1987), who expressed the total drag on an irregular particle as a superposition of drag forces due to normal stress (a.k.a. form drag) and due to tangential stress (friction or skin drag). The form drag is related to the fluid pressure over that surface of the object, which is projected perpendicular to its direction of motion. Therefore, one may define an equivalent sphere with diameter d_n, which has the same projected area as the irregular object. The tangential stress is integrated over the entire surface of the object, thus a surface-equivalent sphere with diameter d_{se} may be used to describe the friction drag. In this way the Stokes' formula for the drag force is written as

$$F = 3\pi\mu v\left(\frac{1}{3}d_n + \frac{2}{3}d_{se}\right).$$
(6.64)

When compared to Equation 6.50, the dynamic shape factor is expressed as

$$\kappa = \frac{1}{3}\frac{d_n}{d_e} + \frac{2}{3}\frac{d_{se}}{d_e}.$$
(6.65)

Alternatively, Equation 6.50 may be written as $F = 3\pi d_n \mu v \kappa_n$, where $\kappa_n = \kappa\, d_e/d_n$, and d_e is the volume-equivalent diameter. Based on experimental data for prisms, Leith (1987) proposed a correlation, which agrees well with experimental data on other geometries, including cylinders, spheroids, and double cones, but overpredicts the shape factor for spheres by about 5%:

$$\kappa_n = 0.357 + 0.684\frac{d_{se}}{d_n} + 0.00154 q_{length} + 0.0104 q_{axis}.$$
(6.66)

Here q_{length} and q_{axis} are the length and axis ratios, respectively, which are defined as follows:

$$q_{length} = \frac{(\text{Length of axis} \parallel \text{to the direction of motion})^2}{\text{Projected area} \perp \text{to the direction of motion}},$$
$$q_{axis} = \frac{\text{Longest axis in the projected area} \perp \text{to the direction of motion}}{\text{Shortest axis in the projected area} \perp \text{to the direction of motion}}.$$
(6.67)

In a series of papers Geller, Mondy, Ingber, and coworkers used a boundary element method to obtain the mobility of irregular particles for linear chains (Geller et al., 1993), branched chains and flakes (Mondy et al., 1996), and long chains (Ingber et al., 1999). This method is a point-kernel technique, which uses superposition of the general solution for the flow generated by point forces on the bounding surfaces of the fluid to satisfy the boundary conditions. For the linear chains of spheroids it was shown that the particle could be modeled as a single prolate spheroid with equal aspect ratio, yielding equivalent results to the second order (Geller et al., 1993). This continues to hold for long chains, simulated up to 700 spheroids (Ingber et al., 1999). For a chain consisting of n identical spheres, each having a radius of R_0, the authors defined the equivalent diameter as $d_e = 2R_0 n^{1/3}$, and

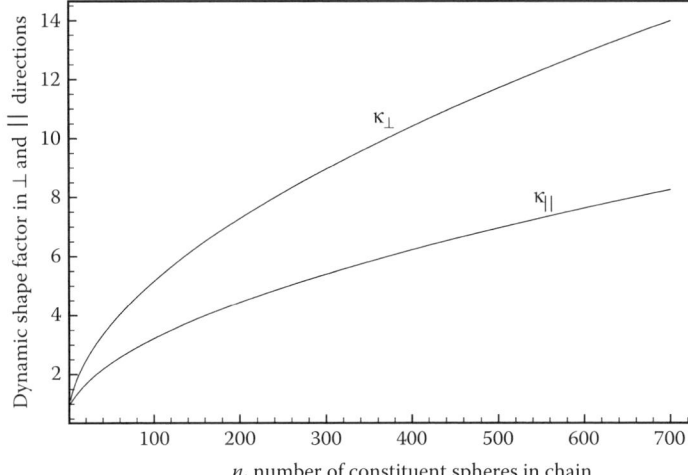

FIGURE 6.7 Dynamic shape factors for a linear chain-agglomerate consisting of n identical spheres, moving parallel (∥) and perpendicular (⊥) with respect to its major axis, as shown in Equations 6.68 and 6.69.

the respective dynamic shape factors for the chain agglomerate moving parallel and perpendicular to its major axis in quiescent fluid were found as

$$\kappa_{\|} = \frac{2}{3} n^{2/3} \left(\ln(2n) - 1.00401 + \frac{0.906526}{\ln(2n)} \right)^{-1}, \tag{6.68}$$

$$\kappa_{\perp} = \frac{4}{3} n^{2/3} \left(\ln(2n) + 0.26269 + \frac{0.38185}{\ln^2(2n)} \right)^{-1}. \tag{6.69}$$

The graphical representation of the above is shown in Figure 6.7. In most practical cases, the dynamic shape factor is not readily available. To compute it, various parameters of the particle are needed which are difficult to obtain experimentally. For a more complete description, including dynamic shape factors for cluster and chain aggregates of particles, the reader is referred to Stöber (1972), for branched chains and flakes, refer to Mondy et al. (1996), while for specifics to uranium, thorium, and plutonium aerosols, Kotrappa (1972) and Kotrappa et al. (1975) are good sources of information.

AEROSOL TRANSPORT

An aerosol forms a dynamic environment. It can be regarded as a particle field, which obeys a multitude of extraneous forces simultaneously acting on it, while being convectively transported by the carrier fluid. The particles, therefore, constantly interact with one another, the carrier fluid, and its boundaries. The collective result of these phenomena is the transport of the particles. The description of how the particles are transported from one spatial location to another, and the changes in the properties of the aerosol field while in transit is of general interest to the aerosol scientist.

MOMENTS OF THE PARTICLE SIZE SPECTRUM

The moments of the particle size distribution occur frequently in the description of the dynamic behavior of aerosols. They are related to important properties of the particles, such as their surface distribution or volume distribution among others, and they have a role in the solution, interpretation, and understanding of the general dynamic equation of aerosol transport. Consider a particle size distribution described by the function $n(v, t)$ such that $dN = n(v, t)\, dv$ is the number of particles in a unit volume of fluid having particle volume of dv about v at time t. Succumbing to mathematical convenience in the description of the particle size, the volume is often preferred to diameter. This is because in particle collision processes, the volumes of the colliding particles are assumed to be conserved. In addition, particles are rarely spherical, albeit it is often assumed that they are. Therefore, when two spherical particles with respective volumes u and v collide and coagulate or agglomerate, the volume of the newly formed particle will be $u + v$. However, the diameter of the new particle, which may be an ellipsoid, will not be a linear combination of the two diameters $(d \neq d_u + d_v)$, and choosing the most appropriate equivalent diameter depends on the process to be described.

In the case of radioactive particles, another property is introduced, s, which is the activity of the particle. Because s does not influence v, but it is conserved in particle collisions, the particle distribution function may be written as $n(v, s, t)$ and the moments can be written in terms of the radioactivity. In the following, the moments of the aerosol distribution are first written in terms of particle volumes, then in terms of particle diameter, and their relations will be pointed out. The significance of writing the moments in terms of radioactivity will be outlined thereafter.

As it was seen in the section on Aerosol Characterization, the aerosol property at a spatial and temporal location $(\bar{r}, t)$ may be defined as

$$q(\bar{r},v,t) = \alpha v^\gamma n(\bar{r},v,t). \tag{6.70}$$

The integral property, therefore, may be written as

$$Q(\bar{r},t) = \int_0^\infty q(\bar{r},v,t)\,dv = \alpha \int_0^\infty v^\gamma n(\bar{r},v,t)\,dv. \tag{6.71}$$

For the case when $\alpha = 1$, the integral aerosol property is related to the moments, when written in terms of the volume variable, as follows:

$$Q^{(\gamma)} = \int_0^\infty v^\gamma n(v)\,dv. \tag{6.72}$$

Here, and in the subsequent development, the spatial and time variables $(\bar{r}, t)$ are dropped for clarity, but they are inherently part of the formalism.

The zeroth moment of the distribution is equivalent to the total particle concentration, $N(t)$:

$$Q^{(0)} = \int_0^\infty n(v)\,dv = N, \tag{6.73}$$

with the dimension of particles per unit volume of carrier fluid.

The first moment of the distribution is equivalent to the total volume concentration or the volume fraction of the aerosol in the carrier fluid, $\phi(t)$:

$$Q^{(1)} = \int_0^\infty v n(v)\,dv = \phi, \tag{6.74}$$

and the mean volume of the particles is

$$\langle v \rangle = \frac{Q^{(1)}}{Q^{(0)}} = \frac{\phi}{N}. \tag{6.75}$$

The volume fraction, $\phi(t)$, is the fraction of the volume that is occupied by particles in a unit volume of the carrier fluid. If the particle density is independent of the volume, the above expression is also valid for the mean mass of the particles.

The 1/3rd moment is related to the mean diameter of the aerosol particles when the particles are spherical:

$$Q^{(1/3)} = \int_0^\infty v^{1/3} n(v) dv = \left(\frac{\pi}{6}\right)^{1/3} \langle d \rangle Q^{(0)}. \tag{6.76}$$

Here, $\langle d \rangle$ is the mean particle diameter, and it is defined as

$$\langle d \rangle = \frac{\int_0^\infty d n(v) dv}{N}. \tag{6.77}$$

Using Equation 6.76, the mean diameter can be expressed as a function of the 1/3rd moment and the total particle concentration as

$$\langle d \rangle = \left(\frac{\pi}{6}\right)^{-1/3} \frac{Q^{(1/3)}}{N}. \tag{6.78}$$

If the particles are spherical, the 2/3rd moment of the distribution can be related to the total surface concentration, to the mean square diameter, and to the mean surface area of the particles in the volume of interest as follows:

$$Q^{(2/3)} = \int_0^\infty v^{2/3} n(v) dv = (36\pi)^{-1/3} \int_0^\infty 4a^2 \pi n(v) dv$$
$$= (36\pi)^{-1/3} \int_0^\infty S n(v) dv = (36\pi)^{-1/3} \sigma. \tag{6.79}$$

Here, a is the particle radius, and we recognize that $4a^2\pi$ is the surface area of the spherical particle, S. Therefore, σ is the total surface concentration of the particles, having an SI unit of m² particle surface per m³ of the carrier fluid. The above equation can be brought to a form that expresses the mean particle surface, $\langle \sigma \rangle$:

$$\langle \sigma \rangle = (36\pi)^{1/3} \frac{Q^{(2/3)}}{Q^{(0)}}, \tag{6.80}$$

and the mean square diameter of the particles, $\langle d^2 \rangle$:

$$\langle d^2 \rangle = \left(\frac{\pi}{6}\right)^{-2/3} \frac{Q^{(2/3)}}{Q^{(0)}}. \tag{6.81}$$

Note that not all moments have an obvious physical significance, and that the order of the moment, γ in Equation 6.72, does not have to be an integer nor a positive number. For example, the harmonic mean diameter can be expressed in terms of the $-1/3$rd moment as follows:

$$\frac{1}{d_H} \equiv \langle d^{-1} \rangle = \left(\frac{\pi}{6}\right)^{1/3} \frac{Q^{(-1/3)}}{Q^{(0)}}. \tag{6.82}$$

By applying the Schwarz inequality to Equation 6.3, an addition relation in the order of the moments can be derived:

$$\left[Q^{(p+s)}\right]^2 \leq Q^{(2p)} Q^{(2s)}, \tag{6.83}$$

where the equality holds for a monodisperse aerosol whose particle volume is constant at a single value. This formula is also useful in approximations where the size distribution is nearly monodisperse, when the coagulation mechanism is a weak function of size distribution, and in establishing limiting conditions.

When the particle diameter is used in the description of the moments instead of the volume, which is often seen in the literature, a formal substitution of d for v in the preceding equations yields the following relations:

The zeroth moment is still equivalent to the total particle concentration, as in Equation 6.73: $M^{(0)} = N = Q^{(0)}$. Here, the notation M refers to the moment written in terms of the particle diameter. The first moment, $M^{(1)}$, is related to the mean diameter:

$$M^{(1)} = \langle d \rangle M^{(0)} = \langle d \rangle N = Q^{(1/3)} \left(\frac{\pi}{6}\right)^{-1/3}. \tag{6.84}$$

The second moment is related to the total surface concentration, σ:

$$M^{(2)} = \frac{\sigma}{\pi} = \frac{(36\pi)^{1/3}}{\pi} Q^{(2/3)}, \tag{6.85}$$

Thus, $\sigma = \pi M^{(2)} = (36\pi)^{1/3} Q^{(2/3)}$, and the mean surface area may be written as

$$\langle \sigma \rangle = \pi \frac{M^{(2)}}{N} = (36\pi)^{1/3} \frac{Q^{(2/3)}}{N}. \tag{6.86}$$

The third moment, which is the last discussed in this chapter, is proportional to the total volume concentration:

$$M^{(3)} = \phi \frac{6}{\pi} = Q^{(1)} \frac{6}{\pi}. \tag{6.87}$$

Note that these relations can be directly derived from the spherical mensuration formulae for volume and surface in terms of the diameter of the sphere. Higher-order moments are also possible to express. They have gained use in the solution of the general dynamic equation for aerosol transport.

In the case of radioactive aerosol particles, it is important to realize that the moments do not have to be expressed in terms of the particle volume. The particle density function or distribution function for radioactive aerosols is often written as $n(v, s, t)$, where s is the radioactivity of the particles having volume v. The activity of the aerosol particle is conserved during the collision process, and the moments may be written similar to Equation 6.72 (dropping the time dependence) as follows:

$$Q_s^{(\gamma)} = \int_0^\infty s^\gamma n(v,s)\,ds. \tag{6.72s}$$

In this way, the zeroth moment is equivalent to the particle distribution without regard to radioactivity, $n(v, t)$, and $Q_s^{(0)}(v)\,dv$ is the number of particles in the range dv about volume v. The first moment, $Q_s^{(1)}(v)\,dv$, is the activity of particles having volume dv about v. Other moments with respect to activity may be obtained similarly.

Aerosol Phase Space

The phase space of the aerosol field is the mathematical dimension where the aerosol transport takes place. The full description of the phase space is equivalent to solving the aerosol transport equation in space and time, a nontrivial task for the analyst. The movement of the particles may be described by their momentum at a spatial location in the carrier fluid. The dependence on time is always inherently understood, and the time variable will be included only in cases when the dynamics requires it. Consider a small packet of particles at a spatial location of $\bar{r}$ having a momentum of $\bar{p}$. Both variables are vector quantities; hence with time included, the phase space has a minimum of six or seven dimensions, depending on the coordinate system used. The momentum is the product of the particle mass and its velocity: $\bar{p} = m\bar{V}$. The velocity is also a vector quantity, and it is the product of the particle speed and the particle direction: $\bar{V} = V\hat{\Omega}$. Here, $\hat{\Omega}$ is the unit vector pointing in the direction of the particle movement. We distinguish the local coordinate system from the directional coordinate system, as illustrated in Figure 6.8. The particle position is described by $\bar{r}$ in the local coordinate system, and the particle direction $\hat{\Omega}$ is specified in the directional coordinate system, which is tied to the particle position. Most often, the local coordinate system is Cartesian while the direction vector, $\hat{\Omega}$, is expressed in spherical coordinates. In this way, the particle direction can be described using only two directional variables, the polar angle, θ, and the azimuthal angle, φ. Using elementary trigonometry, it can be shown that the elemental angle $d\hat{\Omega}$ can be written in terms of $(d\theta\,d\varphi)$.

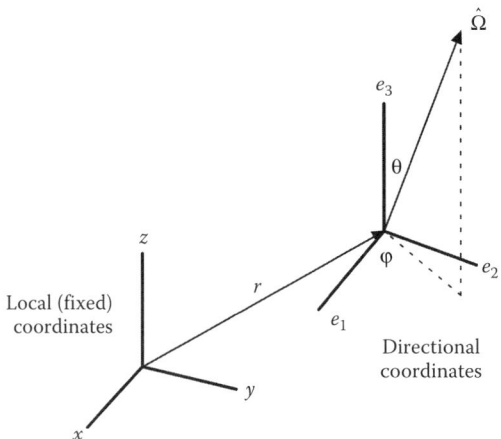

FIGURE 6.8 Particle in a local (or fixed) spatial coordinate system, moving in the direction of $\hat{\Omega}$.

The Physics of Aerosols

Let $n(\bar{r}, \bar{p}) d\bar{r} d\bar{p}$ be the number of particles in volume element $d\bar{r}$ about $\bar{r}$ having a momentum of $d\bar{p}$ about $\bar{p}$. Because $d\bar{p}$ is infinitesimally small, nearly all particles in this packet have the same speed, dV about V, and are moving in the same direction, $d\hat{\Omega}$ about $\hat{\Omega}$. Note that $d\hat{\Omega}$ is a solid angle with a unit of steradian (St), and it represents a cone of directions about $\hat{\Omega}$. Because m, V, and $\hat{\Omega}$ are independent variables, and $d\bar{p} = dm \, dV \, d\hat{\Omega}$, it is often convenient to write that

$$n(\bar{r}, \bar{p}) = n(\bar{r}, p, \hat{\Omega}) = n(\bar{r}, m, V, \hat{\Omega}) \left[\frac{\text{particles}}{\text{cm}_{\text{air}}^3 \, \text{gSt cm/s}} \right], \tag{6.88}$$

which is commonly referred to as angular particle density. The unit in the cgs system is as shown in the square brackets above. When the particles can be identified as being different types, each type having a unique and constant density ρ_i, the momentum becomes $\bar{p} = \rho_i v \bar{V}$, where v is the particle volume, and the angular density can be written as $n_i(\bar{r}, v, V, \hat{\Omega})$. Subscript i identifies the type of the particle whose material density is ρ_i. In the case of radioactive particles, another property, s, which is the activity of the particle, is also introduced: $n_i(\bar{r}, v, s, V, \hat{\Omega})$. Because the presence or absence of the variable s does not affect the generality of the discussion, it is dropped from further equations for clarity. The independent variables in the above formulae represent the phase space element, and they are often abbreviated with $\hat{P}$. Thus,

$$n_i(\bar{r}, v, V, \hat{\Omega}) d\bar{r} \, dv \, dV \, d\hat{\Omega} = n_i(\hat{P}) d\hat{P} \quad \left[\frac{\text{particles}}{\text{cm}_{\text{air}}^3 \text{cm}_{\text{p}}^3 \, \text{St cm/s}} \right] \tag{6.89}$$

is the number of particles in phase space element $d\hat{P} = d\bar{r} \, dv \, dV \, d\hat{\Omega}$ about $\hat{P}$. In many situations, the detailed knowledge of particle direction is not as important as the speed and volume, or the absolute momentum distribution of the particles. Therefore, the integral quantity

$$n(\bar{r}, p) = \int_{4\pi} n(\bar{r}, p, \hat{\Omega}) d\hat{\Omega} \tag{6.90}$$

is introduced, known as the scalar particle density, which represents the particles at position $\bar{r}$, having a momentum p irrespective of their direction of motion. The integration above is performed in all directions. When the particle momentum is not an important metric, the total aerosol particle density is used in the analysis, which is the number of particles per unit volume of air at location $\bar{r}$, irrespective to the particles' momentum and direction:

$$n(\bar{r}) = \int_0^\infty n(\bar{r}, p) dp. \tag{6.91}$$

Note that here $n(\bar{r})$ is independent of particle size because the integration with respect to momentum inherently includes the particle mass, which is a function of the particle size. Hence, Equation 6.91 is equivalent to Equation 6.71 when $\alpha = 1$ and $\gamma = 0$. An interesting insight can be gained when the angular density is written in terms of v and V instead of the momentum, as given in Equation 6.88. It is clear that the total particle density of particle type i is written as

$$n_i(\bar{r}) = \iiint_{4\pi \, V \, v} n_i(\bar{r}, v, V, \hat{\Omega}) \, dv \, dV \, d\hat{\Omega}. \tag{6.92}$$

In order to express the scalar particle density, however, we now have a choice in the order of integration, and there can be several scalar densities. The particle density

$$n_i(\bar{r},v,V) = \int_{4\pi} n_i(\bar{r},v,V,\hat{\Omega})\,d\hat{\Omega}. \tag{6.93}$$

is identical to the scalar density Equation 6.90; only the momentum is written in terms of v and V; hence the particle is now distinguished by its type i. If the integration is continued with respect to the particle speed, we gain the momentum-independent scalar particle density

$$n_i(\bar{r},v) = \int_0^\infty n_i(\bar{r},v,V)\,dV, \tag{6.94}$$

which is just the particle size distribution that was used in deriving the various moments, starting with Equation 6.70, and it is also identical to the number density introduced in Equation 6.1. If the integration of Equation 6.93 is continued with respect to the particle volume instead of its speed, we gain the speed- or momentum-dependent (which may also be called size-independent) particle density:

$$n_i(\bar{r},V) = \int_0^\infty n_i(\bar{r},v,V)\,dv \tag{6.95}$$

The speed- or momentum-dependent forms of the particle density, Equations 6.93 and 6.95 are used when the fundamentals of particle deposition and other kinetic processes are discussed.

An important quantity is the particle angular flux, Ψ, which is defined as

$$\Psi(\bar{r},p,\hat{\Omega})\,dp\,d\hat{\Omega} = Vn(\bar{r},p,\hat{\Omega})\,dp\,d\hat{\Omega} \quad \left[\frac{\text{particles}}{\text{cm}^2 \text{s}}\right], \tag{6.96}$$

and it represents the number of particles crossing a unit surface in unit time whose momentum is dp about p and move in the direction $d\hat{\Omega}$ about $\hat{\Omega}$. It follows that the dimension of Ψ is particles per unit surface per unit time per unit momentum per unit solid angle, and when written in the cgs system, it is [particles/(cm²-s-g-cm/s-St)]. Similar to the derivation used in the case of the angular particle density, we can integrate the angular flux with respect to the direction to obtain the scalar flux:

$$\Phi(\bar{r},p) = \int_{4\pi} \Psi(\bar{r},p,\hat{\Omega})\,d\hat{\Omega} \quad \left[\frac{\text{particles}}{\text{cm}^2 \text{ s g cm/s}} = \frac{\text{particles}}{\text{cm}^3 \text{g}}\right]. \tag{6.97}$$

Finally, the total flux is

$$\Phi(\bar{r}) = \int_0^\infty \Phi(\bar{r},p)\,dp \quad \left[\frac{\text{particles}}{\text{cm}^2 \text{ s}}\right], \tag{6.98}$$

which is the number of particles crossing a unit surface per unit time irrespective of their direction or their momentum. If we define the integral average particle speed as

$$\langle V \rangle = \frac{\int_p Vn(\bar{r},p)\,dp}{\int_p n(\bar{r},p)\,dp}, \tag{6.99}$$

using Equation 6.91 it is easy to show that the scalar flux can be written in terms of the mean particle speed and the total particle density:

$$\Phi(\bar{r}) = \langle V \rangle n(\bar{r}). \quad (6.100)$$

Following the method shown with respect to the particle densities, when the momentum is written in its components, separating the particle speed and particle volume, a speed- or momentum-independent scalar flux, $\Phi(\bar{r}, v)$, and a speed-dependent total flux, $\Phi(\bar{r}, V)$, may be defined. In this way, via integration with respect to particle volume or particle speed, two mean particle speeds may be defined, size averaged and speed averaged, which, in turn can be shown to be identical. Hence, the following relation is obtained:

$$\frac{\Phi(\bar{r},v)}{\Phi(\bar{r},V)} = \frac{n(\bar{r},v)}{n(\bar{r},V)}. \quad (6.101)$$

A note on terminology: The way it is used here, the term *flux* represents a rate quantity in time. The integral of this quantity with respect to time is called *fluence*. Some of the literature, however, uses the term *fluence rate* in lieu of *flux*, and sometimes the word *rate* is dropped for convenience, which may result in confusion.

In the description of particle kinetics, such as deposition, resuspension, and motion by the action of extraneous forces, the particle current bears special significance, as the balance equations are written in terms of the current. Following the method above, we define the angular current as

$$\vec{J}(\bar{r},p,\hat{\Omega}) = \bar{V}n(\bar{r},p,\hat{\Omega}) = \hat{\Omega}\Psi(\bar{r},p,\hat{\Omega}) = \hat{\Omega}Vn(\bar{r},p,\hat{\Omega}). \quad (6.102)$$

This is a vector quantity, pointing in the same direction as the particle movement, $\hat{\Omega}$. Because $\hat{\Omega}$ is a unit vector, $|\vec{J}| = \Psi$. The *net* current is the directional integral of $\vec{J}$ in all angles,

$$\vec{J}(\bar{r},p) = \int_{4\pi} \vec{J}(\bar{r},p,\hat{\Omega}) d\hat{\Omega}, \quad (6.103)$$

and it expresses the particle current regardless of directions. Note that when Equation 6.102 is substituted into Equation 6.103, $\hat{\Omega}$ cannot be taken out of the integral; therefore, $|\vec{J}(\bar{r},p)| \neq \Phi(\bar{r},p)$. The differences among the net current, angular current, and total flux can be demonstrated by considering two identical particles moving along the same line, but in opposite directions, which may be described by the collinear unit vectors $\hat{\Omega}$ and $-\hat{\Omega}$, respectively. According to Equation 6.103, the two particles would cancel to give rise to a net current of zero. However, the angular current, Equation 6.102, is nonzero because one of the particles moves in the direction of $\hat{\Omega}$ while the other in $-\hat{\Omega}$. The total flux, Equation 6.98 or Equation 6.100, is additive in the absolute sense, and will result in a total of two particles. Note that the units of flux and current are identical.

A further important quantity is the flow rate of particles. The angular flow rate is equivalent to that component of the current, which points in the direction of $\hat{e}_n$. Here, $\hat{e}_n$ is a unit vector whose direction is at an angle, θ, with respect to $\hat{\Omega}$. θ can take any value, including zero, which is the case when $\hat{e}_n$ is parallel and pointing in the direction of $\hat{\Omega}$. The concept of the flow rate versus current can be visualized by considering a current of aerosol particles along the direction $\hat{\Omega}$. This may be a stream of particles in a flow field, which crosses a surface whose normal is $\hat{e}_n$. The leakage through this surface in the direction of $\hat{e}_n$ is then a scalar quantity representing the number of particles crossing this surface in the given direction, as follows:

$$j_n(\bar{r},p,\hat{\Omega}) = \hat{e}_n \cdot \vec{J}(\bar{r},p,\hat{\Omega}) = (\hat{e}_n \cdot \hat{\Omega})\Psi(\bar{r},p,\hat{\Omega}). \quad (6.104)$$

We recognize that the dot product $\hat{e}_n \cdot \hat{\Omega}$ is the cosine of the angle between these two vectors, known as the directional cosine, which is often denoted with the Greek letter μ, as $\hat{e}_n \cdot \hat{\Omega} = \cos\theta \equiv \mu$. In this way, the *net* flow rate is

$$j_n(\vec{r},p) = \hat{e}_n \cdot \vec{J}(\vec{r},p) = \int_{4\pi} (\hat{e}_n \cdot \hat{\Omega}) \Psi(\vec{r},p,\hat{\Omega})\, d\hat{\Omega} = \int_0^{2\pi}\!\!\int_{-1}^{1} \mu \Psi(\vec{r},p,\hat{\Omega})\, d\mu\, d\varphi. \qquad (6.105)$$

Here, we took advantage of spherical symmetry, and wrote $\hat{\Omega}$ in terms of the polar and azimuthal angles with the former in terms of the directional cosine. Finally, the total flow rate is obtained as

$$j_n(\vec{r}) = \hat{e}_n \cdot \vec{J}(\vec{r}). \qquad (6.106)$$

The particle flow rate is directly related to the rate of deposition onto a surface whose unit normal is $\hat{e}_n$. This quantity is often called deposition velocity, V_d. In the definition of deposition velocity, it is customary to define the surface normal such that it points into the fluid carrying the particles. Because this direction is opposite the way j_n was defined in Equation 6.104, j_n becomes negative and a negative sign is now needed to compensate this effect:

$$V_d(\vec{r}) = -\frac{\vec{J} \cdot \hat{e}_n}{n_\infty}. \qquad (6.107)$$

Here, n_∞ is the particle density at the outer edge of the boundary layer, which forms at the surface. Further discussion will be provided in the relevant section for deposition.

Particle Deposition and the General Dynamic Equation of Aerosol Transport

Readers who are familiar with the mathematical formalism of radiation transport theory may recognize the similarities with the above description of the aerosol phase space. The difference here is that we are describing microscopic and nanoscopic particles rather than subatomic particles. Just as in the case of radiation transport, the governing equations of aerosol transport follow Boltzmann's transport equation; however, here we write it in the hydrodynamic limit, which is widely known as the General Dynamic Equation (GDE) of aerosol transport. Often, the angular dependence is neglected when the GDE is written. This does not generally lead to any significant error, unless convective transport, diffusion, van der Waals force, and phoretic effects have a nonisotropic distribution, such as near boundaries and surfaces. In the case of particle surface deposition, for example, the equations are written in terms of the angular particle current via the use of particle velocity vectors. When the inertial forces acting on the particle are negligible compared to other forces, the homogeneous continuity equation can be written to describe particle deposition:

$$\frac{\partial n}{\partial t} + \nabla \cdot \vec{J} = 0. \qquad (6.108)$$

The continuity equation is a form of mass balance, but the way it is written above it applies to the number of particles rather than to their mass; therefore it is a particle conservation equation. Here, the current is a superposition of individual currents due to various forces acting on the particles. Among the most often considered processes are the convective, diffusive, phoretic, and extraneous forces:

$$\vec{J} = \sum_k \vec{J}_k, \qquad (6.109)$$

The Physics of Aerosols

with k denoting the kth type of process giving rise to its respective current. In nearly all cases, one can use the definition of the current, given in Equation 6.102:

$$\vec{J}_k = \vec{V}_k n(\vec{r}, p, \hat{\Omega}), \tag{6.110}$$

where $\vec{V}_k$ is the velocity of the particle induced by the kth process, for example, thermophoretic velocity, $\vec{V}_T$, or convective velocity, $\vec{V}_c$, to name only a couple. In the case of diffusion, we can use Fick's law, as shown in Equation 6.23, and write

$$\vec{J}_d = -D\nabla n(\vec{r}, p, \hat{\Omega}). \tag{6.111}$$

When substituted into Equation 6.108, we gain a simplified form of the GDE, where coagulation, condensation, and evaporation, as well as other processes and external sources, are neglected:

$$\frac{\partial n(\vec{r}, v, t)}{\partial t} + \nabla \cdot \left[\vec{V}_c(\vec{r}, v, t) n(\vec{r}, v, t) \right] - \nabla \cdot \left[D(\vec{r}, v, t) \nabla n(\vec{r}, v, t) \right] = 0. \tag{6.112}$$

This equation is suitable to describe deposition due to simultaneous convection and diffusion. The first term on the left-hand side is the rate of change of the number of particles, having a volume of v at time t. The second term is the gain or loss in the number of particles due to convective fluid motion. That is, the fluid can bring or carry away particles of size v from the location of interest. The third term describes the loss or gain due to diffusion. Here, the list of variables in the angular current has been replaced with one more often seen in the literature. It is somewhat less informative than when written fully, but the presence of the velocity vector and the dot product in the above equation is a telltale of angular dependence. In many cases, Equations 6.108 and 6.112 are written in their steady-state form, where the time derivative on the left-hand side is zero, and we have $\nabla \cdot \vec{J} = 0$. This approach is correct for time-independent processes or when the process is averaged over a time period of interest. To obtain the deposition properties of the aerosol for this time period, Equation 6.112 is solved for $n(\vec{r})$ or $n(\vec{r}, v)$, subject to appropriate boundary conditions. The deposition rate, j_A, can be found by integrating the particle flow rate, Equation 6.106, over the surface area of interest, $d\vec{S} = \hat{e}_n \, dA$:

$$j_A = \int_S j_n(\vec{r}) dA = -\int \hat{e}_n \cdot \vec{J}(\vec{r}) dA. \tag{6.113}$$

Here the negative sign is due to the surface normal pointing in the opposite direction of the particle current, as accentuated earlier. Using Equation 6.108, we find

$$j_A = n_\infty \int_S V_d(\vec{r}) dA. \tag{6.114}$$

Because the surface onto which the particles deposit may be an irregular 3D structure, sometimes it may be difficult to evaluate the above integral, despite its apparent simplicity. The most often used boundary conditions in deposition problems are the following: (1) the particle density is zero within a distance of the surface equivalent to the radius of the depositing particle and (2) the particle density is a constant value, n_∞, sufficiently far away. For example, in the case when particles deposit onto a spherical surface of radius R, and when the deposition is due only to diffusion, Equation 6.113 in the steady state can be solved to find

$$j_A = 4\pi R D n_\infty. \tag{6.115}$$

Equation 6.112 is valid not only for deposition, bur for all cases when particle growth and the contribution of external sources are negligible. Particle growth may be due to condensation or evaporation (the latter represents negative growth), but also due to collision of particles, which is known as coagulation. Because of mathematical convenience and because it can occur independently of condensation and evaporation, coagulation is usually handled as a source term, as it will be seen later. For now, we focus on particle growth without coagulation. Let the rate of particle growth due to condensation and evaporation be $I(\bar{r}, v, t)$ such that $I(\bar{r}, v, t)n(\bar{r}, v, t)\,dt\,dv$ is the change in the number of particles having a volume dv about v during dt time interval at location $\bar{r}$. The rate function $I(\bar{r}, v, t)$ can be expressed for the rate of change in the number of particles or for the rate of change in the volume the particles occupy. The form of this function depends on the materials involved and on the environmental conditions, such as temperature distribution and heat transfer. This term can now be included in Equation 6.112 as follows:

$$\frac{\partial n(\bar{r},v,t)}{\partial t} + \frac{\partial}{\partial v}\left[I(\bar{r},v,t)n(\bar{r},v,t)\right] + \nabla \cdot \left[\bar{V}_c(\bar{r},v,t)n(\bar{r},v,t)\right] - \nabla \cdot \left[D(\bar{r},v,t)\nabla n(\bar{r},v,t)\right] = 0.$$

(6.116)

Here, the growth rate is written in terms of the change in the volume of particles. Often, the deposition terms are abbreviated, $R(\bar{r}, v, t)n(\bar{r}, v, t)$, with $R(\bar{r}, v, t)$ being the deposition operator that may include the convective, diffusive, or both terms as shown in Equation 6.115. When independent sources, $S(\bar{r}, v, t)$, are also included, Equation 6.116 may be written as

$$\frac{\partial n(\bar{r},v,t)}{\partial t} + \frac{\partial}{\partial v}\left[I(\bar{r},v,t)n(\bar{r},v,t)\right] + R(\bar{r},v,t)n(\bar{r},v,t) = S(\bar{r},v,t).$$

(6.117)

A number of authors have investigated the nature of the growth term, $I(\bar{r}, v, t)$, for example, Fuchs (1959), Gelbard and Seinfeld (1979), Loyalka and Park (1988), and Lehtinen et al. (1998). For the full description of the aerosol phase space, that is, to gain the full form of the GDE, we need to incorporate the coagulation process into Equation 6.117 and extend the validity of the convective term to include the fluid as well as particle velocities due to phoretic effects:

$$\frac{\partial n(\bar{r},v,t)}{\partial t} + \frac{\partial}{\partial v}\left[I(\bar{r},v,t)n(\bar{r},v,t)\right] + \nabla \cdot \left[\bar{V}(\bar{r},v,t)n(\bar{r},v,t)\right]$$
$$- \nabla \cdot \left[D(\bar{r},v,t)\nabla n(\bar{r},v,t)\right] = S(\bar{r},v,t) + \left(\frac{\partial n(\bar{r},v,t)}{\partial t}\right)_{\text{coag.}}$$

(6.118)

This is one of the more general forms of the aerosol transport equation, cast as particle conservation. The reader may recognize that it is Boltzmann's transport equation as applied at the hydrodynamic limit. Here, $\bar{V}$ denotes the resultant velocity vector due to fluid flow and particle velocities, and the term $(\partial n/\partial t)_{\text{coag}}$ represents the source due to coagulation. The rest of the terms have their meaning as described in the foregoing. Because the coagulation term is an integro-differential equation in its own right, as will be discussed later, Equation 6.118 cannot be solved by simple means, or without introducing simplifying assumptions.

Apart from a few simple cases and simple geometries, the analytical solution of the GDE is intractable and numerical methods of various complexities must be employed. One of the fruitful techniques of solving the GDE is the assumption that over a finite but small spatial interval, far away from surfaces, the particle density is homogeneous, which is known as the "well-mixed hypothesis." Although it has been shown that this technique is inappropriate for large compartments, it is still used in many widely used free and commercially available software. For example,

The Physics of Aerosols

Williams and Loyalka (1991, pp. 256–258) showed that the well-mixed hypothesis significantly overestimates the rate of sedimentation for long times, by up to a factor of 24 for a uniform particle source, and by a factor of 6 for a Dirac delta-function source located at the top of the compartment.

Spatial homogenization, however, can be used successfully when the domain of interest is subdivided into many small interfacing compartments, control volumes, or nodes. A rectangular parallelepiped is commonly used for the control volume for simplifying the calculations. This is a lumped-parameter integral approach where the nodes can communicate with each other. The GDE can be written for each node, and using appropriately chosen boundary conditions at the interfaces, the full GDE can be numerically solved over the entire domain.

To illustrate this concept, consider the convective and diffusive terms in Equation 6.118, both of which involve spatial derivatives. The average particle density over a node of volume υ is

$$n(v,t) = \frac{1}{\upsilon} \int_\upsilon n(\bar{r},v,t)\,d\bar{r} \tag{6.119}$$

Averaging the convective term, and using the Gauss theorem to convert it from volumetric to surface integral, we have

$$\frac{1}{\upsilon}\int_\upsilon \nabla \cdot \left[\bar{V}(\bar{r},v,t)n(\bar{r},v,t)\right]d\bar{r} = \frac{1}{\upsilon}\int_A \left[\bar{V}(\bar{r},v,t)n(\bar{r}_b,v,t)\right] \cdot d\bar{A}, \tag{6.120}$$

where $n(\bar{r}_b, v, t)$ is the particle density at the boundary of the node, $\bar{r}_b$, and A is the surface of the node. If the control volume is sufficiently small, the particle density along its boundary may not appreciably vary, and in the right-hand side of the above equation, we can set $n(r_b, v, t) \cong n(v, t)$. This approximation permits the simplification of Equation 6.120, and the convective flux across the node boundary is written as

$$R_c = \frac{n(v,t)}{\upsilon}\bar{A}\cdot\bar{V}. \tag{6.121}$$

When the control volume is bound on one or more sides by a surface or wall where gravitational deposition occurs, we separate the resultant fluid and particle velocity, $\bar{V}(\bar{r}, v, t)$, to the fluid velocity and the settling velocity, as follows:

$$\bar{V}(\bar{r},v,t) = \bar{U}(\bar{r},t) + \bar{V}_s(v,t). \tag{6.122}$$

If the no-slip condition is employed at the surface of deposition, A_s, then the fluid velocity is assumed to vanish at this boundary, and $\bar{V}(\bar{r}, v, t) = \bar{V}_s(v, t)$, where the latter refers to Equation 6.45. Thus, the deposition operator due to settling can be written as

$$R_s = \frac{n(v,t)}{\upsilon}A_s V_s. \tag{6.123}$$

Averaging the diffusive term, and using Gauss' theorem as above, we have:

$$\frac{1}{\upsilon}\int_\upsilon \nabla \cdot \left[D(\bar{r},v,t)\nabla n(\bar{r},v,t)\right]d\bar{r} = \frac{1}{\upsilon}\int_A D(\bar{r},v,t)\nabla n(\bar{r},v,t)\cdot d\bar{A}. \tag{6.124}$$

For a small control volume, the diffusion coefficient may not vary significantly across the volume of the node. Therefore, this coefficient can be taken out of the above integral and the diffusive transfer across the node boundaries becomes

$$\frac{D(v,t)}{\upsilon} \int_A \nabla n(\bar{r},v,t) \cdot d\bar{A}. \tag{6.125}$$

Also, for small control volumes the gradient of the particle density inside the compartment is not expected to be very large. In addition, the diffusion coefficient for particles in the size range of practical interest (1 nm–100 μm) is in the order of 10^{-2}–10^{-9} cm^2 s^{-1}. When the boundaries of the node do not involve solid surfaces, convective particle transfer across the node interfaces is the dominant transport mechanism. At solid interfaces, however, both convective and diffusive mass transfer may be important. In the case of wall deposition, the diffusive removal can be estimated using boundary layer theory. It is assumed that the particle concentration is zero within one particle radius away from the surface and it is $n(v, t)$ at the edge of the boundary layer. In this way, one can write the gradient of the particle density along the surface normal in terms of the diffusive boundary layer thickness, δ_D, as follows:

$$\hat{e}_n \cdot \nabla n(\bar{r}_s,v,t) \cong \frac{n(v,t)}{\delta_D}. \tag{6.126}$$

Here, $\hat{e}_n$ is the surface normal, $\bar{r}_s$ is the coordinate of the solid surface, $n(\bar{r}_s, v, t)$ is the particle density at the surface, and $n(v, t)$ is the particle density inside the control volume. Substituting into Equation 6.125 and evaluating the integral give the surface deposition by diffusion:

$$R_d = \frac{n(v,t)}{\upsilon \delta_D} A_s D(v,t). \tag{6.127}$$

Diffusion alone is an ineffective mechanism for surface deposition. However, because in the absence of strong resuspension the surface acts as a sink, a strong gradient in the particle density develops within the diffusive boundary layer. Hence, when convective flow takes the aerosol particles close to the surface, the diffusive deposition becomes an important process.

Particle deposition processes have received considerable attention, and there are a large number of correlations based on experimental and theoretical investigations, predicting the rate of particle removal by various deposition mechanisms. The boundary layer thickness also has a rich literature, the details of which are beyond the scope of this work. Suffice to say that determination of the diffusion boundary layer is not a simple task. It is dependent on the particle size, and for aerosols greater than 1 nm, it is much smaller than the momentum boundary layer created by flow of the medium over the surface of interest. Despite the fact that δ_D is particle size dependent, it is often assumed that it is constant. For example, the MAEROS code uses 10^{-5} m for diffusive deposition computations (Gelbard, 1982). The interested reader is referred to the works of Williams and Loyalka (1991) and Friedlander (2000) who give a thorough accounting of particle deposition under various flow regimes and geometries.

The nodal averaging, as outlined in the foregoing, can be extended to include the entire aerosol transport equation or GDE. The procedure involves integrating Equation 6.118 for the ith node, and applying the Gauss theorem. Assuming that each node is coupled to the others through its surface, which in turn may consist of several subareas where particles can leak and where particles can deposit, the number of coupled set of equations obtained is equal to the number of nodes.

Selection and implementation of the numerical scheme to accomplish the aforementioned is not a simple task, as shown by Park (2003). In the CAEROT code (Sajo and Park, 2003b), for example, first the stand-alone coagulation problem is solved for each node, which then forms the coagulation source in Equation 6.118. Subsequently, the computational domain is iteratively swept by a numerical method to solve Equation 6.118, which includes convective transport. This is repeated in each time step until convergence. The time-differencing scheme and the spatial differencing method have substantial influence on the fidelity of the simulation. Although fully implicit time differencing is not expected to diverge, since small time steps introduce more numerical diffusion, it is important that an efficient time step be computed adaptively. Also, simple finite spatial differencing is time consuming and has a weak physical basis. To improve computational efficiency and fidelity, CAEROT uses a semi-Lagrangian approach, in which the aerosol property is transferred along the streamlines, while the node coordinates are fixed relative to the computational domain. Still, the optimal node size may vary as a function of many parameters, including airflow velocity (Sajo et al., 2004).

COAGULATION

Coagulation is the process when two or more particles collide, adhere, and thus form a new particle. We distinguish binary collisions from higher-order collisions. In most cases, a binary collision is assumed, and it is postulated that it is unlikely that three or more particles will collide nearly simultaneously. This is a realistic assumption when the particles are spheroids, but when they are long chains or needles, this may not be always correct. In high particle concentrations, ambient turbulence can rotate, spin, and translate these particles of complex shape in a way that many-body collisions may not be ignored. Nevertheless, most of the theoretical development in aerosol science assumes a binary collision, and we will do the same here to avoid unnecessary mathematical complexity.

In the absence of velocity differentials, that is, when all particles move with the same velocity in the same direction, coagulation does not occur. However, Brownian motion is always present even in the hypothetical case when all other transport mechanisms are absent. Therefore, coagulation can be discussed separately or together with the GDE. Here, we will consider it as a separate phenomenon, but discuss it in the context of dynamic spatial transport. As it is shown in Equation 6.118, the coagulation process appears as a source term in the GDE. In the case of binary collision, a new particle is created from the collision of two smaller particles having respective sizes u and w such that $u + w = v$. From the perspective of the number of those particles whose size is v, or the particle density $n(\bar{r}, v, t)$, this is indeed a source. This process, however, depletes the number of particles whose size is exactly u and w, that is, as $n(\bar{r}, v, t)$ increases by 1, both $n(\bar{r}, u, t)$ and $n(\bar{r}, w, t)$ will decrease by 1 each; one particle is created while two are lost. Therefore, strictly speaking, the number of particles is not conserved by this equation. To alleviate this problem, often the aerosol volume distribution, $q(v, t) = vn(v, t)$, is used as the aerosol property in formulating the GDE because the volumes are conserved.

Because binary collision of particles occurs in discrete steps, an easy way to formulate a mental image of how the particle spectrum changes due to coagulation is to assume that the particle distribution function is discrete, in which all particles fall into a finite number of uniquely identifiable volumes v_i, $i = 1, 2 \ldots j, k \ldots G$. In this way, the rate of change of the number of particles, N_k, having size, v_k, can be written in two parts: a source term corresponding to an influx of particles due to coagulations of smaller particles and a loss term due to particle growth by collision of size v_k with any other size, including v_k.

To express the source term, first focus on two particles of sizes v_i and v_j. Collision of these two particles will give rise to a new particle whose size is v_k. While v_k is fixed, the indices i and j can take any value, and thus they can have a number of combinations, within the constraint that the particle volume conservation $v_i + v_j = v_k$ must be satisfied. This is analogous to the above case where particles with sizes u and w formed a particle of volume v, but with the condition that $w = v - u$. In this way, the conservation of volume can be symbolically written as $v_i + v_{k-i} = v_k$, $k - i$ replacing j.

The number of combinations of how two such particles can form a new one with size v_k depends on i ($i < k$), that is, on the number of unique particle volumes below v_k. Hence, the source term is the sum of these combinations:

$$\frac{1}{2}\sum_{i=1}^{k-1}\beta_{i,k-i}N_i N_{k-i}. \tag{6.128}$$

This formula gives the rate of increase in the number of particles with volume v_k due to collision of particles having smaller volumes i and $k - i$, with the constraint given above. Note that N_i is the particle density, which represents the number of particles having size v_k in a unit volume of the carrier fluid. Factor $\beta_{i,k-i}$ is the coagulation coefficient, which represents the rate of collisions between particles of volumes i and $k - i$. Because this coefficient is symmetric, $\beta_{n,m} = \beta_{m,n}$ and the index of the summation runs to $k - 1$, and the sum must be halved to avoid counting the same type of collision twice. The mathematical form of $\beta_{i,k-i}$ depends on the characteristics of the aerosol phase space, including the particles sizes v_i and v_{k-i}, the flow regime, the pressure and temperature of the carrier fluid, and other parameters if extraneous forces are present. Some of the most important coagulation mechanisms and their coefficients are explained elsewhere in this chapter.

When a particle with volume k collides with another particle having any volume, the number of particles with volume k will be reduced by 1, thus this is a loss. The loss term is the combination of all such possible events, and when the total number of discrete volumes is G, it is described as

$$N_k \sum_{i=1}^{G}\beta_{i,k}N_i. \tag{6.129}$$

Hence, the net rate of change in the number of particles in section k is written as

$$\frac{dN_k}{dt} = \frac{1}{2}\sum_{i=1}^{k-1}\beta_{i,k-i}N_i N_{k-i} - N_k \sum_{i=1}^{G}\beta_{i,k}N_i. \tag{6.130}$$

If the coagulation coefficient is a known function of particle size, the above equation can be iteratively solved for a defined initial condition. The iterative solution is necessitated by the nonlinear nature of Equation 6.130, that is, because $N_{i>k}$ appears on the right-hand side, *a priori* information on all particle densities of sizes greater than v_k, is necessary to solve the equation. In the case when the initial aerosol is monodisperse, the coagulation coefficient is independent of particle size, and Equation 6.130 has an analytical solution (first shown by Smoluchowski in 1917), which does not require detailed information on N_k, only the total number of initial particles must be known, as will be outlined later in this section.

In most practical cases the aerosol size distribution is not discrete but nearly continuous. If the difference in two adjacent discrete particle sizes is denoted by $\Delta v = v_k - v_{k-1}$, and Equation 6.130 is rewritten in the limit of $\Delta v \to 0$ (and $G \to \infty$), we gain a continuous representation in terms of the particle density, $n(v, t)$, as follows:

$$\frac{\partial n(v,t)}{\partial t} = \frac{1}{2}\int_0^v K(u, v - u)n(u,t)n(v - u,t)\,du - n(v,t)\int_0^\infty K(u,v)n(u,t)\,du. \tag{6.131}$$

This formula is often denoted as the Smoluchowski coagulation equation (1917). Here, function $K(u, v)$ is known as the coagulation kernel, and it is defined such that $K(u, v)n(u)n(v)$ is the rate at which particles having volume v collide and adhere to other particles whose volume is u. Note that the symmetry condition $K(u, v) = K(v, u)$ is valid as before. Similar to the case of Equation 6.130, the first term in the right-hand side is the production of particles of volume v via collision of particles

The Physics of Aerosols

having size u with particles of size $v - u$. The second term on the right-hand side is the loss, representing removal of particles whose initial size is v due to collision with particles of size u. Some of the most important coagulation mechanisms and their kernels are explained elsewhere in this chapter.

Equation 6.131 is a nonlinear integro-differential equation, which appears as a source term of the General Dynamic Equation 6.118. This equation can be solved separately from the GDE to find the temporal evolution of the particle size distribution as the result of coagulation, absent of external sources or sinks (such as settling), for example, by the sectional method (Gelbard et al., 1980), the method of moments (Pratsinis, 1988), or by other mathematical transformations (Williams and Loyalka, 1991). In the following, the two most widely used techniques, the sectional method and the method of moments, are briefly outlined.

THE SECTIONAL METHOD OF SOLVING THE COAGULATION PROBLEM

Although the discrete distribution of the aerosol is conceptually appealing, most aerosols, with the possible exception of a monodisperse system, cannot be represented with this simplified model. Because the mathematical form of the coagulation kernel is often convoluted, and because of the complex aerosol size distributions observed in practice, the coagulation Equation 6.131 is intractable in its present form for all but a few cases. A compromise between the discrete method and the continuous analytical expression (6.131) is to approximate the continuous size spectrum with a set of size intervals. In doing so, the particle size spectrum is cast into a finite number of sections or bins. In this way, for example, a particle size distribution between a minimum and a maximum particle diameter, d_{min} and d_{max}, can be subdivided into a predetermined number of sections, within each of which the integral aerosol property of interest, for example, the number of particles, is a known function of size. Figure 6.9 shows an example for such sectionalization between particle diameters 0.01 µm–10 µm. The vertical axis represents the particle density; the horizontal axis

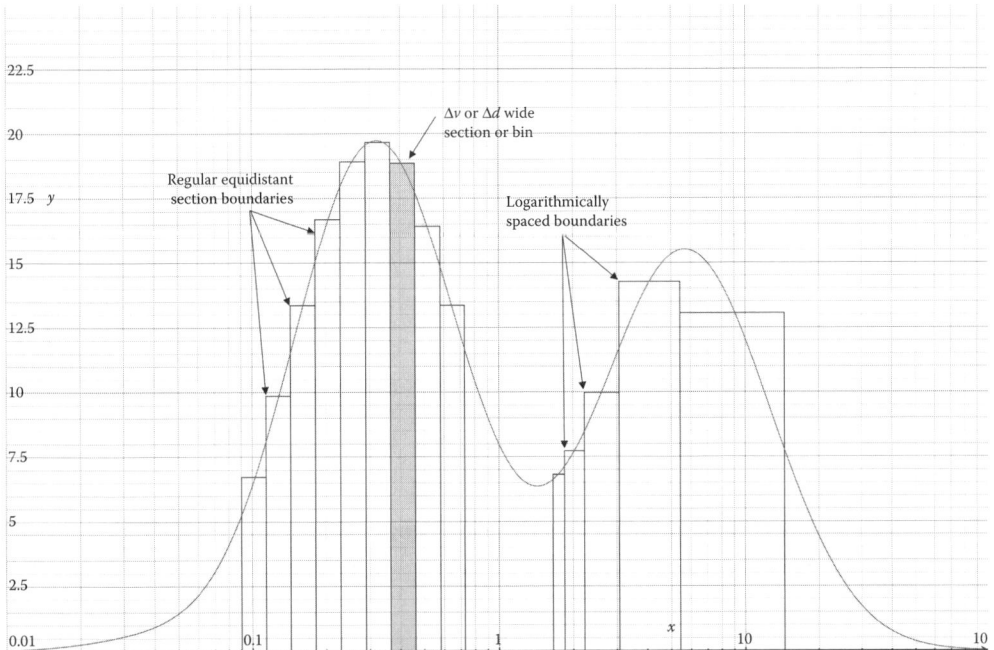

FIGURE 6.9 Continuous particle size distribution and its sectional approximation. Here, the equidistant sectionalization uses equal intervals on the log scale.

represents the particle size. The solid line represents the continuous size distribution, while the histogram about it represents the size bins. Here, it is assumed that within each size bin the number of particles does not change as a function of particle size, that is, the particle size distribution is constant within the bin with respect to size but not with respect to time. Obviously, when more sections are placed in the size distribution, the resolution becomes finer, and the approximation of the continuous line by this sectional method improves.

Using the sectional representation, efficient numerical computations may be set up to calculate the change in the particle size distribution as a function of time due to coagulation alone. In this way, the rate of change of the number of particles (or any integral aerosol property) in a particular section can be written in six parts, corresponding to two source terms and four loss terms (Gelbard et al., 1980). Consider the shaded section in Figure 6.9. Let us call it section k whose bin size is Δv_k with boundaries v_{k-1} and v_k. Here, within this section, the particle density is shown as constant with respect to size at $n(v_k)$, but other appropriate functional forms may also be assumed. The number of particles whose size is in the interval (v_{k-1}, v_k) is then $N_k = n(v_k) \Delta v_k = n(v_k)(v_k - v_{k-1})$. Note that the generalized particle property, $q(v)$, can also be used instead of the particle density, $n(v)$, written analogously to Equation 6.71 but with the boundaries of integration changed to (v_{k-1}, v_k). Also, the size variable of interest may be other than the particle volume, but expressible as a function of the volume, $f(v)$. Conservation equations for a particular section, k, may be written by simply counting the ways particles gain size by coagulating into and out of section k.

A new particle in section k may be created by the collision of two particles in two lower sections having complementary sizes, for example, ones with volumes u and w, such that $v_{k-1} < u + w < v_k$. Thus, the flux of particles (or the integral aerosol property dQ_k/dt) into section k may be written as

$$\left\{\frac{dQ_k}{dt}\right\}_1 = \frac{1}{2}\int_{v_0}^{v_{k-1}}\int_{v_0}^{v_{k-1}} \alpha(u+w)^\gamma C(u,w)\,du\,dw \quad (v_{k-1} < u+w < v_k). \tag{6.132}$$

Here and in the subsequent formulae, factor $C(u, w) = K(u, w) n(u) n(w)$, and v_0 represents the lowest section in the size spectrum. As before, when $\alpha = 1$ and $\gamma = 1$, the aerosol property, Q, becomes the number density of particles, N. Another way particles can appear in section k is due to coagulation of particles from section k with those in lower sections, such that the total size of the particle does not exceed v_k.

$$\left\{\frac{dQ_k}{dt}\right\}_2 = \int_{v_0}^{v_{k-1}}\int_{v_{k-1}}^{v_k} \alpha[(u+w)^\gamma - u^\gamma]C(u,w)\,du\,dw \quad (v_{k-1} < u < v_k \text{ and } u+w < v_k). \tag{6.133}$$

Similarly, a loss of particles from this section can be via coagulation of particles within this section with those of lower sections, such that $u + w > v_k$:

$$\left\{\frac{dQ_k}{dt}\right\}_3 = -\int_{v_0}^{v_{k-1}}\int_{v_{k-1}}^{v_k} \alpha u^\gamma C(u,w)\,du\,dw \quad (v_{k-1} < u < v_k \text{ and } w < v_{k-1} \text{ and } u+w > v_k). \tag{6.134}$$

When particles in section k coagulate with particles within the same section, the loss term can be written in two parts, depending on the destination of the particles, as follows: When the size of the resulting particle is greater than v_k, we have

$$\left\{\frac{dQ_k}{dt}\right\}_4 = -\frac{1}{2}\int_{v_{k-1}}^{v_k}\int_{v_{k-1}}^{v_k} \alpha(u^\gamma + w^\gamma)C(u,w)\,du\,dw \quad (v_{k-1} < (u,w) < v_k \text{ and } u+w > v_k). \tag{6.135}$$

When the resulting particle size remains in section k, that is, when $v_{k-1} < (u, w) < v_k$ and $v_{k-1} < (u + w) < v_k$, the particle volume is conserved and thus for $\gamma = 1$, there is no loss. However, for other cases this may not hold. For example, the number concentration $N(v)$ is reduced, which represents a loss term. For the general case, we write

$$\left\{\frac{dQ_k}{dt}\right\}_5 = -\frac{1}{2}\int_{v_{k-1}}^{v_k}\int_{v_{k-1}}^{v_k} \alpha\left[u^\gamma + w^\gamma - (u+w)^\gamma\right]C(u,w)\,du\,dw \quad (v_{k-1} < (u, w, u+w) < v_k). \tag{6.136}$$

When particles in section k collide with particles of a higher section, the resulting particle is always larger than v_k, and thus it is a loss:

$$\left\{\frac{dQ_k}{dt}\right\}_6 = -\int_{v_k}^{v_G}\int_{v_{k-1}}^{v_k} \alpha\, u^\gamma C(u,w)\,du\,dw \quad (v_{k-1} < u < v_k \text{ and } w > v_k; u+w > v_k). \tag{6.137}$$

Here, v_G represents the last or largest section in the size spectrum. Summing the above equations and converting integrals spanning more than one section into discrete sums yields the full-sectional coagulation equation. A closure relation that permits the expression of the sectional particle flux, dQ_k/dt, in terms of Q_i ($i = 1, 2, \ldots, G$) may be obtained from Equations 6.70 and 6.71 and by assuming that the sectional aerosol property may be written in terms of the size variable and a mean value, $\bar{q}_k(t)$, such that $Q_k(t) = \bar{q}_k(t)[f(v_k) - f(v_k - 1)]$, the discrete form of the sectional particle balance becomes

$$\frac{dQ_k}{dt} = \frac{1}{2}\sum_{i=1}^{k-1}\sum_{j=1}^{k-1}\beta^{(1)}_{i,j,k} Q_i Q_j - Q_k \sum_{i=1}^{k-1}\beta^{(2)}_{i,k} Q_i - \frac{1}{2}\beta^{(3)}_{k,k} Q_k^2 - Q_k \sum_{i=k+1}^{G}\beta^{(4)}_{i,k} Q_i. \tag{6.138}$$

The $\beta^{(m)}$ factors in this formula are the sectional coagulation coefficients, which can be derived from the six components above, and they represent integral values of the coagulation kernels between the appropriate section boundaries (for details, see Gelbard et al., 1980). A summary of the sectional coagulation coefficients is given in Table 6.4. Equation 6.138 represents a system of simultaneous nonlinear differential equations with $k = 1, \ldots, G$. The nonlinearity is provided by the last term, which prevents its reduction to a triangular form. Various numerical methods, particularly higher-order Runge–Kutta rules, can be successfully applied. The number of coefficients necessary to calculate the discrete coagulation equation is $(G^3 + 6G^2 - G)/6$ for an arbitrary discretization of G sections.

In most cases, the particles' size range of interest is so large that having too many sections may impose computational difficulties. A method, which greatly simplifies the set of equations was given by Gelbard and coworkers (1980) in which a geometric constraint is imposed on the selection of the size boundaries, such that $v_k \geq 2v_{k-1}$ ($k = 1, 2, \ldots, G$), that is, the section boundaries are logarithmically placed. This condition ensures that at least one of the particles must originate in section $k-1$ to form a particle in section k, hence the first term in Equation 6.138 reduces to a single summation in the form of $Q_{k-1}\sum_{i=1}^{k-1}\beta^{(1)}_{i,k-1,k}Q_i$. In this way, the number of sections is limited to $G = \ln(v_G/v_0)/\ln(2)$ and the number of required coefficient is reduced to $2G(G+2)$. When the size range of interest is sufficiently large, and when the initial particle size distribution is monotonically decreasing, this method gives a very good compromise between the desired accuracy and computational speed. However, as illustrated in Figure 6.9, such geometric constraints can quickly deteriorate the accuracy of the simulation when large gradients are present in the spectrum. The situation is similar when the distribution is lognormal because for each order of magnitude in the particle size range, $G = 9.96$, and therefore there is a maximum of nine sections in each order of magnitude in the

TABLE 6.4
Sectional Coagulation Coefficients Used in Equation 6.138

Sectional Coagulation Coefficient	Condition
$\beta_{i,j,k}^{(1)} = \int_{v_{i-1}}^{v_i} \int_{v_{j-1}}^{v_j} \dfrac{\theta(v_{k-1} < u+w < v_k)(u+w)K(u,w)}{uw(v_i - v_{i-1})(v_j - v_{j-1})}\, du\, dw$	$2 \leq k \leq G$ $1 \leq i < k$ and $1 \leq j < k$ $\beta_{i,j,k}^{(1)} = \beta_{j,i,k}^{(1)}$
$\beta_{i,k}^{(2)} = \int_{v_{i-1}}^{v_i} \int_{v_{k-1}}^{v_k} \dfrac{[\theta(u+w > v_k)u - \theta(u+w < v_k)v]K(u,w)}{uv(v_i - v_{i-1})(v_k - v_{k-1})}\, du\, dw$	$2 \leq k \leq G$ $i < k$ $\beta_{i,k}^{(2)} \neq \beta_{k,j}^{(2)}$
$\beta_{k,k}^{(3)} = \int_{v_{k-1}}^{v_k} \int_{v_{k-1}}^{v_k} \dfrac{\theta(u+w > v_k)uK(u,w)}{uw(v_k - v_{k-1})^2}\, du\, dw$	$1 \leq k \leq G$
$\beta_{i,l}^{(4)} = \int_{v_{i-1}}^{v_i} \int_{v_{k-1}}^{v_k} \dfrac{uK(u,w)}{uw(v_i - v_{i-1})(v_k - v_{k-1})}\, du\, dw$	$1 \leq k < G$ $i > k$ $\beta_{i,k}^{(4)} \neq \beta_{k,i}^{(4)}$

Note: The factor θ (expression) is zero when the expression within parentheses evaluates to false, and it is 1 when the expression evaluates to true. The highest section number is G.

diameter range. This leads to inefficient sectionalization and poor resolution. Using the geometric constraint, the maximum number of sections within N standard deviations about the geometric mean diameter, $[d_{max}/N\sigma_g, d_{max}N\sigma_g]$, is $G_{max} = 6\ln(N\sigma_g)$.

The formalism seen here can be extended to multicomponent aerosols where a number of different types of particle are simultaneously present. A general technique, based on the sectional approach, to simulate the temporal change of aerosol size and chemical composition due to simultaneous coagulation, interparticle chemical reaction, gas-to-particle conversion, and removal mechanisms, was developed by Gelbard and Seinfeld (1980), and a method to solve the space- and time-dependent multicomponent aerosol problem with application in the marine boundary layer is presented by Gelbard and coworkers (1998).

A computer code to solve the simultaneous coagulation, settling, and condensation problem for a multicomponent aerosol in a well-stirred atmosphere, using the geometric constraint, was developed by Gelbard (1982). A comprehensive model comparison was done by Seigneur and coworkers (1986), which reviewed three types of computer models for simulating aerosol dynamics based on the continuous, sectional, and parameterized approaches. The study found that although the sectional method has isolated disadvantages, it is one of the best performing methods among those considered. Another computer program using the sectional approach, but with arbitrary section boundaries and a number of options to consider interparticle forces, was developed by Sajo and Park (2003a).

THE METHOD OF MOMENTS

As it was pointed out in the foregoing, the way Equations 6.130 and 6.131 are written is not conducive to direct particle conservation because coagulation is a reduction process in the number of particles. Equation 6.70, however, provides a convenient way to generalize Equation 6.131 and relate it to the moments of the size distribution. Let us write Equation 6.70 as

$$q(v, t) = W(v)n(v, t), \tag{6.139}$$

where $W(v) \equiv \alpha v^\gamma$. If Equation 6.131 is multiplied by $W(v)$ and integrated over the particle volume, v, for all sizes, we have

$$\frac{d}{dt}\int_0^\infty W(v)n(v,t)\,dv = \frac{1}{2}\int_0^\infty\int_0^\infty K(u,v)n(u,t)n(v,t)\big[W(u+v) - W(u) - W(v)\big]du\,dv. \quad (6.140)$$

Rearranging this equation, and employing relations 6.70, 6.71, and 6.139 yield

$$\frac{dQ(t)}{dt} = \frac{1}{2}\int_0^\infty\int_0^\infty K(u,v)\big[W(u+v)n(u,t)n(v,t) - q(u,t)n(v,t) - q(v,t)n(u,t)\big]du\,dv. \quad (6.141)$$

This form of the coagulation equation is written in terms of the time rate of change of the integral aerosol property, which can be related to the time rate of change in the various moments. The technique used to obtain Equation 6.141 is known as the moments transformation.

The zeroth moment of the distribution, Equation 6.73, which was found to be equivalent to the total particle concentration, $N(t)$, is obtained by setting $W(v) = 1$. Substituting into Equation 6.141 gives

$$\frac{dQ^{(0)}(t)}{dt} = \frac{dN(t)}{dt} = -\frac{1}{2}\int_0^\infty\int_0^\infty K(u,v)n(u,t)n(v,t)\,du\,dv. \quad (6.142)$$

An elegant analytical solution of this equation can be obtained for the case when the coagulation kernel is not a strong function of particle size, as was proposed by Smoluchowski (1917) for Brownian coagulation of nearly equally sized particles (a.k.a. monodisperse aerosol) in the continuum regime. In this case, for $1 \leq u/v \leq 2$, the kernel is nearly constant at $K = 8kT/3\mu$, and the above equation simplifies to

$$\frac{dN(t)}{dt} = -\frac{1}{2}KN^2(t). \quad (6.143)$$

Integration with respect to time and using the initial condition of $N(t=0) = N_0$, gives a relatively simple solution for the total particle concentration (or integral particle density) as a function of time during the coagulation process:

$$N(t) = \frac{2N_0}{2 + KN_0 t}. \quad (6.144)$$

An initially monodisperse system undergoing pure binary coagulation is a discrete process. In that, particles of size v collide and form particles of size $2v$, which in turn undergo further coagulation to form a series of larger particles having sizes $3v, 4v, \ldots, kv$. In time, this gives rise to a multimodal size distribution, with each mode having a unique volume. By the application of Equation 6.130, it can be shown that an initially monodisperse aerosol in which all particles start out in single mode, $N_1(t=0) = N_0$, and for a coagulation kernel independent of particle size, the particle density in each of the subsequent modes, $k = 1, 2, \ldots$, evolves as follows:

$$N_k(t) = N_0 \frac{(t/\tau)^{k-1}}{(1 + t/\tau)^{k+1}}. \quad (6.145)$$

Here, t/τ is a characteristic dimensionless time, where τ is defined as $\tau = 2/KN_0$. The latter is also known as the half-value time, being the time it takes for the particle number concentration to reduce to one-half of its original value. τ is also proportional to the mean free time between particle collisions, and thus it is related (but not identical) to the relaxation time. The conservation of particles may be written as $N(t) = N_1(t) + N_2(t) + \cdots + N_\infty(t)$. The condition of constant coagulation kernel is satisfied for the case when all particles have the same size. Thus, in a monodisperse system, initially K is indeed constant. But as time passes, an increasing number of particles will be generated whose size may be sufficiently different from the initial size, so as to result in a changed K. Hence, the result shown in Equation 6.145 is generally valid for small values of t/τ, and particularly valid if K is independent of the particle size. Fuchs (1989) demonstrated that for particle size ratio of $u/v < 8$, or in terms of diameters $d_u/d_v < 2$, the coagulation kernel in the transition regime between free molecular flow and the hydrodynamic continuum does not change significantly and it is nearly constant as a function of u/v.

By virtue of Equations 6.144 and 6.145, both N and N_1 are monotonically decreasing functions of time, and all modes, N_k, and $N(t)$ are monotonically decreasing functions of k. However, the modes $k \geq 2$ have regular local maxima at $(t/\tau)_{max} = (k-1)/2$, where their value is independent of all variables, except for the order of the mode, k:

$$N_{k,max} = N_k(t = t_{max}) = 4N_0 \frac{(k-1)^{k-1}}{(k+1)^{k+1}}. \tag{6.146}$$

The first moment of the size distribution, Equation 6.74, is equivalent to the total volume concentration or the volume fraction of the aerosol in the carrier fluid, $\phi(t)$. This is obtained using $W(v) = v$. Because coagulation without compaction preserves the total aerosol volume, independent of the particular progression of the coagulation process in time, absence of external particle sources, and removal mechanisms, the total aerosol volume remains constant and the time rate of change is zero. Substituting $W(v) = v$ into Equation 6.141 cancels all terms inside the integral, and gives the expected result:

$$\frac{dQ^{(1)}(t)}{dt} = \frac{d\phi(t)}{dt} = 0. \tag{6.147}$$

The time rate of change of the second moment can be written similarly. Here, $W(v) = v^2$, and after substitution into Equation 6.141 and rearrangement we find

$$\frac{dQ^{(2)}(t)}{dt} = \int_0^\infty \int_0^\infty K(u,v) q^{(1)}(u,t) q^{(1)}(v,t) \, du \, dv. \tag{6.148}$$

Note that the factor of 1/2 has disappeared, and so did the second-order aerosol property $q^{(2)}$. Here, $q^{(\gamma)}$ is defined as in Equation 6.70 but with $\alpha = 1$. As in the case of Equation 6.143, if the coagulation kernel is a weak function of particle size, the above equation simplifies to

$$\frac{dQ^{(2)}(t)}{dt} = K\left[Q^{(1)}(t)\right]^2 = K\phi^2(t), \tag{6.149}$$

and we find that the time rate of change of the second moment is proportional to the square of the first moment, that is, to the square of the total volume concentration. Following this line of argument, it is possible to show that higher moments can be written in terms of lower-order moments. Using the binomial theorem to expand $W(u+v)$ in Equation 6.141, we obtain a generalized equation for the temporal rate of change of the γth moment,

$$\frac{dQ^{(\gamma)}(t)}{dt} = \frac{1}{2} \int_0^\infty \int_0^\infty K(u,v) \sum_{i=1}^{\gamma-1} \binom{\gamma}{i} q^{(\gamma-i)}(u) q^{(i)}(v) \, du \, dv, \tag{6.150}$$

and in the particular case when the coagulation kernel does not change appreciably with particle size, as above, we have

$$\frac{dQ^{(\gamma)}(t)}{dt} = \frac{K}{2}\sum_{i=1}^{\gamma-1}\binom{\gamma}{i}Q^{(\gamma-i)}(t)Q^{(i)}(t). \qquad (6.151)$$

It is readily seen that Equations 6.143 and 6.147 are merely special cases of Equation 6.151 for $\gamma = 0$ and $\gamma = 1$, respectively. The progression of the summation in negative direction is interpreted as $\sum_{i=1}^{-k} a_i = \sum_{i=0}^{k-1} -a_{-i}$. For monodisperse aerosols, Equation 6.83 may be applied using the equality sign, and the above formula can be simplified to

$$\frac{dQ^{(\gamma)}(t)}{dt} = K\left[Q^{(\gamma/2)}(t)\right]^2 (2^{\gamma-1} - 1). \qquad (6.152)$$

The above development forms the basis of the moments method of solving the coagulation equation, which is an alternative to the sectional method. In doing so, the Smoluchowski equation is transformed into a set of ordinary differential equations written either for the integral aerosol property or for the moments. However, depending on the particle size distribution and on the aerosol flow regime (and on the corresponding functional form of the coagulation kernel), these equations present various degrees of difficulties. For most practical forms of $K(u,v)$, Equation 6.141 and its moments expansions (6.150) yield more unknowns than the available number of equations; therefore it is not closed. A closure relation may be gained by assuming a mathematical form of the aerosol scalar density (or size distribution), $n(v,t)$, which is conducive to analytical or numerical integration of the resulting equations. This technique is known as "kernel approximation." In particular, the lognormal (Equation 6.6) and the gamma distributions (Equation 6.9) have been used by many authors to write the first three or four moments (e.g., $Q^{(0)}$, $Q^{(1)}$, $Q^{(1/3)}$, and $Q^{(2/3)}$ or $M^{(0)}$, $M^{(1)}$, and $M^{(3)}$ corresponding to $\gamma = 0$, 1, and 2) as a closed set of equations. Williams (1986) showed that in this way some of the resulting equations could be solved analytically. In some other cases, a combination of techniques is used. For example, in the free molecular flow regime where Brownian coagulation is dominant, Pratsinis (1988) approximated the coagulation kernel and brought it to a form that permits analytic integration by assuming that *a priori* information is available for the initial particle size distribution.

These techniques suffer from the limitation that the initial form of the particle density is fixed; not only the actual particle density function may be different from the one assumed; the accuracy of the method is bound by the number of unknowns that the preselected size distribution entails—a restriction, which is not present when the sectional method is used. Other techniques attempt to resolve this constraint by assuming that the moments of the size distribution or the coagulation kernel can be approximated by a series. For example, Barrett and Jheeta (1996) postulated that the logarithm of the moments could be written in terms of a finite-order polynomial, as follows:

$$\ln Q^{(\gamma)} = \sum_{i=0}^{p} a_i \gamma^i. \qquad (6.153)$$

The unknown coefficients, a_i, can be obtained from the equations written for the $p + 1$ moments. Conceptually, this technique stems from the representation discussed with respect to Equation 6.72, in that a higher (unknown) moment can be approximated by a series of lower (known) moments. For the moments $Q^{(0)}$ and $Q^{(2)}$ using values $p = 2$, 3, and 4, the authors showed that in the case of Brownian coagulation, the order of the polynomial has no substantial effect, and their results are in good agreement with those of other researchers using different techniques.

Using the moments and one or more of the balance equations, such as the general dynamic equation or one of its simplified forms, it is possible to express some of the moments without explicit reliance and *a priori* assumptions on the shape or mathematical form of the particle size distribution. This was shown by Friedlander (2000) for the case of condensation with respect to the zeroth and second moments, $M^{(0)}$ and $M^{(2)}$, but the concept is applicable to other phenomena as well.

COAGULATION KERNELS

Collision and agglomeration is due to relative motion between the particles, which is caused primarily by Brownian motion, gravitational settling, and turbulence. Other processes of secondary importance include thermophoretic and diffusiophoretic effects, acoustic waves, and other forces that give rise to velocity differentials between the particles. Depending on the forces acting on the particles, and on the particle size relative to that of the carrier fluid molecules, the aerosol particles can move with the fluid as a continuum, experience slip, or take on significantly different velocities. All of these processes greatly depend on the particle-fluid flow regime, and some of them can coexist. In addition, because the particle size spectrum may span several orders of magnitude, the dominant flow regimes affecting the various parts of the size spectrum may be substantially different. Therefore, the coagulation problem and the selection of the appropriate coagulation kernel for the problem at hand is a difficult task. The following discussion is organized according to these processes and their dominant flow regimes. Among the many different types of coagulation kernels, only those are considered here that occur most frequently in practice: various regimes of Brownian motion, Fuchs's method, gravitational coagulation, and turbulent coagulation. There are a variety of other types of coagulation kernels, including Williams' unified theory of coagulation, whose details may be found in Williams (1988), and, for example, in Williams and Loyalka (1991).

BROWNIAN COAGULATION—DIFFUSION REGIME

Thermally driven motion of fluid molecules and particles suspended therein gives rise to random collisions. When the particle is large compared to the mean free path of the carrier fluid's molecules or atoms, the Knudsen number Kn < 1 and the particle–fluid system forms a continuum. In this case, the predominant process for relative particle motion is diffusion, and hydrodynamic calculations of the drag force coupled with solving the diffusion equation give acceptable results for the collision frequency. Consider a particle with radius b whose center is attached to a polar coordinate system. In this way, from the perspective of this particle, the diffusion process by which other particles deposit onto this one is isotropic. The rate at which particles of radius a diffuse toward a particle of radius b can be written using Equation 6.30 in polar coordinates as follows:

$$\frac{\partial C(r,t)}{\partial t} = D_0 \left[\frac{\partial^2 C(r,t)}{\partial r^2} + \frac{2}{r} \frac{\partial C(r,t)}{\partial r} \right]. \tag{6.154}$$

The boundary conditions are similar to those found in deriving Equation 6.115, namely $C(r = a + b) = 0$ and $C(r = \infty) = C_\infty$. That is, the particle concentration is zero within a distance of the surface of particle b equivalent to the radius of the depositing particle, a, and the particle concentration is a constant value, C_∞, sufficiently far away. Here, D_0 is the relative diffusion coefficient. It can be shown that $D_0 = D_a + D_b$, and using Equation 6.34, we have

$$D_0 = \frac{kT}{6\pi\mu} \left(\frac{1}{a} + \frac{1}{b} \right). \tag{6.155}$$

The solution of Equation 6.154 is straightforward, and it is left to the reader. Using Fick's law, Equation 6.23, or the deposition rate derived in Equation 6.113, the flow rate of particles a to particle b can be expressed as

$$j = 4\pi(a+b)^2 D_0 \left[\frac{\partial C(r,t)}{\partial r} \right]_{r=a+b}. \tag{6.156}$$

Employing the solution of Equation 6.154 yields

$$j = 4\pi D_0 C_\infty (a+b)\left(1 + \sqrt{\frac{\tau}{t}}\right), \tag{6.157}$$

where τ is the characteristic time, proportional to the diffusion time over a distance of $(a+b)$, $\tau = (a+b)^2/\pi D_0$. In ambient atmospheric conditions, the diffusion time over the range of 1 μm is approximately 10^{-2} s. Therefore, for practical time scales, the term $\sqrt{\tau/t} \approx 0$ and $j = 4\pi D_0 C_\infty (a+b)$. The coagulation kernel is the rate at which particles collide due to diffusion per unit flux, and it may be obtained by dividing the flow rate with the ambient particle concentration:

$$K(a,b) = 4\pi D_0 (a+b) = \frac{2kT}{3\mu}\left(\frac{1}{a} + \frac{1}{b}\right)(a+b). \tag{6.158}$$

Often, the form of the coagulation kernel that uses particle volumes is more useful. If u and v represent the volumes of particles whose radii are a and b, respectively, we have

$$K(u,v) = \frac{2kT}{3\mu}\left(2 + \left(\frac{u}{v}\right)^{1/3} + \left(\frac{v}{u}\right)^{1/3}\right). \tag{6.159}$$

A useful approximation of the above formula may be obtained by noticing that when $1 \leq u/v \leq 2$,

$$(u/v)^{1/3} + (v/u)^{1/3} \approx 2; \quad \text{thus in this range, } K(u,v) \cong 8kT/3\pi.$$

BROWNIAN COAGULATION—SLIP FLOW REGIME

When the particle size is in the range of the mean free path, that is, when $Kn \approx 1$, Cunningham's slip correction factor can be used to extend the result obtained for the diffusion regime. This is justified because Cunningham's factor accounts for the deviation from Stokes' law, or the change in the drag coefficient with increasing Knudsen number, as seen earlier in this chapter. Modifying the diffusion coefficient, Equation 6.34, to take into account the slip condition, for a particle of radius a,

$$D_a = \frac{kT}{6\pi\mu a} C_a, \tag{6.160}$$

and Equation 6.158 is rewritten as follows:

$$K(a,b) = \frac{2kT}{3\mu}\left(\frac{C_a}{a} + \frac{C_b}{b}\right)(a+b). \tag{6.161}$$

This may be written in terms of particle volumes, where u and v represent the volumes of particles whose radii are a and b, respectively, as follows:

$$K(u,v) = \frac{2kT}{3\mu}\left[C_a\left(\frac{v}{u}\right)^{1/3} + C_b\left(\frac{u}{v}\right)^{1/3} + C_a + C_b\right]. \tag{6.162}$$

Although Cunningham's slip correction is valid in the range of Kn > 1, as Williams and Loyalka (1991) point it out, for Kn ≫ 1, which is the free molecular flow regime, Equation 6.161 or Equation 6.162 would degenerate into an incorrect form. The range between Kn < 1 and Kn > 1 is a transition regime between the continuum and free molecular flow regimes, and it was studied, among others, by Fuchs (1989).

Brownian Coagulation—Free Molecular Flow Regime

When the particle is smaller than the mean free path, that is when Kn > 1, the particle dimensions are increasingly comparable to the surrounding gas molecules, and they will experience periods of free travel in between collisions. This is a marked departure from the continuum regime where the granular nature of the carrier fluid is not experienced by the particles. In contrast, in this flow regime, the particles' interactions with the air molecules determine their motion, which is, however, still random. Under this condition, the number of collisions can be obtained from the Maxwell–Boltzmann distribution, Equation 6.15.

The collision rate between aerosol particles of radii a and b was shown by Williams (1971) to follow statistical mechanical averaging of the Maxwellian:

$$R_{ab} = \iint f_a(\bar{v}_a) f_b(\bar{v}_b) |\bar{v}_a - \bar{v}_b| \sigma(|\bar{v}_a - \bar{v}_b|) dv_a\, dv_b. \tag{6.163}$$

Here, $f(\bar{v})$ is the Maxwellian written for the velocity vectors $\bar{v}_a$ and $\bar{v}_b$ for particles a and b, respectively, such that, for example, for particle a,

$$f_a(v) = n_a \left(\frac{m_a}{2\pi kT}\right)^{3/2} e^{-(m_a v^2 / 2kT)}. \tag{6.164}$$

The function $\sigma(|\bar{v}_a - \bar{v}_b|)$ is the cross section for collision with a relative velocity $\bar{v} = \bar{v}_a - \bar{v}_b$. In the case of rigid spheres, $\sigma(v) = (a+b)^2 \pi$. Using the reduced velocity, $\bar{V} = (m_a \bar{v}_a + m_b \bar{v}_b)/(m_a + m_b)$, Equation 6.163 is evaluated to yield

$$R_{ab} = n_a n_b (a+b)^2 \left[8\pi kT \frac{m_a + m_b}{m_a m_b}\right]^{1/2}. \tag{6.165}$$

The coagulation kernel is obtained by dividing the collision rate with $(n_a\, n_b)$:

$$K(a,b) = (a+b)^2 \left[8\pi kT \frac{m_a + m_b}{m_a m_b}\right]^{1/2}. \tag{6.166}$$

If the colliding particles are of the same material, the masses m_a and m_b can be converted to density and particle radii, and the coagulation kernel becomes

$$K(a,b) = \left(\frac{6kT}{\rho}\right)^{1/2} (a+b)^2 \left(\frac{1}{a^3} + \frac{1}{b^3}\right)^{1/2}. \tag{6.167}$$

In terms of particle volumes, where as previously, u and v represent the volumes of particles whose radii are a and b, respectively:

$$K(u,v) = \left(\frac{6kT}{\rho}\right)^{1/2}\left(\frac{3}{4\pi}\right)^{1/6}(u^{1/3}+v^{1/3})^2\left(\frac{1}{u}+\frac{1}{v}\right)^{1/2}. \tag{6.168}$$

FUCHS' METHOD FOR ALL BROWNIAN REGIMES

Fuchs (1989) developed a technique, which covers the entire range of Knudsen numbers, including the transition regime. His assumption was that the particle is surrounded by a fictitious jump distance, δ, and the diffusion theory, shown in Equation 6.154, is only valid for distances $r > a + b + \delta_{ab}$, where δ_{ab} is the combined jump distance of the two particles a and b. Within the distance between the particle surface and $r = a + b + \delta_{ab}$, Fuchs assumed that the particles obey the kinetic theory of gases in vacuum. Obtaining the diffusion current at the $r = a + b + \delta_{ab}$ interface and the kinetic current within the distance $r < a + b + \delta_{ab}$, he found that

$$K(a,b) = \frac{4\pi(a+b)(D_a+D_b)}{\left((a+b)/(a+b+\delta_{ab})\right) + \left((4(D_a+D_b))/((a+b)\bar{v}_{ab})\right)}. \tag{6.169}$$

Here, the diffusion coefficients D_a and D_b include Cunningham's correction factor, as in Equation 6.160, and $\bar{v}_{ab} = (\bar{v}_a^2 + \bar{v}_b^2)^{1/2}$ with $\bar{v}$ being the average thermal velocity such that $\bar{v}_a = (8kT/\pi m_a)^{1/2}$, as seen in the discussion pertaining to the Maxwellian, Equation 6.15. Further, Fuchs derived that the combined jump distance is

$$\delta_{ab} = \left(\delta_a^2 + \delta_b^2\right)^{1/2} \tag{6.170}$$

with

$$\delta_a = \frac{(2a+L_a)^3 - (4a^2+L_a^2)^{3/2}}{6aL_a} - 2a, \tag{6.171}$$

$$L_a = \frac{8D_a}{\pi\bar{v}_a}. \tag{6.172}$$

The coefficients δ_b and L_b follow a similar pattern as given in Equations 6.171 and 6.172.

GRAVITATIONAL COAGULATION

As seen earlier with respect to the Stokes' velocity, Equation 6.45, particles under gravitational settling will attain different terminal velocities, which is proportional to their mass and size. This gives rise to differential velocities, thus faster particles may collide with slower ones that are within their trajectory. This is a surprisingly complex problem because as one particle approaches the other, the particle trajectory tends to curve due to interparticle forces. When such forces are neglected, the collision rate can be obtained by expressing the net current of incident particles on one another. Consider a particle with radius a, having a velocity of $\bar{v}_a$, incident on a field of particles with radii b, having concentration $C_{b,\infty}$ and moving in the same direction as particle a, but with velocity $\bar{v}_b$. Using the formalism introduced in the description of the aerosol phase space, in particular Equations 6.102

and 6.103, the current for particles a and b becomes the product of the collision area, the relative particle velocity, and the particle concentration as follows:

$$\vec{J} = \pi(a + b)^2 \left(\bar{v}_a - \bar{v}_b\right) C_{b,\infty}. \tag{6.173}$$

The coagulation kernel is the ratio of $|\vec{J}/C_{b,\infty}|$, which can be easily obtained from the above equation. Using Stokes' law for the settling velocity, Equation 6.45, with Cunningham's slip correction, $v_s = mgC/3\pi d\mu$, the settling velocities for particles a and b may be obtained and substituted into Equation 6.173. Hence, the coagulation kernel can be written as

$$K(a,b) = \frac{g}{6\mu}(a + b)^2 \left|\frac{m_a C_a}{a} - \frac{m_b C_b}{b}\right|. \tag{6.174}$$

The absolute value is necessary because unlike the current, which is a vector quantity (see Equation 6.102), the coagulation kernel is a scalar, which depends only on the magnitude of the velocity differential. If the particles are of the same material composition with density ρ, the above equation simplifies to

$$K(a,b) = \frac{2\rho g}{9\mu} \pi(a + b)^2 \left|C_a a^2 - C_b b^2\right|. \tag{6.175}$$

This may also be written in terms of particle volumes when the Cunningham factors are written in terms of volume, denoted as C_u and C_v (u and v correspond to particles a and b, respectively):

$$K(u,v) = \frac{\rho g}{6\mu} \left(\frac{3}{4\pi}\right)^{1/3} \left(u^{2/3} + v^{2/3}\right) \left|C_u u^{2/3} - C_v v^{2/3}\right|. \tag{6.176}$$

Among other investigators, Pruppacher and Klett (1974) attempted to account for the interparticle forces, which are neglected in the above development, via introducing the collision efficiency, ε. Based on the recognition that the velocity of a falling particle deviates from the vertical as it passes near another particle and that the particle trajectory may cross the Stokes' streamlines, they derived an improved relationship for the coagulation kernel in terms of particle velocities, v_a and v_b, as follows:

$$K(a,b) = \pi(a + b)^2 (v_a - v_b)\varepsilon, \tag{6.177}$$

where the collision efficiency is

$$\varepsilon \cong \frac{1}{2}\left(\frac{a}{a+b}\right)^2. \tag{6.178}$$

The derivation of the collision efficiency assumes that the flow regime falls under Stokes' law, and the particle motion is dominated by the fluid flow. Thus, the particle velocities v_a and v_b, can be obtained from Equation 6.45 with or without Cunningham's correction. A further assumption is that there is a significant size difference between the colliding particles, such that the smaller particle does not alter the flow about the larger particle. This set of assumption imposes a limitation on the applicability of Equation 6.177 to small values of (a/b). Williams and Loyalka (1991) noted that Equation 6.178 is reasonably accurate up to $a/b = 0.5$.

SIMULTANEOUS BROWNIAN AND GRAVITATIONAL COAGULATION

In practical cases, most particle size distributions are polydisperse, and an appreciable quantity of particles appears across both ends of the spectrum. Under this condition, coagulation by both gravitational settling and Brownian motion may become important. The conventional handling of simultaneous Brownian and gravitational coagulation is the arithmetic addition of the individual collision kernels, which is often referred to as sum kernel:

$$K_B(a,b) + K_G(a,b) = \frac{2kT}{3\mu}\left(\frac{1}{a} + \frac{1}{b}\right)(a+b) + \frac{2g\rho\pi}{9\mu}(a+b)^2|a^2 - b^2|. \tag{6.179}$$

Here, the first term, K_B, is the kernel due to Brownian motion with Kn < 1, Equation 6.158, and the second term, K_G, represents the kernel of gravitational settling, Equation 6.175, without slip correction.

This simple method of summation does not account for the enhanced collision rate due to diffusive processes in the wake of settling particles and due to the relative motion of particles. Thus, the exact kernel, K_{BG}, is greater than the sum kernel, $K_B + K_G$. In considering these effects, Simons and coworkers (1986) rigorously derived the rate of deposition of particles having a radius b onto a particle of radius a as a result of simultaneous Brownian diffusion and relative velocity due to gravitational settling. For diffusion-dominated flow, they showed that the contribution of the gravitational settling to the total coagulation kernel is twice of what is predicted by the sum kernel. A correction factor for the sum kernel was introduced, γ, equivalent to the ratio of the true kernel and the sum kernel, $K_{BG} = \gamma(K_B + K_G)$, in terms of an infinite sum of the ratio of modified Bessel functions I and K of half-integer order, as follows:

$$\gamma = \frac{4\pi}{\beta(4+\beta)}\sum_{n=0}^{\infty}(-1)^n(2n+1)\frac{I_{n+(1/2)}(\beta/2)}{K_{n+(1/2)}(\beta/2)}. \tag{6.180}$$

Here, β is a dimensionless factor, which is proportional to the relative importance of diffusion versus gravitational settling, and it is equivalent to $4K_G/K_B$:

$$\beta = 4\pi\frac{\rho g}{3kT}ab|a^2 - b^2|. \tag{6.181}$$

The largest discrepancy between the exact and the sum kernels occurs at particle diameters for which $\beta \approx 6$, yielding a local maximum of about 1.27 in γ. When gravitational settling is important (as measured by the relative particle velocity), β is large, whereas in diffusion-dominated regimes, β is small. However, because of the absolute value in the above equation, a large β does not necessarily mean that the absolute size of one of the colliding particles is large. For example, the (a, b) pairs of (0.4 μm, 3.0 μm) and (1.25 μm, 2.25 μm) both yield $\beta = 100$, whereas $\beta = 0$ for all particle sizes when $a = b$. For $\beta = 0$, $\gamma = 1$, and as the relative dominance of Brownian coagulation decreases, β increases, and at $\beta = 6$, the $\gamma(\beta)$ function attains a maximum of about 1.27. The γ function gradually decreases thereafter as β further increases, smoothing into 1.0 at very large arguments, as shown in Figure 6.10.

The numerical evaluation of the $\gamma(\beta)$ function is hindered by the gradually slowing convergence of the sum in Equation 6.180 for large orders, n, which are necessary for $\beta > 100$. For each successive ratio of the Bessel functions, the difference $|\Sigma I_{v+1}/K_{v+1} - \Sigma I_v/K_v|$, where $v = 1/2 \ldots n + 1/2$,

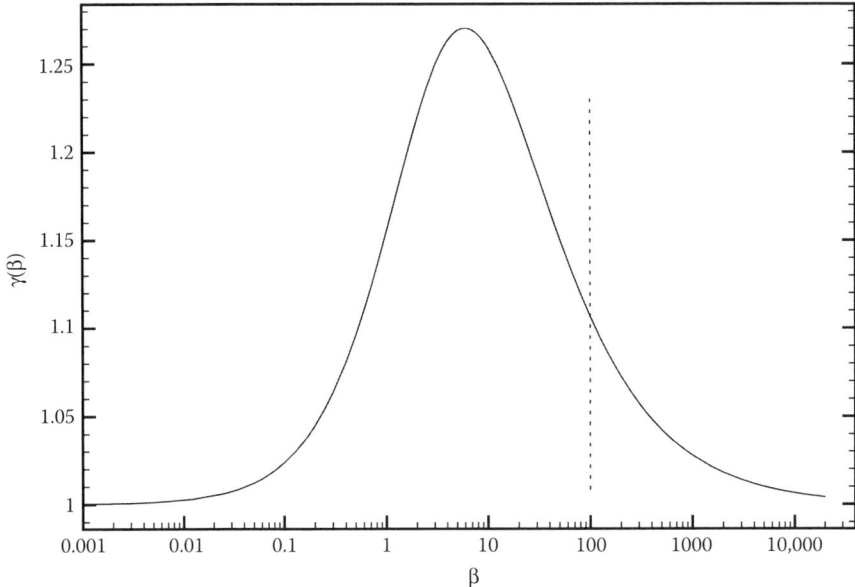

FIGURE 6.10 γ(β) function, Equation 6.180 which adjusts the value of the sum kernel for combined Brownian and gravitational settling.

becomes progressively smaller, while I_v/K_v precipitously decreases for large arguments. Owing to the finite size of the computer's internal number representation, at large arguments and large orders, a disastrous loss of digits occurs, and this sum becomes very difficult to evaluate by conventional means. Simons (1986) obtained results up to β = 100, and using multiple-precision arithmetic, Sajo (2008) extended the range to β = 20,000. Beyond this value, γ(β) is effectively unity. A practical approximation to Equation 6.180, which avoids using special functions, is as follows:

$$\gamma(0 < \beta \leq 5.5) = -0.17812 e^{-\beta} - 0.02038 e^{-2\beta} + 0.02755\beta - 0.002650\beta^2 + 1$$
$$\gamma(5.5 < \beta \leq 1110) = 0.2043 \ln(\beta) - 0.01497 \ln^2(\beta) + 11.5122\beta^{-2/3} - 29.8050\beta^{-1}$$
$$+ 32.8900\beta^{-4/3} - 14.5431\beta^{-5/3} + 0.01121\beta^{1/3} + 0.1304, \quad (6.182)$$
$$\gamma(1110 \leq \beta \leq 20,000) = 1 + 2.8430\beta^{-2/3}.$$

COAGULATION BY TURBULENT DIFFUSION

Most flow regimes in practice entail turbulence. This means that the fluid flow is characterized by chaotic motion and by the formation of transient vortices at many different length scales. Larger eddies transfer their energy to smaller ones and in this way a hierarchy of eddies is formed until the cascade reaches the level of turbulent microscale. Aerosol particles carried by this turbulent flow, thus experience an enhanced rate of collision. The dimensionless Re number is an indicator of whether the flow is turbulent or laminar. The energy dissipation rate of the turbulent eddies, ε_T (with units in m²/s³), can be estimated based on the root-mean-square turbulent velocity, $\bar{u}$, and length scale of the energetic eddies, L, as follows:

$$\varepsilon_T \approx \frac{\bar{u}^3}{L}, \quad (6.183)$$

and for isotropic eddies, Taylor (1935) derived that

$$\frac{\varepsilon_T}{15\nu} = \overline{\left(\frac{\partial u_x}{\partial x}\right)^2}. \quad (6.184)$$

Here $\nu = \mu/\rho$ is the kinematic viscosity, u_x is the x-directional component of the absolute velocity, and the large overbar represents statistical averaging. For example, in confined atmospheres, such as glove boxes or laboratory environments, the length scale of energetic eddies is in the order of 3 m, and the root-mean-square of the turbulent velocity is about 0.1 m/s. Using Equation 6.183, this gives $\varepsilon_T \approx 3.33E - 4$ m^2/s^3.

The turbulent microscale can be estimated as

$$\ell \approx \left(\frac{\nu^3}{\varepsilon_T}\right)^{1/4}, \quad (6.185)$$

thus the microscale corresponding to the above example, under standard pressure and temperature in air, is 1775 µm.

Turbulence is often discussed in terms of low-momentum diffusive motion and high-momentum convective motion. Large aerosol particles can often attain high momentum, and when their density is substantially higher than that of the air, their inertia can become sufficiently high to escape one turbulent eddy and enter another one. When the particle is smaller than the length scale of the eddies, they become entrained by them, and turbulent diffusion will be the dominant factor in their rate of collision. Because the microscale of turbulence in air are in the order 100 µm and above, most particles that are suspended in the air for more than a few seconds or minutes are smaller than this scale (the sedimentation velocity of a 100-µm particle is about 25 cm/s). These particles experience a stochastic movement throughout the fluid transported by isotropic turbulence; therefore, the classical diffusion treatment can be invoked to describe their motion.

Turbulent diffusion is usually described via the turbulent diffusion coefficient, D_T, which—for mathematical convenience—is analogous to the molecular diffusivity, albeit it depends on the flow conditions and it is not a property of the fluid or the aerosol. This approach assumes that a relation exists between the turbulent flux and the gradient, similar to that between the flux and the gradient for diffusive mass transport, as given by Fick's law, Equation 6.23. Therefore, the diffusion equation for the aerosol undergoing turbulent motion may be written in spherical coordinates as

$$\frac{1}{r^2}\frac{d}{dr}\left[r^2(D_0(r) + D_T(r))\frac{d}{dr}C(r)\right] = 0. \quad (6.186)$$

Here, D_0 is the relative Brownian diffusion coefficient, Equation 6.155, and the variable r is the distance between the centers of the aerosol particles. Clearly, $r \geq a + b$, and the equality is valid in the case of touching particles. Using Levich's work (1962), Williams and Loyalka (1991) demonstrated that the solution of Equation 6.186 leads to the coagulation kernel of

$$K(a,b) = 4\pi D_0 (a+b) g(\chi), \quad (6.187)$$

where the dimensionless correction factor $g(\chi)$ and its argument χ are as follows:

$$g(\chi) = \left[1 - \chi\left(\frac{\pi}{2} + \tan^{-1}\chi\right)\right]^{-1}, \quad (6.188)$$

$$\chi = (a+b)\left[\beta_0\left(\frac{\varepsilon_T}{\nu}\right)^{1/2}\frac{1}{D_0}\right]^{1/2}, \qquad (6.189)$$

with $\beta_0 = 0.15$. In the case when χ is small, Brownian particle motion is dominant, and in the limit when $\chi \to 0$, Equation 6.187 degenerates into Equation 6.158, which is Brownian coagulation in the diffusion regime. For large values of χ, turbulence becomes dominant, and in the limit of $\chi \to \infty$, Equation 6.187 becomes

$$K(a,b) = 1.8\pi(a+b)^3\left(\frac{\varepsilon_T}{\nu}\right)^{1/2}. \qquad (6.190)$$

Thus, Equations 6.158 and 6.190 provide the limiting cases for the combined Brownian and turbulent coagulation.

SIMULTANEOUS COAGULATION MECHANISMS

As it was seen above, combination of the coagulation kernels are not obvious in the case when several simultaneous collision processes are superimposed. Simple summation of kernels, even if they represent independent processes, may not yield correct results. When there is no theoretical or practical formulas establishing the relations of the kernels, such as in the cases of gravitational and Brownian or Brownian and turbulent coagulation, often a geometric addition of the collision rates is considered. In this way, for example, a simultaneous Brownian, gravitational, and turbulent coagulation can be expressed as

$$K_{\text{Simul}}(u,v) = K_B + \left[K_T^2 + K_G^2\right]^{1/2}. \qquad (6.191)$$

Other summation methods, similar to the above expression, and general numerical techniques are enumerated by McDonald (1988).

REFERENCES

Aitchison, J. and Brown, J.A.C. *The Lognormal Distribution*, Cambridge University Press, Cambridge, UK, 1957.

Allen, M.D. and Raabe, O.G. Re-evaluation of Millikan's oil drop data for the motion of small particles in air, *J. Aerosol Sci.* 13:537–547, 1982.

Allen, M.D. and Raabe, O.G. Slip correction measurements of spherical solid aerosol particles in an improved Millikan apparatus, *Aerosol Sci. Tech.* 4:269–286, 1985.

Barrett, J.C. and Jheeta, J.S. Improving the accuracy of the moments method for solving the aerosol general dynamics equation, *J. Aerosol Sci.* 27:1135–1142, 1996.

Benhenni, M., Yousfi, M., Bekstein, A., Echwald, O. and Merbahi, N. Analysis of ion mobility and diffusion in atmospheric gaseous mixtures from Monte Carlo simulation and macroscopic laws, *J. Phys. D: Appl Phys.* 39:4886–4893, 2006.

Cunningham, E. On the velocity of steady fall of spherical particles through fluid medium. *Proc. Royal Soc.*, 83:357–365, 1910.

Davies, E.J. and Schweiger, G. *The Airborne Microparticle. Its Physics, Chemistry, Optics, and Transport Phenomena*, Springer-Verlag, Berlin, 2002.

DeCarlo, P.F., Slowik, J.G., Worsnop, D.R., Davidovits, P., and Jimenez, J.L. Particle morphology and density characterization by combined mobility and aerodynamic diameter measurements. Part 1: Theory, *Aerosol Sci Tech.* 38:1185–1205, 2004.

Einstein, A. *Investigation on the Theory of the Brownian Movement*, Dover, New York, 1956.

Friedlander, S.K. Smoke, dust, and haze. *Fundamentals of Aerosol Dynamics*, 2nd ed., Oxford University Press, New York, 2000.

Fuchs, N.A. *Evaporation and Droplet Growth in Gaseous Media*, Pergamon, New York, 1959.
Fuchs, N.A., Daisley, R.E., and Fuchs, M. *The Mechanics of Aerosols*, Dover Publications, New York, 1989.
Fuller, E.N., Schettler, P.D., and Giddings, J.C. A new method for the prediction of binary gas-phase diffusion coefficients, *Ind. Eng. Chem.* 58:19–27, 1966.
Gelbard, F.M. *MAEROS User Manual*, NUREG/CR-1391 SAND 80–0822, Sandia National Laboratories, Albuquerque, NM, 1982.
Gelbard, F.M., Fitzgerald, J.W., and Hoppel, W.A. A one-dimensional model to simulate multicomponent aerosol dynamics in the marine boundary layer 3. Numerical methods and comparisons with exact solutions, *J. Geophys. Res.* 103:16119–16132, 1998.
Gelbard, F.M. and Seinfeld, J.H. Exact solutions of the general dynamic equation for aerosol growth by condensation, *J. Colloid Interface Sci.* 78:173, 1979.
Gelbard, F.M. and Seinfeld, J.H. Simulation of multicomponent aerosol dynamics, *J. Colloid Interface Sci.* 78:485–501, 1980.
Gelbard, F.M., Tambour, Y., and Seinfeld, J.H. Sectional representations for simulating aerosol dynamics, *J. Colloid Interface Sci.* 76:541–556. 1980.
Geller, A.S., Mondy, L.A., Rader, D.J., and Ingber, M.S. Boundary element method calculations of the mobility of nonspherical particles—1. Linear chains, *J. Aerosol Sci.* 24:597–609, 1993.
Gilliland, E.R. Diffusion coefficients in gaseous systems, *Ind Eng. Chem.* 26:681, 1934.
Goldstein, S. *Modern Developments in Fluid Dynamics*. Clarendon Press, New York, 1938.
Hutchins, D.K., Harper, M.H., and Felder, R.L. Slip correction measurements for solid spherical particles by modulated dynamic light scattering, *J. Aerosol Sci. Technol.* 22:202–218, 1995.
Ingber, M.S., Womble, D.E., Geller, A.S., Rader, D.J., and Mondy, L.A. Boundary element method calculations of the mobility of nonspherical particles—3. Parallel implementation for long chains, *J. Aerosol Sci.* 30:127–130, 1999.
Jennings, S.G. The mean free path in air. *J. Aerosol Sci.* 19:159–166, 1988.
Jovanovic, J.V., Vrhovac, S.B., and Petrovic, Z.L. Application of Blanc's law at arbitrary electric field to gas density ratios. *Eur. Phys. J. D*, 28:91–99, 2004.
Junge, C.E. *Air Chemistry and Radioactivity*, Academic Press, New York, 1963.
Kim, J.H., Mulholland, G.W., Kukuck, S.R., and Pui, D.Y.H. Slip correction measurements of certified PSL nanoparticles using a nanometer differential mobility analyzer for Knudsen number from 0.5 to 83, *J. Res. Natl. Inst. Stand. Technol.* 110:31–54, 2005.
Kotrappa, P. Shape factors for aerosols of coal, UO_2, and ThO_2, in respirable size range. Chapter 16, in T. T. Mercer, P. Morrow, W. Stöber, Eds., *Assessment of Airborne Particles*. Charles C. Thomas Publisher, Springfield, IL, 1972.
Kotrappa, P., Sundararajan, A.R., Bhanti, D.P., and Menon, V.B. Dynamic shape factors for PuO_2 aerosols useful in autoradiographic particle size analysis. *Health Phys.* 29:701–704, 1975.
Kunkel, W.B. Magnitude and character of errors produced by shape factors in Stokes' law estimates of particle radius, *J Appl. Phys.* 19:1056, 1948.
Lehtinen, K., Kulmala, M., Vesala, T., and Jokiniemi, J. Analytical method to calculate condensation rates of a multicomponent droplet, *J. Aerosol Sci.* 29, 1041, 1998.
Leith, D. Drag on non-spherical objects. *J. Aerosol Sci. Technol.* 6, 153–161, 1987.
Levich, V.G. *Physicochemical Hydrodynamics*. Prentice-Hall, Englewood Cliffs, NJ, 1962.
Loyalka, S.K. and Park, J.W. Aerosol growth by condensation and generalization of Mason's formula. *J. Colloid Interface Sci.* 125:712, 1988.
Mandelbrot, B.B. *The Fractal Geometry of Nature*. W.H. Freeman, New York, 1977.
Mathur, G.P. and Thodos, G. Self diffusivity of substances in gaseous and liquid states, *AIChE J.* 11:613, 1965.
Mathur, G.P. and Saxena, S.C. A new method for calculation of diffusion coefficients of multicomponent gas mixtures, *J. Pure Appl. Phys.* 4:266, 1966.
McDonald, B.H. Assessing numerical methods used in nuclear aerosol transport models. In: E. del la Loggia and J. Royen (eds), *Proceedings of Water Cooled Reactor Aerosol Code Evaluation and Uncertainty Assessment*, EUR-11351 EX, Commission of the European Communities, Luxembourg, 1988.
Millikan, R.A. The isolation of an ion, measurement of its charge, and the correction of Stokes's Law, *Science* 32:349–397, 1910.
Millikan, R.A. Coefficients of slip in gases and the law of reflection of molecules from the surfaces of solids and liquids, *Phys. Rev.* 21:217–238, 1923.
Mondy, L.A., Geller, A.S., Rader, D.J., and Ingber, M.S. Boundary element method calculations of the mobility of nonspherical particles—2. Branched chains and flakes, *J. Aerosol Sci.* 27:537–546, 1996.

Oberbeck, A. Über stationäre Flüssigkeitsbewegungen mit Berücksichtigung der inneren Reibung, *J. Reine Angew. Math.*, 81:62–80, 1876.

Oseen, C. *Neuer Methoden und Ergebnisse in der Hydrodynamic*, Academische Verlag, Leipzig, Deutchland, 1927.

Park, H. Development and implementation of fine-structure aerosol spectrum coagulation kernels and deposition mechanisms using advanced nodal methods in the CAEROT code. MS Thesis, Louisiana State University, Nuclear Science Center, 2003.

Perrin, J. *Brownian Movement and Molecular Reality*, Taylor & Francis, London, 1910.

Perry, R.H. and Green, D.W., Eds. *Perry's Chemical Engineers' Handbook*, 8th ed., McGraw-Hill, New York, 2007.

Pratsinis, S.E. Simultaneous nucleation, condensation, and coagulation in an aerosol reactor, *J. Colloid Interface Sci.*, 20:101–111, 1988.

Proudman, I. and Pearson, J.R.A. Expansion at small Reynolds numbers for the flow past a sphere and a circular cylinder, *J. Fluid Mech.* 2:237–262, 1957.

Pruppacher, H.R. and Klett, J.D. *Microphysics of Clouds and Precipitation*. Reidel Publishing, New York, NY, 1974.

Reid, R.C., Prausnitz, J.M., and Sherwood, T.K. *The Properties of Gases and Liquids*. McGraw-Hill, New York, 1977.

Reist, P.C. *Aerosol Science and Technology*, 2nd ed., McGraw-Hill, New York, 1993.

Sajo, E. and Park, H. Simulation of single species aerosol coagulation and deposition using the sectional method with arbitrary size boundaries, *Health Phys.* 84:S159, 2003a.

Sajo, E. and Park, H. Computational model to predict size-specific aerosol transport in space and time, *Health Phys.*, 84:S159–S160, 2003b.

Sajo, E., Park, H., and Scott, L.M. Validation of the CAEROT code: 3D aerosol transport and deposition in confined atmospheres, *Health Phys.*, 86:S210–S211, 2004.

Sajo, E. Evaluation of the exact coagulation kernel under simultaneous Brownian motion and gravitational settling, *J. Aerosol Sci. Tech.* 42:134–139, 2008.

Sandler, S.I. and Mason, E.A. Kinetic theory deviations from Blanc's law of ion mobilities. *J. Chem Phys.* 48:2873–2895, 1968.

Seigneur, C., Hudischewskyj, A.B., Seinfeld, J.H., Whitby, K.T., Whitby, E.R., Brock, J.R., and Barnes, H.M. Simulation of aerosol dynamics: A comparative review of mathematical models, *J. Aerosol Sci. Tech..* 5:205–222. 1986.

Simons, S., Williams, M.M.R., and Cassell, J.S. A kernel for combined Brownian and gravitational coagulation, *J. Aerosol Sci.* 17:789–793, 1986.

Smoluchowski, M. von. Versuch einer mathematischen Theorie der Koagulationskinetik kolloider Losungen, *Z. f. Phys. Chemie.* 92:129–168, 1917.

Stokes, G.G. On the effect of internal friction of fluids on the motion of pendulums, *Cambridge Philos. Soc. Trans.* 1(8):18–32, 1851.

Stöber, W. Dynamic shape factors of non-spherical aerosol particles. Chapter 14, in: T.T. Mercer, P.E. Morrow, and W. Stöber, Eds., *Assessment of Airborne Particles*. Charles Thomas Publisher, Springfield, IL, 1972.

Sutton, O.G. *Mathematics in Action*. G. Bell and Sons, London, 1957.

Takata, N. A method to calculate mobilities for ions in gas mixtures. *J. Phys. D: Appl Phys.* 18:1795–1802, 1985.

Taylor, G.I. Statistical theory of turbulence. *Proc. Royal Soc. (London)*, A.151:421, 1935.

Williams, M.M.R. *Mathematical Methods in Particle Transport Theory*. Butterworths, London, 1971.

Williams, M.M.R. Some topics in nuclear aerosol dynamics. *Prog. Nuc. Energy* 17:1–52, 1986.

Williams, M.M.R. A unified theory of aerosol coagulation. *J. Phys. D: Appl. Phys.* 21:875–886, 1988.

Williams, M.M.R. and Loyalka, S.K. *Aerosol Science: Theory and Practice with Special Applications to the Nuclear Industry*. Pergamon Press, Oxford, 1991.

Whitby, K.T., Husar, R.B., and Liu, B.Y.H. The aerosol distribution of the Los Angeles smog. *J. Colloid Interface Sci.* 39:177–204, 1972.

Whitby, K.T. The physical characteristics of sulfur aerosol. *Atmos. Environ.* 12:174–178, 1978.

Whitby, K.T. Aerosol and ozone formations in the Columbus, Ohio urban plume on July 29 and 30, 1980. Publication No. 447, Particle Technology Laboratory, Department of Mechanical Engineering, University of Minnesota, Minneapolis, MN, 1981.

Willeke, K. Temperature dependence of particle slip in a gaseous medium. *J. Aerosol Sci.* 7:381, 1976.

Xiong, C. and Friedlander, S.K. Morphological properties of atmospheric aerosol aggregates. *Proc. Natl. Acad. Sci.* 98:11851–11856, 2001.

7 Behavior of Radioactive Aerosols and Gases

Mark D. Hoover

CONTENTS

Introduction 135
Physical Forms of Airborne Radioactive Materials 136
    Dusts 136
    Fumes 136
    Smokes 136
    Mists 137
    Vapors 137
    Gases 137
    Multiple or Mixed Physical Forms 137
Factors Affecting the Dispersion and Disposition of Aerosols and Gases 138
    Aerodynamic Equivalent Diameter 139
    Thermodynamic Equivalent Diameter 141
    Comparison of Aerodynamic and Thermodynamic Effects
        as a Function of Particle Size 141
    Relative Importance of Resuspension from Surfaces 142
Understanding and Interpreting the Lognormal Aspects of Airborne
    Particle Size Distributions 143
Modeling Exposure Pathways and Their Significance 147
    A Key-Parameter Equation for Exposure Modeling 148
    Monte Carlo Simulation of Uncertainty for Prospective and Retrospective
        Assessments of Airborne Radioactivity 151
An Example Monte Carlo Estimation of Radiation Dose to a Worker from
    an Accidental Release of Plutonium from a Radioactive Waste Drum 152
        Influence of Initiating-Event Probabilities and Other Conditional Probabilities
            for Accident Consequences 152
Conclusion 154
References 154

INTRODUCTION

Aerosols and gases of radioactive materials share the physical and chemical forms and dispersion behavior of nonradioactive aerosols and gases. The physical form (gas, particle, size, density, shape) of an airborne material determines its aerodynamic and thermodynamic behavior and influences the choice of sampling and analytical techniques. The chemical form of a material determines its solubility, translocation, and reactivity in biological systems and the environment. Chapter 6 of this book has described important mathematical aspects and principles of the physics of aerosols. The current chapter presents general concepts and information on the behavior of radioactive aerosols and gases and the associated considerations for air sampling. Topics include the physical forms of airborne radioactive materials, factors affecting the dispersion and disposition for aerosols and gases, issues for understanding and interpreting the lognormal aspects of airborne particle size distributions, mathematical modeling approaches for assessment and control of exposures to airborne radioactivity, and an example of Monte Carlo simulation of uncertainty in aerosol release and behavior parameters for prospective and retrospective assessments of airborne radioactivity.

PHYSICAL FORMS OF AIRBORNE RADIOACTIVE MATERIALS

Consistent with historically useful classifications of general aerosol categories (cf., Lapple, 1961; Dennis, 1976; Hinds, 1999), radioactive materials can be in the form of dusts, fumes, smokes, mists, vapors, gases, or combinations of these forms. The term "aerosol" refers here to a suspension of particles (solid or liquid) in a gas. Both the particulate and the gas components are of concern for sampling airborne radioactive materials. Although some definitions of "aerosol" limit the term to suspensions that are "stable," a practical case can easily be made for the broader definition that generally includes any quasi-stable particle suspension (Dennis, 1976). Within the broader definition, the fact that an aerosol may be changing with time becomes an important consideration for determining when, where, and how to conduct and interpret the results of air sampling. It should be expected that air samples taken at different times may have different composition and different relevance. As noted in the following discussion, the physical form of an airborne material depends on both its inherent physicochemical properties, such as its melting and boiling points, as well as on the conditions of its creation, evolution, and dispersion.

Dusts

Dusts are aerosols of solid particles produced by mechanical disruption or disintegration of solid materials or by dispersion of previously pulverized materials. Particles with aerodynamic diameter greater than 50 μm have high gravitational settling velocities (more than 7 cm/s) and are sometimes called "inertials." The aerodynamic diameter range of atmospheric dusts is usually relatively large, approximately 0.1–20 μm. Airborne dusts can result from handling of radioactive powders, as well as from drilling, sawing, maintenance, or disruption of radioactively contaminated materials or structures. Dusts may contain other substances of health concern such as asbestos, beryllium, or silica.

Fumes

Fumes are released when solid materials undergo a change of state by sublimation or evaporation, followed by condensation. Most evaporation and condensation processes produce relatively small particles. Fumes range in size from approximately 0.01–0.1 μm in diameter. The amount of fuming increases with temperature. Chemical reactions may be involved. For example,

$$\text{Pb (metal)} \rightarrow \text{Pb (vapor)} \quad (7.1)$$

and

$$2\text{Pb (vapor)} + O_2 \rightarrow 2\text{PbO (fume)} \quad (7.2)$$

Smokes

Smokes are airborne combustion products of any size. If radioactive material is contained in the material being combusted, then the radioactivity may be incorporated into the evolving smoke. Smoke can be released from deliberate or inadvertent combustion of materials, equipment, buildings, industrial facilities, or vehicles and their contents. Burning plastics can release toxic and corrosive smoke, as well as gases and vapors.

MISTS

Mists are aerosols of liquids dispersed in air as droplets. Dispersion can be by splashing, spraying, or nebulization from special devices. Dispersion of liquids can result in exposure to the droplets themselves, to dusts from drying of solids that were dissolved or suspended in the droplets, or to vapors that evaporate from the droplets. Bursting of bubbles from liquid surfaces can release airborne droplets. As noted by Dennis (1976), mists with droplet diameter between 5 and 40 µm are sometimes referred to as "fogs"; atmospheric aerosols of droplets produced by chemical reactions or condensation are referred to as "smog"; and the term "haze" is used to describe aerosols potentiated by discharge of man-made pollutants or by exudation of resinous materials from vegetation, forest fires, or volcanic eruptions.

VAPORS

Vapors are condensable gases from evaporation of liquid materials or sublimation from solid materials. Some liquids are relatively volatile and vaporize quickly at room temperature. Others have a low vapor pressure and persist for long periods as liquids. Historical methods for determining the heats of sublimation and diffusion behavior for metals have involved measurements of airborne and condensed radioisotopes following their incorporation into and subsequent release from metal samples (Kornev and Zubkovsky, 1956). Vapors can exist at concentrations up to their saturation vapor pressure at the prevailing temperature. Supersaturated vapors typically condense to form a liquid particle aerosol or fog or coatings on the surfaces of preexisting particles. The particle coating process can lead to variations in particle composition as a function of particle size (i.e., the mass fraction of condensed material in particles of smaller diameter will be higher than in particles of larger diameter).

GASES

Gases expand and contract according to the ideal gas law as a function of temperature and pressure, have relatively low density and viscosity compared to solid and liquid states of matter, and readily diffuse to distribute themselves throughout a container or the environment. Radioactive gases include tritium, radon, argon, krypton, xenon, iodine, and ruthenium tetroxide. The greater molecular weight of molecules such as radon, as compared to air, can result in higher concentrations of the heavier gases in poorly mixed, low-lying areas.

MULTIPLE OR MIXED PHYSICAL FORMS

Radioisotopes can be present in multiple physical forms and can change physical form, depending on conditions of chemistry and temperature. For example, tritium (the radioactive form of hydrogen) can be present as tritiated hydrogen gas (HT), as a liquid or vapor of tritiated water (HTO), or in solid form (as tritiated organic materials of wide ranging chemical makeup or as tritiated inorganic materials such as metal oxides). Tritium released as a gas can become incorporated in liquids or solids, and thereby be retained in or on building materials and equipment for extended periods. Carbon-14 can similarly be present in multiple forms (e.g., ^{14}C-labeled carbon dioxide, liquid organic compounds, or solids). The noble gases radon, krypton, and xenon may be dissolved and transported in liquids; their release from solids such as soils, rock, concrete, or metals can be impaired by coatings or by the physical integrity of the solid structure.

Examples of radioactive materials encountered (or not encountered) in multiple physical forms are plentiful in the published history of accidents involving nuclear reactors. As summarized by Chamberlain (1991), radioactive gases and the relatively volatile fission products tellurium, iodine, and cesium were released during the Windscale facility accident on October 10, 1957, when a fire in the graphite moderator resulted in a highly oxidizing environment. In contrast, when an inadvertent loss of coolant at Three Mile Island on March 29, 1979, caused the temperature of the reactor to rise to over 2000°C, radioactive gases xenox and krypton were released to the environment, but the expected release of gaseous iodine-131 and iodine-129 did not occur. The low volatility of the iodine was attributed to the high pH of the cooling water (about 8), the reducing potential of the hydrogen present in the reactor containment building (formed when the hot zirconium fuel clad underwent an oxidation reaction with steam), and the possible presence of silver from melted control rods. Chamberlain also noted that releases of tellurium and ruthenium were lower than expected during the Chernobyl accident on April 26, 1986, apparently because those refractory elements were not oxidized to their more volatile forms. The release of the volatile ruthenium tetroxide had been an important component of the releases during the fire at Windscale.

FACTORS AFFECTING THE DISPERSION AND DISPOSITION OF AEROSOLS AND GASES

Figure 7.1 illustrates a model of important pathways for dispersion and disposition of radioactive aerosols and gases into the air, onto and between surfaces throughout the workplace, onto clothing and skin, and into the breathing zones of people. Gases are generally dispersed with the prevailing air currents. Physical relationships between the point of release, air ventilation inlets and outlets, and worker location are important. Near the source of release for a material or in the early phase of a release, large particles can comprise a significant component of the airborne concentration. Farther from the source or at later time periods, larger particles may have already settled by gravitational forces and the remaining aerosol may have a significantly smaller size distribution. Time-dependent factors of resuspension from surfaces can result in continued airborne concentrations long after the initial release.

Figure 7.2 illustrates mechanisms by which particles can be removed from the air and deposited on workplace surfaces, in ventilation filtration and air cleaning systems, and in the respiratory tract. These mechanisms include gravitational settling and impaction (aerodynamic properties), interception (a basic matter of particle size), Brownian diffusion (a thermodynamic property), electrostatic attraction (dependent on both the charge on the particle or droplet and the charge on the collection surface), and thermal diffusion (the net tendency for gases and particles to migrate away from warmer surfaces where temperatures and the associated dispersion by diffusion are higher, and toward cooler surfaces where temperatures and associated dispersion by diffusion are lower).

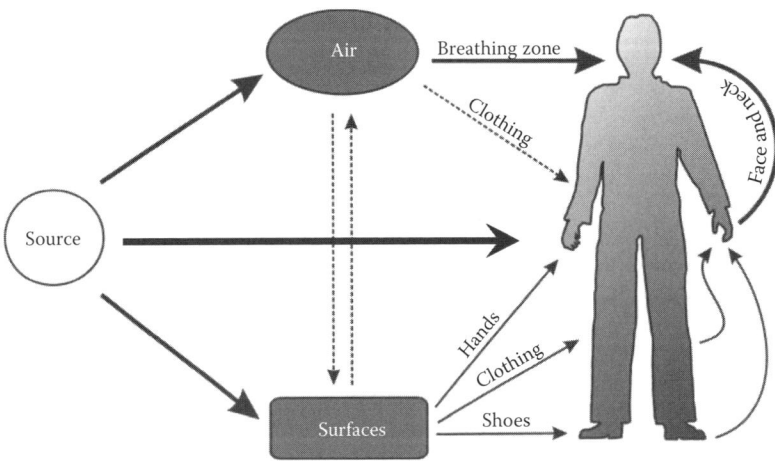

FIGURE 7.1 Illustration of important pathways for exposure of people to aerosols and gases dispersed from a source. (Adapted from Day, G.A. et al., *Ann Occup Hyg* 51: 67–80, 2007. With permission.)

Behavior of Radioactive Aerosols and Gases

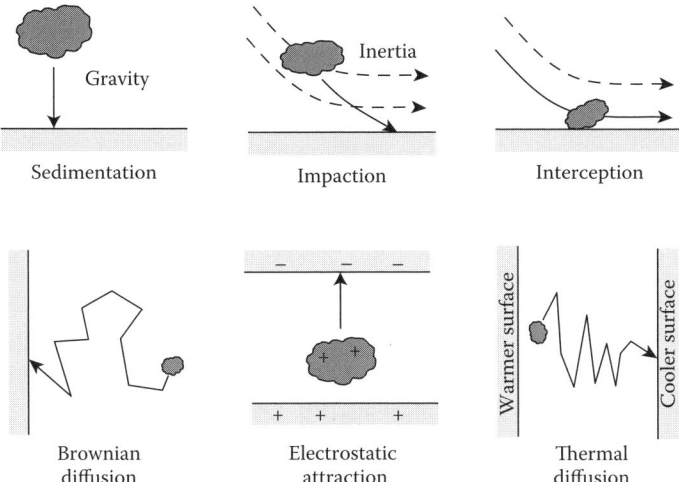

FIGURE 7.2 Fundamental mechanisms of particle collection in the environment, in air filtration and air cleaning systems, and in the human respiratory tract.

Figure 7.3 illustrates the relative importance of gravitational (aerodynamic) settling and diffusion (thermodynamic behavior) as a function of physical particle size. For example, a 1-μm diameter sphere of unit density (1 g/cm³) settles approximately 30 μm in one second and will diffuse a distance of about 7 μm in the same time period. If the particle size is reduced to 0.1 μm, the particle would settle less than 1 μm in 1 s, while diffusing more than 30 μm. Additional discussion about the concepts and comparative importance of the aerodynamic and thermodynamic properties of particles are presented in the following sections.

AERODYNAMIC EQUIVALENT DIAMETER

To simplify considerations of particle behavior, particularly in the respiratory tract and for particle sampling and filtration, it is convenient to treat two particles of different size, shape, and density as aerodynamically equivalent if they have the same terminal settling velocity in air. For convenience,

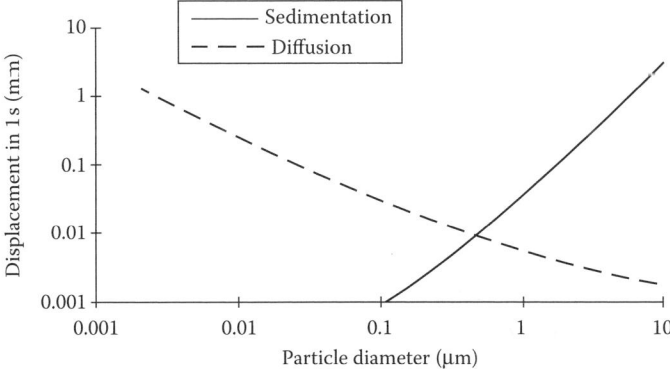

FIGURE 7.3 Comparison of the relative importance of particle motion from the gravitational sedimentation (vertical displacement) and diffusion mechanisms (root mean square distance the particle travels as a result of Brownian motion) as a function of particle diameter for unit density spheres. Note that logarithmic scales are required on both axes to address orders-of-magnitude differences in the degrees of displacement and particle diameters of interest. (Adapted from Raabe, O.G., *Internal Radiation Dosimetry*, Raabe, O.G., Ed., Medical Physics Publishing, Madison, WI, pp. 111–142, 1994.)

the aerodynamic equivalent diameter of a particle is defined as the diameter of a unit-density sphere whose terminal settling velocity in air is the same as that of the particle in question. As was noted in Equation 6.44 in Chapter 6 of this book, the terminal settling velocity, v_s, of a spherical particle of physical diameter d_p can be expressed as

$$v_s = \frac{mg}{3\pi d_p \mu} \tag{7.3}$$

where m is the particle mass, g is the acceleration of gravity, and μ is the viscosity of air. If the mass term, m, is replaced by the product of the particle density, ρ, and the particle volume, $\pi d_p^3 \rho/6$, and the value of 9.8 m/s² is used for the acceleration of gravity, g, and the value for the viscosity of air at sea level (1.85×10^{-5} kg/m/s) is used for μ, then the following equation provides an informative estimate of the terminal settling velocity v_s (m/s) for a spherical particle with diameter d_p (µm) and density ρ (g/cm³):

$$v_s = 2.94 \times 10^{-5} d_p^2 \rho \tag{7.4}$$

Thus, the approximate terminal settling velocities for spherical, unit-density particles of diameter 1, 10, 100, and 1000 µm are 2.94×10^{-5}, 2.94×10^{-3}, 0.294, and 29.4 m/s, respectively. (Note that the 100-fold increase of settling velocity for a 10-fold increase in particle diameter is due to the fact that terminal settling velocity as expressed in Equation 7.3 is directly proportional to particle mass [i.e., proportional to particle diameter cubed] and inversely proportional to particle diameter.) Thus, aerodynamically large particles are likely to be removed from the air by gravitational settling, while aerodynamically small particles will remain airborne for longer periods. Note, however, that air currents can cause larger particles to remain airborne longer than would be expected based on simple calculations of gravitational settling time.

The influence of particle density and particle shape can both be important when using air sampling devices to estimate aerosol concentration as a function of particle size. For example, if light-scattering techniques are used to characterize the airborne particle number concentration as a function of physical (or optically equivalent) particle size for aerosols of unknown composition, the assumption of unit density can result in two errors: for particles with density greater than 1 g/cm³, there may be underestimation of the aerodynamic diameter of the detected particles and underestimation of the mass or radioactivity associated with the detected particles, and for particles with density less than 1 g/cm³ there may be overestimation of the aerodynamic diameter of the detected particles and overestimation of the mass or radioactivity associated with the detected particles.

As a first approximation (i.e., neglecting the slip correction that applies mostly to particles smaller in physical diameter than 1 µm and shape-factor correction that applies to particles that are prolate rather than spherical), the aerodynamic equivalent diameter of a particle, d_{ae}, has the following relationship to its density, ρ, and its physical diameter, d_p:

$$d_{ae} = d_p \rho^{1/2} \tag{7.5}$$

Thus, a uranium oxide (U_2O_3) particle with a density of 8.3 g/cm³ will have an aerodynamic diameter that is 2.88 (the square root of 8.3 is 2.88) times greater than its physical diameter. Referring back to the linear dependence of terminal settling velocity on particle density shown in Equation 7.4, the terminal settling velocities for uranium oxide particles of diameter 1, 10, 100, and 1000 µm will be approximately 2.44×10^{-4}, 0.0244, 2.44, and 244 m/s, respectively, which are greater by a factor of 8.3 (the particle-density factor, and therefore the particle-mass factor) than the terminal settling velocities noted earlier in this chapter for unit-density particles of those same diameters.

As a first approximation, the influence of particle shape on aerodynamic diameter can be considered to be relatively small. For example, Stöber (1972) demonstrated that d_{ae} for a doublet of 1-µm diameter spheres of unit density is only 1.19 µm, and d_{ae} for linear triplet is only 1.28 µm. Even fibers with a length-to-diameter ratio between 10 and 100 will behave as if their aerodynamic diameter is relatively similar (i.e., within a factor of 3 or 4) to their diameter, rather than proportional to their length. Stöber attributes this behavior to the fact that chain aggregate particles and fibers settle prevailingly in a position where they face the direction of relative air motion with their maximum cross section. Thus, the mass of a chain aggregate particle containing many times the mass of a single particle can be underestimated by simple measurement of aerodynamic diameter. (See Chapter 6 of this book for exact methods of determining of d_{ae} as a function of particle physical size, density, and shape.)

THERMODYNAMIC EQUIVALENT DIAMETER

The thermodynamic equivalent diameter of a particle of interest is defined as the diameter of a spherical particle that has the same diffusion coefficient in air as the particle of interest. Whereas aerodynamic diameter is approximately equal to the physical diameter of a particle times the square root of its density, the thermodynamic diameter of a particle is approximately equal to its physical size.

Because diffusion is a thermodynamic process, thermodynamic equivalent diameter is used to characterize particle size in the range where particle motion is dominated by diffusion. Diffusion of a particle in any given gaseous medium depends on both the temperature and the particle size. Molecules of air are in a constant state of motion and collide with one another at speeds that depend on their temperature. For air at a temperature of 25°C and a pressure of 760 mm Hg, the mean free path (mean distance traveled between air-molecule collisions) is 0.067 µm. Particles with diameters much greater than the mean free path of the surrounding gas are said to be in the "continuum" regime where their surfaces are constantly being impacted by air molecules from all directions. In contrast, particles whose diameters are smaller than the mean free path of the air are said to be in the "free-molecular" regime where the impact by air molecules is intermittent and the effect of being struck by a moving air molecule overwhelms the effects of gravity on the particle. A "transition" regime exists between the continuum and free-molecular regimes. Under free-molecular conditions, particle motion is controlled by the bombarding air molecules and the particles move about in random directions. This motion is called Brownian motion and the process of particle movement is called diffusion.

COMPARISON OF AERODYNAMIC AND THERMODYNAMIC EFFECTS AS A FUNCTION OF PARTICLE SIZE

Figures 7.4a and 7.4b compare total deposition in the human respiratory tract as a function of particle size and particle density in the thermodynamic and aerodynamic regimes. As detailed in ICRP Publication 66, Annexe D, section D.4.3, paragraph D33 (ICRP, 1994), the total deposition factor, N, equals the quadratic addition of the thermodynamic deposition, n_{th}, and the aerodynamic deposition, n_{ae}:

$$N = (n_{th}^2 + n_{ae}^2)^{1/2} \qquad (7.6)$$

The fact that both n_{th} and n_{ae} are simultaneously small for particles in the size range of 0.3 µm results in the existence of the minimum-particle-deposition regime, which has important implications for particle deposition efficiency on filter media, deposition losses in aerosol sampling lines, as well as for deposition in the human respiratory tract.

When thermodynamic diameter is used as the metric of particle size (Figure 7.4a), deposition of very small particles in the respiratory tract is the same for all particles of the same thermodynamic diameter, regardless of particle density, while deposition of larger particles is clearly dependent on particle density. Conversely, when aerodynamic diameter is used as the metric of particle size (Figure 7.4b) deposition of large small particles in the respiratory tract is the same for all particles of the same aerodynamic diameter, while deposition of very small particles of the same

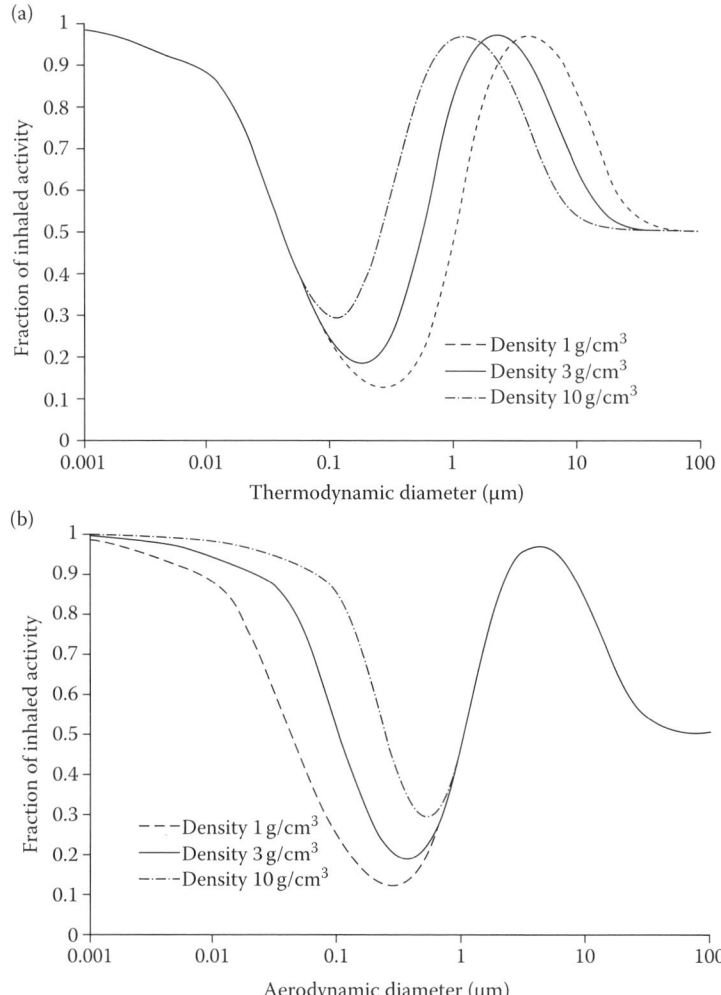

FIGURE 7.4 Comparisons of the influence of thermodynamic diameter (a) and aerodynamic diameter (b) for monodisperse spheres as a function of three different densities on the fraction of inhaled material deposited in the respiratory tract as predicted by the Human Respiratory Tract Model (ICRP, 1994) for a reference person doing light work. Note that thermodynamic diameter is not an appropriate measure of particle size for diameters greater than 1 μm (where inertial effects dominate particle deposition) and that aerodynamic diameter is not an appropriate measure of particle size for diameters less than 0.1 μm (where diffusion dominates particle deposition). (See ICRP, 2002 for additional related presentations.)

aerodynamic diameter is dependent on particle density (because high-density particles will have smaller physical size than low-density particles of the same aerodynamic diameter).

Relative Importance of Resuspension from Surfaces

The relative importance of resuspension of particles from surfaces as compared to exposure associated with direct release of an aerosol from a process or other point of release can be highly variable. Resuspension may become important at later times if surfaces that became contaminated during the initial event are later disturbed by activities such as foot traffic, mechanical handling, or poorly selected cleaning methods such as dry sweeping.

Resuspension of particles from a surface is generally an inefficient process. The ratio of particles-per-unit-volume-of-air above a surface (e.g., particle number or mass per m³) to particles-per-unit-area on the surface (e.g., particle number or mass per m²) is typically in the range of only 10^{-4}–10^{-6}, depending on the diameter of the particles, the elapsed time since their initial deposition, the nature and moisture content of the surface, and the amount of energy applied. An extensive body of experimental data on airborne releases of radioactive aerosols and gases from a variety of handling processes and accident conditions has been reviewed and summarized in the United States Department of Energy (DOE) *Handbook on Airborne Release Fractions/Rates and Respirable Fractions for Nonreactor Nuclear Facilities* (DOE, 1994b).

The importance of airborne particle number (regardless of origin) depends on the level of concern for dispersion and inhalation exposure to an individual particle. If the mass of an individual particle is on the order of a picogram, then on the order of 10^{12} particles are associated with a gram of mass, and billions of particles are associated with micrograms of mass. Thus, subjecting a gram of material to forces that can induce resuspension might release a mass of between of 10^{-4} and 10^{-6} g, which would be equivalent to between 10^8 and 10^6 particles. In the case of highly radioactive materials such as ^{238}Pu, even low rates of resuspension can result in airborne concentrations of health significance (see, e.g., Scott et al., 1997; Scott and Fencl, 1999). If individuals must be protected from inhaling only a few particles, then airborne concentrations must be kept low by engineering controls, or (as a last resort) by respiratory protection having a high assigned protection factor.

UNDERSTANDING AND INTERPRETING THE LOGNORMAL ASPECTS OF AIRBORNE PARTICLE SIZE DISTRIBUTIONS

Normally distributed particle sizes arise when a multitude of small additive or subtractive factors influence a process or when the particles are nearly all the same size. Examples are diameters of ball bearings or latex spheres of the same nominal diameter. Lognormally distributed particle sizes arise when multiplicative factors act on a variable that would otherwise have a single magnitude. Examples are the breakup of large objects into smaller objects or certain types of agglomeration of smaller objects into larger objects. This is intuitively so because the original source material or agglomerates of fragmented material can range in size to many orders of magnitude larger than the smallest particle size, because large amounts of energy are required to break large particles into smaller and smaller fragments, and because particle diameters cannot be smaller than the fundamental diameters of their molecules or atoms.

Unless particles are specially made to be uniformly sized (e.g., in a laboratory or a special manufacturing process), aerosols tend to have a lognormal size distribution. This means that the distributions of particle count, surface area, or mass are statistically normal when plotted against the logarithm of particle diameter, rather than against particle diameter itself. Environmental sampling data for chemicals and other materials have also been shown to be lognormally distributed (see, e.g., Esmen and Hammad, 1977).

A fortuitous feature of the lognormal distribution is that the relationships among the diameters of interest for aerosol science have been mathematically described by Hatch and Choate (1929) and can be conveniently calculated based on the two characteristic parameters of the lognormal distribution, which are the count median diameter, d_g, and the geometric standard deviation, σ_g.

For the count mode diameter, d^*:

$$d^* = d_g \exp(-1.0 \ln^2 \sigma_g) \tag{7.7}$$

For the count mean diameter, $\bar{d}$:

$$\bar{d} = d_g \exp(0.5 \ln^2 \sigma_g) \tag{7.8}$$

For the diameter of average surface area, d_a:

$$d_a = d_g \exp(1.0 \ln^2 \sigma_g) \qquad (7.9)$$

For the surface area median diameter, d'_a:

$$d'_a = d_g \exp(2.0 \ln^2 \sigma_g) \qquad (7.10)$$

For the surface area mean diameter, $\bar{d}_a$:

$$\bar{d}_a = d_g \exp(2.5 \ln^2 \sigma_g) \qquad (7.11)$$

For the diameter of average mass (or volume), d_m:

$$d_m = d_g \exp(1.5 \ln^2 \sigma_g) \qquad (7.12)$$

For the mass (or volume) median diameter, d'_m:

$$d'_m = d_g \exp(3.0 \ln^2 \sigma_g) \qquad (7.13)$$

For the mass (or volume) mean diameter, $\bar{d}_m$:

$$\bar{d}_m = d_g \exp(3.5 \ln^2 \sigma_g) \qquad (7.14)$$

Figure 7.5 illustrates relationships of the lognormal aerosol particle size distribution among the count median diameter (*CMD*, d_g), count mode diameter (d^*), count mean diameter ($\bar{d}$), diameter of average mass (d_m), surface area median diameter (*SAMD*), and mass median diameter (*MMD*) for an aerosol with an activity median aerodynamic diameter (*AMAD*) of 5 μm, a geometric standard deviation (σ_g) of 2.5, and a particle density of 3.0 g/cm³. These distribution parameters are the reference

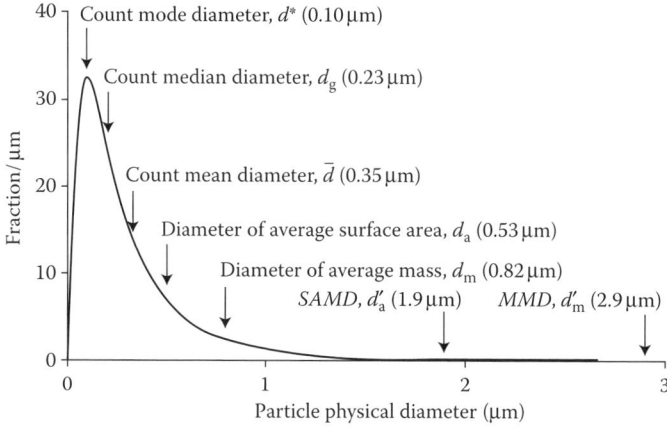

FIGURE 7.5 Illustration of the lognormal aerosol particle size distribution relationships among diameters of interest for aerosol characterization and aerosol behavior. For this example, the activity median aerodynamic diameter (*AMAD*) is 5 μm, the geometric standard deviation (σ_g) is 2.5, and the particle density is 3.0 g/cm³. These distribution parameters are the reference values recommended by the ICRP (1994) for aerosols encountered in occupational exposure settings.

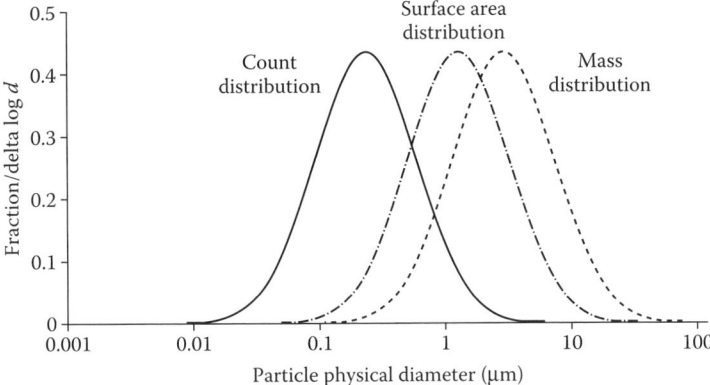

FIGURE 7.6 Comparison of frequency distribution functions on a logarithmic scale for the count, surface area, and mass distribution relationships as a function of particle diameter for the lognormally distributed aerosol presented in Figure 7.5. Note that use of a logarithmic scale on the abscissa shows that the distributions are normally distributed in log space.

values recommended by the ICRP (1994) for aerosols encountered in occupational exposure settings. The value of 3.0 g/cm³ for density is recommended because it is typical of many natural materials. Note that an *AMAD* of 1 μm (with σ_g 2.5 and particle density 3.0 g/cm³) is the recommended assumption for assessment of exposures in the general environment (ICRP, 1994).

Figure 7.6 uses a logarithmic scale for the abscissa to compare the count distribution, surface area distribution, and mass distribution as a function of particle diameter for the lognormally distributed aerosol that was shown in Figure 7.5. Note that use of the logarithmic scale on the abscissa shows that the distributions are normally distributed in log space.

Figure 7.7 is a log-probability plot that further illustrates the relationships among diameters for the lognormal distribution. Once again, the aerosol in this example has the same distribution parameters as the aerosol shown in Figures 7.5 and 7.6. Note that the percent-less-than-stated-size for the particle count data is *linear* for a logarithmic presentation of diameter on the abscissa. Curves are also shown for percent-less-than-stated-size as a function of surface area and as a function of mass. All three data sets are linear in log space. If a plot of percent-less-than-stated-size versus the logarithm of particle diameter is not linear, then the distribution is not lognormal. Distributions that are not lognormal can result when the observed aerosol is composed of multiple modes that may or may not be individually lognormal. Deviation from the lognormal can also result when the observed aerosol has been depleted of its small-particle fraction by diffusion or depleted of its large-particle fraction by gravitational settling.

Because the small and large particle-size components of an aerosol can be important to understanding basic behavior issues for aerosols, it is useful to be aware of and able to assess and interpret the degree of dispersion of particle size as it relates to the various metrics of particle diameter and to the geometric standard deviation for the particle size distribution. Table 7.1 compares the properties of the normal and lognormal distributions and facilitates understanding of how the percent of aerosol in a given size range is determined. The z-score is used to denote the normal standard (probit) variable. (1 z-unit is 1 probit, and 1 probit is 1 standard deviation unit.) Relationships that are familiar from the normal distribution are as follows:

50% of the values lie between ±0.675σ
68% between ±1σ
90% between ±1.64σ
95% between ±1.96σ
97.5% between ±2.24σ

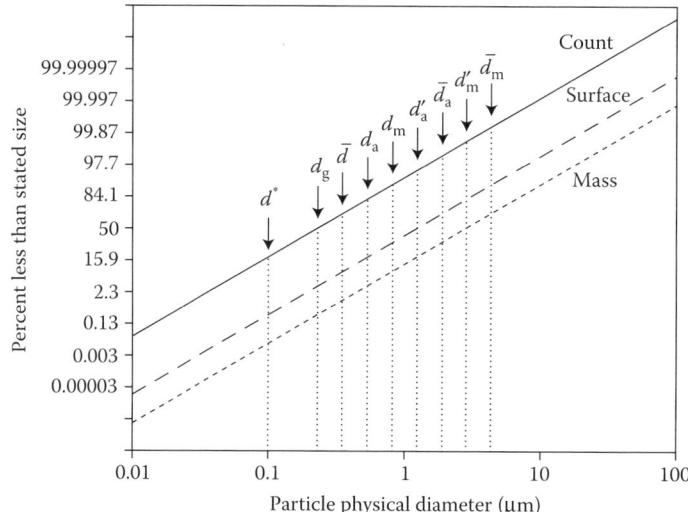

FIGURE 7.7 Log-probability plot illustrating the relationships among diameters for the lognormally distributed aerosol presented in Figure 7.5. Note that the percent less than stated size for particle count data is linear for a logarithmic presentation of diameter on the abscissa. For this distribution, the count mode diameter (d^*) is 0.10 μm, the count median diameter (d_g) is 0.23, the count mean diameter ($\bar{d}$) is 0.35 μm, the diameter of average surface area (d_a) is 0.54 μm, the diameter of average mass (d_m) is 0.82 μm, the surface area median diameter (d'_a) is 1.25 μm, the surface area mean diameter ($\bar{d}_a$) is 1.90 μm, the mass median diameter (d'_m) is 2.89 μm, the mass mean diameter ($\bar{d}_m$) is 4.39, and the mass median aerodynamic diameter (*AMAD*) is 5.0 μm.

99% between ±2.58σ
99.9% between ±3.29σ
99.99% between ±3.89σ

As shown in Table 7.1, these relationships are readily transformed for use with the lognormal distribution. For example, because 95% of a normally distributed population is contained within ±1.96σ above and below the median particle diameter (d_{50}), 95% of a lognormally distributed population is contained between diameter $d_{50}/\sigma_g^{1.96}$ and diameter $d_{50} \cdot \sigma_g^{1.96}$. Thus, for example, in

TABLE 7.1
Comparative Properties of the Normal and Lognormal Distributions, Where z is Used to Denote the Normal Standard (Probit) Variable

Normal Distribution	Lognormal Distribution
$z = \dfrac{x - \mu}{\sigma}$	$z = \dfrac{\ln x - \ln \mu}{\ln \sigma}$
$d_{0.135} = \mu - 3\sigma$	$d_{0.135} = d_{50}/\sigma^3$
$d_{2.27} = \mu - 2\sigma$	$d_{2.27} = d_{50}/\sigma^2$
$d_{15.9} = \mu - 1\sigma$	$d_{15.9} = d_{50}/\sigma$
$d_{50} = \mu$	$d_{50} = d_{50}$
$d_{84.1} = \mu + 1\sigma$	$d_{84.1} = d_{50}\,\sigma$
$d_{97.7} = \mu + 2\sigma$	$d_{97.7} = d_{50}\,\sigma^2$
$d_{99.865} = \mu + 3\sigma$	$d_{99.865} = d_{50}\,\sigma^3$

Note: 1 z-unit is 1 probit, and 1 probit is 1 standard deviation unit.

Behavior of Radioactive Aerosols and Gases 147

a distribution with $d_{50} = 3$ µm and $\sigma_g = 2$, 95% of the population is included between $3\ \mu m/2^{1.96} = 0.77$ µm and $3\ \mu m \times 2^{1.96} = 11.7$ µm. Such information can guide in the selection of appropriate sampling devices, respiratory protection, and air cleaning or filtration systems.

Note that the "tails" of the aerosol distribution may be significant for assessing the consequences of an aerosol release. A coarse aerosol with a median diameter in the range of 10's or 100's of µm may (depending on the geometric standard deviation) have a respirable component that is of concern for inhalation. The small particle "tail" of the size distribution can strongly influence the degree of concern for long-distance dispersion and inhalation of the aerosol. The large particle "tail" or the intermediate particle size may comprise what is deposited on surfaces. What is available for subsequent analysis from surface wiping or other sampling of contaminated surfaces can influence the degree of understanding of the aerosol. What is available for resuspension can influence the residual risk from the aerosol release, and may be far less of a concern (because of an inherently large particle size) than the original aerosol release.

Another calculation of interest is to estimate the activity median aerodynamic diameter of an aerosol for which only 5% of the particles have a diameter in the "respirable" size range (i.e., aerodynamic diameter less than 10 µm). For example, assuming a geometric standard deviation of 2, and using the following equation:

$$\frac{d_{50}}{d_5} = \frac{d_{95}}{d_{50}} = \sigma_g^{1.64} \tag{7.15}$$

$$d_{50} = d_5\, \sigma_g^{1.64} = 10 \times 2^{1.64} = 31\ \mu m \tag{7.16}$$

it is seen that the *AMAD* can be larger than 30 µm and still contain a significant fraction of particles in the respirable size range.

A further calculation of interest is to estimate the respirable (sub-10-µm) fraction of an aerosol of a given *AMAD* and σ_g. For example, if the radioactivity in an aerosol is lognormally distributed with an *AMAD* of 25 µm and a σ_g of 2.5, then the standard normal variable, z is

$$\begin{aligned} z &= \frac{\ln 10 - \ln 25}{\ln 2.5} \\ &= -1 \end{aligned} \tag{7.17}$$

The z-score of ±1 probit corresponds to the range from 16% to 84%. Therefore, 16% of the radioactivity in this aerosol is associated with particles having aerodynamic diameter less than 10 µm. For other values of z, the associated probability can be obtained from tabular values of the standard normal variable or from standard spreadsheet formulas.

MODELING EXPOSURE PATHWAYS AND THEIR SIGNIFICANCE

In an ideal world, a comprehensive exposure pathway model such as illustrated in Figure 7.1 would be identified and understood for transport of particles from a source such as a production or handling operation to air, to surfaces, and to workers or other individuals. The pathway model would be accompanied by a comprehensive job-exposure matrix describing the temporal history of work jobs and associated exposure conditions for each potentially exposed individual (e.g., Figure 7.8). Exposure characteristics would be based on a combination of direct measurements and process knowledge for individual job activities. There would be an understanding of the particle size distribution of any potentially airborne particles and the likelihood that such particles might deposit in the various regions of the respiratory tract. There would also be a scheme for selecting an appropriate level of control based on information about "determinants of exposure" such as the amount of material

	Exposure period 1	Exposure period 2	Exposure period 3	...
Job 1	Time(J_1, P_1)	Time(J_1, P_2)	Time(J_1, P_3)	...
Job 2	Time(J_2, P_1)	Time(J_2, P_2)	Time(J_2, P_3)	...
Job 3	Time(J_3, P_1)	Time(J_3, P_2)	Time(J_3, P_3)	...
⋮	⋮	⋮	⋮	⋱

FIGURE 7.8 Conceptual design of a comprehensive job-exposure matrix to document aerosol exposure characteristics as a function of time for each worker (Hoover et al., 2007). Exposure characteristics can be based on a combination of direct measurements and process knowledge for individual job activities. Creation of a meaningful job-exposure matrix requires credible assessment of the detailed conditions and behavior of the particles and gases of interest under all circumstances.

handled, its dustiness, flammability, reactivity, and so on, and the potential toxicity of a material due to its particle size, mass, number, surface area, functionality, or other biologically relevant properties.

To support decision making about measurement, exposure, and control, there would be straightforward equations (see example below) or comprehensive algorithms that would relate the conditions of work activities and worker exposure to estimated health effects from individual work tasks throughout a process life cycle and supply chain. There would also be an understanding of the inherent uncertainty or variability associated with key aspects of exposure situations and of how the various possible measures of exposure (e.g., average, cumulative, peak, or temporal patterns) relate to the potential acute or chronic health consequences of exposure (e.g., irritation, sensitization, carcinogenicity, or other effects or toxicity to the skin or eyes, respiratory tract, central nervous system, or other target organs).

Prospectively, the pathway model, job-exposure matrix, and relational equation approach would be used to design safe work facilities and practices. *Contemporaneously*, the modeling and exposure assessment approach would be used to monitor and verify that exposures are under control. *Retrospectively*, the approach could be used in health surveillance to reevaluate and understand any dose–response relationships between exposures to nanoparticles and unexpected health effects, if sufficient information about the material characteristics and behavior were available.

A KEY-PARAMETER EQUATION FOR EXPOSURE MODELING

Although very complicated mechanistic models can be applied, the use of a relatively straightforward "key-parameter" equation as presented in the following discussion illustrates that assessing and managing the dispersion and exposure behavior of radioactive particles and gases can be fundamentally viewed and understood as a linear sequence of contributing factors (Hoover et al., 1997). The equation presented is generally applicable to assessment of occupational or environmental exposures by inhalation or skin contact to any type of potentially toxic material.

The equation is based on historical decision analysis techniques and methods for estimating potential associated health effects to humans from accidental releases of radioactive materials. For example, application of a variation of this calculation approach was used in the U.S. Department of Energy handbook on *Airborne Release Fractions/Rates and Respirable Fractions for Nonreactor Nuclear Facilities* (DOE, 1994b) to document a wide range of experimental results from spilling, pouring, ignition, and other disruption and handling activities involving uranium, plutonium and other nuclear materials. Similarly, a variation of this exposure assessment approach was used in

Behavior of Radioactive Aerosols and Gases

the U.S. Department of Energy Safety Notice on *Decision Analysis Techniques* (DOE, 1995) to estimate the inhalation dose to a worker from a fire involving plutonium-contaminated rags in a glovebox enclosure. More recently, the equation has been used to describe exposure assessment considerations for nanoparticles in the workplace (Hoover et al., 2007).

The equation comprises the following simple linear relationships for an informative exposure assessment:

$$\text{Risk of Health Effect} = \frac{(MAR \cdot DR \cdot ARF \cdot RF \cdot LPF \cdot BR \cdot T \cdot DCF)}{V} \quad (7.18)$$

where

- *MAR* is the amount of material-at-risk (i.e., Bq, Ci, nanograms, milligrams, grams, kilograms, or megagrams).
- *DR* is the damage ratio (i.e., the fraction of the material that is disrupted or subject to dispersion during the work activity). The *DR* term takes into account the fact that not all of the material that may be present in a process is subject to disruption. A consistent approach must be used to associate a *DR* with a corresponding *MAR*. For example, material in storage would not have the same *DR* factor for mechanical disruption and dispersion as material being processed in a manufacturing step.
- *ARF* is the airborne release fraction (or rate for a continuous release). The *ARF* term takes into account the fact that only a small fraction of the disrupted material may be actually dispersed into the air by the process or event of interest, and that the airborne release fraction will depend on the physical form of the material (perhaps 0.00002 for a vitrified source material in a closed metal container, 0.0001 for a solid, 0.001 for a liquid, and 0.01 for a powder). The *Handbook on Airborne Release Fractions/Rates and Respirable Fractions for Nonreactor Nuclear Facilities* (DOE, 1994b) is a useful compilation of experimental data on *ARF* values for release of radioactive aerosols and gases from a variety of handling processes and accident conditions. If a variation of Equation 7.8 is used for assessment of dermal exposure, the release fraction can be associated with those particles that could present themselves for skin exposure.
- *RF* is the respirable fraction (i.e., particles smaller than 10 μm aerodynamic diameter), a factor considered in the model when concern is only for particles that may deposit in the alveolar region of the respiratory tract. The magnitude of the *ARF* and *RF* values are scenario-dependent, which provides practitioners with opportunities to reduce the magnitude and consequence of an aerosol release by controlling environmental factors such as humidity, temperature, conditions of material aging or storage, and methods of equipment or facility operation and housekeeping. The *ARF* and *RF* values can also be interdependent, with the DOE handbook (1994b) noting that the highest *RF* values are often associated with the smallest *ARF*s. In addition, depending on the degree of refinement of the exposure assessment model, *RF* can be adjusted based on actual particle size information about the inhaled aerosol to relate the *uptake* of inhaled material in the human body by deposition in the various regions of the respiratory tract to the *intake* of airborne material by breathing. Figure 7.3 applied the particle-size-dependent relationships of the ICRP Human Respiratory Tract Model (ICRP, 1994) to illustrate how deposition in the respiratory tract varies with particle size. The relative health consequences of particle deposition in the head airways, conducting airways, and alveolar region of the respiratory tract depend on the type of radioactivity and the biological behavior of the material after deposition. Chapter 10 in this book can be consulted regarding the concerns for biological effects of radioactivity in the human body, including the effects following translocation from the point of exposure to other organs. Note that if a variation of Equation 7.8 is used for assessing dermal exposure,

considerations can be given to particles that might penetrate the stratum corneum of human skin and reach the epidermis or the dermis. Different considerations can be given to larger particles that may only have potential to enter damaged skin.

- *LPF* is the leak path factor (e.g., 0.05 for situation in which only 5% of the aerosol is allowed to reach the exposed individuals because 95% of the airborne particles released by the dispersion event are captured by an engineered control system or by the use of personal protective equipment and; 1.0 for a situation in which no controls are applied). The *LPF* term can be used in a prospective design mode to identify a level of control that will be required or adequate to reduce potential exposures to a safe level, and (wherever possible) to a level that is as low as reasonably achievable (ALARA).
- *BR* is the breathing rate (e.g., 0.02 m^3/min for the ICRP reference man [ICRP, 1975] involved in light work activity). Note that some work activities involve greater levels of exertion than the standard assumption of light exercise; other activities require less. Note also that *BR* can influence the ratio of uptake to intake for an aerosol of a given size in the respiratory tract and that breathing at a higher rate of intake does not necessarily result in a proportionally higher uptake (cf., ICRP, 1994, 2002).
- *T* is the time duration of exposure (i.e., minutes or other suitable time units). Time periods selected for the model should be consistent with the work activity and with the temporal properties of any warning system for alerting workers to the need to leave a work area or take other protective measures. If statutory or administrative occupational exposure limits are available, the modeling results can be compared to those limits and time periods used in the model can be consistent with determining compliance with those limits (e.g., 8-h time-weighted averages or 15-min short-term exposure limits).
- *DCF* is the dose conversion factor (e.g., injury or illness per unit of delivered dose of radioactive or other material of interest to the body or affected organ of the exposed individual). For many toxic materials, *DCF* is the most problematic factor in the equation (i.e., What are the health consequences of concern and what are the associated dose–response relationships?). However, in the case of radiation exposure assessment and control, the unifying *DCF* approach of "latent cancer fatalities per unit of radioactivity inhaled" and the associated extensive understanding of the relationships between the physicochemical properties and the biological behavior of inhaled radioactive materials that lead to cancer have provided a harmonized basis for the development and implementation of material-specific radiation protection programs. For materials that cause other potential health outcomes (e.g., inflammation, pulmonary fibrosis, neurological effects, cardiopulmonary effects, or other injury or disease), it is necessary to seek unifying bases for relating the probability of potential health effects to the various potential measures of exposure. Depending on the material of interest and the associated health outcomes of concern, an entire array of *DCF* relationships and values may be needed to address the full range of general and organ-specific health outcomes that may result from exposure to the material. Thus, multiple applications or summations of the equation can be used to address the range of health effects that might be associated with exposures to a given material. This is particularly true in light of the fact that the effects may vary by intensity and duration of exposure, physical form of the toxicant, environmental conditions, and other considerations.
- Lastly, *V* is the effective volume into which the aerosol materials of concern are dispersed (i.e., m^3 or other volume units, with appropriate considerations for the rate of dilution air exchange or exhaust). A variety of approaches can be used to estimate the appropriate dilution volume. For a release of material at or near floor level or ground level, it may be reasonable to assume an initial dilution into a hemisphere, with the diameter of the hemisphere dependent on the dynamics of the dispersion event. Appropriate spatial and temporal relationships for dilution and dispersion of the aerosol cloud and location of the workers or other receptors of interest can be incorporated into the model. For example, a decay

constant, λ, such as 0.02/min fractional removal rate (1.2 room changes/h), might be used for a radioactive waste-handling building. A simple Gaussian plume model is often used to estimate the dispersion of the aerosol cloud as it moves from the point of release.

MONTE CARLO SIMULATION OF UNCERTAINTY FOR PROSPECTIVE AND RETROSPECTIVE ASSESSMENTS OF AIRBORNE RADIOACTIVITY

Although Equation 7.18 can readily be used to make a single-point calculation involving the estimated material-at-risk, damage ratio, airborne release fraction, respirable fraction, exposure time, and other exposure parameters, it is informative to select realistic *ranges and probability distribution functions* for each of the key factors and to calculate a resulting probability distribution function for the consequences of an exposure scenario.

The uncertainty modeling can be done using a Monte Carlo forecasting and risk analysis program (e.g., Crystal Ball® [Decisioneering, 2002], which runs under Microsoft Windows® in the Excel® spreadsheet program). In a Monte Carlo approach, each trial involves a random selection of values from the respective probability distribution functions of the model parameters. All the trials together provide a probability distribution function for worker doses and consequences from each accident or event. A typical simulation run can involve 10,000 trials.

The selection of appropriate ranges and probability distribution functions for the model parameters is an essential part of the process. The ranges and distributions can be selected based on reasonable assumptions that have some basis in known physical processes. For example, the simple "triangular distribution" has the following features: (1) it has zero probability for the minimum expected value, which is followed by (2) a linearly increasing probability that is maximum for the most likely expected value, which is followed by (3) a linearly decreasing probability that has a value of zero for the maximum expected value of the parameter.

Examples of distributions available in simulation programs such as Crystal Ball are shown in Figure 7.9. Distributions cover the full range of options including point estimates; uniform distributions in which all values in a range are equally likely; the relatively simple triangle distribution with a defined estimate of the minimum, most-likely, and maximum values; standard functions with "tails" such as the normal or lognormal distributions; more specialized functions such as the Weibull or Beta distributions; and specially formulated "custom" distributions based on specific user knowledge. As cautioned in the *Preparation Guide for U.S. Department of Energy Nonreactor Nuclear Facility Safety Analysis Reports* (DOE, 1994a), it is important to avoid approaches or models that are unnecessarily complicated.

Care should similarly be taken in selecting parameter values from data sources such as the DOE aerosol release handbook (DOE, 1994b) to avoid over-interpretation of the data (i.e., "pencil

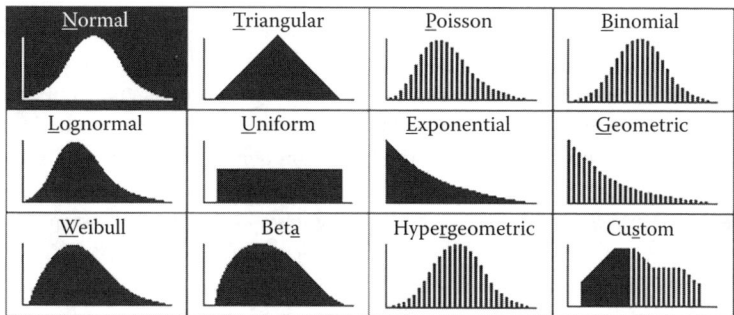

FIGURE 7.9 Examples of statistical distributions that can be used in Monte Carlo simulation. (Data from Decisioneering, *Crystal Ball Version 5.2 User Manual: Forecasting and Risk Analysis for Spreadsheet Users*, Decisioneering, Inc., Denver, CO, 2002.)

sharpening"). The stated purpose of the handbook is to provide a compendium and analysis of experimental data from which *ARFs* and *RFs* may be derived. As specifically recommended in the aerosol release handbook, the estimated dispersion terms for accidents at nonreactor nuclear facilities are intended to be used *to better understand* the potential bounding hazards of activities involving radioactive materials. Although the summary of data in the handbook is quite extensive and contains a good deal of useful interpretation, the authors of the handbook note that:

> It is generally not productive to attempt to use the experimental data cited in this handbook to develop assumed statistical distributions of values for probabilistic assessments. The overall collection of data available for a wide variety of stresses will not support fine statistical resolution as a technically meaningful activity, and this handbook specifically rejects citation as a defensible basis for such attempts.

Taking into account admonitions not to over-interpret the available handbook data, the experimental data provided can be evaluated and used in a responsible manner by selecting experimental results that are relevant to the accident conditions being considered, and by using simple ranges of values that are meaningful for the accident situations. Although it is possible to find data that are relevant to some situations of interest, the handbook authors state that this is not possible for all accident situations. This lack of a complete knowledge base points to the ongoing need to develop, compile, and validate additional information about the dispersion behavior of all types of material in general and radioactive materials in particular.

AN EXAMPLE MONTE CARLO ESTIMATION OF RADIATION DOSE TO A WORKER FROM AN ACCIDENTAL RELEASE OF PLUTONIUM FROM A RADIOACTIVE WASTE DRUM

Table 7.2 and Figure 7.10 present an example of the input parameters, output values, and results for a Monte Carlo simulation of the statistical distribution of radiation dose (rem) that may result to a worker as the result of an accidental release of plutonium aerosol from dropping of a radioactive waste drum. The simulation was conducted by Hoover et al. (1997) using the Crystal Ball forecasting and risk analysis program (Decisioneering, 2002) and the exposure model described in Equation 7.18. An important message is that use of the full distribution estimates of the contributing parameter values gives a much more comprehensive picture of the possible consequences than can be derived from single-point estimates of the median value or the highest value alone. Ramachandran (2005) can be consulted for an example of Monte Carlo modeling for exposure of workers in an industrial hygiene setting.

INFLUENCE OF INITIATING-EVENT PROBABILITIES AND OTHER CONDITIONAL PROBABILITIES FOR ACCIDENT CONSEQUENCES

In conducting a prospective or retrospective evaluation of aerosol behavior and consequences, it is important to assess and understand the influence of any initiating-event probabilities and other conditional probabilities for accident consequences. In the estimated dose consequences described in the example above for breach of a drum containing radioactive waste, it is assumed that (1) the drum was dropped and (2) the drum lid and drum liner were breached. For a complete understanding of the likelihood and consequence of such events, the probability of the initiating event (the drum drop itself) would be taken into account. Such a probability (e.g., anticipated number of drops per thousand drums handled, or probability per year based on an assumed schedule of work) will depend on factors such as the design of the fork lift or other handling device, the distance traveled, and the level of training of the operator. If an event has already occurred, prospective estimates of anticipated exposure consequences would take advantage of any available knowledge about the actual physical conditions of the release and would use the modeling approach to augment available measurement data for aerosol concentration or other indications of actual event consequences.

TABLE 7.2
Example of Input and Output Parameters for a Monte Carlo Simulation of the Statistical Distribution of Radiation Dose (rem) That May Result to a Worker due to an Accidental Release of Plutonium Aerosol from Dropping of a Radioactive Waste Drum

Simulation Input Parameter	Distribution or Value
Material available for release (*MAR*) in a radioactive waste drum (Ci)	Triangular distribution with minimum: 0, likeliest: 6.3, maximum: 81
Damage ratio (*DR*) for drum drop (assuming breach of the lid and liner)	Triangular distribution with minimum: 0.01, likeliest: 0.1, maximum: 0.25
Airborne release fraction (*ARF*)	Triangular distribution with minimum: 3×10^{-5}, likeliest: 8×10^{-5}, maximum: 5×10^{-4}
Respirable fraction (*RF*)	0.1
Leak path factor (*LPF*)	1.0 (no control)
Breathing rate (*BR*) (m³/min)	0.02
Initial volume of aerosol cloud (*V*) (m³)	500 (6.1-m [20-ft] diameter hemisphere)
Decay constant for cloud dilution (λ) (fraction/min)	0.02 (equivalent to 1.2 air changes per hour)
Time in the aerosol cloud (*T*) (min)	Triangular distribution with minimum: 0.4, likeliest: 0.5, maximum: 10
Dose conversion factor (*DCF*) (rem/Ci inhaled)	5.1×10^8 (for potential use in calculating health consequences)
Equation for projected dose (rem)	$MAR \cdot DR \cdot ARF \cdot RF \cdot LPF \cdot BR \cdot T \cdot (1 - \exp(-\lambda \cdot T))/(\lambda \cdot V)$
Statistical Output Parameter	**Value**
Number of simulation trials	10,000
Mean projected dose (rem)	5.12
Median projected dose (rem)	2.67
Mode projected dose (rem)	0.45
Standard deviation (rem)	7.00
Variance	49.04
Skewness	3.31
Kurtosis	19.12
Coefficient of variability	1.37
Range minimum (rem)	0.0088
Range maximum (rem)	88.70
Range width (rem)	88.69
Mean standard error (rem)	0.07

Note: The simulation was conducted by Hoover et al. (1997) using the Crystal Ball forecasting and risk analysis program (Decisioneering, 2002).

In conducting both retrospective and prospective analyses, it is also important to note that the mere occurrence of an "initiating event" for an accident does not guarantee that a release of radioactive particles or gases has occurred or will occur. In other words, dropping a waste drum does not guarantee that the drum will be breached. There is a conditional probability that the drum will release material after being dropped. That probability will depend on parameters such as the design of the drum, the distance of the drop, and the characteristics of the surface onto which the drop is made. For example, in a series of tests done at the Rocky Flats Plant (Bearly et al., 1988; Rocky Flats, 1993), failure rates based on dropping of drums from a height of 17 ft were found to be 0.33 probability of a lid failure and 0.077 probability of a liner failure. Thus, the overall conditional probability of release of radioactivity as a result of dropping a waste drum from a height of 17 ft is only $0.33 \times 0.077 = 2.5 \times 10^{-2}$. If prospective modeling is being done for ongoing or anticipated

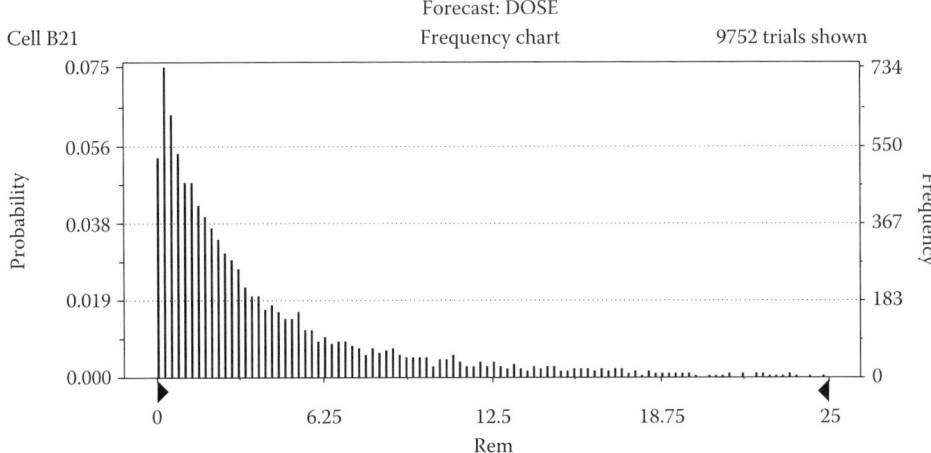

FIGURE 7.10 Example statistical distribution from a Monte Carlo simulation of the radiation dose that may result to a worker as the result of an accidental release of plutonium aerosol from dropping of a radioactive waste drum for assumed accident-related parameters as described in Table 7.2. The simulation assumes that the drum lid and drum liner are dislodged or otherwise breached as a result of the drop. After 10,000 trials the minimum predicted dose was 0.0088 rem, the maximum predicted dose was 88.9 rem, and the median value of predicted dose distribution was 2.7 rem. The display range for presentation of the distribution plot is from 0 to 25 rem and comprises 9752 of the 10,000 trials. The simulation was conducted by Hoover et al. (1997) using the Crystal Ball forecasting and risk analysis program (Decisioneering, 2002).

activities, such conditional-event probabilities can be used in conjunction with the probability distribution functions for consequences to plan appropriate steps for mitigation or other response.

CONCLUSION

In conjunction with other information in this book, an understanding of the factors that influence the behavior of radioactive particles and gases can help in the anticipation of dispersion and exposure situations that might be encountered in the workplace and the general environment, the recognition and evaluation of those situations, and the design of appropriate air sampling and response strategies to mitigate and control the consequences of those situations. A copy of the AEROSAMP spreadsheet for making the aerosol calculations described in this chapter can be obtained from the author.

REFERENCES

Bearly, L.A., Paynter, J.K., and Lombardi, E.F., *Full-Scale Drop-Impact Tests with DOT Specification 7A Waste Containers*, WPS 88-001, Rockwell International, Golden, CO, December 1988.

Chamberlain, A.C., *Radioactive Aerosols*, Cambridge University Press, Cambridge, 1991.

Day, G.A., Dufresne, A., Stefaniak, A.B., Schuler, C.R., Stanton, M.L., Miller, W.E., Kent, M.S., Deubner, D.C., Kreiss, K., and Hoover, M.D., Exposure pathway assessment at a copper–beryllium alloy facility, *Ann Occup Hyg* 51: 67–80, 2007.

Decisioneering, *Crystal Ball Version 5.2 User Manual: Forecasting and Risk Analysis for Spreadsheet Users*, Decisioneering, Inc., Denver, CO, 2002.

Dennis, R., Ed., *Handbook on Aerosols*, Report TID-26608, U.S. Department of Energy, Washington, DC, 1976.

DOE, *Preparation Guide for U.S. Department of Energy Nonreactor Nuclear Facility Safety Analysis Reports*, DOE-STD-3009-94, U.S. Department of Energy, Washington, DC, July 1994a.

DOE, *Airborne Release Fractions/Rates and Respirable Fractions for Nonreactor Nuclear Facilities*, DOE-HDBK-3010-94, DOE Handbook, U.S. Department of Energy, Washington, DC, December 1994b.

DOE, *Decision Analysis Techniques*, Safety Notice Issue No. 95-01, Office of Nuclear and Facility Safety, Office of Operating Experience Analysis and Feedback, U.S. Department of Energy, Washington, DC, August 1995.

Esmen, N. and Hammad, Y., Lognormality of environmental sampling data, *Environ Sci Health* A12: 29–41, 1977.

Hatch, T. and Choate, S.P., Statistical description of the size properties of non-uniform particulate substances. *J. Franklin Inst.* 207: 369–387, 1929.

Hinds, W.C., *Aerosol Technology—Properties, Behavior, and Measurement of Airborne Particles*, 2nd ed., John Wiley & Sons, New York, 1999.

Hoover, M.D., Newton, G.J., and Farrell, R.F., Monte Carlo simulation of potential radiological doses to workers during postulated accidents involving radioactive waste, *Trans Am Nucl Soc* 77: 493–494, 1997.

Hoover, M.D., Stefaniak, A.B., Day, G.A., and Geraci, C.L., Exposure assessment considerations for nanoparticles in the workplace, Chapter 5 in: *Nanotoxicology: Characterization, Dosing, and Health Effects*, Monteiro-Riviere, N.A. and Tran, C.L., Eds., Informa Healthcare, New York, pp. 71–83, 2007.

ICRP, *Reference Man: Anatomical, Physiological and Metabolical Characteristics*, ICRP Publication 23, International Commission on Radiological Protection, Elsevier Science, Oxford, 1975.

ICRP, *Human Respiratory Tract Model for Radiological Protection*, ICRP Publication 66, International Commission on Radiological Protection, Elsevier Science, Oxford, *Ann ICRP* 24(1–3), 1994.

ICRP, *Guide for the Practical Application of the ICRP Human Respiratory Tract Model*, Supporting Guidance 3. International Commission on Radiological Protection, Elsevier Science, Oxford, *Ann ICRP* 32(1/2), 2002.

Kornev, Yu.V. and Zubkovsky, S. L., A new method for studying the process of sublimation of metals, *Atomic Energy* 2(4): 427–431, 1956. DOI: 10.1007/BF01489632.

Lapple, C.E., Characteristics of particles and particle dispersoids, *Stanford Res Inst J* 5: 95, 1961.

Raabe, O.G., Characterization of radioactive airborne particles, in *Internal Radiation Dosimetry*, Raabe, O.G., Ed., Medical Physics Publishing, Madison, WI, pp. 111–142, 1994.

Ramachandran, G., *Occupational Exposure Assessment for Air Contaminants*, CRC Press, Boca Raton, FL, 2005.

Rocky Flats, *Final Safety Analysis Report Building 664*, EG&G Rocky Flats, Inc., Golden, CO, July 31, 1993.

Scott, B.R., Hoover, M.D., and Newton, G.J., On evaluating respiratory tract intake of high-specific-activity alpha-emitting particles for brief occupational exposure, *Radiat Prot Dosim* 69: 43–50, 1997.

Scott, B.R. and Fencl, A., Variability in PuO_2 intake by inhalation: Implications for worker protection at the U.S. Department of Energy, *Radiat Prot Dosim* 83(3): 221–232, 1999.

Stöber, W., Dynamic shape factors of nonspherical aerosol particles, Chapter 14 in *Assessment of Airborne Particles—Fundmentals, Applications, and Implications to Inhalation Toxicity*, Mercer, T.T., Morrow, P.E., and Stöber, W., Eds., Charles C. Thomas Publisher, Springfield, MA, pp. 249–289, 1972.

8 Filtration

Mark D. Hoover

CONTENTS

Introduction 157
Filtration Fundamentals 158
    Air Filters are not Sieves 158
    "Pore Size" is Only a Guide 158
    Filters Require Proper Strength, Support, and Sealing 160
Selection of Filter Media for Air Sampling 160
    Physical Characteristics of Filter Media 160
    Pressure Drop Characteristics 164
    Particle Collection Efficiency 165
    Radiation Shielding in the Filter Matrix 166
        Filters for α-Spectroscopy 166
    Considerations of "Front" and "Back" Filter Surface 168
    Analytical and Radiochemistry Issues 169
    Transportation and Storage of Filter Samples 169
    Justifying a Change in the Selection of Filter Media 169
    Avoiding Unexpected Changes in Filter Media 170
Filtration for Nuclear Air and Gas Treatment 170
Building Filtration and Air Cleaning 171
    Filtration Considerations 172
Air-Cleaning Considerations 173
Filtration for Respiratory Protection 173
Legal Requirements for Respiratory Protection 174
    APFs for Different Types of Respiratory Protection 176
    Classes of Filter Media for Air-Purifying Respirators 176
    Clarification into Dust Masks, Surgical Masks, and Other Media 177
Conclusion 178
References 178

INTRODUCTION

Filtration is the most widely used method for collecting samples of radioactive aerosols. Methods and equipment range from low-volume, miniature lapel samplers (1 L min^{-1} or less) for collecting aerosols in the breathing zone of individual workers to high-volume samplers (sampling rates up to about 60 m^3 h^{-1}) for short- or long-term sampling in the workplace or environment. Many filter media are available for use in collection of airborne particles (*cf.*, Davies, 1973; Brock, 1983; Liu et al., 1983; Brown, 1993; Spurny, 1998; Lee and Mukund, 2001; Lippmann, 2001). Filtration materials include various membranes, as well as cellulose, glass, quartz, and plastic fibers. Sintered structures of metals or mineral particles have been used for filtration at high temperatures.

    Filtration fundamentals and key issues for filter selection and use for air sampling are described below. Selected issues and considerations of value to the air sampling practitioner are also included

FILTRATION FUNDAMENTALS

AIR FILTERS ARE NOT SIEVES

A common misconception is that filters act as sieves and that there is a direct relationship between the "pore size" of a filter and the minimum particle size that can be collected. In reality, filters capture particles by a combination of physical processes, which include direct interception, impaction, Brownian diffusion, electrical attraction, and gravitational sedimentation (Figure 8.1). Filtration efficiency is lowest in the particle size range where the combined effects of diffusion, interception, and inertial impaction are minimal (Figure 8.2). The most penetrating particle size (MPPS) (Figure 8.3) is typically in the range of approximately 50–500 nm (0.05–0.5 μm). Particles above the MPPS are collected predominantly by impaction and interception and particles below the MPPS are collected predominantly by Brownian diffusion.

The exact value of MPPS depends on the characteristics of the filter (e.g., fiber diameter and packing density) and typically decreases slightly as face velocity (the velocity at which air enters the filter) increases. The general basis for the decrease in MPPS with increasing face velocity is that particles slightly smaller than the initial MPPS now lack the required residence time for collection by diffusion, while particles slightly larger than the initial MPPS now have sufficient inertia to be collected.

If only very large or very small particles are of concern, then even filters with low efficiency at the MPPS may be adequate for some applications. For filters of high efficiency, a negligible reduction in collection efficiency is present at the MPPS, and the primary concern for filtration efficiency relates to unintended leakage around the filter or through areas that may have been damaged by improper handling.

"PORE SIZE" IS ONLY A GUIDE

It is instructive for the air sampling practitioner to be aware that the concept of "pore size" for any type of filter is based on historical evaluations of filter media for liquids via methods such as a "bubble-point" test (*cf.*, Brock, 1983); manufacturer reports of pore size are *not* based on direct

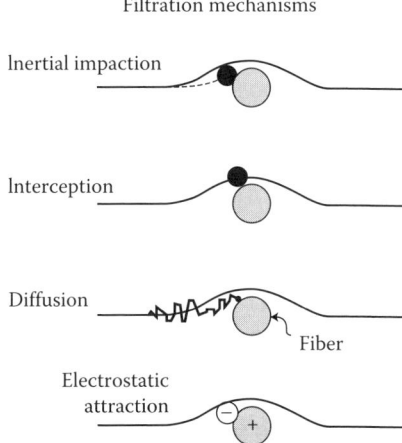

FIGURE 8.1 Illustration of mechanisms that contribute to collection of airborne particles on filters. Particles can also be deposited by gravitational sedimentation.

Filtration

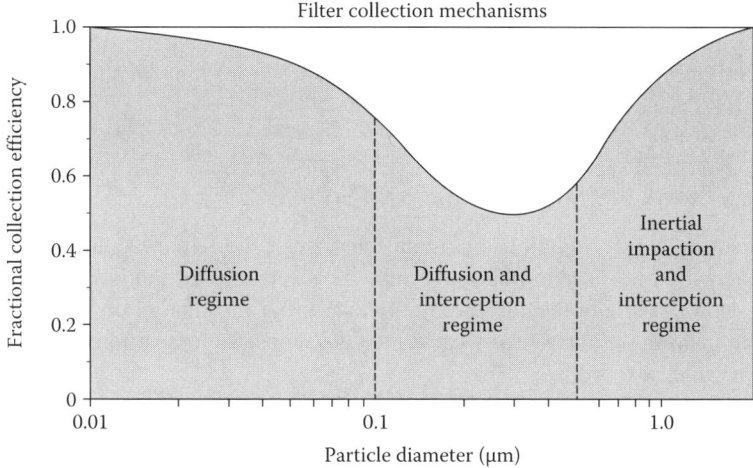

FIGURE 8.2 Generic illustration of filtration efficiency as a function of particle diameter. Filtration is least efficient in the particle size range where the combined effects of diffusion, interception, and inertial impaction are minimal.

measurements of physical filter morphology and dimensions, such as through electron microscopy. A bubble-point test is conducted by mounting a filter in a holder, placing a layer of water on the top surface of the filter, slowly increasing the pressure of air applied to the underside of the filter, and noting the pressure at which the appearance of bubbles is observed on the water side. Filters that require higher pressures to induce the appearance of bubbles are assigned smaller pore sizes. As noted by Brock (1983) and others, discrepancies routinely exist between empirical test results and theoretical predictions (e.g., based on capillary theory). Thus, the concept of pore size is useful as a guide, but only as a rough guide, to the relative expected performance of different filter media.

Other techniques for determining pore size include airflow methods, gas diffusion across a wet filter medium, the mercury intrusion method (which also provides information on void volume and pore size distribution), water intrusion-pressure determination, and the bacterial-retention method.

Because collection occurs by a complex combination of mechanisms, filters with nominal pore sizes larger than 1 μm can be very efficient collectors of submicrometer particles. As demonstrated

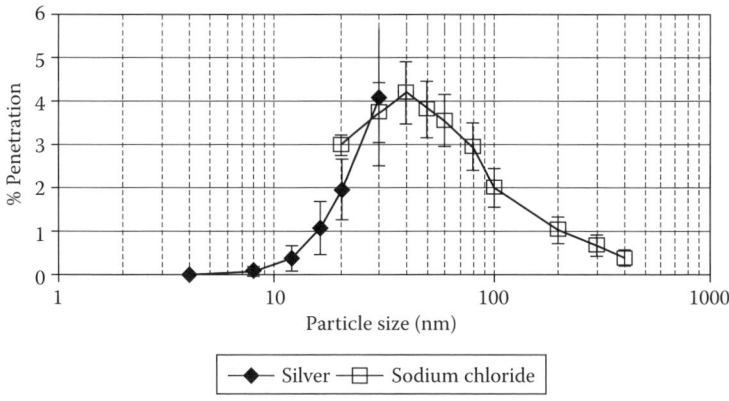

FIGURE 8.3 Example of data collected using a combination of test aerosols to determine the MPPS and penetration percent for an air filter with overall collection efficiency greater than 95%. Each data point represents the mean and standard deviation from the evaluation of five samples of the same filter media. (Data from Rengasamy, S., et al., *J. Occup. Environ. Hyg.* 5, 556–564, 2008a.)

by Lindeken et al. (1964), membrane filters show no serious degradation of collection efficiency until the pore diameters exceed 5 µm. As described further below, filters with pore size as large as 5 µm are often preferred because they have lower pressure drops than smaller pore size filters; yet they retain high efficiency values.

FILTERS REQUIRE PROPER STRENGTH, SUPPORT, AND SEALING

The filter should be strong enough to maintain integrity at the required sampling flow rates and during handling activities. Fragile filters that may be acceptable for laboratory use should not be used in field conditions where they are easily broken during installation or removal. Loss of filter material during handling with tweezers can cause unacceptable loss of material for radiological, chemical, or gravimetric analysis.

A backup support (e.g., a porous plate or disk of cellulose, metal, or other construction, fine screen; etc.) that produces negligible pressure drop should be used behind the filter to prevent filter distortion or breakage. The support should be sturdy enough to avoid distortion under the expected conditions of filter pressure drop, even if the filter is laden with dust.

The sample holder should provide adequate structural support while not cutting, crimping, or otherwise damaging the filter. The composition, construction, and surface texture of the filter holder and associated sealing mechanism should prevent sampled air from bypassing the filter; facilitate changing of the filter; facilitate decontamination; and be resistant to thermal, mechanical, chemical, or radiation damage under the expected conditions of use.

If a gasket is used to seal the filter to the backing plate, the gasket should be in contact with the filter along the entire circumference to ensure a good fit. Periodic inspections of the gasket should be performed to detect degradation and to eliminate buildup of dust or filter material, which could result in sampled air bypassing the filter.

SELECTION OF FILTER MEDIA FOR AIR SAMPLING

Table 8.1 summarizes the type of information that is useful for selecting an appropriate filter for sampling airborne radioactive particles. The table had its genesis in research to provide a technical basis for selection and use of filter media in continuous air monitors (CAMs) for α-emitting radionuclides (Hoover and Newton, 1991, 1992). That work included development and qualification of the new option of a Teflon™ membrane filter with a high-contrast, black-fiber backing for use in α-CAMs. Contents of the table were subsequently expanded for inclusion as Informative Annex D in the American National Standard for *Sampling and Monitoring Releases of Airborne Radioactive Substances from the Stacks and Ducts of Nuclear Facilities* (ANSI N13.1, 1999).

Table 8.1 includes examples of a variety of membrane, track-etch, microglass fiber, and coarse fiber filters. Representative scanning electron micrographs of filter types from the example table are shown in Figure 8.4. Inclusion in the table does not constitute an endorsement of any particular manufacturer or filter type. Conversely, the absence of any particular filter from the table does not constitute a rejection of that medium. For general sampling applications, information is provided on durability, flow resistance, and efficiency. Information on collection of radon decay products and energy resolution for α-spectroscopy are included for α-CAM applications.

PHYSICAL CHARACTERISTICS OF FILTER MEDIA

The physical characteristics of filter media are related to their composition and method of fabrication and can vary dramatically (Figure 8.4).

Cellulose membrane filters are formed by "casting" a liquid film onto a moving plate, with a "doctor blade" that levels and controls the thickness of the film (*cf.*, Brock, 1983). The film consists

TABLE 8.1
Characteristics of a Selection of Filters Evaluated for Use in Sampling Radioactive Particles

Filter Type	Filter Composition (and Durability)	Typical Flow Rate (L min^{-1} cm^{-2} psi^{-1})[a]	FWHM of the Po-218 PEAK (keV)[b]	Relative Radon Progeny Counts in the Pu ROI[c]	Relative Radon Progeny Collection Efficiency[d]	Filter Efficiency Range (%)[e]
Millipore type SMWP (5.0 μm pore size) Millipore Corp, Bedford, MA	Mixed esters of cellulose acetate and cellulose nitrate (fragile; electrostatic; and both sides similar)	3.2	670	1	1	98.1–>99.99
Millipore type AW19 (5.0 μm pore size) Millipore Corp	Homogeneous, microporous polymers of cellulose esters formed around a cellulose web (rugged and both sides similar)	3.2	470	0.57	0.99 ± 0.01	99.93–>99.99
Durapore SVLP (5.0 μm pore size) Millipore Corp	Polyvinylidene fluoride (rugged and both sides similar)	2.8	790	1.55	0.67 ± 0.01	—
Fluoropore FSLW (3.0 μm pore size) Millipore Corp	Polytetrafluoro-ethylene bonded to polypropylene high-density fibers (rugged; front is membrane; back is fiber; and sides are barely distinguishable by naked eyes)	4.6	350	0.47	1.04 ± 0.02	98.2–>99.98
Fluoropore FMLB (5.0 μm pore size) Millipore Corp	Polytetrafluoroethylene bonded to polypropylene high-density fibers (rugged; front is membrane; back is fiber; and sides are distinguishable by naked eyes–high contrast backing)	12	460	0.67	0.96 ± 0.04	
Versapor 3000 (3.0 μm pore size) Gelman Sciences, Ann Arbor, MI	Acrylic copolymer on a nylon fiber support (rugged and both sides similar)	5.0	590	0.94	0.75 ± 0.02	99.7–>99.99
Gelman type A/E (~1.0 μm pore size) Pall-Gelman, East Hills, NY	Borosilicate glass fiber without binder (breakable during handling and both sides similar)	5.0	≥1000	1.31	0.92 ± 0.01	99.6–>99.99
Whatman EPM 2000 Whatman LabSales, Hillsboro, OR	Borosilicate glass microfiber without binder (breakable during handling and both sides similar)	4.0	≥1000	1.48	1.00 ± 0.03	—

continued

TABLE 8.1 (continued)
Characteristics of a Selection of Filters Evaluated for Use in Sampling Radioactive Particles

Filter Type	Filter Composition (and Durability)	Typical Flow Rate (L min^{-1} cm^{-2} psi^{-1})[a]	FWHM of the Po-218 PEAK (keV)[b]	Relative Radon Progeny Counts in the Pu ROI[c]	Relative Radon Progeny Collection Efficiency[d]	Filter Efficiency Range (%)[e]
Whatman–41 Whatman LabSales	Cotton cellulose filter paper (rugged; currently used primarily for liquid filtration; and both sides similar)	5.0	≥1500	1.65	0.42 ± 0.01	43– >99.5
Nuclepore (0.6 μm pore size) VWR Scientific, Pleasanton, CA	Polycarbonate membrane (rugged; thin; very electrostatic; currently used primarily for liquid filtration; and the collection side recommended by the manufacturer is the shiny side)	0.8	500	0.89	0.85 ± 0.02	53– >99.5
Millipore type AABP (0.8 μm pore size) Millipore Corp	Mixed esters of cellulose (fragile; electrostatic; and the collection side is darker)	1.4	520	0.91	1.05 ± 0.01	99.999– >99.999

Source: Adapted from Hoover, M.D. and Newton, G.J., *Annual Report of the Inhalation Toxicology Research Institute for 1990–1991*, LMF-134, National Technical Information Service, Springfield, VA, pp. 16–19, 1991; Hoover, M.D. and Newton, G.J., *Annual Report of the Inhalation Toxicology Research Institute for 1991–1992*, LMF-138, National Technical Information Service, Springfield, VA, pp. 5–7, 1992.

Note that a version of this table has also been included as an informative annex in the American National Standard for Sampling and Monitoring Releases of Airborne Radioactive Substances from the Stacks and Ducts of Nuclear Facilities (ANSI N13.1, 1999).

[a] Flow rate determined under 5 psi (35 kPa) vacuum pressure drop at 620 mm Hg ambient atmospheric pressure (Albuquerque, NM).

[b] FWHM is the typical full-width-at-half-maximum of the Polonium-218 peak obtained with a 2.5-cm-diameter filter and a 2.5-cm-diameter solid-state detector with a 0.5-cm separation distance during sampling of room air at the Inhalation Toxicology Research Institute in Albuquerque, NM.

[c] Radon progeny background counts in the Pu ROI for the filter of interest, divided by similar counts obtained simultaneously on a Millipore SMWP filter.

[d] Total radon progeny background counts on the filter of interest, divided by similar counts obtained simultaneously on a Millipore SMWP filter. Mean and standard uncertainty for five replicate tests.

[e] The range of filter efficiency values given generally correspond to a particle diameter range of 0.035–1 μm, a pressure drop of 1–30 cm Hg, and a face velocity range of 1–100 cm s^{-1}. Values are from Liu et al. (1983), Liu (1992), Hoover et al. (1997a), and Hoover et al. (1997b).

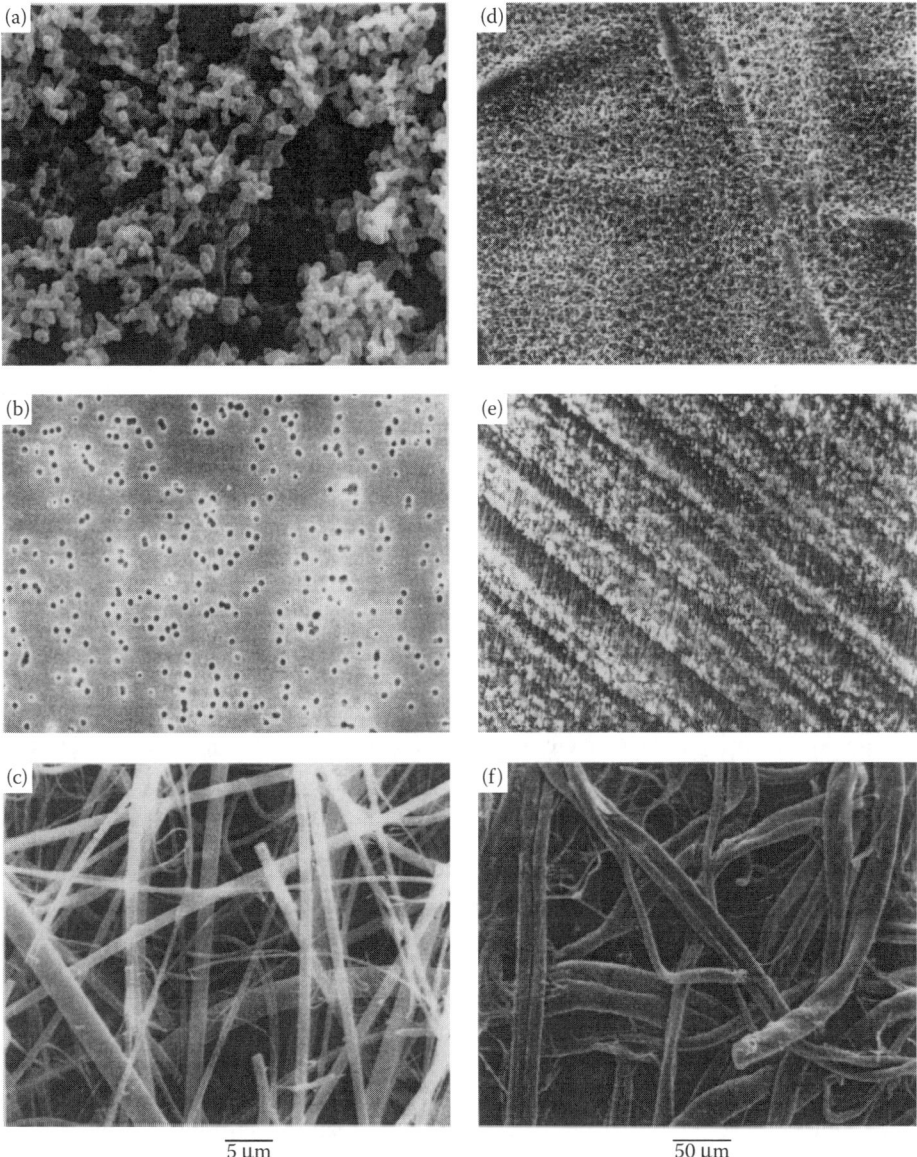

FIGURE 8.4 Typical scanning electron photomicrographs at 1600× magnification of (a) a Millipore 5-µm pore size, unsupported, mixed-cellulose membrane filter, (b) a Nuclepore 0.8-µm pore size track-etch filter, and (c) a Whatman EPM-2000 microglass fiber filter, and at 200× magnification of (d) a Gelman Versapor fiber-supported cellulose membrane filter, (e) a Fluoropore Teflon™ membrane filter, and (f) a Whatman-41 cotton cellulose fiber filter.

of dissolved cellulosic materials in a solvent that is subsequently evaporated, allowing the cellulosic material to dry into the desired porous array of randomly attached, bead-like structures. The mixed cellulose ester membrane filter illustrated in Figure 8.4a shows the resulting porous membrane structure that provides a tortuous path for airflow. The plate side of membrane filters tends to be slightly flatter than the surface that was facing upward during production. At a lower magnification, Figure 8.4d shows a fiber-supported cellulose membrane filter. Inclusion of the fiber support alters the "smoothness" of the filter surface, especially on the nonplate side.

Teflon membrane filters are made by controlled stretching of dense Teflon film. As illustrated in Figure 8.4e, the stretched film has the appearance of an array of ultrafine fibrils between ridges of less-stretched membrane. As noted below in the section "Filters for α-Spectroscopy," these fibrils provide a very flat surface for excellent front-surface collection of airborne particles.

Composite membrane filters are made by depositing a membrane onto a porous or fibrous support. Use of a fibrous support provides structural integrity to fragile membranes such as mixed cellulose esters, and is especially helpful for providing a desirable "stiffness" to filters for ease of handling. Unsupported Teflon membrane filters can be difficult to handle because they are especially flimsy and tend to wrap themselves around the tweezers during handling. They also readily accumulate an electrostatic charge, which can cause them to move erratically in the proximity of other objects such as the filter holder or the counting or weighing equipment. Integration of a fibrous backing into the filter construction [such as the high-density polyethylene fibers that comprise the entire backing of the Millipore Fluoropore FMLB filter (Millipore Corp., Bedford, MA)] reduces the electrostatic charge of the filters and makes them easier to handle. The handling quality of other types of Teflon membrane filters has been improved by incorporation of a support ring of polyethylene fibers around the outer back edge of the filter.

Track-etch filters are made by exposing a solid membrane (typically polycarbonate film) to a source of nuclear radiation, allowing the intensity of the radiation source and the time of exposure to determine the number of randomly spaced damage sites that are induced in the membrane, and then etching the membrane in a chemical solution for a length of time suitable to enlarge the damage sites to pores of the desired size. Greater irradiation creates a greater number of pore sites; longer etching creates larger pores. Each pore is "straight," but the pores have different angles relative to each other because of the directional nature of the radiation entering the membrane. Thus, the pores appear round on the front surface where the radiation entered, and "woody" on the back surface where the radiation has penetrated through and exited the membrane. The track-etch example in Figure 8.4b is a Nuclepore filter with 0.6 μm pore size (VWR Scientific, Pleasanton, CA).

Fibrous filters are beds of randomly oriented fibers such as microglass as illustrated by the Whatman EPM-2000 filter (Whatman LabSales, Hillsboro, OR) shown in Figure 8.4c. Fiber diameter, packing density, and overall depth of the fiber bed influence the performance of fibrous filters. At a lower magnification, Figure 8.4f shows the historically important Whatman-41 cotton cellulose fiber filter. As shown in Table 8.1 and discussed further below, collection efficiency of the finer, microglass filter is substantially superior to that of the coarse cotton fiber filter. In addition, although the Whatman-41 has an advantage of being easily dissolved for chemical analyses, it has a collection efficiency that decreases dramatically at low flow rates. Compared to membrane fibers and track-etch filters, fibrous filters are often referred to as "depth" filters.

Electrostatic or "electret" filters can be created by placing an electrostatic charge on the fibers within the filter. This improves collection efficiency without an increase in fiber packing density or pressure drop. Electrostatic formulations are more typically used in filters for building filtration or respiratory protection; performance characteristics of filters for air sampling typically depend on their "mechanical" properties. The collection efficiency of some (but not all) electrostatic filters can degrade with time to the underlying mechanical collection efficiency, especially if used at high humidity or exposed to certain chemical vapors, gases, or aerosols (*cf.*, Barrett and Rousseau, 1998; Martin and Moyer, 2000; Moyer and Bergman, 2000).

PRESSURE DROP CHARACTERISTICS

Typical flow rates per unit filter area per unit pressure drop (L min^{-1} cm^{-2} psi^{-1}) presented in Table 8.1 were determined using a flow meter calibrated by a standard traceable to the National Institute of Standards and Technology. The reported characteristic values for each filter type were calculated from measurements made in Albuquerque, NM (620 mm Hg ambient atmospheric pressure) with an applied filter pressure drop of 5 psi (35 kPa) vacuum. Conversion of the pressure drop characteristic into other units such as face velocity per unit pressure drop (e.g., cm s^{-1} per psi) (*cf.*, Liu et al., 1983) can easily be made from the data presented.

The units of L min^{-1} cm^{-2} psi^{-1} presented here were chosen for user convenience in determining whether a given filter of a given cross section is likely to provide a suitable flow rate at a reasonable applied pressure drop. For example, the characteristic value of 3.2 L min^{-1} cm^{-2} psi^{-1} for the Millipore SMWP filter indicates that application of a 2-psi pressure drop to a filter with a 5-cm^2 cross-sectional area (nominal filter diameter 25 mm) will yield a flow rate of 32 L min^{-1}. Thus, the SMWP is readily suitable for applications requiring a flow rate of 1 or 2 cfm (28–56 L min^{-1}).

In comparison, the Fluoropore 5-µm pore size Teflon membrane filter has an even better characteristic pressure drop value of 12 L min^{-1} cm^{-2} psi^{-1}, indicating that it can be operated at the same flow rate as the SMWP with an applied pressure drop that is lower by nearly a factor of 4. Conversely, the characteristic pressure drop value of only 0.8 L min^{-1} cm^{-2} psi^{-1} associated with the 0.6-µm pore size Nuclepore track-etch filter indicates that operation at the same flow rate as the SMWP would require an applied pressure drop that is higher by a factor of 4; use of this small-pore filter would be impractical at more than a few L min^{-1} unless a larger filter cross-sectional area was employed.

The calculations illustrated above are predicated on the assumption of a linear relationship between flow rate, cross-sectional area, and pressure drop (i.e., compliance with Darcy's law of resistance). As noted by Davies (1973), Darcy's law applies to low-resistance, high-efficiency filters for which the airflow is both viscous and incompressible. Deviations from the linear relationship occur if the air is rarified by large pressure drop as it passes through the filter, if the geometry of the filter matrix is compressed or otherwise altered by increasing pressure drop, or if chemical reactions or the accumulation of dust alter the fundamental resistance of the filter media. Therefore, use of reference information from filter manufacturers is a starting point, and it is prudent for air sampling practitioners to follow up with their own evaluations and verification of the pressure drop characteristics of any candidate or applied filter media.

PARTICLE COLLECTION EFFICIENCY

To limit the uncertainty in air sampling results, filters that are used for sampling airborne radioactive particles should have a minimum efficiency of 95%. Efficiency values should be applicable to the conditions of use. In particular, the collection efficiency depends on the filter airflow face velocity (Liu et al., 1983).

If the published or manufacturer's data on filter collection efficiency are not available for the particle sizes of interest, then the efficiency should be determined by the user. This can be done by placing a highly efficient membrane or glass fiber filter behind the filter of interest and then comparing the mass or radioactivity penetrating to the backup filter to the total mass or radioactivity collected on both filters (*cf.*, Hickey et al., 1991). If a filter with efficiency lower than 95% is required to meet the overall sampling objectives (e.g., requirements for low-pressure drop for extended sampling durations or requirements for dissolution of the filter for radiochemistry), then a correction for efficiency should be made. Because filter efficiency is a function of airflow rate, care should be taken to maintain a sample flow rate that is adequate to achieve the desired collection efficiency.

As reported in Table 8.1, filter efficiencies published by Liu et al. (1983) ranged from >99.999 at all particle sizes and flow rates for the Millipore type AA, 0.8-µm pore size membrane filter to <50% for the Whatman-41 cotton cellulose filter at low flow rates and small particle sizes. To confirm the particle collection efficiency of the new, high-contrast backed Fluoropore filter, Hoover and Newton (1992) along with Dr. B.Y.H. Liu (University of Minnesota, Minneapolis, MN) conducted penetration tests using the method mentioned in Liu et al. (1983). He reported collection efficiency for the 5-µm pore size filter to be 98.3% greater than 99.99% over the size range 0.03 µm to 1.0 µm diameter. He also confirmed the collection efficiency for the standard 3-µm pore size Fluoropore filter to be 99.90% greater than 99.99% over that same size range.

In addition, a plutonium dioxide aerosol was used by Hoover and Newton (1992) at the Inhalation Toxicology Research Institute to conduct five aerosol penetration tests of 2.5-cm-diameter Fluoropore filters operated at 28.3 L/min. The radioactivity of plutonium dioxide aerosol (physical diameter 0.3 μm, aerodynamic diameter 1.0 μm, and geometric standard deviation 1.6) collected on each filter was compared to the radioactivity that penetrated to a backup filter. Efficiency was 99.94 ± 0.03%, which confirmed the excellent performance of the new filter.

Table 8.1 also includes collection efficiency for radon decay products as determined by simultaneous sampling of room air on all filters, followed by simultaneous radioactivity counting using the ZnS(Ag) method. This efficiency evaluation can readily be made by users under their actual conditions of filter use (*cf.*, Lindeken, 1961).

RADIATION SHIELDING IN THE FILTER MATRIX

As noted in ANSI N13.1 (1999), if penetration of radioactive material into the collection media or self-absorption of radiation by the material collected would reduce the count rate by more than 5%, a correction factor should be used. A dual filter method can also be used to measure the reduction of counting efficiency by energy absorption in the filter medium (Hickey et al., 1991). Evaluation of self-absorption in the material collected may require separate radiochemical analyses.

Concerns for penetration of particles into the filter matrix are a function of the type of filter, the type of radiation, and the radiation counting method being used. Shielding by the filter media is seldom a concern for detection of γ-radiation. Although energy degradation concerns are greatest for α-emitting radionuclides, even glass microfiber filters such as the Gelman A/E glass (Pall Gelman, East Hills, NY) collect particles near enough to the filter surface that radioactivity counting results from the ZnS(Ag) method are as accurate as radiochemical results. Higby (1984) demonstrated that absorption of α-radiation emitted from airborne particles collected on glass fiber filters does not constitute a major source of error in estimating concentrations of airborne α-emitting radionuclides, but excellent resolution in α-spectroscopy requires use of membrane-type filters, which are front-surface collectors. Additional information on issues for α-spectroscopy is provided in the following section.

Filters for α-Spectroscopy

An active area of work has related to the selection of appropriate filter media for α-spectroscopy, including use in α-CAMs. The three major requirements for a filter for α-spectroscopy are (1) that it collects particles on its surface so that the α-energy spectrum is not degraded by burial of particles in the filter matrix, (2) that its collection efficiency exceeds 95% (and ideally exceeds 99%) so that the results are not biased by sample losses, and (3) that it has a low-pressure drop so that samples can be collected at a reasonably high flow rate of 25–60 L min^{-1} (1–2 cfm).

Although the Millipore SMWP filter had historically been widely recommended by CAM manufacturers, concerns were raised about the fragile nature of that filter (Hoover and Newton, 1991). Recognizing that several newer filter types had become available, Hoover and Newton (1991) sought a more durable membrane filter that would have a reasonably low-pressure drop and excellent surface collection efficiency to provide a high resolution for α-spectroscopy. Performance of the filter options that were considered is described in Table 8.1.

Filter performance characteristics reported in Table 8.1 for α-spectroscopy resolution were based on total collection efficiency for radon decay products and on detection of the 6.0-MeV α-emission of polonium-218 (a naturally occurring decay product of ambient radon-222, which causes interference in instruments used to detect plutonium or uranium isotopes). Figure 8.5 illustrates the differences in α-energy spectra obtained with a very efficient front-surface-collecting filter (the Fluoropore) as compared to the standard Millipore SMWP and to the Whatman-41 fiber filter. Clearly, the Fluoropore filter provides better energy separation than the SMWP or the fiber filter.

Filtration

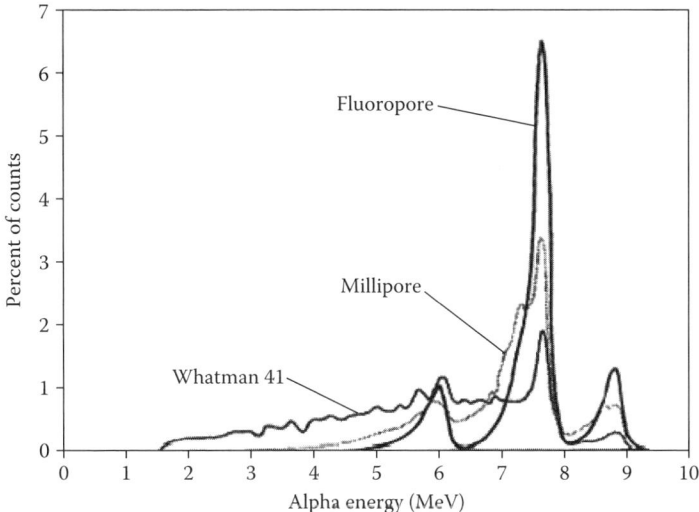

FIGURE 8.5 Illustration of the influence of filter type on the quality of the radon progeny energy spectrum in an α-CAM. The peak on the left is from Bi-212 (6.08 MeV) and Po-218 (6.00 MeV). The peak in the middle is from Po-214 (7.68 MeV). The peak on the right is Po-212 (8.78 MeV). The lower-energy tail of the peak from Bi-212 and Po-218 extends into the region where the α-emissions of Pu-239 (5.2 MeV) and Pu-238 (5.5 MeV) are found. The Millipore Fluoropore Teflon membrane filter provides superior resolution compared to the Millipore SMWP mixed cellulose ester filter. The Whatman-41 cotton fiber filter is not useful for spectroscopy because of the poor resolution it provides.

The fiber filter is not suitable for use in α-CAMs because penetration of collected particles into the filter matrix causes severe broadening of the radon progeny peaks.

The total collection efficiency for radon decay products reported in Table 8.1 is of interest because it influences the amount of background subtraction required in the plutonium region of interest (ROI). Efficiency results for collection of radon progeny were normalized to the amount collected on the Millipore SMWP. The broadening of the α-energy peak for each filter was quantified by two additional factors included in Table 8.1: (1) the full-width-at-half-maximum resolution of the polonium-218 peak (the lowest of the three energy peaks shown in Figure 8.5) and (2) the fraction of the radon decay product counts that occur in the plutonium ROI. There are important differences between the filters in the extent to which the energy spectra of the radon decay product peaks extend into the plutonium ROI (4.3–5.6 MeV). Again, counts in the plutonium region were normalized to the counts obtained on the Millipore SMWP filter during simultaneous sampling. This provides a straightforward comparison of which filter is more suitable for CAM use.

As shown in Table 8.1, resolution for α-spectroscopy of the polonium-218 α-emission at 6.0 MeV ranges from as low as 350 keV (full-width-at-half-maximum) for the Fluoropore 3-μm PTFE membrane filter to >1500 keV for the Whatman-41 cotton cellulose fiber filter. As noted above, the poor resolution associated with the Whatman-41 filter makes filters unsuitable for use in CAMs that are employed to detect plutonium or uranium in the presence of ambient radon decay products. The Versapor 3000 filter (Pall-Gelman, East Hills, NY) consists of an acrylic copolymer on a nonwoven nylon fiber, which provides a lower pressure drop and has a performance similar to that of the Millipore SMWP. The Durapore 5-μm pore size polyvinylidene fluoride membrane filter from Millipore can also be considered, although it provides a poorer spectral quality. All types of fiber filters are unacceptable due to poor spectral quality from burial of particles in the fiber bed. Accumulation of a layer of dust on a filter surface does improve its surface collecting properties by reducing irregularities in the surface, but performance of fiber-type filters in spectroscopic applications is poorer than that of membrane-type filters. Small-pore filters are unacceptable because of their high-pressure drop.

The Fluoropore filter provides the best resolution and thus the least interference with plutonium detection. Spectral performance of the 3-μm pore-size Fluoropore filter is slightly better than that of the 5-μm pore-size Fluoropore filter, but the larger pore size version is an excellent option because it provides a significantly lower pressure drop (less than half) with only a third less spectral quality. As noted above in the section "Filtration Fundamentals," it is important to note that filters are not sieves, and that particles much smaller than the pore size are collected by diffusion and impaction mechanisms. Thus, larger pore filters are preferred in many applications, such as in α-CAMs, because they retain good particle collection efficiency with a lower pressure drop, which allows longer sampling times before the sampling flow rate is decreased by pressure buildup across the filter.

Alpha-CAM users can select from a number of suitable filters, as long as they calibrate their instruments with the selected filter. Filters other than those illustrated in Table 8.1 may also be selected if they can be shown to provide suitable ruggedness, spectral quality, efficiency, and pressure drop. Moore et al. (1993) also provide a useful examination of factors such as filter selection and filter diameter that affect α-particle detection in CAMs.

Considerations of "Front" and "Back" Filter Surface

If the performance characteristics of the front and back surfaces of the filter are not within 5% of each other for the intended purpose of the sample, there should be a readily accomplished means of identifying the appropriate surface for particle collection.

Note that there is little visible distinction between the Teflon membrane collection side and the fiber support side of original versions of the Fluoropore filter that are made with white support fibers, rather than black support fibers. Collection on the fibrous back side of the Fluoropore filters leads to an unacceptably broad α-energy spectrum. In fact, the radon progeny collection characteristics of the support side of the filter are similar to the Whatman-41 filter shown in Figure 8.5. At the request of α-CAM users, the manufacturer developed the color-coded distinction between the front and back sides of this filter so that such sampling errors can now be avoided.

The small-pore membrane filter (Millipore AABP) listed in Table 8.1 has a black collection surface, which makes it easy for the operator to reliably load the correct (dark) side of the filter into the sampling chamber. Conversely, the Teflon membrane filter (Millipore FMLB) listed in Table 8.1 has black support fibers on the *back* side, which means that "the black side is the back side" and that the "gray" side (not the black side) is the collection side. Sampling procedures should clearly document any specific handling requirements related to the collection side for each filter being used. The details of those procedures should be adequately addressed during the training of technicians and analysts.

Even if performance characteristics of the front and back surfaces of a given filter are not different, it is prudent, and frequently necessary (e.g., when samples are collected for electron microscopy or α-spectroscopy), to know which side has been used for collection. Some practitioners write the capital letter "P" with an indelible marker near an edge on the collection side of the filter. The letter P is useful because it can be literally interpreted to mean "particle" side and (more importantly) because it has a unique, right-facing orientation; even if the ink were to leak through to be visible on the back side of the filter, it would be readily evident to the user that the letter is only correctly oriented when viewed from the collection side. Alternative labeling options (to avoid possible confusion of "P" with "b" or "9") are to use the letter "R" (for right side) or the letter "N" (for near side). Some manufacturers provide filters with a printed grid on one side, which (depending on the nature of the filter and the discretion of the user) can be used to designate either the front or the back.

As a more elegant solution, some filters are prepackaged (or mounted by the user) into a cardboard or other type of carrier that supports and remains with the filter, identifies its collection side, provides space for sample number, location, date and time, or other user-written or barcode information, and (in most cases) has a clipped corner or other physical feature that prevents improper insertion or orientation of the collection side in the sampling head.

ANALYTICAL AND RADIOCHEMISTRY ISSUES

If a filter and the collected particle sample are to be separated for an analytical method that requires the particles to remain intact, the user should select a filter medium that can be easily dissolved by a method that will not alter, dissolve, or otherwise interfere with the particles of interest. If particles are to be separated from the filter by methods such as ultrasonic agitation, then it is desirable that the filter media remain intact during that process. If dissolution of both the filter media and the particles is acceptable (or required), then more aggressive digestion methods can be used. Low-pressure drop, cellulose filters are commonly used, and samples can be easily reduced to ash or dissolved for analysis by analytical chemistry or radiochemistry.

TRANSPORTATION AND STORAGE OF FILTER SAMPLES

Filters are sometimes placed into paper envelopes or other packaging for transport or storage before or after radioactivity counting. Care should be taken to account for potential sample loss during packaging and transport. Particles are generally well retained on filter media by van der Waals forces (i.e., the relatively weak attraction between neutral atoms and molecules arising from polarization induced in each by the presence of the other). The potential for loss of material from the filter increases with increasing mass of the collected material. When a filter sampling program is established, or when the conditions of sample collection or mass loading are significantly altered, users should consider the potential impact of sample loss on counting results. It should not be assumed that lack of visible loss of particulate material from the filter is confirmation of the absence of loss. Simple evaluations of sample loss under the conditions of filter use can be made by counting or wiping and counting the total container contents as well as the filter.

Some plastic filter containers (e.g., the so-called "petri-slides") have a secure-fitting lid that includes an inner lip that presses against the perimeter of the collection side of the filter to prevent the filter from touching the lid or otherwise moving within the container in ways that would induce loss of material from the filter. Such containers also include convenient space for sample identification and labeling.

Filters that are only to be counted for γ-radiation can be secured between layers of a sticky tape. For gross-α counting, a particle-laden filter can be placed (collection side up) onto a backing of a sticky tape and secured with a covering layer of ZnS(Ag)-coated plastic scintillation media. Long-term storage of filter samples in such a manner for archival purposes, however, is not always feasible, especially for high-specific-activity, α-emitting radionuclides such as plutonium-238. Radiation damage to the filter, packaging tape, or plastic container may allow release of radioactivity. If long-term retention of filters is needed for programmatic purposes, precautions and care should be taken when handling samples that are aged.

JUSTIFYING A CHANGE IN THE SELECTION OF FILTER MEDIA

There are a number of situations, such as changes in mission or operational conditions, which may warrant reconsideration of the type of filter media being used in an existing air sampling program. As noted in ANSI N13.1 (1999), decisions on changing the filter media should be based on a number of considerations, including potential loss of continuity between historical and future sampling results, potential impacts on vacuum system performance, requirements for retesting of the aerosol collection or counting system, requirements for revision and approval of documentation, retraining requirements for workers, and potential impacts on secondary uses of the filter samples, such as periodic chemical analyses for process control. Although some filter media date back many decades, their continued use is not justified simply because of historical precedents. Similarly, a change in filter media is not justified simply because a new alternative exists. As new filter types become available, comparisons such as those illustrated in Table 8.1 can be made to ensure that appropriate filter types are selected for sampling radioactive aerosol particles.

AVOIDING UNEXPECTED CHANGES IN FILTER MEDIA

It is prudent to be aware that filter media delivered from the stockroom, the supplier, or the manufacturer are not always what is intended or required. Packaging and labeling errors sometimes occur after receipt of filters by the user organization as a result of poor procedures for storing and issuing the media; although rare, filters are sometimes improperly packaged and labeled by the manufacturer or supplier. On rare occasions, critical performance characteristics of filter media *as manufactured* have been known to occur.

An example of an unexpected change in filter composition and performance was observed for the Versapor 3000 media during the research on filter media for α-CAMs conducted by Hoover and Newton (1991, 1992) and Hoover et al. (1997a, 1997b). In the course of that work, the Versapor 3000 was found to have a rugged construction (homogeneous microporous cellulose ester polymers formed around a cellulose web) that made it stronger than the Millipore SMWP, provided a favorable pressure drop performance similar to the SMWP, and resulted in a superior spectral quality approaching that of the Fluoropore FMLB filter. The Versapor 3000 also readily dissolves in nitric acid, making it useful for applications involving radiochemistry. Unfortunately, the availability of the cellulose web support material ceased following implementation of Kyoto Protocol restrictions on the use of fluorocarbons. An alternate support web was substituted by the manufacturer, which resulted in similar strength and similar flow rate performance, but a surface that is less smooth and therefore less suitable for α-spectroscopy, and a composition that is less dissolvable for radiochemistry applications. Thus, the current version of the filter, which is designated as the Versapor 3000T, has features that are somewhat useful for α-spectroscopy, but less ideal than those of the original version.

As noted in previous sections of this chapter, incorrect or altered filter media can sometimes be identified by changes in filter appearance, pressure drop, spectral results, or dissolution behavior. Expert groups, such as the Air Monitoring Users Group (AMUG, www.amug.us), play an important role in fostering the exchange of information about such quality assurance issues among air monitoring practitioners, researchers, regulators, and manufacturers.

FILTRATION FOR NUCLEAR AIR AND GAS TREATMENT

Air sampling practitioners will find it useful to understand the major considerations of filtration for air and gas treatment in nuclear power and nuclear weapons facilities, including differences in terminology and testing procedures compared to air sampling. Air sampling measurements by the practitioner provide important input to selection and validation of appropriate filtration systems.

The American Society of Mechanical Engineers (ASME) AG-1 *Code on Nuclear Air and Gas Treatment* (ASME, 2003) addresses systems and equipment used in nuclear facilities to capture and control radioactive aerosols and gases. The U.S. DOE *Nuclear Air Cleaning Handbook* (DOE, 2003) specifically addresses those issues for nuclear applications in DOE and National Nuclear Security Administration (NNSA) facilities.

As described in the DOE handbook and summarized in the following paragraphs, commercially available filters are divided into three distinct categories based on how they operate to remove suspended particulate matter from the air passing through them.

Heating, ventilation, and air conditioning (HVAC) filters are the largest category and include media composed of highly porous beds of resin-bonded glass or plastic fibers with diameters ranging from 1 to 40 μm. The fibers act as targets for collecting airborne dust. As their name indicates,

HVAC filters are widely used for air cleaning in mechanical ventilation systems. They are almost all single-use, disposable items and are used in all sectors of the nuclear industry, including as prefilters to reduce the amount of coarse dust reaching more efficient filters located downstream.

High-efficiency particulate air (HEPA) filters are single-use, disposable, extended-medium, dry-type filters with (1) a minimum particle removal efficiency of no less than 99.97% for 0.3-µm particles, (2) a maximum resistance, when clean, of 1.0 in water gauge when operated at 1000 cfm, and (3) a rigid casing that extends the full depth of the medium. HEPA filters are widely used throughout all phases of the nuclear industry. The filtering medium of HEPA filters is thinner and more compressed, and contains smaller diameter fibers than HVAC filters. Filters of identical construction and appearance, but having a filtering medium with a retention efficiency of 99.9995% for 0.1 µm particles, are referred to as ultralow penetration aerosol (ULPA) filters.

Industrial cleanable cloth filters are the third category of commercial air filters. As the designation indicates, these filters have built-in mechanisms (such as a mechanical shaker or a reverse-flow and air-pulse feature) for periodically cleaning the filtering surfaces of accumulated dust. Unlike the first two types, industrial cleanable cloth filters rely on building a thick layer of dust on the surface of the cloth to provide a high-efficiency filtering medium. Cleanable cloth filters are used in the nuclear industry for ore processing and refining and for similar tasks involving high concentrations of coarse mineral dusts. The category of industrial cleanable cloth filters also comprises special types of particulate filters for chemical and combustion operations, including deep beds of sand in graded granular sizes, deep beds of glass fibers, and stainless steel membranes formed from compressed and sintered granules or fibers. Stainless-steel membrane filters operate like industrial cleanable cloth filters in that they depend on a dust layer for high-efficiency particle removal and must be cleaned periodically, usually by reverse compressed air jets.

Similar to considerations for air sampling, it is essential that the filter media for nuclear air-cleaning applications be compatible with the aerosol of interest and the environmental conditions, and that appropriate steps be taken to prevent damage to the filter media or leakage. As noted in the DOE handbook, factors to be considered in the design of an air-cleaning system to provide satisfactory working conditions for personnel and to prevent the release of radioactive or toxic substances to the atmosphere include

- Nature of the contaminants to be removed (e.g., radioactivity, toxicity, corrosivity, particle size and size distribution, particle shape, and viscosity)
- Heat (e.g., process heat and fire)
- Moisture (e.g., sensible humidity process vapors and water introduced from testing)
- Radiation (e.g., personnel exposure and material suitability considerations)
- Other environmental conditions to be controlled
- Upset or accident or accident hazard considerations

As specified in the ASME AG-1 Code, the test aerosol for certification of HEPA filters for nuclear applications is thermally generated dioctyl phthalate (DOP). Because of concerns for possible reduced collection efficiency of small particles at higher flow rates, AG-1 currently restricts the face velocity of HEPA media to not more than 2.5 cm s^{-1} (5 ft min^{-1}). Alderman et al. (2008) evaluated the collection efficiency of several AG-1 ASME-certified, deep-pleat, nuclear-grade HEPA filters and found little reduction in collection efficiency at face velocities up to 4.5 cm s^{-1}. MPPS values were generally found to be in the range of 110–130 nm, and decreased slightly with increasing face velocity.

BUILDING FILTRATION AND AIR CLEANING

Air sampling practitioners will also find it useful to understand the major considerations for more general applications of building filtration and air cleaning. The National Institute for Occupational Safety and Health (NIOSH) (2003) *Guidance for Filtration and Air Cleaning Systems to Protect*

Building Environments from Airborne Chemical, Biological, or Radiological Attacks provides an informative summary of building filtration and air-cleaning issues, along with a comprehensive list of key references. It is a companion to the NIOSH (2002) *Guidance for Protecting Building Environments from Airborne Chemical, Biological, or Radiological Attacks*. As described further below, filtration generally refers to removal of liquid or solid particles, and air cleaning generally refers to removal of gases and vapors.

FILTRATION CONSIDERATIONS

The American Society of Heating, Ventilating, and Air-Conditioning Engineers (ASHRAE) has developed a standard test method (current version: ANSI/ASHRAE 52.2-2007) to describe and rate air filters according to their collection efficiency over three particle size ranges: Range 1 (0.3–1 μm), Range 2 (1–3 μm), and Range 3 (3–10 μm). In a position statement on bioterrorism, the National Air Filtration Association (NAFA, 2003) discusses the use of these tests in preparing for bioterrorism events, and notes that the standard provides a tool for selecting an appropriate filter for a specific application. As summarized in Table 8.2, the collection efficiency of a filter in each of the three ranges can be used to assign the filter a Minimum Efficiency Reporting Value (MERV) on a scale of 1–20. Although higher MERV values indicate higher collection efficiencies, the MERVs *are not* the mathematical result of an efficiency formula; interpreting the meaning of an MERV requires

TABLE 8.2
ANSI/ASHRAE Standard 52.2 MERVs for Air Filters

	Average Particle Collection Efficiency in Specified Size Ranges		
MERV	Range 1 0.3–1 μm	Range 2 1–3 μm	Range 3 3–10 μm
1	—	—	<20%
2	—	—	<20%
3	—	—	<20%
4	—	—	<20%
5	—	—	20–35%
6	—	—	35–50%
7	—	—	50–70%
8	—	—	>70%
9	—	<50%	>85%
10	—	50–65%	>85%
11	—	65–80%	>85%
12	—	>80%	>90%
13	<75%	>90%	>90%
14	75–85%	>90%	>90%
15	85–95%	>90%	>90%
16	>95%	>95%	>95%
17	>99.97%	—	
18	>99.99%	—	
19	>99.999%	—	
20	>99.9999%	—	

Source: Adapted from ASHRAE, *Method of Testing General Ventilation Air-Cleaning Devices for Removal Efficiency by Particle Size*, ANSI/ASHRAE Standard 52.2, American Society of Heating, Refrigerating, and Air Conditioning Engineers, Atlanta, GA, 2007.

consulting the table of MERV definitions. Because filter collection efficiency increases for mechanical filters (but not for electrostatic filters) as the filter loads (*cf.*, NIOSH, 2003), collection efficiency in each size range for MERV determination is a composite efficiency based on testing of clean and incrementally loaded filters.

Practical considerations include ensuring that air is not allowed to flow around, rather than through, a filter. The guidance notes that electrostatic filters (composed of polarized fibers) may provide an economical alternative for obtaining higher collection efficiency without the higher pressure drops of higher MERV filters. Caution is advised, however, because the collection efficiency of electrostatic filters can degrade with time to the underlying mechanical collection efficiency, especially if used at high humidity or exposed to certain chemical vapors, gases, or aerosols (*cf.*, Martin and Moyer, 2000; Moyer and Bergman, 2000).

AIR-CLEANING CONSIDERATIONS

As noted in the NIOSH (2003) guidance, choosing the appropriate sorbent or sorbents for a gaseous or vapor airborne contaminant is a complex decision that should involve consultation with a qualified professional and consideration of many factors. Important factors include

- *Specificity of the sorbent for the contaminant.* Not all sorbents work for all contaminants. For example, natural zeolites are hydrophilic and are effective for organic solvents and for low-molecular-weight halides such as chlorinated fluorocarbons (CFCs). They do not have an affinity for nonpolar molecules. Synthetic zeolites can be made to be hydrophilic if they are alumina-rich or hydrophobic if they are silica-rich. Activated carbon is nonpolar, and therefore effective for organic vapors. However, activated carbon is not effective for volatile, low-molecular-weight gases such as ammonia and formaldehyde. Activated carbon can be impregnated with various materials for use with specific chemical contaminants. For example, impregnation with copper/silver salts enables collection of phosgene, chlorine, and arsine; and impregnation with phosphoric acid enables collection of ammonia.
- *Compatibility of pore size.* Porous sorbents cannot adsorb molecules larger than their pore size.
- *Contact time.* The agent must spend sufficient time in the vicinity of the sorbent to be captured and controlled by either physical adsorption or chemisorption. This involves considerations such as velocity of airflow, depth of the collection layer, and available surface area of the sorbent. Thinner or less porous layers require lower flow rates to provide adequate residence time for collection by diffusion.
- *Sorbent capacity.* Breakthrough can occur if the collection capacity of the sorbent is exceeded. Calculations of the expected lifetime for the sorbent should be made based on knowledge of the concentration of the agent being presented or measurements of breakthrough should be made. Fatah et al. (2000) provide information about detection methods that may be used to detect concentrations or breakthrough of chemical, biological, or radiological agents. A mathematical model for predicting chemical breakthrough (Wood and Snyder, 2007) is available on the NIOSH website.
- *Competition for sorption sites.* Silica gel and alumina are effective for trapping polar compounds, but they have high affinity for water. Breakthrough may occur under wet or humid conditions.

FILTRATION FOR RESPIRATORY PROTECTION

Air sampling practitioners will also find it useful to understand the major considerations related to filtration for respiratory protection. An important underlying concept for respiratory protection is the assigned protection factor (APF), which is defined as the minimum anticipated protection provided by a properly functioning respirator or class of respirators to a given percentage of properly fitted and trained users (*cf.*, NRC, 2006). Consistent with this definition, a respirator with an APF of 10 would be acceptable for use in work conditions where the airborne concentration of an aerosol of concern is not more than 10 times the allowable concentration. As described further below, there are a number of specific legal requirements for selection and use of respiratory protection with different degrees of APF. Similar to the role of air sampling for process and building filtration, results of air sampling measurement by the practitioner contribute to decision-making about the appropriate selection and use of respiratory protection.

LEGAL REQUIREMENTS FOR RESPIRATORY PROTECTION

Legal requirements for respiratory protection in nuclear and other radiation-related facilities operated under the auspices of the U.S. Nuclear Regulatory Commission (NRC) are provided in 10 CFR 20 (NRC, 2006). Of particular interest are Subpart H on *Respiratory Protection and Controls to Restrict Internal Exposure in Restricted Areas* and Annex A on *Assigned Protection Factors for Respirators*. Table 8.3 presents the APF values that are currently assigned by the NRC to the various types of respirators. The NRC stipulates that if the licensee assigns or permits the use of respiratory protection equipment to limit the intake of radioactive material, then (1) the licensee shall use only respiratory protection equipment that is tested and certified by the NIOSH except as otherwise noted; and (2) if the licensee wishes to use equipment that has not been tested or certified by NIOSH, or for which there is no schedule for testing or certification, the licensee shall submit an application to the NRC for authorized use of this equipment except as provided in this part. The NRC further requires that the application include evidence that the material and performance characteristics of

---

TABLE 8.3
Assigned Protection Factors for Respirators as Presented in Appendix A to the 10 CFR 20 Standards for Protection against Radiation

	Operating Mode	APF[a]
I. Air-Purifying Respirators [Particulate[b] Only][c]		
Filtering-facepiece disposable[d]	Negative pressure	[d]
Facepiece, half[e]	Negative pressure	10
Facepiece, full	Negative pressure	100
Facepiece, half	PAPRs	50
Facepiece, full	PAPRs	1000
Helmet/hood	PAPRs	1000
Facepiece, loose fitting	PAPRs	25
II. Atmosphere-Supplying Respirators (Particulate, Gases, and Vapors[f])		
1. Air-line respirator		
Facepiece, half	Demand	10
Facepiece, half	Continuous flow	50
Facepiece, half	Pressure demand	50
Facepiece, full	Demand	100
Facepiece, full	Continuous flow	1000
Facepiece, full	Pressure demand	1000
Helmet/hood	Continuous flow	1000
Facepiece, loose fitting	Continuous flow	25
Suit	Continuous flow	[g]

TABLE 8.3 (continued)
Assigned Protection Factors for Respirators as Presented in Appendix A to the 10 CFR 20 Standards for Protection against Radiation

	Operating Mode	APF[a]
2. Self-contained breathing apparatus (SCBA)		
Facepiece, full	Demand	100[h]
Facepiece, full	Pressure demand	10,000[i]
Facepiece, full	Demand, recirculating	100[h]
Facepiece, full	Positive–pressure, recirculating	10,000[i]
III. Combination Respirators		
Any combination of air-purifying and atmosphere-supplying respirators	APF for type and mode of operation as listed above	

Source: NRC, *Standards for Protection against Radiation, Code of Federal Regulations*, Title 10—*Energy*, Chapter I—*Nuclear Regulatory Commission*, Part 20, Revised, January 1, 2006, www.access.gpo.gov, U.S. Government Printing Office, Washington, DC, 2006.

[a] These APFs apply only in a respiratory protection program that meets the requirements of this part. They are applicable only to airborne radiological hazards and may not be appropriate to circumstances when chemical or other respiratory hazards exist instead of, or in addition to, radioactive hazards. Selection and use of respirators for such circumstances must also comply with Department of Labor regulations. Radioactive contaminants for which the concentration values in Table 8.1, Column 3 of Appendix B to Part 20 are based on internal dose due to inhalation may, in addition, present external exposure hazards at higher concentrations. Under these circumstances, limitations on occupancy may have to be governed by external dose limits.

[b] Air-purifying respirators with APF < 100 must be equipped with particulate filters that are at least 95% efficient. Air-purifying respirators with APF = 100 must be equipped with particulate filters that are at least 99% efficient. Air-purifying respirators with APFs > 100 must be equipped with particulate filters that are at least 99.97% efficient.

[c] The licensee may apply to the Commission for the use of an APF greater than 1 for sorbent cartridges as protection against airborne radioactive gases and vapors (e.g., radioiodine).

[d] Licensees may permit the use of this type of respirator by individuals who have not been medically screened or fit-tested on the device provided that no credit be taken for their use in estimating intake or dose. It is also recognized that it is difficult to perform an effective positive or negative pressure preuse user seal check on this type of device. All other respiratory protection program requirements listed in § 20.1703 apply. An APF has not been assigned for these devices. However, an APF equal to 10 may be used if the licensee can demonstrate a fit factor of at least 100 by use of a validated or evaluated, qualitative or quantitative fit test.

[e] Under-chin type only. No distinction is made in this Appendix between elastomeric half-masks with replaceable cartridges and those designed with the filter medium as an integral part of the facepiece (e.g., disposable or reusable disposable). Both types are acceptable so long as the seal area of the latter contains some substantial type of seal-enhancing material such as rubber or plastic, the two or more suspension straps are adjustable, the filter medium is at least 95% efficient, and all other requirements of this Part are met.

[f] The APFs for gases and vapors are not applicable to radioactive contaminants that present an absorption or submersion hazard. For tritium oxide vapor, approximately one-third of the intake occurs by absorption through the skin so that an overall protection factor of 3 is appropriate when atmosphere-supplying respirators are used to protect against tritium oxide. Exposure to radioactive noble gases is not considered a significant respiratory hazard, and protective actions for these contaminants should be based on external (submersion) dose considerations.

[g] No NIOSH approval schedule is currently available for atmosphere-supplying suits. This equipment may be used in an acceptable respiratory protection program as long as all the other minimum program requirements, with the exception of fit testing, are met (i.e., § 20.1703).

[h] The licensee should implement institutional controls to assure that these devices are not used in areas immediately dangerous to life or health (IDLH).

[i] This type of respirator may be used as an emergency device in unknown concentrations for protection against inhalation hazards. External radiation hazards and other limitations to permitted exposure such as skin absorption shall be taken into account in these circumstances. This device may not be used by any individual who experiences perceptible outward leakage of breathing gas while wearing the device.

the equipment are capable of providing the proposed degree of protection under anticipated conditions of use; and that this must be demonstrated either by licensee testing or on the basis of reliable test information.

Legal requirements for use of respiratory protection against radioactive and other toxic or hazardous materials in general industry are provided by OSHA in 29 CFR 1910.134 (OSHA, 1998). OSHA stipulates that (1) a respirator shall be provided to each employee when such equipment is necessary to protect the health of such employee and (2) the employer shall provide the respirators that are applicable and suitable for the purpose intended.

APFs for Different Types of Respiratory Protection

The four types of respiratory protection described in the NRC and OSHA standards are

1. *Air-purifying respirators (APRs)*, which use an air-purifying filter, cartridge, or canister to remove specific air contaminants from the ambient air being inhaled by the user;
2. *Powered air-purifying respirators (PAPRs)*, which use a blower to provide filtered air to the user via a facepiece, hood, or helmet;
3. *Supplied-air respirators (SARs), also known as airline respirators*, which provide filtered air on demand, continuously, or via a pressure demand or other positive-pressure mode to the user from a source of breathing air that is not designed to be carried by the user; and
4. *Self-contained breathing apparatus (SCBA)*, which supply breathing air to the user from a source that is carried by the user.

Differences currently exist between some APF values assigned by the NRC (and listed here in Table 8.3) and values assigned by OSHA. Those differences arose in 2006 when OSHA revised its respiratory protection standard to add definitions and requirements for APFs (OSHA, 2006). Notably, OSHA now assigns an APF of 50 to full-facepiece negative-pressure APRs, full-facepiece PAPRs, and full-facepiece demand SCBA, whereas NRC still assigns an APF of 100 to those types of respirators. In making the revisions, OSHA stated in particular that proper respirator selection using APFs is an important component of an effective respiratory protection program. Accordingly, OSHA concluded that the APFs provided in the revised standard are necessary to protect employees who must use respirators for protection from airborne contaminants. As described both technically and historically in the OSHA (2006) documentation, the final APFs were issued after thoroughly reviewing the available literature, including chamber simulation studies and workplace protection factor studies, comments submitted to the record, and hearing testimony. The final APFs adopted by OSHA provide employers with critical information to use when selecting respirators for employees exposed to atmospheric contaminants found in general industry, construction, shipyards, longshoring, and marine terminal workplaces.

Classes of Filter Media for Air-Purifying Respirators

Considerations for determining the appropriate APF for a given type of respirator include both the manner in which the respirator fits (e.g., fit-factor issues to prevent leakage) and the intrinsic filtration efficiency of filter media. Commercially available filters for use in APRs are designated as nine classes according to three categories each for the following two criteria:

- Resistance to degradation
  - *N* indicating not resistant to the presence of oil particles in the work environment
  - *R* indicating oil resistant
  - *P* indicating oil proof

Filtration

- Efficiency
  - *95* indicating greater than 95% efficient
  - *99* indicating greater than 99% efficient
  - *100* indicating greater than 99.97% efficient

Thus, a class P99 filter can be expected to filter more than 99% of particles at ~0.3 μm diameter at 85 L min^{-1} flow rate, and is suitable for use in an environment where oil particles such as lubricants, cutting oils, or glycerine are present. The footnotes associated with Table 8.3 illustrate some of the efficiency requirements stipulated by the NRC for particulate air filters that are acceptable for use with different types of respirators (e.g., air-purifying respirators with an APF = 100 must be equipped with particulate air filters that are at least 99% efficient). The footnotes also contain other important caveats such as considerations for total body exposure (not just inhalation exposure) for airborne radioactive contaminants such as tritium oxide vapor (i.e., where approximately one-third of the bodily intake of tritium occurs by absorption through the skin so that an overall protection factor of only 3 is appropriate when atmosphere-supplying respirators are used to protect against tritium oxide).

An extensive listing of respiratory protection information can be found on the OSHA website at www.osha.gov/SLTC/respiratoryprotection. Guidance for the selection and use of respirators are also provided in the *American National Standard Practices for Respiratory Protection* (ANSI/AIHA, Z88.2-1992). Details of NIOSH responsibilities and activities in the area of respiratory protection, including information on the regulatory requirements and procedures for approval of respiratory protective devices by NIOSH, can be found on the NIOSH National Personal Protection Technology Laboratory (NPPTL) website: www.cdc.gov/niosh/npptl/topics/respirators. The *Respirator Selection Logic* published by NIOSH (2005) is particularly informative. The selection logic includes recommendations for use of suitable chemical sorbent cartridges or canisters in addition to the appropriate type of particulate filter when exposure to gas or vapor contaminants is expected. Considerations of selection of appropriate gas or vapor sorbent media for worker protection are also relevant to considerations for selection of media for conducting air sampling.

Recent work by NIOSH and others to demonstrate the efficiency of filter media for nanoparticles is relevant to filtration issues for radon decay products and other ultrafine radioactive aerosols (*cf.*, Rengasamy et al., 2007, 2008a, 2008b, 2009a, 2009b; Shaffer and Rengasamy, 2009). As was illustrated in Figure 8.3, filter efficiency performance and MPPS issues for ultrafine particles are consistent with classical physical principles of diffusion, impaction, characteristics of the filter media, and airflow conditions.

CLARIFICATION INTO DUST MASKS, SURGICAL MASKS, AND OTHER MEDIA

As a final consideration for understanding filtration-related aspects of respiratory protection, it is useful to clarify some issues regarding "dust masks," "surgical masks," and other media that are sometimes mistaken for "approved" respiratory protection methods.

Dust masks are not respirators. The term "dust mask" applies to devices that resemble filtering facepiece respirators (FFRs), but have not been approved by NIOSH for respiratory protection against particulate exposure. Dust masks are often used around the home or garden or inadvertently in the workplace by users who are not formally trained in procedures for respiratory protection. Although most home improvement and hardware stores do provide FFRs that have been certified by NIOSH and are clearly labeled as NIOSH-approved, many of the commercial products found in such stores are simple dust masks. In a recent evaluation of the particle collection efficiency of commercially available dust masks, Rengasamy et al. (2008a) found the collection efficiency of some masks to be less than 25%.

Surgical masks are not respirators. The historical development and use of surgical masks was to protect the sterility of the surgical field from contamination by fluids or other particles dispersed from the mouths or noses of the surgeons and other operating room personnel. Surgical masks are not designed to seal tightly to the face. Various types of surgical masks are worn by healthcare personnel to protect against body fluid splashes to the nose and mouth. Some surgical masks have been "cleared" by the Food and Drug Administration (FDA) based on a review by FDA of the manufacturer's test data and proposed claims (FDA, 2004). Three categories of FDA-cleared surgical masks are based on high, moderate, or low filtration efficiencies as determined under specified test conditions. Rengasamy et al. (2009a) conducted a detailed evaluation of five models of FDA-cleared surgical masks and found that filtration efficiency of the masks varied widely. As described by the FDA in their 2008 document on *Personal Protective Equipment and Patient Care* (FDA, 2008), a new category of surgical masks known as the *surgical N95 respirator* is a NIOSH-approved N95 FFR which also meets FDA-required tests for fluid resistance and differential pressure.

Common fabrics such as cotton/polyester shirt material, cotton handkerchief material, and washcloths or other toweling materials are not respirators. The particle-collection efficiency of common materials have, however, been evaluated by investigators such as Cooper et al. (1983). Particle-collection efficiency of common fabric materials is typically very low and leakage around the fabric can be significant. Although of unreliable efficiency, common fabrics may provide some level of protection in areas or in emergencies where respirators are either not available or not routinely used.

Respirator filters are not surrogate air sampling filters. Although examination of respirator filter media has sometimes been considered as a possible method for estimating potential inhalation exposure of the respirator user, there does not appear to be any published evidence of how the material collected on a respirator filter might compare with actual airborne concentrations for an aerosol of concern.

CONCLUSION

Selection of an appropriate filter for the intended conditions of use should be based on careful considerations including strength and durability, effective "pore size" of the filter as it affects both collection efficiency for the expected particle size and filter resistance to airflow and pressure drop at the intended cross-sectional area of use, expected changes in pressure drop due to dust loading, background radioactive content of the filter, radiation self-absorption within the filter, surface-collection or depth-collection properties of the filter when α-spectroscopy is used, chemical compatibility and solubility for analytical processes, and cost. Users are cautioned to be selective in their choice of filter media. Periodic reconfirmation of filter characteristics and performance should be conducted to ensure that the intended and appropriate filter media are being used. Changes in objectives or conditions should also be noted so that appropriate reevaluations of filter selection and use can be made. Applying an informed approach to selection and use of media for air sampling, building filtration, and respiratory protection can ensure that air measurements are relevant to airborne agents of concern and that control technologies are properly selected and used.

REFERENCES

Alderman, S.L., Parsons, M.S., Hogancamp, K.U., and Waggoner, C.A., Evaluation of the effect of media velocity on filter efficiency and most penetrating particle size of nuclear grade high-efficiency particulate air filters, *J. Occup. Environ. Hyg.* 5(11), 713–720, 2008.

ANSI, *Sampling and Monitoring Releases of Airborne Radioactive Substances from the Stacks and Ducts of Nuclear Facilities*, ANSI N13.1, American National Standards Institute, New York, 1999.

ANSI/AIHA, *American National Standard Practices for Respiratory Protection*, ANSI/AIHA Z88.2-1992, American National Standards Institute, New York, 1992.

ASHRAE, *Method of Testing General Ventilation Air-Cleaning Devices for Removal Efficiency by Particle Size*, ANSI/ASHRAE Standard 52.2, American Society of Heating, Refrigerating, and Air Conditioning Engineers, Atlanta, GA, 2007.

ASME, *Nuclear Air and Gas Treatment*, ASME AG-1, American Society of Mechanical Engineers, New York, 2003.

Barrett, L.W. and Rousseau, A.D., Aerosol loading performance of electret filter media, *Am. Ind. Hyg. Assoc. J.* 59, 532–539, 1998.

Brock, T.D., *Membrane Filtration: A User's Guide and Reference Manual*, Science Tech Publishers, Madison, WI, 1983.

Brown, R.C., *Air Filtration: An Integrated Approach to the Theory and Applications of Fibrous Filters*, Pergamon Press, Oxford, 1993.

Cooper, D.W., Hinds, W.C., Price, J.M, Weker, R., and Howell, S.Y., Common materials for emergency respiratory protection: Leakage tests with a Manikin. *Am. Ind. Hyg. Assoc. J.* 44, 720–726, 1983.

Davies, C.N., *Air Filtration*, Academic Press, London, 1973.

DOE, *Nuclear Air Cleaning Handbook*, Slawski, J., Fretthold, J.K., Hargan, M.R., and Zavadoski, R.W., Eds., DOE-HDBK-1169-2003, U.S. Department of Energy, Washington, DC, 2003.

Fatah, A.A., Barrett, J.A., Arcilesi Jr, R.D., Ewing, K.J., Lattin, C.H., and Helinski, M.S., *Guide for the Selection of Chemical Agent and Toxic Industrial Material Detection Equipment for Emergency First Responders*, NIJ Guide 100-00, National Institute of Justice, U.S. Department of Justice, Washington, DC, 2000. Available at http://www.ncjrs.org/pdffiles1/nij/184449.pdf. Accessed September 04, 2009.

FDA, Guidance for Industry and FDA Staff: Surgical Masks—Premarket Notification [501(k)] Submissions; Guidance for Industry and FDA., U.S. Food and Drug Administration, Washington, DC, 2004. Available at http://www.fda.gov/MedicalDevices/DeviceRegulationandGuidance/GuidanceDocuments/ucm072549.htm. Accessed November 03, 2009.

FDA, *Personal Protective Equipment [PPE] and Patient Care*, U.S. Food and Drug Administration, Washington, DC, 2008. Available at http://www.fda.gov/cdrh/ppe/masksrespirators.html. Accessed September 04, 2009.

Hickey, E.E., Stoetzel, G.A., and Olsen, P.C., *Air Sampling in the Workplace*, NUREG-1400, U.S. Nuclear Regulatory Commission, Washington, DC, 1991.

Higby, D.P., *Effects of Particle Size and Velocity on Burial Depth of Airborne Particles in Glass Fiber Filters*, PNL-5278, Battelle Pacific Northwest Laboratory, Richland, WA, 1984.

Hoover, M.D., Fencl, A.F., and Newton, G.J., *Performance Qualification Tests of the Millipore AW-19 Membrane Filter for Air Sampling*, ITRI-970501, Lovelace Respiratory Research Institute, Albuquerque, NM, 1997a.

Hoover, M.D., Fencl, A.F., and Newton, G.J., *Laboratory Test Results on Collection Efficiency of the Gelman Versapor 3000 Filter for 0.3 µm Aerodynamic Diameter Particles*, ITRI-970502, Lovelace Respiratory Research Institute, Albuquerque, NM, 1997b.

Hoover, M.D. and Newton, G.J., Technical bases for selection and use of filter media in continuous air monitors for alpha-emitting radionuclides, in *Annual Report of the Inhalation Toxicology Research Institute for 1990–1991*, LMF-134, National Technical Information Service, Springfield, VA, pp. 16–19, 1991.

Hoover, M.D. and Newton, G.J., Update on selection and use of filter media in continuous air monitors for alpha-emitting radionuclides, in *Annual Report of the Inhalation Toxicology Research Institute for 1991–1992*, LMF-138, National Technical Information Service, Springfield, VA, pp. 5–7, 1992.

Lee, K.W. and Mukund, R., Filter collection, in *Aerosol Measurement: Principles, Techniques, and Applications*, 2nd ed., Baron, P.A. and Willeke, K., Eds., John Wiley & Sons, New York, 2001.

Lindeken, C.L. Use of natural airborne radioactivity to evaluate filters for alpha air sampling, *Am. Ind. Hyg. Assoc. J.* 22(4), 232–237, 1961.

Lindeken, C.L., Petrock, F.K., Phillips, W.A., and Taylor, R.D., Surface collection efficiency of large-pore membrane filters, *Health Phys.* 10, 495–499, 1964.

Lippmann, M., Filters and filter holders, in *Air Sampling Instruments for Evaluation of Atmospheric Contaminants*, 9th ed., Cohen, B.S. and McCammon, C.S., Eds., American Conference of Governmental Industrial Hygienists, Cincinnati, OH, 2001.

Liu, B.Y.H., Personal communication. University of Minnesota, Minneapolis, MN, 1992.

Liu, B.Y.H., Pui, D.Y.H., and Rubow, K.L., Characteristics of air sampling filter media, in *Aerosols in the Mining and Industrial Work Environments*, Ann Arbor Science, Ann Arbor, MI, pp. 989–1038, 1983.

Martin, S.B. and Moyer, E.S., Electrostatic respirator filter media: Filter efficiency and most penetrating particle size effects, *App. Occup. Environ. Hyg.* 15, 609–617, 2000.

Moore, M.E., McFarland, A.R., and Rodgers, J.C., Factors that affect alpha particle detection in continuous air monitors. *Health Phys.* 65, 69–81, 1993.

Moyer, E.S. and Bergman, M.S., Electrostatic N95 respirator filter media efficiency degradation resulting from intermittent sodium chloride aerosol exposure. *Appl. Occup. Environ. Hyg.* 15, 600–608, 2000.

NAFA, *NAFA position statement on Bio-Terrorism—Using ASHRAE Standard 52.2 in Preparedness for Bio-Terrorism*, National Air Filtration Association, Virginia Beach, VA, 2003. Available at http://www.nafahq.org/LibaryFiles/Articles/Article004.htm. Accessed September 04, 2009.

NIOSH, *Guidance for Protecting Building Environments from Airborne Chemical, Biological, or Radiological Attacks*, National Institute for Occupational Safety, Health Department of Health and Human Services, Publication No. 2002-139, Cincinnati, OH, 2002. Available at http://www.cdc.gov/niosh/docs/2002-139/. Accessed September 04, 2009.

NIOSH, *Guidance for Filtration and Air-Cleaning Systems to Protect Building Environments from Airborne Chemical, Biological, or Radiological Attacks*, National Institute for Occupational Safety and Health, Department of Health and Human Services, Publication No. 2003-136, Cincinnati, OH, 2003. Available at: http://www.cdc.gov/niosh/docs/2003-136/. Accessed September 04, 2009.

NIOSH, *Respirator Selection Logic 2004*, National Institute for Occupational Safety and Health, Department of Health and Human Services, Publication No. 2005-100, Cincinnati, OH, October 2004. Available at http://www.cdc.gov/niosh/docs/2005-100/. Accessed September 04, 2009.

NRC, *Standards for Protection against Radiation, Code of Federal Regulations*, Title 10—*Energy*, Chapter I—*Nuclear Regulatory Commission*, Part 20, Revised, January 1, 2006, www.access.gpo.gov, U.S. Government Printing Office, Washington, DC, 2006.

OSHA, *Occupational Safety and Health Standards, Personal Protective Equipment, Respiratory Protection*, 29CFR1910.134, Washington, DC, 1998.

OSHA, *Assigned Protection Factors; Final Rule*, Federal Register, 71:50121–50192, 2006.

Rengasamy, S., Verbofsky, R., King, W.P., and Shaffer, R.E., Nanoparticle penetration through NIOSH-approved N95 filtering facepiece respirators, *J. Int. Soc. Res. Prot.* 24, 49–59, 2007.

Rengasamy, S., King, W.P., Eimer, B., and Shaffer, R.E., Filtration performance of NIOSH-approved N95 and P100 filtering-facepiece respirators against 4–30 nanometer size particles, *J. Occup. Environ. Hyg.* 5, 556–564, 2008a.

Rengasamy, S., Eimer, B.C., and Shaffer, R.E., Nanoparticle filtration performance of commercially available dust masks, *J. Int. Soc. Res. Prot.* 25, 27–41, 2008b.

Rengasamy, S., Miller, A., Eimer, B.C., and Shaffer, R.E., Filtration performance of FDA-cleared surgical masks, *J. Int. Soc. Res. Prot.* 26, 54–70, 2009a.

Rengasamy, S., Eimer, B.C., and Shaffer, R.E., Comparison of nanoparticle filtration performance of NIOSH-approved and CE-marked particulate filtering facepiece respirators, *Ann. Occup. Hyg.* 53(2), 117–128, 2009b.

Shaffer, R.E. and Rengasamy, S., Respiratory protection against airborne nanoparticles. *J. Nanopart. Res.*, DOI 10.1007/s11051-009-9649-3, published online May 23, 2009.

Spurny, K.R., Ed., *Advances in Aerosol Filtration*, Lewis Publishers, Boca Raton, FL, 1998.

Wood, G.O. and Snyder, J.L., Estimating service lives of organic vapor cartridges III: Multiple vapors at all humidities, *J. Occup. Environ. Hyg.* 4(5), 363–374, 2007. An online version of the model is available at http://www.cdc.gov/niosh/npptl/multivapor/multivapor.html. Accessed October 09, 2009.

9 Behavior of Radon and its Decay Products

Phillip Jenkins

CONTENTS

Introduction 181
Uranium Decay Series and ^{222}Rn 182
    Solution to Decay Equations Using a Recurrence Formula 184
    Potential α-Energy Concentration 185
    Radon Sample in a Closed Container or Defined Volume of Air 186
    Effect of Deposition on Radon Progeny Concentration 187
    Radon Progeny Buildup on a Filter 190
Thorium Decay Series and ^{220}Rn 193
    Solution to Decay Equations Using a Recurrence Formula 194
    Potential α-Energy Concentration 196
    Thoron Sample in a Closed Container or Defined Volume of Air 197
    Effect of Deposition on Thoron Progeny Concentration 198
    Thoron Progeny Buildup on a Filter 198
Actinium Decay Series and ^{219}Rn 201
    Solution to Decay Equations Using a Recurrence Formula 202
    Potential α-Energy Concentration 204
    Actinon Sample in a Closed Container or Defined Volume of Air 204
    Effect of Deposition on Actinon Progeny Activities 204
    Actinon Progeny Buildup on a Filter 205
References 208

INTRODUCTION

It is important that users of continuous air monitors (CAMs) and other air samplers understand the behavior of radon and radon decay products (hereafter called "radon progeny") for two reasons: (1) radon and radon progeny can be a significant source of radiation dose to workers, and documentation of occupational exposure to these sources may be required and (2) radon progeny collected on the filter of a CAM or of another air sampling device can interfere with the measurement of long-lived α-particle emitters and thus must be measured accurately in order to subtract the interference from the signal due to the long-lived α-emitters. Increases in the count rate of a CAM caused by an increase in radon progeny concentration could cause the CAM to alarm, initiating an evacuation of the work area. It is important, therefore, to be able to identify correctly changes in radon and radon progeny concentrations in order to avoid unnecessary disruptions due to evacuations and also creating unnecessary concern and anxiety in the work force. At the same time, depending on the situation, the increase in radon progeny itself indicated by the CAM may be the radiation of interest and may also be a valid reason for evacuation or the initiation of ventilation and/or respiratory protection. As for other air sampling devices, counts from radon decay products may, if not properly accounted for, lead to erroneously high results with perhaps similar consequences regarding unnecessary work force disruptions and anxiety.

Radon is a chemical element and a member of the family of noble gases, which means that it is chemically inert for all practical purposes. With 86 protons, it is the most massive of the noble gases. Of interest here are the three isotopes of radon that are members of naturally occurring ^{222}Rn decay series (hereafter in this chapter, the word "radon" is used to refer only to the specific radionuclide ^{222}Rn) is a member of the Uranium Decay Series, which is headed by ^{238}U. ^{220}Rn, historically called "thoron," is a member of the Thorium Decay Series, which is headed by ^{232}Th. ^{219}Rn, historically called "actinon," is a member of the Actinium Decay Series, which is headed by ^{235}U.

Uranium can be found in most soil and rock. Because ^{238}U comprises a large percentage of natural uranium, radon is the most abundant of the three isotopes considered here. Thorium is also found in soil and rock and is prevalent in some areas. With a half-life of 55.6 s, however, usually only the thoron that is formed near the surface of soil or other thorium-bearing materials can escape to the air. Still, thoron can be a health concern and an interference to CAM measurements. This is particularly true where thorium is specifically processed or stored.

Because ^{235}U comprises only 0.7% by weight (2.2% by activity) of natural uranium, actinon is found in low abundance in nature compared to ^{222}Rn. As long as the natural abundance is not changed, it is of little concern. However, there are situations where the natural abundances have been disturbed by human activities, making actinon and/or its decay products measurable in air, possibly an interference for CAMs and even at concentrations large enough to be of concern for personnel protection. Therefore, the user of a CAM should be aware of the radiological properties of actinon; hence, they are included in the discussion here.

Eisenbud and Gesell (1997) describe sources of radon, thoron, and actinon. Evans published an excellent discussion of the behavior of radon progeny (Evans, 1969). The treatment here extends the work of Evans to include thoron and actinon progeny, as well as the use of a recurrence formula that makes the calculations of concentration simpler and more efficient.

URANIUM DECAY SERIES AND ^{222}Rn

The Uranium Decay Series consists of 14 radionuclides headed by ^{238}U, which has a half-life of 4.468×10^9 years (Browne, 1986). All of the decay products of ^{238}U, even ^{234}U with a half-life of 2.454×10^5 years (Browne, 1986), have existed long enough that they are in secular equilibrium with ^{238}U. Therefore, for every becquerel (or curie) of ^{238}U on Earth, there is also somewhere on Earth a becquerel (or curie) of each of the radioactive decay products of ^{238}U (discounting materials that may have left Earth as a result of various space programs). The entire Uranium Decay Series is listed in Table 9.1.

Uranium is ubiquitous in nature and can be found to some degree in most rock and soil. Therefore, thorium (^{230}Th and ^{234}Th) and radium (^{226}Ra) are also found in most rock and soil. Because of differences in chemical forms, some limited separation can occur due to groundwater movement and other natural processes. But because radon is a noble gas, a significant portion of it is free to migrate away from its parent ^{226}Ra. A sufficient quantity of radon migrates out of rock and soil to produce a radon concentration in outdoor ambient air that ranges from 4 to 19 Bq m^{-3} (0.1–0.5 pCi L^{-1}) (Eisenbud and Gesell, 1997) and averages about 15 Bq m^{-3} (0.4 pCi L^{-1}) in the United States at ground level. Soil gas containing an elevated concentration of radon can migrate into buildings or residences where it is not dispersed as readily as it would be outdoors, which can lead to an elevated concentration indoors. Further, buildings where residues of ^{226}Ra exist due to past or present processing activities have the potential for an elevated concentration of radon.

When radon separates from its parent ^{226}Ra, radon and its short-lived decay products form a decay series, which is a subgroup of the entire uranium series. This subgroup is depicted in Figure 9.1. The short-lived decay products of radon are ^{218}Po, ^{214}Pb, ^{214}Bi, and ^{214}Po; hereafter, called "radon progeny." This subgroup effectively ends with ^{210}Pb, which has a half-life of 22 years. This half-life is much larger than those of radon and radon progeny. Therefore, when radon and radon progey decay completely, the resulting activity or concentration of ^{210}Pb is very much smaller than those of

Behavior of Radon and its Decay Products

TABLE 9.1
Uranium Decay Series[a]

Nuclide	Half-Life	Mode of Decay
^{238}U	4.468×10^9 years	α
^{234}Th	24.1 days	β
^{234m}Pa	1.17 min	β
^{234}U	2.454×10^5 years	α
^{230}Th	7.54×10^4 years	α
^{226}Ra	1600 years	α
^{222}Rn	3.825 days	α
^{218}Po	3.05 min[b]	α
^{214}Pb	26.8 min	β
^{214}Bi	19.7 min[b]	β
^{214}Po	164 μs	α
^{210}Pb	22.3 years	β
^{210}Bi	5.013 days	β
^{210}Po	138.4 days	α
^{206}Pb	Stable	

[a] Browne (1986) unless noted otherwise.
[b] Friedlander (1964).

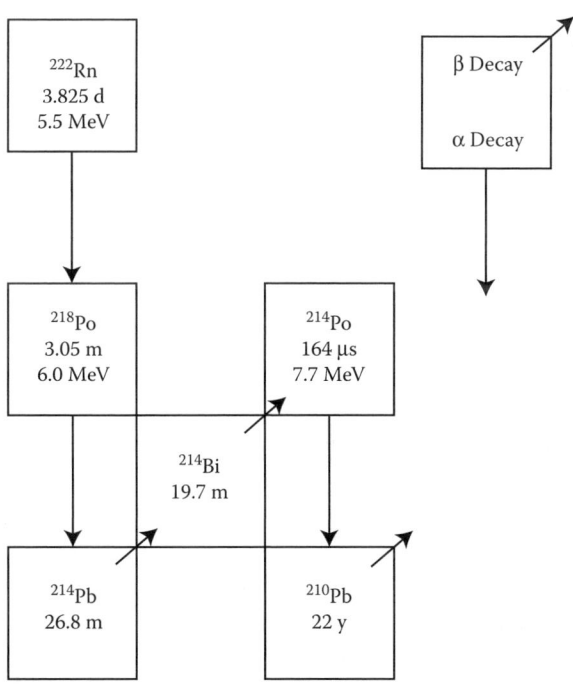

FIGURE 9.1 ^{222}Rn Decay Series.

the radon and radon progeny and in most cases the resulting ^{210}Pb and the remainder of the Uranium Decay Series are of no concern. Exceptions to this can occur in situations where ^{210}Pb and the remainder of the series can build up on filters or other surfaces over long periods of time, but such situations are beyond the scope of this chapter.

SOLUTION TO DECAY EQUATIONS USING A RECURRENCE FORMULA

The equations that describe the decay and ingrowth of radon and radon progeny are often referred to as the "Bateman equations." (Bateman, 1910). The equations permit the user to calculate the activity or air concentration at time t of the various chain members from the decay of the initial quantity of the parent radon. These equations can be used for any set of linear serial transformations and are often applied to the serial decay of a chain of radionuclides, such as the Uranium Decay Series. The equations as published by Bateman tend to grow large and tedious beyond the second transformation or, in this case, the second radionuclide in the decay series. A recurrence formula that can be used to make the Bateman equations more efficient was published by Hamawi (Hamawi, 1971) and discussed by Scherpelz and DesRosiers (1981). Use of the recurrence formula simplifies the equations significantly and makes them more efficient for use in computer programs, spreadsheets, internal codes used by radiation monitors, and so on. The use of the recurrence formula for modeling the decay and ingrowth of radon progeny was discussed by Jenkins (Jenkins, 2002).

Initially considered here are the equations that express the concentrations of radon and radon progeny in air as a function of time with no losses (e.g., ventilation or plateout of particulate radon progeny onto surfaces) except for radioactive decay. Because ^{214}Po has an extremely short half-life, it is customary to assume that it is always in secular equilibrium with ^{214}Bi (i.e., the activity of ^{214}Po is always equal to that of ^{214}Bi). As far as the mathematics are concerned, it can be assumed that the α-particle emitted by ^{214}Po immediately follows the β-decay of ^{214}Bi and that ^{214}Bi decays directly to ^{210}Pb emitting a β-particle and an α-particle in the process. Therefore, no terms related to ^{214}Po appear in the equations. Also, because ^{210}Pb has a long half-life compared with those of the radon progeny, it is assumed in the equations that the decay chain ends with ^{214}Bi.

The application of the recurrence formula leads to the following process. First, the factors designated as f_{ii} are defined, which are merely the exponential terms describing the radioactive decay of each individual radionuclide.

$$f_{00} = \exp(-\lambda_0 t), \tag{9.1}$$

$$f_{11} = \exp(-\lambda_1 t), \tag{9.2}$$

$$f_{22} = \exp(-\lambda_2 t), \tag{9.3}$$

$$f_{33} = \exp(-\lambda_3 t), \tag{9.4}$$

where λ_0, λ_1, λ_2, and λ_3 are the decay constants for ^{222}Rn, ^{218}Po, ^{214}Pb, and ^{214}Bi, respectively (s^{-1}), and t is time (s).

The f_{ij} factors are defined as follows:

$$f_{01} = \frac{(f_{00} - f_{11})}{(\lambda_1 - \lambda_0)} \text{ (for decay of initially present } ^{222}\text{Rn to } ^{218}\text{Po)}, \tag{9.5}$$

$$f_{12} = \frac{(f_{11} - f_{22})}{(\lambda_2 - \lambda_1)} \text{ (for decay of initially present } ^{218}\text{Po to } ^{214}\text{Pb)}, \tag{9.6}$$

$$f_{23} = \frac{(f_{22} - f_{33})}{(\lambda_3 - \lambda_2)} \text{ (for decay of initially present } ^{214}\text{Pb to } ^{214}\text{Bi)}, \quad (9.7)$$

$$f_{02} = \frac{(f_{01} - f_{12})}{(\lambda_2 - \lambda_0)} \text{ (for decay of initially present } ^{222}\text{Rn to } ^{214}\text{Pb)}, \quad (9.8)$$

$$f_{13} = \frac{(f_{12} - f_{23})}{(\lambda_3 - \lambda_1)} \text{ (for decay of initially present } ^{218}\text{Po to } ^{214}\text{Bi)}, \quad (9.9)$$

$$f_{03} = \frac{(f_{02} - f_{13})}{(\lambda_3 - \lambda_0)} \text{ (for decay of initially present } ^{222}\text{Rn to } ^{214}\text{Bi)}. \quad (9.10)$$

The recurrence nature of this process can easily be seen from this sequence of equations. Once all the "f factors" have been defined, then solving for the concentrations of radon and radon progeny is simple.

$$C_0 = C_{0,0} f_{00}, \quad (9.11)$$

$$C_1 = C_{0,0} \lambda_1 f_{01} + C_{1,0} f_{11}, \quad (9.12)$$

$$C_2 = C_{0,0} \lambda_1 \lambda_2 f_{02} + C_{1,0} \lambda_2 f_{12} + C_{2,0} f_{22}, \quad (9.13)$$

$$C_3 = C_{0,0} \lambda_1 \lambda_2 \lambda_3 f_{03} + C_{1,0} \lambda_2 \lambda_3 f_{13} + C_{2,0} \lambda_3 f_{23} + C_{3,0} f_{33}, \quad (9.14)$$

where $C_{0,0}$, $C_{1,0}$, $C_{2,0}$, and $C_{3,0}$ are the initial concentrations (or activities) of ^{222}Rn, ^{218}Po, ^{214}Pb, and ^{214}Bi, respectively (Bq m^{-3} or Bq) and C_0, C_1, C_2, and C_3 are the concentrations (or activities) of ^{222}Rn, ^{218}Po, ^{214}Pb, and ^{214}Bi, respectively (Bq m^{-3} or Bq), as a function of time.

POTENTIAL α-ENERGY CONCENTRATION

A concept that has been in use since about 1957 is the collective concentration in air of the radon progeny (Evans, 1969). When radon and radon progeny are inhaled, the resulting radiation dose to the lung is primarily due to the radon progeny and not to radon itself. Further, the dose to the lung is primarily from the emissions of α-particles, with the β-particles and γ rays contributing a small percentage of the total dose. Finally, significant fractions of the radon progeny that are inhaled are retained in the lung and, due to their short half-lives, they decay before they can be expelled from the lung. Because of these factors, a unit called the "Working Level (WL)" was proposed that expresses the potential alpha energy concentration (PAEC) in air of radon progeny. The WL unit is defined as any combination of the short-lived decay products of radon in 1 L of air that will result in the ultimate emission of 1.3×10^5 MeV of α-particle energy (USNRC, 2008). At the time that the WL unit was proposed, the occupational maximum permissible concentration (MPC) of radon in air, with the radon progeny in secular equilibrium, was 3700 Bq m^{-3} (100 pCi L^{-1}). The WL unit was defined such that it was approximately equal to the PAEC of the radon progeny at the occupational MPC at that time. The International System (SI) unit for PAEC is J m^{-3}. One WL is equivalent to 20.8 µJ m^{-3}. The PAEC can be calculated from the individual radon progeny concentrations by either of the following two equations:

$$\text{PAEC}(\mu\text{J m}^{-3}) = 0.000578 C_1 + 0.00285 C_2 + 0.00210 C_3, \quad (9.15)$$

where C_1, C_2, and C_3 are the concentrations of ^{218}Po, ^{214}Pb, and ^{214}Bi, respectively (Bq m^{-3}) and

$$\text{PAEC(WL)} = 0.00103C_1 + 0.00507C_2 + 0.00373C_3, \qquad (9.16)$$

where C_1, C_2, and C_3 are the concentrations of ^{218}Po, ^{214}Pb, and ^{214}Bi, respectively (pCi L^{-1}).

The equilibrium equivalent concentration, EEC, is defined as the activity concentration of radon that would result *if* the radon progeny were in secular equilibrium with the radon and have the same PAEC as the nonequilibrium mixture to which the EEC refers. This quantity is expressed in the unit Bq m^{-3} (or pCi L^{-1}). The EEC can be derived from the individual concentrations of the radon progeny using the following formula:

$$\text{EEC} = 0.105C_1 + 0.516C_2 + 0.379C_3. \qquad (9.17)$$

The equilibrium factor, F, is the ratio of the EEC to the actual radon gas concentration;

$$F = \frac{\text{EEC}}{C_0}. \qquad (9.18)$$

The concentration of each of the radon progeny can be expressed as a fraction of the concentration of radon;

$$F_i = \frac{C_i}{C_0}. \qquad (9.19)$$

This ratio is frequently referred to in the radon literature. For example, an oft-quoted mean indoor ratio for all the progeny relative to the radon concentration is 1.0/0.5/0.3/0.2 where each ratio, beginning with radon itself is found using Equation 9.19 (NCRP, 1984).

The equilibrium factor (Equation 9.18) can be calculated from the following:

$$F = 0.105F_1 + 0.516F_2 + 0.379F_3. \qquad (9.20)$$

The dose from radon progeny can be related to the potential α-energy exposure, which is the integral over time of the PAEC. The traditional unit of potential α-energy exposure is the Working Level Month (WLM). The potential α-energy exposure in WLM is calculated by multiplying the PAEC in WL by the number of hours worked in an environment with that value of PAEC and dividing by 170 working hours per month. The current occupational Derived Air Concentration (USNRC, 2008) for exposure to radon and radon progeny in equilibrium is 1110 Bq m^{-3} (30 pCi L^{-1}) for radon and 1/3 WL for radon progeny. Therefore, the annual occupational limit for potential α-energy exposure is 1/3 WL times 12 working months, or 4 WLM. The SI unit for potential α-energy exposure is J h m^{-3}. One WLM is equivalent to 3.6 mJ h m^{-3}.

RADON SAMPLE IN A CLOSED CONTAINER OR DEFINED VOLUME OF AIR

One method of measuring radon is by capturing a grab sample of filtered air containing radon in a scintillation cell and quantifying the resulting α-emissions from ^{222}Rn, ^{218}Po, and ^{214}Po after secular equilibrium is established. While the measurement method is outside the scope of this chapter, the behavior of radon and its progeny in a closed container, such as a scintillation cell, is instructive here. It is assumed that the sampled air containing radon is filtered to prevent radon progeny or any other α-emitting radionuclides that might be present in the air from entering the container. Therefore, $C_{1,0}$, $C_{2,0}$, and $C_{3,0}$ in Equations 9.11 through 9.14 are zero, thus simplifying these equations. (Note that it is still necessary to calculate all of the values of f_{ij} even though some of them are not used in Equations 9.11 through 9.14 under this assumption.)

Mathematically, this is similar to following a defined volume of air as it moves through a work area, if it is assumed that (1) the initial concentrations of radon progeny in the volume of air are zero, (2) there is no loss of radon progeny due to deposition, (3) there is no loss due to dilution, and (4) any loss due to radon or radon progeny migrating out of the volume are compensated by radon or radon progeny migrating into the volume from surrounding air. Actually, the primary interests here are the relative concentrations of radon and radon progeny; hence, loss due to dilution can be ignored if it is assumed that dilution would affect all of the radionuclides equally.

The decay and ingrowth over time of the radon and radon progeny are shown in Figure 9.2. On the y-axis are shown the activities of radon and radon progeny expressed as a fraction of the original activity of radon in a container or a defined volume of air. It can be seen from Figure 9.2 that ^{218}Po, with a half-life of only 3.05 min, comes into secular equilibrium with the radon quickly. After about 3 h, the next radionuclide in the series, ^{214}Pb, comes into secular equilibrium with the radon, and after about 4 h all of the radon progeny are in secular equilibrium with the radon. The concept of PAEC is not applicable to a closed container, as the radon progeny are likely deposited on the inside surface of the container and not in the air; however, it is applicable to a defined volume of air. Therefore, the equilibrium factor, F, is shown in Figure 9.2 where it is seen to increase to a value of 1 after about 4 h. After this time, the radon progeny remain in secular equilibrium with the radon and decrease in activity with the half-life of radon.

EFFECT OF DEPOSITION ON RADON PROGENY CONCENTRATION

Radon may enter a room or work area from several different sources, such as from residues from current or past operations or from natural emanation from soil or building materials. For simplicity, here it is assumed that radon, without progeny, enters a defined volume of air from a single source. As this defined volume of air moves through a work area, a portion of the radon progeny becomes deposited on various surfaces and is removed from the air. As in the previous subsection, loss of radon or radon progeny due to ventilation is not considered here.

Consider a defined volume of air containing some initial concentration of radon with no progeny. As this volume of air travels through a work area, the radon progeny concentrations increase toward secular equilibrium as described in the previous section. If there are no losses of the radon progeny

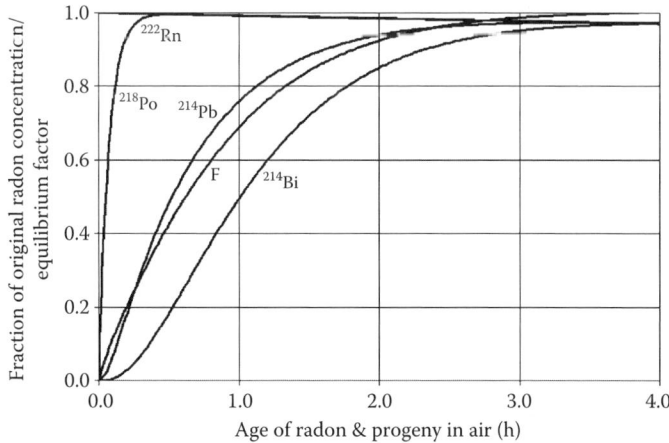

FIGURE 9.2 Radon in a defined volume of air or closed container, showing ingrowth of radon progeny with time, no deposition.

due to deposition on surfaces, then the concentrations of the radon progeny are described by Figure 9.2. However, it is more realistic to assume that radon progeny deposit on surfaces depending on several factors such as the surface areas the air encounters as it travels through the work area and the concentration of aerosols in the air. The equations presented in the previous section can be applied to this situation with only minor modifications. To simulate this, a rate of deposition is added to the decay factors for the radon progeny in Equations 9.1 through 9.10. In other words, in these equations each λ_i, $i = 1$–3, is replaced by ω_i, where

$$\omega_i = \lambda_i + \delta_i \tag{9.21}$$

and δ_i is the deposition rate in the same unit of reciprocal time as λ_i, such as (s^{-1}). Note that the λ_i values in Equations 9.12 through 9.14 are unchanged as these denote terms of production rates of the radon progeny not loss rates.

For the purposes here, example values of deposition rate (δ_i) are assumed in order to demonstrate how the radon progeny and equilibrium factor behave. It is customary to speak in terms of the "age of the air." In this context, the age of the air is either the time since pure radon has been introduced into the air from some source of ^{226}Ra or the time since the air has passed through a filter that removed the radon progeny from the air. Therefore, it is really the "age of the radon progeny" that is considered here.

Figure 9.3 shows the behavior of the radon progeny with time if a deposition rate of 1 h^{-1} is assumed for all of the radon progeny. In this case, the concentration of ^{218}Po quickly achieves a value of 92.8% of the radon concentration before decreasing with the half-life of radon. The concentration of ^{214}Pb increases to 55.7% of that of the radon concentration in a little over 2 h before decreasing with the half-life of radon. The concentration of ^{214}Bi increases to 37.7% of that of the radon in about 3 h and then decreases with the half-life of radon. The equilibrium factor increases to a value of 0.54 in about 3 h and remains at that value.

If the assumed deposition rate for all of the radon progeny is increased to 10 h^{-1}, then the result obtained is as shown in Figure 9.4. In this case, the concentration of ^{218}Po quickly achieves a value of 57.5% of the radon concentration before decreasing with the half-life of radon. The concentration of ^{214}Pb increases to 7.7% of that of the radon concentration in about 45 min before decreasing with the half-life of radon. The concentration of ^{214}Bi increases to 1.3% of that of the radon in a little less than 1 h and then decreases with the half-life of radon. The equilibrium factor increases to a value of 0.106 in a little less than 1 h and remains at that value.

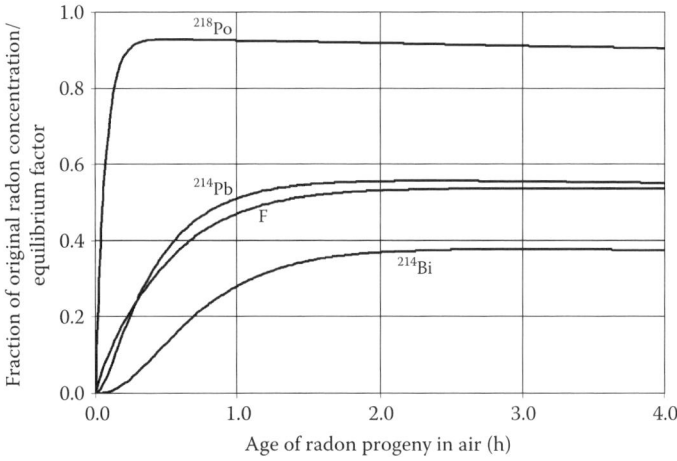

FIGURE 9.3 Radon progeny in air, deposition rate of 1 h^{-1} for all radon progeny.

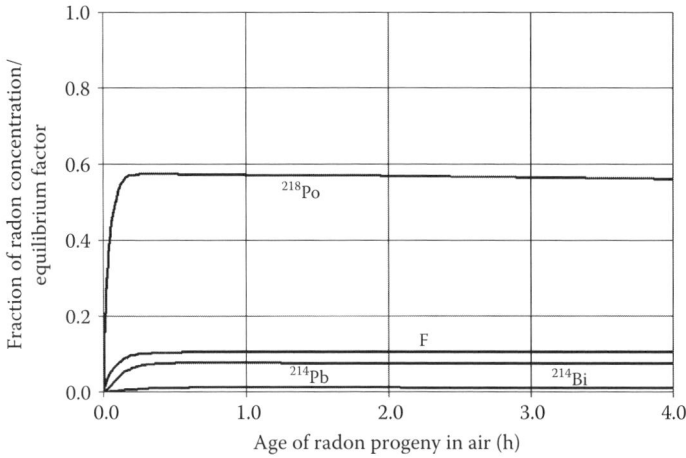

FIGURE 9.4 Radon progeny in air, deposition rate of 10 h^{-1} for all radon progeny.

In a situation where the concentration of aerosol in the air is small, the fraction of ^{218}Po that is not attached to an aerosol (the unattached fraction) may be significant and therefore the deposition rate for ^{218}Po may be greater than that for the other radon progeny. Figure 9.5 shows the result of assuming that the deposition rate of ^{218}Po is 10 h^{-1} and the deposition rate of ^{214}Pb and ^{214}Bi is 1 h^{-1}. In this case, the concentration of ^{218}Po behaves in the same manner as shown in Figure 9.4, increasing quickly to achieve a value of 57.5% of the radon concentration before decreasing with the half-life of radon. However, the concentration of ^{214}Pb increases to 34.5% of that of the radon concentration in about 2 h before decreasing with the half-life of radon. The concentration of ^{214}Bi increases to 23.3% of that of the radon in a little less than 3 h and then decreases with the half-life of radon. The equilibrium factor increases to a value of 0.333 in a little less than 3 h and remains at that value.

If realistic values of deposition rate are known for a given situation, they can be applied to the equations described above to produce a description of the behavior of the radon progeny in that situation. Further, the modeling can be increased in complexity by considering multiple sources of radon entering the airstream at different points in time. Such complex modeling is beyond the scope of this chapter, but could be achieved using the equations presented here.

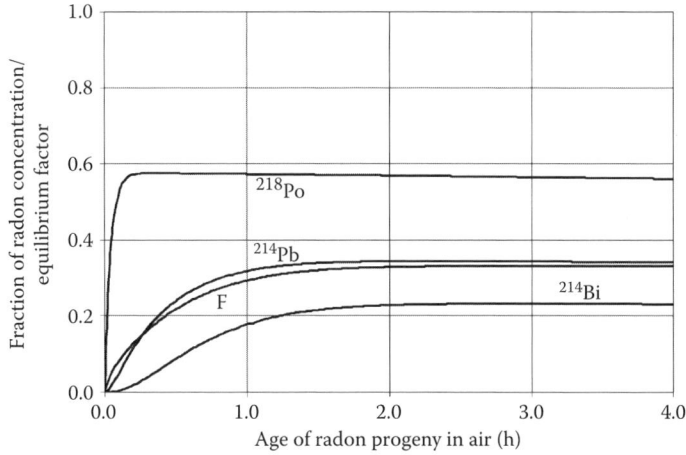

FIGURE 9.5 Radon progeny in air, deposition of 10 h^{-1} for ^{218}Po and 1 h^{-1} for ^{214}Pb and ^{214}Bi.

RADON PROGENY BUILDUP ON A FILTER

Of interest here is the collection of radon progeny on the filter of a CAM or other air samplers and the manner in which the radon progeny behave during the collection process. The equations that describe the buildup of the radon progeny on a filter are similar to Equations 9.1 through 9.14 presented above using the recurrence formula; however, the integrated form of the "f" factors is now used. Here they are called "h" factors. The integrated form is necessary, because the collection rate (concentration × flow rate) must be integrated over the sampling time to produce the collected activity at the end of the sampling time. In this case, it is necessary to account for decay and buildup of radon progeny on the filter during the sampling time. If the material being collected were not radioactive or had a very long half-life, such as ^{238}U, then the integral would simply be (concentration × flow rate × time). Note that since radon is not collected on the filter, only the factors for the radon progeny are included. The initial "h_{ii}" factors are defined as follows:

$$h_{11} = \frac{[1 - \exp(-\lambda_1 t_s)]}{\lambda_1}, \tag{9.22}$$

$$h_{22} = \frac{[1 - \exp(-\lambda_2 t_s)]}{\lambda_2}, \tag{9.23}$$

$$h_{33} = \frac{[1 - \exp(-\lambda_3 t_s)]}{\lambda_3}, \tag{9.24}$$

where λ_1, λ_2, and λ_3 are the decay constants of ^{218}Po, ^{214}Pb, and ^{214}Bi, respectively (s^{-1}) and t_s is the duration of the collection on the filter (s).

Then applying the recurrence formula, the "h_{ij}" factors are calculated as follows:

$$h_{12} = \frac{(h_{11} - h_{22})}{(\lambda_2 - \lambda_1)}, \tag{9.25}$$

$$h_{23} = \frac{(h_{22} - h_{33})}{(\lambda_3 - \lambda_2)}, \tag{9.26}$$

$$h_{13} = \frac{(h_{12} - h_{23})}{(\lambda_3 - \lambda_1)}. \tag{9.27}$$

Then the activities of the radon progeny as a function of collection time, t_s, are calculated as follows:

$$A_1 = F\varepsilon C_1 h_{11}, \tag{9.28}$$

$$A_2 = F\varepsilon(C_1 \lambda_2 h_{12} + C_2 h_{22}), \tag{9.29}$$

$$A_3 = F\varepsilon(C_1 \lambda_2 \lambda_3 h_{13} + C_2 \lambda_3 h_{23} + C_3 h_{33}), \tag{9.30}$$

where A_1, A_2, and A_3 are the activities on the filter of ^{218}Po, ^{214}Pb, and ^{214}Bi, respectively (Bq), C_1, C_2, and C_3 are the concentrations in air ^{218}Po, ^{214}Pb, and ^{214}Bi, respectively (Bq m^{-3}), F is the pump flow rate (m^3 s^{-1}), and ε is the collection efficiency.

Behavior of Radon and its Decay Products

For any specific "age" of the radon progeny in air, the collection and ingrowth of the radon progeny on the filter can be calculated using these equations. The situations considered in the previous section are used here as examples.

The basis of the calculations here are

1. Initial ^{222}Rn concentration of 100 Bq m^{-3} (2.7 pCi L^{-1})
2. Pump flow rate of 0.0472 m^3 s^{-1} (100 cfm)
3. Collection efficiency of 100% ($\varepsilon = 1$)
4. Radon progeny "age" of 1 h

First, consider the situation of no loss of radon progeny due to deposition, where the relative radon progeny concentrations are shown in Figure 9.2. For this example, the 1-h radon progeny concentrations are 99.3, 75.8, and 49.4 Bq m^{-3} for ^{218}Po, ^{214}Pb, and ^{214}Bi, respectively. The activities of the radon progeny on the filter are shown in Figure 9.6 as a function of sample collection time in hours. Also included in Figure 9.6 are the total α-activity (the sum of ^{218}Po and ^{214}Bi/^{214}Po activities) and the total β-activity (the sum of ^{214}Pb and ^{214}Bi activities). The activities on the filter reach maximum values after about 4 h of collection time and remain at constant values afterward as long as the concentrations of the radon progeny in the air remain constant.

For the situation where the deposition rate for the radon progeny is 1 h^{-1} (the corresponding relative concentrations are shown in Figure 9.3) the 1-h concentrations of the radon progeny are 92.5, 51.1, and 28.1 Bq m^{-3} for ^{218}Po, ^{214}Pb, and ^{214}Bi, respectively. The activities of the radon progeny on the filter are shown in Figure 9.7. The curves in Figure 9.7 appear similar to those in Figure 9.6, except that smaller maximum values of the activities are reached.

For the situation where the deposition rate for the radon progeny is 10 h^{-1} (the corresponding relative concentrations are shown in Figure 9.4) the 1-h concentrations of the radon progeny are 57.3, 7.7, and 1.3 Bq m^{-3} for ^{218}Po, ^{214}Pb, and ^{214}Bi, respectively. The activities of the radon progeny on the filter are shown in Figure 9.8. Again, the curves in Figure 9.8 appear similar to those in Figure 9.7, except that even smaller maximum values of the activities are reached.

Finally, for the situation where the deposition rate for ^{218}Po is 10 h^{-1} and for ^{214}Pb and ^{214}Bi is 1 h^{-1}, the corresponding relative concentrations are shown in Figure 9.5, the 1-h concentrations of the radon progeny are 57.7, 31.9, and 17.7 Bq m^{-3} for ^{218}Po, ^{214}Pb, and ^{214}Bi, respectively. The activities of the radon progeny on the filter are shown in Figure 9.9. As expected, the curve for ^{218}Po is the same

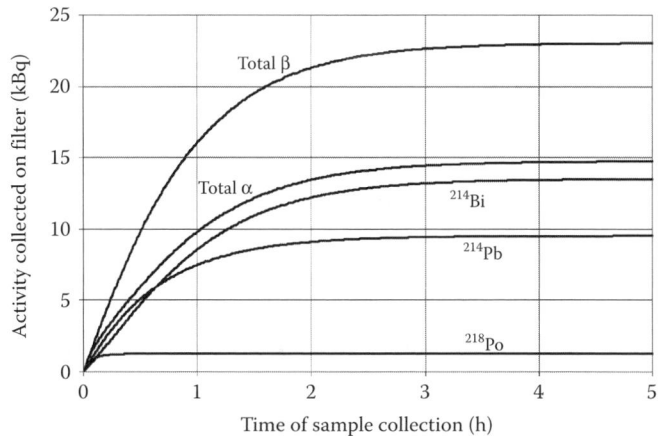

FIGURE 9.6 Collection and ingrowth of radon progeny on filter (scenario from Figure 9.2).

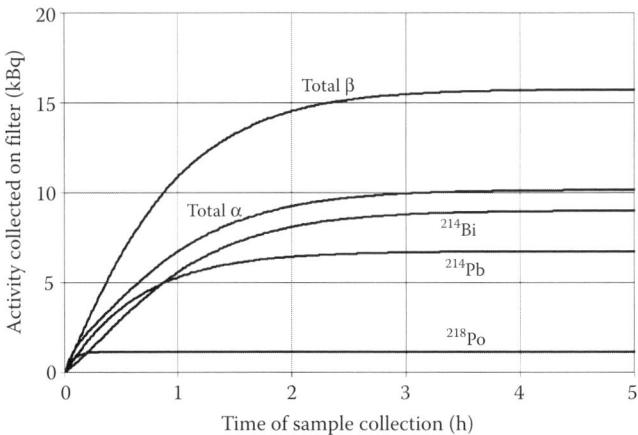

FIGURE 9.7 Collection and ingrowth of radon progeny on filter (scenario from Figure 9.3).

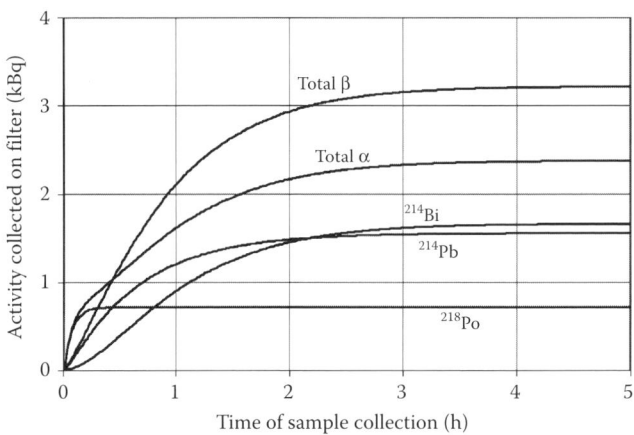

FIGURE 9.8 Collection and ingrowth of radon progeny on filter (scenario from Figure 9.4).

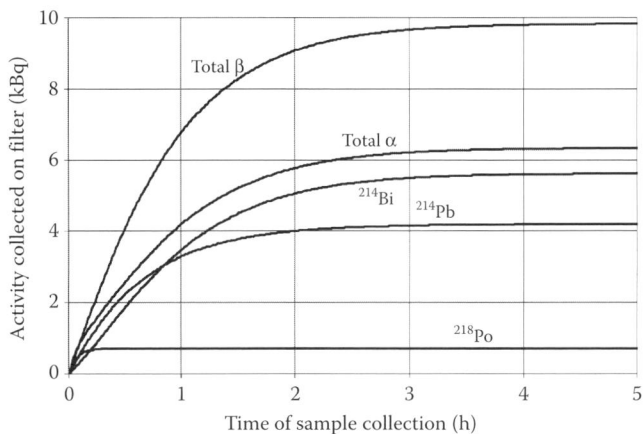

FIGURE 9.9 Collection and ingrowth of radon progeny on filter (scenario from Figure 9.5).

as in Figure 9.8, but the maximum activities of the other two radon progeny are larger in Figure 9.9 due to less loss from deposition.

Using the method demonstrated here, one could produce the values for activity of the radon progeny, as well as total α- and β-activities, collected on a filter for whatever values of the various parameters (radon concentration, deposition rates, age of radon progeny, pump flow rate, and collection efficiency) are of interest.

Throughout this discussion, it was assumed that the radon concentration remains constant. In reality, the radon concentration likely does not remain constant due to changes in meteorological conditions, such as barometric pressure, temperature, soil moisture, wind speed, and perhaps operational factors. It is beyond the scope of the discussion here to consider modeling the behavior of radon progeny in air with the radon concentration varying with time; however, it should be clear that the radon progeny always tend toward an equilibrium condition with radon, following the equations that are presented here with the change in radon concentration reflected in a similar change in the radon progeny concentrations with some time lag. Although not shown here, if the radon is removed from the air, for example, by flushing the scintillation cell considered above with nitrogen, the radon progeny decay to insignificant levels in about 4 h; essentially the reverse of the ingrowth of the radon progeny in the first place. So, in the case of a drop in radon concentration in air, the concentrations of the radon progeny may be greater than that of the radon, and the equilibrium factor can actually be greater than 1, for a period of time until the radon progeny fully respond to the drop in radon concentration.

THORIUM DECAY SERIES AND ^{220}Rn

The Thorium Decay Series consists of 11 radionuclides headed by ^{232}Th, which has a half-life of 1.405×10^{10} years (Browne, 1986). All of the decay products of ^{232}Th have relatively short half-lives, the longest being that of ^{228}Ra (5.75 years) (Browne, 1986) and therefore are in secular equilibrium with ^{232}Th. The entire sequence of the Thorium Decay Series is listed in Table 9.2.

In the Radon Decay Series, the four radionuclides following radon have half-lives significantly smaller than that of radon and therefore can establish a condition of secular equilibrium with the

TABLE 9.2
Thorium Decay Series[a]

Nuclide	Half-Life	Mode of Decay
^{232}Th	1.405×10^{10} years	α
^{228}Ra	5.75 years	β
^{228}Ac	6.13 h	β
^{228}Th	1.913 years	α
^{224}Ra	3.66 days	α
^{220}Rn	55.6 s	α
^{216}Po	0.150 s	α
^{212}Pb	10.64 h	β
^{212}Bi	1.0092 h	α36%, β64%
^{212}Po	298 ns	α
^{208}Tl	3.053 min	β
^{208}Pb	Stable	

[a] Browne (1986).

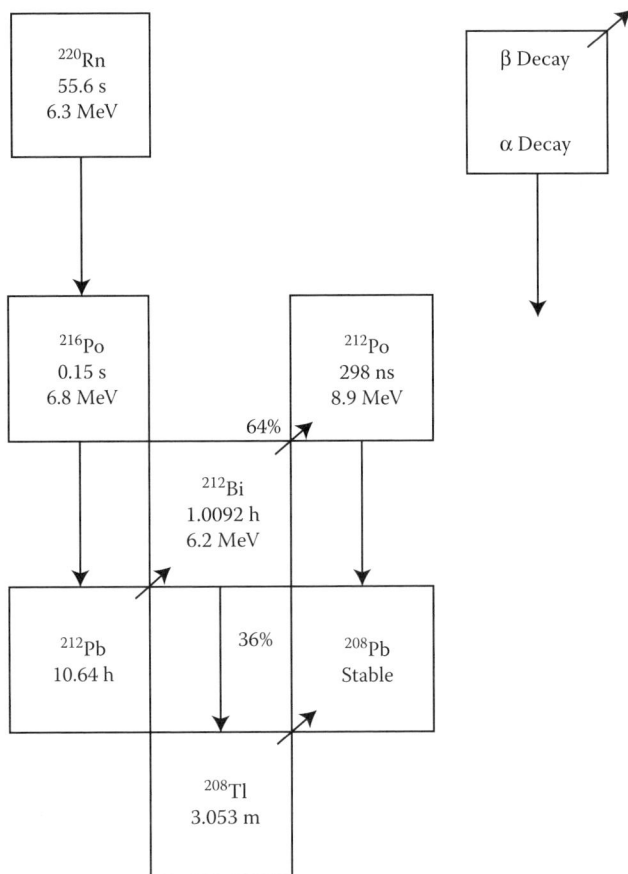

FIGURE 9.10 ^{220}Rn (thoron) Decay Series.

radon. The dynamics of the Thoron Decay Series, shown in Figure 9.10, are very different from that. The half-life of thoron is only 55.6 s. The half-life of its immediate decay product (^{216}Po) is 0.145 s; hence, it establishes a condition of secular equilibrium with thoron. The next radionuclide in the series (^{212}Pb) has a half-life of 10.64 h. It is impossible for ^{212}Pb to establish a condition of secular equilibrium with thoron. However, the progeny of ^{212}Pb have smaller half-lives and establish a condition of transient equilibrium with ^{212}Pb.

Further, the Thoron Decay Series splits into two branches due to the fact that ^{212}Bi decays 64% of the time by β-particle emission and 36% of the time by α-particle emission. Following the emission of a β-particle from ^{212}Bi, the next radionuclide in the series (^{212}Po) has such a short half-life (3×10^{-7} s) that one can consider ^{212}Po always to be in secular equilibrium with ^{212}Bi. The activity or concentration of ^{212}Po is always 64% of that of ^{212}Bi. Mathematically, one could assume that ^{212}Bi emits an α-particle 100% of the time; 36% of the time the α-energy is 6.2 MeV and 64% of the time the α-energy is 8.9 MeV (the α-particle that actually is emitted by ^{212}Po).

SOLUTION TO DECAY EQUATIONS USING A RECURRENCE FORMULA

In the same manner as discussed above for radon decay products, the recurrence formula can be applied to the decay and buildup of thoron and thoron decay products. In this case, there are six radionuclides to consider. However, it is not necessary to consider ^{212}Po in the recurrence formula, because its activity or concentration is always 64% of that of ^{212}Bi, as was explained.

First define the f_{ii} factors that are the exponential terms describing radioactive decay.

$$f_{00} = \exp(-\lambda_0 t), \qquad (9.31)$$

$$f_{11} = \exp(-\lambda_1 t), \qquad (9.32)$$

$$f_{22} = \exp(-\lambda_2 t), \qquad (9.33)$$

$$f_{33} = \exp(-\lambda_3 t), \qquad (9.34)$$

$$f_{44} = \exp(-\lambda_4 t), \qquad (9.35)$$

where λ_0, λ_1, λ_2, λ_3, and λ_4 are the decay constants for ^{220}Rn, ^{216}Po, ^{212}Pb, ^{212}Bi, and ^{208}Tl, respectively (s^{-1}) and t is time (s).

Then applying the recurrence formula, define the following f_{ij} factors:

$$f_{01} = \frac{(f_{00} - f_{11})}{(\lambda_1 - \lambda_0)} \text{ (for decay of initially present } ^{220}\text{Rn to } ^{216}\text{Po)}, \qquad (9.36)$$

$$f_{12} = \frac{(f_{11} - f_{22})}{(\lambda_2 - \lambda_1)} \text{ (for decay of initially present } ^{216}\text{Po to } ^{212}\text{Pb)}, \qquad (9.37)$$

$$f_{23} = \frac{(f_{22} - f_{33})}{(\lambda_3 - \lambda_2)} \text{ (for decay of initially present } ^{212}\text{Pb to } ^{212}\text{Bi)}, \qquad (9.38)$$

$$f_{34} = \frac{(f_{33} - f_{44})}{(\lambda_4 - \lambda_3)} \text{ (for decay of initially present } ^{212}\text{Bi to } ^{208}\text{Tl)}, \qquad (9.39)$$

$$f_{02} = \frac{(f_{01} - f_{12})}{(\lambda_2 - \lambda_0)} \text{ (for decay of initially present } ^{220}\text{Rn to } ^{212}\text{Pb)}, \qquad (9.40)$$

$$f_{13} = \frac{(f_{12} - f_{23})}{(\lambda_3 - \lambda_1)} \text{ (for decay of initially present } ^{216}\text{Po to } ^{212}\text{Bi)}, \qquad (9.41)$$

$$f_{24} = \frac{(f_{23} - f_{34})}{(\lambda_4 - \lambda_2)} \text{ (for decay of initially present } ^{212}\text{Pb to } ^{208}\text{Tl)}, \qquad (9.42)$$

$$f_{03} = \frac{(f_{02} - f_{13})}{(\lambda_3 - \lambda_0)} \text{ (for decay of initially present } ^{220}\text{Rn to } ^{212}\text{Bi)}, \qquad (9.43)$$

$$f_{14} = \frac{(f_{13} - f_{24})}{(\lambda_4 - \lambda_1)} \text{ (for decay of initially present } ^{216}\text{Po to } ^{208}\text{Tl)}, \qquad (9.44)$$

$$f_{04} = \frac{(f_{03} - f_{14})}{(\lambda_4 - \lambda_0)} \text{ (for decay of initially present } ^{220}\text{Rn to } ^{208}\text{Tl)}. \qquad (9.45)$$

Once all the f factors have been defined, the concentrations of thoron and thoron progeny are solved using the following equations:

$$C_0 = C_{0,0} f_{00}, \quad (9.46)$$

$$C_1 = C_{1,0} f_{11} + C_{0,0} \lambda_1 f_{01}, \quad (9.47)$$

$$C_2 = C_{2,0} f_{22} + C_{1,0} \lambda_2 f_{12} + C_{0,0} \lambda_1 \lambda_2 f_{02}, \quad (9.48)$$

$$C_3 = C_{3,0} f_{33} + C_{2,0} \lambda_3 f_{23} + C_{1,0} \lambda_2 \lambda_3 f_{13} + C_{0,0} \lambda_1 \lambda_2 \lambda_3 f_{03}, \quad (9.49)$$

$$C_4 = C_{4,0} f_{44} + 0.36[C_{3,0} \lambda_4 f_{34} + C_{2,0} \lambda_3 \lambda_4 f_{24} + C_{1,0} \lambda_2 \lambda_3 \lambda_4 f_{14} + C_{0,0} \lambda_1 \lambda_2 \lambda_3 \lambda_4 f_{04}], \quad (9.50)$$

$$C_5 = 0.64 C_3, \quad (9.51)$$

where, $C_{0,0}$, $C_{1,0}$, $C_{2,0}$, $C_{3,0}$, and $C_{4,0}$ are the initial concentrations (or activities) of ^{220}Rn, ^{216}Po, ^{212}Pb, ^{212}Bi, and ^{208}Tl, respectively (Bq m^{-3} or Bq) and C_0, C_1, C_2, C_3, C_4, and C_5 are the concentrations (or activities) of ^{220}Rn, ^{216}Po, ^{212}Pb, ^{212}Bi, ^{208}Tl, and ^{212}Po, respectively (Bq m^{-3} or Bq), as a function of time.

POTENTIAL α-ENERGY CONCENTRATION

The concept of equilibrium does not apply to thoron progeny as it does to radon progeny; hence, there is no equilibrium factor for thoron progeny. However, the concept of PAEC does apply to thoron progeny. The PAEC for thoron progeny is defined in the same manner as for radon progeny. One WL of thoron progeny is any combination of the short-lived decay products of ^{220}Rn in 1 L of air that will result in the ultimate emission of 1.3×10^5 MeV of potential α-particle energy (USNRC, 2008). The PAEC can be calculated from the individual thoron progeny concentrations by either of the following two equations:

$$\text{PAEC}(\mu\text{J m}^{-3}) = 0.0692 C_2 + 0.00657 C_3, \quad (9.52)$$

where C_2 and C_3 are the concentrations of ^{212}Pb and ^{212}Bi, respectively, (Bq m^{-3}) and

$$\text{PAEC}(\text{WL}) = 0.123 C_2 + 0.0117 C_3, \quad (9.53)$$

where C_2 and C_3 are the concentrations of ^{212}Pb and ^{212}Bi, respectively (pCi L^{-1}).

Note that the very short-lived thoron progeny (^{216}Po and ^{212}Po) are not included in Equations 9.52 and 9.53 as they contribute insignificantly to the total PAEC. Further, comparing these equations to Equations 9.15 and 9.16 shows that for a given concentration, thoron progeny produce a larger PAEC than radon progeny. This is primarily due to the half-life of ^{212}Pb (10.64 h), which is larger than the half-lives of any of the radon progeny and also partly to the energy of the α-particle emitted by ^{212}Po (8.9 MeV), which is larger than that emitted by ^{214}Po (7.7 MeV).

The dose from thoron progeny can be related to the potential α-energy exposure, which is the integral over time of the PAEC. The potential α-energy exposure in WLM is calculated by multiplying the PAEC in WL by the number of hours worked in an environment with that value of PAEC and dividing by 170 working hours per month. The occupational Derived Air Concentration (USNRC, 2008) for exposure to thoron and thoron progeny in equilibrium is 333 Bq m^{-3} (9 pCi L^{-1}) for thoron or 1 WL for thoron progeny (USNRC, 2008). Therefore, the annual occupational limit for potential α-energy exposure to thoron progeny is 1 WL times 12 working months, or 12 WLM. One WLM is equivalent to 3.6 mJ h m^{-3} in SI units.

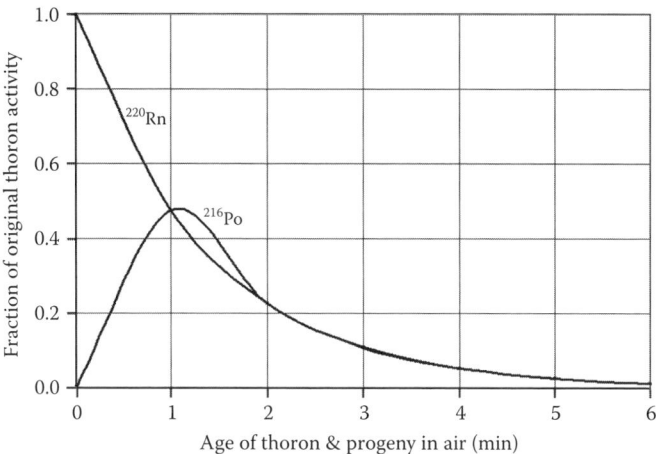

FIGURE 9.11 Thoron in closed container or defined volume of air showing ingrowth of ^{216}Po during the first six minutes, no deposition.

THORON SAMPLE IN A CLOSED CONTAINER OR DEFINED VOLUME OF AIR

A sample of pure thoron in a closed container (such as a scintillation cell) or defined volume of air decays rapidly to insignificant levels. Figure 9.11 shows the decay of thoron and the ingrowth and subsequent decay of ^{216}Po. This is analogous to the decay and ingrowth of radon and radon progeny shown in Figure 9.2, as no loss due to deposition is assumed. However, because thoron and ^{216}Po decay rapidly, Figure 9.11 includes only the first 6 min of the process. As shown in Figure 9.12, ^{212}Pb reaches a maximum value of 0.144% of the initial thoron activity in 9 min, and then decays with its half-life of 10.64 h. ^{212}Bi reaches a maximum value of 0.113% of the initial thoron activity in about 4 h and then decays in transient equilibrium with ^{212}Pb. As discussed above, the activity of ^{212}Po is always 64% of the activity of ^{212}Bi. ^{208}Tl reaches a maximum value of 0.041% of the initial thoron activity in about 4 h and subsequently decays approaching a state of transient equilibrium with ^{212}Pb.

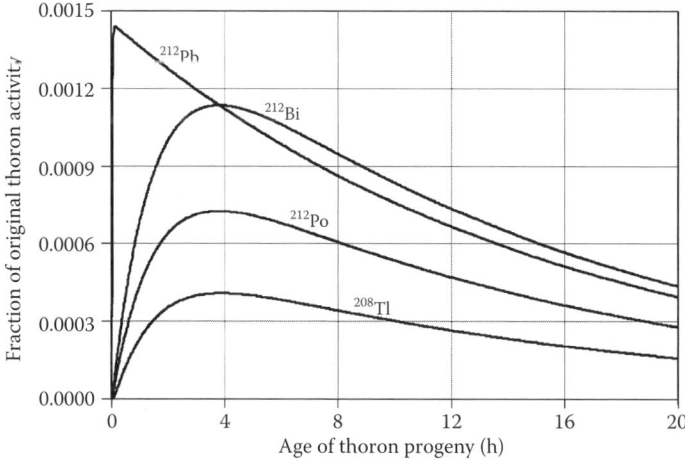

FIGURE 9.12 Ingrowth and decay of thoron progeny in a closed container or defined volume of air, no deposition.

EFFECT OF DEPOSITION ON THORON PROGENY CONCENTRATION

Deposition of thoron progeny can be modeled by adding a deposition rate as discussed in the section for radon progeny. The effect of assuming a deposition rate of 1 h^{-1} on the thoron progeny is shown in Figure 9.13. In this case, ^{212}Pb reaches a maximum value of 0.133% of the initial thoron activity in 5 min, and then decays with its half-life of 10.64 h. ^{212}Bi reaches a maximum value of 0.027% of the initial thoron activity in 45 min and then decays in transient equilibrium with ^{212}Pb. Again, the activity of ^{212}Po is always 64% of the activity of ^{212}Bi. ^{208}Tl reaches a maximum value of 0.009% of the initial thoron activity in 50 min and subsequently decays approaching a state of transient equilibrium with ^{212}Pb.

THORON PROGENY BUILDUP ON A FILTER

Of interest here is the collection of thoron progeny on a filter and the manner in which the thoron progeny behave during the collection process. The equations that describe the buildup of the thoron progeny on a filter are similar to Equations 9.31 through 9.51 presented above, using the recurrence formula; however, the integrated form of the "f" factors is now used. As before, they are referred to as "h" factors. Note that since thoron is not collected on the filter, only the factors for the thoron progeny are included here. The initial "h_{ii}" factors are defined as follows:

$$h_{11} = \frac{[1 - \exp(-\lambda_1 t_s)]}{\lambda_1}, \tag{9.54}$$

$$h_{22} = \frac{[1 - \exp(-\lambda_2 t_s)]}{\lambda_2}, \tag{9.55}$$

$$h_{33} = \frac{[1 - \exp(-\lambda_3 t_s)]}{\lambda_3}, \tag{9.56}$$

$$h_{44} = \frac{[1 - \exp(-\lambda_4 t_s)]}{\lambda_4}, \tag{9.57}$$

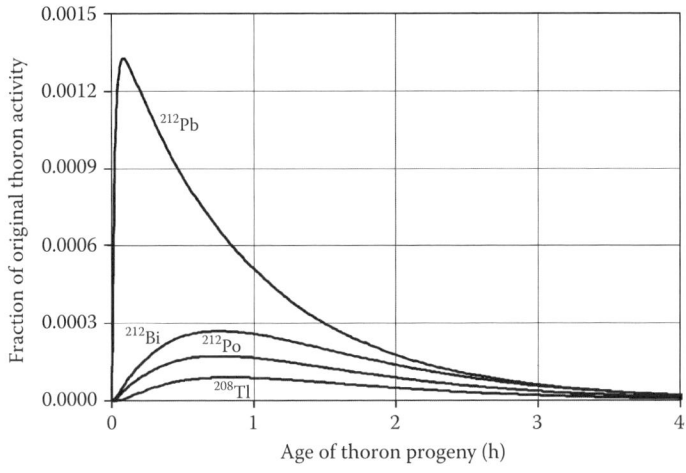

FIGURE 9.13 Thoron progeny in air, deposition rate of 1 h^{-1}.

Behavior of Radon and its Decay Products

where t_s is the duration of the collection on the filter (s) and λ_1, λ_2, λ_3, and λ_4 are the decay constants of ^{216}Po, ^{212}Pb, ^{212}Bi, and ^{208}Tl, respectively (s^{-1}).

Then applying the recurrence formula, the "h_{ij}" factors are calculated as follows:

$$h_{12} = \frac{(h_{11} - h_{22})}{(\lambda_2 - \lambda_1)}, \tag{9.58}$$

$$h_{23} = \frac{(h_{22} - h_{33})}{(\lambda_3 - \lambda_2)}, \tag{9.59}$$

$$h_{34} = \frac{(h_{33} - h_{44})}{(\lambda_4 - \lambda_3)}, \tag{9.60}$$

$$h_{13} = \frac{(h_{12} - h_{23})}{(\lambda_3 - \lambda_1)}, \tag{9.61}$$

$$h_{24} = \frac{(h_{23} - h_{34})}{(\lambda_4 - \lambda_2)}, \tag{9.62}$$

$$h_{14} = \frac{(h_{13} - h_{24})}{(\lambda_4 - \lambda_1)}. \tag{9.63}$$

The activities of the thoron progeny as a function of collection time, t_s, are calculated as follows:

$$A_1 = F\varepsilon C_1 h_{11}, \tag{9.64}$$

$$A_2 = F\varepsilon (C_1 \lambda_2 h_{12} + C_2 h_{22}), \tag{9.65}$$

$$A_3 = F\varepsilon (C_1 \lambda_2 \lambda_3 h_{13} + C_2 \lambda_3 h_{23} + C_3 h_{33}), \tag{9.66}$$

$$A_4 = F\varepsilon (0.36 C_1 \lambda_2 \lambda_3 \lambda_4 h_{14} + 0.36 C_2 \lambda_3 \lambda_4 h_{24} + 0.36 C_3 \lambda_4 h_{34} + C_4 h_{44}), \tag{9.67}$$

$$A_5 = 0.64 A_3, \tag{9.68}$$

where A_1, A_2, A_3, A_4, and A_5 are the activities on the filter of ^{216}Po, ^{212}Pb, ^{212}Bi, ^{208}Tl, and ^{212}Po, respectively (Bq), C_1, C_2, C_3, and C_4 are the concentrations in air of ^{216}Po, ^{212}Pb, ^{212}Bi, and ^{208}Tl, respectively (Bq m^{-3}), F is the pump flow rate (m^3 s^{-1}), and ε is the collection efficiency.

In a similar manner as was discussed for radon progeny, the basis of the calculations here are

1. Initial ^{220}Rn concentration of 100 Bq m^{-3} (2.7 pCi L^{-1})
2. Pump flow rate of 0.0472 m^3 s^{-1} (100 cfm)
3. Collection efficiency of 100% ($\varepsilon = 1$)
4. Thoron progeny "age" of 1 h

First, consider the situation of no loss of thoron progeny due to deposition, where the relative thoron progeny concentrations are shown in Figure 9.12. For this example, the 1-h thoron progeny concentrations are 0.136, 0.0685, and 0.0234 Bq m^{-3} for ^{212}Pb, ^{212}Bi, and ^{208}Tl, respectively. The

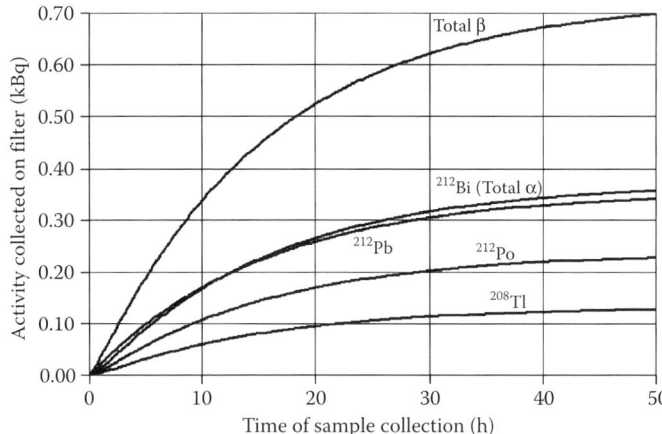

FIGURE 9.14 Thoron progeny collection and ingrowth (scenario from Figure 9.12).

activities of the thoron progeny on the filter are shown in Figure 9.14 as a function of sample collection time in hours. Also included in Figure 9.14 are the total α-activity (which is the same as the ^{212}Bi activity) and the total β-activity (the sum of ^{212}Pb and ^{208}Tl activities). The activities on the filter continue to increase beyond the 50 h of collection time shown in Figure 9.14, reaching maximum values in about 5 days and remain at constant values afterward as long as the concentrations of the thoron progeny in the air remain constant.

For the situation where the deposition rate for the radon progeny is 1 h^{-1} (the corresponding relative concentrations are shown in Figure 9.13) the 1-h concentrations of the thoron progeny are 0.0512, 0.0258, and 0.00879 Bq m^{-3} for ^{212}Pb, ^{212}Bi, and ^{208}Tl, respectively. The activities of the thoron progeny on the filter are shown in Figure 9.15. The curves in Figure 9.15 appear similar to those in Figure 9.14, except that smaller maximum values of the activities are reached.

Using the method demonstrated here, one could produce the values for activity of the thoron progeny, as well as total α- and β-activities, collected on the filter for whatever values of the various parameters (thoron concentration, deposition rates, age of thoron progeny, pump flow rate, collection efficiency) are of interest.

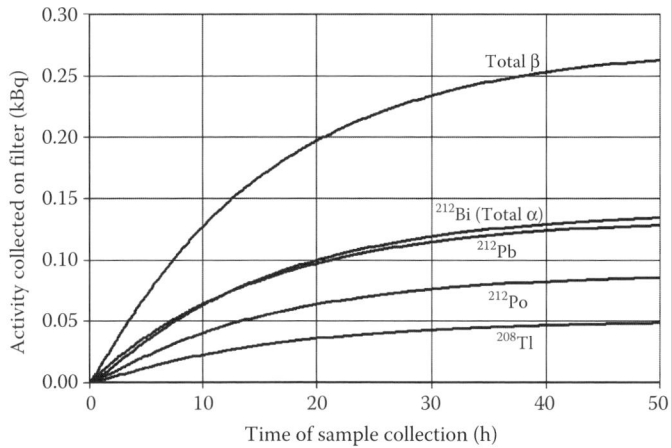

FIGURE 9.15 Thoron progeny collection and ingrowth (scenario from Figure 9.13).

Throughout this discussion, it was assumed that the thoron concentration remains constant. In reality, the thoron concentration likely varies due to changes in meteorological conditions, such as barometric pressure, temperature, soil moisture, wind speed, and perhaps operational factors. Although not shown here, if the thoron is removed from the air, for example, by flushing a scintillation cell used for a thoron grab sample with nitrogen, the thoron progeny decay to insignificant levels much more slowly than do radon progeny due to the half-life of ^{212}Pb. Some measurement techniques utilize this fact by following the decline in α-particle emissions after the scintillation cell is flushed or after the sampling pump is turned off. The radon decay products decay to insignificant levels after 4 h, after which counts from α-particles following a half-life of 10.64 h can be attributed to thoron progeny.

ACTINIUM DECAY SERIES AND ^{219}Rn

The Actinium Decay Series consists of 11 radionuclides headed by ^{235}U, which has a half-life of 7.037×10^8 years (Browne, 1986). One of the decay products of ^{235}U, ^{231}Pa, has a long half-life of 3.276×10^4 years (Browne, 1986). Its decay product, ^{227}Ac, has a half-life of 21.77 years (Browne, 1986). All of the decay products of ^{235}U are in secular equilibrium. The entire Actinium Decay Series is listed in Table 9.3.

Because of the small natural abundance of ^{235}U compared with ^{238}U and ^{234}U, the actinium series including ^{219}Rn (actinon) is usually of little importance. However, there are places where the natural abundances have been disturbed by human activities, making actinon and/or its decay products measurable in air, possibly an interference for CAMs and even at concentrations large enough to be of concern for personnel protection. Therefore, the user of a CAM should be aware of the radiological properties of actinon; hence they are included in the discussion here.

The dynamics of the Actinon Decay Series, shown in Figure 9.16, are both similar and different from those of the Radon and Thoron Decay Series. As in the Thoron Decay Series, actinon has a very short half-life (3.96 s) and decays by α-particle emission to another α-particle emitter, which also has a very short half-life (^{215}Po, 1.78 ms). Hence, like thoron and its immediate decay product, actinon and its immediate decay product decay rapidly. In the Thoron Decay Series, the dynamics are controlled by ^{212}Pb with a half-life of 10.64 h. In the Actinon Decay Series, however, the controlling

TABLE 9.3
Actinium Decay Series[a]

Nuclide	Half-Life	Mode of Decay
^{235}U	7.037×10^8 years	α
^{231}Th	1.063 days	β
^{231}Pa	3.276×10^4 years	α
^{227}Ac	21.77 years	β
^{227}Th	18.718 days	α
^{223}Ra	11.43 days	α
^{219}Rn	3.96 s	α
^{215}Po	1.78 ms	α
^{211}Pb	36.1 min	β
^{211}Bi	2.14 min	α
^{207}Tl	4.77 min	β
^{207}Pb	Stable	

[a] Browne (1986).

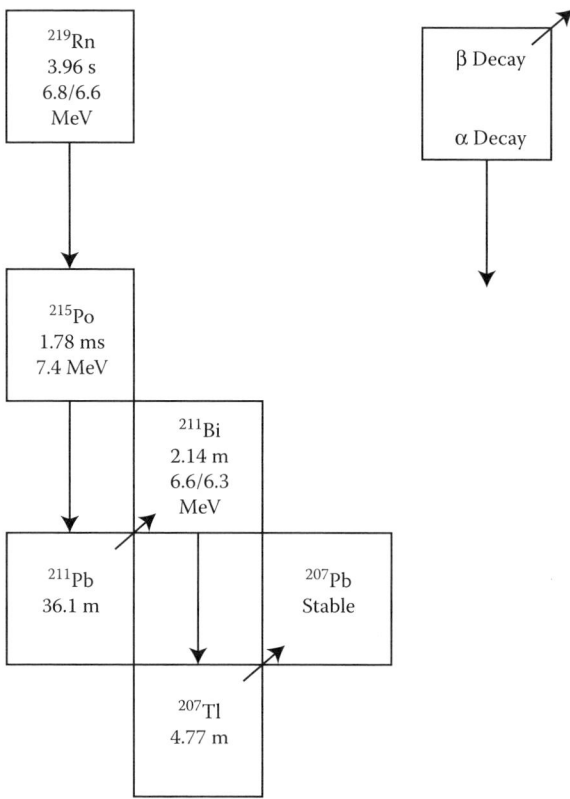

FIGURE 9.16 ^{219}Rn (actinon) Decay Series.

radionuclide is ^{211}Pb with a half-life of only 36.1 min. Hence in that respect, the actinon decay products are somewhat similar to radon decay products. In a scintillation cell, for example, when the cell is flushed, the gross α-count rate from both radon decay products and actinon decay products diminish to background at about the same rate. When using gross α-counting and not α-spectroscopy, it is easy to mistake actinon decay products for radon decay products.

SOLUTION TO DECAY EQUATIONS USING A RECURRENCE FORMULA

In the same manner as discussed for radon decay products, the recurrence formula can be applied to the decay and buildup of actinon and actinon decay products. Here, there are five radionuclides to consider.

First, define the f_{ii} factors that are the exponential terms describing radioactive decay.

$$f_{00} = \exp(-\lambda_0 t), \tag{9.69}$$

$$f_{11} = \exp(-\lambda_1 t), \tag{9.70}$$

$$f_{22} = \exp(-\lambda_2 t), \tag{9.71}$$

$$f_{33} = \exp(-\lambda_3 t), \tag{9.72}$$

$$f_{44} = \exp(-\lambda_4 t), \tag{9.73}$$

where λ_0, λ_1, λ_2, λ_3, and λ_4 are the decay constants for ^{219}Rn, ^{215}Po, ^{211}Pb, ^{211}Bi, and ^{207}Tl, respectively (s^{-1}) and t is time (s).

Then applying the recurrence formula, define the following f_{ij} factors:

$$f_{01} = \frac{(f_{00} - f_{11})}{(\lambda_1 - \lambda_0)} \text{ (for decay of initially present } ^{219}\text{Rn to } ^{215}\text{Po)}, \tag{9.74}$$

$$f_{12} = \frac{(f_{11} - f_{22})}{(\lambda_2 - \lambda_1)} \text{ (for decay of initially present } ^{215}\text{Po to } ^{211}\text{Pb)}, \tag{9.75}$$

$$f_{23} = \frac{(f_{22} - f_{33})}{(\lambda_3 - \lambda_2)} \text{ (for decay of initially present } ^{211}\text{Pb to } ^{211}\text{Bi)}, \tag{9.76}$$

$$f_{34} = \frac{(f_{33} - f_{44})}{(\lambda_4 - \lambda_3)} \text{ (for decay of initially present } ^{211}\text{Bi to } ^{207}\text{Tl)}, \tag{9.77}$$

$$f_{02} = \frac{(f_{01} - f_{12})}{(\lambda_2 - \lambda_0)} \text{ (for decay of initially present } ^{219}\text{Rn to } ^{211}\text{Pb)}, \tag{9.78}$$

$$f_{13} = \frac{(f_{12} - f_{23})}{(\lambda_3 - \lambda_1)} \text{ (for decay of initially present } ^{215}\text{Po to } ^{211}\text{Bi)}, \tag{9.79}$$

$$f_{24} = \frac{(f_{23} - f_{34})}{(\lambda_4 - \lambda_2)} \text{ (for decay of initially present } ^{211}\text{Pb to } ^{207}\text{Tl)}, \tag{9.80}$$

$$f_{03} = \frac{(f_{02} - f_{13})}{(\lambda_3 - \lambda_0)} \text{ (for decay of initially present } ^{219}\text{Rn to } ^{211}\text{Bi)}, \tag{9.81}$$

$$f_{14} = \frac{(f_{13} - f_{24})}{(\lambda_4 - \lambda_1)} \text{ (for decay of initially present } ^{215}\text{Po to } ^{207}\text{Tl)}, \tag{9.82}$$

$$f_{04} = \frac{(f_{03} - f_{14})}{(\lambda_4 - \lambda_0)} \text{ (for decay of initially present } ^{219}\text{Rn to } ^{207}\text{Tl)}. \tag{9.83}$$

Once all the "f factors" have been defined, the concentrations of actinon and actinon progeny can be calculated from the following:

$$C_0 = C_{0,0} f_{00}, \tag{9.84}$$

$$C_1 = C_{1,0} f_{11} + C_{0,0} \lambda_1 f_{01}, \tag{9.85}$$

$$C_2 = C_{2,0} f_{22} + C_{1,0} \lambda_2 f_{12} + C_{0,0} \lambda_1 \lambda_2 f_{02}, \tag{9.86}$$

$$C_3 = C_{3,0} f_{33} + C_{2,0} \lambda_3 f_{23} + C_{1,0} \lambda_2 \lambda_3 f_{13} + C_{0,0} \lambda_1 \lambda_2 \lambda_3 f_{03}, \tag{9.87}$$

$$C_4 = C_{4,0} f_{44} + C_{3,0} \lambda_4 f_{34} + C_{2,0} \lambda_3 \lambda_4 f_{24} + C_{1,0} \lambda_2 \lambda_3 \lambda_4 f_{14} + C_{0,0} \lambda_1 \lambda_2 \lambda_3 \lambda_4 f_{04}, \tag{9.88}$$

where, $C_{0,0}$, $C_{1,0}$, $C_{2,0}$, $C_{3,0}$, and $C_{4,0}$ are the initial concentrations (or activities) of ^{219}Rn, ^{215}Po, ^{211}Pb, ^{211}Bi, and ^{207}Tl, respectively (Bq m^{-3} or Bq) and C_0, C_1, C_2, C_3, and C_4 are the concentrations (or activities) of ^{219}Rn, ^{215}Po, ^{211}Pb, ^{211}Bi, and ^{207}Tl, respectively (Bq m^{-3} or Bq), as a function of time.

POTENTIAL α-ENERGY CONCENTRATION

It is assumed here that the PAEC for actinon progeny can be defined in the same manner as for radon or thoron progeny; where one WL of actinon progeny is any combination of the short-lived decay products of ^{219}Rn in 1 L of air that will result in the ultimate emission of 1.3×10^5 MeV of potential α-particle energy. The PAEC can be calculated from the individual actinon progeny concentrations by either of the following two equations:

$$\text{PAEC}(\mu \text{ J m}^{-3}) = 0.00328C_2 + 0.000195C_3, \quad (9.89)$$

where C_2 and C_3 are the concentrations of ^{211}Pb and ^{211}Bi, respectively, (Bq m^{-3}) and

$$\text{PAEC(WL)} = 0.00584C_2 + 0.000346C_3, \quad (9.90)$$

where C_2 and C_3 are the concentrations of ^{211}Pb and ^{211}Bi, respectively (pCi L^{-1}).

Note that the very short-lived decay product of actinon (^{215}Po) is not included in Equations 9.89 and 9.90 as it contributes insignificantly to the total PAEC. Further, comparing these equations to Equations 9.15 and 9.16 shows that for a given concentration, actinon progeny produce a somewhat smaller PAEC than radon progeny.

The dose from actinon progeny can be related to the potential α-energy exposure, which is the integral over time of the PAEC. The potential α-energy exposure in WLM is calculated by multiplying the PAEC in WL by the number of hours worked in an environment with that value of PAEC and dividing by 170 working hours per month. One WLM is equivalent to 3.6 mJ h m^{-3}. There is no published occupational Derived Air Concentration for exposure to actinon and actinon progeny.

ACTINON SAMPLE IN A CLOSED CONTAINER OR DEFINED VOLUME OF AIR

A sample of pure actinon in a closed container (such as a scintillation cell) or defined volume of air decays extremely rapidly to insignificant levels. Figure 9.17 shows that the actinon decays through six orders of magnitude in about 1.3 min. The first decay product of actinon, ^{215}Po, has such a small half-life that it virtually immediately comes into secular equilibrium with the actinon. Figure 9.17 is analogous to Figure 9.2 for radon and Figure 9.11 for thoron, as no loss owing to deposition is assumed. As shown in Figure 9.18, ^{211}Pb reaches a maximum value of 0.18% of the initial actinon activity in about 1 min, and then decays with its half-life of 36.1 min. ^{211}Bi reaches a maximum value of 0.154% of the initial actinon activity in 9 min and then decays in transient equilibrium with ^{211}Pb. ^{207}Tl reaches a maximum value of 0.132% of the initial actinon activity in 20 min and subsequently decays in transient equilibrium with ^{211}Pb.

EFFECT OF DEPOSITION ON ACTINON PROGENY ACTIVITIES

Deposition of actinon progeny can be modeled by adding a deposition rate as discussed in previous sections for radon and thoron progeny. The effect of assuming a deposition rate of 1 h^{-1} on the actinon progeny is shown in Figure 9.19. In this case, ^{211}Pb reaches a maximum value of 0.177% of the initial actinon activity in about 1 min, and then decays with its half-life of 36.1 min. ^{211}Bi reaches a maximum value of 0.133% of the initial actinon activity in 7 min and then decays in transient equilibrium with ^{211}Pb. ^{207}Tl reaches a maximum value of 0.098% of the initial actinon activity in 16 min and subsequently decays in transient equilibrium with ^{211}Pb.

Behavior of Radon and its Decay Products

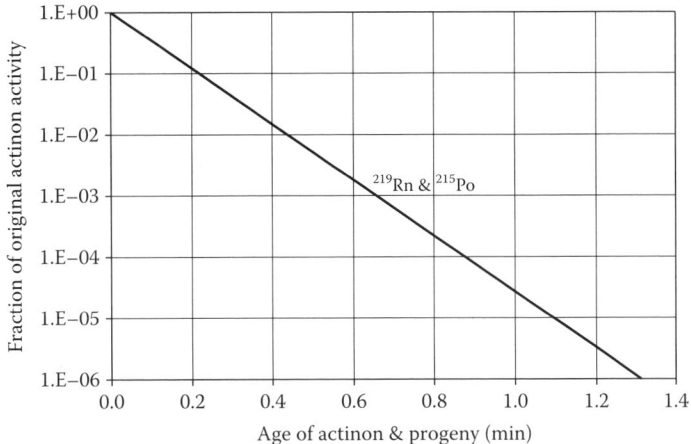

FIGURE 9.17 Actinon with ^{215}Po in secular equilibrium in closed container or defined volume of air during the first 1.4 minutes, no deposition.

ACTINON PROGENY BUILDUP ON A FILTER

Of interest here is the collection of actinon progeny on the filter of a CAM and the manner in which the actinon progeny behave during the collection process. The equations that describe the buildup of the actinon progeny on a filter are similar to Equations 9.69 through 9.88 presented above using the recurrence formula; however, the integrated form of the "f" factors is now used. As before, they are referred to as "h" factors. Note that since actinon is not collected on the filter, only the factors for the actinon progeny are included here. The initial "h_{ii}" factors are defined as follows:

$$h_{11} = \frac{[1 - \exp(-\lambda_1 t_s)]}{\lambda_1}, \tag{9.91}$$

$$h_{22} = \frac{[1 - \exp(-\lambda_2 t_s)]}{\lambda_2}, \tag{9.92}$$

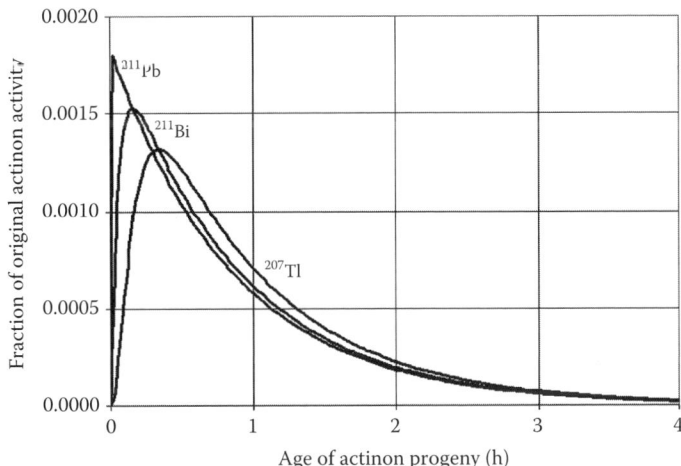

FIGURE 9.18 Ingrowth and decay of actinon progeny in a closed container or defined volume of air, no deposition.

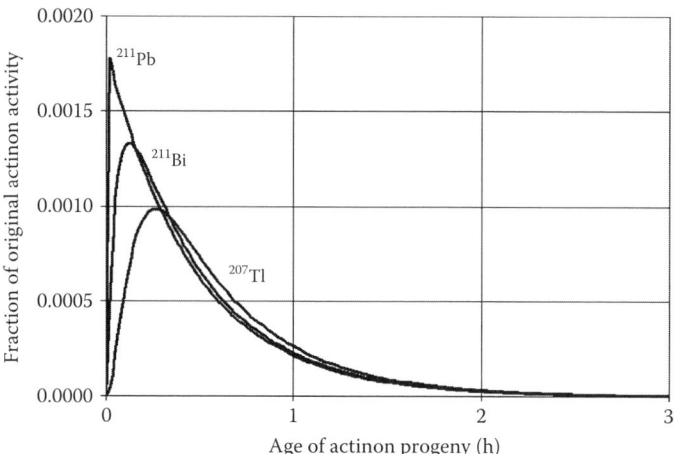

FIGURE 9.19 Actinon progeny in air, deposition rate of 1 h^{-1}.

$$h_{33} = \frac{[1 - \exp(-\lambda_3 t_s)]}{\lambda_3}, \tag{9.93}$$

$$h_{44} = \frac{[1 - \exp(-\lambda_4 t_s)]}{\lambda_4}, \tag{9.94}$$

where $\lambda_1, \lambda_2, \lambda_3,$ and λ_4 are the decay constants of ^{215}Po, ^{211}Pb, ^{211}Bi, and ^{207}Tl, respectively (s^{-1}) and t_s is the duration of the collection on the filter (s).

Then applying the recurrence formula, the "h_{ij}" factors are calculated as follows:

$$h_{12} = \frac{(h_{11} - h_{22})}{(\lambda_2 - \lambda_1)}, \tag{9.95}$$

$$h_{23} = \frac{(h_{22} - h_{33})}{(\lambda_3 - \lambda_2)}, \tag{9.96}$$

$$h_{34} = \frac{(h_{33} - h_{44})}{(\lambda_4 - \lambda_3)}, \tag{9.97}$$

$$h_{13} = \frac{(h_{12} - h_{23})}{(\lambda_3 - \lambda_1)}, \tag{9.98}$$

$$h_{24} = \frac{(h_{23} - h_{34})}{(\lambda_4 - \lambda_2)}, \tag{9.99}$$

$$h_{14} = \frac{(h_{13} - h_{24})}{(\lambda_4 - \lambda_1)}. \tag{9.100}$$

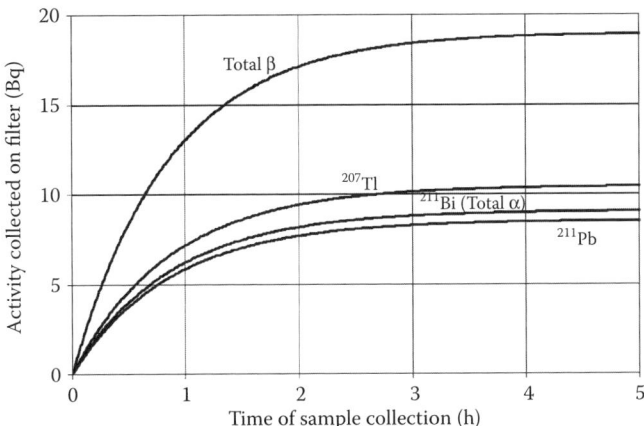

FIGURE 9.20 Actinon progeny collection and ingrowth (scenario from Figure 9.18).

Then the activities of the actinon progeny as a function of collection time, t_s, are calculated as follows:

$$A_1 = F\varepsilon C_1 h_{11}, \tag{9.101}$$

$$A_2 = F\varepsilon(C_1\lambda_2 h_{12} + C_2 h_{22}), \tag{9.102}$$

$$A_3 = F\varepsilon(C_1\lambda_2\lambda_3 h_{13} + C_2\lambda_3 h_{23} + C_3 h_{33}), \tag{9.103}$$

$$A_4 = F\varepsilon(C_1\lambda_2\lambda_3\lambda_4 h_{14} + C_2\lambda_3\lambda_4 h_{24} + C_3\lambda_4 h_{34} + C_4 h_{44}), \tag{9.104}$$

where A_1, A_2, A_3, and A_4 are the activities on the filter of ^{215}Po, ^{211}Pb, ^{211}Bi, and ^{207}Tl, respectively (Bq), C_1, C_2, C_3, and C_4 are the concentrations in air of ^{215}Po, ^{211}Pb, ^{211}Bi, and ^{207}Tl, respectively (Bq m^{-3}), λ_1, λ_2, λ_3, and λ_4 are the decay constants of ^{215}Po, ^{211}Pb, ^{211}Bi, and ^{207}Tl, respectively (s^{-1}), F is the pump flow rate (m^3 s^{-1}), and ε is the collection efficiency.

In a similar manner as was discussed in previous sections for radon and thoron progeny, the basis of the calculations here are

1. Initial ^{219}Rn concentration of 100 Bq m^{-3} (2.7 pCi L^{-1})
2. Pump flow rate of 0.0472 m^3 s^{-1} (100 cfm)
3. Collection efficiency of 100% ($\varepsilon = 1$)
4. Actinon progeny "age" of 1 h

First consider the situation of no loss of actinon progeny due to deposition, where the relative actinon progeny concentrations are shown in Figure 9.18. For this example, the 1-h actinon progeny concentrations are 0.0579, 0.0615, and 0.0708 Bq m^{-3} for ^{211}Pb, ^{211}Bi, and ^{207}Tl, respectively. The activities of the actinon progeny on the filter are shown in Figure 9.20 as a function of sample collection time in hours. Also included in Figure 9.20 are the total α-activity (which is the same as the ^{211}Bi activity) and the total β-activity (the sum of ^{211}Pb and ^{207}Tl activities). The activities on the filter reach maximum values in about 5 h and remain at constant values afterward as long as the concentrations of the actinon progeny in the air remain constant.

For the situation where the deposition rate for the actinon progeny is 1 h^{-1} (the relative concentrations for which are shown in Figure 9.19) the 1-h concentrations of the actinon progeny are 0.0213, 0.0227, and 0.0261 Bq m^{-3} for ^{211}Pb, ^{211}Bi, and ^{207}Tl, respectively. The activities of the actinon progeny

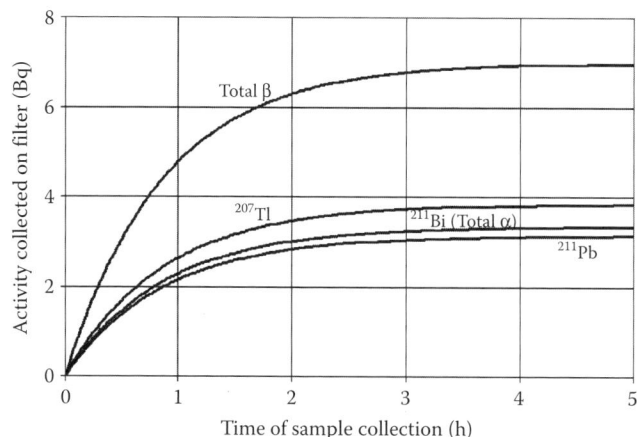

FIGURE 9.21 Actinon progeny collection and ingrowth (scenario from Figure 9.19).

on the filter are shown in Figure 9.21. The curves in Figure 9.21 appear similar to those in Figure 9.20, except that smaller maximum values of the activities are reached.

Using the method demonstrated here, one could produce the values for activity of the actinon progeny, as well as total α- and β-activities, collected on the filter of a CAM for whatever values of the various parameters (actinon concentration, deposition rates, age of actinon progeny, pump flow rate, and collection efficiency) are of interest.

Throughout this discussion, it was assumed that the initial actinon concentration remains constant. In reality, the initial actinon concentration likely does not remain constant due to changes in meteorological conditions, such as barometric pressure, temperature, soil moisture, wind speed, and perhaps operational factors.

REFERENCES

Bateman, H. The solution of a system of differential equations of radioactive decay. *Proc. Camb. Phil. Soc.* 1910, 15, 423–427.
Browne, E., Firestone, R. B., and Shirley, V. S., Eds. *Table of Radioactive Isotopes*, John Wiley & Sons, New York, NY, 1986.
Eisenbud, M. and Gesell, T. *Environmental Radioactivity*, 4th ed., Academic Press, San Diego, CA, 1997.
Evans, R. D. Engineer's guide to the elementary behavior of radon daughters. *Health Phys.* 1969, 17(2), 220–252.
Friedlander, G., Kennedy, J. W., and Miller, J. M. *Nuclear and Radiochemistry*, John Wiley & Sons, New York, NY, 1964.
Hamawi, J. M. A useful recurrence formula for the equations of radioactive decay. *Nucl. Tech.* 1971, 11, 84–88.
Jenkins, P. H. Equations for modeling of grab samples of radon decay products. *Rad. Saf. J.* 2002, 83(Suppl. 1), S48–S51.
Evaluation of Occupational and Environmental Exposures to Radon and Radon Daughters in the United States, NCRP Report No. 78, National Council On Radiation Protection and Measurements, 1984.
Scherpelz, R. I. and DesRosiers, A. E. A modification to a recurrence formula for linear first-order equations. *Health Phys.* 1981, 40, 904–907.
U.S. Nuclear Regulatory Commission, 2008, Standards for Protection against Radiation, *Code of Federal Regulations*, Title 10, Part 20, U.S. Government Printing Office, Washington, DC, 2008.

10 Internal Dosimetry of Inhaled Radioactive Aerosols

Charles A. Potter

CONTENTS

Introduction 209
Morphology and Deposition 210
Clearance and Absorption 213
Dose to the Lung 214
Gastrointestinal Tract 215
Systemic Metabolism 216
Derived Air Concentration 217
Summary 218
References 218

INTRODUCTION

The air around us is full of particles. With each breath we take, particles enter the nose and mouth. Particles may deposit at any point along their pathway by diffusion or deposition. They may impact on outer surfaces and never enter the body. If particles do enter, they may deposit on a multitude of surfaces beginning with the interior of the nose or mouth. The particles pass into the esophagus and into the bronchi where pathways branch off to different locations. As they continue their entry through the respiratory tract, their pathway gets even more tortuous as the tubules get smaller and smaller, until they terminate in the pulmonary alveoli.

In the alveoli, metabolic exchange of oxygen and carbon dioxide occurs with the bloodstream across capillary walls. Absorption into the bloodstream may happen here or at any point along the path where particles deposit. However, particles do not have to be absorbed out of the respiratory tract into an adjacent compartment within the body (such as the lymph system or bloodstream)—they may also be cleared. Tiny hairs called cilia are continually moving particles up and out of the lung. This process competes with the process of absorption. Particles composed of insoluble compounds may likely be cleared back to the top of the throat where upon these particles may be swallowed and traverse into the gastrointestinal tract beginning with the esophagus.

Just because a particle enters the gastrointestinal or "GI" tract does not mean that its fate is excretion. Some fraction of particles cleared into the GI tract will be absorbed into the bloodstream from the stomach or the small intestine. The absorbed material, either from the respiratory tract or the GI tract, will have dissolved separating the individual constituent compounds into more readily transportable species: ions, atoms, and complexes. These atoms or other species do not stay in the bloodstream for a long period of time. They may deposit in individual organs and tissues typically with a 6-h translocation half-time.

Organs or tissues in which these elements deposit are element-specific and are dependent on the metabolism of that organ or tissue. Iodine is well-known for its affinity for the thyroid. Alkaline earth elements such as calcium, radium, and strontium tend to mimic natural calcium and deposit in bone. Actinides such as uranium may be translocated to the kidneys and in the case of plutonium, taken up in the bone or liver where it is effectively retained. There is typically more than one organ or tissue in any metabolic model. The atoms, after remaining in the organ or tissue for some time, may dissociate from that organ or tissue back into the bloodstream.

In the bloodstream, there are still several pathways to follow. The atoms may be reabsorbed into another organ or tissue. From there, their fate is the same as described in the previous paragraph either to primary (principal) or secondary target organs. Alternatively, the atoms may be destined for excretion. For example, the atoms may be removed as the blood passes through the kidneys. From the kidneys, the atoms move along the renal pathway to the bladder and ultimately are excreted in urine. They may also be cleared from the bloodstream in the liver or into the intestines where they ultimately will be excreted in the feces.

If the particles/atoms involved are radioactive, they may impart a radiation dose during the residence time in the body. This occurs when the unstable atoms decay and release energy. The radioactive decay process competes with all of the deposition, absorption, clearance, and translocation processes described above. Some decay products are also radioactive subject to characteristic decay times and transformation to yet other species. Radioactive decay releases α-, β-, x-ray, or γ-radiation depending on the radionuclide. α- and β-radiation deposits energy locally, meaning that the source and target tissues are typically the same. However, γ-photons emitted isotropically in a particular organ can conceivably irradiate all other organs and tissues in the body to some extent. This results in an absorbed dose to affected organs and tissues, as well as an associated equivalent dose.

Determination of a whole-body detriment from internalized radioactivity and the associated radiation exposure is a nontrivial exercise. It is accomplished by defining an "effective dose" denoted E. This type of dose is estimated by determining an equivalent dose to each organ and tissue in the body, weighting those doses using "tissue weighting factors," and summing the weighted doses as

$$E = \sum_T w_T H_T, \qquad (10.1)$$

where w_T is the tissue weighting factor for tissue T and H_T the absorbed dose to target tissue T.

Dose from internal emitters is not received instantaneously, as it is from external sources. Rather, dose is imparted gradually as radioactive material remaining in the system decays and emits radiation. Therefore, a conventional time period over which dose is accumulated must be defined. This accumulated dose is called "committed effective dose" and denoted $E(\tau)$ where τ is the time over which the dose is accumulated. Convention for occupational workers is that the working life begins at age 20 and ends at age 70, providing a dose accumulation period of 50 years. Because τ is defined as 50 years, the notation is modified accordingly so that the committed effective dose may then be denoted $E(50)$.

The purpose of this chapter is to describe the mechanisms by which radioactive material present as airborne particles is taken into the body, distributed among organs and tissues, and ultimately causes dose to the body. Most of the information provided will be on the respiratory tract and how material deposits and is cleared or absorbed, as well as how dose to the lung is imparted. Some information on systemic metabolism is provided where the term "systemic" refers to material absorbed into the bloodstream and distributed. Finally, the calculation of "derived air concentration," the airborne concentration guideline, will be described.

MORPHOLOGY AND DEPOSITION

There are two respiratory tract models available, both developed during the 1990s. The "Human Respiratory Tract Model for Radiological Protection" was published in 1994 by the International Commission on Radiological Protection (ICRP) as ICRP Publication 66. In 1997, the publication "Deposition, Retention, and Dosimetry of Inhaled Radioactive Substances" was published by the

Internal Dosimetry of Inhaled Radioactive Aerosols

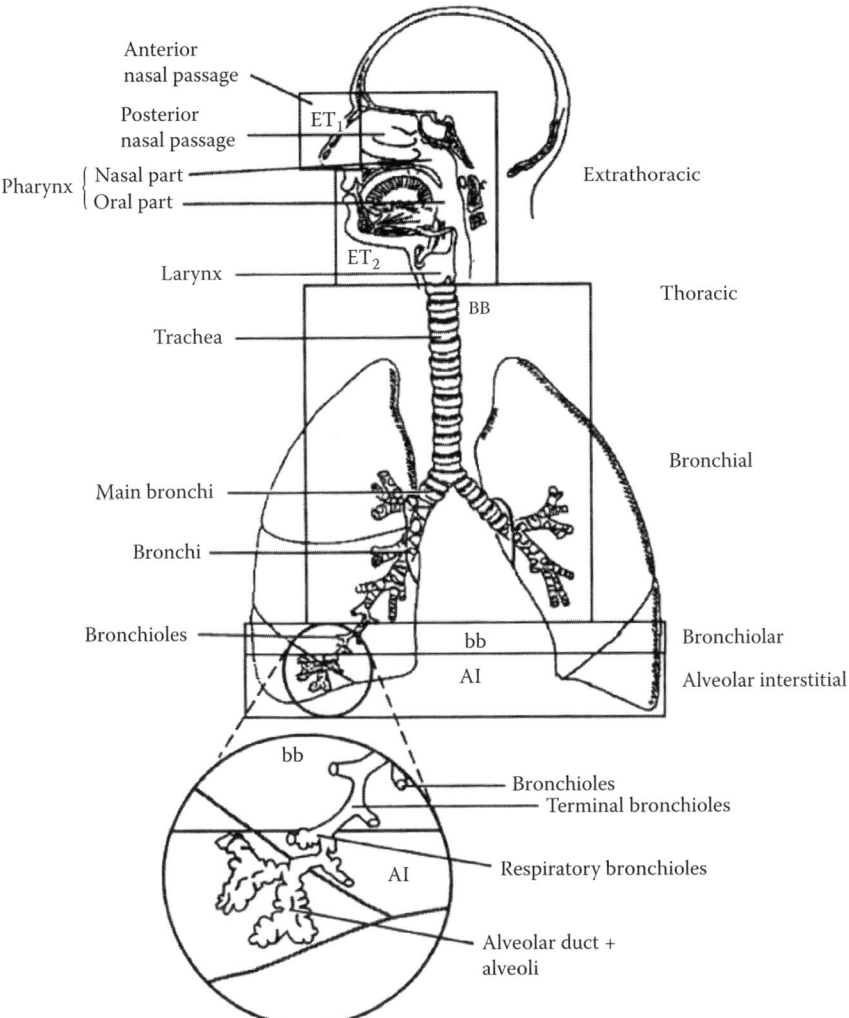

FIGURE 10.1 Morphology of the human respiratory tract as defined in ICRP Publication 66. (From International Commission on Radiological Protection. *Human Respiratory Tract Model for Radiological Protection*. Oxford: Pergamon Press; ICRP Publication 66, *Ann. ICRP* 24(1–3); 1994a. With permission.)

National Council on Radiation Protection and Measurements (NCRP) as NCRP Report No. 125. While both are worthy of note, the internal dosimetry community has embraced that promulgated by the ICRP. The preface to the NCRP Report No. 125 states, "… The NCRP recommends the adoption of ICRP Publication 66 for calculating exposures for radiation workers and the public …" (NCRP, 1997). Therefore, this chapter will use morphology, deposition, and clearance terms as defined in ICRP Publication 66.

The respiratory tract, shown in Figure 10.1 can be divided into two main regions. The upper respiratory tract that includes nasal passages, pharynx, and larynx—everything above the thorax—is called the *extrathoracic* region. The trachea and lungs represent what is called the *thoracic* region. The extrathoracic region is further divided into two subregions: extrathoracic region 1 (ET1) and extrathoracic region 2 (ET2). The ET1 region contains only the anterior nasal passage. The ET2 region includes the nasal and oral parts of the pharynx, mouth, and larynx. Deposition occurs in both regions with the ET2 region clearing to the GI tract and deposition in the ET1 region clearing to the environment.

The thoracic region is broken down into three subregions. The highest of these is the bronchial (BB) region. This includes bronchi comprising generations 0–8, which includes the trachea (0), bronchi (1), and larger bronchioles (2–8). The bronchiolar (bb) region contains bronchioles of generations 9–15 and is designated bb. The deepest region of the lung is designated alveolar–interstitial (AI). This region contains bronchioles generations 16–26 and the alveoli where gaseous exchange with the blood occurs. Deposition occurs in all three of these regions, and that deposition clears to the region above (AI → bb, bb → BB, etc.). Clearance will be more fully discussed in the section on absorption below.

These respiratory tract regions and subregions can be compared to a series of air filters. Upon inhalation, air flows through ET1, followed by ET2, then BB, then bb, then AI. Particulates may be deposited in any of the regions or remain in the inhaled air. They then have another chance at deposition in each of the regions as the air is exhaled. Each of these filters has a filtration efficiency and a volume that is a function of physiological parameters. The filtration efficiency is basically the quotient of the difference between filter input and output and the filter input itself. Volumes are dependent on tidal volume (amount of air breathed in or out during normal respiration) and functional residual capacity (amount of air left in the lungs after expiration of a tidal breath) of the lung itself and varies based on activity level, for example, sitting, light exercise, and so on.

Deposition mechanisms are dependent on particle size. There are two mechanisms by which particles impinge on surfaces of the respiratory tract: aerodynamic processes and thermodynamic processes. Aerodynamic processes where particles are carried by airflow dominate with particles greater than 1 μm in diameter. Thermodynamic processes dominate when particles are smaller than 0.1 μm where particles move by Brownian motion. All particles exhibit both of these mechanisms. Deposition fractions, fractions of the intake in each filter (lung region/subregion) are calculated using the filter efficiency and volume described above.

Airborne particulates are rarely (if ever) monodisperse, that is, the particle sizes follow a distribution described as lognormal defined by a mean particle size and a geometric standard deviation. These characteristics of particle size distributions are taken into account in the deposition charts

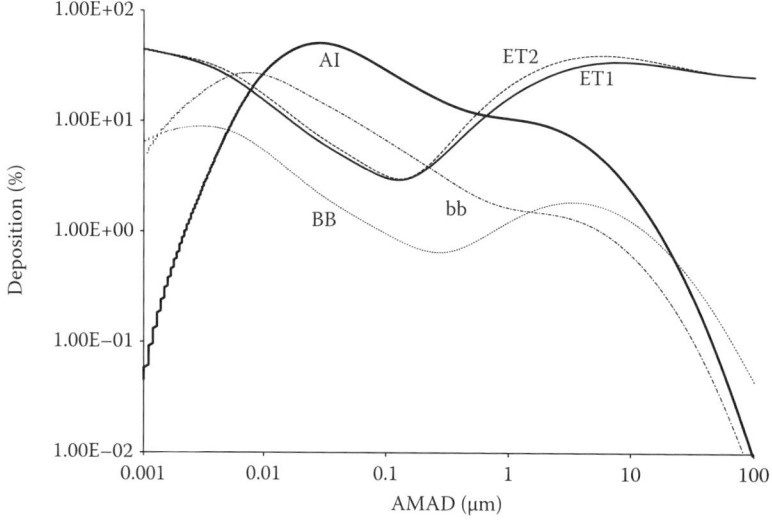

FIGURE 10.2 Regional deposition of particulates in lung regions. AMAD refers to the activity median aerodynamic median, that is, particle size, of the distribution (data created using LUDEP). (*Note:* LUDEP 2.0, Personal Computer Program for Calculating Internal Doses Using the ICRP Publication 66 Respiratory Tract Model. National Radiological Protection Board, Chilton, UK.)

provided in ICRP Publication 66 as shown in Figure 10.2. Default parameters used in the calculation of dose coefficients (ICRP, 1995) are based on Reference Man (ICRP, 1975). In this case, Reference Man is a nonsmoker, nose breather, and performs light work consisting of 5.5 h of light exercise (breathing 1.5 m^3/h) and 2.5 h of sitting (breathing 0.54 m^3/h) in an 8-h day. He is exposed to a polydisperse aerosol of a 5 μm particle size. ICRP Publication 66 allows for variation of all these characteristics as well as sex of the worker.

CLEARANCE AND ABSORPTION

Clearance and absorption are competing processes. The clearance process describes physical movement of particles from one respiratory tract region to another. The body typically tries to remove particles from the respiratory tract using cilia, hairs that line each lung tube. The cilia move particles from deeper to shallower regions, up the tracheal stem, to the pharynx where they may be swallowed. Absorption, on the other hand, is the process where the particles break down and the constituents are absorbed into the bloodstream. Both processes occur simultaneously in most compartments. It is assumed that the rate at which material is either cleared or absorbed out of a lung compartment (or any organ or tissue compartment) is proportional to the amount of material in that compartment.

Clearance from one region to another is described by clearance half-times. What is likely a much more complicated mathematical process is simplified by defining two or three compartments in each subregion, each with a different clearance half-time. As can be seen in Figure 10.3, the BB region is divided into three subregions BB1, BB2, and BBseq. BB1 and BB2 represent this biphasic

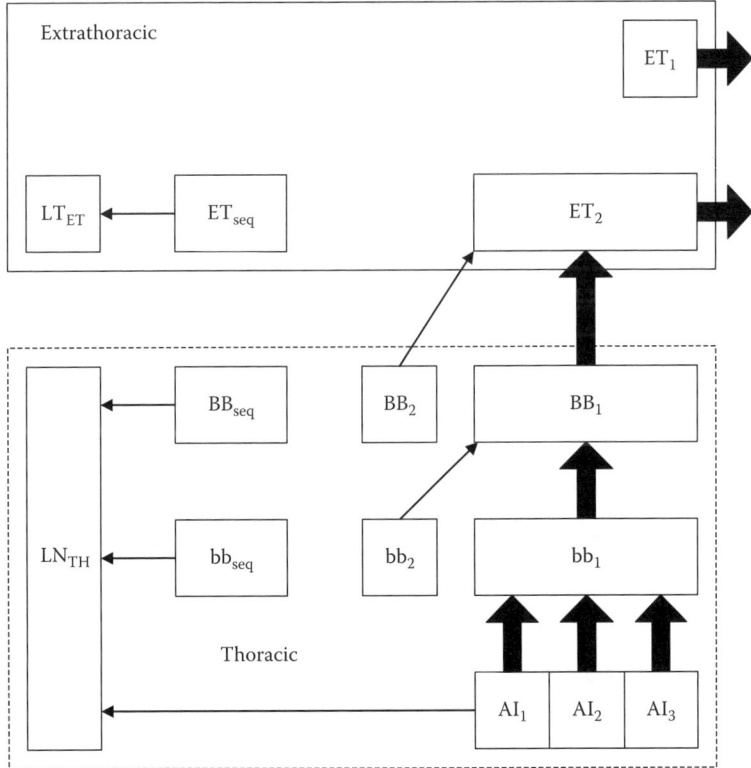

FIGURE 10.3 Respiratory tract clearance pathways for particulates. (From International Commission on Radiological Protection. *Human Respiratory Tract Model for Radiological Protection*. Oxford: Pergamon Press; ICRP Publication 66; *Ann. ICRP* 24(1–3); 1994a. With permission.)

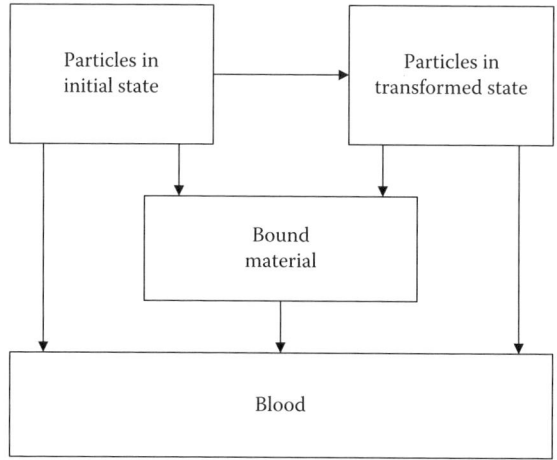

FIGURE 10.4 Respiratory tract absorption from lung tissues into bloodstream. (From International Commission on Radiological Protection. *Human Respiratory Tract Model for Radiological Protection.* Oxford: Pergamon Press; ICRP Publication 66; *Ann. ICRP* 24(1–3); 1994a. With permission.)

clearance process to the ET2 compartment (BBseq will be discussed later). The bb region has a similar breakdown. The AI region is broken down into three compartments, again representing a more complicated clearance pattern out of the subregion. Clearance occurs in most compartments with the exception of the lymph nodes for which there is only absorption.

Absorption occurs out of every compartment except ET1, which only clears to the environment. The absorption of a particular element is dependent on the solubility of the compound. The full absorption model shown in Figure 10.4 is applied to each compartment (i.e., bb1, bb2, as opposed to bb subregion) from which there is absorption, which generally includes initial state, transformed state, and bound state material. The initial state is simply the undissolved particulate. The transformed state denotes some chemical transformation that may occur in lung fluid. Particles in the transformed state may still clear (or transfer) either to a metabolically adjacent compartment, or be absorbed. The bound state describes material in the lung tissue that does not clear, but may remain embedded for a relatively long time and that presumably imparts considerable dose to the lung.

It is the expectation of the authors of ICRP Publication 66 that individual parameters for different compounds will be defined through experiments and those parameters applied to the lung model. However, since data of this type was not available for the publication, default parameters were provided. Three absorption classes were defined: F for fast, M for moderate, and S for slow. Associated absorption half-times for the initial and transformed states were provided for each class (class F material is not transformed). The bound compartment was not used in this default system, but its possible use for insoluble elements such as plutonium is intriguing.

Some material is deposited in the lung tissue such that it is "sequestered" and does not clear. Compartments for this sequestered material are designated with the subscript "seq." The material deposited therein clears to the regional lymph nodes where it is absorbed into the bloodstream. Material in the lymph nodes is absorbed through the same absorption processes as the absorption mechanisms for the lung tissue.

DOSE TO THE LUNG

Dose to any target tissue from any source tissue is calculated in the same way. First, the number of disintegrations (transformations) in the source tissue is determined. This value is then multiplied by the "specific effective energy" for the target and source, and that value is corrected by unit conversion factors. The specific effective energy is defined by ICRP Publication 30 as "the energy (MeV), suitably modified for quality factor [radiation weighting factor], imparted per gram of a target tissue (T) as a consequence of the emission of a specified radiation (i) from a transformation occurring in source tissue (S)." (ICRP, 1979) These values can be summed over the radiations for a specific radionuclide. Included in this value is the "absorbed fraction," which ICRP Publication 30 defines as "the fraction of energy emitted as a specified radiation type in a specified source tissue which is absorbed in a specified target tissue." (ICRP, 1979) Absorbed fractions for particles (α and β) are typically unity where the source and target tissues are the same and zero otherwise.

Particles deposited in respiratory tract tissue break down in the lung fluid and are eventually absorbed, as previously described. Unlike other organs and tissues, the radionuclides retained in the lung prior to absorption are not considered to be distributed throughout the lung tissue; they undergo the processes of the tissues and subregions in which they are located. Bound material is considered to be radionuclide chemically bound to the lung, while sequestered particles are just that intact particles sequestered in a particular subregion. Source tissues are defined accordingly. Respiratory tract regions are also considered in this process.

ET_1, which describes deposition clearing very quickly from the anterior nasal passage, contains only one source tissue which is "surface deposit." Since only one tissue is represented and ET_1 has no transformed or bound state, it is less important exactly what the term "surface deposit" means. Alternatively, ET_2 contains three source tissues: surface fluid, sequestered particles, and bound material. Surface fluid includes radionuclide in the initial and bound states in the ET_2 compartment, while sequestered particles contains material in the initial and bound states in the ET_{seq} compartment. The LN_{ET} includes initial state, transformed state, and bound material in the extrathoracic lymph nodes.

The BB and bb regions have quick and slow clearance compartments in addition to sequestered particles and bound materials. The quick clearance compartments represent a 2 μm "gel" layer of mucous over the cilia and the slow clearance compartments a 4 μm "sol" layer of mucous in which the cilia are immersed. The quick clearance compartments are represented by BB_1 and bb_1, while the slow clearance compartments are represented by BB_2 and bb_2. The mucous and sequestered compartments include both initial and transformed states of radionuclide in those compartments. The bound compartments stand alone. Tissue considerations are not made in the description of the AI region. One source tissue is defined including all three compartments, both initial and transformed states, and bound material.

Target tissues considered are identified with each deposition subregion, but are described by particular target tissues. The target tissue in ET_1 is the keratinized epithelium of the skin in the anterior part of the nose. ET_2 represents the stratified squamous epithelium of the main extrathoracic airways. Ciliated epithelium of the bronchi and bronchioles are represented by the BB and bb regions, respectively. The AI region describes alveolar–interstitium and lymph nodes are represented by LN_{ET} and LN_{TH}. Absorbed fractions between each source and target tissue are defined and, as expected, are energy and particle (e.g., α, β) dependent. Self-absorption of radiation energy within inhaled particles is not considered. The inclusion of self-absorption would reduce the calculated dose to the lung, and consequently the effective dose, so its exclusion results in conservative dose estimates.

GASTROINTESTINAL TRACT

The discussion to this point has surrounded the respiratory tract and associated tissues. We now turn to metabolism of material outside of the respiratory tract. As described above, material can be both physically cleared into the extrathoracic region and swallowed as well as absorbed into the bloodstream. We first will look at the cleared material and the GI tract's part in the process.

The GI tract model divides the tract into four major parts: stomach, small intestine, upper large intestine (ascending and transverse colon), and lower large intestine (transverse and descending colon, sigmoid colon, and rectum). Material either passes through the GI tract or is absorbed into the bloodstream. Absorption is defined by an f_1 parameter, which is essentially the fraction of material entering the GI tract that is absorbed into the bloodstream. If the f_1 parameter is equal to unity, all material is assumed to be absorbed directly from the GI tract. Therefore, it can be seen that inhaled material has two opportunities for absorption: one directly through the lung tissue and a second opportunity via the GI tract.

As with respiratory tract compartments and systemic organs/tissues, the rate of translocation of material from one organ in the GI tract to another is proportional to the amount of material in the organ. While this may be realistic for other organs and tissues, it does not accurately represent the bolus flow of material through the GI tract. That being said, over time the model can be used more accurately to calculate transformations in contents of the GI compartments as time increases past intake.

The walls of each GI tract organ are the target tissues. Dose to the walls of the stomach, small intestine, upper large intestine, and lower large intestine comes from radiation emitted from radionuclides in the contents of the respective compartments. Self-absorption of radiation in the bolus is considered by assuming that only half of the nonpenetrating radiation emitted in the contents reaches the intestinal walls. Also included in dose considerations are x- and γ-radiations that provides dose to organs outside of the GI tract as well as dose received by the compartments of the GI tract from x- and γ-radiation emitted from other organs and tissues.

SYSTEMIC METABOLISM

Once the material is in the bloodstream, organs and/or tissues where it deposits are dependent on the body's metabolic processes for the particular element. Health physicists are familiar with radium's tendency to be attracted to the bone matrix and iodine's tendency to be absorbed into the thyroid. Elements may deposit in one organ, be retained there for some period of time, be released back into the bloodstream, be absorbed by another organ, and continue this type of process before being eventually excreted.

The model for uranium depicted in Figure 10.5 shows 19 compartments. Uranium is absorbed into blood plasma from compartments of the respiratory and/or GI tract. From there it participates in a recycling process with red blood cells or migrates into four general regions: skeleton, soft tissue, liver, or kidneys. As can be seen, these general regions are broken into compartments that describe either different physiological tissues, as with bone (skeleton), or that describe different clearance rates and mechanisms, as with the liver. The other compartments represent urinary and fecal excreta.

A review of the removal rates from organs and tissues shows the physiological vs. mathematical differences of the compartments within the organ/tissue regions. Soft tissue is represented by three clearance half-times: 2 h (ST0), 20 days (ST1), and 100 years (ST2). Turnover in the liver is represented by short-term removal (7 days) described as Liver 1 and longer-term removal (10 years) described as Liver 2. Turnover in the kidney also is represented by a urinary path with a 7-day clearance half-time and an "other kidney tissue" path with a 5-year clearance half-time. Conversely, bone tissues are represented by cortical and trabecular surfaces and volumes. Bone surface turnover is relatively quick represented with a 5-day clearance half-time. Exchangeable bone volume is represented by a relatively short 30-day clearance half-time while nonexchangeable trabecular bone volume and nonexchangeable cortical bone volume are represented by 3.85- and 23-year clearance half-times, respectively. Urinary bladder and GI tract contents are represented by their respective models, while urine and feces are compartments purely for mathematical use of internal dosimetrists, allowing them to derive expected contents in a urine or fecal sample for a particular time after intake. It should be noted that except for pathways leading to excretion, uranium is metabolized back into the bloodstream where it is available for further translocation into the same or a different organ or tissue.

Internal Dosimetry of Inhaled Radioactive Aerosols

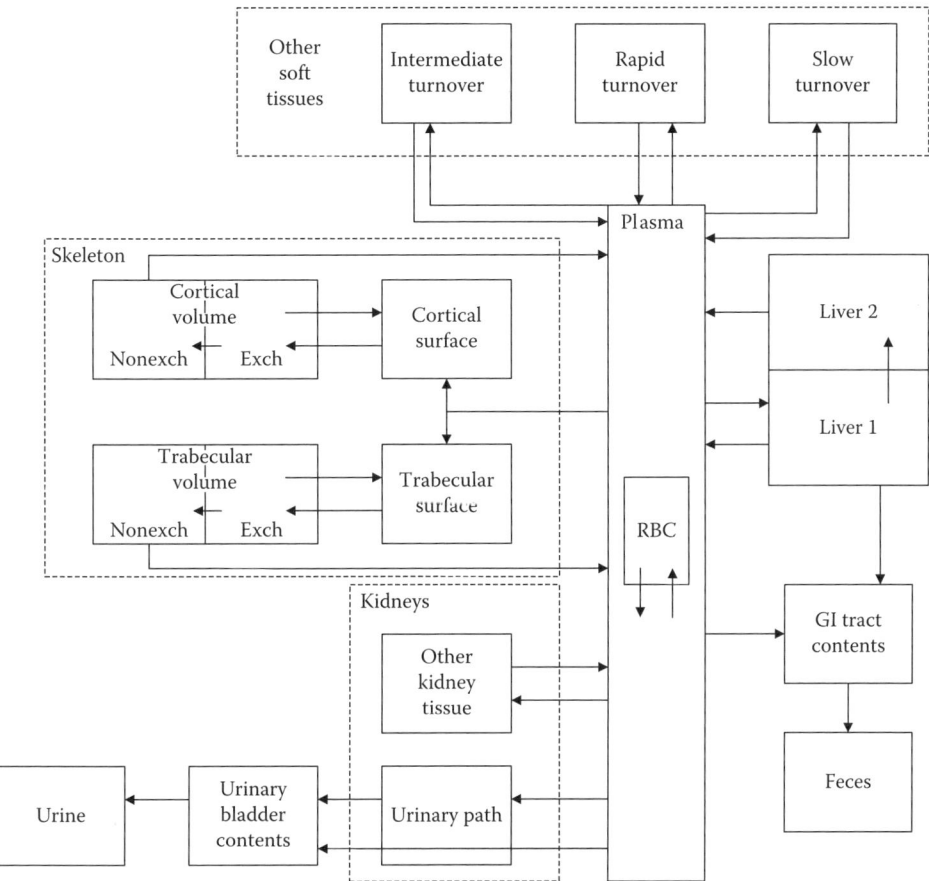

FIGURE 10.5 Metabolic model for uranium. (From International Commission on Radiological Protection. *Age-dependent Doses to Members of the Public from Intake of Radionuclides: Part 3 Ingestion Dose Coefficients*. Oxford: Pergamon Press; ICRP Publication 69; *Ann. ICRP* 25(1); 1995. With permission.)

While uranium is deposited in an organ or tissue, it will decay and its radiations will be emitted. Radioactive progeny may also decay in the same organ or tissue and provide its radiations also. In most cases, the energy from the emitted α- and β-particles are assumed to be absorbed in the organ or tissue in which they are emitted. However, emitted photons will provide dose to neighboring organs and tissues, and those contributions are included in calculations of dose. For simplicity, internally deposited radioactive material is assumed to be immediately and uniformly distributed throughout the volume of the organ/tissue in question.

DERIVED AIR CONCENTRATION

When considering air monitoring or respiratory protection, as well as when evaluating air monitoring results, the occupational health physicist typically uses the "derived air concentration" or DAC as a guide. Determining the value of the DAC is a three-step process. First, a dose coefficient or dose conversion factor must be determined. These values are radionuclide, and intake pathway dependent, and their derivation is nontrivial. They represent the dose received per unit intake of radionuclide, and typical units are sieverts per becquerel (Sv/Bq). Once the dose coefficient is obtained, an "annual limit on intake" or ALI can be calculated. The quotient of the ALI and the amount of air breathed by Reference Man in a working year (250 days, 8 h/day of light activity) is the DAC.

Dose coefficients are calculated by assuming a unit intake and calculating the dose to each organ and tissue represented. Effective (whole-body) dose is calculated by taking the dose to each organ and tissue, multiplying by appropriate tissue weighting factors, and summing those values. Tissue weighting factors are provided for a series of organs and tissues and represent the relative contribution of the organ/tissue's dose to the total detriment to the whole body. Also considered when calculating internal dose is "commitment" over an extended time period with the realization that internal dose is not always delivered in a short period of time. Committed effective dose is calculated by determining the number of transformations as described above in the organ or tissue over a 50-year time period (ICRP, 1990). The dose coefficient is then the value calculated for committed effective dose per unit intake assumed. Intake of an initially pure parent is assumed and radioactive progeny are included in this calculation.

The ALI is easily calculated by determining the quotient of the dose limit (0.05 Sv or 5 rem for whole body, 0.5 Sv or 50 rem for organ/tissue) and the dose coefficient. Under the current system used in the United States, the ALI would be the lowest of the values calculated for committed effective dose and committed equivalent dose to each organ or tissue. Regulatory bodies using recommendations described in ICRP Publication 60 (ICRP, 1990) have somewhat different requirements. Under those recommendations, deterministic (organ/tissue) effects are not considered in the determination of an ALI, and the reference dose used is 0.02 Sv.

As stated above, once the ALI is determined, the DAC is calculated by determining the quotient of the ALI and the amount of air breathed by Reference Man in a year. For light work, the default condition for calculation of occupational DACs, the amount of air breathed in a year is 2400 m^3 (ICRP, 1994a).

SUMMARY

This chapter described the processes in which inhaled material deposits in the respiratory tract, is translocated through clearance or absorption to other organs and tissues, and ultimately causes dose and equivalent dose to those organs and tissues. Also discussed is the determination of dose coefficients, annual limits on intake, and derived air concentrations for use by the operational internal dosimetrist and operational health physicist.

REFERENCES

International Commission on Radiological Protection. *Report of the Task Group on Reference Man*. Oxford: Pergamon Press; ICRP Publication 23; 1975.

International Commission on Radiological Protection. *Limits for Intakes of Radionuclides by Workers*. Oxford: Pergamon Press, ICRP Publication 30; *Ann. ICRP* 2(3/4); 1979.

International Commission on Radiological Protection. *Recommendations of the International Commission on Radiological Protection*. Oxford: Pergamon Press; ICRP Publication 60; *Ann. ICRP* 21(1–3); 1990.

International Commission on Radiological Protection. *Human Respiratory Tract Model for Radiological Protection*. Oxford: Pergamon Press; ICRP Publication 66; *Ann. ICRP* 24(1–3); 1994a.

International Commission on Radiological Protection. *Dose Coefficients for Intakes of Radionuclides by workers, Replacement of ICRP Publication 61*. Oxford: Pergamon Press; ICRP Publication 68; *Ann. ICRP* 24(4); 1994b.

International Commission on Radiological Protection. *Age-dependent Doses to Members of the Public from Intake of Radionuclides: Part 3 Ingestion Dose Coefficients*. Oxford: Pergamon Press; ICRP Publication 69; *Ann. ICRP* 25(1); 1995.

National Council on Radiological Protection and Measurements. *Deposition, Retention, and Dosimetry of Inhaled Radioactive Substances*. Bethesda, MD: NCRP; Report No. 125; 1997.

Part III

Fundamentals of Sampling System Design and Operation for Airborne Radioactivity

11 Basic Air Sampling Equipment

Mark L. Maiello

CONTENTS

Introduction 221
Components of the Sampling Train 222
    Inlet Port 222
    Filter Holders 223
        Bypass Leakage 223
        Electrostatic Losses 224
    Tubing 224
        Characteristics 224
        Wall Losses 224
        Contraction Fittings 225
    Flow Measurement 226
        Rotameters 226
        Mass Flow Meters 230
        Critical Orifices and Venturi Meters 232
    Pumps 233
        Introduction 233
        Terminology 234
        Pump Characteristics 234
        Types of Pumps 235
        Rotary Vane Pumps 236
        Gear Pumps 237
        Lobe Pumps 237
        Diaphragm Pumps 237
        Linear Pumps 239
        Piston Pumps 239
        Rocking Piston Pumps 239
        Oil versus Oil-Less Pumps 239
        Miniature Pumps 240
        Choice of Pumps 240
        Backpressure 240
        Pulsation Control 242
References 242

INTRODUCTION

Sampling of radioactivity in air consists of what appears to be straight forward engineering, but is in fact, a quite sophisticated technology supported by detailed and well-documented research and applied practice.

The airborne radioactivity must be captured using a trap of some sort. The process of capturing the radioactivity requires sampling air by drawing it to the trap using tubing and a motive force—an air pump. Once captured, the radioactivity must be analyzed to quantitatively determine the magnitude of the radioactivity captured. However, this value of radioactivity is only that which was captured. In the effluent or ambient air from which the sample was obtained, there is a total amount of radioactivity. The relationship between the two must be bridged to obtain a value of radioactivity that can be used to determine whether a health hazard exists. At this point, despite the relatively straightforward engineering of sampling, somewhat complicated procedures and professional judgments must be invoked to obtain a usable value of radioactivity.

In theory, the usable value the analyst desires is not the total amount of radioactivity in the atmosphere, although in some cases that knowledge is useful. The usable value is the concentration of the radioactivity—the activity per unit volume. This is usually expressed in $\mu Ci\ L^{-1}$ or $mCi\ L^{-1}$ and/or in the various prefixed versions of $Bq\ L^{-1}$ or $dpm\ L^{-1}$. Radiation dose to humans is time dependent and related to the airborne radioactive concentration. The longer one breathes a radioactive contaminant, the larger the dose to the organ systems of the body will become. Since breathing can be consigned a volume rate, for example, $20\ L\ min^{-1}$, the concentration of the radioactive contaminant becomes a useful value for determining the hazard to exposed individual. By calculating the total amount of radioactivity inhaled (activity per unit volume multiplied by breathing rate), calculations and modeling techniques can be invoked to determine the radiation dose to the exposed individual. This is the reason for all radiation and radioactivity measurements.

These procedures and judgments used to determine the air concentration involve assessments of capture efficiency, corrections for various influences that cannot be controlled, such as the pressure and temperature at the measurement location, and perhaps the use of computer programs to eliminate interfering signals from unwanted radioactivity, as from natural radio-elements like Radon-222.

In this chapter, we will discuss the fundamental engineering features of radioactive air sample collection keeping in mind that the methods presented in the later sections of the book discuss specific procedures, corrections, and cautions that must be applied or at least considered to achieve the desired measurement result.

COMPONENTS OF THE SAMPLING TRAIN

To capture radioactivity, one can construct an air sampling train comprised of an inlet port, a tubing, contraction fittings, a trap (filter or other media), a flow rate measurement device, and an air pump. Although simple in theory, air sampling systems require diligence and care to operate properly and to maintain. Particulate losses can occur in all parts of the system upstream of the trap. In the case of radioiodine, surface reactions can occur that result in loss of sample. Such losses should be quantified, and the final results should be corrected to account for them.

An understanding of the various components of an air sampling train is essential for accurate airborne radioactivity measurements. The following sections of this chapter provide a brief introduction to these various air sampling components.

INLET PORT

In many but certainly not all instances, the inlet port of an air sampling system is a filter holder containing a filter appropriate to the contaminant of interest. An example of a filterless inlet port is that used in an effluent duct or "stack" sampling. In this case, the relatively high air velocity in the duct makes isokinetic particulate sampling necessary using a filter downline of a special inlet port. Not all radioactive air sampling are intended to measure air particulates but even those systems that do not often employ a filter at the inlet port to protect the balance of the sampling train against aerosol or dust contamination.

Filter Holders

The general requirements of filter holders as described by Marshall and Stevens (1980) are the following:

- Provides sufficient support to prevent damage to the filter
- Is well sealed to eliminate air leaks
- Does not cause inertial effects on air particles resulting in preferential collection of certain particle sizes
- Minimizes turbulence to reduce wall deposition
- Obtains a representative sample of the ambient aerosol at the sampling position at least up to 10 µm in particle diameter
- Is designed so that it can be decontaminated (this mitigates against cross-contamination from repetitive samples)
- Allows easy retrieval and replacement of the filter

Filter holders are comprised of many materials: nylon, polyvinyl chloride/polystyrene, metal, and metal with an anodized aluminum finish, among others. The filter holders or "cassettes" are basically a mounting for the fragile filters. They provide a support, often in the form of a fine metal screen, to maintain the rigidity of the filter against the force of the impinging air flow stream. The lateral supports of the cassette protect the filter from physical damage that may be imparted by the ambient environment. The cassettes may be *open face* such that the sampling area of the filter is unprotected or *closed face* such that only a small-diameter opening in the cassette face allows air to penetrate to the filter. The open face configuration is used when the filter is sampling air directly from the environment. Closed face configuration allows the filter cassette to be connected to other upstream components of the sampling train such as a prefilter or a stack sampling line. In this set up, the small diameter opening serves as a connecting port.

Depending on their design and materials of construction, filter holders may suffer from phenomena that can have detrimental influences on sampling results.

Bypass Leakage

A filter holder supports a filter on a backup screen inside a base that also incorporates an outlet fitting for downline connection to the remainder of the sampling train. The filter is held in place along the perimeter with a ring (open face configuration) or by inserting a cap similar to the base into the ring. The cap incorporates a fitting for connection to the inlet tube (closed face). This arrangement is sometimes referred to as *press-fit* because the pieces must be connected snuggly to hold the filter in place without air leaks (bypass leakage) occurring along the seam where the two pieces couple. This is the design of filter holders of 25 mm and 37 mm diameter of acrylic-copolymer construction commonly used in the United States. Leakage will cause particulate matter ordinarily captured on the filter to bypass collection resulting in an underestimated measurement of the aerosol.

It has been recommended that press-fit cassettes be assembled using a tool such as a mechanical press (Baron, 2003). A press can provide a consistent amount of force, mitigating against a leaky coupling that could occur with manual sealing of the cassettes. However, coupling the two pieces together using too much force can damage the plastic, which may also produce a leak. Covering the seal around the cassette perimeter with shrink bands or tape will not prevent bypass leakage.

Bypass leakage can be detected visually by observing the filter edges. A good seal will show a clean edge around the filter perimeter, although this finding is subject to a false-negative conclusion. A bypass leak may reveal itself by dust streaks on the filter edge or by an incomplete impression mark made by the cap or ring. Further investigation may reveal dust deposits on the cassette wall where the base and cap or ring join.

Another cause of leaks is the use of incompatible backup pads and filters. These circumstances require leak testing before putting a cassette batch into use. This test is described in Baron et al. (2002).

Electrostatic Losses

Some filter-holder materials are nonconductive, for example, PVC/polystyrene and nylon, and therefore may hold an electrostatic charge. Charging may occur when cassettes are rubbed against other surfaces (triboelectric charging). The charge is retained if the cassette is not grounded. Aerosols may become charged as well. The classic examples are the radionuclides in the Rn-222 chain of decay such as Po-218 and Pb-214. Depending on several factors, including the voltage on the collecting surface and the relative humidity of the air, electrostatic fields may collect charged aerosols such as radon progeny with an effective flow rate easily approaching 1 LPM (Maiello, 1986). Also see Maiello and Harley (1989, 1990) for a calculation of charged radon progeny fractions. Therefore, electrostatic forces of attraction and repulsion may affect aerosol sample collection to a significant degree. Whatever is collected on the surface of the cassette rather than on the filter is lost to the radioactive counting system ultimately producing a negatively biased result (Baron and Deye, 1990).

To avoid electrostatic losses, the cassettes should preferably be made of a conductive material such as metal or at the least, a static dissipative material such as plastic incorporating graphite (graphite spray coatings are also available). Another method to overcome this loss is to collect the material on the cassette surface for counting. But such collection may not be convenient or totally effective. A design making the cassette and the filter into one unit that can be efficiently counted is also another solution to this problem (Puskar et al., 1992).

Tubing

Characteristics

Sample line tubing is constructed of many materials. A few characteristics are important to consider for radioactive and general air sampling:

- Chemical interaction with the effluent material should not occur. This is especially important when sampling for radioiodine because polyvinylchloride, copper, and Buna-N will chemically interact with iodine causing losses. Polyethylene, aluminum, Teflon®, carbon, and stainless steels are preferred tubing materials (Kabat, 1983).
- Electrostatic plate-out must be avoided. Much like the problem associated with plastic filter heads, loss of particulate matter due to electrostatic forces can be considerable. Liu (1985) reports this phenomenon occurring in polyethylene and polytetrafluoroethylene (Teflon) tubing. Flexing of the tubing may produce electrostatic buildup in an analogous fashion to piezoelectric charging. Therefore, conductive tubing that can dissipate charge buildup is preferred, for example, polyvinylchloride tubing that appears to behave much like metal tubing.
- Mechanical rigidity against vacuum forces must be sufficient. The tube walls must not collapse when the pump creates the low pressure necessary to draw sample air into the sampling system. If it is necessary to bend the sample tube, the inside tube diameter must not change by more than 15% (ANSI/HPS, 1999). There are materials commercially available such as Tygon® and polypropylene-based tubing such as Norprene® that exhibits good rigidity, but the conductivities of these materials should be checked with the manufacturers to avoid the electrostatic problem mentioned above.

Wall Losses

Other more fundamental forces contribute to the loss of particulates in sampling tubes. These include Brownian diffusion, gravity, and turbulent diffusion. A means to quantify loss is by

Basic Air Sampling Equipment

measuring the concentration of the effluent at the tube entry (C_{entry}) and concentration of the effluent at the tube exit (C_{exit}). This is defined as penetration P:

$$P = \frac{C_{exit}}{C_{entry}}. \tag{11.1}$$

With a laminar flow stream entering a straight vertical tube from the top (exit port at the bottom), $P = 1$. However, if the entry is at the bottom, and the particle settling velocity is greater than the air velocity along the tube axis, $P = 0$, owing to the effects of gravity. For a straight, horizontal tube with a nonturbulent air stream, penetration will be affected by gravity as well. Penetration also depends on the tube length, internal diameter, and particle settling velocity.

For a tube with a 90° bend (horizontal to vertical bend), the penetration under laminar conditions becomes dependent on the area of particle deposition at the bend and the internal tube diameter (Vincent, 1989).

However, it is generally acknowledged that the combination of flow rate and tube diameter of most air sampling systems results in turbulent rather than laminar flow (ANSI/HPS, 1999). If the Reynolds number (Re) of a system exceeds 2200, the flow is considered turbulent. An equation for Re is

$$Re = \frac{4\rho q}{\pi \mu d_t}, \tag{11.2}$$

where ρ is the gas density (kg m^{-3}), q is the volumetric flow rate through the tube (m^3 s^{-1}), μ is the dynamic viscosity of the gas (Pascal-s), d_t is tube diameter (m).

Under turbulent conditions, penetration through a straight horizontal tube can be calculated using

$$P = \exp\left[-\frac{L\pi v_e d_t}{q}\right],$$

where L is the straight length of the tube and v_e is the effective depositional velocity of the particles (the vector sum of the velocities that are in the radial direction away from the center of the tube axis toward the tube walls: turbulent inertial deposition and Brownian diffusion velocity, and the downward directed velocity: gravitational settling terminal velocity).

Penetration at a bend is related to the Stokes number (Stk). The Stk is calculated as follows:

$$Stk = \frac{CU_m D_a^2 \rho_w}{9\mu d_t}, \tag{11.3}$$

where C is the Cunningham slip correction for particles (dimensionless), D_a is the aerodynamic particle diameter (m), U_m is the mean velocity of the gas in the tube (m s^{-1}), and ρ_w is the density of water at 4°C (1000 kg m^{-3}).

When Re is 6000 $\leq$ Re $\leq$ 10,000, the penetration can be correlated with Stk as

$$P = 10^{-0.963 Stk}. \tag{11.4}$$

Owing to the complexity of most sampling systems, the total penetration (or loss) of particulate matter is estimated using numerical methods. DEPOSITION is a numerical code accepted by the U.S. Nuclear Regulatory Commission for use by its licensees (Anand et al., 1993) to demonstrate the characteristics of a sampling transport system.

Contraction Fittings

Although it should be discouraged, it is often necessary to transition from one tube diameter to another within a sampling train. Whenever possible, tube diameters should be chosen to mate securely with filter cassette connectors, flow rate meters, pump connectors, and other components of the sampling train. If it becomes necessary to transition to another diameter, contraction fittings are used. However, these fittings can be another source of particle loss. Transitioning from an inlet diameter d_i (corresponding to an inlet area A_i) to an outlet diameter d_o (corresponding to an outlet area A_o) means that the flow rate through the fitting is forced through some contraction half-angle (θ). Such angles may range from 10° to 90°. It has been found that losses to the fitting walls, WL = 1 – P, increase with increasing half-angle for a given area ratio A_o/A_i, flow rate, and particle size (Muyshondt et al., 1996). Experimentally obtained wall losses correlate well with the following equation:

$$\text{WL} = \frac{1}{1 + (X/ae^{b\theta})^c}, \tag{11.5}$$

where $X = \text{Stk}(1 - A_o/A_i)$, $a = 3.14$, $b = -0.0185$, and $c = -1.24$ (these constants were found by the method of least squares).

At some values of $\text{Stk}(1 - A_o/A_i)$ and θ, for example, $\text{Stk}(1 - A_o/A_i) < 0.25$ and $\theta = 12°$, losses in the contraction fitting are <5%. Losses become significant quite rapidly as $\text{Stk}(1 - A_o/A_i)$ increases for a given θ. When Stk is large, aerosol trajectories become straight lines. The contraction fitting will not influence them so that the wall losses reach an asymptotic maximum defined by

$$\text{WL} = \left(1 - \frac{A_o}{A_i}\right). \tag{11.6}$$

FLOW MEASUREMENT

Rotameters

All parts of the sampling train are important, but the measurement of air flow rate is critical because it is used to calculate the contaminant air concentration. The flow rate device must be able to withstand long periods of use without significant changes to its operating characteristics. For example, the clogging effects of entrained dust must be minimal over long sampling periods, if this will effect the flow reading or else cleaning will be required. Accuracy is another matter that must be considered. These devices are usually calibrated to an intermediate or primary standard (instruments that can be certified to be accurate to ±2% or better based on a volume standard). This guarantees that under the conditions of calibration, the readout of the flow meter is accurate within a specified degree of uncertainty. The calibration must not be affected by field use or else recalibration will be required. In any event, calibration is commonly performed and documented on no longer than an annual basis. More information is supplied in the section on calibration.

Although many devices are available to measure flow rate, the most common and the one used in the methods of this book is the rotameter (see Figure 11.1). Its low cost, adaptability to many sampling trains, and simplicity of use contributed to its popularity.

Rotameters are essentially a tapered tube that is wider at the top than it is at the bottom. A float is placed within the tube such that it does not contact the tapered tube walls. The air stream can flow around the float. The float can be any number of shapes including spherical or cylindrical.

The float is allowed to seek a level that is essentially a balance between the suspending force of the sampled air stream and the mass of the float. The mass of the float and gravity apply a downward force on the float while air velocity pressure, mass of the air displaced by the float (buoyancy), and aerodynamic drag on the float act upward on it.

There will be a pressure reduction across the float (less above it; more below). Other factors, such as turbulent flow in the tapered tube, play a role in the stability of the float. Early designs

Basic Air Sampling Equipment

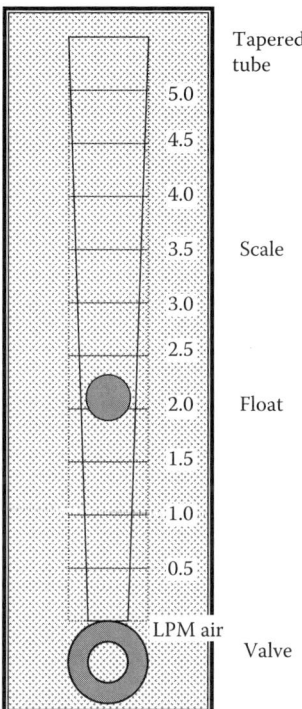

FIGURE 11.1 Rotameter.

incorporated slots in the body of the float to help achieve stability. This would cause the float to rotate (thus the term "rotameter"). It is common now to incorporate such slots or "flow guides" into the tapered tube design. A scale is placed next to or on both sides of the tapered tube so that the final position of the float can be related to the air flow rate in the sampling line.

Rotameters are made from a variety of materials. Typical, general use devices use polycarbonate for the meter body and tube, stainless steel, glass, or aluminum for the float, and stainless steel for fittings and other metal parts. Specialty rotameters use materials suitable such as Teflon for working with ultrapure gas or corrosive flow lines. The flow rate range for a single rotameter is usually a factor of 10, but many sizes exist allowing the investigator to choose one suitable for the work at hand. Some rotameters can measure a minimum of 0.5 LPM; other models up to 2800 LPM. There are even interchangeable models that permit the flow tube to be swapped. Using such a "kit," a range from 0.002 to 22 LPM can be achieved. The air pressure at which rotameters operate safely also varies with model from about 35–200 psi.

Another varying feature of rotameters is the readout accuracy. Manufacturers' claims typically vary between ±2% and ±10%. Because these devices are not highly accurate nor is it always easy to interpolate between the relatively few scale markings typically provided, calibration has heightened importance.

The equation of air flow through rotameters is described by Fisher (1952). A general equation is given by Caplan (1985) as

$$Q = C\sqrt{\frac{1}{2}\pi g \left(\frac{D_t^2 - D_f^2}{D_f}\right)}\sqrt{V_f(\rho_f - \rho_a)\frac{1}{\rho_a}}, \tag{11.7}$$

where Q is the volume flow rate through the rotameter (cm³ s⁻¹), C is the discharge coefficient (dimensionless), g is the gravitational acceleration (cm s⁻²), D_t is the tube diameter (cm), D_f is the float diameter (cm), V_f is the float volume (cm^3), ρ_f is the density of the float (g cm^{-3}), and ρ_a is the density of the air (g cm^{-3}).

Re is defined as Re = $(\rho D_f V_t)/\mu$, where V_t is the terminal velocity of the sphere in an infinite volume of fluid (air). C is a function of air viscosity and the Re may be taken as a constant for the conditions encountered for most applied (occupational) air sampling work. Note that $\rho_f \gg \rho_a$ rendering $\rho_f - \rho_a$ a constant also. The other terms in the equation that are constant are D_f, V_f, g, and π allowing for the combination into a new constant K.

The equation can be rewritten to reflect this as

$$Q = K(D_t^2 - D_f^2)\sqrt{\frac{1}{\rho_a}}. \tag{11.8}$$

The scale reading "R" on the rotameter is equal to $(D_t^2 - D_f^2)$, where D_f is also a constant. Therefore, the above equation becomes

$$Q = KR\sqrt{\frac{1}{\rho_a}} \tag{11.9}$$

and

$$R = \frac{Q}{K}\sqrt{\rho_a}. \tag{11.10}$$

Note that the reading is proportional to the square root of the air density, which will change with the altitude and temperature of the sampling station. However, as the reader can see, the correction will not be large because of the square root proportionality.

The overarching rule of air sampling calibration is that it should be performed with the sampling system operating as it would be in the field (with all sampling components connected), and under field atmospheric conditions (temperature, pressure, and perhaps humidity and particle density simulated in the calibration procedure as closely as possible).

One may make a correction to determine the flow rate or total volume of air sampled at "industrial hygiene standard conditions." These are defined as a temperature of 25°C and a pressure of 760 mm Hg (standard conditions are defined for the pure sciences as 0°C and 760 mm Hg or 1 atm). Alternatively, one may correct a sampling result based on a calibration done under laboratory conditions to compensate for the ambient conditions in the field (air pressure and temperature) to determine the true air flow or total volume of air sampled at the sampling station.

Boyle's gas law states that for a given mass of gas held at constant temperature, pressure is inversely proportional to the volume. The law of Charles and Gay-Lussac states that for a given mass of air held at constant pressure, volume is directly proportional to the temperature. To summarize

$$\frac{PV}{T} = \Re \quad \text{(for a fixed mass of gas)}, \tag{11.11}$$

where $\Re = 8.314$ J deg^{-1} mol^{-1} (the universal gas constant; where T is expressed as degrees Kelvin or K) and $\Re_a = 287.0$ J kg^{-1} deg^{-1} (the gas constant for dry air; 1 mole of which weighs 28.97 g).

Therefore, the ideal gas law for air is

$$P = \rho_a \Re_a T. \tag{11.12}$$

It is seen that for a T equal to the calibration temperature and an air pressure in the field less than the calibration pressure P, the volume of ambient air expands. Thus, the air is less dense. Therefore, the force exerted on the rotameter float will be less at a given air velocity. At an altitude, it will not rise as high as it would have at sea-level conditions for the same flow rate. This can be summarized in the following equation:

$$R_a = R_c \sqrt{\frac{\rho_a}{\rho_c}}, \tag{11.13}$$

where R_a is the rotameter scale reading at an altitude, R_c is the calibration reading, ρ_a is the air density at an altitude (g cm^{-3}), and ρ_c is the calibration air density (g cm^{-3}).

If the ambient air pressure is above that at calibration conditions (ambient T equal to the calibration conditions), the increased air density provides more force and the float rises farther than it would under the calibration conditions for equal flow rates. This is analogous to converting flow rates observed in the field at altitudes greater than sea level into standard condition flow rates (sea level at some specified temperature). For example, if one samples at an altitude greater than sea level but desires a flow rate of 3 *standard* liters per minute, one would have to set the rotameter float at a higher setting than 3 to compensate for the lower air density that, as indicated above, would not cause the float to rise as high as at sea level. The formula is similar to Equation 11.13 with one important change under the square root sign:

$$R_a = R_s \sqrt{\frac{\rho_c}{\rho_a}}, \tag{11.14}$$

where R_a is the flow rate setting at altitude, R_s is the desired standard conditions flow rate, ρ_a is the air density at altitude (g cm^{-3}), and ρ_c is the calibration air density (g cm^{-3}).

Changes in ambient temperature effect air density as well. When T is higher in the field than at calibration, air density decreases due to expansion of the atmosphere. When T is lower in the field relative to that at calibration, the air density will be higher. Therefore, as in Equation 11.13, a higher rotameter reading will be indicated at lower temperatures for the same air velocity as at calibration. If we wished to obtain a standard condition volume flow rate, the rotameter would be set lower than the desired flow using Equation 11.14.

Of course, P and T change simultaneously in the field. If P is inversely related to volume and volume is directly proportional to temperature, the following correction to obtain *actual* air flow rate through the rotameter is needed:

$$R_a = R_c \sqrt{\frac{P_a}{T_a} \times \frac{T_c}{P_c}}. \tag{11.15}$$

To obtain a desired rotameter flow rate for standard occupational conditions

$$R_a = R_s \sqrt{\frac{P_c}{T_c} \times \frac{T_a}{P_a}}. \tag{11.16}$$

The question becomes when to apply these corrections. A suggested action level is when ambient pressure and temperature can induce a 5% or greater deviation from the calibrated flow rates of the rotameter. Note from the ideal gas law (Equation 11.12) that air density is proportional to air pressure (at constant T) and that both pressure and density decrease exponentially with altitude (characteristic of compressible fluids like air). Ideally, the pressure and density are reduced to e^{-1} or 37% of the pressure and density at sea level when an altitude of 8 km is reached. However, the dynamic meteorology of the atmosphere does not make the exponential relationship exact. The mean decrease in temperature with altitude in the troposphere is about 6.5°C km^{-1}. Again, this is an average that may not be the case for any particular sampling time (Williamson, 1973). Assuming average conditions at 8 km, the ratio of the pressure at altitude to that at sea level where the rotameter was calibrated would be 0.37. The temperature at altitude would be (6.5 × 8) degrees cooler as that at calibration (assumed to be 25°C at sea level) or −27°C. Using the absolute Kelvin scale of temperature, this is a ratio of calibration temperature to ambient temperature of (298 K/246 K) or 1.2. Multiplying 0.37 and 1.2 and taking the square root (Equation 11.15) yields a correction to the calibrated flow in a rotameter used under these temperature and pressure conditions of 0.67, which is certainly of a magnitude to justify applying the correction. Considering that actual temperature and pressure profiles can easily vary from the textbook predictions, it is important to measure both when using a rotameter under conditions that are suspected to be significantly different from those under which it was calibrated. Further, one must consider the ultimate use of the air sampling data. Researchers collecting nonoccupational scientific data may desire to maintain error terms below 5%. And, collecting air sampling data for comparison between locations under varying conditions of pressure and temperature may require conversion to some defined standard conditions. When considering measurements in support of occupational dose estimation, calibration should be referenced to the altitude that the sampler is used because regulatory pollutant concentration limits are based on the actual volume of air breathed and not a volume corrected to another set of conditions, for example, sea-level conditions.

The above discussion is concerned only with the effects on the rotameter reading brought about by pressure and temperature changes in the ambient atmosphere. It is possible to induce nonstandard conditions or conditions differing from calibration conditions, within the sampling train. As pointed out by Craig (1971), it is the usual practice to place the rotameter downstream from the air sampling trap or device. This protects the rotameter from contaminants in the air stream. However, if the sampler induces a pressure drop downstream, it will mean that the rotameter is operating at an air pressure less than ambient. For example, when sampling a very dusty atmosphere, a layer of dust will progressively build up on a filter, clogging pores and producing a pressure drop in the balance of the sampling train. As the pressure in the line decreases, the flow rate and thus the rotameter reading will change. Even if one attempted to hold the flow rate constant by continually resetting the rotameter back to the starting flow rate, the reading R, as we saw above, translates to a different actual flow rate with every reduction of pressure in the sampling line.

To compensate for this sampling line pressure drop effect on the rotameter, several techniques can be employed:

- Calibration of the rotameter must be performed with the sampling train operating just as it would in the field.
- Low flow rates should be used for filtration-sampling of dusty atmospheres to preclude variable pressure drops in the sampling line.
- Include a pressure monitor just upstream of the rotameter to detect pressure drops caused by the sampling mechanism or by line leaks.

Mass Flow Meters

To avoid the pressure and temperature corrections associated with rotameters, one can employ the mass flow meter. Rotameters measure the rate of air volume passing a point in the sampling line. Mass flow meters measure the mass of air being transported in the sampling line. As we saw in the section "rotameters," volume (and therefore volume rates) can be changed by varying the ambient air pressure and temperature. Air mass, on the other hand, is invariant. If one fills a balloon with gas, the volume will change if one decreases the outside pressure (the balloon will expand) or if the outside temperature decreases (the balloon will shrink). However, in both cases, assuming the balloon does not leak, the mass of air, that is, the number of air molecules, remains the same.

Mass flow meters operate by measuring a change of electrical resistance between two points (refer to Figure 11.2). A capillary tube is used to siphon off a small fraction of the flowing gas from the main stream inside the flow meter. The resistance change is induced by introducing heat to the measured gas at an upstream point in the capillary tube and at another point downstream. Heat will be carried

Basic Air Sampling Equipment

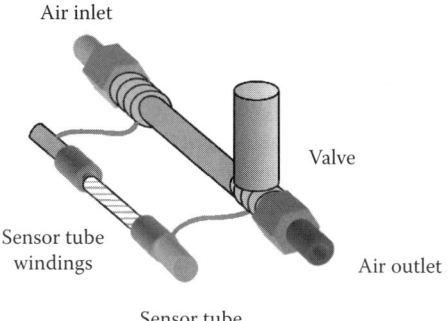

FIGURE 11.2 Mass flow meter.

to the downstream point within the meter by the gas flow. Temperature transducers change electrical resistivity with changes in temperature. They are used to detect the upstream and downstream temperature differential. The transducers are part of a time-tested electrical circuit known as the Wheatstone bridge that can accurately measure the difference in resistivity (Wolf, 1973).

The Wheatstone bridge (Figure 11.3) requires power to supply current to four resistors arranged on one side each of a box-shaped circuit. The current in the circuit is therefore split between each pair of resistors. One pair has known resistances R_1 and R_2. The other pair consists of a variable resistance R_3 and an unknown resistance R_x. In the mass flow meter, the unknown resistance is that induced by the temperature difference on the flowing gas. A current measuring device resides at the center of the box where it can monitor the current flowing between each side of the circuit. When no current flows between each side (achieved by adjusting the variable resistor automatically), it means that the ratio of resistances R_1 and R_2 equal the ratio of resistances R_3 and R_x. This also means that the unknown resistance can be determined since

$$\frac{R_1}{R_2} = \frac{R_3}{R_x} \tag{11.17}$$

and

$$R_x = R_3 \frac{R_2}{R_1}. \tag{11.18}$$

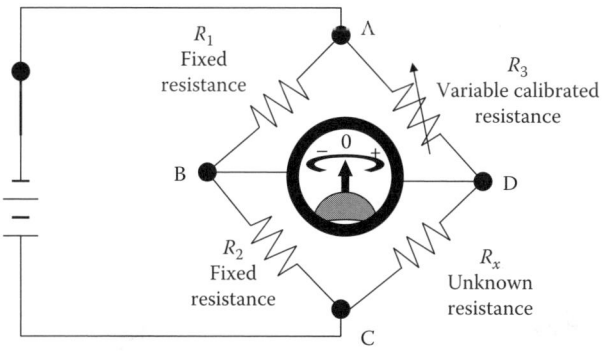

FIGURE 11.3 Wheatstone bridge circuit.

The mass flow readout can be calibrated to interpret different values of R_x as different gas mass flow rates. The measurement is ultimately based on resistivity and not on any parameter of the gas. Therefore, there is little dependence on line pressure or temperature changes. The deviations in accuracy of mass flow meters are typically 0.10% deg^{-1} and 0.02% psi^{-1} pressure change (Swearingen, 1999). Another advantage of the mass flow meter is that it can provide a recordable electronic signal of the measured air flow. Typical flow ranges are, like rotameters, handled by different models. Flow rates less than 1 or 2 standard cm^3 of air per minute or higher than 1000 standard liters per minute are typically available through commercial manufacturers.

Critical Orifices and Venturi Meters

The critical orifice is a flow rate controller with one setting that is dependent on the diameter of the orifice (Figure 11.4). A thin circular plate with an orifice cut into its center so that the hole is sharp edged constitutes the critical orifice device. The device works by constricting the air flow through the orifice and measuring the change in pressure as the air streamlines are forced together to traverse the orifice. Under normal atmospheric conditions, the pressure in the *vena constricta* or narrowest part of the flow stream must be greater than 47% of the upstream pressure so that a stable flow rate is achieved unaffected by a further reduction in downstream pressure and if invariant conditions upstream are maintained. The difference between the *vena contricta* pressure and that upstream may be directly measured with a manometer, but commonly no field measurements are made except to initially calibrate the critical orifice. Once calibrated for flow rate, the orifice is installed into the sampling line. Different flow rates require replacement with an orifice of different diameter.

The requirement for the large pressure drop (47 kPa) across the orifice is a disadvantage because a high-power pump is needed. Further, the orifice must be machined carefully to achieve an accurate air flow. In practice, because the air flow in the *vena constricta* should achieve the speed of sound, the orifice can be damaged by routine use, requiring frequent inspections. The orifice must also be kept clean. This can be accomplished by employing a filter in the air sampling train upstream of the orifice. Critical orifices must be calibrated using a bubble flow meter or other calibration device to determine if the claimed nominal flow rate is accurate. Leaks in the air sampling system, particularly

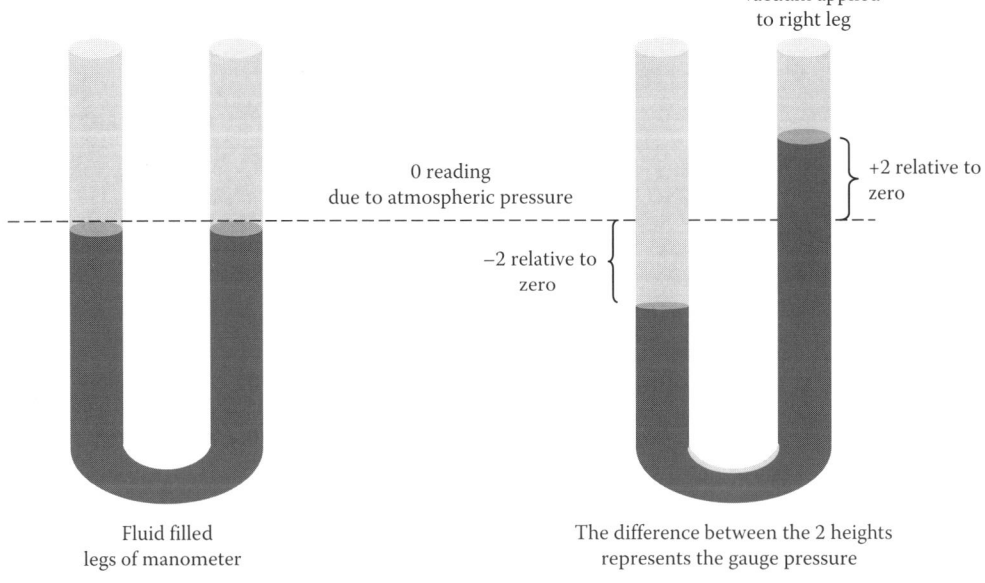

FIGURE 11.4 Manometer.

Basic Air Sampling Equipment

at fittings, will compromise the use of a critical orifice. These disadvantages are offset by the advantages of reliability (no moving parts), small size, relatively low cost, and, unlike the rotameter, critical orifices (and the related venturi meter) can be made so that they are not dependent on the pressure at the inlet. These qualities make the device attractive for field use. Critical orifices can be made in many diameters to produce flow rates from 0.1 to several L min^{-1} as long as suitable air pump power is available.

The venturi meter, though more expensive to construct than the critical orifice, is claimed to be a more accurate flow controlling device (Wang and Zhang, 1999). Instead of a sharp-edged plate, the venturi meter consists of a converging inlet, a throat, and a diverging outlet that provides the resistance to the air flow to create the needed difference in upstream and downstream pressures (P_1 and P_2, respectively) to allow the flow rate to be measured.

The mass flow rate (kg s^{-1}) through the venturi meter increases as the ratio of P_1 to P_2 decreases. The maximum occurs when the pressure ratio reaches a critical point of 0.53. The critical pressure of the gas (P_c) occurs at this point which is in the throat of the venturi. When $P_2 \leq P_c$, the throat pressure = P_c and mass flow rate is maximum. When $P_2/P_1 < P_c/P_1$, the mass flow is independent of the downstream pressure P_2 and will remain constant despite any changes to P_2. The gas velocity in the throat will equal the speed of sound and the throat is "choked."

A theoretical analysis of the critical air flow conditions results in the following equation for volumetric flow rate Q which includes a term C_d (coefficient of discharge) to account for air flow friction (Wang and Zhang, 1999):

$$Q = C_d A_t \sqrt{\frac{\gamma P_1}{\rho_1} \left[\frac{2}{\gamma + 1} \right]^{\frac{\gamma+1}{\gamma-1}}}, \tag{11.19}$$

where A_t is the critical orifice throat area (m^2), γ is the specific heat ratio of the gas, P_1 is the upstream pressure (Pa), and ρ_1 is the upstream gas density (kg m^{-3}).

When the air temperature is 293 K, P_1 is 1.013×10^5 Pa, ρ_1 is 1.19 kg m^{-3}, and $\gamma = 1.41$, the throat diameter (d_t) (mm) of the critical orifice can be determined using Equation 11.19 to obtain

$$d_t = \sqrt{\frac{Q}{157.6 C_d}}, \tag{11.20}$$

The discharge coefficient is determined experimentally. In optimized designs of the venturi meter, it is >0.9. Thus, the flow rate Q may be determined for the throat diameter d_t.

Optimally designed venturi meters with critical pressure drops as low as 9 kPa can be achieved. This is important because a high-powered pump is not necessary to achieve the required critical flow condition.

PUMPS

Introduction

Pumps may be used to create a partial vacuum that moves air molecules through the sampling line; thus, they are often referred to as "vacuum pumps." The sampling line is not a closed system being open at the input side to accept effluent or ambient air and also open at the exhaust side of the pump. As opposed to the creation of a sustainable vacuum by evacuating a sampling bottle or test chamber, the vacuum created in a sampling train is sustainable only while the pump operates and is immediately lost once the pump is turned off. Obviously, the purpose of pump action for continuous sampling, as in a sampling train, is to create an adequate air flow rate. An adequate flow rate is one that is estimated to collect at least the minimum detectable radioactivity per unit of air volume within a "reasonable" time period, for example, commensurate with period of effluent release and that can be measured with the detection equipment at hand. Grab sampling is associated with evacuated sampling containers that are opened to the atmosphere when ready to collect an air sample (see the methods section on measuring gamma-ray emitting gases).

An excellent source of information on vacuum science is the *Vacuum and Pressure Systems Handbook* (Gast, 2007) that can be found online at www.gastmfg.com.

Terminology

Since this is a discussion about creating a reduced air pressure in the sampling line, it makes sense that much of pump terminology is associated with this one parameter. Pressure is measured and stated by either including or not including atmospheric pressure in the final value.

Gauge pressure is the air pressure within the sampling train only and does not include the atmospheric air pressure in the final reported value. Gauge pressure is the pressure relative to the atmospheric pressure. Therefore, it may be positive if higher than atmospheric pressure or negative if lower than atmospheric pressure. Gauge pressure is designated in the old English system of units as psig (pounds per square inch gauge).

Absolute air pressure, designated analogously as psia, is the sum of the gauge pressure and the atmospheric pressure. Atmospheric pressure, which varies temporally and with altitude, is a mean of 14.7 psia at sea level. If we create a vacuum in a sample collection bottle of −6 psig, we could also report it as (−6 psi + 14.7 psi) = 8.7 psia.

The classic measurement device of gauge pressure is the manometer (refer to Figure 11.4). This device is a "U"-shaped tube with left and right legs open to the atmosphere. The tube is partially filled with mercury which, because atmospheric air pressure is equal in both legs, will be of equal heights. This is the zero mark of the manometer. When a vacuum pump is attached to one leg (say the right leg), the mercury in the right leg will be pulled upward as air is evacuated from that leg and the mercury rises to fill the void. The height attained by the mercury in the right leg and the associated depression of the mercury in the left leg (both measured in inches or millimeters) is measured on a scale. The difference between the two is a measure of the gauge pressure. Note that atmospheric pressure is zeroed out by measuring the difference in both legs.

The analogous measurement of absolute pressure is done with a barometer. This device consists of a glass column open at one end over a reservoir of mercury. The mercury will seek a height in the column based on the force of the atmospheric pressure on the reservoir. Higher atmospheric pressure forces the mercury higher up the column and *vice versa*. The measurement is reported as the height of the mercury column in mm Hg or in Hg. This is direct reading of ambient air pressure. Thus, a barometric reading is a reading of absolute pressure.

Air pressure may also be measured in inches of mercury, that is, "in Hg." Mean sea level air pressure is equivalent to 29.92 in Hg. Other units include the atmosphere (atm) where 1 atm = 14.7 psia. In fact, there are a confusing number of pressure units including inches of H_2O (in H_2O), millimeters of Hg (mm Hg, also known as 1 Torr), pounds per square foot, kilograms per square centimeter, and Newtons per square meter (also known as the Pascal or "Pa"). A summary of free air capacities of selected vacuum pumps is shown in Table 11.1.

Pump Characteristics

Pumps can be rated in many ways. One parameter that is often used is *vacuum rating*. This is the vacuum that the pump can create in a closed system when operating at sea-level conditions of 29.92 in Hg. Since the latter can change with altitude, the *adjusted vacuum rating* is often used:

$$\text{AVR} = P_{\text{atm}} \times \frac{\text{NVR}}{P_{\text{std}}},$$

where AVR is the adjusted vacuum rating, P_{atm} is the ambient atmospheric pressure, NVR is the nominal vacuum rating (provided by the pump manufacturer), and P_{std} is the standard atmospheric pressure.

TABLE 11.1
Free Air Capacities of Selected Vacuum Pumps

Vacuum Pump	Maximum Vacuum (in. Hg)	Free Air Capacity (0 in. Hg ambient pressure) (CFM)
Diaphragm	23.5–29.0[a]	0.5–3.5
Linear	5.0–12.1	0.4–8.5
Rotary vane (oil)	10.0–28.0	1.3–55
Rotary vane (oil less)	15.0–27.0	1.3–55
Gear	0.004	0.9
Piston	27.5–28.5	1.3–10.5
Rocking piston	25.5–29.0[a]	1.2–2.7
Lobed	15	—

Source: Adapted from Gast, *Vacuum and Pressure Systems Handbook*, Gast Manufacturing, Inc., Benton Harbor, MI, available at www.gastmfg.com/pdf/vacpresshadbk.pdf, accessed 12/27/07, 2007.
Note: The maximum vacuum theoretically possible is 29.92 in Hg (14.7 psi) at sea level or "perfect vacuum."
Gear pump information from www.coleparmer.ca "vacuum gear pump."
[a] 29.0 in. Hg vacuum achievable with a 2-stage version of the pump.

All variables must be in consistent units of air pressure.

For example, a pump with a nominal vacuum rating of 28 in Hg will have an adjusted vacuum rating, in Denver, Colorado (24.8 in Hg atmospheric pressure), of

$$\text{AVR} = 24.8 \times \frac{28}{29.92}, \tag{11.21}$$

AVR = 23.2 in Hg, which is only 83% of the manufacturer's rating.

Another rating perhaps more applicable to air sampling is the *air removal rate*. The *free air or open capacity* is simply the air removal rate through the pump without connections to any sampling equipment. The air removal rate will decrease once a "vacuum load" is placed on the pump. For example, if the pump is removing air from a chamber, the air removal rate will decrease as the pressure in the chamber decreases. However, for open systems like continuous air samplers, the air removal rate gives an indication of the air flows that can be achieved by the pump.

It should be noted that the maximum vacuum that many pumps used in air sampling can attain is not much more than about 95% of "perfect" vacuum.

Types of Pumps

Pumps contain chambers that are alternately filled and evacuated with air through mechanical action. The mechanical action drives a parcel of air out of the chamber. This evacuation of air in the pump chamber creates a low-pressure zone relative to atmospheric pressure. If the chamber can be opened to the ambient atmosphere, the low pressure automatically causes filling of the chamber with another parcel of air and the sequence is repeated. The input side of the pump is the source of the air that is driven through the pump. The exhaust side of the pump dumps the compressed air to the ambient atmosphere. If this pump were connected to a sampling train, the vacuum would be translated through the train drawing air completely through it.

Think of a piston like that in a car engine. It resides in a cylinder through which it glides. On the down stroke, the air in the cylinder beneath the piston is compressed creating a partial vacuum above the cylinder. The compressed air can be channeled away through a one-way exhaust valve that only opens on the down stroke. On the up stroke, an inlet valve opens allowing ambient air to fill the cylinder that is under partial vacuum. The process is repeated causing air to be driven from the inlet to the exhaust. This is the case in an air sampling train. The effect is to continually draw parcels of air through the train. If the pump were connected to a closed system such as a sampling canister, the source of inlet air would eventually be diminished as the air in the canister was drawn away.

The pump family can be divided into two broad categories: *positive displacement* and *nonpositive displacement, dynamic* types. Positive displacement pumps move air by repetitively creating a known volume that fills with air and is later decreased in size to force the air out. The piston pump mentioned above is an example. Dynamic pumps, also termed fans or blowers, induce momentum (and kinetic energy) to air and channel it away to do useful work. This is usually accomplished by using impellers or blades turning at high speed. Units are termed "fans" if they provide about 0.5 psi pressure on the outlet side, while they are termed "blowers" if they provide pressure above approximately 5 psi. Since these devices are generally not applicable to radioactive air sampling, they will not be discussed further.

Positive displacement pumps are further divided into *rotary* and *reciprocating* types. These are fairly self-explanatory. The mechanisms of rotary pumps use rotating action to create the motive force. Examples in common use for air sampling include vane pumps, gear pumps, and lobe pumps. Reciprocating pumps use the repeating linear motion of pistons or piston-like mechanisms to produce air movement. Common types include the diaphragm and piston pumps.

Rotary Vane Pumps

This pump is configured as a cylinder inside which a rotor turns eccentrically, that is, off-axis. The off-center centrifugal rotation of the camshaft brings it into periodic contact with the inner walls of the cylinder. Vanes that either slide in and out with the eccentric movement (sliding vane) or that are hinged to the camshaft so that they ride between it and the cylinder walls (hinged vane) move to produce or decrease air spaces. These air spaces are repetitively opened for filling with air and then closed to force the air out (Figure 11.5). One advantage of this design is that inlet and outlet valves (recall the automobile piston example above) are not required. The moving vanes completely control

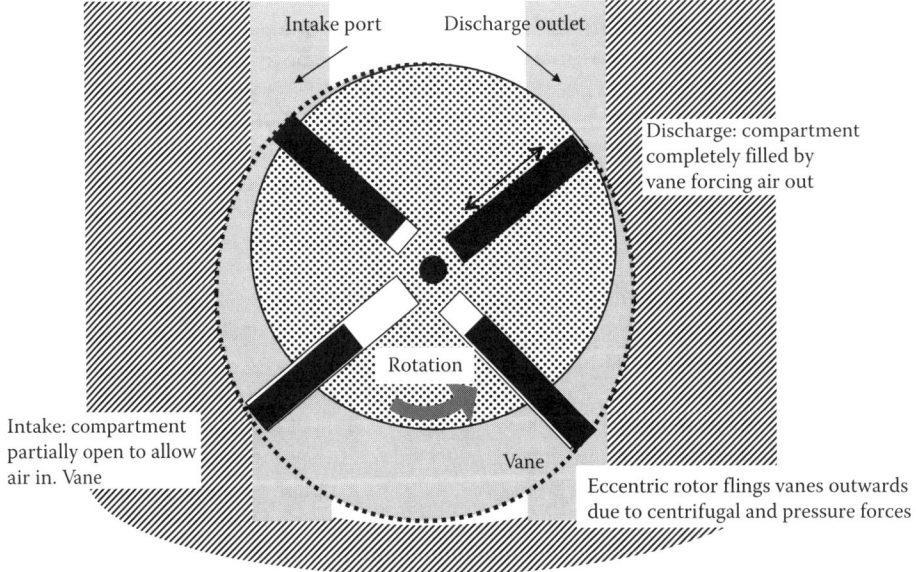

FIGURE 11.5 Rotary vane pump.

Basic Air Sampling Equipment

the air flow from inlet to outlet port. Another advantage is the pulse-free air flow produced by these pumps (see diaphragm pumps described below). Little vibration is created that can be transmitted to other sampler components. Rotary vane pumps are capable of very high free air flow rates.

Gear Pumps

An *internal* gear pump also uses an eccentric cylindrical geometry of two gears: an outer gear (rotor) and an inner or idler gear. The rotor is in contact with the semicylindrical casing of the pump. As the gears rotate, the eccentric configuration will cause the teeth of the idler to come out of the mesh with the rotor. These voids can then be filled with air through the pump intake. The rotation will eventually bring the teeth into the mesh, decreasing the voids and compressing the air toward the pump outlet (Figure 11.6). A crescent-shaped blade located at the out-of-mesh position makes contact with the teeth of the idler gear to prevent air from flowing backward from outlet to inlet.

An *external* gear pump uses two gears that mesh without an eccentric configuration. This arrangement places two identical gears central to the inlet and outlet of the pump. They are arranged one above the other so that they counter rotate to paddle air through the pump. Air is compressed where the teeth mesh and expelled where they come out of contact. On the opposite side of the mesh area, the teeth of both gears contact the interior of the pump casing to prevent air leakage (Figure 11.7).

Lobe Pumps

A lobe pump is much the same as an external gear pump. However, the gears are replaced with bilobed, trilobed, or multilobed impellers. This geometry allows continuous sealing between the lobes. Special timing gears keep the impellers properly synchronized (Figure 11.8).

Diaphragm Pumps

The principle of a diaphragm pump is much simpler than the others discussed above. Instead of a sliding seal used in the rotary pumps described above, this reciprocating pump uses the short stroke of a connecting rod to rapidly move a flexible diaphragm across a small chamber (Figure 11.9). The alternate up and down strokes are coordinated with the opening and closing of outflow and intake valves. When the air is expelled from the chamber by forcing it through the open outflow valve (intake valve closed), a vacuum is created, which is filled when the diaphragm is relaxed and

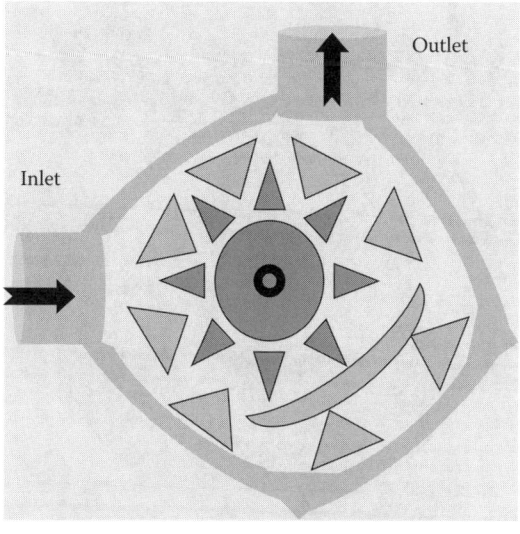

FIGURE 11.6 Internal gear pump.

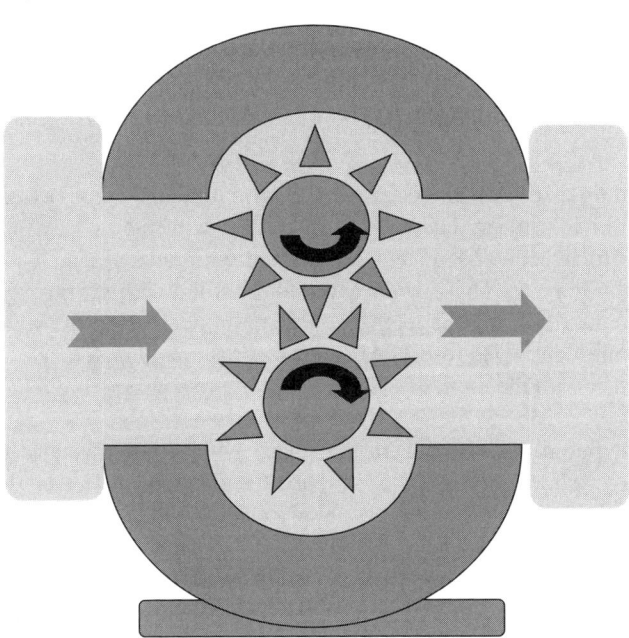

FIGURE 11.7 External gear pump.

the intake valve is opened (outflow valve closed). Diaphragm pumps are relatively quiet but the reciprocating motion of the electrically powered connecting rod will impart a vibration to the pump housing that can be transmitted to other components in the sampling train. The rapid pulsation of the air flow will also be apparent especially when reading the rotameter. However, there are means to mitigate the latter effect.

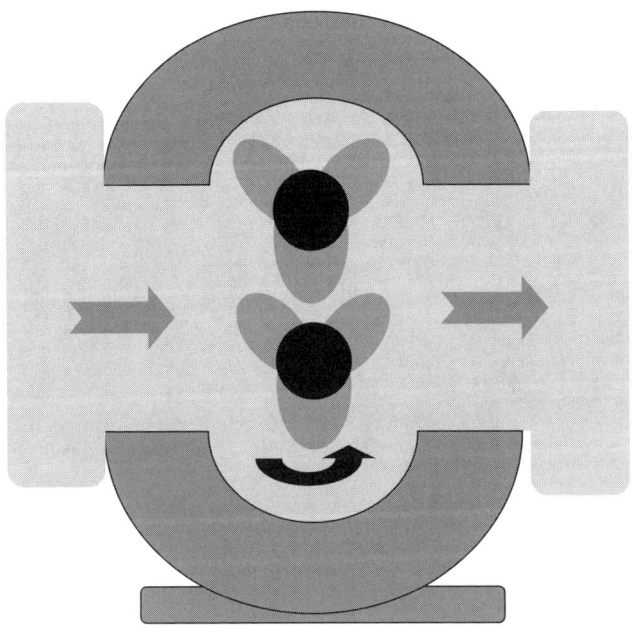

FIGURE 11.8 Lobe pump.

Basic Air Sampling Equipment

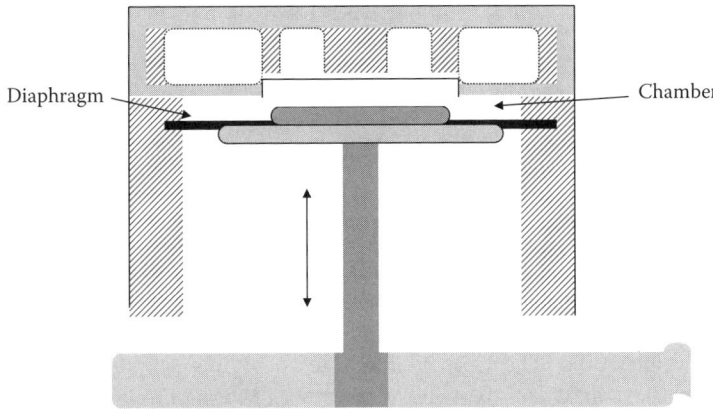

FIGURE 11.9 Diaphragm pump.

Linear Pumps

A linear pump is a close relative of the diaphragm pump. In fact, the principle is the same. The difference lies in the method used to impart the reciprocal motion to the diaphragm. The linear pump eschews a purely mechanical means of oscillation for an electromagnetic method. The result is extremely quiet operation and smooth air flow. However, the air flow that can be achieved is relatively less than other types of pumps but is probably still adequate for many air sampling methods.

Piston Pumps

This pump also uses a reciprocating motion to create vacuum. Instead of a flexible diaphragm, this pump produces vacuum using a traditional piston arrangement. As with the diaphragm pump, intake and outflow valves coordinated to the movement of the piston are required. Piston pumps can create a high vacuum but they also exhibit relatively high vibration and noise levels.

Rocking Piston Pumps

This design rigidly fixes a flexible cup to a connecting rod. As seen in the diaphragm pump, there is an "eccentric" connected to the rotating main axel of the pump motor that imparts a reciprocating motion to the connecting rod. As the rod moves upward, the cup expands. This maintains a seal against the vacuum cylinder walls and compensates for the motion of the rod. The cup can be made of durable Teflon eliminating the need for lubricating oil. The advantages include small-size, low-weight and relatively quiet operation.

Oil versus Oil-Less Pumps

Pumps may be lubricated with oil or use internal working surfaces, such as Teflon, that obviate the need for oil. Since pumps are generally located at the end of the sampling line, oil will not generally enter the sampling line. However, during pump operation, oil may be released to the nearby environment through the pump exhaust. Oil released into a working environment, especially one where relatively high amounts of radioactivity are in use, could represent a radioactive source of air and surface contamination and thus becomes an occupational health physics issue. Some pump exhausts can be fitted with filters to decrease aerosol emissions. These filters could become a radioactive waste over time. In fact, the pump itself will become internally contaminated with repetitive use especially if particulate radioactive aerosols of long half-life were sampled. Both the waste oil and the pump itself can be expensive to dispose of as radioactive wastes.

Oil-less pumps will not generate a radioactive oil aerosol but, if the sampling train trapping mechanism for the contaminant is not 100% efficient, the exhaust air flow may contain some radioactivity. Again, the pump could become contaminated over time. An exhaust filter may be necessary depending on the air sampling environment.

Oil-lubricated pumps provide a higher vacuum (more suction) because oil is an efficient seal against internal air leaks.

Miniature Pumps

Pumps that are not intended for personal air sampling, particularly of the diaphragm and rocking piston types, can be miniaturized to very small weights and sizes. Weights for miniature pumps of 15 g–2.5 kg are typical depending on the type of electrical motor. By comparison, a typical personal air sampler may weigh 600 g. The free air flow rates of miniature pumps are reduced as well. Examples on the market range approximately from 0.9 LPM to approximately 7 LPM. Some personal air samplers can achieve 5 LPM or higher. The use of miniature pumps for radioactive air sampling is certainly possible especially if low flow rates are desirable. The low weight and size make them suitable for compact systems and for unusual mounting arrangements. However, one must be careful not to tax these pumps with excessive backpressures. Long running times without proper cooling will also curtail the lifetime of these and most pumps. If other features are desired such as programmable start times, constant flow control, battery power, and automatic restart after a change in flow rate, a personal air sampler may be a better choice. However, operating in constant flow mode of personal air samplers can reduce the maximum flow rate to less than 1 LPM.

Choice of Pumps

Radioactive air sampling is driven by the need to measure a certain minimum amount (or concentration) of radioactivity with a chosen degree of statistical confidence. The minimum amount could be, for example, an order of magnitude or more lower than the regulatory air concentration for a particular radionuclide. Like most occupational or environmentally oriented ventures, air sampling is influenced temporally; for example, there may be minimal time that the air sampling practitioner can spend sampling before a result is required. Or, the sampler may need to monitor a workplace practice over a relatively long period such as an eight-hour shift. The total volume of air sampled is key to making the measurement meaningful. For grab sampling, a pump that can evacuate the sampling container in reasonable amount of time is always desirable.

The choice of pump is affected by the amount of time needed for sampling. Essentially, a decision is required to determine if the flow rate of the pump is adequate to acquire the required volume of air, and therefore the minimal amount of radioactivity that can be detected by analysis, in the allotted amount of time. For most applications like personal air sampling, the manufacturer has already made the choice for you. But if you are building a sampling train from scratch, the choice of pump is on your shoulders.

Other issues affect the choice of pump. For example, one may desire a lightweight unit for field use. This is one reason personal air samplers are sometimes used for long-term environmental air sampling. Another issue is vibration. Diaphragm and piston pumps will vibrate and this motion can be translated to other system components. Mechanical vibrations can induce unwanted motions to delicate table-mounted equipment like glass bubblers. Noise is also a consideration. Low noise levels may be desired in certain work environments while at remote air sampling stations it is not a concern. Backpressure is another consideration (see below).

Refer to Table 11.1 for a summary of the pumps described above. Most of the pumps discussed above can adequately handle typical air sampling needs.

Backpressure

The term "backpressure" (sometimes referred to as "pressure drop") refers to the total resistance to air movement inherent in the air sampling system. It is also another measure of pump performance.

Basic Air Sampling Equipment

TABLE 11.2
Pump and Filter Backpressure Comparison

Maximum Pump Flow Rate for 8 h Duration (LPM)	Maximum Pump Backpressure (in H_2O)	Filter Backpressure (in H_2O)
5	8	30
4	18	25
3	23	18

Source: Adapted from Roberson, R. What is backpressure? Tech Corner, available at www.sensidyne.com, accessed 12/27/07, 2007.

The total resistance is comprised primarily of friction forces. Friction will occur when air molecules collide with other air molecules and with the walls of the sampling system. Thus, the walls of the tubing and the pump are significant sources of friction. As the collision rate increases, frictional forces increase. Thus, backpressure increases at higher air sampling flow rates.

Perhaps the most important source of friction is the sampling medium. This may be a filter or a collection vessel filled with reactive media, or both. If the collection medium is a filter only, air movement will slow at the point of resistance, that is, at the filter cassette or head. Behind the filter, downline from the cassette, the air may move with less resistance since the frictional forces are only comprised of tube-wall friction. As a result, a partial vacuum or low-pressure area will be created behind the filter head. Filters with low porosity will create higher backpressure than filters with larger pores. As particulates accumulate on the filter, the vacuum behind the filter will increase, increasing the backpressure. A smaller-diameter filter will also produce higher backpressure than a larger diameter filter.

Thus, backpressure is inversely proportional to the air pressure inside the sampling line. The creation of a low-pressure region behind the filter is a measure of the backpressure. A manometer can be used to monitor this effect and, in fact, is a recommended procedure to determine if sampling lines are also suffering from leaks (Craig, 1971).

Pumps can be specified according to backpressure and flow rate. This is especially true of personal air samplers that run on battery power. If the backpressure is too high for the pump, the battery power can be drained before the desired sampling time, for example, an 8-h workshift, is achieved. This pump specification must be compared with flow rate/backpressure specifications for filters. An example of such a determination follows (Roberson, 2007).

Assume that our pump has the following specifications (refer to Table 11.2):

At 5 LPM, the pump will process a significant amount of air but it will be totally inadequate to handle the backpressure. In fact, some pumps will automatically shut down under these conditions to prevent internal damage. At lower flow rates, the pump is better suited to handling the reduced backpressures caused by the filter. The backpressure for this filter is specified as 18 in water at 3 LPM and 25 in water at 4 LPM. The pump can sustain (for 8 h) a backpressure less than that produced by the filter only at 3 LPM. There appears to be enough difference to accommodate any increase in backpressure from filter particulate loading during sampling.

For radioactivity measurements, this flow rate must be used to calculate the lower limit of detection and the critical measurement level (defined in a later chapter) of the radiation measurement system used to analyze the filter. This result is compared with the occupational inhalation limit, that is, the Derived Air Concentration (DAC), for the radionuclide in question. If the lower limit of detection and the critical level are not significantly below the DAC, the sampling and/or measurement system may not be adequate for regulatory and occupational health purposes. To alleviate this, the investigator may change the pump to achieve a higher flow rate capable of sustaining an 8-h sampling period under the filter backpressure. Or, if the filter can be changed to one of larger diameter or larger pore size, lower backpressures and thus higher flow rates can be achieved. The latter will directly determine if the DAC can be comfortably measured. Lastly, improvements in the radiation measurement system can lower the detection limits. Sometimes a combination of all these procedures may be needed.

Pulsation Control

Most pumps except for the rotary vane and linear types will create varying degrees of pulsating air flow in the sampling train. Under many circumstances, this can be ignored, but if the pulsation makes the flow rate measurement difficult to make (by oscillating the float in the rotameter) or imparts unwanted mechanical vibrations to the sampling system or nearby equipment, then the pulsation should be damped. This can be conveniently done by inserting an exhaust muffler or pulse dampener into the sampling train downline of the flow rate meter. This may be a tightly sealed glass jar, metal, or plastic chamber with appropriate input and outlet ports such that the air in the sampling line can expand. Mufflers can also collect moisture in the air. This may be important if the sampling trap or reaction vessel, for example, a gas diffusion bubbler, is using a liquid solution to capture the radioactivity. The bubbling action increases the contact time of the chemical solution with the airborne radioactivity. It may also create a mist that can be pulled into the sampling train toward critical components such as the flow measurement device and the pump. A column of silica gel can be installed upstream of these components and the muffler to capture water in the air stream.

REFERENCES

Anand, N.K., McFarland, A.R., Wong, F.S., and Kocmoud, C.J., *DEPOSITION: Software to Calculate Particle Penetration through Aerosol Transport Systems*, U.S. Nuclear Regulatory Commission Report NUREG/GR-006, U.S. Government Printing Office, Washington, DC, 1993.

ANSI/HPS, *Sampling and Monitoring Releases of Airborne Radioactive Substances from the Stacks and Ducts of Nuclear Facilities*, American National Standard/Health Physics Society N13.1-1999, Health Physics Society, McLean, VA, 1999.

Baron, P. and Deye, G., Electrostatic effects in asbestos sampling I: Experimental measurements, *Am. Ind. Hyg. Assoc. J.*, 51, 51, 1990.

Baron, P., Khanina, A., Martinez, A.B., and Grinshpun, S.A., Investigation of filter bypass leakage and a test for aerosol sampling cassettes, *Aerosol Sci. Technol.* 36, 857, 2002.

Baron, P., Aerosol sampling: Minimizing particle loss from cassette bypass leakage, in *NIOSH Manual of Analytical Methods*, National Institute for Occupational Safety and Health, Cincinnati, OH, 2003.

Caplan, K.J., Rotameter corrections for gas density, *Am. Ind. Assoc. J.*, 46, B–10, 1985.

Craig, D.K., The interpretation of rotameter air flow, *Health Phys.*, 21, 328, 1971.

Fisher, K., How to predict calibration of variable-area flow meters, *Chem. Eng.*, June, 180–184, 1952.

Gast, *Vacuum and Pressure Systems Handbook*, Gast Manufacturing, Inc., Benton Harbor, MI, available at www.gastmfg.com, accessed 12/27/07, 2007.

Kabat, M.J., Deposition of airborne radioiodine species on the surface of metals and plastics, in *Proceedings of the 17th DOE Air Cleaning Conference*, CONF-820833, M.W. First, Ed., U.S. Department of Energy, Washington, DC, 1983.

Liu, B.Y., Piu, D., Rubow, K.L., and Szymanski, W.W., Electrostatic effects in sampling and filtration, *Ann. Occup. Hyg.*, 29, 251–269, 1985.

Maiello, M., The development of an environmental gamma-ray and radon detector using thermoluminescent dosimetry and the teflon electret, PhD dissertation, New York University, University Microfilms International, Ann Arbor, MI, 1986.

Maiello, M. and Harley, N., Determining the charged fractions of ^{218}Po and ^{214}Pb using an environmental γ-ray and Rn detector, *Health Phys.*, 57, 51, 1989 with *Erratum Health Phys.*, 59, 365, 1990.

Marshall, M. and Stevens, D.C., The purposes, methods and accuracy of sampling airborne particulate radioactive materials, *Health Phys.*, 39, 409, 1980.

Muyshondt, A., McFarland, A., and Anand, N.K., Deposition of aerosol particles in contraction fittings, *Aerosol Sci. Technol.*, 24, 205, 1996.

Puskar, M.A., Fergon, S.M., Harkins, J.M., and Hecker, L.H., Gravimetric determination of airborne dust by using a filter cartridge inside a closed face, 37 mm polystyrene cassette, *Am. Ind. Assoc. J.*, 53, 692, 1992.

Roberson, R., What is backpressure? Tech Corner, available at www.sensidyne.com, accessed 12/27/07, 2007.
Swearingen, C., Choosing the best flowmeter, *Chem. Eng.*, July, 62–68, 1999.
Vincent, J.H., *Aerosol Sampling—Science and Practice*, John Wiley & Sons, Chichester, UK, pp. 207–216, 1989.
Wang, X. and Zhang, Y. Development of a critical air flow venturi for air sampling, *J. Agric. Eng. Res.*, 73, 257, 1999.
Williamson, S.J., *Fundamentals of Air Pollution*, 1st ed., Addison-Wesley, Reading, MA, chaps. 3 and 4, 1973.
Wolf, S., *Guide to Electronic Measurements and Laboratory Practice*, 1st ed., Prentice-Hall, Englewood Cliffs, NJ, pp. 153–156, 1973.

12 Calibration of Air Samplers and Monitors

James T. Voss and Jeffrey J. Whicker

CONTENTS

Introduction	246
General Calibration Procedures	246
Calibration of Radiation Detector Systems	246
Precalibration Inspection	246
Precalibration Electronic Testing	247
Calibration of Radiation Detectors with Radioactive Reference Sources	247
Calibration of Air Sampling Flow Rates	249
Additional Important Considerations	252
Traceability	252
Considerations for Environmental Conditions	252
Uncertainty	254
Auditability	255
Quality Assurance Program	256
Conclusion	258
References	258
Additional Sources of Information	258
Appendix A: Generic Calibration Procedure	260
James T. Voss and Jeffrey J. Whicker	
Purpose	260
Scope	260
Hazard Analysis	260
Identified Hazards and Mitigation	260
References	260
Equipment	260
Procedure	260
Inspection and Repair	260
Electrical Testing	261
Pulser Calibrations (Useful for Some Count Rate-Based CAMs)	261
Efficiency Calibration (Detector Calibration)	261
Performance (Alarm) Check	261
Flow Calibration	262
Documentation	262
Example Calibration Record	262
Appendix B: Multiple-Frame-of-Reference Method for Rotameter Correction Factors	264
Mark D. Hoover	
Introduction and Objectives of the Method	264
Operation	265
Frames of Reference	265

The Rotameter Equation 266
The Ideal Gas Law 266
The Scale Factor Equation 267
Example Calculations 269
Other Influences 270
Recommendations to Air Sampling Practitioners 270
References 270

INTRODUCTION

A detailed evaluation of the concentrations of radioactive materials in air is required for effective protection of workers and the public (*cf*., DOE, 2007; EPA, 2007; NRC, 2009). For this, we must be confident that the air sampling equipment used provides accurate measurements. The primary objectives of calibration are to ensure that instruments used for air sampling and monitoring are working and that their measurement responses are within acceptable limits when compared to reference standards. Specific to air sampling and monitoring, the calibration process requires comparing measurements of airflow rates and the response of nuclear counting instruments to recognized reference standards. In addition, the instrument responses are adjusted to match the expected response over the measurement range of the instruments when compared under the conditions of use.

This chapter describes terminology and specific techniques related to the calibration of air sampling and monitoring equipment, and important considerations for calibration. A discussion of the interpretation of airflow rates and volumes reported between air sampling facilities is included with emphasis on the consideration of whether to express the results in terms of standard volumes or ambient volumes.

GENERAL CALIBRATION PROCEDURES

The calculation of concentrations of radionuclides in air (C_A) requires measurements of the airflow rate (F), sampling times (T_s), net count rate (R_N), and detector efficiency (ε), as shown in Equation 12.1. These variables require assessment of their accuracy and precision.

$$C_A = \frac{R_N}{F \cdot T_s \cdot \varepsilon} \tag{12.1}$$

where C_A is in units of activity of unit volume such as Bq m^{-3}, F is units of volume of air per unit time such as m^{-3} min^{-1}, T_s is in units of time such as min, R_N is counts per minute or cpm, ε is the ratio of cpm to disintegrations per minute or dpm.

Specifically, the instruments that measure these key variables are detectors that measure the radioactivity level (net count rate divided by the detector efficiency) and the instruments used to measure sample volume (flow rate sampling time), and each requires calibrations. The error associated with the measurement of time is so small that it is generally ignored.

CALIBRATION OF RADIATION DETECTOR SYSTEMS

Accurate measurement of the amount of radioactivity collected on a filter or other medium or the amount contained in a volume of a chamber requires that the nuclear counter used be calibrated. This involves several steps ranging from precalibration inspection and testing of an instrument through final testing against a reference radioactive source (see the Appendix A for a general calibration procedure).

Precalibration Inspection

Before calibrating an instrument received either from the field or manufacturer, inspect the instrument for damage that would prevent it from operating acceptably, verify instrument response using

Calibration of Air Samplers and Monitors

a radioactive check source, and confirm that the target airflow rate can be achieved. Combined, these observations and test results describe the "as-found" condition of the equipment and they should be documented. The as-found readings may be used to determine the validity of the instrument's indications while in service between calibrations. As-found readings should be taken by exposing the detector to a radioactive source and recording the instrument response or the detector efficiency. As-found readings should also be recorded for the airflow sensor to verify that the readings are within expectations.

Precalibration Electronic Testing

Precalibration testing of the electronics of an instrument should be performed to verify meter/display linearity. The signal used to perform electronic calibration should mimic the pertinent characteristics of the detector signal (e.g., pulse width, pulse height). Precalibration testing may include adjusting pulse width, setting high voltage (HV), amplifier gain, thresholds, and windows (*note*: these can be interrelated). Improper settings of these can result in inaccurate energy assignment (linearity), missed counts due to excessive dead times, pulses that are improperly placed outside the upper and lower thresholds of energy windows, or having pulses not processed due to improper widths or amplitudes. α and β particulate air monitoring instruments may use single or multichannel analyzers. These instruments may require the use of a pulse generator or a radioactive source with known radiation energies. The instrument response (energy and count rate) can be tested by introducing pulses of varying frequency and amplitude from the generator and checking the energy assignment and measured count rates of the pulses. When used, pulser testing, should be conducted over the expected range of radiation energies and count rates expected. Adjustments to the instrument response (HV, gain, dead time corrections, pulse width, etc.) can be made to ensure that the instrument response matches that expected. Precalibration testing for alarming air monitors should include verification of radiation and failure alarms, alarm delay time, and alarm relays. The tests described here will require the expertise of an electronics specialist or air sampling technician with the proper experience.

Calibration of Radiation Detectors with Radioactive Reference Sources

The calibration of radiation detectors should generally be done annually or more frequently under changing environmental conditions, or when an instrument fails a daily or weekly performance check (see the example of a failed chi-square test in Figure 12.1). The general procedure is to compare the measurement response of an instrument to the response of either a reference instrument (one calibrated to a standard reference source) or to the radioactive source strength of the reference source (IAEA, 1999). If needed, the instrument under test can be electronically adjusted to match the expected response or a calibration factor can be used to mathematically adjust the measurement to match the expected response. An equation for the calibration factor (F) is shown in Equation 12.2, where R is the conventional true reference value of the quantity and M is the measurement indication given by the instrument.

$$F = \frac{R}{M} \qquad (12.2)$$

Radiation detector systems, such as real-time air monitoring instruments, should be calibrated under an appropriate set of environmental conditions and with radioactive sources that exhibit the radiation type and energies of the airborne radioactivity to be monitored. Table 12.1 shows a set of standard conditions under which calibrations might occur (though the calibrations should be done under the conditions of instrument use), and Table 12.2 identifies a selection of sources that should be used for calibration. To determine the choice of radioactive source strength to calibrate with, consider that analog readout instruments may have either linear or logarithmic scales, and

RADIOLOGICAL CONTROL ALPHA/BETA COUNTING INSTRUMENT RECEIPT TEST	Facility: TA-21 DP West	FC ERD&D

Instrument
- Counting Interval T_g (min): 1.0
- ID No.: 6810
- Model: 2929
- Detector ID No.: 6810
- Calibration Expiration Date: 11/22/06
- Location: RCT office

Radioactive Standard
- ID No.: ER-DU-001
- Isotope: D38
- Activity (dpm): 11,000
- Date Determined: 2/15/06

Reason for Test
- ◉ Receipt
- ○ Failed First Receipt Test

Background Determination (cpm)

Background Counting Interval T_b (min): 1.0

$$R_B = \frac{N_B}{T_b} = \frac{}{1.0} = 0.00$$

Chi-Squared (χ^2) Test

Lower	Upper
5.60	42.95

Average Number of Counts per Interval

$$\overline{X} = \frac{\Sigma X}{20} = 5{,}829.45 \text{ counts/interval}$$

Chi-Squared (χ^2) Test

No. of Counts	X	$(X - \overline{X})$	$(X - \overline{X})^2$
1	5,784	−45.45	2,065.7
2	5,391	−438.45	192,238.4
3	5,829	−.45	.2
4	5,900	70.55	4,977.3
5	5,792	−37.45	1,402.5
6	5,811	−18.45	340.4
7	5,771	−58.45	3,416.4
8	5,805	−24.45	597.8
9	5,879	49.55	2,455.2
10	5,906	76.55	5,859.9
11	5,888	58.55	3,428.1
12	5,809	−20.45	418.2
13	5,882	52.55	2,761.5
14	5,786	−43.45	1,887.9
15	5,804	−25.45	647.7
16	5,980	150.55	22,665.3
17	5,988	158.55	25,138.1
18	5,794	−35.45	1,256.7
19	5,894	64.55	4,166.7
20	5,896	66.55	4,428.9
Totals Σ	116,589	0.00	280152.95

Chi-Squared Calculation

$$\chi^2 = \frac{\Sigma (X - \overline{X})^2}{\overline{X}} = 48.06$$

This χ^2 value must fall within the limits above.

Chi-Squared Test Passed? ○ Yes ◉ No

Standard Deviation

$$\sigma = \sqrt{\frac{\Sigma (X - \overline{X})^2}{n-1}} = 121.43$$

UCL = 6,194 counts/interval

LCL = 5,465 counts/interval

Efficiency Calculation

$$\text{Measured Efficiency} = \frac{\frac{\overline{X}}{T_g} - R_B}{\text{Source Activity}} = .53 \text{ cpm/dpm}$$

Calibration Efficiency E_C (%/100) = .329

Uncertainty In Calibration Efficiency (%) = 37.92

Efficiency Test Passed? ◉ Yes ○ No

Test and Log Completed By
- Name (print): RCT
- Signature:
- Date: 2/15/06

Log Validated By
- Name (print):
- Signature:
- Date:

A-6002-162 (02/98)

FIGURE 12.1 Example receipt test report and daily operational test log for a radiological control alpha/beta counting instrument. Note that the instrument in this example has failed the chi-squared test.

after adjustment, the response of the instrument should be checked with source strengths between approximately 20% and 80% of scale. If the instrument is designed to auto range (automatically set the range of response based on signal strength), a single calibration activity may be used but after adjustment, the response of the instrument should be checked at approximately 20% and 80% of scale.

Calibration of Air Samplers and Monitors

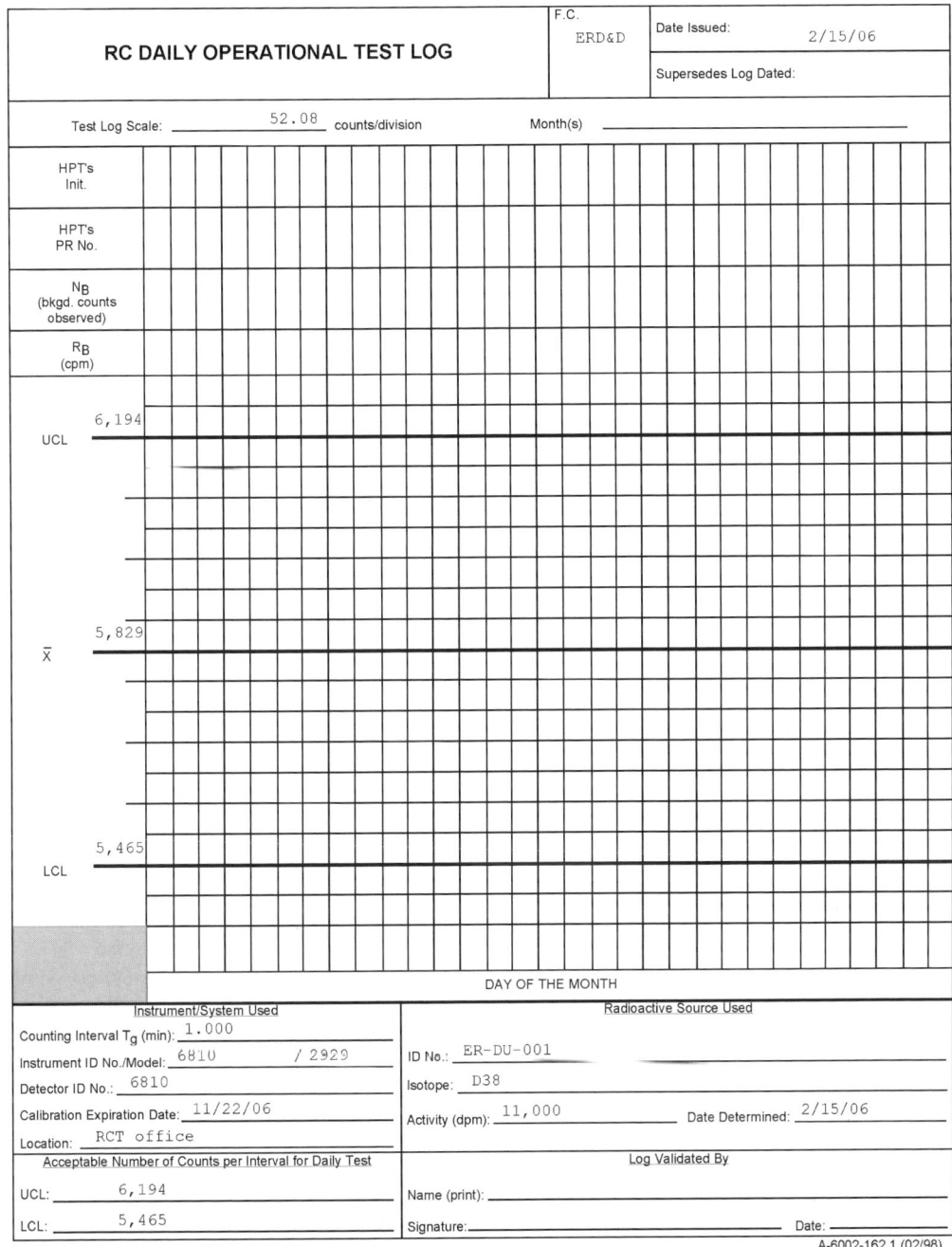

FIGURE 12.1 (Continued)

CALIBRATION OF AIR SAMPLING FLOW RATES

As shown in Equation 12.1, concentration measurements require an accurate measurement of the rate air is sampled to compute the volume of air sampled. Therefore, the devices used to measure the sampling rate should be calibrated with a reference instrument annually (more often if conditions change or less often if stability is documented) (ANSI, 2009). Airflow meter readings should be within ±15% of the conventionally true value of the actual flow rate (ANSI, 1989). Note that if the

TABLE 12.1
Standard Conditions for Calibration

Influence Quantities	Acceptable Range for Standard Test Conditions
Warmup time	
Electronic devices	≤10 min
Air or gas circuit	≥60 min
Relative humidity	Ambient ±10%, not to exceed 75%
Ambient temperature	20°C to 24°C
Atmospheric pressure	70 to 106 kPa
Line voltage	Nominal 1%
Frequency	60 Hz ± 0.5 Hz
AC power-supply wave form	Sinusoidal with total harmonic distortion ≤5%
Background ambient photon radiation (external)	<2.5% of full scale of range or decade under test, but nominally should not exceed 50 μrad/h (0.5 μGy/h), referenced to air
Nonionizing electromagnetic field of external origin	<50% of the lowest value that causes interference
Magnetic induction of external origin	Less than twice the induction due to the earth's magnetic field
Controls	Set for normal operation per manufacturer's recommendations
Contamination by radionuclides	Negligible
Reference point	Effective center

TABLE 12.2
Examples of Radiation Sources for Calibration

Source	Half-Life (Years)	Radiation Type	Energy (MeV)
^{239}Pu	2.41×10^4	Alpha	5.155
^{230}Th	7.54×10^4	Alpha	4.770
^{14}C	5730	Beta	0.156
^{147}Pm	2.62	Beta	0.224
^{99}Tc	2.14×10^{10}	Beta	0.292
^{36}Cl	3.01×10^{10}	Beta	0.710
^{204}Tl	3.78	Beta	0.763
^{210}Pb-Bi + progeny	20.4	Beta	0.016
			0.061
			1.161
^{90}Sr-^{90}Y	27.7	Beta	0.546
			2.284
^{238}U + progeny	4.51×10^{10}	Beta	0.100
			0.191
			2.281
^{137}Cs	30.2	Photon	0.662
^{60}Co	5.27	Photon	1.173
			1.332

Calibration of Air Samplers and Monitors

instrument uses a mass flow meter, appropriate corrections for ambient pressure and temperature are required, as these affect air density and thus the volume of air sampled.

All flow meters should be calibrated against devices that are traceable to NIST (National Institute of Standards and Technology) or another National/International Standards Laboratory. Before calibrating the airflow-rate sensor, the air "in-leakage" (i.e., the flow through the instrument not passing through the collection media) should be verified to be less than 5% (ANSI, 1999) of the nominal flow rate. Flow meters should be calibrated with a filter or device installed that produces a pressure drop equivalent to the pressure drop of the filter type to be used during normal operations.

During calibration, the standard flow meter should be inline with the filter so that it measures air going through the filter. The entire flow path through the air sampler should be tested for in-leakage prior to calibration to prevent measuring air that has entered the system from leakage paths. Rotameters and mass flow meters should be calibrated at the expected pressure, temperature, and average flow rate anticipated during use, and should also be calibrated at one point above and one point below the anticipated sampling flow rate (e.g., 75% and 125% of the anticipated sampling flow rate). These tests should cover the anticipated range of flow rates for the instrument.

Critical flow venturis may be used as an airflow control device in some equipment (ASTM, 2005). Validating correct operation of a critical flow venturi involves steps similar to any airflow calibration. One exception is the need to verify flow across a wide range since the critical flow venturi, by its nature, is not intended to operate over a range of airflow but at a single set point. Critical venturi flowmeters should be calibrated with a sufficient pressure differential across the meter to ensure that the velocity at the throat of the meter is sonic. The absolute pressure at the entrance of the critical flow meter should be within required boundaries of the average absolute pressure anticipated at that location. The temperature at the entrance of the critical flowmeter during calibration should be within predefined temperature tolerance limits for the average temperature anticipated at that same location during sampling.

Many reference instruments used to calibrate air samplers and monitor pumps and airflow meters are themselves calibrated under specific environmental test conditions such as at room temperature and at sea level. Because the impact temperature and ambient atmospheric pressure (a strong influence) have on the volume of gas sampled (see the Appendix B), one must be aware of the limitations of the reference instruments and make necessary corrections, as suggested by Equation 12.3.

$$Q_A = Q_S \left(\frac{P_A}{P_S}\right)\left(\frac{T_S}{T_A}\right) \tag{12.3}$$

Here Q_A is the ambient sampling flow rate (m^3 s^{-1}), Q_S is the standardized sampling flow rate, that is, calibrated at sea level (m^3 s^{-1}), T_S is the standardized temperature in Kelvin, and T_A is the average ambient temperature under which the measurements were made, also in Kelvin. P_A and P_S are the atmospheric pressure (e.g., in Pascals) at ambient and standard conditions, respectively.

Even after calibration to a reference instrument, calibrated flow meters that are used under environmental conditions different from those present during calibration must be corrected for. For example, when a rotameter is used at a pressure or temperature it was not calibrated for, Equation 12.4 should be applied to correct for flow rate (see the Appendix B).

$$Q_O = Q_C \sqrt{\frac{T_O}{T_C} \times \frac{P_C}{P_O}} \tag{12.4}$$

Here Q_O is the sampling flow rate under operating conditions, Q_C is the sampling flow rate under calibration conditions, T_O is the temperature in Kelvin under operating conditions, and T_C is the temperature in Kelvin under calibration conditions, P_O and P_C are the atmospheric pressure (e.g., in Pascals) at operating and calibration conditions, respectively.

The consideration of whether to express air concentration results in terms of standard volumes [based on conditions of standard temperature and pressure (STP)*] or ambient volumes (based on pressure and temperature during the sample collection) is determined by the purpose of the measurements. In most cases, air concentrations are made for the purpose of worker protection, therefore the concentrations should be calculated using ambient volumes. There are occasions when comparison of the concentration measurements with regulatory guides is appropriate. This may require use of volumes measured under STP conditions, in which case the results could be stated in terms of STP (e.g., EPA, 2007). Procedures for the calibration of airflow rate measurement systems should be stated in sufficient detail to allow proper interpretation of airflow rates and volumes reported by the equipment (i.e., ambient or standard volumes).

ADDITIONAL IMPORTANT CONSIDERATIONS

There are a number of considerations and concepts related specifically to the calibration of air sampling and monitoring instruments and the documentation of the calibration. These include traceability, repeatability, compensation for environmental radiations and conditions, auditability, and the establishment and maintenance of a solid quality assurance program. Calibration laboratories should have established (documented) procedures for the calibration of air sampling equipment.

TRACEABILITY

Traceability is that property of the result of a measurement or the value of a standard whereby it can be related to stated references, usually national or international standards, through an unbroken chain of comparisons all having stated uncertainties. The concept of traceability is often expressed by the adjective *traceable*. The unbroken chain of comparisons is called a *traceability chain*. Figure 12.2 is an example of a source certificate showing traceability to NIST.

The calibration program should include a system for selecting, using, calibrating, controlling, and maintaining measurement standards (radioactive sources) and other radioactive reference materials used as measurement standards. Calibration laboratories should establish the traceability of their own measurement standards and measuring instruments to the appropriate base unit(s) of measurement included in the International System of Units (SI) (e.g., mass, length, temperature, time, electrical current, etc.) by means of an unbroken chain of calibrations or comparisons. The link to fundamental SI base units may be achieved by reference to the NIST or other national measurement standards. National measurement standards may be primary standards, based on fundamental physical constants and quantities, or they may be secondary standards, which are tested and calibrated to primary reference standards. If calibration services from other laboratories are used, traceability of measurements should be assured by having the laboratory demonstrate competence, measurement capability and traceability to primary standards. All calibration certificates should contain the measurement result of the standard, for example, disintegrations s^{-1} (Bq), including the measurement uncertainty and/or a statement of compliance tracing the measurement to an identified metrological specification such as one from the International Organization for Standardization (ISO).

CONSIDERATIONS FOR ENVIRONMENTAL CONDITIONS

It is incumbent on the calibration laboratory to monitor, control, and record the environmental conditions, as required by the relevant specifications, methods and procedures or where they influence the quality of the results. Due attention should be given to dust, electromagnetic disturbances,

---

* STP has variable definitions often dependent on the profession one is working in. One should define STP in any publication of data before comparing with ambient or other conditions.

Calibration of Air Samplers and Monitors

Gamma Reference Source

Source no.	OG 104
Drawing	VZ-1311
Dimensions of active surface	⌀ 47 mm
Overall dimensions	⌀ 56.5 mm × 25.4 mm
Construction	The radionuclidic mixture is homogeneously distributed onto the active area of the plastic foil. The activated foil is covered on both sides with a paper label and a plastic foil.

Nuclide	Gamma-ray energy [MeV]	Activity [Bq]	Emission rate [s^{-1}]
Americium-241	0.060	3.41E03	1.22E03
Cadmium-109	0.088	2.13E04	7.72E02
Cobalt-57	0.122	5.96E02	5.10E02
Cerium-139	0.166	8.65E02	6.91E02
Mercury-203	0.279	2.32E03	1.89E03
Tin-113	0.392	2.97E03	1.93E03
Strontium-85	0.514	3.80E03	3.74E03
Caesium-137	0.662	2.68E03	2.28E03
Yttrium-88	0.898	6.36E03	5.97E03
Cobalt-60	1.173	3.21E03	3.20E03
Cobalt-60	1.333	3.21E03	3.21E03
Yttrium-88	1.836	6.36E03	6.31E03

Reference date	1 June 2006 at 12.00 GMT
Leakage and contamination test	Wipe test according to ISO 9978
Date of wipe test	21 June 2006
Measuring method	The activity was measured with a gamma spectrometer system consisting of a calibrated high purity germanium detector and a multichannel analyser.
Traceability	Additional to the direct traceability to the PTB through the DKD this product complies with the requirements for traceability to NIST specified in the American National Standard "Traceability of Radioactive Sources to the NIST and Associated Instrument Quality Control (ANSI N42.22-1995)". As a requirement of the ANSI N42.22-1995 QSA Global GmbH participates in the NEI/NIST Measurements Accurance Program of the Nuclear Power Industry.
Uncertainty	The relative uncertainty of the activity is 3%. The reported uncertainty, determinded according to the DKD-3 report is based on the standard uncertainty multiplied by a coverage factor of k = 2, providing a level of confidence of 95%. (Ref. NIST Technical Note 1297/"Guide to the Expression of Uncertainty in Measurement" ISO Guide, 1995).
Radioactive impurities	At the time of calibration the following radioactive impurities were detected: Co-56<1 Bq; Co-58<1 Bq; Zn-65<1 Bq; Ag-110m<1 Bq; Cs-134<3 Bq
Quality assurance system	The quality assurance system of QSA Globel GmbH was certified by Lloyd's Register Quality Assurance (LRQA) according to ISO 9001, issue 2000. Isotrak products meet the requirements of 10CFR50 Appendix B in the USA.

FIGURE 12.2 Example of a calibration source certificate showing traceability to NIST.

interfering radiations (particularly ^{220}Rn and ^{222}Rn and their respective progeny), humidity, electrical supply, temperature, and sound and vibration levels. Calibrations and any testing supporting calibration results should be halted when the environmental conditions jeopardize the accuracy of the results of the tests and/or calibrations. The calibration laboratory should ensure that the environmental conditions do not invalidate the results or adversely affect the required quality of any measurement. Laboratory facilities including but not limited to energy sources, lighting and environmental conditions should be such as to facilitate correct performance of the tests and/or calibrations (see Table 12.1). If the environmental conditions cannot be changed, then the measures taken to compensate for those environmental conditions should be documented. Efforts should be taken to ensure good housekeeping in the calibration laboratory.

Particular care should be taken when calibrations are undertaken at field (or temporary measurement sites) or locations other than a permanent laboratory facility. The technical requirements for accommodation of the environmental conditions and a description of the conditions that can affect the results of tests and calibrations should be documented.

UNCERTAINTY

Consideration of uncertainty should be made when evaluating the results from a calibration procedure. The IAEA report No. 16 on calibration of radiation instruments discusses how to determine and report uncertainties (IAEA, 1999) which is relevant to calibrations of air sampling instrumentation because calculations of air concentrations are most often based on radiation count rate meters. Guidance is also provided for propagation of errors through calculation of calibration factors. The following is a brief summary of the standard approach to calculating the uncertainty for multiple measurements from radiation count data and the propagation of error.

The mathematical formulae of the mean, $\bar{X}$, and the standard deviation, $s(x)$, of a repeated set of measurements are calculated using Equations 12.5 and 12.6, respectively.

$$\bar{X} = \frac{1}{n}\sum_{i=1}^{n} x_i \tag{12.5}$$

$$s(x) = \sqrt{\frac{1}{n-1}\sum_{i=1}^{n}(x_i - \bar{x})^2} \tag{12.6}$$

For independent count data from radiation detectors, either Poisson or Gaussian distribution statistics apply, and the estimated standard deviation is calculated as the square root of the measured number of counts (x) that has been summed over a counting interval (t) (Knoll, 2000) where

$$N = \sum_{t=0}^{t} x, \tag{12.7}$$

$$\sigma^2 = N, \tag{12.8}$$

and

$$s = \sqrt{N}. \tag{12.9}$$

When calculating radioactivity levels, detector efficiencies, calibration factors, and so on, can be important to propagate the uncertainties of the measurements through the calculation shown in Equation 12.10.

$$\sigma_u^2 = \left(\frac{\partial u}{\partial x}\right)^2 \sigma_x^2 + \left(\frac{\partial u}{\partial y}\right)^2 \sigma_y^2 + \left(\frac{\partial u}{\partial z}\right)^2 \sigma_z^2 + \cdots \tag{12.10}$$

Standard textbooks, such as Knoll (2000), provide solutions for Equation 12.10 for counting measurements, but essentially, the random nuclear counting errors in the sample and background add in quadrature in the error in the net count. Since the net count and all other factors used for calculation of air concentration are related as quotients or products, it is the relative error that adds in quadrature for all other error terms. The calibration error, due to uncertainty in the accuracy of the calibration standards and to error introduced in the conduct of the calibration procedures, should also be added as appropriate in the determination of total uncertainty.

The precision of the calibration measurements can also be an indicator or the quality of the measurements. Precision is the quality of repeatability of measurement data; the similarity of successive independent measurements of a single magnitude generated by repeated applications of a process under specified conditions.

AUDITABILITY

Calibration programs need to be of high quality to assure that accurate measurements are being performed. For this, assessments and audit shall be performed. The process of "auditing" applies to evaluations of management systems, for example, procedures, qualification of personnel, technical basis documents, quality assurance documents; whereas, "assessment" applies to the actual performance of the calibration processes, for example, through interlaboratory comparisons, random instrument checks, and so forth.

One of the key aspects of a calibration program that applies to both audits and assessments is record keeping. In essence, all records should be kept in such a way as to provide transparency and full disclosure on how an instrument was calibrated, the conditions of the testing, and the results. These records should be easily retrievable. What are the records required? For an individual Air Sampler Calibration the following records are generally needed:

- Type of device under calibration
- Model # of device under calibration
- Serial # of device under calibration
- Name and ID of the owner of the device under calibration
- Calibration procedure ID
- Environmental factors such as temperature, humidity, barometric pressure
- As-found readings
- Final readings
- Equipment used to perform the calibration, with model and serial #s and their last date of calibration and their calibration due dates
- Name of person performing the calibration
- Unique ID of the person performing the calibration
- Date and time the calibration was performed
- Length of time the calibration is valid for
- Any limitations on the useable range or environmental conditions under which the air sampler is expected to operate acceptably

In addition, similar information for all of the equipment used to perform the calibration is also required, but those records do not have to be directly attached to the air sampler calibration record, though these records should be readily available.

Qualification records of the individuals performing the calibrations are also required. Those records are generally kept in a central file but can be linked back to specific air sampler calibrations. The following information is generally needed.

- Name of the person performing the air sampler calibration;
- Unique ID of the person performing the air sampler calibration;

- Method used to qualify the individual such as experience, training, education, and so on;
- Date the individual was qualified to perform the calibration;
- Periodic requalification of the individual if necessary.

QUALITY ASSURANCE PROGRAM

Quality control is a system for verifying and maintaining a desired level of quality in the calibration program by careful planning, use of proper equipment, continued inspection, and corrective action as required. Calibration laboratories should have quality control procedures for monitoring the validity of tests and calibrations undertaken. The resulting data should be recorded in such a way that trends are detectable and, where practicable, statistical techniques, for example, uncertainty evaluations, parametric and nonparametric testing for differences among means, and Chi-square tests (Gilbert, 1987; Knoll, 2000), shall be applied to the reviewing of the results and identifying occasions where calibration sources or equipment is operating outside acceptable limits.

This monitoring should be planned and reviewed and may include, but not be limited to the following:

- Regular use of certified reference materials and/or internal quality control using secondary reference materials
- Participation in interlaboratory comparison or proficiency-testing programs
- Replicate tests or calibrations using the same or different methods
- Retesting or recalibration of the individual air samplers
- Correlation of results for different characteristics of the air samplers

Quality control data should be analyzed and, where they are found to be outside predefined criteria, planned actions should be taken to correct the problem and to prevent incorrect results from being reported. A quality assurance manual should describe how the quality assurance criteria are satisfied and are integrated into the overall calibration laboratory system. The following 10 elements can form the basis for a good quality assurance program.

1. *Management Program*
   - Establish an organizational structure, functional responsibilities, levels of authority, and interfaces for those managing, performing, and assessing the work.
   - Establish management processes, including planning, scheduling, and providing sufficient resources for the work.
2. *Personnel Training and Qualification*
   - Train and qualify personnel to be capable of performing their assigned calibration assignments.
   - Provide continuing training to personnel to maintain their job proficiency.
   - The personnel responsible for the opinions and interpretation included in test reports should, in addition to possessing the appropriate qualifications, training, experience, and satisfactory knowledge of the testing, also have:
     - relevant knowledge of the technology used for the manufacturing of the items, materials, products, and so on. tested, or the way they are used or intended to be used, and of the defects or degradations which may occur during or in service;
     - knowledge of the national or international standards and/or general requirements expressed in the governmental regulations that have required use of the measurement equipment, for example, to assure that a regulatory limit such as an airborne radionuclide concentration is not exceeded; and,
     - an understanding of the significance of deviations found with regard to the normal use of the items, materials, products, and so on.

3. *Quality Improvement*
   - Establish and implement processes to detect and prevent quality problems associated, directly or indirectly, with instrument performance.
   - Identify, control, and correct items, services, and processes that do not meet established requirements.
   - Identify the causes of problems and work to prevent recurrence.
   - Review any and all quality-related information to identify laboratory, instrument, or calibration items, services, and processes needing improvement.
   - Establish the philosophy that quality improvement requires a continuous effort to produce better calibration services.
4. *Documents and Records*
   - Prepare, review, approve, issue, use, and update documents used to prescribe processes, specify requirements, or establish design (see below).
   - Specify, prepare, review, approve, and maintain records.
5. *Work Processes*
   - Perform work consistent with technical standards, administrative controls, and other hazard controls adopted to meet regulatory or contract requirements, using approved instructions, procedures, or other appropriate means.
   - Identify and control laboratory test equipment and items to be calibrated to ensure their proper use.
   - Maintain laboratory test equipment and items to be calibrated to prevent their damage, loss, or deterioration.
   - Calibrate and maintain equipment used for process monitoring or data collection, for example, computers and related software.
6. *Design*
   - Design tests, test equipment, calibration equipment, and the calibration process using sound engineering/scientific principles and appropriate standards.
   - Verify or validate the adequacy of design products using individuals or groups other than those who performed the work.
   - Verify or validate work before approval and implementation of the calibration process.
7. *Procurement*
   - Procure items and services that meet established requirements for the calibration work.
   - Evaluate and select prospective suppliers on the basis of specified criteria.
   - Establish and implement processes to ensure that approved suppliers continue to provide acceptable items and services.
8. *Inspection and Acceptance Testing*
   - Inspect and test specified items, for example, electronic test equipment or environmental chambers, services, and processes using established acceptance and performance criteria.
   - Calibrate and maintain laboratory test equipment used for inspections and tests of air sampling equipment.
9. *Management Assessment*
   - Laboratory managers should periodically assess their management procedures to identify and correct problems that hinder the organization from achieving the objectives of the calibration laboratory.
10. *Independent Assessment*
    - Plan and conduct independent assessments to measure service quality, to measure the adequacy of work performance, and to promote improvement.
    - Establish sufficient authority, and freedom from line management, for the group performing independent assessments.

- Ensure persons who perform independent assessments are technically qualified and knowledgeable in the areas to be assessed.
- Laboratory intercomparison is the organization, performance, and evaluation of calibrations on the same or similar air samplers by two or more laboratories in accordance with predetermined conditions. Participation in a suitable program of interlaboratory comparisons should be implemented where possible.

CONCLUSION

Calibration of air samplers and monitors is composed of the actual procedure that records instrument response relative to reference standards (of both the radiation detector and the associated air sampling equipment, e.g., pumps, rotameters) and the quality control program that assures that the calibration is performed so as to meet recognized standards of excellence. This marriage of procedure and quality is required to assure that the radioactive air sampling measurements achieve an accuracy that is acceptable for comparison to regulatory requirements and that can be used to estimate releases to the environment. Eventually, no matter the difficulties inherent in doing so, such measurements are often extrapolated to radiation protection quantities, including human dose. With such consequences, air sampling equipment must be well calibrated and the system of calibration, including the quality assurance measures described in this chapter, strictly adhered to and documented.

REFERENCES

American National Standards Institute (ANSI). 1999. *Sampling and Monitoring Releases of Airborne Radioactive Substances from the Stacks and Ducts of Nuclear Facilities.* ANSI N13.1.

American National Standards Institute (ANSI). 1989. *Performance Specifications for Health Physics Instrumentation—Occupational Airborne Radioactivity Monitoring Instrumentation.* ANSI N42.17B.

American National Standards Institute (ANSI). 2009. *Radiation Protection Instrumentation Test and Calibration—Air Monitoring Instruments.* ANSI N323C.

American Society of Mechanical Engineers (ASTM). 2005. *Measurement and Fluid Flow in Pipes using Orifice, Nozzle, and Venturi.* ASTM MFC-3M-2004.

Department of Energy (DOE). 2007. *Code of Federal Regulations.* Washington, DC: U.S. Government Printing Office; 10 CFR Part 835. Available at http://www.hss.energy.gov/HealthSafety/WSHP/Radiation/rule.html verified August 2008.

Environmental Protection Agency (EPA). 2007. *National Primary and Secondary Ambient Air Quality Standards.* 10 CFR Title 40 Part 50.

Gilbert, R.O. 1987. *Statistical Methods for Environmental Pollution Monitoring.* New York: Van Nostrand Reinhold.

International Atomic Energy Agency (IAEA). 1999. *Calibration of Radiation Protection Monitoring Instruments.* Report No. 16. Vienna, Austria: IAEA.

Knoll, G.F. 2000. *Radiation Detection and Measurement.* Hoboken, NJ: John Wiley & Sons.

Nuclear Regulatory Commission (NRC). 2009. *Code of Federal Regulations.* Washington, DC: U.S. Government Printing Office; 10 CFR Part 20. Available at http://www.access.gpo.gov/cgi-bin/cfrassemble.cgi?title=200910, verified April 2010.

ADDITIONAL SOURCES OF INFORMATION

American National Standards Institute (ANSI). 1976. *Glossary of Terms in Nuclear Science and Technology.* ANSI N1.1.

American National Standards Institute (ANSI). 1974 (Reaffirmed 1991). *Specification and Performance of On-Site Instrumentation for Continuously Monitoring Radioactivity in Effluents.* ANSI N42.18.

American National Standards Institute (ANSI). 2000. *Performance Specifications for Tritium Monitors.* ANSI N42.30.

Hinds, W.C. 1982. *Aerosol Technology: Properties, Behavior, and Measurement of Airborne Particles.* New York: John Wiley & Sons.

International Standards Organization (ISO). 1990. ISO 25-General requirements for the competence of calibration and testing laboratories (ISO publications are available from the ISO Central Secretariat, Case Postale 56, 1 rue de Varembé, CH-1211, Genève 20, Switzerland/Suisse. ISO publications are also available in the United States from the Sales Department, American National Standards Institute, 11 West 42nd Street, 13th Floor, New York, NY 10036, USA.)

IEC 50(394)-1995. *International Electrotechnical Vocabulary*, Chapter 394, Nuclear Instrumentation, Instruments.

National Bureau of Standards (NBS). 1989. *Calibration Services User's Guide*. Special Publication 250.

National Institute of Standards and Technology (NIST). 2001. *NIST Handbook* 150, 2001 Edition, *Procedures and General Requirements*. NIST.

VIM. 1993. *International Vocabulary of Basic and General Terms in Metrology*. International Organization for Standardization.

APPENDIX A: GENERIC CALIBRATION PROCEDURE

James T. Voss and Jeffrey J. Whicker

Purpose

This procedure provides guidance and instructions to assist a knowledgeable instrument technician in doing a routine adjustment and calibration of a Continuous Air Monitor (CAM). This procedure is written as a generalized example. CAM manufacturers will have model specific calibration procedures that should be used.

Scope

This calibration procedure is specific to a CAM used for α-radiation detection, but it could be modified for other instruments. Calibration of other detectors or use at other facilities may require different procedures. This procedure provides instructions for doing routine performance checks and calibrations to assure proper functioning of the CAM and includes a quantitative detector efficiency determination for radiation measurement.

Hazard Analysis

Identified Hazards and Mitigation

1. The internal surfaces of this CAM may be contaminated. Swipe the internal surfaces and perform an α/β count. If positive (e.g., greater than an acceptable limit), use latex gloves while decontaminating the CAM.
2. The electroplated calibration source can cause radioactive contamination if improperly handled. The electroplated source should be held around the outside edge and the electroplated surface should not be subjected to abrasion.
3. The CAM includes high voltage but low current supplies that are used for operation of the detector. The instrument operates on 110 V AC and access to the interior in an energized state constitutes a potential shock hazard.

References

1. CAM-specific technical manual
2. Others as appropriate

Equipment

1. Pulsers
2. Oscilloscope
3. Digital volt meter and probe (DVM)
4. Clean filter papers
5. Appropriate radioactive check sources
6. Basic instrument tool kit

Procedure

Inspection and Repair

1. Inspect the instrument for broken displays or indicators, loose connectors, and any other physical damage. Repair as needed. If the detector face needs cleaning, common solvents

Calibration of Air Samplers and Monitors

such as methanol may be used, but ensure that the solvents used will not damage the detector.

2. Test the CAM for air in-leakage according to procedure for this testing. Record the results on the Calibration Record in Attachment 1 (see below).

Electrical Testing

1. Voltage checks: Power up the CAM and check background levels (e.g., to ensure no contamination on the detector) and/or connect an oscilloscope to the preamplifier output. If the signal-to-noise ratio on the preamp output is high (e.g., greater than 100 mV) replace the detector. If the detector replacement does not cure the problem, reinstall the old detector and troubleshoot the preamp.
2. Verify that the low voltages on the pre-amplifier are set appropriately. If not, repair as needed.
3. Verify that the pulse height analyzer (PHA) voltages are set appropriately. If not, repair as needed.

Pulser Calibrations (Useful for Some Count Rate-Based CAMs)

1. Connect the pulser to the preamp board.
2. Connect an oscilloscope to the preamp output and adjust the amplitude height on the oscilloscope. Set the pulse width for 1 μs on the oscilloscope and use the pulse-width potentiometer to adjust the width as needed.
3. Set the pulser to a high rate, for example, 10,000 pulses/min. On the Calibration Record, record the initial CAM response relative to the pulse input rate.
4. Repeat steps using a lower pulse rate, say 100. The count rates of the CAM should match within better than 10% of the correct values. Record these final readings on the Calibration Record.

Efficiency Calibration (Detector Calibration)

1. Place the radioactive standard source into the CAM holder and begin count cycle.
2. Record monitor response (count rate, concentration, etc.)
3. Remove the source from the CAM and close the detector door.
4. Collect a background of sufficient length. The background should be sufficiently low and within established bounds. If the background is higher, troubleshoot and repair as needed.
5. Using the final readings obtained in Step 2 above, determine CAM efficiency using the following formula:

$$\text{Alpha efficiency (\%)} = \frac{100 \times \{\text{count rate of source (cps)} - \text{background (cps)}\}}{\text{Radioactive source activity (Bq)}}$$

The average efficiency should be within acceptable limits. If not, repair as needed.

6. Record the CAM efficiency on the Calibration Record. Include information regarding the instrument, radioactive source, and counting efficiency information on the Calibration Record.

Performance (Alarm) Check

1. Insert the designated radioactive sources (can use both high and low activity sources if the CAM has preset low and high alarm levels) for doing the alarm test of the CAM.
2. Record the alarm response (yes—alarm actuates or no-alarm failed; and indicate the type of alarm, that is, low-level alarm or high-level alarm). Confirm on the Calibration Record that the alarm test was done. Enter the results of the test on the Calibration Record.

Flow Calibration

1. Connect the calibrated flow meter in line with the CAM and pump.
2. Start the pump.
3. Using a calibrated bubble-meter or other calibrated flow-rate measurement device, take readings at the low and high ends and at the mid-point of the normal air flow operating range.
4. Record the readings from the calibrated flow meter and the CAM indicator.
5. Calculate the percent difference and compare with acceptable tolerances.
6. Record the results on the Calibration Record.

DOCUMENTATION

1. Verify that the Calibration Record is complete. Sign and date the Record.
2. Attach a new calibration sticker to the instrument with your name, calibration date, calibration void date, and the new efficiency factor (%).
3. Update the instrument record in the RIP database per Reference 3.
4. File the documentation package.

EXAMPLE CALIBRATION RECORD

Figure A1 shows an example calibration report.

EXAMPLE CALIBRATION REPORT

EQUIPMENT TO BE CALIBRATED: AIR SAMPLER TYPE:

CALIBRATION PROCEDURE:

Manufacturer: Serial #:   Calibration date:

Calibration equipment information
Source ID #:   Source isotope: Source activity:
Reference air flow meter ID #:

Environmental conditions during calibration
Temperature: % R/H: Barometric pressure:

As-found readings
Background count rate: Count rate with calibration source:
Detector efficiency: Calibration source peak channel #:
Air flow rate indication: Reference air flow meter indication:
Detector signal to noise: Amplifier gain setting:
HV setting: Alarm check OK:

FIGURE A1 Example calibration report.

Sampler air inleakage OK:

Electronics testing with pulser:

Pulser CPM	Initial Reading	Final Reading
10,000		
1000		
100		
40		

Equipment adjustments

Repairs:

Replace parts:

Adjust amplifier gain:

Adjust HV setting:

Equipment calibration (as-left readings)

Background count rate: Count rate with calibration source:

Detector efficiency: Calibration source peak channel #:

Air flow rate indication: Reference air flow meter indication:

Detector signal to noise: Amplifier gain setting:

HV setting: Alarm check OK:

Sampler air inleakage OK:

State any limitations of the calibration:

Calibrated by:

Reviewed by:

FIGURE A1 (Continued)

APPENDIX B: A MULTIPLE-FRAME-OF-REFERENCE METHOD FOR ROTAMETER CORRECTION FACTORS

Mark D. Hoover

INTRODUCTION AND OBJECTIVES OF THE METHOD

Rotameters (also known as variable-area meters) are widely used and relatively inexpensive industrial hygiene tools for monitoring and controlling the flow air during air sampling. Unfortunately, they are seldom operated at the temperature and pressure under which they were calibrated. In particular, rotameters are typically located downstream of a sampling device, which causes the pressure inside the rotameter to be lower than the ambient atmosphere from which the sample is being drawn (Figure B1). In addition, sampling is sometimes conducted of atmospheres with specific gravity that is different than the specific gravity of the atmosphere used for calibration (e.g., use of a rotameter calibrated with air to sample from an atmosphere of argon).

Sampling errors occur when the rotameter equation and the ideal gas law are not used, or are improperly used, to correct calibration conditions to the conditions of use. Depending on the temperature, pressure, and specific gravity conditions, errors can exceed the 5% level recommended for acceptable data quality without a correction (Hickey et al., 1993). The selection or substitution of incorrect rotameter components and the occlusion or fouling of the tube or float can also result in errors.

The objectives of this appendix are to

1. Describe a multiple-frame-of-reference approach to promote understanding and minimize the possibilities for errors in the calibration and use of rotameters for air sampling.
2. Identify and clarify the governing equations for operation of rotameters, including
   a. The Rotameter Equation
   b. The Ideal Gas Law
   c. The Scale Factor Equation

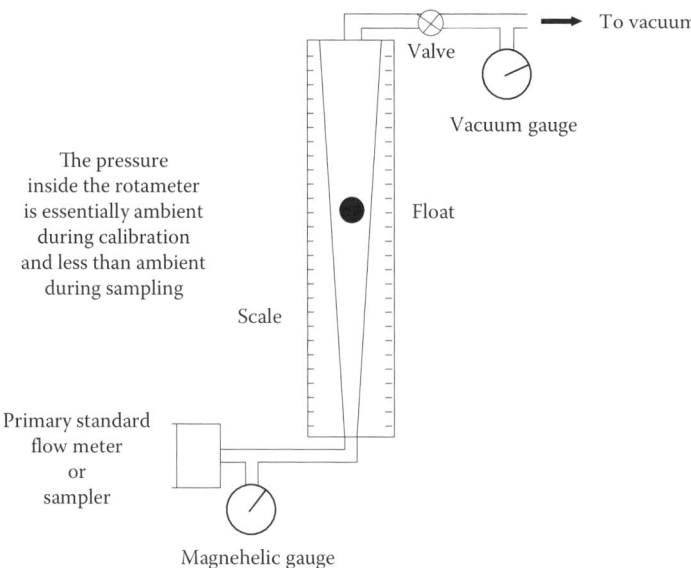

FIGURE B1 Typical rotameter setup for calibration with a low-pressure-drop, primary-standard flow meter or operation with an air sampler.

Calibration of Air Samplers and Monitors

3. Illustrate proper use of the equations to relate rotameter flow rates among the frames of reference as a function of calibration and operating conditions.
4. Summarize other good practice to avoid common errors in the use of rotameters.

OPERATION

Rotameters are smoothly tapered tubes with a graduated numerical scale and a visible float. The term "rotameter" originated from the visible rotation of certain floats during active operation.

Floats come in a variety of materials and densities (e.g., glass, sapphire, brass, stainless steel, monel, titanium, tantalum, etc.), shapes (spherical, elliptical, plumb bob, spool, hat-shaped, etc.), and diameters. The underlying principle of operation of the rotameter under conditions normally applicable to air sampling is the Bernoulli Principle, which states that an increase in the velocity of a gas (i.e., an increase in the velocity pressure) is accompanied by a corresponding decrease in the local static pressure of the gas. (This is the principle that causes "lift" on an airplane wing.) During operation of the rotameter, the float is subject to a downward force from gravity and an upward lift from the Bernoulli effect as air accelerates in an upward direction around the float between the cross-sectional area of the float and the wall of the tube. (Note that the float is *not* "dragged upward" by viscous interactions with the moving air. The viscosity of gases such as air is not sufficient to influence the position of the float.)

The force of gravity on the float is constant, but the magnitude of lift on the float depends on the velocity achieved by the air as it passes between the float and the wall of the tube. At small flow rates the float will rise to a position near the bottom of the tube where the annular region between the float and the wall of the tube is small enough to provide an air velocity that is sufficient to induce the needed lift. At higher flow rates the float will rise until the annular region between the float and the wall of the tube is proportionally larger and ultimately sufficient to provide the air velocity that is associated with the needed lift.

The height of the float provides a visual confirmation that flow is occurring and a direct indication of the magnitude of the flow rate. More than one float can be included in a single rotameter to extend the useful operating range (e.g., a low-density float such as glass to provide on-scale readings at low flow rates, and a high-density float such as tantalum to provide on-scale readings at higher flow rates). The density of the float has a square root influence on the performance of the rotameter. Thus, a rotameter with a range of 0–10 L/min with a glass float (density 2.4 g/cm^3) will have a range of 0–26 L/min when used with a tantalum float (density 16.6 g/cm^3). (The square root of the quotient 16.6/2.4 is 2.6.)

Appropriate correction equations for the influences of temperature, pressure, and specific gravity in rotameters have been available for many years (e.g., Schoenborn and Colburn, 1939; Rhodes, 1941; Craig, 1971). However, these equations are frequently misunderstood, misused, or ignored.

FRAMES OF REFERENCE

To assist in the proper calibration and use of rotameters, it is instructive to consider the relationships among the physical conditions in the following five separate frames of reference:

1. Calibration temperature and pressure (CTP), which correspond to the temperature, pressure, and specific gravity of the gas for which a calibration relationship was determined between volumetric flow rate of the calibration gas and the height of the float.
2. Operating temperature and pressure (OTP), which correspond to the temperature, pressure, and specific gravity of the gas inside the rotameter during operation.
3. Ambient temperature and pressure (ATP), which correspond to the temperature and pressure in the environment from which the gas sample is being taken or, alternatively, into which the gas sample is being delivered.

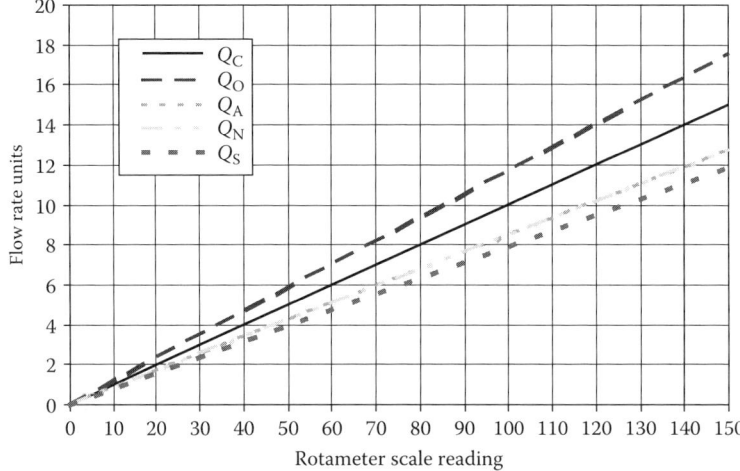

FIGURE B2 Illustration for the example conditions listed in Table B1 of how identical rotameter scale readings can correspond to different flow rates in the five frames of reference. (Note that the conditions for Q_A and Q_N are the same in this example so the curves for those conditions overlap.)

4. Normal temperature and pressure (NTP), which is traditionally 20°C (293 K) and 1 atm (760 mm Hg).
5. Standard temperature and pressure (STP), which is traditionally 0°C (273 K) and 1 atm (760 mm Hg).

Figure B2 and Table B1 provide an example of the relationships among flow rates in these frames of reference as a function of rotameter calibration and operating conditions.

THE ROTAMETER EQUATION

The basic rotameter equation involves the following square root relationship between the flow rate Q_C required to raise the float to a given height under calibration conditions and the flow rate Q_O required to raise the float to that same height under operating conditions:

$$Q_O = Q_C \sqrt{\frac{P_C}{P_O} \cdot \frac{T_O}{T_C} \cdot \frac{SG_C}{SG_O}} \tag{B.12.1}$$

where P_C and P_O are the calibration and operating pressures, T_C and T_O are the calibration and operating temperatures, and SG_C and SG_O are the calibration and operating specific gravities of the gas.

Lift on the float, as noted above, is due to the Bernoulli effect and, therefore, viscosity is not a factor in the rotameter equation for air sampling applications. Note that the Rotameter equation is only used to convert between calibration and operating conditions, or, conversely, between operating conditions and calibration conditions.

THE IDEAL GAS LAW

The gases typically encountered in health physics and industrial hygiene settings (air, nitrogen, argon, carbon monoxide, helium, etc.) behave as ideal gases. Therefore, the Ideal Gas Law can be

TABLE B1
Input and Output Data for the Example Shown in Figure B2 of How Identical Rotameter Scale Readings Can Correspond to Different Flow Rates in the Five Frames of Reference

Input Data:	CTP	OTP	ATP	NTP	STP
Gas specific gravity, SG	1	1			
Temperature, T (K)	293	293	293	293	273
Pressure, P (mm Hg)	760	552	760	760	760
Rotameter calibration scale factor (scale units/flow unit):				10	
Maximum rotameter scale reading:				150	
Flow rate value for the calculation of flow rate relationships:				10	

Output Data:	Rotameter Scale Reading	Flow Rate Relationships				
		CTP	OTP	ATP	NTP	STP
Case 1	100	*10.00*	11.73	8.52	8.52	7.94
Case 2	85	8.52	*10.00*	7.26	7.26	6.77
Case 3	117	11.73	13.77	*10.00*	10.00	9.32
Case 4	117	11.73	13.77	10.00	*10.00*	9.32
Case 5	126	12.59	14.78	10.73	10.73	*10.00*

	Rotameter Scale Reading	Calculated Data for Calibration Curves				
		Flowrates				
		Q_c	Q_o	Q_a	Q_n	Q_s
Scale minimum:	0	0	0	0	0	0
Scale maximum:	150	15.0	17.6	12.8	12.8	11.9

Note: For example, rotameter manufacturer and model, method for reading float level, etc.: add as appropriate.

Abbreviations: CTP is calibration temperature and pressure, OTP is operating temperature and pressure, ATP is ambient temperature and pressure, NTP is normal temperature and pressure, and STP is standard temperature and pressure.

used to convert the flow rate under operating conditions (Q_O) into the flow rate under any of the other conditions of interest (Q_S, Q_N, or Q_A). For example,

$$Q_A = Q_O \left(\frac{T_A}{T_O} \cdot \frac{P_O}{P_A} \right) \tag{B.12.2}$$

The equations to convert between Q_O, Q_S, Q_N, and Q_A using appropriate combinations of the Rotameter Equation and the Ideal Gas Law are shown in Table B2.

THE SCALE FACTOR EQUATION

Modern rotameters are designed to have a linear relationship between the volumetric flow rate of the gas and the height of the float. The Scale Factor Equation can be expressed in any consistent set of units as

$$Q = RR \cdot SF \tag{B.12.3}$$

where Q is the volumetric flow rate of the gas in the frame of reference of concern, RR is the rotameter scale reading corresponding to the height of the float, and SF is the appropriate scale factor relationship between Q and RR in the frame of reference of concern.

TABLE B2
Equations That Define the Relationships among Flow Rates for an Identical Rotameter Reading (Float Height) in the Five Frames of Reference[a,b,c]

To ⇒ From	Calibration Conditions	Operating Condition	Ambient Conditions	Normal Conditions	Standard Conditions
Calibration conditions	$Q_c = Q_c$	$Q_o = Q_c \sqrt{\dfrac{P_c}{P_o} \cdot \dfrac{T_o}{T_c} \cdot \dfrac{SG_c}{SG_o}}$	$Q_A = Q_c \sqrt{\dfrac{P_c}{P_o} \cdot \dfrac{T_o}{T_c} \cdot \dfrac{SG_c}{SG_o}} \cdot \dfrac{P_o}{P_A} \cdot \dfrac{T_A}{T_o}$	$Q_A = Q_c \sqrt{\dfrac{P_c}{P_o} \cdot \dfrac{T_o}{T_c} \cdot \dfrac{SG_c}{SG_o}} \cdot \dfrac{P_o}{P_N} \cdot \dfrac{T_N}{T_o}$	$Q_s = Q_c \sqrt{\dfrac{P_c}{P_o} \cdot \dfrac{T_o}{T_c} \cdot \dfrac{SG_c}{SG_o}} \cdot \dfrac{P_o}{P_s} \cdot \dfrac{T_s}{T_o}$
Operating conditions	$Q_c = Q_o \sqrt{\dfrac{P_o}{P_c} \cdot \dfrac{T_c}{T_o} \cdot \dfrac{SG_o}{SG_c}}$	$Q_o = Q_o$	$Q_A = Q_o \cdot \dfrac{P_o}{P_A} \cdot \dfrac{T_A}{T_o}$	$Q_N = Q_o \cdot \dfrac{P_o}{P_N} \cdot \dfrac{T_N}{T_o}$	$Q_s = Q_o \cdot \dfrac{P_o}{P_s} \cdot \dfrac{T_s}{T_o}$
Ambient conditions	$Q_c = Q_A \cdot \dfrac{P_A}{P_o} \cdot \dfrac{T_o}{T_A} \sqrt{\dfrac{P_o}{P_c} \cdot \dfrac{T_c}{T_o} \cdot \dfrac{SG_o}{SG_c}}$	$Q_o = Q_A \cdot \dfrac{P_A}{P_o} \cdot \dfrac{T_o}{T_A}$	$Q_A = Q_A$	$Q_N = Q_A \cdot \dfrac{P_A}{P_N} \cdot \dfrac{T_N}{T_A}$	$Q_s = Q_A \cdot \dfrac{P_A}{P_s} \cdot \dfrac{T_s}{T_A}$
Normal conditions	$Q_c = Q_N \cdot \dfrac{P_N}{P_o} \cdot \dfrac{T_o}{T_N} \sqrt{\dfrac{P_o}{P_c} \cdot \dfrac{T_c}{T_o} \cdot \dfrac{SG_o}{SG_c}}$	$Q_o = Q_N \cdot \dfrac{P_N}{P_o} \cdot \dfrac{T_o}{T_N}$	$Q_A = Q_N \cdot \dfrac{P_N}{P_A} \cdot \dfrac{T_A}{T_N}$	$Q_N = Q_N$	$Q_s = Q_N \cdot \dfrac{P_N}{P_s} \cdot \dfrac{T_s}{T_N}$
Standard conditions	$Q_c = Q_s \cdot \dfrac{P_s}{P_o} \cdot \dfrac{T_o}{T_s} \sqrt{\dfrac{P_o}{P_c} \cdot \dfrac{T_c}{T_o} \cdot \dfrac{SG_o}{SG_c}}$	$Q_o = Q_s \cdot \dfrac{P_s}{P_o} \cdot \dfrac{T_o}{T_s}$	$Q_A = Q_s \cdot \dfrac{P_s}{P_A} \cdot \dfrac{T_A}{T_s}$	$Q_N = Q_s \cdot \dfrac{P_s}{P_N} \cdot \dfrac{T_N}{T_s}$	$Q_s = Q_s$

[a] Only the rotameter equation is required for conversions between Q_c and Q_o.
[b] Only the ideal gas law is required for conversions among Q_o, Q_A, Q_N, and Q_S.
[c] Both the rotameter equation and the ideal gas law are required for conversions between Q_c and either Q_A, Q_N, or Q_S.

Calibration of Air Samplers and Monitors

For example, if a rotameter has an *SF* of 1 scale unit per L/min, then an *RR* of 10 corresponds to a flow rate of 10 L/min. The scale factor provided by the manufacturer can only be used if the rotameter is operated at the same conditions under which it was calibrated. Under actual conditions of operation, a rotameter reading of *RR* (corresponding to a flowrate of Q_C under the conditions of calibration) is the result of a flowrate of Q_O under the conditions of operation. As described in the examples below, relationships of *SF* to flow rates in the other frames of reference must be determined using the equations in Table B2. A Microsoft Excel® spreadsheet to perform the calculations can be obtained from the author.

Example Calculations

For the example shown in Table B1 and Figure B2, the numerical value of a flow rate of interest (10 flow-rate units in this case) was entered into the spreadsheet, and the resulting flow-rate relationships were calculated for each of the following five cases:

1. The value of interest is the flow rate at CTP,
2. The value of interest is the flow rate at OTP,
3. The value of interest is the flow rate at ATP,
4. The value of interest is the flow rate at NTP, and
5. The value of interest is the flow rate at STP.

The example in Table B1 and Figure B2 involves a rotameter that is calibrated in air at sea-level pressure and operated to sample the ambient air at sea level with a sampling device (e.g., a filter) that provides a pressure drop of 4 psi (208 mm Hg). A temperature of 293 K is assumed for the calibration, operating, and ambient conditions. Traditional values of temperature and pressure are used for the NTP and STP conditions. The example further assumes that the calibration scale factor (scale units/flow unit) for the rotameter is 10. Thus, if the flow units are L/min, a scale reading of 100 under the conditions of calibration corresponds to flow rate of 10 L/min.

> *Case 1* illustrates the flow rates at OTP, ATP, NTP, and STP that will result if the rotameter is adjusted to a scale reading 100. This case is important because it corresponds to the common error of assuming that the manufacturer's calibration scale can be used to set the desired sampling flow rate for collection of an ambient air sample. Because of the reduced pressure induced by the sampling device, the internal operating pressure in the rotameter is only 760 – 208 = 552 mm Hg. At that condition, an operating gas flow rate of 11.73 L/min occurs at the scale reading of 100. Similarly, because the ambient pressure from which the sample is taken is 760 mm Hg, the ambient flow rate is only 8.52 L/min. Therefore, setting the rotameter to a scale reading of 100 will result in an ambient sampling rate that is 15% lower than is indicated by the original calibration curve. This is an error to be avoided.
>
> *Case 2* illustrates the flow rates at CTP, ATP, NTP, and STP that will result if the rotameter is adjusted to provide a flow rate of 10 L/min at OTP. This case is included for completeness, but is essentially of academic interest because (as noted in ANSI, 1999, 2009) the primary concern of the air sampling practitioner is to correctly achieve a flow rate of interest at ATP. Nevertheless, the relative relationships among the five frames of reference are consistent across all cases, and it is useful to understand that (under the conditions of this example) the rotameter scale reading will only be 85 when the operating flow rate is 10 L/min, and that the corresponding ATP flow rate will be only 7.26 L/min.
>
> *Case 3* illustrates the flow rates at CTP, OTP, NTP, and STP that will result if the rotameter is correctly adjusted to provide a flow rate of 10 L/min at ATP. This is the case of greatest interest to air sampling practitioners because it reveals the rotameter reading that must be set to correctly collect an ambient air sample at the desired flow rate. For the current example, the rotameter scale reading must be set at 117 to provide an ambient flow rate of

10 L/min. Note that although the conversion equations can be used to establish this target setting, it is prudent to initially and periodically verify by ambient volumetric calibration that the desired flow rate is actually being achieved in practice.

Cases 4 and 5 are only occasionally of interest to correctly set a rotameter to provide a flow rate that is expressed in NTP or STP units. Sampling to achieve a given flow rate under NTP or STP conditions is not an objective when conducting air sampling for worker or environmental protection, but may be of interest for conducting a chemical material balance during process sampling.

OTHER INFLUENCES

The following will influence the reliability of results when using rotameters:

1. Use of different fittings and tubing than those used for calibration can yield different pressure drops.
2. Use of different floats can give different readings.
3. Use of multiple rotameters of differing flow ranges can give incorrect results if the wrong rotameter is installed for a given application.
4. Use of rotameters designed to accommodate modular, exchangeable tube inserts can give incorrect results if the wrong insert is installed.
5. Occlusion, fouling, or chemical erosion of the tube or float can alter results.

RECOMMENDATIONS TO AIR SAMPLING PRACTITIONERS

1. Understand the relationships among the following five frames of reference: calibration, operating, ambient, normal, and standard conditions.
2. Define, justify, and document the physical conditions in each frame of reference for each air sampling application.
3. Understand how rotameters work.
4. Use the Rotameter Equation to equate the flow rate under calibration conditions to the flow rate under operating conditions.
5. Use the Ideal Gas Law to equate the flow rate under operating conditions to the flow rate under ambient, normal, or standard conditions.
6. Use the Scale Factor Equation to relate scale units to flow rate units in the various frames of reference.
7. Use a calibrated volumetric or mass flow meter to periodically verify sampling flow rates under the conditions of operation.
8. Anticipate, evaluate, and control potential errors associated with changes in temperature, pressure, and specific gravity; selection or substitution of incorrect rotameter components; and occlusion, fouling, or erosion of the tube or float.

REFERENCES

ANSI, *Sampling and Monitoring Releases of Airborne Radioactive Substances from the Stacks and Ducts of Nuclear Facilities*, ANSI N13.1, American National Standards Institute, New York, 1999.

ANSI, *Radiation Protection Instrument Test and Calibration—Air Monitoring Instruments*, ANSI N323C, American National Standards Institute, New York, 2009.

Craig, D.K., The interpretation of rotameter air flow readings, *Health Phys.* 21: 328–332, 1971.

Hickey, E.E., Stoetzel, G.A., Strom, D.J., Cicotte, G.R., Wiblin, C.M., and McGuire, S.A. *Air Sampling in the Workplace*, NUREG 1400, U.S. Nuclear Regulatory Commission, Washington, DC, 1993.

Rhodes, T.J., *Industrial Instruments for Measurement and Control*, McGraw-Hill, New York, 1941.

Schoenborn Jr, E.M. and Colburn, A.P., The flow mechanism and performance of the rotameter, *Trans. Am. Inst. Chem. Eng.* 35: 359–389, 1939.

13 Principles of Air Sampler Placement in the Workplace

Jeffrey J. Whicker

CONTENTS

Background 271
Aerosol and Gas Transport is Dominated by Ventilation-Induced Airflow 272
Airflow Studies for Testing Ventilation and Sampler Placement 273
Conducting Airflow Studies for Improved Air Quality 274
Conducting Airflow Studies for Sampler Placement 276
    Evaluation of Placement of Retrospective Air Samplers 278
    Evaluation of Placement of Continuous Air Monitors 278
    Optimization of Air Sampling and Monitor Placement 279
Further Considerations for Room Airflow Testing 280
    Representative Facility Conditions 280
    Representative Aerosol Tracer Release 280
Conclusions 282
References 282

BACKGROUND

Air sampling programs should be designed to meet the following objectives: (1) where needed, provide real-time monitoring of air concentrations and alarm if the levels exceed the preset and/or desired warning limits, (2) estimate exposure levels when appropriate, and of importance to this chapter (3), test the effectiveness of aerosol confinement and ventilation (Marshall and Stevens, 1980). Protection of workers from dispersible radioactive material requires that ventilation-driven airflow is used and that it is sufficient for containment and control. Airflow is a highly effective tool for preventing dispersible material from escaping containment structures such as glove boxes and chemical hoods, and if released into a room, directed air flow can prevent the unwanted spread of the radioactive material into other areas of the facility resulting in unwarranted spread of contamination and worker exposures.

Ventilation-driven airflow patterns within the room dominate aerosol and gas transport and significantly impact the level of hazard to workers. The temporal and spatial dispersion patterns of accidentally released radioactive aerosols and gases in rooms are highly complex. Yet these complex dispersion patterns, both at the room scale and at the smaller scale of worker breathing zones (BZs), determine both worker exposure and the measurement of air concentrations by air samplers and real-time monitors. There can be steep concentration gradients, especially during the early phases of a release, and radioactive aerosols and gases can take a number of minutes to disperse to locations for alarming air monitors (Scripsick et al., 1979; Whicker et al., 1997). This can lead to delays in detection for worker warning, and if the dilution of the released material is significant enough, poor detection capability (Crites, 1994).

Well-designed room ventilation has proved to be very effective in controlling dispersible contaminants and the placement and number of air sampling instrumentation relative to airflow patterns is important for worker protection. Yet ventilation-driven airflow patterns in a room are not easily predicted from simple analysis of the ventilation design and room layout. Therefore, measurement of airflow, aerosol, and gas dispersion patterns in a room is required to determine the proper numbers and placement of air samplers and monitors. This contributes to high levels of worker protection through fast and sensitive detection of released materials (Whicker et al., 2003). Therefore, it is important to make measurements of airflow and aerosol and gas dispersion patterns in rooms that contain significant inhalation hazards. Both the U.S. Department of Energy (DOE) and the U.S. Nuclear Regulatory Commission (NRC) recognize the importance of airflow studies for protection of workers (NRC, 1992; DOE, 2007). Each of these regulatory agencies discuss placement of air samplers and additional guidance on airflow studies is provided in the NRC document NUREG-1400 (Hickey et al., 1993). While NUREG-1400 provides very valuable suggestions in this regard, this chapter provides an enhanced discussion on airflow testing for maintaining good air quality and for air sampler placement. Some recent applications from studies of indoor airflow, dispersion studies of aerosols and gases in rooms and the interplay between worker exposure and air monitoring are also discussed.

AEROSOL AND GAS TRANSPORT IS DOMINATED BY VENTILATION-INDUCED AIRFLOW

Aerosol transport is governed by several fundamental mechanisms (Hinds, 1982). Figure 13.1 shows a simplified schematic of aerosol dispersion from a point source. After release into a workroom, an aerosol is transported in the x-direction by ventilation-driven airflow. At the same time, the aerosol cloud disperses in the x, y, and z directions due to small-scale turbulence (turbulent diffusion) and Brownian diffusion. Downward migration through gravitational settling will also occur for heavier particles.

Unless the particles are quite large (nonrespirable), studies conducted in ventilated spaces have shown ventilation-induced convective flow and associated turbulent diffusion are the primary mechanisms of aerosol transport (Hinds, 1982; Whicker et al., 2000a). For example, a spherical 1 μm particle with unit density will have a settling velocity of 0.03 cm s^{-1}, a Brownian average velocity of 0.0008 cm s^{-1}, and a velocity of about 2–3 cm s^{-1} from small-scale turbulence. Comparing these velocities to an average room air velocity of about 15 cm s^{-1} found in a nuclear facility (Whicker et al., 2000a,b), it is easy to conclude that forced-convective flow, derived from each room's ventilation system, drives transport of aerosol in these rooms, especially for respirable particles with aerodynamic diameters <10 μm. Ventilation-driven airflow dominates transport of gases as well, though the Brownian motion component is much higher (e.g., ~2 cm s^{-1} for tritium gas).

Therefore, the ventilation system, combined with the room configuration, sets the boundary conditions for airflow patterns through their effects on the fundamental quantities of velocity, direction, and turbulence of the airflow. Each of these fundamental quantities contributes to the effectiveness of confinement of the aerosol, and if released into an occupied room, the levels of exposure.

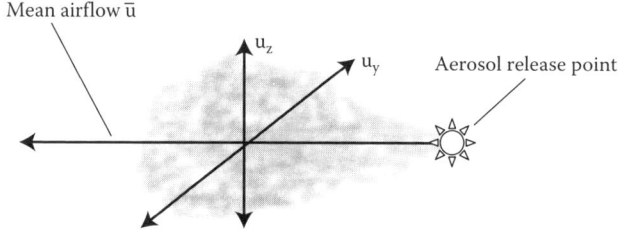

FIGURE 13.1 Illustration of aerosol dispersion at a single point.

AIRFLOW STUDIES FOR TESTING VENTILATION AND SAMPLER PLACEMENT

Inhalable forms of radioactive material are worked with every day in laboratories worldwide, which necessitate numerous engineered barriers, administrative controls, and protective air sampling programs to ensure a safe working environment. Given the overwhelming importance of forced convection from ventilation-driven airflow for transporting aerosols and gas, it is critical to perform tests to verify the desired level of function of the ventilation-induced airflows that support these controls and programs.

There are several general services ventilation-driven airflow provides for protecting nuclear workers that can be tested. First, airflow studies can be used to assess the effectiveness of ventilation-based containment. Negative pressure containment systems such as that found in hoods, glove boxes, and in some cases, between adjoining rooms, are used extensively for confining radioactive material inside the containment. Confinement in these cases is assured by pulling air into an area with lower relative pressure, and airflow studies can be used to test for this desirable flow. Second, room ventilation is generally designed to rapidly dilute and remove any airborne material released into the room. Airflow studies can be designed to test the ability of room ventilation to remove airborne radioactive material from the BZs of workers and from the general room air after an accidental release. Third, measurement of ventilation-driven airflow patterns and its relation to placement of air monitoring equipment in a room help to determine the optimal placement of samplers. Each of these is discussed later in this chapter.

Airflow studies can be characterized as either *qualitative* or *quantitative*. Each type of study has its own advantages and disadvantages. Decisions of which combination of qualitative or quantitative techniques used depend on numerous factors. These factors include the hazard level, distribution of possible release locations, room and ventilation complexity, cost of the studies, and the purpose of the airflow study (e.g., confirm containment, exposure assessment, etc.).

Qualitative airflow studies are tests that are done by releasing a tracer into a room and then visually tracking the movement of the tracer through the room. Several types of tracers can be used including: (1) smoke generators and tubes, (2) helium-filled balloons, and (3) neutrally buoyant bubbles. The airflow patterns can be recorded on worksheet drawings with narrative descriptions or by using photographs or videotapes. These studies are considered qualitative because measurements of tracer movement are determined visually, concentration measurements are not made, and larger tracers (balloons and bubbles) do not respond to small-scale turbulence that is important for dispersion of smaller particles. Therefore, uncertainty with qualitative techniques is likely to be larger than that associated with quantitative airflow studies. However, done properly, qualitative airflow studies are straightforward and economical, and they provide valuable estimates of direction and velocity of airflow.

For quantitative airflow studies, nontoxic tracer aerosols or gases are released at a number of room locations and concentrations of the tracer are then measured at another set of selected points in the room. The release locations are selected such that are they represent the room's potential release points, and the sampling locations represent potential air sampling locations. Measurements of the tracer concentrations can be resolved in time and space, and therefore provide accurate measurements not only of speed and direction but also of aerosol dilution. These types of studies can provide information on potential worker exposures and the adequacy of the air monitor/sampler location through comparisons of time-dependent concentrations at a worker's BZ relative to monitor/sampler locations.

Another type of quantitative airflow study involves direct measurement of airflow as opposed to measurement of tracer concentration. This is accomplished using either thermal or sonic anemometers (Wasiolek et al., 1999). These measurements not only provide information on the velocity and direction of the airflow, but also provide information on the turbulence. However, these are point measurements in space and therefore multiple spatial measurements are needed. For example, to measure the linear velocity of air flowing into a chemical hood requires a series of measurements across the entire face of the hood.

CONDUCTING AIRFLOW STUDIES FOR IMPROVED AIR QUALITY

Before testing, criteria for acceptable airflow should be established. These criteria will be based on four general goals for ventilation-induced airflow. Another set of airflow tests are used to evaluate placement of air samplers and monitors. In this section, measurements for the main four general goals of ventilation and the techniques for testing if the ventilation-driven airflows provide adequate worker protection are described.

Four general goals of ventilation-induced airflow: The first goal of ventilation is to provide a pressure differential between rooms so that the lowest relative pressure is in the areas with the highest potential for aerosol release. This pressure differential provides directional airflow that greatly diminishes or prevents the spread of the radioactive aerosol beyond the initial point or room of escape. Second, high-velocity, directed airflow is used to prevent an aerosol or gas from escaping the confinement structure. Examples of this include chemical hoods or local exhaust ventilation used for machining. To be effective at confinement, the local airflow velocity into an open containment structure should follow published guidelines (ACGIH, 1992). Beyond airflow velocity, the National Institute of Health (NIH, 1996) explored other factors which influence the effectiveness of local exhaust ventilation and concluded:

a. Room velocities and turbulence in the area of the local exhaust ventilation should be low and not interfere with the airflow into the exhaust,
b. Exhaust hoods located in corners of rooms perform better than those in the center of a wall,
c. High-velocity jets of air impinging into the work area in front of the hood decrease capture efficiency of the hood and should be avoided, and
d. To avoid interference from competing flows, each room exhaust should be separated by at least 4 ft.

Third, if a radioactive material escapes confinement and is released into the room air, the goal of the room ventilation is to rapidly move the aerosol out of the worker's BZ, dilute it with clean air, and rapidly exhaust the aerosol from the room. Whicker et al. (1997) showed that steep concentration gradients exist for extended periods around the release point. Thus, a worker collocated with the release point can receive most of their exposure in the first few minutes. In contrast, Flynn et al. (1996) showed that with proper airflow, the concentration in the BZ can be dramatically reduced. Therefore, airflow studies should measure direction and velocity of the air at workstations and the level of dilution in the room.

Fourth, airborne radioactive material in a room should be rapidly exhausted from all areas in a room. Clearance times can be estimated from the number of room air exchanges per hour. However, it is important to note that the room air exchange rate is an average for the whole room, yet within the room there can be zones of stagnant airflow. A release in one of these stagnant areas can result in concentrated aerosol that slowly mixes into the rest of the room. This could lead to high exposures, thus showing the importance of identifying stagnate areas.

Measurements to Test the Four Goals of Airflow: The first goal of ventilation is to provide successive negative pressure differentials between areas, with the least relative pressure in the area with the greatest potential for a release. This is often the room in which the material is stored and worked with. Secondary ventilation zones, with greater relative pressure could be adjacent hallways or rooms. Measurements of the differential pressure between ventilation zones can be monitored to ensure the desired pressure differentials and a simple puff of smoke near connecting doorways or zones can ensure proper airflow direction.

The second goal is to provide local exhaust containment. To meet this objective, a high velocity jet of air is drawn from the room and into a containment structure, such as a chemical fume hood. To ensure the adequate containment of toxins at the source, airflow velocity into the capture hood has to be sufficiently high to be successful (ACGIH, 1992). To test this, quantitative velocity measurements can be made using a calibrated thermal anemometer. This test consists of a series of air velocity measurements that are made across the face of the ventilation hood to calculate a mean velocity that is compared to published guidelines.

The third goal of the ventilation design is to rapidly remove aerosol from BZs. For testing of the airflow patterns in potential BZs nearest the release point (potentially the greatest exposed worker), puffs of smoke could be generated into the local BZ at potential release locations and the velocity and direction of the smoke in the BZ noted. The smoke should rapidly clear from the BZ by either being drawn into the nearby local exhaust or by mixing into the rest of the room and then being exhausted from the room through the general room exhausts. Though this approach would be relatively easy to implement, this qualitative technique would only provide limited information on the rate of dispersal, for example, time until the aerosol is no longer visible in the BZ. A quantitative approach would be to measure the concentration of a tracer aerosol in the BZ over time or the concentration integrated over time in units of Bq min m^{-3} (the product of an average concentration and sample time which is eventually converted into DAC-hour for exposure calculation).

The fourth goal of room ventilation is to rapidly remove accidentally released airborne material from the room. On average, the reduction in aerosol concentration with time can be predicted from the given air exchange rate. Thus, at 15 room air exchange rates per hour, one complete air change occurs every 4 min. Therefore, an aerosol should be effectively removed (<5% remaining) after about 20 min. The time-dependent concentration in the room can be described as shown below

$$C(t) = C_o \, e^{-Qt/V} \tag{13.1}$$

where $C(t)$ is the air concentration at time t, C_o is the initial air concentration, Q is the flow rate of the room (e.g., m^3 min^{-1}), V is the volume of the room in m^3. After an acute release of a tracer becomes mixed in the room, the tracer concentration decreases exponentially. The rate constant of $-Q/V$ (fraction lost per unit time) is the measured room air exchange rate.

The time-dependent concentration model in Equation 13.1 assumes a perfectly and instantaneously mixed aerosol in the room, which is not true at the beginning of the release. A more likely time profile for the concentration for a short-term release is provided in Figure 13.2a. This plot shows that the concentration at a location will increase as the aerosol cloud arrives, reaches a peak, and then decreases as the ventilation clears the room of the aerosol. To estimate the room air exchange rate, concentration data can be plotted with time on a semi-log plot and the slope is calculated based on the extended tail of the function as shown in Figure 13.2b. The room air exchange rate is the slope of this tail and is measured in units of fraction of room volume per time. Other mixing models have been used to estimate the BZ concentration assuming first-order kinetics and a multicompartment model with mixing parameters (Rodgers et al., 1998).

The importance of having a sufficient room air exchange rate should not be underestimated. Figure 13.3 illustrates this point. Analysis of the 15-min integrated occupant exposure as a function of room air exchange rate shows that the relative exposure (relative to 1 air change per hour or ACH) decreases with ACH (Figure 13.3a). Therefore, increases in air exchange rate result in significant increases in relative protection (Figure 13.3b). Results from Whicker et al. (2002) also suggest that mixing of aerosol/gas is significantly faster in rooms with higher ACH and that precise placement of air sampling and monitoring equipment is less important in rooms with good mixing qualities. Ventilation design must balance this with other factors such as contamination control, collective dose for high occupancy rooms, thermal comfort of workers, and cost to operate the HVAC system (Awbi, 1991).

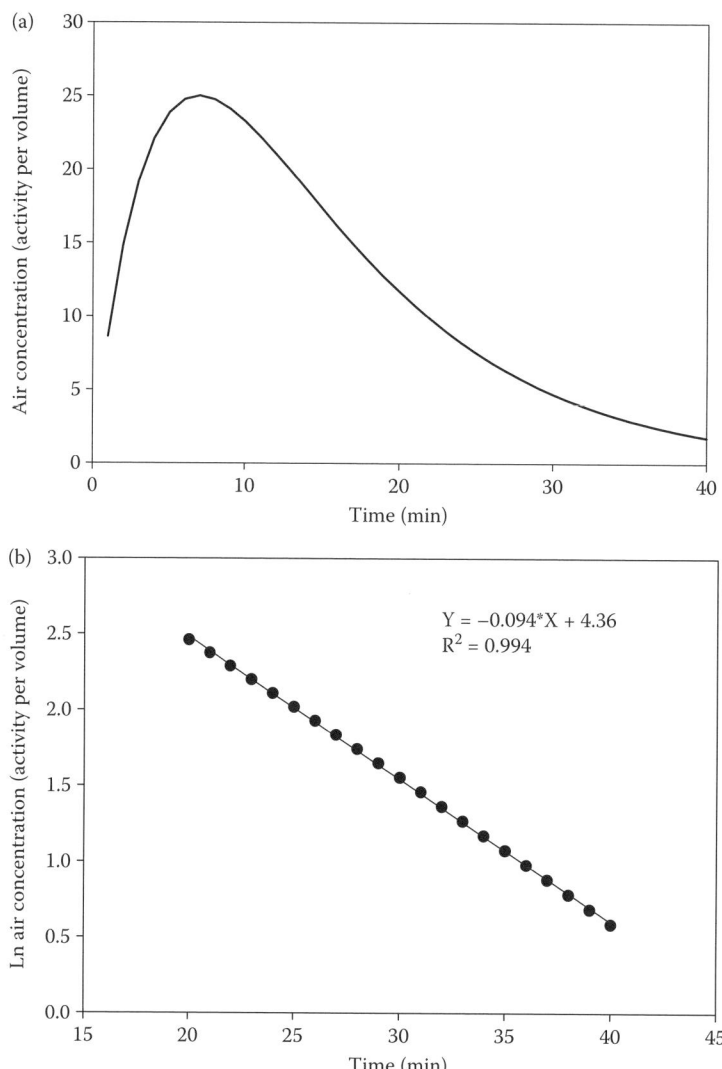

FIGURE 13.2 The air concentration through time for a hypothetical short release with clearing through room ventilation (a). (b) Shows the regression of the natural logarithm (ln)-transformed concentration through time for the tail (>20 min) of the release shown in (a). The slope of the (b) represents the estimated number of air exchanges per minute.

CONDUCTING AIRFLOW STUDIES FOR SAMPLER PLACEMENT

Using information derived from airflow studies, facility ventilation designs, air sampling measurement equipment, and monitoring strategies can be optimally configured to provide the best level of worker protection.

Sensitive air sampling and monitoring instruments can provide a high level of worker protection if the equipment is positioned properly within the room relative to the airflow patterns between the release and sampling locations. Data on dispersion patterns of aerosols and gases in ventilated rooms are essential to ensuring that the air monitors are located in sufficient quantities and in positions that provide fast and sensitive detection (Whicker et al., 1997). Improper placement of sensors in a room will result in significantly longer detection times and lower detection sensitivity, which result in much higher occupant exposures. Further, because of the significant cost associated with

Principles of Air Sampler Placement in the Workplace

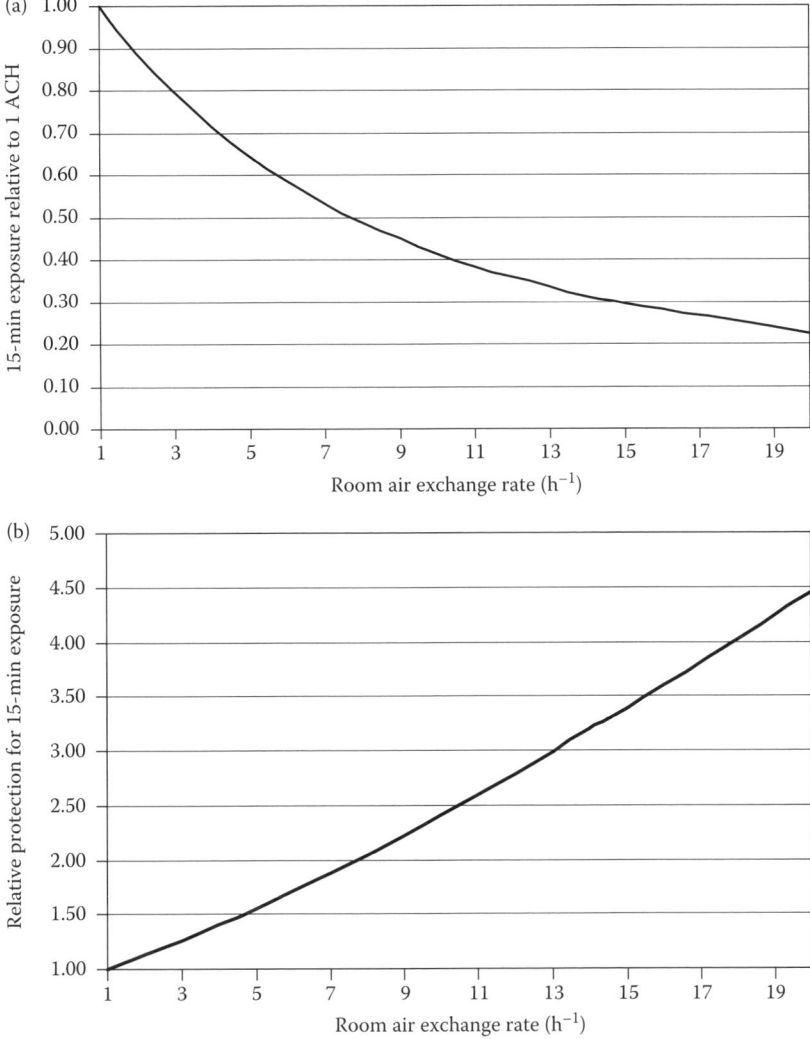

FIGURE 13.3 15-min exposures relative to a room air exchange rate of one as a function of the number of room air exchanges per hour (a). This relationship shows the importance of having higher air exchange rates for protecting workers as the relative protection increases with the air exchange rate (b).

these monitors, the number and location must be optimized to provide a balance between cost and human safety (Whicker et al., 2003).

Placement of air sampling and monitoring equipment should consider the amount of released material reaching the sampler location, that is, the amount of dilution, and the time required for the material to reach the sampling location. While qualitative tests can provide simple information such as whether the air samplers are placed downwind of the potential release locations, these tests cannot be used to measure dilution or transport rates in an absolute sense.

Sampling of general room air can be done either in real-time or retrospectively, that is, confirmation of the release after it has occurred. For real-time monitoring it is important that the monitors be positioned in the room such that the airflow transports the released aerosol to the monitor quickly and with little dilution. For "retrospective" samplers that produce results after a preset sampling period, timeliness of detection is less important, but dilution of the released aerosol should be minimized to achieve the desired sensitivity. For these reasons, optimal placement of all air sampling instrumentation needs to be assessed.

EVALUATION OF PLACEMENT OF RETROSPECTIVE AIR SAMPLERS

Placement of air samplers may be determined using an airflow study that estimates the dilution of aerosol that occurs between the release point location and other receptor locations in the room (Scripsick et al., 1979). Receptor locations could be workers' BZs or potential air sample locations elsewhere in a room. These dilution measurements are useful to determine if the sensitivity of the air sample measurements are sufficient to adequately monitor containment effectiveness, to help determine the location of a release and to help confirm and estimate the level of a worker's exposure. Depending on the physical form of the accidentally released material or the expected size of the released particles, this could be done using either tracer gas or particles. For this, the tracer could be released from representative locations in the room and the aerosol concentrations at the BZ and at numerous other locations in the effected room would be measured. The ratio of the measured tracer concentration at the sampler location divided by the concentration near the release point is a good measure of the dilution. This is a relative measure of the amount of sample collected at a potential sample location compared to the release location. For example, a ratio of 0.1 would imply that the aerosol was diluted by a factor of 10 between the release and the sampling locations. The greater the dilution factor (DF), the more sensitive the sampling must be for a release from that location. With multiple measurements from multiple release locations, one can determine the best locations to provide full coverage. Historical air sampling data can also be useful for evaluating dilution.

EVALUATION OF PLACEMENT OF CONTINUOUS AIR MONITORS

Evaluating the transport rate of released material and the placement of air samplers is primarily important for real-time monitors where fast detection is critical for worker protection. Workers who may be in the aerosol cloud are not alerted to the release until the cloud reaches the air monitor and is detected. Therefore, evaluation of the proper placement of Continuous Air Monitors (CAMs) should include measurements of both dilution and the transport time to the CAMs from potential release locations. Tracer studies using real-time detectors can be used to quantify these transport times. These measurements are made by releasing a tracer from a set of representative release locations at a known time and then quantifying the moment the tracer is detected at the sampling location.

One metric that has been proposed for evaluating placement of CAMs is the DF. This metric has the advantage that it is dose-based by combining air concentration and exposure time (Whicker et al., 1997). The DF is calculated as

$$DF(t) = \frac{\int_0^t C_{\text{release}}(dt)}{\int_0^t C_{\text{receptor}}(dt)} \quad (13.2)$$

where C_{release} is the instantaneous concentration in the local area of the release point, C_{receptor} is the instantaneous concentration at some specific receptor location in the room, and t is the integration time. Equation 13.2 provides useful information on the amount of dilution between the release point and other locations in a room, but DF varies dramatically with the selection of t, the time length of the integration. Generally for an acute, puff-type release, DF is quite large at the beginning of the release (concentration is high at the release point and low at the receptor) and decreases to single value as the aerosol thoroughly mixes and is exhausted from the room. The final steady-state value of DF, that is, where t is long, is useful for placement of retrospective air samplers where fast detection is not important, as discussed earlier.

For placement of CAMs, fast detection is important. Evaluation of CAM placement can be done using the Exposure Fraction (EF) as shown in Equation 13.3 below.

$$\mathrm{EF}(t) = \frac{\int_0^{\mathrm{TA}} C_{\mathrm{BZ}}(\mathrm{d}t)}{\int_0^{\mathrm{Tbkg}} C_{\mathrm{BZ}}(\mathrm{d}t)} \tag{13.3}$$

A worker's exposure can be calculated by integrating the concentration in the BZ of the worker through time until the aerosol cloud reaches a receptor location prompting an immediate alarm that stops the worker's exposure. The time to alarm (TA) is assumed to be the length of time it takes a released aerosol to reach a receptor location and be detected. Thus, the numerator in Equation 13.3 represents the worker exposure, assuming an alarm occurs. The numerator is divided by the concentration integrated until the aerosol is flushed from the room through ventilation and the concentration essentially returns to background (assuming a discrete release period). The denominator represents the workers dose if there were no alarming CAMs in the room. The ratio in Equation 13.3 is then the exposure with alarm divided by the exposure without an alarm and is a measure of the level of protection that a CAM at a position would provide for a release.

OPTIMIZATION OF AIR SAMPLING AND MONITOR PLACEMENT

Calculations using EF for numerous sampling locations and for various release locations and exposure scenarios can provide the data needed to optimize the number and placement of CAMs in a room (Whicker et al., 2003). The general technique is to select a number of sampling locations to be evaluated and place tracer samplers at these locations. Tracer aerosol or gas can then be released from a representative set of potential release locations, and the concentration with time measured at each evaluated sampling location and in a hypothetical BZ of a worker, who is often assumed to cause the release and be within a few feet of the release point. EFs are calculated for each combination of monitor and release location, and the optimal selection of numbers and locations of samplers can be calculated based on the lowest EF when averaged over all release locations in the study. Whicker et al. (2003) analyzed measurements of EF for a series of test releases in a plutonium laboratory (with room size of about 12 m × 18 m × 5 m) and compared relative protection between the current placement strategy and alternative strategies for placement and number of CAMs (Figure 13.4). The results show that adding additional CAMs in different locations could

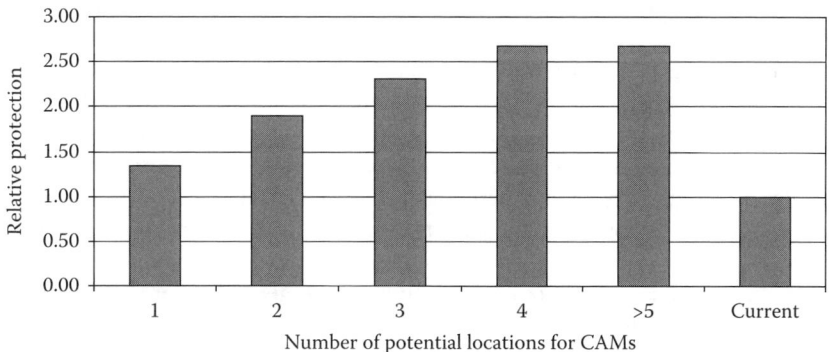

FIGURE 13.4 Data from Whicker et al. (2003) showing projected dose savings as a function of the number of potential CAM locations relative to the dose savings from the one CAM in the room at the time of the test. Data show that relative protection increases (more than a factor of 2) when three additional CAMs are installed. No additional protection is realized with the addition of four or more CAMs.

significantly improve worker protection, but that after additions up to four CAMs, no additional protection is achieved. Data of this type can provide critical guidance toward an optimized balance between cost of the continuous air monitoring program and worker protection.

FURTHER CONSIDERATIONS FOR ROOM AIRFLOW TESTING

REPRESENTATIVE FACILITY CONDITIONS

It is important that the conditions under which the airflow testing is conducted be representative of the operating conditions. A limitation of airflow studies is that the results may not be used to extrapolate to workroom conditions outside those existing during measurements. The U.S. DOE recognized the dynamic nature of facility ventilation and other factors that can influence airflow patterns. Therefore, in nuclear facilities DOE requires testing of airflow every three years or more or whenever a facility changes in a way that could significantly impact airflow patterns (DOE, 2007).

There are certain variables that have been shown to significantly affect airflow patterns and aerosol dispersion in a room (Vavasseur et al., 1986; Awbi, 1991; Buchanan et al., 1995; Whicker et al., 1999, 2002). These variables include

1. Release location
2. Removal or addition of large structures especially those close to supply vents
3. Ventilation rates
4. Changes in supply diffuser configuration
5. Sources of heating or cooling
6. Characteristics of the aerosol contaminant (e.g., particle size) and the release (e.g., energetic versus passive, etc.)
7. Local ventilation operation (e.g., fans, room exhaust)

Therefore, the conditions of these relevant factors should be noted prior to the airflow tests, and a current map of the room made. Conditions existing during measurements should represent those during normal operation.

REPRESENTATIVE AEROSOL TRACER RELEASE

The tracer and the mechanism for the release should closely simulate accidental release conditions. Important concerns are particle size and release energy. The tracer (particle or gas) should have similar aerodynamic properties to the material of concern. Because of the concern for sampling during accident conditions for protection of personnel in the room, the tracer should behave similarly to respirable particles (i.e., an aerodynamic diameter <10 µm). The energy of the tracer release should also be similar to accidental releases. For example, slow leaks and puffs are quite different phenomena that should be mimicked carefully. Differences in the energy of the tracer release can lead to different dispersion characteristics and distort the results. The temperature of the tracer aerosol can also affect the dispersion pattern and should be considered.

Smoke Generators: Airflow studies using tracers require that aerosol or gas be generated, released, and its transport measured. Many measurements of airflow in facilities today are based on releases from smoke generators (Fritz et al., 2006). Smoke generators produce large amounts of visible smoke. Other generators produce smaller amounts of aerosol, but have greater flexibility in choosing the characteristics of the aerosol (i.e., size, liquid or solid, and amount). Theatrical smoke generators are often used for visual tests because of their high output that makes the aerosol visible for longer-time frames. Measurement of the aerosol transport can range from observation and making notes on a map to more quantitative measurements of concentrations that are resolved in time and space.

It is important to generate aerosol that is representative of the aerosol expected in an actual release, so the instrumentation used to generate the smoke should generate aerosol that is not thermally buoyant, not jetted out of the generator with great velocity, and the particles should be nontoxic and represent inhalable particle sizes, if appropriate. For example, several smoke generators from two manufactures were tested for these characteristics (Whicker et al., 2000b). In this study, a Campbell Scientific Sonic Anemometer was used to measure velocities and temperatures. The optical size of the smoke particles was measured using a Met One optical particle sizer (refer to Figures 13.5a–c for some of the results from this study, specifically results from a smoke generator that provided a favorable tracer). As shown in Figure 13.5a, the background total velocity fluctuated between 3 and 17 cm s^{-1} before the generator was turned on. When the generator was activated there was little detectable additional flow. This slow type of release is a good simulation of releases that are relatively passive such as a glove box, glove puncture, or low-energy puffs from containment

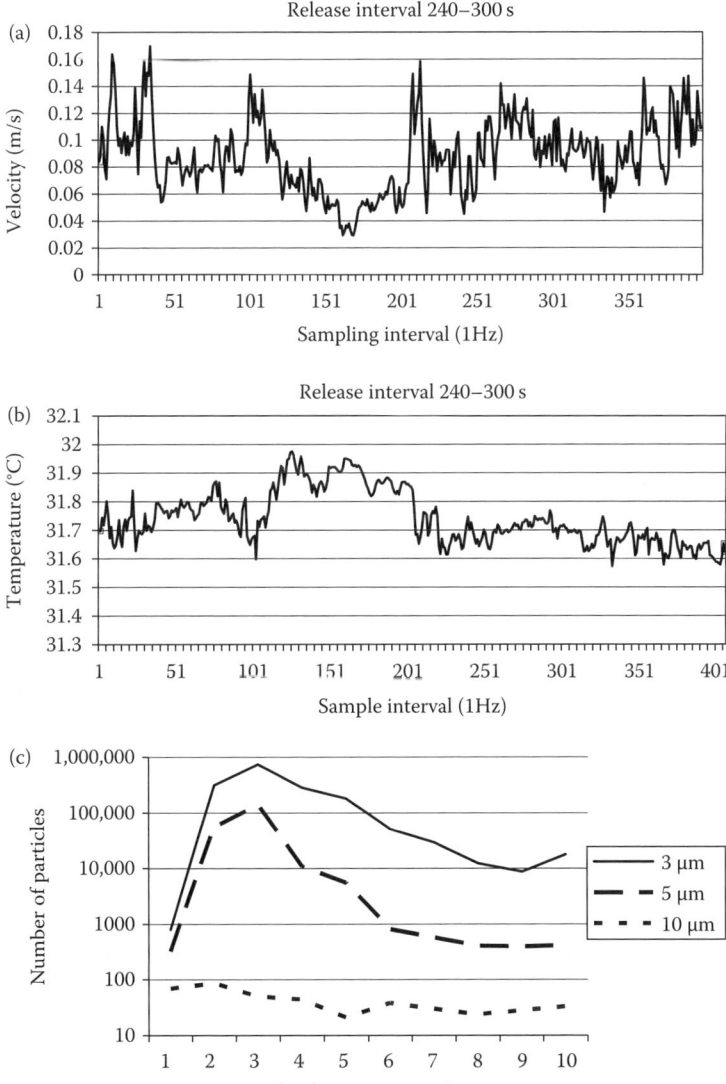

FIGURE 13.5 Results from evaluation of a smoke generator showing little jetting of airflow from nozzle (a), no increase in temperature (b), and particle sizes predominantly <10 μm (c).

structures, which are common types of releases in many nuclear facilities. Figure 13.5b shows that there was little detectable change in the temperature so this aerosol generator could be considered for studies where the releases are expected to be approximately at room temperature. The measured particle size of the smoke several minutes after generation is shown in Figure 13.5c. The particle optical sizes are estimates of the physical diameters. It is clear from this graph that most of the generated smoke particles have diameters less than 10 μm and so, fall in the respirable range.

CONCLUSIONS

In conclusion, airflow studies are a critically important part of the safety program at facilities where inhalation hazards exist. A good facility design provides structural confinement and local exhaust ventilation in order to minimize the potential escape of radioactive material, and if confinement is lost, the room ventilation should be designed to rapidly dilute and clear the material released from workers' BZs and exhaust the contaminant from the room air. Further, effectively designed air sampling and monitoring programs with properly positioned equipment provides confirmation of containment and warns workers of confinement loss. An effective combination of engineered confinement, facility ventilation, and air sampling/monitoring can provide for the elimination or reduction of worker inhalation risk. The described airflow studies are an important mechanism to ensure that the ventilation is working as designed and that workers are sufficiently protected.

REFERENCES

ACGIH. American Conference of Governmental Industrial Hygienists: Range of capture velocities. In *Industrial Ventilation: A Manual of Recommended Practice*, 21st ed. Vol. 3, p. 6. American Conference of Industrial Hygienists, Cincinnati, OH, 1992.

Awbi, H.B. *Ventilation of Buildings*. E&FN SPON, New York, 1991.

Buchanan, C.R., Chung, I.P., and Dunn-Rankin, D. A numerical study of contaminant mixing. *J Instit Environ Sci*, September/October, 15, 1995.

Crites, T.R. Alpha air monitor alarm sensitivity: Operational experience. *Radiat Protect Dosim*, 53, 65, 1994.

DOE, U.S. Department of Energy. *Radiation Protection Program Guide for Use with Title 10, Code of Federal Regulations, Part 835, Occupational Radiation Protection, Section 10: Air Monitoring*. DOE G 441.1-1B, Department of Energy, Washington, DC, 2007.

Flynn, M.R., Lackey, B.D., and Muthedath, P. Experimental and numerical studies on the impact of work practices used to control exposures occurring in booth-type hoods. *Am Ind Hyg Assoc J*, 57, 469, 1996.

Fritz, B.G., Khan, F., and Mendoza, D.P. Evaluation of airflow patterns following procedures established by NUREG-1400. *Operational Radiation Safety*, 91, S18–S23, 2006.

Hickey, E., Stoetzel, G.A., Strom, D.J., Cicotte, G.R., Wiblin, C.M., and McGuire, S.A. Air sampling in the workplace. *U.S. Nuclear Regulatory Commission Report NUREG-1400*; U.S. Nuclear Regulatory Commission, Washington, DC, 1993.

Hinds, W.C. *Aerosol Technology: Properties, Behavior, and Measurement of Airborne Particles*. John Wiley & Sons, New York, 1982.

Marshall, M. and Stevens, D.C. The purposes, methods and accuracy of sampling for airborne particulate radioactive materials. *Health Phys*, 39, 409, 1980.

National Institutes of Health (NIH). *Methodology for Optimization of Laboratory Hood Containment*, Vols. I and II. Bethesda, MD, 1996. Available at http://des.od.nih.gov/farhad/cover.htm [verified May 2006].

NRC, U.S. Nuclear Regulatory Commission. Air sampling in the workplace. *Regulatory Guide 8.25*. U.S. Government Printing Office, Washington, DC, 1992.

Rodgers, J.C, Whicker, J.J., and Voss, J.T. Comparison of continuous air monitor utilization: A case study. *Rad Prot Manage*, May/June, 56, 1998.

Scripsick, R.C., Stafford, R.G., Beckman, R.J., Tillery, M.I., and Romero, P.T. Evaluation of a radioactive aerosol surveillance system. In *Advances in Radiation Protection Monitoring, Proceedings of an International Symposium on Advances in Radiation Protection Monitoring*. IAEA-SM-229/62, International Atomic Energy Agency, Stockholm, Sweden, 1979.

Vavasseur, C., Muller, J.P., Aubertin, G., and Lefevre, A. Application of tracer gas methods to the measurements of ventilation parameters in nuclear power plants and various industrial sectors. In *Ventilation '85*, Goodfellow, H.D., Ed., Elsevier Science Publications, Amsterdam, the Netherlands, 75, 1986.

Wasiolek, P.T., Whicker, J.J., Gong, H., and Rodgers, J.C. Room airflow studies using sonic anemometry. *Indoor Air*, 9, 125–133, 1999.

Whicker, J.J., Baker, G., and Wasiolek, P.T. Quantitative measurement of airflow inside a nuclear facility. *Health Phys*, 79, 712, 2000a.

Whicker, J.J., Rodgers, J.C., Fairchild, C.I., Scripsick, R.C., and Lopez, R.C. Evaluation of continuous air monitor placement in a plutonium facility. *Health Phys*, 72, 734, 1997.

Whicker, J.J., Rodgers, J.C., and Moxley, J.S. A quantitative method for optimized placement of continuous air monitors. *Health Phys*, 85, 599, 2003.

Whicker, J.J., Rodgers, J.C., Wasiolek, P.T., Ammerman, C., Lopez, R., and Moore, M.E. Implications of room ventilation and containment design for minimization of worker exposure to plutonium aerosols. *Los Alamos National Laboratory Report LAUR-99-5968*, 1999.

Whicker, J.J., Scripsick, R.C., Day, G.A., Wannigman, D.L., and Creek, K.L. Airflow studies at the Los Alamos National Laboratory Beryllium Facility. *Los Alamos National Laboratory Report LAUR-00-5164*, 2000b.

Whicker, J.J., Wasiolek, P.T., and Tavani, R.A. Influence of room geometry and ventilation rate on airflow and aerosol dispersion: Implications for worker protection. *Health Phys*, 82, 52, 2002.

14 The Practice of Continuous Air Monitoring for Alpha-Emitting Radionuclides

John C. Rodgers

CONTENTS

Overview of the Practice 285
Application 288
Contributing Factors to CAM Sensitivity 290
    Sampling Rate and Placement 290
    Interference from Radon and Progeny 291
    Deposition on a Filter 293
    Decision Level and the Background Correction Algorithm 294
Interference Effects and their Control 297
Precision, Accuracy, and Bias (for α-CAM Alarms) 298
Apparatus Description: Room CAMs and Personal Lapel CAMs 305
Alpha-CAM Placement in the Workplace 307
Calibration and Gain Control 308
Calculation of Exposure in DAC-h 309
Setting and Testing CAM Alarm Levels 310
Summary 311
References 312

OVERVIEW OF THE PRACTICE

Alpha continuous air monitoring practice refers to a process whereby an air sample is continuously collected into a suitable instrument and onto a filter, and simultaneously, in real-time (or near real-time), analyzed as the sample gets collected. The instrument used in this practice will be referred to as the "alpha-CAM" or simply "CAM." The major components of a CAM are

- Special sample inlet
- Holder for the filter
- Alpha-radiation detector and preamplifier positioned coaxially over the filter
- Flow measuring device
- Data accumulation and storage electronics
- Embedded computer with its associated firmware and spectroscopy software
- Audible alarm for worker notification

Autonomous data analysis in the instrument is based on count data obtained by direct detection of characteristic α-radiation emitted by the radioactive substances collected on a filter. It is thus unlike "continuous sampling" or intermittent "grab sampling," in which radiological air samples are collected for later analysis off-line. Both continuous sampling and continuous air monitoring share many features, including in most circumstances, the need to eliminate radon daughter interference. Therefore, they are sometimes referred to as interchangeable, but the distinguishing features of continuous air monitoring are real-time radionuclide-specific detection, counting, and analysis.

Continuous air monitoring for α-emitting radionuclides differs in other important features from continuous air sampling. Continuous air monitoring requires the design of the sampling inlet to be such that incoming flow is directed around an α-detector (typically a silicon solid-state detector attached to a charge-sensitive preamplifier) positioned coaxially with the filter disc, and closely spaced with the filter so that the trajectories of α-particles from collected radioactive particles on the filter up to the detector are kept as short as possible (Figure 14.1). Another difference is the need to accumulate net count data (gross α-count minus radon progeny interference) in relatively short time intervals for rapid analysis of specific radionuclides of concern. Perhaps the most important

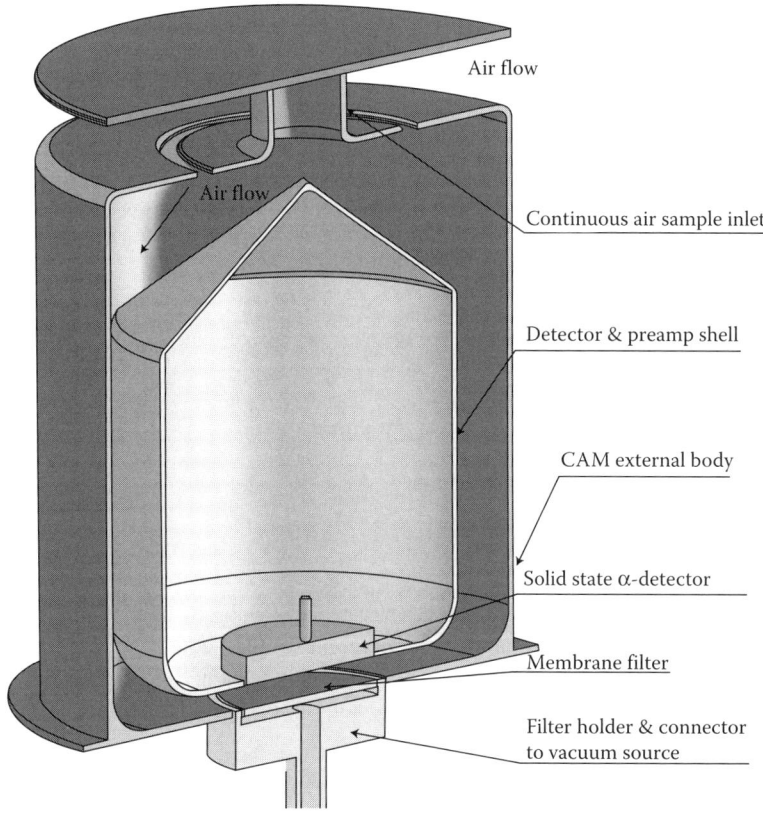

FIGURE 14.1 Cross-section schematic view of an alpha-CAM: Airflow is initiated and continuously maintained through a filter holder attached to a vacuum source and a flow meter. Air enters through a radial inlet and an opening at the top of the CAM body, flows around and down through a narrow gap between the body and an inner preamplifier/detector shell that isolates the detector and electronics from the flow, and then through the filter where particulate matter is filtered from the airstream. A silicon α-detector is positioned coaxially with the filter, separated by a narrow air gap (5–7 mm). The diameter of the active area of the detector and the exposed surface of the filter are identical to ensure maximum counting efficiency. The aerodynamic design of the CAM inlet, body, and internal detector/preamp shell, especially of the inlet openings and the deflection cone at the top and transition zone at the base, are such that at the specified sampling flow rate (typically 56.6 LPM), a representative sample is aspirated and a uniform deposit of particulate matter having ADs in the range from near diffusion size to 15 μm over the face of the filter is obtained.

and distinguishing feature of a CAM is its alarm capability when detection of a significant accumulation of specific radionuclides of concern occurs. The triggering of an alarm also produces a permanent, time-stamped record of the event.

A critical first stage in the function of a CAM is the aspiration of a fully representative sample of particles through the inlet stage of the instrument. That the sample should be representative has implications for inlet design. A simple tube inlet projecting out from one side of the CAM body into the air is susceptible to bias against inclusion of the large particle components of the sample due to inertial effects (Vincent et al., 1986). Similarly, if an inlet tube creates a defined jet of air into or across the sample filter, large particle losses on CAM internal surfaces can create significant sample bias due to size-selective losses (Biermann and Valen, 1983). Even the use of a transport tube to extract and transport a sample from a location of interest to the CAM will incur a risk of substantial sample loss and bias (Whicker et al., 1999). Internal losses can occur on surfaces beyond the inlet. As can be seen in Figure 14.1, the sharply divergent flow around the preamplifier/detector housing in the CAM head configuration could result in removal of large particles by inertial effects. Hence this transition region, as well as the inlet, must be designed such that unbiased transport of all particles is accomplished so that all of the particles aspirated are collected, and a uniform deposit of all sizes of particles, from diffusion size through 10–15 μm aerodynamic equivalent diameter (AED), will occur. That is to say, larger diameter particles should not be deposited disproportionately around the edges of the sample filter where detection efficiency is low, nor concentrated in the center or one side such that rapid burial can occur. Historically, many alternative α-CAM instrument designs have failed acceptance tests due to unrepresentative sampling or nonuniform sample deposition (Biermann and Valen, 1983; McFarland et al., 1990; Rodgers and Kenney, 1988).

A further critical element of CAM design is the utilization of Multichannel Analyzer (MCA) technology to accumulate count data binned according to pulse height (i.e., proportional to α-energy). Count data are thereby stored at the end of every count cycle as a histogram of α-energy counts in each of the MCA channels (typically, 256 channels). The binned dataset is referred to as an α-spectrum. Spectroscopic analysis of the resultant count versus energy spectra requires extraction of radionuclide specific net α-activity in the presence of α-emissions from particulate radon progeny in the air. In the absence of a highly effective design of these techniques, the contributions of radon progeny to the α-count rate in a targeted region of interest (ROI) in the spectrum can overwhelm attempts to detect low-level actinide radionuclide concentrations of concern. Older CAM technologies based on single-channel analyzer approaches have proven unreliable in this regard (Crites, 1994), since no single correction factor will account for dynamic changes in α-spectrum characteristics as monitoring conditions change.

The practice of α-CAM surveillance is used to provide a timely alarm that will alert workers to immediately exit an area or don respiratory protection equipment. The presence of particulates of α-emitters such as uranium, thorium, americium, or plutonium that may suddenly become elevated as a result of loss of control or containment must be rapidly detected. In most applications, only a few strategically placed CAMs are needed to achieve monitoring and alarm goals. To complement the limited number of CAMs, continuous air sampling using simple filter holders connected to a vacuum source, that is, filter air sampler (FAS) units, are often used to provide samples for retrospective dosimetry at far more numerous fixed positions near each work station. Here the sample can be analyzed by radiochemical techniques after radon progeny interference has been allowed to decay away. The result of continuous air sampling alone is an estimated accumulated exposure integrated over the time of sampling, which is often much longer than the duration of the release event. While spatial resolution is higher, the elements of time of the event and its duration are missing so that only an approximate exposure to an individual can be provided. Simultaneous monitoring with CAMs can help fill in this important piece of the exposure scenario. Bioassay can provide further evidence.

α-CAM procedures presented here do not address radiological stack effluent monitoring applications where sample extraction and transport through nozzles and sample lines are needed, although the monitor itself may be an α-CAM instrument. Particulate radiological stack monitoring is described elsewhere in this book and in an American National Standard (ANSI) entitled "Sampling and Monitoring Releases of Airborne Radioactive Substances from the Stacks and Ducts of Nuclear Facilities" (ANSI/HPS, N13.1-1999). The present chapter does not address the special requirements of sampling for radioactive gases and vapors such as that would be required for radioiodine monitoring or monitoring for radioactive gases, for example, radioactive isotopes of hydrogen (tritium and tritiated water vapor), krypton, or xenon. Some methods for these are described elsewhere in this book.

The practitioner should keep in mind that regulatory requirements for α-continuous air monitoring have been developed (Hickey et al., 1993; USDOE, 2007a) and applied in many facilities. These are generally specifications of the conditions in which early warning of significantly elevated levels of airborne α-emitting radionuclides should be provided in the event of an accidental loss of control or containment, and what levels are significant.

The following discussion includes considerations of instrument design and performance, how they are assessed, and how α-CAM installation (especially placement in the workspace) can influence monitoring performance.

APPLICATION

Real-time continuous air monitoring, in addition to FAS, is required when it is necessary to detect and provide warning of airborne radioactivity concentrations that warrant immediate action to terminate inhalation of airborne radioactive material (Hickey et al., 1993; USDOE, 2007a). This helps to assure that intakes of radionuclides are maintained as low as reasonably achievable (ALARA).

Real-time detection and air monitoring in the workplace accomplishes other objectives as well, including assessing and logging airborne radiological conditions as they change over time. This aids in detecting the gradual buildup of airborne radioactive materials. Documenting exposure conditions as they evolve provides valuable timeline data for worker dose assessment and can help identify locations in a workroom where a release is likely to have occurred. From the regulatory perspective, the basis for determining whether or not ALARA goals have been met, and for interpreting radiological air monitoring data, is by comparison with the derived air concentration (DAC) for the radionuclide of concern in a particular workplace and with time-integrated exposure expressed as the product of concentration and duration of exposure, DAC-h. The derivation of an air concentration point of reference is based on limiting annual dose, but it begins with establishing a limit on annual intake that would constrain radiological dose to exposed workers. Regulatory bodies such as the Nuclear Regulatory Commission (NRC) and the Department of Energy (DOE) have established annual limits on intake (ALI) specific to each radionuclide of concern that correspond to the regulatory dose limits for individual organs of the body or the whole body itself. The ALI (in units of µCi or Bq) is the smaller value of intake (inhalation or ingestion) of a given radionuclide in a year by an adult reference man, which would result in a committed effective dose equivalent (CEDE) of 5 rem (0.05 Sv) or of 50 rem (0.5 Sv) to an individual organ or tissue. The ALI of a specific radionuclide can be calculated from the annual dose limit (in rem or Sv) divided by its applicable dose conversion factor (rem/µCi, Bq/Sv) depending on which ICRP methodologies are used for regulatory purposes. The examples and calculations given in this chapter use NRC regulations implementing ICRP 26/30 methodologies. Since it is easier to measure and control concentrations of radionuclides in air rather than measure worker intake, each ALI is used to calculate a corresponding DAC. The DAC for any radionuclide is that concentration in air that if occupationally exposed to for a work year (50 weeks of 40 h each, for a total of 2000 h) would result in an intake of the ALI for inhalation. The normal breathing rate for light work is 2400 m^3/year; so the DAC is the ALI quantity (units of radioactivity) divided by the annual volume breathed (units of m^3). For example, the inhalation ALI for soluble ^{239}Pu is 6×10^{-3} µCi (2.2×10^2 Bq); hence,

$$1 \text{ DAC} = (6 \times 10^{-3})/(2400)/(10^6 \text{ mL/m}^3) \sim 3 \times 10^{-12} \text{ µCi/mL (111 Bq/mL)} \qquad (14.1)$$

In the case of an instrument such as the CAM where continuous sample collection is occurring even when the sampled activity is being counted, the method for estimating the air concentration (or DAC) is by recording a net count at the beginning of an interval of time, then recording the net count again at the end of that interval, and finally from the difference in net counts (and using other variables being measured in the CAM, including the volume of air sampled and sample collection time), calculating the concentration in the volume of air sampled that would have produced the observed count difference. Obviously, even in the absence of accumulating radioactivity of concern, there will be random differential results arising from the Poisson fluctuations in background counting and correction procedures: sometimes positive, sometimes negative. Ideally, the long-term average of these fluctuating results will be near zero for a blank filter (see Figure 14.4). Absent a near-zero average, any bias in the background subtraction procedure (algorithm) can be assessed from an evaluation of these long-term averages. The lower limit of detection (L_D) can be calculated from the standard deviation in the average for the blank (Curie, 1984).

When estimating the accumulated exposure over a period of monitored time, the product of the DAC and the hours exposed or integrated exposure expressed in units of DAC-h applies. This is often a more significant quantity than the DAC at a particular moment in time, and can be easily measured and logged by an α-CAM. The net radionuclide α-activity on the CAM filter sample at the end of each count integration interval, total volume of air sampled in the interval, and time over which the filter has been in use are all continuously measured or determined by the embedded microcomputer-based operating system. The quotient of the first two quantities (net activity and volume, converted into DAC units by dividing the result with the DAC factor for the radionuclide of interest) multiplied by the time is the DAC-h. Note that no matter at what point in time in a monitored period the radioactivity is first detected, and for how long within that period the radioactivity remains elevated, the DAC-h is the same: for example, an exposure of 8 DAC-h could be the result of an exposure to 1 DAC for 8 h on the third day of a week-long sampling period or the result of an exposure to 80 DAC for 0.1 h on the first day. The DAC-h directly addresses the quantity of radiological protection concern, which is potential worker exposure, that is, 1 DAC-h corresponds to 2.5 mrem CEDE. This is the exposure to contaminated air moving through the monitored location of the CAM. Individual workers may well experience a higher or lower personal exposure. For example, an individual may be persistently upwind of a release that is detected by a CAM.

Measuring concentration (expressed in units of DAC), as done in the continuous sampling and detection procedure of a CAM, requires the determination of successive differences in counts as additional volumes of air are drawn through the filter. Such differences are frequently negative simply due to the inherently random (Poisson) nature of radioactive decay. Utilizing the integrated DAC-h exposure instead of concentration provides an advantage because it is always positive and trending upward with each increment of the target radionuclide collected, reaching a plateau when, and if, target radionuclide air concentrations drop to background. For example, assume a puff release of α-emitters has occurred over a relatively brief interval when a glovebox seal was punctured, resulting in an elevated air concentration that rises to a peak value and then declines, all in about one hour. Suppose the concentration averages about 1 DAC before being completely cleared from a room, that is, 1 DAC-h exposure has occurred. Then collected radionuclide activity on the filter will ramp up, leading to a maximum count rate logged in the CAM database during that one hour. The accumulation rate is determined by the sampling rate and by fluctuations in air concentration at the monitored location. Beginning at that point in time at which the last of the dissipating cloud has been sampled, each successive count of the CAM filter will result in approximately the same net activity (dpm) being recorded, but along with that both a longer monitoring time (h) and larger total volume (cm^3) of air passed through the filter will be recorded. From the definition of the DAC-h, (dpm*h/cm^3/DAC factor) one can see that as a result of the coordinated increase in hours and increase in sample volume, and a nearly constant count, the DAC-h result will stay approximately constant at 1 DAC-h. To continue with a numerical case, suppose the sampling rate is 28.3 lpm (1 CFM) resulting in 1698 L sampled, and the activity on the filter corresponds to what would accumulate from an exposure to an average of 1 DAC for 1 h [about 5 pCi (0.19 Bq)], then after another hour of sampling after the air is cleared, the filter activity will be determined to be the same, the time multiplier would now be 2 h, and also the sampled air volume term in the denominator would also now be a factor of two larger (3896 L) so that the exposure estimate reported would remain at one DAC-h. The interpretation of a time history of DAC-h is hence relatively unambiguous even if the concentration has fluctuated considerably due to turbulent mixing in a room. As a result, by close observation of the trend in DAC-h, rather than DAC or count rate, a CAM user would not have to reenter a contaminated workroom to reset the CAM following an alarm in order to determine if the ventilation system had cleared the air of contaminants. Once the air is clear, the DAC-h levels will plateau in the presence of a fixed amount of activity on the filter.

If large quantities of an easily dispersed α-emitting radionuclide are planned to be used in a confined workspace, an assessment should be made to optimize the number and placement of α-CAMs in the work area such that the greatest level of protection can be achieved within the constraints of the physical layout of the room, existing ventilation patterns, availability of power, communication, vacuum lines, and costs. It should be borne in mind that optimal placement could achieve higher levels of protection than simply adding more CAMs. Technical references such as the NRC Report, "Air Sampling in the Workplace" (Hickey et al., 1993), as well as many sources in the technical literature (Rodgers et al., 1998; Wicker et al., 1997; Wicker et al., 1999) can provide detailed numerical guidance in the determination of how these factors apply in specific cases. See Chapter 13 by Whicker in this book for an overall discussion of CAM placement.

CONTRIBUTING FACTORS TO CAM SENSITIVITY

SAMPLING RATE AND PLACEMENT

In a typical laboratory setting, CAM monitoring is conducted at approximately 2 CFM (944 cm^3 s^{-1}) in order to achieve good sensitivity. For example, if a puff release of ^{239}Pu created a momentary respirable air concentration in the vicinity of a CAM of about 30 DAC, α-activity on the filter in 2 min at this sampling rate would produce a count rate of about 45 cpm (assuming a detection efficiency of 30%). Such a rate would be readily detected, and an alarm would be generated by the CAM. The corresponding exposure, on the order of 1 DAC-h (30 DAC × 2/60 h), would be considered good sensitivity. In some older CAM designs, the inlet design, size of the filter and detector, choice of filter medium, and capacity of house vacuum (or the vacuum pump characteristics) may limit sample extraction rates to 1 CFM (472 cm^3 s^{-1}) or less. Lower sampling rates may also develop during a long period of continuous monitoring or in dusty environments due to the effects of buildup of inert aerosols on a filter. Small particle aerosols, such as those found in smoke or produced by internal combustion engines are particularly effective in causing filter clogging (Brown, 1993). Lower sampling rates are acceptable within limits as long as worker protection objectives can still be met. However, a carefully considered low flow threshold should be established that results in a maintenance alert to the operators.

Modern CAM design utilizes both a larger diameter detector and a filter (47 mm vs. 25 mm), and thus it is compatible with a higher sample flow rate, for example, 2 CFM versus 1 CFM (944 vs. 472 cm^3 s^{-1}). The resultant sample activity will be twice as large for any given air concentration. Moreover, the larger detector/filter geometry results in higher counting efficiency (32% vs. 20% for a 25 mm geometry). This results in 60% more counts from the same sample quantity. The detector and filter diameters *must* be the same to achieve this higher performance. The increased count per unit air concentration significantly improves the precision of the measurement.

The CAM user should be aware that due to the very low concentration limits associated with transuranic α-emitters, the number of particles associated with a transuranic DAC can be exceedingly small. For example, only six particles per 1000 m^3 (0.006 particle/m^3) of air correspond to 1 DAC of ^{238}Pu, assuming the activity median aerodynamic diameter typical of plutonium laboratories (5 μm

AMAD) (Dorrian and Bailey, 1995). By implication, there is a low probability of capturing even a single particle in a given cubic meter of air sampled. Hence, in situations where low-level releases are more apt to occur or where there expected to be rapid dilution during aerosol transport from a release event over to CAM locations in the room, there is likely to be a low probability for rapid detection of low DAC concentrations, and this would be made worse by low sampling rates. Therefore, it is prudent not to allow sampling flow to fall to low levels. Careful placement, multiple CAM unit deployment, and maintaining adequate sampling rate, all play a role in measurement optimization. Improved sensitivity for detection in the presence of background interference is also a factor.

INTERFERENCE FROM RADON AND PROGENY

The sensitivity of the α-CAM methods is highly dependent on the accuracy of the background compensation method used, since if the process is inaccurate, leading to over- or undercorrection of background interference, the detection threshold could be at a higher level of counts. The background issue for CAM monitoring is that typically radon, both ^{220}Rn and ^{222}Rn, and their particulate decay products (radioactive isotopes of polonium, bismuth, and lead) are often found in elevated concentrations in indoor environments. Concentrations of these radionuclides in ambient air follow a diurnal cycle of rise and decline, with occasional significant deviations caused by changing meteorological conditions (wind, barometric pressure, and precipitation). Certain of the decay products in the radon decay chain are energetic α-emitters. As these decay products accumulate on the sample filter, their α-emissions contribute to a broad spectrum of detected counts at energies that completely overlap the characteristic energies of the target radionuclides (i.e., in the ROI of 3.2–5.6 MeV in the α-energy spectrum, Figure 14.2).

The dispersion in energies seen in the α-spectrum is the result of α-particle interactions with the filter medium, inert dust particles, and air in the gap that must exist between the filter and the α-detector. As a result, even though high-resolution energy-discriminating solid-state α-detectors are used in α-CAMs, the ability to distinguish between counts contributed by radon decay products and those from target uranium or transuranic radionuclides is confounded by degraded energy α-counts from

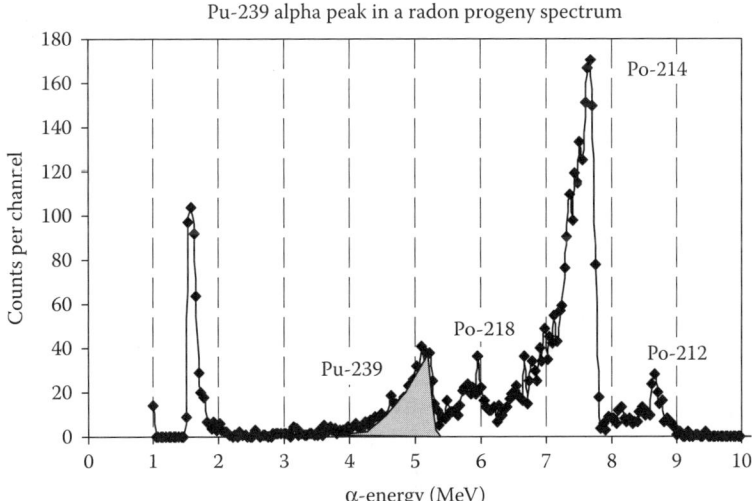

FIGURE 14.2 A typical CAM-generated α-energy peak spectrum showing the presence of α-emitting radon progeny, including ^{218}Po and ^{212}Bi at about 6 MeV, ^{214}Po at 7.68 MeV, and ^{212}Po at 8.78 MeV. A ^{239}Pu peak corresponding to approximately 30 cpm is found with center at 5.15 MeV. There is a clear evidence of the low-energy tails of these background radionuclides overlapping the ^{239}Pu ROI between about 3.2 MeV and 5.6 MeV, in the shaded area. The diamonds are the count data.

radon progeny. The problem is compounded by the fact that target radionuclide concentrations of concern are in the range of a few pCi m^{-3} (hundredths of a Bq m^{-3}) while radon progeny backgrounds may be in the range of a few hundred to a few thousand pCi m^{-3} (tens of Bq m^{-3}). Overall, background count rates are typically several times the count rates associated with 8 DAC-h activity accumulations on the sample filter, which all but preclude gross α-counting as a detection method.

In Figure 14.2, it can be seen that the count rates in those channels associated with the low-energy tails of radon progeny α-peaks are lower than that in the peak channels for these same radionuclides. Yet even these lower-energy background α-contributions can be at rates higher than those arising in the same region from the initial collection of radionuclides of concern. For example, consider the case of an 8 DAC-h ^{239}Pu exposure in an 8 h work day: the DAC for soluble ^{239}Pu is 6.66 dpm m^{-3}, so a CAM sampling at 2 CFM (0.057 m^3 min^{-1}) for an 8 h day with a counting efficiency of ~0.3 cpm dpm^{-1} (typical for a 47 mm diameter filter and detector geometry) will produce a count rate of 54 ± 7 cpm. In contrast, if ^{222}Rn (gas) is present at typical concentrations in the range of 2×10^3 dpm m^{-3} to 4×10^3 dpm m^{-3} (1–2 pCi l^{-1}, 37–74 Bq m^{-3}), then the levels of short-lived ^{218}Po ($T_{1/2}$ = 3 min) (particulate) in transient equilibrium with radon can be present in the range of 1000–2000 dpm m^{-3}, depending on the degree of equilibrium (typically, 50% indoors). Hence, sampling at 2 CFM could result in filter levels of approximately 200 cpm of ^{218}Po after 15 or 20 min of sampling, or about three to four times the plutonium count rate. With an α-emission energy of 6 MeV, approximately 20–40% of those counts will occur in the low-energy tail overlapping the transuranic ROI beginning at about 5.6 MeV and extending down to about 3.2 MeV. A smaller percentage of the higher-energy radon progeny α-emitters (i.e., ^{214}Po and ^{212}Po at 7.68 and 8.78 MeV, respectively) typically contribute smaller percentages of counts to this region, but interference can be enhanced with significant accumulation of these on the filter during long periods of sampling coupled with the broadening of peaks caused by the effect of dust collection. One would anticipate a larger interference of these overlapping counts with transuranic α-emitters with peak energies closer to 6 MeV, such as ^{241}Am (5.49 MeV) and ^{238}Pu (5.50 MeV), than with ^{239}Pu (5.15 MeV). An essential strategy for dealing with the effects of interference overlap is to periodically *reset* the MCA used to accumulate a multichannel count versus energy spectrum, that is, clear all accumulated counts in each channel and continue the accumulation of an α-spectrum from the zero baseline. There are then, two clock cycles in CAM operation: one is the accumulated time over which air has flowed through the filter ("filter time"), depositing whatever radioactive particles are present during the long monitoring interval onto the filter, and the other is the shorter "count cycle time" between periodic resets of the MCA when α-spectra are generated. There is an unavoidable trade-off between maintaining a short count cycle that would help reduce uncompensated overlap counts and having a long enough count cycle so that the statistics of the background correction algorithms applied to spectrum data are acceptable.

On-filter in-growth of progeny radionuclides from collected parent radionuclides is of particular significance in the case of the ^{220}Rn decay chain, when present. A decay product in this chain, ^{212}Po, can both be collected from sources in the air and grow in from ^{212}Bi parents on the sample filter, yielding a source (35.9% abundance) of 6.05 MeV α-emissions. See the section "Interference Effects and their Control" below for a further discussion of radon progeny interference in terms of counting statistics.

In environments where the radon progeny are attached to ambient dust particles they will collect on the filter along with the dusts. In special facility HVAC design conditions where all incoming ventilation air is HEPA filtered, a significantly higher fraction of these freshly formed radon progeny will remain unattached to particulates, and in this condition can be removed during sampling by a properly designed, small pore diameter screen positioned in the inlet (McFarland et al., 1992). The removal process is driven by the high diffusion rate of nanometer-size particles, which causes them to rapidly contact the pore walls. This is in contrast with the low diffusion rate for larger particles that pass through unimpeded. If the unattached fraction is low, this will not have a large effect on radon progeny interference reduction. In certain CAM inlet designs there are mechanisms to remove a portion of not only the unattached, but also portions of the attached radon progeny using impactor techniques such as direct impaction of a high-volume (>10 CFM) jet of sampled air onto a grease-coated planchet directly below the jet (Alexander, 1966) where α-emissions are detected by scintillation techniques, multistage virtual impaction techniques (Yule, 1978), or a combination of a cyclone preseparator and virtual impactor (Rodgers, 2007). "Virtual" impaction is achieved by causing the jet of air to impinge on an enclosed column of low-volume slowly moving air, which results in the separation of large particles into the low-volume flow while the fine particles (including much of the radon progeny particles) continue with the high-volume deflected portion. Only the low-volume portion is filtered and counted. In addition to sometimes poor separation of the target transuranic particulates from the radon progeny contaminated dusts, another disadvantage of these impaction techniques is that in many cases the detection principle is gross α-counting, which does not yield energy distribution data. This can be a problem in some applications where larger particles carry radon progeny with them to the collection substrate.

DEPOSITION ON A FILTER

Of several mechanisms contributing to the peak broadening effect described above, the effect of structure of the filter medium used is perhaps most significant. The use of thick fibrous filter media such as cellulose filters (as opposed to thin, smooth membrane filters) is a well-known contributor to peak broadening and therefore should be avoided (Moore et al., 1993), although the collection efficiency for very small particle fractions may be higher. Radon progeny are largely associated with very small-sized particles that allow them to penetrate deep into a fibrous mat. α-Emissions from such distributions can experience considerable energy losses during transit to the surface. For example, α-peak analysis of room air sampled with a membrane filter such as Millipore Fluoropore 5 μm will typically exhibit a ^{218}Po peak with a full-width-at-half-maximum (FWHM) of 325 keV, while the FWHM of this α-peak using Millipore Type A glass fiber is 410 keV. As illustrated and explained in Figure 14.3, peak width, as expressed by FWHM, provides a useful measure of peak spread. The

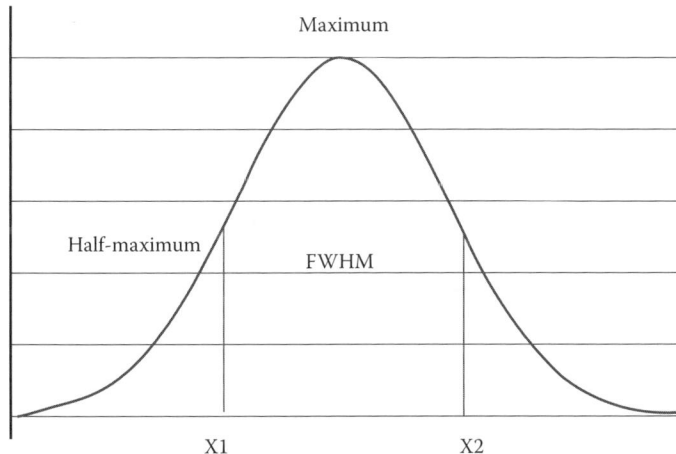

FIGURE 14.3 FWHM: Assuming that the above plot is a simplified display of an α-spectrum peak, the FWHM is a measure of the difference of the two extreme values of the independent variable (maximum X2 and minimum X1) such as the channel number or α-energy in keV, at which the dependent variable (such as α-count) is equal to half of its maximum value. Hence, the FWHM provides a measure of the energy spread of an α-peak. The spread of an α-peak is indicative of the combined effects of the inherent resolution of the detector, and α-energy losses in the air gap between detector and filter, in the radioactive particles themselves, due to burial by inert dusts and burial in the filter medium.

added 85 keV width may or may not be of significance in counting, depending in part on whether or not peak width is a critical factor for peak identification and counting.

Elevated concentrations of dusts and smoke particles can affect peak shape as well. If radon progeny are attached to large particles when being sampled, the adsorption effects on α-emissions can increase interference by creating broader background peaks.

The presence of a prior deposition of dust aerosols, which can create a porous mat of fiber-like structures (dendrites), has been thought to have the potential for creating conditions whereby low-angle α-emissions from radioactive particles subsequently deposited in the mat could result in significant energy losses before the α-particle arrives at the detector. However, a detailed study of the influence of dust loading on α-energy resolution and burial by Huang et al. (2002) has shown that the large aerodynamic diameter (AD) particles associated with an aerosol consisting of mainly transuranic particulates tend to deposit on the surface of preexisting dust layers on a filter and thus α-emissions are likely to be relatively unaffected by the previous deposits. In other words, under certain circumstances prior dust loading only increases the "thickness" of the filter. Specifically, Huang et al. found using gold particles labeled with radon (^{220}Rn) progeny as surrogates of transuranic α-emitters, that a thin deposit of gold particles (0.06 mg cm^{-2}) deposited on a thick layer of salt dust (7.98 mg cm^{-2}) exhibited relatively little broadening of the ^{214}Po radon progeny peak (477 ± 35 keV vs. 400 ± 32 keV for the reference peak), whereas a well-mixed thick layer of gold and salt dust (2.08 mg cm^{-2}) exhibited considerable absorption and dispersion effects (1320 ± 60 keV) (Huang et al., 2002). In an accidental release scenario it is expected that the freshly deposited transuranic particulate will likely deposit in a thin layer resulting in a relatively well-defined α-energy peak spectrum, even when the filter is partially loaded with aerosols from earlier sampling.

DECISION LEVEL AND THE BACKGROUND CORRECTION ALGORITHM

Even with sharply defined α-peaks, a technical challenge remains for α-spectrum data analysis because a method for systematically and effectively removing counts contributed by α-emissions in the low-energy background region that mask the contributions from radionuclides of concern (e.g., ^{239}Pu) must be employed. If not done properly, the correction process can lead to a net count result that is undercorrected, causing an overestimate of true target activity. Conversely, overcorrection can lead to loss of true counts resulting in negative net counts, thus effectively desensitizing the instrument. For example, if as a result of incorrect background correction methods the net transuranic count is consistently returned as a negative −10 ± 3 counts, then in order to achieve a detection threshold of positive 10 counts, there would have to be approximately twice as much activity present compared with a method that more nearly corrected background to zero. Therefore, the correction method must be very accurate to avoid errors of both the first and second kinds (false positive and false negative). As a rule, the threshold for the chronic alarm should be set far enough above the expected minimum detection level of most measurements, so that there is likely to be less than one false-positive alarm every four months (and even fewer, when a large number of α-CAMs are installed). At the same time, the background correction method should be accurate enough that the attainable level will be at concentrations low enough such that adequate early alarms can be reliably generated while avoiding potentially serious false-negative outcomes.

In statistical terms, the decision level, DL (alternatively referred to as the Critical Level, L_C), is that net measurement outcome at or above which a decision is made that target radionuclides are present in the sample, while accepting with low probability (typically 5%) that this decision is false positive (claiming activity is present when in fact the target activity is zero). This is a qualitative, binary decision: "activity present" or "no activity present." Note that the DL is not the same as the "minimum detectable activity" or MDA. The quantitative MDA takes into consideration the probability of not concluding that significant activity is present when it actually *is* present, that is, the "false-negative" error. The MDA, therefore, is used as a determinant of whether or not a measurement procedure (which would include sampling rate of the measurement, monitoring device design, α-spectrum analysis methodology, etc.) provides adequate quantitative detection capability.

To illustrate the DL versus MDA distinction, consider the different procedures used to determine each. If, for example, the measurement of plutonium on a CAM filter yields N_s net counts after determining and subtracting N_b background interfering counts from the gross signal S, then the counting error associated with the measurement is attributable to the variance in the gross count of the sample plus variance due to the background count because $N_s = S - N_b$. Hence, the standard deviation of the net count is $\sigma_s = (N_s + 2*N_b)^{1/2}$, where N_b occurs twice in the counting process: once with the gross count and once with the background or "blank" count itself. For net positive counts near the qualitative threshold for detection, the decision that has to be made is whether or not the true target radionuclide count is actually zero, which is to say, if $N_s = 0$. The answer to that question requires a determination of the standard deviation of the net count of the filter sample when it is known that the target radionuclide is absent, σ_0, and a selection of a desired level of confidence so that there will be only a small chance that an error will be made by concluding that target activity is present when the true activity is zero. Typically, the confidence level is chosen so that the probability of falsely concluding that a blank observation represents a "real" signal is $\alpha = 5\%$. Choosing this means that $DL = L_C$ such that the fraction $1 - \alpha$ corresponds to the correct decision "not detected" with 95% probability. Typically, $\sigma_0 = \sigma_b$ for low levels of activity. A constant, k_α, derived from the standardized normal distribution corresponding to the probability level of $1 - \alpha = 95\%$ is used as a multiplier of the standard deviation of the background ("blank") to determine the DL:

$$DL = k_\alpha * (2*N_b)^{1/2} \text{ counts,} \tag{14.2}$$

where $k_\alpha = 1.645$ at 95% confidence level and $(2*N_b)^{1/2} = \sigma_b$ the standard deviation of the blank count.

When the net counting result is equal to or greater than the DL, one would say that one has detected a positive amount of activity equal to or greater than the DL with a "false-positive" probability of 0.05, or 5%. Refer to Figure 14.4 where the 95% confidence interval of the blank count distribution is illustrated. Before reporting a *quantitative* measurement result based on any given net count, consideration must be given to the possibility of erroneously reporting *no* detected activity when in fact it is present. A count level that includes consideration of the probability of such "false-negative" errors is the Quantitative Determination Level (L_D), which is a level of counts beyond the DL such that there is an equally small chance of this second type of error: $L_D = L_C + k_\beta * \sigma_b$. If the same confidence level of avoiding errors of this type is chosen, $\beta = 5\%$ (i.e., $k_\beta = 1.645$, at the 95% confidence level), and the count times for the sample and background are the same, then $k_\beta = k_\alpha = 1.645$, and then, $L_D = k_\alpha(2*N_b)^{1/2} + k_\beta(2*N_b)^{1/2} = 4.65*(N_b)^{1/2}$.

Referring again to Figure 14.4, it is important to recognize that the Critical or DL, L_C, should not be used as the Determination Level L_D, since to do so would be to equate $\beta = 0.5$, illustrated by a distribution having the same standard deviation as the blank, but centered at L_C. It may be tempting to equate the two if the focus of attention is on avoiding false (positive) alarms. However, this choice would mean, in effect, that there is a 50% chance that when the *true* net count is at the critical level, measurement will find a smaller net count leading to the erroneous conclusion that the activity is zero. The consequence of committing this type of error takes the form of what may be referred to as a "lost opportunity for alarm." This is illustrated in Figure 14.4 by the shaded area on the left side of the distribution centered on L_C. By selecting $\beta = 0.05$ ($k_\beta = 1.645$, at the 95% confidence level), as shown in the right-most distribution, both types of errors have the same low statistical probability of occurrence, highlighted in color at the overlap of these distributions.

The quantitative determination level as well as the DL is based on measurements of the blank for the measurement process, that is, whatever interference that might be contributing counts to N_b is included in the blank count. In the CAM application, N_b is the sum of the counts associated with the

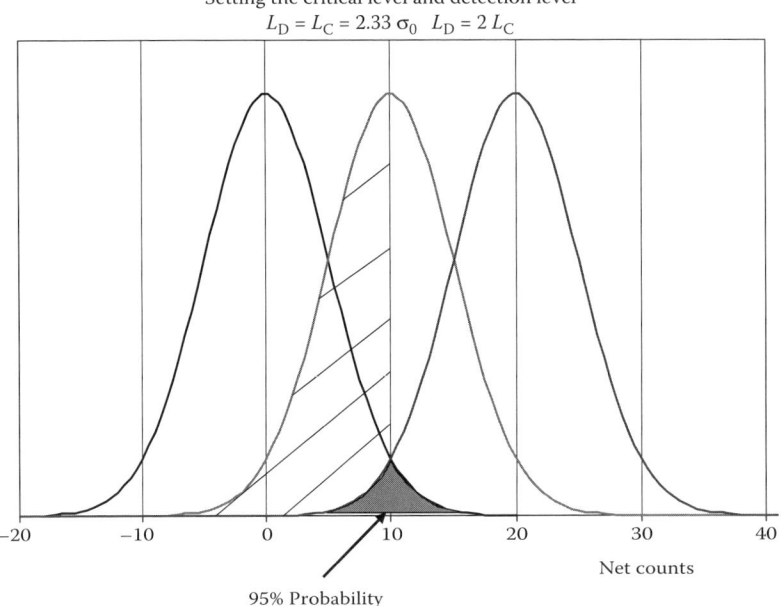

FIGURE 14.4 The normal probability distribution on the left represents the distribution of counts associated with the net "blank" count from a CAM measurement (i.e., no target radionuclides are present). The mean is zero since the activity is zero, but owing to interference effects from background being subtracted from gross counts in the transuranic ROI, there is a distribution of outcomes on either side of zero. The 95% probability interval is associated with a confidence level of 0.05 for the L_C. In this example, $L_C = 10$ net counts. The probability of a false-positive error (concluding activity is present when it is not) is then only 5% at this count level. The center distribution is one that would result if the L_D is set equal to the L_C. The shaded area to the left of the center represents the probability (50%) of a false-negative determination (i.e., falsely concluding there is no net activity) if the net count is in the vicinity of L_C. The proper setting for L_D is shown in the right-most distribution ($L_D = 2L_C$), so that the probability of a false-negative determination has the same value as that for a false-positive error, or only 5%. The shaded area illustrates the symmetrical 95% probability intervals.

overlap of the low-energy tails from the higher-energy radon progeny peaks. The standard deviation of the blank can be determined prior to applications where contaminants may be present by obtaining a long series of background counts for representative integration times. For reporting the MDA, the L_D based on statistical analysis of a large set of CAM blank measurements can be converted to units of radioactivity by dividing L_D (counts) by spectrum integration time and counting efficiency (cpm/dpm). Hence, the MDA for the measurement process should be established by obtaining repeated count measurements of the radionuclide of concern under representative background conditions, that is, using the data from a CAM collecting background air containing only a typical range of concentration of radon progeny interference and inert dusts collected on the filter medium of choice. The standard deviation of the resultant net target (plutonium) counts from this measurement process, σ_b, is then applied to calculate the MDA:

$$\text{MDA} = \frac{4.65 * \sigma_b}{\varepsilon * t_c} \text{dpm}, \qquad (14.3)$$

where ε is the detector efficiency (cpm/dpm), σ_b is the standard deviation of the standard blank counts, and t_c is the corresponding standard count time (min). (For a more detailed explanation, see Currie, 1984.)

The sensitivity of an α-CAM is usually expressed in exposure terms that is, DAC-h sensitivity. The DAC-h, as previously described, is the time integral of exposure to a given transient room air concentration condition, where the radioactivity concentration is expressed in DAC by dividing with the DAC factor for the target radionuclide. A typical regulatory sensitivity requirement for α-CAMs is 8 DAC-h under conditions of relatively low radon background. Many modern α-CAM designs can achieve much lower sensitivity while still maintaining an acceptably low false alarm rate. A facility health physicist or ALARA coordinator may therefore determine a lower operating alarm threshold taking into account the MDA for the facility CAMs. Or, the alarm point may have to be set higher if elevated radon background conditions cause unacceptably high false alarm rates among the many CAMs in that facility.

The upper range of an α-CAM DAC or DAC-h measurement is limited in principle by the limits of the detector electronics to resolve incoming α-pulse events or by associated analyzer dead time. For most practical applications, this upper limit is not of much consequence due to the extremely low concentration limits allowed for α-emitting radionuclides in the workplace, although it can be important even when elevated levels of airborne α-emitting radionuclides are present, such as under circumstances where α-CAMs are used to monitor the safety of workers who are protected by personal protective equipment (PPE) such as respirators. Alarm thresholds are then set at appropriate levels reflecting the protection factors of the PPE in use. Even then upper range limitations are unlikely to be a concern.

INTERFERENCE EFFECTS AND THEIR CONTROL

Sources of interference in α-air monitoring by CAMs include not only the radon progeny collected on the sample filter during the monitoring period between filter changes, but also the electronic pulse noise introduced into detector electronics from power lines or radio frequency (RF) sources such as radio transmitters, for example, walkie-talkies, and noise created by light scattering into the detector, for example, exposure to laser light from nearby laboratory equipment. The interference from RF noise can be addressed by shielding the CAM electronics or by restrictions on the use of RF devices in the laboratory. Light sensitivity can also be controlled with surface coatings on CAM body components, including the detector itself, and limitations on colocation of α-CAMs and lasers. Electronic noise induced by external activities (radio operation or laser operation) increases the number of false alarms in a random fashion and may again be best controlled by restrictions of use of certain devices in specific laboratory areas. The effect of radon progeny interference is a systematic increase in the likelihood of a false-positive alarm as a consequence of incomplete background interference correction in the ROI, as discussed above. Electronic noise tends to affect a broad range of apparent energies and is most likely a random effect that can be identified by distorted α-spectra or high count rates in the lowest energy regions of the spectra.

As radon background interference goes up, the standard deviation of net background counts in the target ROI will tend to go up as well, which is reflected in a higher MDA for the procedure under those conditions. If a given false-positive alarm rate is to be maintained under conditions of increased background, the alarm set point will have to be higher. For example, consider the case of an α-CAM having a background standard deviation of 0.7 DAC-h when the radon levels are nominally 1 pCi L^{-1} (37 mBq L^{-1}), and 1.5 DAC-h when background is elevated at 4 pCi L^{-1} (148 mBq L^{-1}) (the conversion from net counts to DAC-h assumes a flow rate of 2 CFM and an integration period of 15 min per measurement). If the desired threshold for alarm (set point) is 8 DAC-h, then at background levels of 1 pCi L^{-1} (37 mBq L^{-1}), the set point would be 11.4 standard deviations above background levels. But at 4 pCi L^{-1} (148 mBq L^{-1}), it is only 5.3 standard deviations above. While the probability of a false alarm on any one measurement in this latter condition is small, in a large set of measurements, the probability of alarm becomes more significant. The particulars of this example, and tables for estimating probability of false alarm under varying conditions, are discussed in great detail in Hoover et al. (1995) (also see the Appendix in Chapter 28 for air sampling of Pu). If, rather than a conservative set point of 8 DAC-h, a much lower value of 2 DAC-h was attempted, the risk of false alarm is quite high, being only 1.3 standard deviations above the background, and less than the MDA. The point here is that knowledge of the statistics of blank measurements for a given CAM application is key to the success of setting up critical operating parameters. A high false alarm rate is an unacceptable operating condition in a facility due to the high costs incurred by disruptions of work activity and it may have a deleterious effect on employee attitudes toward response to alarms.

PRECISION, ACCURACY, AND BIAS (FOR α-CAM ALARMS)

Since precision and accuracy are sometimes confused with each other, it is helpful to recall that *precision* refers to the degree of agreement in a series of measurements of the same quantity—in other words, the clustering of values around their own average. It might be expressed as the relative percent difference between replicates. *Accuracy* is the closeness of a measurement to the true value. It can be expressed as the relative percent difference between observed and conventionally true (or known) values. Figure 14.5 illustrates the difference between these terms. While curve "A" of this figure represents a more accurate determination of a parameter with a mean value of 10 than does curve "B," neither of these would be considered accurate if the true mean value is 30 (Curve "C"). While a CAM-reported concentration value may not be considered representative of the concentration actually breathed by workers, it nonetheless must be as sensitive and accurate a measurement

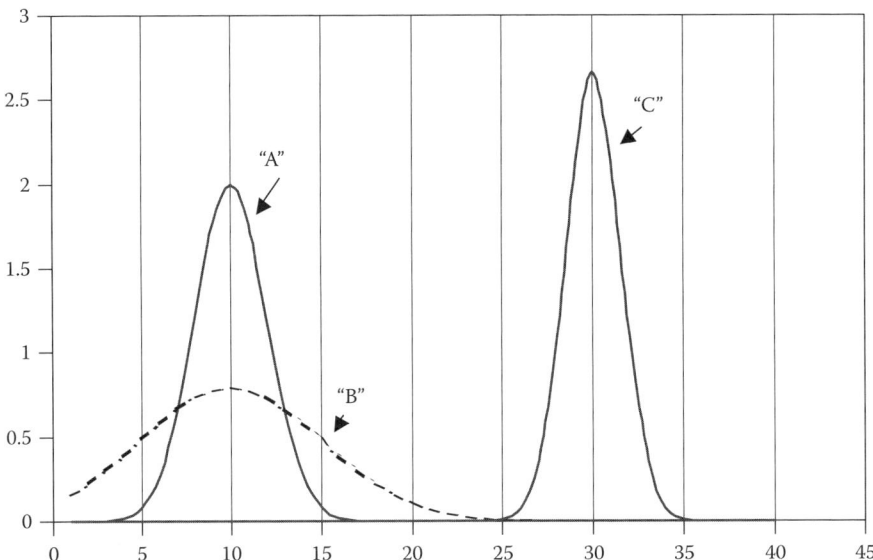

FIGURE 14.5 Hypothetical data curves "A," "B," and "C" illustrate the difference between precision and accuracy. Curve "A" represents a relatively narrow data distribution compared with curve "B" having the same mean, and hence the corresponding dataset would be said to be a more precise determination than that of curve "B." However, if the true mean is three times larger, represented by curve "C," then the datasets in neither curve "A" nor "B" would be considered accurate. A systematic bias in the measurement process, such as an incorrect calibration of the detector, or air in-leakage in the sampling train leading to systematic underestimate of the sample volume could lead to lack of accuracy. Imprecision represented in curve "B" might be the result of noise in the detector electronics or insufficient counting times in the measurement process. Improvements in the design, calibration, and operation of a CAM system might have the desired effect of removing bias and imprecision in the measurements made, and thus result in datasets such as curve "C," but this in itself would not assure that the mean of "C," μ_C, represents a quantitative determination if $\mu_C < L_D$.

of *potential alarm conditions* as possible. In such a context, the relationship between precision, accuracy, and bias can be understood as follows. As explained in a preceding section, each net count measurement in a continuous series of sample counting steps (integrations) represents the difference between a gross count of the ROI in the α-spectrum of a particular radionuclide, for example, the 5.15 MeV region where the α-peak of ^{239}Pu is found, and the simultaneous counts contributed to this ROI by radon progeny. The net ^{239}Pu count is then the difference between the gross count and the background contributions. Therefore, there are random measurement uncertainties (unpredictable deviations) from the true value of *both* of these components at any time step due to the inherent statistical nature of radioactive decay described by Poisson statistics. These variations lead to the spread in the data around the mean value. In addition to random effects, there can be systematic bias contributed by application of an algorithm for radon progeny background estimation that systematically over- or underestimates interference, by calibration uncertainties due to operator error in calibrating the detector efficiency or flow meter, and even by the effect of a CAM inlet design that systematically biases collection efficiency for a portion of the target aerosol, for example, if particles above a certain size were effectively removed from the sample by deposition on internal surfaces ahead of the filter (McFarland et al., 1990).

Various combinations of bias and precision faults can lead to low accuracy:

- *Large bias + low precision = low accuracy*: A poorly parameterized or applied background correction algorithm and/or calibration of critical components (bias), combined with inadequate spectrum count accumulation in the peak and background regions leading to poor precision, results in poor accuracy in detecting alarm conditions.
- *Low bias + low precision = low accuracy*: Even if the background algorithm is applied properly and the component calibrations are correct with little bias, poor counting statistics (low precision) from inadequate spectrum counts will still lead to poor accuracy.
- *Large bias + high precision = low accuracy*: Again, if the correction of the gross count is systematically either excessive or underestimated by the compensation algorithm or calibrations are deficient, then even when the precision of the determination of background and target counts are high, the net result will be inaccurate because it will be systematically biased negative or positive.

To illustrate with a quantitative example, in the case of one type of α-CAM design [25 mm diameter detector and filter, 1 CFM sampling rate (472 cm^3 s^{-1}), and 0.2 cpm dpm^{-1} efficiency] the background compensation is based on a multiple ROI ratio compensation algorithm (Hoover and Newton, 1992). The strategy of using multiple ROIs begins with establishing count summation regions in portions of the leading and tail edges of α-energy spectrum peaks. These counts and count ratios of individual ROI counts are applied in a predictive mathematical model to produce an estimated net Pu count, here it is labeled Pu(ROI-1):

$$\text{Pu(ROI-1)} = K_{Pu} * \text{ROI-2} * \left(\frac{\text{ROI-3}}{\text{ROI-4}}\right), \tag{14.4}$$

where K_{Pu} is the empirical calibration constant for Pu.

As seen in Figure 14.6, the net Pu ROI counts are predicted from counts summed in the three higher-energy regions: near 6 MeV, in the tail of the 7.68 MeV peak, and on the leading edge of the 7.68 MeV peak. The constant K_{Pu} is determined in a separate calibration process using a calibrated plutonium source and representative radon background conditions. Since this method depends on establishing the empirical constant K_{Pu} under certain conditions (typically lab conditions), it is susceptible to bias effects if conditions in the field are very different, and if short count times are employed. The consequences can be systematic under- or overcorrection.

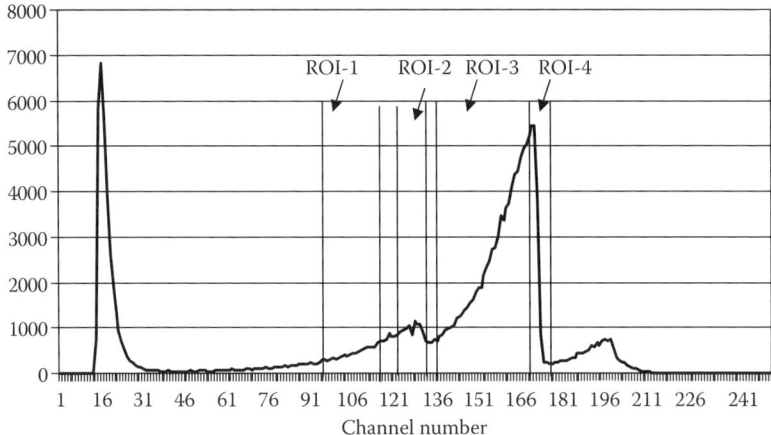

FIGURE 14.6 α-Energy spectrum illustrating the ROI method. ROI-1 is the transuranic ROI where counts from $^{238-239}$Pu would be expected to accumulate, ROI-2 is the region where the 6 MeV α's from the radon progeny accumulate, ROI-3 is the low-energy-tail region of accumulation of 7.68 MeV α's from the radon progeny, while ROI-4 is the high-energy region of this peak. Counts in the several ROIs can be applied in a calculation to yield a corrected net cpm in ROI-1.

An alternative algorithm based on fitting an exponential function to the low-energy radon progeny "tail" counts from the higher-energy α-peaks interfering with the ^{239}Pu ROI further illustrates how background correction strategies influence the net transuranic count (Hoover et al., 1995).

Figure 14.7 illustrates a 256-channel α-spectrum collected with a CAM utilizing such an exponential tail algorithm. The spectrum shows radon progeny peaks at 8.78, 7.68, and 6.0 MeV. The strategy of this algorithm is to process the α-spectrum to identify channel points between peaks with the least count ("valley points"), which delineate one peak from another, and then fit a two-parameter exponential function between the count at the valley point and a baseline point at the far low-energy region of the spectrum.

The valley points between peaks are identified by an "X" in this figure. The region between X_1 and X_0 is the plutonium ROI. In the tail fitting algorithm, an exponential equation of the form

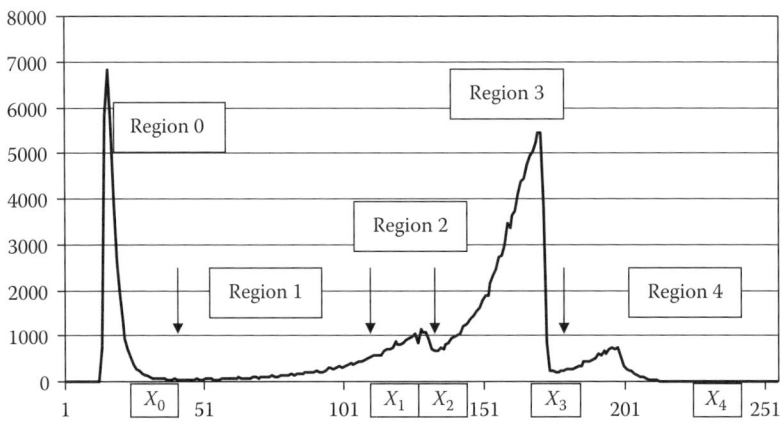

FIGURE 14.7 A 256-channel α-energy spectrum illustrating the location of critical valley points X_0, X_1, X_2, X_3, and X_4, which define count summation regions and points for fitting exponential tails of the higher-energy peaks.

$y = e^{mX+b}$ (where y is the count, m is the slope, and b is the intercept) is fit between successive valley points (X_3 to X_0, X_2 to X_0, etc.). The contributions to counts in the Pu ROI from each of the higher-energy radon progeny peaks, as determined by the successive exponential functions, are then subtracted from the gross Pu count. The net Pu count is therefore the residual from this subtraction process. Then the Gaussian portion of each peak is removed (Canberra Albuquerque, 2003). The calculation of net Pu counts takes the form

$$\text{Net Pu CPM} = \text{Gross Pu ROI count} - T_{8.78} - T_{7.68} - T_{6.0}, \quad (14.5)$$

where T_i are the total tail counts in the Pu ROI from each radon progeny.

Equation 14.5 illustrates that if the exponential function fit of the low-energy-tail regions of each higher-energy peak is a good fit to the data and if the Pu ROI count is solely due to background, the net Pu cpm will average very close to zero. Referring back to Figure 14.2, for example, the large contribution of radon progeny counts to the spectrum relative to the expected contribution of ^{239}Pu at about 4 DAC-h exposure can be seen. While the peak is well defined in this spectrum, the broadening of peaks with dust accumulation and the increase in background peak areas when radon background is high make low-level peak identification much more difficult.

The "exponential tail" compensation algorithm concept tracks the changing spectrum peak shapes of the radon progeny as sampling conditions evolve since the valley points are not defined by fixed channel numbers, but are searched for each cycle. It is not based on a fixed calibration factor and count ratios as in the case of the multi-ROI ratio method. However, the "exponential tail" method is also sensitive to the distortion effects of heavy aerosol loading of the filter which may smear out the valley points to low precision in short integration times.

A sequence of 10-min sampling intervals of a filter collecting normal background levels of radon progeny (tenths of pCi L^{-1}) is shown in Figure 14.8. The resultant net plutonium counts for the 10-min case, expressed as equivalent DAC-h, have an average value of -0.40 ± 0.02 DAC-h, indicating a slight overcompensation bias in the algorithm. The 30-min integration procedure case illustrates that with better counting statistics and higher precision the bias nearly disappears (Avg. = $5 \times 10^{-3} \pm 0.72$ DAC-h). Extrinsic factors such as inert dust burial of successive depositions of radon progeny can, of course, result in peak distortions that disrupt the normal application of background correction algorithms and introduce biased results even for long integration (high-precision) counts.

The question of accuracy is best answered by tests with Pu test aerosols in controlled atmospheres. For example, in tests with plutonium aerosol delivered directly to the inlet of the Alpha Sentry CAM head sampling at 2 CFM, and employing 5 min integration, the coefficient of variance

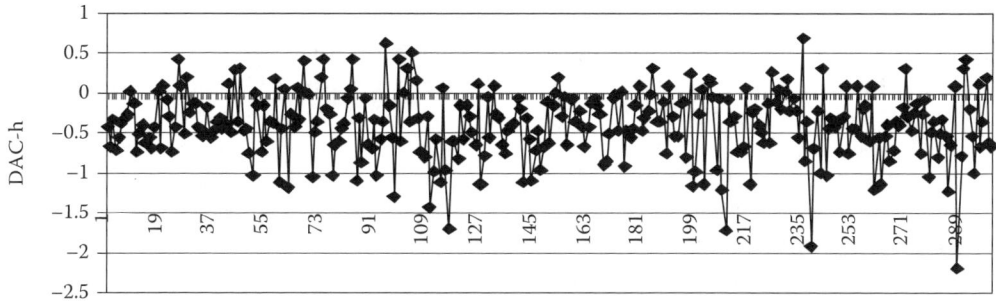

FIGURE 14.8 Week-long procedure blank determinations (ambient air sampled containing the radon progeny but no Pu or Am) using the "exponential tail" algorithm to calculate net Pu DAC-h with a 10-min integration time. The negative average DAC-h result in the 10-min integration data indicates a slight overcorrection bias in the method with relatively few counts per channel collected.

TABLE 14.1
Standard Deviation of Reported/Estimated Plutonium Background (DAC-h)

Effective ^{222}Rn Concentration (pCi/L)	5-min Averaging[a]	15-min Averaging[b]	30-min Averaging[c]
0.02	0.6	0.3	0.2
1.0	1.3	0.7	0.5
4.0	2.5	1.5	1.0

[a] Average standard deviation values based on laboratory tests at Inhalation Toxicology Research Institute (ITRI) involving a 5-min averaging period in the Canberra Alpha Sentry CAM. Values are proportional to the square root of the radon concentration.
[b] Standard deviation values calculated for a 15-min averaging period, assuming Poisson statistics. Increasing the counting time by a factor of 3 reduces the reported standard deviation by a factor of 1/sqrt(3).
[c] Standard deviation values calculated for a 30-min averaging period, assuming Poisson statistics. Increasing the counting time by a factor of 6 reduces the reported standard deviation by a factor of 1/sqrt(6).

(COV) of the response of the instrument was a low 6.5%. The reported activity was 96% ± 6% of the actual activity delivered as determined by an independent counting method (Hoover and Newton, 1992). The level of background radon interference and the size of the Pu exposure are reflected in the response of the instrument. As further illustrated in Table 14.1, as the level of radon progeny increases, the variability in the outcome (standard deviation) also increases. With this level of accuracy, the expected false-positive alarm rate for an 8 DAC-h alarm threshold should be negligible at radon progeny backgrounds less than about 4 pCi L^{-1} (148 Bq m^{-3}) and only 2×10^{-3} per year at 4 pCi L^{-1}.

Recent CAM designs are moving toward higher sampling rates and more sophisticated spectrum analysis algorithms based on peak fitting. An important example of the effectiveness of the application of peak-fitting spectral analysis algorithms is shown in Figure 14.9. The method used is based on background correction by Marquardt–Levenberg nonlinear least-squared fitting of a mathematical peak function to each peak identified in the spectrum, and removing their contribution to the Pu ROI on a channel-by-channel basis. A peak shape function has been identified, which accurately represents the physics of α-particle interactions in α-detectors. It represents the convolution of a normalized Gaussian function with parameter σ and a left-side exponential with parameter τ:

$$Y(x) = \frac{A\sigma\sqrt{2\pi}}{2} \times \tau \exp\left[(x - C)\tau + \frac{\sigma^2\tau^2}{2}\right] \times \text{erfc}\left[\frac{1}{\sqrt{2}}\left(\frac{(x - C)}{\sigma} + \sigma\tau\right)\right] \quad (14.6)$$

where A is the peak amplitude, C is the peak channel number, x is the channel number, and erfc[ ] is the complementary error function, and σ and τ are fitting parameters of the Gaussian and exponential.

A convolution is defined mathematically as an integral that expresses the degree of overlap of one function as it is integrated over another (it blends one function with another). In this case, it is the blending of a Gaussian function (a symmetrical peak shape) with a left-side exponential (a left-facing tail).

The resultant peak function (Equation 14.6), first proposed by L'Hoir (1984), has been shown to well describe the asymmetric peak shape of α-interactions in silicon (Bortels and Collaers, 1987). In this peak-fitting method, the function is fit to each of the α-spectrum peaks by nonlinear least-squares method (often referred to as the Levenberg–Marquardt method), whereby the fitting parameters of the model are adjusted to minimize the squared difference between the model prediction and the data at each point in the peak ROI range (Press et al., 1986). The principle of this method is

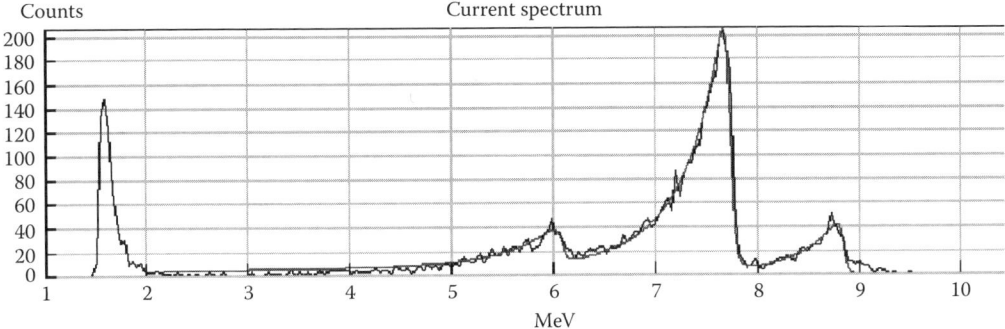

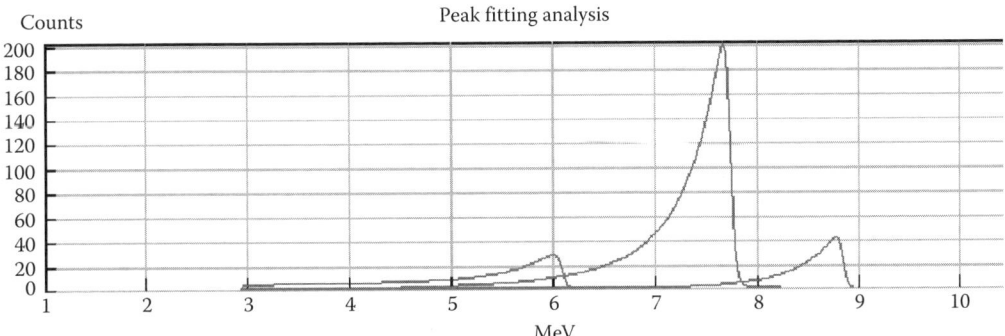

FIGURE 14.9 Peak-fitting algorithm applied to a 256-channel α-spectrum. Shown in the upper graph are the raw counts per channel data (jagged line) overlaid by a fitted peak function for all the peaks (smooth curve). Shown in the lower graph are the individual fitted peaks at 6.0, 7.68, and 8.78 MeV. Peak fitting is the result of applying a nonlinear least-squares-fitting procedure using a mathematical model of an α-peak (the convolution of a Gaussian peak and a left-sided exponential).

easily stated but difficult to implement: for a given set of data, propose a model of the data, and then draw a model curve that fits as closely as possible to the data points. Put mathematically, the concept is to select a set of parameters **p** of a model $y = f(x, \mathbf{p})$ such that the deviation between the model predictions **y** and the measurement dataset **U** is as small as achievable by some fitting procedures. The idea of the closeness of the fit can be understood in terms of the collective "distance" between a set of N data points and a corresponding set of model predicted values, where the individual distances may be either positive or negative. But the squared differences are absolute, and they are the proper basis for the measure. It is the sum of squared differences that is the criterion of goodness of the fit between model and data: the smaller the sum of squared differences, the better the fit. The task then becomes to minimize parameter χ^2, which can be defined as

$$\chi^2 = \sum_{i=1}^{N} (y_i(\mathbf{p}) - \mathbf{U}_i)^2. \tag{14.7}$$

The set of parameters $\mathbf{p}_o$ that minimizes χ^2 is the target of a complex search process requiring an iterative solution of a set of equations like Equation 14.7 and their derivatives, starting from a suboptimal initial parameter set and moving toward the optimal set one step at a time, since the equations are clearly nonlinear and complex. A number of methods to accomplish this are known, but the Levenberg–Marquardt method is one of the more powerful and efficient ones available.

The overlaps of the tails from all fitted high-energy peaks into the ROI of lower-energy peaks are systematically removed by subtraction, including the transuranic peak ROI. In this case, therefore, the net transuranic peak count is determined as the gross integral peak count (sum of counts under the fitted model to a transuranic peak if found and fitted) minus the tail contributions of each higher-energy peak (if any). In background conditions, subtraction of overlapping tail counts from a lower-energy peak ROI often results in scattered residual counts (plus or minus), with no resolvable peak result, that is, there is no sufficiently defined peak to attempt a fit. In this case, a "non-detect" condition exists and a zero count may be returned. While this constitutes "left-censoring" of data, it may be considered a defensible practice in this case where accurate detection is paramount. As seen in Figure 14.10, the intercomparison of the "exponential tail" and "peak shape" fitting procedures on the same dataset reveals that the peak-fit analysis results in a much lower detection limit, and fewer opportunities for a false-positive alarm under dusty conditions when the spectral peak shapes are distorted by dust-burial effects.

It is incumbent on the CAM user to carefully evaluate the expected performance of candidate α-CAMs. Consideration must be given to the effect on performance that the interference conditions discussed above might have. Use and effects of alternative algorithms should be judged. Finally, one must estimate how the accuracy of the measurement affects realistically attainable alarm thresholds with an acceptable false alarm rate.

The accuracy of the final α-CAM exposure or concentration measurement (and hence alarm condition detection) is also affected by uncertainty associated with volumetric flow determinations obtained from internal α-CAM air-flow metering. Flow meter calibration accuracy of a few percent is obtainable by commonly used external reference calibration equipment and procedures. However, a potentially large contribution to inaccuracy in the volumetric determination is in-leakage of air into the flow due to filter seal leaks, especially when pressure drop builds across the filter as it loads with dusts. Such a systematic bias in the sample volume can lead to significant underestimation of concentration and hence false-negative α-CAM alarm responses. This potential problem can be addressed by user training that stresses the importance of proper filter mounting during filter changes, and a rigorous maintenance and testing program that includes checks on filter in-leakage. Flow rates can be continuously monitored as reported by the flow meter. By implementing a

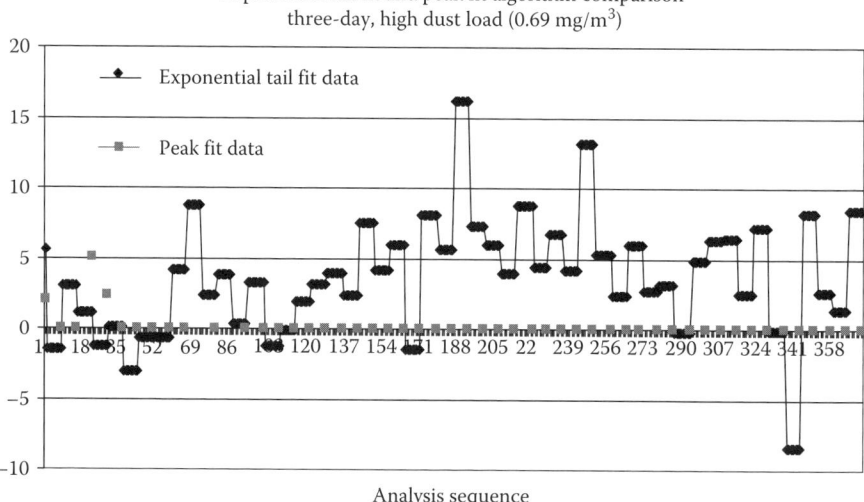

FIGURE 14.10 Comparison of background correction utilizing the "exponential tail" fitting (diamonds) and the "peak"-fitting procedures (squares). The data are from a study conducted in the salt-laden atmosphere underground at the Waste Isolation Pilot Plant (WIPP); hence there are significant salt dust burial effects on α-spectra being fit. (Adapted from Rodgers, J., *Comparison of Background Compensation Algorithms in Environmental Monitoring: Data from LANL-WIPP Investigations of ECAM Performance Underground at WIPP*. Presentation, Department of Energy Air Monitoring Users Group Meeting, Pojoaque, NM, Los Alamos National Laboratory Report LA-UR-03–6693, July 28, 1994.)

maintenance alarm function (not the radiation alarm that has been discussed so far), a CAM user can be alerted to a filter loading condition that could result in loss of detection efficiency, sensitivity and accuracy due to reduced sample collection rate, and thus signal the need for a filter change. When under control, volume uncertainty is not a significant issue.

Uncertainty in the calibration of detector efficiency is another contributor to inaccuracy in CAM measurement and alarm response. The primary determination of CAM α-detector efficiency can readily be accomplished with high accuracy using NIST traceable uniformly plated α-sources having sufficient activity. The efficiency determined by this means represents the ideal "geometric" efficiency for the detector—filter configuration, with only the effects of energy absorption of intervening air included. However, systematic bias can enter into the detection process if a portion of the activity collected on the sample filter becomes lost from contributing to the count due to burial into the filter substrate or under later deposits of inert dust particles (Moore et al., 1993) or if radioactive particles are not uniformly deposited on the filter. In addition to outright absorption, the effects of inert dusts or filter fibers can include serious distortions of the α-spectrum leading to inaccurate background correction: both over- and undercorrection of the gross count can occur. Although most transuranic labs have HEPA filtered air and high air exchange rates, positioning a CAM next to aerosol-generating devices such as carbon vane vacuum pumps can cause these problems. Fibrous filter media such as fiberglass or cellulosic filters are prone to sample burial. The use of smooth, hard surface filter media such as membrane filters can largely prevent burial bias as long as the dust load in the air is not large. The use of such surface collection media also results in better defined α-energy spectra (peaks having smaller FWHM), which significantly improves the performance of background algorithms generally. With the use of proper filter media and attention to filter loading, detection efficiency is not a significant issue.

APPARATUS DESCRIPTION: ROOM CAMs AND PERSONAL LAPEL CAMs

The α-CAM apparatus consists of a sample collection head housing a filter holder, an α-detector positioned closely opposite the filter, electronics for α-signal processing and analysis (typically, a preamplifier, an amplifier, an embedded PC card, and an MCA), a flow meter, and an alarm annunciator (flashing light and/or horn). A user interface with controls, a readout/display, and a means for data logging and added alarm functionality may also be part of the sampling head or a separate component that can be positioned immediately outside the monitored area. Current α-CAM technology relies on wired or wireless networking to establish communication between an array of α-CAMs in a facility and the facility operations center and emergency response technicians. In this case, a networked computer becomes the user interface. Detailed incident information is made available immediately on computer resources throughout the facility.

To implement the networked approach, an embedded single-board computer in each CAM supports a web server application in firmware that uploads current CAM data (all critical CAM function status, α-spectrum, alarm status, and more) in a form accessible by web browsers connecting to the network. Then data communications through a high-speed Local Area Network (LAN) and optionally, RF broadcasts in a Virtual LAN, Ethernet, or by satellite links can be easily accomplished (Canberra Albuquerque, 2003). A new industry standard open network protocol termed "Rad Net" has been widely adopted by nuclear instrument manufactures and users to facilitate networking by greatly reducing network overhead, reducing costs and enabling flexibility in future enhancements such as email generation and pager systems. The protocol is intended to become an American National Standard (see www.radnetusergroup.org).

As indicated in Figure 14.1, air flows into the head through a radial inlet at the top, and down around a cylindrical or conical detector container to the filter positioned directly opposite, and closely spaced, to the detector. The filter holder (which may be a special cartridge or card) is preloaded with membrane-type filter paper before inserting in the head. Successful spectroscopy of α-emissions from the sample requires well-defined energy peaks. As previously noted, α-particle interactions with a fibrous mat introduces variable energy losses resulting in undesirable peak broadening (Moore et al., 1993). Typically, a membrane filter means that the filter medium has low porosity and higher pressure drop. However, as discussed by Lee and Ramamurthi (1993) and previously mentioned, the development of a class of composite filters consisting of a thin Teflon (PTFE) membrane attached to a polypropylene fiber backing has created a filter capable of high flow/cm^2, but exhibiting surface collection properties of membrane filters. Annex D of ANSI/HPS N13.1-1999 discusses some of the characteristics of filters used in air monitoring. Fluoropore 5 μm filters have flow rates of up to 59 L/min/cm^2 as well as an α-energy peak resolution of 350 keV FWHM compared to 1500 keV FWHM for Whatman 41 cellulose fiber filters (based on the ^{218}Po peak) and so are an excellent choice for α-CAM use.

The flow meter for α-CAMs may be a direct mass flow metering device, or a more complex elbow meter device with a mass flow sensor (McFarland et al., 1994). The CAM elbow meter includes a combination of an elbow pressure drop generator and a shunt-type mass flow sensor for providing an output which gives the mass flow rate of a gas that is nearly independent of the density of the gas. For air, the output is also approximately independent of humidity. The 90-deg elbow creates a pressure (p) differential across the two legs of the elbow meter. A resultant change in pressure Δp, proportional to the flow through the elbow, moves a small flow through the associated mass flow sensor. The voltage output of the sensor is calibrated using a reference flow calibrator set to provide readings in actual CFM (ACFM) (Canberra Albuquerque, 2003). Use of an elbow meter has the considerable advantage that there is a minimum constriction in flow for operation and no constraints on the tube entrance length for proper function.

The inlet stage of a CAM demands careful design to insure that a representative sample of particles in the free air stream outside is successfully extracted and transported down to the CAM filter. The high density of transuranic elements means that particulates of these radionuclides have ADs much larger (~3×) than their corresponding physical diameter (Elder et al., 1974). As a consequence, there is an increased likelihood of particle loss in the inlets of α-CAMs due to impaction on inner surfaces (McFarland et al., 1990). The effect of loss of large AD particles on the detection performance of α-CAMs is substantial: the activity of 10 μm AD particles is 100 times that of 1 μm particles. Careful CAM inlet design as described by McFarland et al. (1990) assures that a high percentage of particles having ADs ≥10 μm AD that are aspirated into the inlet are transported to the sample filter and that the collected particulate is uniformly deposited over the full surface of the filter. Aerosol wind tunnel studies (McFarland et al., 1992) have demonstrated over 80% penetration of 10 μm AD particles from the inlet to the filter. Uniform sample deposits on the filter are feasible. Other α-CAM head designs, especially those utilizing a simple "clam-shell" design where sample particles enter from one side and collect on a filter holder in the opposite side have much lower penetration efficiency (Biermann and Valen, 1983; McFarland et al., 1990). The α-CAM user should be satisfied that a selected CAM head inlet has good sample collection properties for particles of large AD in order to respond properly to transuranic particle releases.

Another format of the α-CAM apparatus is the personal lapel CAM. In this design, the individual user wears a low-volume, battery-operated air pump in a belt pack along with data processing electronics and a small sampling head is supported on the user's lapel connected to the pump by a flexible tube. In this form, it is the region near the breathing zone (BZ) of the user that is monitored, rather than the general room air. The output of the data analyzer can be communicated to the LAN by means of wireless broadcasts from the belt pack. This is in principle different from traditional *personal air sampling* whereby an integrated filter sample is taken over a work period and later analyzed off-line.

While BZ monitoring can be quite effective in alerting the user to a release in the immediate vicinity, and especially in applications where the sampling head is attached inside PPE, it is not without its own limitations. Owing to the low sampling rate and small size of the lapel sampling head, the BZ sampler may not extract a representative sample of the radioactive particulate present. In a review by Whicker (2004) of the relationships between BZ sampling and bioassay determination of internal dose, it was found that BZ measurements often do not correlate well with bioassay results. This discrepancy was explained by the findings that BZ sampling results can be significantly affected by measurement conditions such as the orientation of the sampler with respect to the source, where the head is worn (right or left lapel), the design of the sampling head as it reflects sampling efficiency of all particle sizes, particle size distribution, local air velocities and directions, and the sharp concentration gradients in and around the BZs of workers under accident conditions (Wicker, 2004). In some work conditions, a combination of room and BZ–CAMs may be beneficial.

ALPHA-CAM PLACEMENT IN THE WORKPLACE

The successful application of α-CAMs for worker protection requires more than the selection of an accurate, calibrated, and well-maintained instrument. It also requires proper placement in various work environments where it is needed to achieve high levels of worker protection. The need for proper placement is called for by regulation and federal guidance (Hickey et al., 1993; USDOE, 2007b) but is often overlooked or given passing attention in practice. The best evidence of this was developed in a study of certain DOE facilities where worker protection was dependent on extensive use of α-CAMs (Crites, 1994). Shown in Table 14.2 are summary results, where the air concentration exposure (DAC-h) was estimated from retrospective FAS monitoring data collected in rooms where CAMs were operational. The CAM instrumentation in use at these facilities was claimed to have an operational sensitivity of 20–45 DAC-h, based on such factors as airflow patterns, detector efficiency, and filter absorption loss. It is especially noteworthy that in a very large percentage (71%) of the instances where even substantial releases occurred (>500 DAC-h), there was *no* CAM alarm. Part of the difficulty implicated in these data and noted by Crites (1994) is CAM design and placement. If for example, the airborne particulate was not being adequately sampled (especially large particles), there would have been little or no Pu particulates collected to generate significant counts, or if the compensation algorithm was biased such that the tendency was to overcorrect radon background, the CAMs would not alarm even when exposures were quite elevated. However, it was concluded that monitor placement could have had a major role in this poor instrument performance. That placement is a significant contributing factor to the lack of CAM alarm response was confirmed in later studies in one of these facilities (Whicker et al., 1997). The exclusive deployment of CAMs at corner ventilation registers rather than in interior locations closer to points of release was found to consistently reduce the probability for rapid and effective detection and alarm. The extremely high concentration gradients that typically accompany a puff release from glove box containment failure, for example, glove puncture, are often dissipated into low concentrations when exiting the registers. Also detrimental was the use of long transport tubes to the CAM sampling heads (Whicker et al., 1999). Poor particulate extraction efficiency and losses to the walls of the tube during transport reduces the probability of rapid CAM response even further.

TABLE 14.2
DOE Plutonium Processing and Handling Facility α-CAM Alarm Response History

DOE Laboratory Pu Exposure Condition (DAC-h)	Number of Release Event Challenges to Room CAMs	Number of CAM Alarms Recorded	Percent Success (%)
2–10	127	19	15
10–50	107	13	12
50–100	37	6	16
100–500	33	5	15
>500	14	4	29

Source: Crites, T., *Radiation Protection Dosimetry*, 53:65–68, 1994.

Identifying room airflow patterns by the release of tracer particles or by using computational fluid dynamics modeling studies (Rodgers et al., 1998; Wasiolek, 1999) can aid in the selection of the correct number and proper placement of CAMs in a workroom so that prompt and reliable alarms will be generated. The critical parameters of such studies that must be considered in determining CAM placement are lag times in contaminant plume passage from potential release locations to CAM location options, and peak concentration at those locations. The degree of protection afforded to workers by the CAM alarm will be compromised if the CAMs are not placed in such a way that they will intercept at least a portion of the radioactive particulates released in an accident in a very short time postrelease. Optimal CAM placement within the workroom obtained by such studies has been shown to provide much shorter lag times compared to locations at corner ventilation registers, which up to the time of the Crites studies had been the norm. See Chapter 13 by Whicker for a thorough discussion of sampler placement.

CALIBRATION AND GAIN CONTROL

The only external reference calibrations possible and required for α-CAMs are the efficiency calibration of the α-detector (cpm/dpm) and the volumetric flow calibration for the flow meter, as previously discussed. The geometric efficiency (cpm/dpm) is obtained by a counting procedure using NIST-traceable α-standards. Proper operation of the ROI or peak-fitting algorithm for background correction should be based on well-known characteristics of ambient radon progeny readily collected on a calibration filter and may require data entries by the user. In some older CAM designs, it is assumed that there is a fixed association between α-energy and MCA channel numbers, so that once the user has selected certain channel numbers to correspond to various ROIs they remain the same throughout the monitoring cycle. In practice, however, the effects of the accumulation of particulates on a CAM filter lead to systematic energy losses in α-emissions reaching the detector. The result is a shift in the association of peak characteristics (ROI and peak centroid) relative to channel numbers in the α-spectrum. In clean room conditions the shift may not be large or consequential. But if particulate generating conditions exist (e.g., carbon dusts emitted by carbon-vane pumps, chemical fumes, etc.), significant shifts can occur, leading to degradation of background correcting algorithms. To counter this effect, some CAM algorithms include a step in which a fiduciary background peak in the α-spectrum is identified and used to periodically recalibrate the α-energy spectrum. For example, if the gain (or equivalently, the slope) of the CAM α-spectrum is fixed (e.g., channel 256 corresponds to 10 MeV), a linear association between peak energy Y and channel number can be described by an energy versus channel number equation of the form

$$Y \text{ (MeV)} = \textbf{Slope}\text{(MeV/channel)}* \textbf{channel} + \textbf{Intercept} \text{ (MeV)}. \tag{14.8}$$

where Slope = 10/256 = 0.039 MeV/channel, and Intercept is empirically determined.

Then if the ^{214}Po radon progeny peak at 7.68 MeV is selected as the fiduciary peak, any other peak channel location can be found utilizing the intercept value derived from Equation 14.8 for the Y = 7.68 MeV case. If the nominal (clean filter) channel number associated with this peak is 173, the intercept in Equation 14.8 would be 0.92 MeV. Suppose, to continue the example, a peak shift of lower four channels (from channel 173 to channel 169) has occurred as a result of accumulated dusts, and was detected by a peak search of the spectrum. Then by solving for the new intercept value in Equation 14.8, one would find a new intercept of 1.08 MeV, thus recalibrating the spectrum relative to the fiduciary peak. By this strategy, a continuously adjusted internal calibration can be maintained without user intervention. To be successful, the strategy requires both that the MCA is very linear and the peak search is accurate.

Volumetric flow calibration is accomplished with the use of a reference-calibrated flow meter. It is important to assure that the calibration meter is inserted into the flow train ahead of the filter (to avoid filter-induced pressure drop effects on volume determination) and that the calibration is referenced to ACFM at the elevation of the CAM application, and not Standard CFM (SCFM), since regulatory concentration limits are based on volume of air breathed at the work location, not at sea level.

CALCULATION OF EXPOSURE IN DAC-h

At each end of each sample collection and counting step in the continuous monitoring process of a CAM, an α-energy spectrum is analyzed for the presence of the target radionuclides, and a background corrected net count is generated. From net counts in the target ROI, A_{net}, the DAC-h (and DAC if desired) implied by the counting parameters, including sample collection time and integration time, can be computed. For example, the plutonium DAC-h is calculated as

$$\text{Exposure (DAC-h)} = \frac{(T_i * A_{net}(T_i))}{(T_c * \varepsilon * Z_{DAC} * K_{DAC\text{-}hr} * V)}, \quad (14.9)$$

where T_i is the time since the filter was changed (in hours) at the ith count, T_c is the time over which the spectrum was collected (in hours), V is the volume of air (in liters) that has gone through the filter since it was last changed, which is the sum of sampled volumes at each time step, ε is the efficiency of the detector (cpm dpm^{-1}), Z_{DAC} is the DAC factor for the target radionuclide (μCi mL^{-1}), such as 3×10^{-12} μCi mL^{-1} for ^{239}Pu, and $K_{DAC\text{-}h}$ is the unit conversion factor [1.332×10^{11} mL *(μCi * h * L)$^{-1}$].

Note: the equation may be used with metric DAC values in MBq m^{-3} if the volume is again measured in liters and the unit conversion factor is changed to 3.6×10^6.

Determination of air concentration from CAM data depends on measuring incremental changes in filter activity as sampling continues. The filter is not changed of course; hence, the activity of the previous count $A_{net}(T_{i-1})$ must be subtracted from the current activity measured $A_{net}(T_i)$ to yield the measure of the accumulated counts just in the current cycle. Then the concentration is calculated from

$$\text{Concentration (DAC)} = \frac{T_i * [A_{net}(T_i) - A_{net}(T_{i-1})]}{(T_c^2 * \varepsilon * Z_{DAC} * K_{Conc} * V)}, \quad (14.10)$$

where T_i is the integration time between the ith count and the $(i-1)$-th count, K_{Conc} is the unit conversion factors to yield concentration in proper units, and other terms are as previously defined for the DAC-h calculation.

The one-sigma counting error in the net count, $A_{net}(T)$ at time T is given by

$$\text{Err}(A_{net}) = 1.65 * [A_{net}(T) + 2.0 * (\Sigma(\text{Tail}_i))]^{1/2}, \quad (14.11)$$

where $\Sigma(\text{Tail}_i)$ represents the sum of the overlapping tail contributions from each of the higher-energy radon progeny to the counts in the ROI of the target radionuclide, such as ^{239}Pu.

The overall uncertainty (expressed in fractional percent) including the contribution to uncertainty from the other measured terms can be estimated from the sum of squared uncertainties in each term:

$$U(\text{total}) = (U_{count}^2 + U_{volume}^2 + U_{calibration}^2 + U_{time}^2 + \cdots)^{1/2}. \quad (14.12)$$

These and other calculations are typically made internally in the CAM computer firmware automatically and reported to the user at the user interface unit (or LAN broadcast) and logged in data files for later retrieval.

To clarify the above formulas it should be noted that detection and collection of the sample occur simultaneously without a filter change. In contrast to the case where an air sample with a fixed volume is collected for a certain period of time and then moved off-line to a counting device, the CAM sample is collected over a long period at a certain *rate* which might change over time as the sample filter becomes loaded. Hence, the CAM instrument logs *total sample volume* (denominator of Equations 14.9 and 14.10) *and total time from filter change* (numerator) to track the quantity of air sampled and the sampling duration. This can be seen as corresponding to an average sampling rate term in the denominator of both equations. It also establishes the integrated time (hours) of exposure recorded by the CAM. Continuing with the counting process, the CAM filter is counted continuously, but counts are summed in finite time steps using an MCA, as has been described above. Individual count sums are accumulated in discrete energy bins to produce an α-energy spectrum (a histogram). At the end of each count cycle, the spectrum is read from the MCA and passed to the data analysis computer for determination of net transuranic or uranium counts, A_{net}. Then the MCA is reset (all channel counts cleared to zero) and another count started. The count time (denominator) converts the observed count into count rate. Since any amount of α-emitter activity collected between the time a fresh filter was loaded into a CAM and the most current count interval will be counted and recounted as time goes on, the observed count rate represents the *cumulative exposure* of the CAM up to that time. The unit of exposure measure, the DAC-h, is thus the cumulative exposure to airborne radioactivity expressed relative to the DAC for the particular radionuclide in question. (Note that based on the method by which the DAC is derived by regulatory agencies, the DAC corresponds to a dose or dose equivalent, as expected for a measure of worker exposure.) In the expression for concentration, an estimate of an accumulating activity due to drawing contaminated air through the filter during the most recent counting interval can be obtained only by subtracting successive counts of activity on the filter, and from that, a concentration measurement derived. As a result of the fact that radioactive decay is subject to Poisson statistics, successive counts of nominally the "same" quantity of radioactive material will result in counts in the range

$$A(T) = N \pm N^{1/2}, \qquad (14.13)$$

where N is the net number of counts in time T, and hence the estimated concentration estimate (derived from differences in successive counts) will oscillate up and down (possibly plus and minus) even if the true activity on the filter remains constant. The overall uncertainty in DAC will then include error propagated from two separate measurements of net activity $A(T)$, which can be relatively large. Uncertainty in DAC-h by contrast is only based on the most recent determination of activity. Again, the units of concentration are typically given in units of the DAC for the radionuclides in question.

SETTING AND TESTING CAM ALARM LEVELS

Since the role of the α-CAM is to provide early warning of a release, it is necessary to set one or more suitable thresholds for alarm. The threshold can be set to an alarm either at a certain concentration or at a given level of exposure (DAC-h). Since the quantitative detection of a release at near-MDA levels may require counting for a fairly long period of time (5–15 min), it is desirable to have a quick response determination based simply on a rapid count estimation strategy in the transuranic ROI, which does not require long count times, but which has a large threshold for alarm so that the probability of a false-positive alarm will be very low. Such an alarm is usually referred to as the "acute alarm." The count time for this alarm condition can be much shorter than what would be required to acquire a statistically significant α-spectrum for more detailed analysis (minutes or less).

The alarm based on detection of a low-level chronic release is referred to as the "chronic alarm." In this case, there must be sufficient counts in all elements of the spectrum to allow for accurate peak detection and peak fitting to occur. Spectrum integration times of 5–30 min may be required. As previously noted, dusty, high radon level conditions may make it difficult to achieve reliable low-level exposure estimates. The chronic alarm setting is typically based on limiting exposure and thus is a DAC-h alarm, although a concentration alarm threshold may also be used. Selection of a suitable DAC-h alarm threshold, L_{alarm}, requires careful evaluation of CAM sensitivity as expressed by the DL (expressed in units of DAC-h). Then the alarm logic for a given exposure condition at time T is as follows: if Exposure $(T) > L_{\text{alarm}}$, and if $L_{\text{alarm}} > $ DL, then activate the alarm. The second condition is required, since if the user selected alarm level is less than the current DL, a reliable alarm cannot be set. The alarm threshold could then be automatically adjusted upward in the CAM's internal operating system.

Based on the defining characteristics of these two types of alarm, it is evident that it should be possible to test the acute alarm function of a CAM simply by inserting a high activity source of the target radionuclide into the CAM instead of a filter. The acute alarm should quickly sound. However, the same test will not work for testing the chronic alarm at some desired level of sensitivity (e.g., 8 DAC-h) since it is based on spectrum analysis of a sample filter when a representative level of radon, dusts, and so on are present. This type of alarm function test is usually only performed as part of a type test or acceptance test by a facility when CAMs are purchased. If the concern of the alarm test is only to determine whether or not the alarm system is functional, the acute alarm test should suffice.

SUMMARY

Alpha-CAM technology has been developed and has evolved over a period of several decades in response to the need for an instrument that is continuously sampling workplace air that might, under accident conditions, become contaminated with α-emitting radionuclide particulates so that an early warning of the release of hazardous quantities could be provided to workers in time to evacuate or don protective gear.

The performance of α-CAMs can be optimized through instrument design, especially the design of the flow path from the CAM inlet to the filter, such that a large fraction of particles having an AED of 10 µm or less are efficiently collected and uniformly deposited on the sample filter (McFarland et al., 1990). Collection of the large AD particle components in the sample will greatly increase the probability of prompt detection due to the inherently high activity of such particles. Improvements of the background compensation algorithms that more accurately and reliably remove extraneous counts due to radon progeny α-emissions will increase sensitivity and reduce false alarms. Different background compensation strategies may perform better than others. An instrument with MCA capabilities offers more opportunities for improved background correction than does one with only single-channel analysis (or gross α-counting).

The user should be aware that selecting commercial α-CAMs based on different inlet design and background correction techniques may have both cost and worker safety implications. For example, while more sophisticated CAM technology may increase installation costs, too many false alarms may erode worker confidence in alarms, leading to what might otherwise be avoidable exposure (with attendant health care costs), and may also cause excessive lost work time costs due to frequent room evacuations. Hence, cost savings in instrument selection may be offset by hidden operational costs.

The function of optimized placement of CAMs in a workplace is to provide prompt, accurate, and reliable detection of elevated concentrations of actinide α-emitters in the air breathed by the workers. Detection accuracy takes precedence over accuracy in determining worker dose directly from CAM data in the monitoring process. Detection is complemented by independent air concentration measurements around a work area for purposes of dose assessment in the event of an accidental release detected by CAMs. For the latter purpose, a large number of FAS monitors are deployed around a laboratory space as close to the BZs of work stations as feasible. A critical flow venturi can be utilized with a FAS to very effectively control sampling flow rate (Parulian et al., 1996). Then, an accurate volume can be assigned to each sample. Sample analysis of FAS filters is carried out off-line using α-spectroscopy to identify the radionuclide of interest after radon progeny interference has been allowed to decay. An accurate estimate of exposure is then assigned based on the data from the nearest FAS monitors to the worker locations. The room CAMs can indicate when the release actually occurred. This is true even in situations where the airborne concentrations were not high enough to cause an alarm, but were still high enough to be detected (>DL) and were at levels below the L_D (which should be at or near the alarm set point).

Considerable progress has been made over many years meeting the challenges associated with proper implementation of CAM technology, as has been described in the foregoing.

REFERENCES

Alexander, J.M., A continuous monitor for prompt detection of airborne plutonium, *Health Physics*, 12:533, 1966.

American National Standard (ANSI), *Sampling and Monitoring Releases of Airborne Radioactive Substances from the Stacks and Ducts of Nuclear Facilities*, ANSI/HPS N13.1–1999, Health Physics Society, McLean, VA, 1999.

Biermann, A. and Valen, S., CAM particle deposition evaluation, in: Griffith, R. V., Ed., *Hazards Control Department Annual Technology Review*, Livermore, CA: Lawrence Livermore National Laboratory Report, UCRL-50007-83, pp. 79–82, 1983.

Bortels, G. and Collaers, P., Analytical function for fitting peaks in alpha-particle spectra from Si detectors, *Applied Radiation and Isotopes*, 38(10):831–837, 1987.

Brown, R.C., *Air Filtration*, Pergamon Press, Oxford, UK, 1993.

Canberra Albuquerque. *Multi-Head NetCAM Users Manual*, Vsn. 1, Albuquerque, NM, 2003.

Crites, T., Alpha air monitor alarm sensitivity: Operational experience. *Radiation Protection Dosimetry*, 53:65–68, 1994.

Currie, L.A., *Lower Limit of Detection*, USNRC NUREG/CR-4007, 1984.

Dorrian, M.D. and Bailey, M.R., Particle size distributions of radioactive aerosols measured in workplaces, *Radiation Protection Dosimetry*, 60:119–133, 1995.

Elder, J., Gonzales, M., and Ettinger, H., Plutonium aerosol size characteristics, *Health Physics*, 27:45–53, 1974.

Hickey, E., Stoetzel, G., Strom, D., Cicotte, G., Wiblin, C., and McGuire, S., *Air Sampling in the Workplace*. U.S. Nuclear Regulatory Commission (NRC) Guide NUREG-1400, 1993.

Hoover, M.D. and Newton, G.J., *Statistical Limitations in the Sensitivity of Continuous Air Monitors for Alpha-emitting Radionuclides*. Annual Report of the Inhalation Toxicology Research Institute, LMF-138, Albuquerque, NM, 1992.

Hoover, M.D., Newton, G.J., Yeh, H.C., Seiler, F.A., and Boecker, B.B. Evaluation of the Eberline Alpha-6 Monitor for use in the Waste Isolation Pilot Plant: Report of Phase II. Inhalation Toxicology Research Institute Report, January 31, 1990

Hoover, M.D., Newton, G.J., Fencl, A.F., and Marcinkovich, M., *Independent Evaluation of Los Alamos National Laboratory Continuous Air Monitor Instrumentation at the Inhalation Toxicology Research Institute*. ITRI Report ITRI-951102, Albuquerque, NM, 1995.

Huang, S., Schery, S., Alcantara, R., Rodgers, J., and Wasiolek, P., Influence of dust loading on the alpha-particle energy resolution on continuous air monitors for thin deposits of radioactive aerosols, *Health Physics*, 83(6):884–891, 2002.

Lee, K.W. and Ramamurthi, M., Filter collection, in: Willeke, K. and Baron, P., Eds., *Aerosol Measurement, Principles, Techniques, and Applications*, Van Nostrand Reinhold, New York, NY, 1993.

L'Hoir, A., Study of the asymmetrical response of silicon surface barrier detectors to MeV light ions. Application to the precise analysis of light ions energy, I. Helium ions. *Nuclear Instruments and Methods in Physics Research*, 233:336–345, 1984.

McFarland, A., Ortiz, C., and Rodgers, J., Performance evaluation of continuous air monitor (CAM) sampling heads. *Health Physics*, 58(3):275–281, 1990.

McFarland, A., Rodgers, J., Ortiz, C., and Moore, M. A continuous sampler with background suppression for monitoring alpha-emitting aerosol particles. *Health Physics*, 62(1):400–406, 1992.

McFarland, A., Rodgers, J., Ortiz, C., and Nelson, D., Elbow Mass Flow Meter. U.S. Patent 5337603, August 16, 1994.
Moore, M., McFarland, A., and Rodgers, J., Factors that affect alpha particle detection in continuous air monitor applications. *Health Physics*, 65:69–81, 1993.
Parulian, A., McFarland, A., and Rodgers, J., A constant-flow air sampler for workplace environments. *Health Physics*, 71:870–878, 1996.
Press, W., Flannery, B., Teukolsky, S., and Vetterling, W., Modeling of data, in: *Numerical Recipes, The Art of Scientific Computing*, Chapter 14, Cambridge University Press, New York, NY, 1986.
Rodgers, J., *Comparison of Background Compensation Algorithms in Environmental Monitoring: Data from LANL-WIPP Investigations of ECAM Performance Underground at WIPP*. Presentation, Department of Energy Air Monitoring Users Group Meeting, Pojoaque, NM, Los Alamos National Laboratory Report LA-UR-03-6693, July 28, 1994.
Rodgers, J., Environmental Continuous Air Monitor Inlet with Combined Preseparator and Virtual Impactor, U.S. Patent 732477, June 19, 2007.
Rodgers, J. and Kenney, J., *A Critical Assessment of Continuous Air Monitoring Systems at the Waste Isolation Pilot Plant*, Environmental Evaluation Group Report EEG-38, March, 1988.
Rodgers, J., Whicker, J., and Voss, J., Comparison of continuous air monitor utilization, *Radiation Protection Management*, 15(3):56–64, May/June 1998.
U.S. Department Of Energy (USDOE), *Code of Federal Regulations*, Title 10, Part 835, Occupational Radiation Protection, Subpart E, Monitoring of Individuals and Areas, 72FR31904, June 8, 2007a.
U.S. Department of Energy (USDOE), Section 10, *Air Monitoring Guide* DOE G441.1-1B, 2007b.
Vincent, J.H., Stevens, D.C., Mark, D., Marshall, M., and Smith, T.A., On the aspiration characteristics of large-diameter thin-walled aerosol sampling probes at yaw orientations with respect to the wind. *Journal of Aerosol Science*, 17:211–224, 1986.
Wasiolek, P., Whicker, J., Gong, H., and Rodgers, J., Room airflow studies using sonic anemometry. *Indoor Air*, 9:125–133, 1999.
Whicker, J., Relationship of air sampling measurements to internal dose: A review. Presentation, *37th Annual Midyear Meeting of the Health Physics Society,* "Air Monitoring and Internal Dosimetry," 2/8/2004–2/11/2004, Augusta, GA.
Whicker, J., Rodgers, J., Fairchild, C., Scripsick, R., and Lopez, R., Evaluation of continuous air monitor placement in a plutonium facility. *Health Physics*, 72:734–743, 1997.
Whicker, J., Rodgers, J., and Lopez, R. Assessment of the need for transport tubes when continuously monitoring for radioactive aerosols. *Health Physics*, 77:322–327, 1999.
Yule, T.J., An on-line monitor for alpha-emitting radionuclides, *IEEE Transactions on Nuclear Science*, NS-25:762, 1978.

15 Principles of Sampling Airborne Radioactivity from Stacks

John Glissmeyer

CONTENTS

Introduction	315
Probe Placement	316
General	316
Mixing	316
Mixing Criteria	317
Mixing Demonstration Method	318
Configurations Tested	318
Use of Previously Tested Configurations or Scale Models	324
Generic Mixing Tests	324
Sampling Train	327
Nozzles	327
Alignment	328
Isokinetic Concept	328
Transmission Ratio	329
Modeling Aspiration Efficiency	331
Transport in the Sampling System	332
Straight Tubes	333
Bends	335
Sample Collection	335
Sample Controls	336
Stack Flow Measurement	336
Sample Flow Measurement	337
Leak Tightness	337
Materials	338
References	338

INTRODUCTION

There are many instances where air samples must be collected from process or effluent streams. In many ways, the air sampling process is similar to sampling from calm or outdoor air. The difference here is that the air is moving along a confined path in one direction. The particular challenge is where and how to extract the sample from the moving air. The remainder of the sampling process is generally covered elsewhere in this book.

The overall success or performance of the sampling system is largely a function of the characteristics of the radionuclide-bearing constituents, the path through the sampling train, and the efficiency of the sample collector or detector. In order of greatest to lowest sampling difficulty are nonreactive gases, vapors, reactive vapors, and particles. The sample transmission for reactive vapors is usually determined by the reactivity of the particular effluent molecules. The sample transmission for particles is determined by their aerodynamic size. Most of this chapter focuses on particle sampling. Exceptions to the concepts for the other constituent types will be pointed out throughout the chapter.

PROBE PLACEMENT

GENERAL

Stack sampling principles are applicable to extracting gaseous samples from ducts, pipes, and discharge stacks or vents. The air sampled is typically extracted from the moving airstream with a probe of some type. Sampling probes and the sampling system components should be located where they can be accessible for inspection and maintenance and where these activities can be safely performed within acceptable environmental conditions of temperature, weather exposure, and dose rates. Inspection and test ports of adequate size should be located just upstream of the probe to facilitate any inspection and testing activities. Figure 15.1 shows a typical arrangement of probe, test or inspection ports, and an optional flow sensor. The inspection ports should be located somewhat upstream of the sampling nozzle so that the nozzle edges can be observed for particle buildup and alignment. The inspection ports can double as test ports when testing for satisfactory sampling conditions and calibration of the optional flow sensor. The flow sensor can be located coplanar with the sampling nozzle as long as it is offset from the nozzle. The nozzle should be located within the center half of the cross-sectional area of the stack.

MIXING

The sampling probe should be located where the potential contaminants of concern are well mixed in the bulk airflow. Therefore, the sample stream will contain the contaminants in the

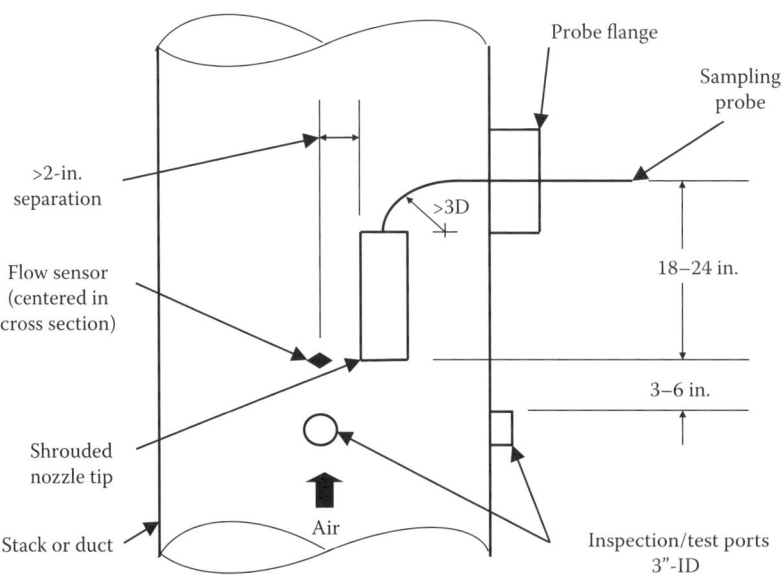

FIGURE 15.1 Arrangement of sampling probe, inspection ports, and flow sensor.

Principles of Sampling Airborne Radioactivity from Stacks

same concentration as in the bulk flow. This is the case regardless of the number of nozzles on the sampling probe. Well-mixed bulk airflow enables the use of single-nozzle probes.

The standard for air sampling from stacks (ANSI/HPS N13.1-1999 *Sampling and monitoring releases of airborne radioactive substances from the stacks and ducts of nuclear facilities*) provides criteria to define mixing that must be demonstrated with a series of objective tests.

It is generally assumed that mixing must occur between the sampling probe and the nearest upstream fan or junction of streams feeding the stack, whichever is closest to the probe. The worst-case situation is where a contaminated airstream joins the effluent airstream along one side of a straight section of stack or duct.

MIXING CRITERIA

The ANSI/HPS N13.1-1999 standard for air sampling provides numerical criteria and methods for determining if satisfactory mixing has occurred upstream of the sampling probe. They are summarized as follows.

1. Uniform air velocity—It is important that the gas momentum across the stack cross section where the sample is extracted be well mixed or uniform. Consequently, the velocity is measured at several points in the stack at the location of the sampling nozzle. The uniformity is expressed as the variability of the measurements about the mean. This is expressed using the relative coefficient of variation (COV), which is the standard deviation divided by the mean and expressed as a percentage. As the COV value becomes lower, the velocity becomes more uniform. The acceptance criterion is that the COV of the air velocity must be ≤20% across the center two-thirds of the area of the stack.
2. Angular flow—Sampling nozzles are usually aligned with the axis of the stack. If the air travels up the stack in cyclonic fashion, the air velocity vector approaching the nozzle could be misaligned with the sampling nozzles enough to impair the extraction of particles. Consequently, the flow angle is measured in the stack at the elevation of the sampling nozzle. The average air velocity angle must not deviate from the axis of the sampling nozzle by more than 20°.
3. Uniform concentration of tracer gases—A uniform contaminant concentration in the sampling plane enables the extraction of samples that represent the true concentration. This is demonstrated using a tracer gas to represent gaseous effluents. The acceptance criteria are that (a) the COV of the measured tracer gas concentration is ≤20% across the center two-thirds of the sampling plane* and (b) at no point in the sampling plane does the concentration vary from the mean by >30%.
4. Uniform concentration of tracer particles—Uniformity in contaminant concentration at the sampling elevation is further demonstrated using tracer particles large enough to exhibit inertial effects. Particles of 10-μm aerodynamic diameter (AD)† are used by default unless it is known that larger particles are present in the airstream. The acceptance criterion is that the COV of particle concentration is ≤20% across the center two-thirds of the sampling plane.

---

* Cross section of the duct or stack where the sampling nozzle inlet is located.
† Aerodynamic diameter—The diameter of a unit density sphere with the same settling velocity in still air as the actual particle. For spherical particles, the AD is approximately the actual particle diameter multiplied by the square root of the particle's density $AD = D_p \sqrt{\rho_p}$, where D_p is the particle diameter and ρ_p is the particle density.

Mixing Demonstration Method

There are three alternative approaches to demonstrate conformance with the mixing criteria for probe locations.

- Perform the testing on the constructed stack.
- Copy the geometry of a stack that has already been tested and then perform the validation steps on the constructed stack.
- Perform the compliance tests on a scale model of the stack, followed later by the validation test on the constructed stack.

Four types of tests are performed for a single operating configuration of the stack

Velocity uniformity	Minimum two repeat runs
Flow angle	Minimum two repeat runs
Gas tracer mixing	Minimum six repeat runs per effluent stream
Particle tracer mixing	Minimum one run per effluent stream and one repeat

Significant changes to a stack compared to the originally tested operating configuration will require additional test runs. Flow controls may also need to be calibrated periodically (usually annually).

The preparations for testing include the following:

- Quality assurance plan appropriate for the application.
- Test plan.
- Procedures.
- Tracer generating equipment.
- Tracer injection ports in each stream (or just upstream of the stack for a worst-case simulation). The particle or gas tracer is injected at the centerpoint of the stream. The gas tracer is also injected at four other points in the cross section within 20% of a hydraulic diameter* (HD) of the walls (or corners in a rectangular duct).
- Instrumentation calibrated as appropriate for the individual tests,
- Test ports near the elevation of the air sampling probe as illustrated in Figure 15.1.
- Access platforms to the tracer injection and test ports of a size large enough for the work.

The particle tracer test on the constructed stack requires the most preparation for access. In the direction of the transects through the two test ports, a platform would be needed to allow the insertion and positioning of a rigid probe with a particle counter attached to the end of the probe outside the stack.

Although favorable outcomes are expected from the onsite tests, there is some risk if the testing provides negative results. Retrofitting can be expensive, although the most obvious recourse might be to move the sampling probe up the stack to achieve more appropriate mixing conditions.

Configurations Tested

Most of the above testing would be eliminated if the stack in question has the same geometry as one already tested. Rodgers et al. (1996) described one of the first demonstrations of mixing on

---

* The hydraulic diameter is equal to the actual diameter of a circular duct. It is equal to $(2HW)/(H + W)$ of a rectangular duct where H and W are the height and width of the duct cross section.

an actual stack using the modern methods adopted in the ANSI/HPS-1999 standard. Following that, most such tests have been performed at U.S. Department of Energy facilities and are documented in reports. Some of these configurations are described here and the data are contained in the referenced reports. Most of the configurations shown were tested as satisfactory according to the criteria in the "Mixing Criteria" section above. These tests should expedite the use of other geometrically similar installations, subject to the similarity constraints given below in the section "Modeling."

Figures 15.2 through 15.4 show some variations in the configuration where fans discharge directly into a vertical stack at a 45° angle. Tracer injections were at or near the fan discharges. Tests were performed on the actual stacks or full-scale models as in the case of Figure 15.4. In Figure 15.2, there are dampers between the injection ports and the stack. There were no dampers in the configurations of Figures 15.3 or 15.4. Dampers close to the stack and the injection point can affect the mixing test results. In Figure 15.4, there are two sampling probes in series.

Figure 15.5 shows a vertical stack where the airstreams from two fans meet just before a 45° entry to the stack. This was tested on a scale model and the results were validated on the actual stack.

Figure 15.6 shows an unusual arrangement where several fans discharge into separate plenums, which then join prior to a 90° bend at the stack base. The turbulence from this junction promoted mixing. This is an example of a test performed with a scale model with the results validated on the actual stack.

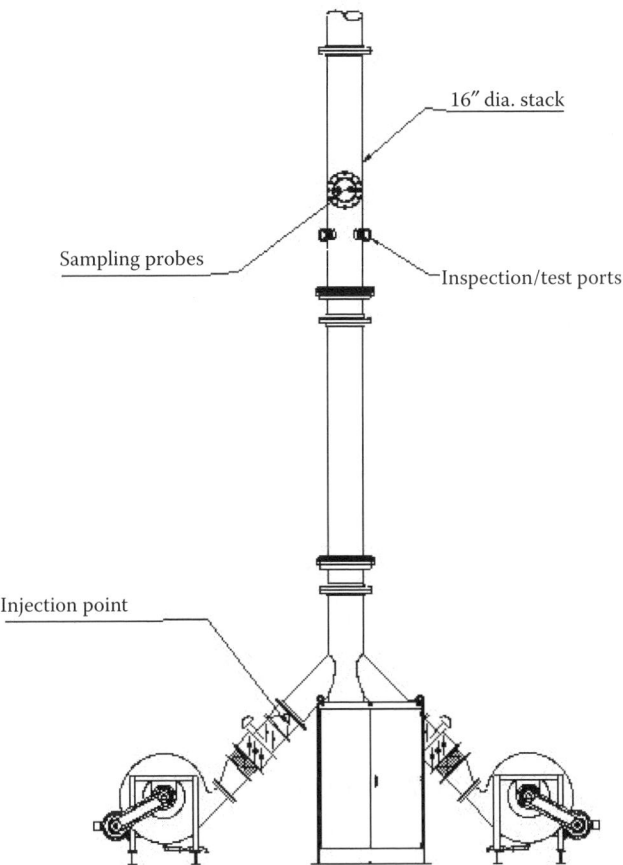

FIGURE 15.2 A stack with two fans with dampers discharging at an upward angle. (Adapted from Glissmeyer, J.A. and Maughan, A.D. *Qualification Tests for the Air Sampling System at the 296-Z-7 Stack*. PNNL-13687, Pacific Northwest National Laboratory, Richland, WA, 2001b.)

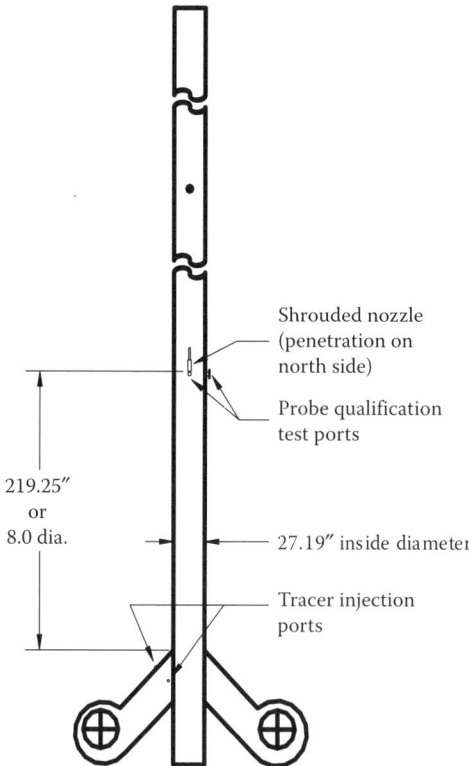

FIGURE 15.3 A stack with two fans without dampers discharging at an upward angle. (Adapted from Glissmeyer, J.A. and Maughan, A.D., 1999, *Project W420 Air Sampler Probe Placement Qualification Tests for Four 6-inch Diameter Stacks: 296-A-25, 296-B-28, 296-S-22, and 296-T-18*. PNNL-12016, Pacific Northwest National Laboratory, Richland, WA.)

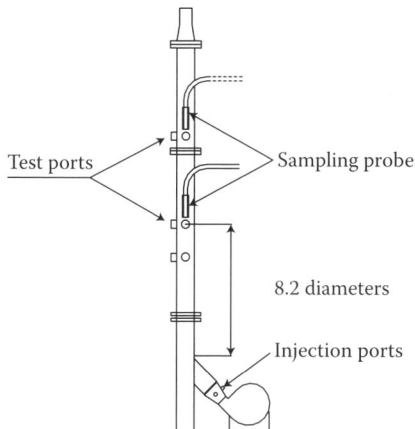

FIGURE 15.4 A stack with a single fan discharging at an upward angle. (Adapted from Glissmeyer, J.A. and Maughan, A.D., 1999, *Project W420 Air Sampler Probe Placement Qualification Tests for Four 6-inch Diameter Stacks: 296-A-25, 296-B-28, 296-S-22, and 296-T-18*. PNNL-12016, Pacific Northwest National Laboratory, Richland, WA.)

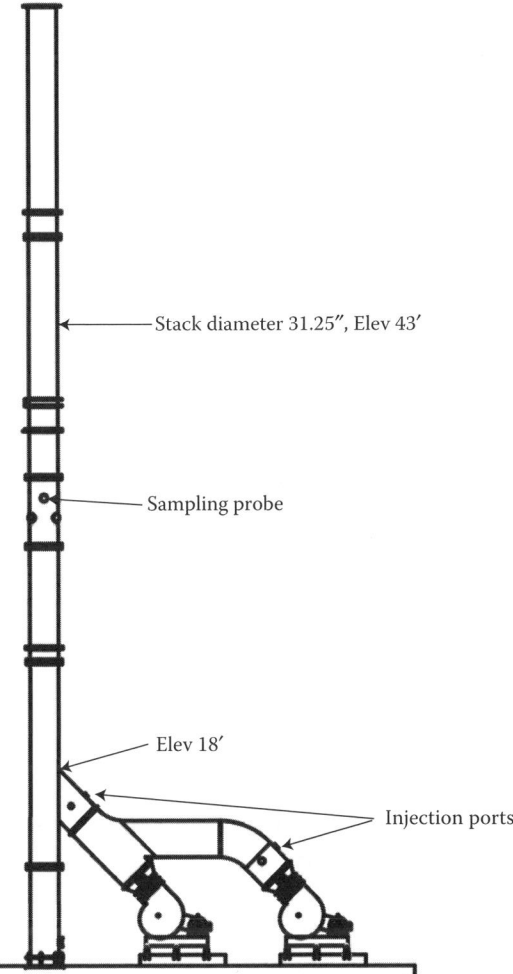

FIGURE 15.5 A stack configuration with two fans discharging via a common manifold. (Adapted from Glissmeyer, J.A. and Maughan, A.D., 1998b, *Airborne Effluent Monitoring System Certification for New B-Plant Ventilation Exhaust Stack*. PNNL-12017, Pacific Northwest National Laboratory, Richland, WA.)

Figure 15.7 is an example of four fans discharging into a common plenum. The air then passes through two bends before reaching the stack (Ballinger et al., 2004). These scale model results were compared with those from a computational fluid dynamics model by Barnett et al. (2003).

Figure 15.8 is an example of an instance where the test results did not meet the mixing criteria (Ballinger et al., 2004). The stack included an axial vane fan in the horizontal duct run. Tracer injection was just downstream of the fan and the measurement and sampling ports were located just below the scaffolding shown in the figure. The gas tracer test failed to meet the criterion; hence particle tracer tests were not performed. The length of straight duct was insufficient for mixing and there were no airflow direction changes (such as bends) to create the large-scale turbulence that promotes mixing.

Figure 15.9 depicts a vertical stack that was tested as a scale model. The model was assembled in the horizontal direction for convenience in testing. This reorientation does not affect the results. The stack flow comes from four separate ducts discharging directly into the bottom of the stack. Were it not for a static mixer (see below for a description), satisfactory mixing would not have occurred in the stack.

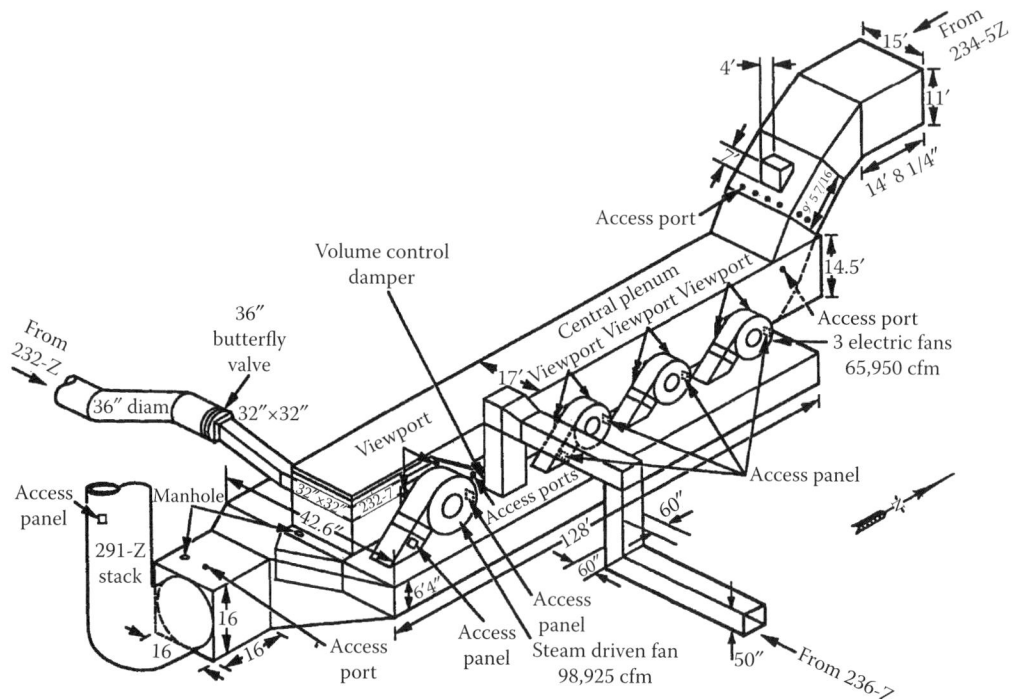

FIGURE 15.6 Downdraft fans feeding parallel plenums which converge at a bend. (Adapted from Glissmeyer, J.A., Maughan, A.D., and Jarvis, T.T., *Qualification Tests for the New Air Sampling System at the 296-Z-1 Stack*. PNNL-14057, Pacific Northwest National Laboratory, Richland, WA, 2002.)

FIGURE 15.7 Four fans manifolded together feeding a stack via two bends. (Adapted from Ballinger, M.Y., et al. *Health Phys.* 86(4):406–415, 2004.)

Principles of Sampling Airborne Radioactivity from Stacks 323

FIGURE 15.8 A horizontal stack powdered by an in-line axial fan. (Adapted from Ballinger, M.Y., et al. *Health Phys.* 86(4):406–415, 2004.)

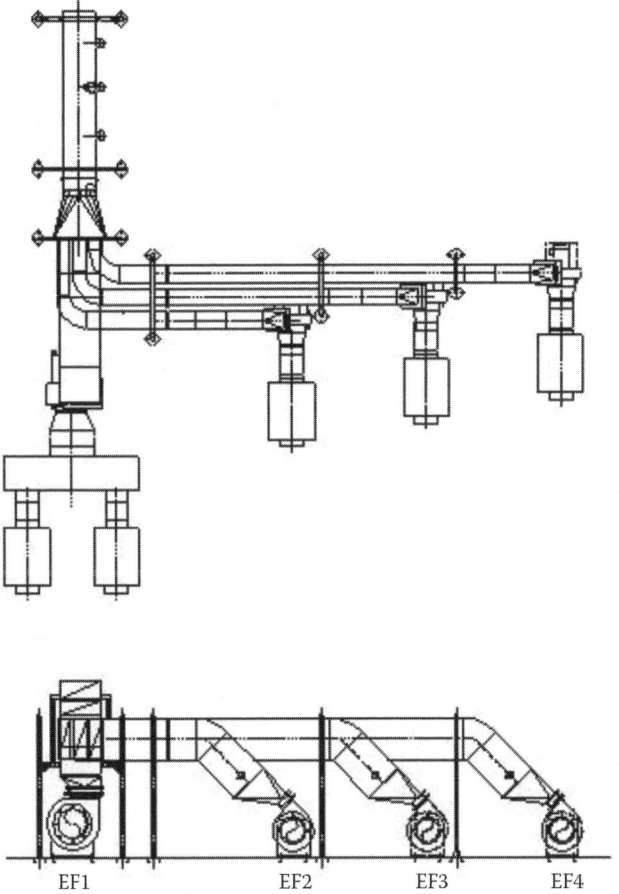

FIGURE 15.9 Plan and end views of a scale model where four effluent streams combine at the base of the stack. (Adapted from Glissmeyer, J.A. *Assessment of the 296-S-21 Stack Sampling Probe Location*. PNNL-16014, Pacific Northwest National Laboratory, Richland, WA, 2006.)

USE OF PREVIOUSLY TESTED CONFIGURATIONS OR SCALE MODELS

As described in some of the above examples, data from a previously tested stack or from a scale model can be used to demonstrate conformance of a planned air monitoring system. To apply data from a previously tested system or a scale model the following additional criteria must be met:

- The surrogate and its sampling location must be geometrically similar to the actual stack.
- The product of the surrogate's mean air velocity times the HD must be within a factor of six of the product of the same parameters for the actual stack.
- The Reynolds number* for the prototype and surrogate stacks must be >10,000.

Furthermore, the surrogate results are considered applicable if, in subsequent tests at the sampling probe location on the full-scale stack,

- The velocity profile and flow angle in the constructed stack meets the ANSI N13.1-1999 criteria
- The velocity uniformity COV for the constructed and surrogate stacks agrees within 5% COV.

GENERIC MIXING TESTS

If one is at the point of designing a new stack and air sampling system, it is recommended that design features that enhance mixing be included downstream of the fans, effluent control equipment, and all additions to the bulk flow. Elements that enhance mixing do so by causing large-scale turbulence in the stream and may include one or more 90° turns in the duct, converging airstreams, a mixing box, or a commercial static mixer. Turning vanes and flow straigtheners have the undesirable effect of inhibiting mixing for a considerable distance.

Mixing is promoted by large-scale turbulence such as eddies occurring at a change in the direction of airflow. McFarland et al. (1999a) investigated the types of flow direction changes in round ducts that promote mixing. Different types of static mixers were also tested that included vanes, duct perimeter rings that partly obstruct the flow, and wedge-shaped obstructions. Static mixers can provide mixing if a change in the flow direction is not possible, albeit at some cost in pressure drop. A typical static mixer is shown in Figure 15.10. It is an arrangement of concentric sets of fixed vanes. The vanes in the two sets direct the air passing through them in opposite directions, causing significant turbulence which accomplishes the mixing.

McFarland et al. (1999b) developed a "Generic Mixer." This unit included a configuration tested for mixing and placement of the sampling nozzle. Seo et al. (2006) performed similar tests for square and rectangular ducts. Similar tests were performed for three configurations of generic mixing boxes by Han et al. (2007).

Figure 15.11 illustrates some of the generic configurations tested by McFarland et al. (1999a,b), Seo et al. (2006), and Han et al. (2007). Table 15.1 summarizes some of the key results. The number of dimensionless duct diameters, that is, length/hydraulic diameter (L/HD) needed to achieve a gas tracer COV of 10% following the changes in the flow direction or mixers is listed. A COV value less than 10% was chosen so that the results would be conservative. Also, the L/HD length is generally that of

---

* The Reynolds number is dimensionless and is a general measurement of turbulence in a channel or pipe. It is calculated as $Re = \rho_{air} DU/\mu_{air}$, where ρ_{air} is the air density, D is the tube diameter, U is the mean velocity through the tube, μ_{air} is the air viscosity. Where the value is greater than 2100, the flow is said to be turbulent. For values <1000, the flow is laminar. The transition region lies between.

Principles of Sampling Airborne Radioactivity from Stacks 325

FIGURE 15.10 A commercial static mixer.

the worst-case tracer test result. Han et al. (2007) tested with both gas and particle tracers, whereas the other tests used only gas tracers. The velocity COV was about equal to, or less than, the tracer COV.

The mixing in a system comes at the expense of increased energy to move the air. Table 15.1 lists pressure coefficients, C_p, for the various duct elements for comparison. The larger the C_p, the greater the energy cost for accomplishing the increased mixing.

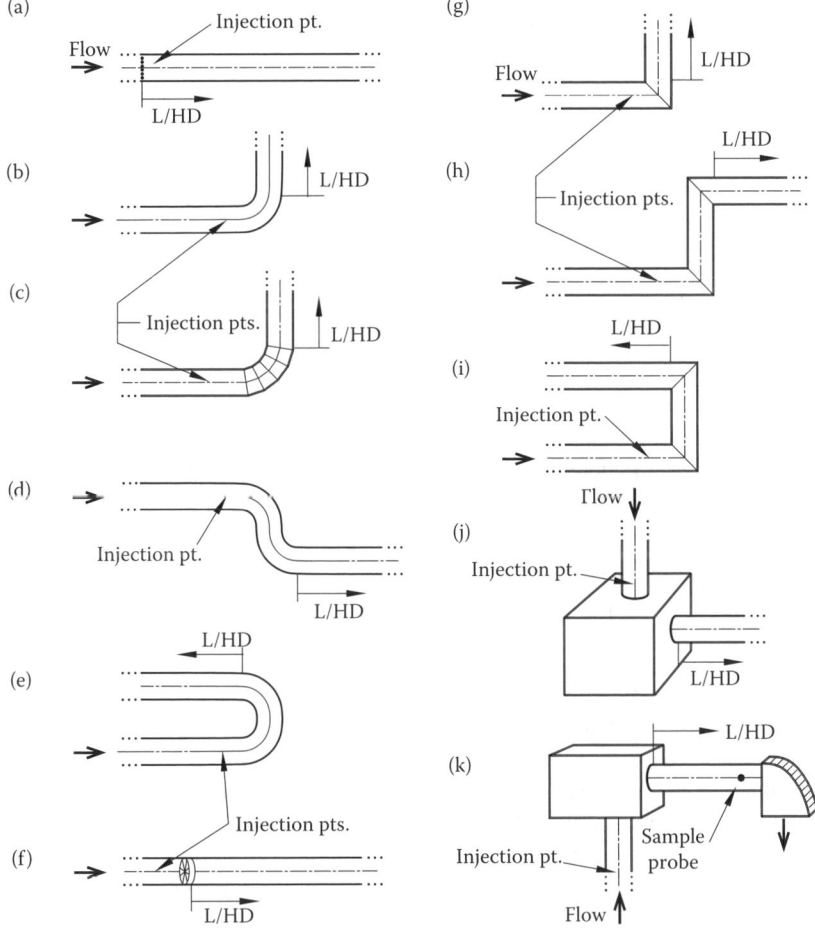

FIGURE 15.11 Tested mixing elements in ducts and stacks.

TABLE 15.1
Summary of Key Results for the Generic Configurations Tested

Duct Feature	Approximate L/HD to Achieve 10% COV for Tracer Gas[a]	Pressure Coefficient[b]	Reference
Straight round duct, Figure 15.11a	≈40	—	Seo et al. (2006)
90° bend in the round duct, Figure 15.11b	≈11	0.6	McFarland et al. (1999a)
90° five-gore bend in the round duct, Figure 15.11c	≈11	0.6	McFarland et al. (1999a)
Two 90° bends in S-configuration, in the round duct, ≈Figure 15.11d	7	1.7	McFarland et al. (1999a)
Two 90° bends in U-configuration, in the round duct, Figure 15.11e	≈12	1.0	McFarland et al. (1999a)
Straight duct with commercial mixers, Figure 15.11f	4	2.1	McFarland et al. (1999a)
Generic mixer box with no internal elements, Figure 15.11k	4	—	McFarland et al. (1999a)
Generic mixer with internal elements, Figure 15.11k	2.5	4.5	McFarland et al. (1999b)
Straight square duct, Figure 15.11a	≈40	—	Seo et al. (2006)
90° mitered bend, in the square duct, Figure 15.11g	≈13	0.84	Seo et al. (2006)
Two 90° mitered bends in S-configuration, in the square duct, Figure 15.11h	6	2.20	Seo et al. (2006)
Two 90° mitered bends in U-configuration, in the square duct, Figure 15.11i	≈13	1.26	Seo et al. (2006)
Generic mixer box (LH-GTP configuration), Figure 15.11j	4	0.65	Han et al. (2007)

Source: Adapted from Gast Manufacturing Inc., *Vacuum Pressure Handbook* available at www.gastmfg.com/pdf/vacpresshadbk.pdf

[a] 10% was chosen to ensure that the criteria of 20% COV would be achieved if users fabricate parts of similar design.

[b] The pressure coefficient is defined as $C_p = \dfrac{\Delta P}{(1/2)\rho_{air}U^2}$, where ΔP is the pressure drop across the mixing element, ρ_{air} is the air density, and U is the mean velocity through the tube.

In the round duct with a single 90° bend, the 10% COV was achieved in about 11 duct diameters, and with two bends in an "S" arrangement, only seven duct diameters were needed. With two bends in a "U" arrangement, 12 diameters were needed. It was estimated that about 40 diameters would be needed in the straight duct. With a static mixer like that shown in Figure 15.10, three duct diameters were needed for 10% COV.

Seo et al. (2006) showed that mixing results for a square duct were very similar to the round duct. Such was not the case for a rectangular duct. At 9.5 L/HD downstream of a single 90° bend, the gas tracer COV for the rectangular duct (with an aspect ratio of 3:1) was 62% compared to 28% for the square duct. For velocity, the rectangular duct had a COV of 29% versus 6.4% for the square duct.

Han et al. (2007) demonstrated that mixer boxes achieved 10% COV for velocity, gas and particle tracers in a short 4 L/HD, provided a 90° turn in the duct is acceptable. The 4 L/HD compares very favorably to the 11 L/HD needed with a plain 90° bend. The pressure coefficient with the mixer boxes was similar to the plain bends and much lower than with the commercial mixers.

Flow straighteners* are sometimes used to even out the air velocity in the duct cross section. Turning vanes are often used in duct elbows or stack entrances to minimize large-scale turbulence and the resultant pressure drop as the air makes the turn. Flow straighteners and turning vanes inhibit

---

* A device that provides uniform and parallel (laminar) flow of air. It consists of a disc-shaped arrangement or "honeycomb" of tightly spaced flow paths. The honeycomb material can be metal or plastic.

Principles of Sampling Airborne Radioactivity from Stacks

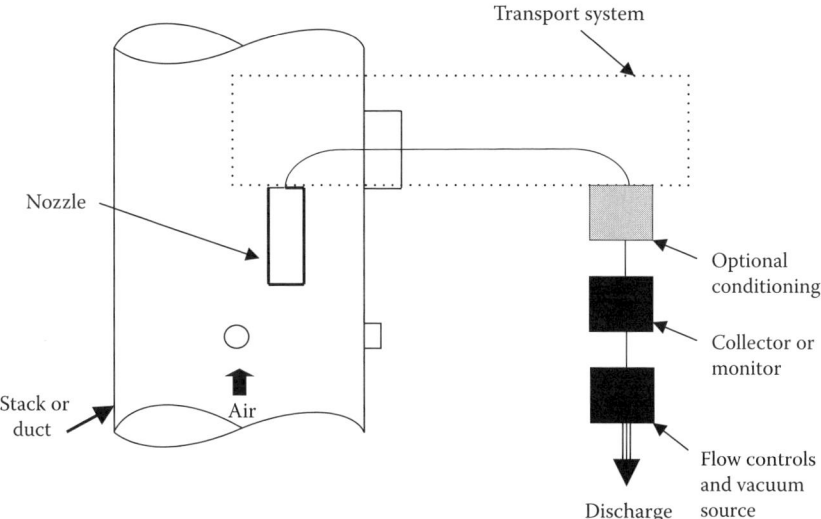

FIGURE 15.12 An idealized sampling system.

mixing. Where such features are used, the mixing must be complete prior to or after the straightener and turning vane and upstream of the sampling probe. Seo et al. (2006) showed that for a mitered 90° bend in a square duct, the COV for tracer gas increased from 24% to 86% (at L/HD = 9.5) when turning vanes were added inside the bend.

SAMPLING TRAIN

Figure 15.12 shows the idealized components of a stack sampling train. These include

1. Nozzle(s).
2. Transport line made up of straight and bent sections of tubing, and sometimes expansions, contractions, splitters, and mixers.
3. An optional conditioning system, which is only needed in certain situations. This includes a dilutor, drier, concentrator, humidifier, and diffusion denuder.*
4. A sampler collector. This is typically a filter paper or continuous air monitor. An absorber or adsorber cartridges for certain gas and vapor constituents is also used as needed.
5. The sample flow controls. This includes the vacuum source, flow meters, flow controls, and the stack flow measurement system.

These components typically are arranged in series; however, there are many instances where the sampling system is made up of primary and secondary sampling trains. The primary train extracts and transports a bulk sample stream. One or more secondary trains obtain subsamples from the primary train. For example, separate secondary trains can obtain sample streams for a continuous air monitor, radioiodine sampling, noble gases, or tritium.

Nozzles

The first part of the sampling train is usually one or more nozzles mounted on a probe. The nozzles serve to separate the sample stream from the bulk flow. Because the bulk flow must be shown to be

---

* A device used to discriminate between gases and aerosols in a sampling system. Denuders are often constructed of multiple wetted parallel plates arranged inside a cylindrical shell.

well mixed, there is no need to have more than one nozzle. Single-nozzle probes are easier to fabricate and usually perform better at transmitting the extracted particles than the multinozzle variety. There has been no demonstration that multiple nozzles produce better results than a single one. For the same sample flowrate, a single nozzle will have a larger internal diameter than multiple nozzles. ANSI/HPS N13.1-1999 requires certain demonstrated performance for transmitting the sample constituents regardless of the type of probe and the number of nozzles.

For those who prefer multiple nozzles, the stack cross section is divided into the same number of equal areas as there are nozzles. There is no need to have more than six nozzles, and one should be located in the centroid of each of the equal areas. See for example, U.S. Environmental Protection Agency (EPA) Method 1 for guidance on arranging the equal areas (40CFR60, Appendix A, Method 1, 2000).

ALIGNMENT

The axis of the nozzles should parallel the axis of the airstream. This occurs if the airstream is not cyclonic at the point of sampling. The EPA provides a simple method for determining the mean flow angle, which must be less than 20° relative to the axis of the stack (40CFR60, Appendix A, Method 2, 2000).

ISOKINETIC CONCEPT

The concepts of isokinetic sampling and stack sampling have long been associated. It is a concept intended to explain the efficiency with which particles in the sampling stream are separated from the bulk flow. It is easiest to understand using the diagrams shown in Figure 15.13 that the sides of a nozzle of some diameter are aligned parallel to the direction of flow in a laminar stream.

The objective of isokinetic sampling is to achieve the same concentration of particles of all sizes inside the nozzle as there is in the free stream. In Figure 15.13a, the air velocity inside the nozzle

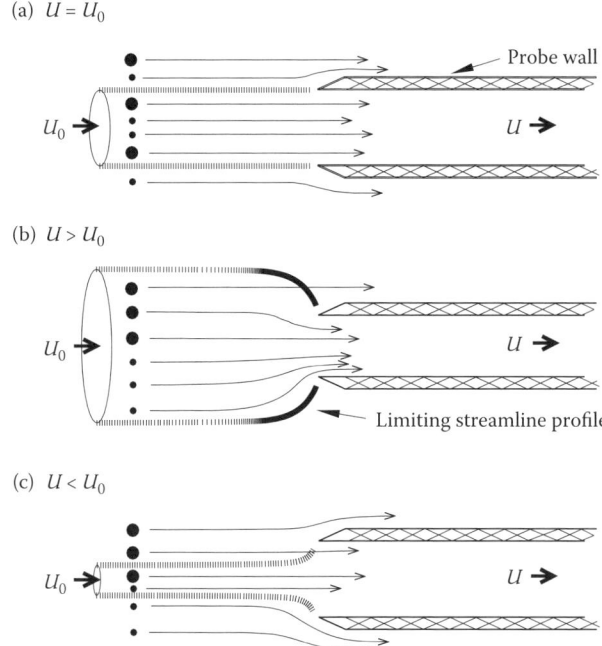

FIGURE 15.13 Illustration of (a) isokinetic, (b) super-isokinetic, and (c) sub-isokinetic sampling.

Principles of Sampling Airborne Radioactivity from Stacks

opening equals the free stream velocity, and this state is called "isokinetic." In this state, the airstream should enter the nozzle indicated in the figure and its limiting or critical diameter equals the inlet diameter of the nozzle. Depicted are particles of two sizes. The particle of smaller diameter is influenced entirely by drag forces and follows the airstream. The larger diameter is influenced in part by drag forces, but also has sufficient inertia to deviate from the airstream. In the isokinetic state, particles of both sizes that approach the nozzle inside the critical volume should enter the nozzle, because the airstream neither converges nor diverges at the nozzle.

In Figure 15.13b, the air velocity in the nozzle is greater than that of the free stream, and the critical volume approaching the nozzle is larger than the nozzle inlet. This state is called "superisokinetic." The smaller particles influenced mostly by drag follow the airstreams into the nozzle and all of them in the critical volume enter the nozzle. The larger particles near the edge of the critical volume may possess enough inertia to cross the airstream and not enter the nozzle. Consequently, not all of those particles that should enter the nozzle do, and the sample stream is somewhat depleted in these larger particles.

In Figure 15.13c, the air velocity in the nozzle is less than that of the free stream, and this state is called "subisokinetic." The critical volume approaching the nozzle has a smaller diameter than the nozzle. The smaller particles follow the streamlines because they are influenced mostly by drag. Those outside the critical volume but inside the projected nozzle diameter follow the streamlines and do not enter the nozzle. In the same region, the larger particles with inertia cross the streamlines and enter the nozzle. Thus, the sample stream is somewhat enriched in large particles relative to what would represent the particles in the critical volume.

In past practice, all that was necessary for ideal stack sampling was to achieve and maintain the isokinetic state. However, in real stack sampling situations, the airflow is very turbulent and there are really no smooth streamlines. Particles interact with the edge of the nozzle and the surfaces inside and outside the nozzle in nonideal ways. So the picture in Figure 15.13 is not fully accurate. Nonetheless, the concepts are in common usage and are useful tools to partly understand the dynamics of air sampling. In practice, the state of isokinesis is not sufficient in itself to ensure ideal sampling and is not necessary.

TRANSMISSION RATIO

The design and use of a sampling nozzle can have a significant impact on the quality of a sample. There are two basic factors by which a nozzle can produce a nonrepresentative sample: (1) operation in such a manner that the aspiration efficiency, A_e, is not unity and (2) losses on the internal walls of the nozzle, Wl. These two terms, that is, aspiration efficiency and wall losses, are defined as

$$A_e = \frac{c_i}{c_\infty} \tag{15.1}$$

and

$$Wl = \frac{c_i - c_e}{c_i}, \tag{15.2}$$

where c_i is the concentration at the nozzle inlet plane, c_∞ is the aerosol concentration in the free stream, and c_e is the concentration at the nozzle exit plane.

The effects of aspiration ratio and wall losses are manifested in the transmission ratio, T_r, which is the ratio of aerosol concentration at the nozzle exit plane to aerosol concentration in the free stream:

$$T_r = \frac{c_e}{c_\infty}. \tag{15.3}$$

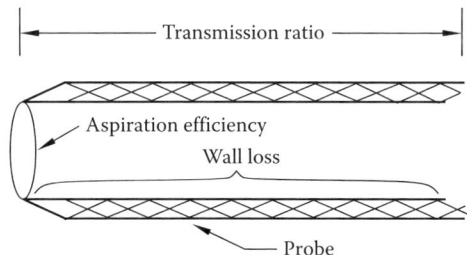

FIGURE 15.14 Illustration of a sampling nozzle and relationships of aspiration efficiency, wall loss, and transmission ratio.

The transmission ratio, aspiration ratio, and wall losses are related by

$$T_r = A_e(1 - Wl). \tag{15.4}$$

Figure 15.14 illustrates the parts of the nozzle pertaining to aspiration efficiency, wall loss, and transmission ratio. The transmission ratio encompasses these three parameters providing a measure of the amount of aerosol that actually penetrates from the free stream to the exit plane of the nozzle.

Glissmeyer and Ligotke (1995) illustrated the point that sampling isokinetically does not guarantee the best particle sample. They performed tests with two probes with shrouded nozzles (McFarland et al., 1989) and tapered-single nozzle probes at two airspeeds and four particle sizes. All of these probes consisted of a single nozzle, a 90° bend, and a filter holder as depicted in Figure 15.15. Figures 15.16 and 15.17 show the overall concentration ratios (concentration measured/concentration in free stream) measured at the collection filter.

The tapered probe (F) was operated isokinetically and the other (E) was similar in design, but with a larger nozzle, and was operated subisokinetically. The modern shrouded nozzle is not operated isokinetically, but is optimized to provide fairly level transmission ratios for a broad range of particle size, free stream velocity, and nozzle flowrate. Two commercially available shrouded probes, C and D, were operated at 57 lpm. (Probe C was optimized for a low velocity range of 2.5–8.5 m s^{-1} and Probe D was optimized for a range of 8–16 m s^{-1}). Note that the two shrouded probes had the most stable and highest concentration ratios at the velocity for which it was optimized. The isokinetic probe (F) has the lowest overall concentration ratio and the subisokinetic probe (E) a little higher.

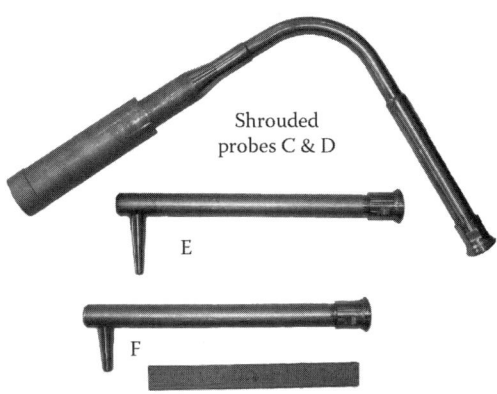

FIGURE 15.15 Tested sampling nozzles and probes.

Principles of Sampling Airborne Radioactivity from Stacks

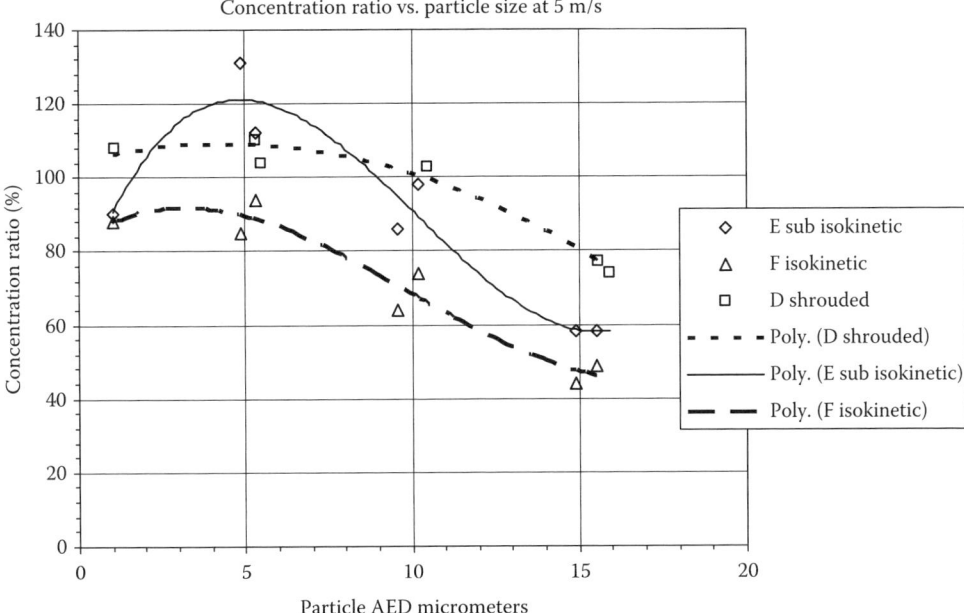

FIGURE 15.16 Comparison of data and polynomial fits of performance results for sampling nozzles at 5 m/s.

MODELING ASPIRATION EFFICIENCY

There have been many correlations of experimental data that allow the system designer to estimate the aspiration efficiency of sharp-edged nozzles. The model of Gong et al. (1995) has been used to describe the performance of the shrouded nozzle; however, experimental data from the nozzle manufacturers can provide the most useful information.

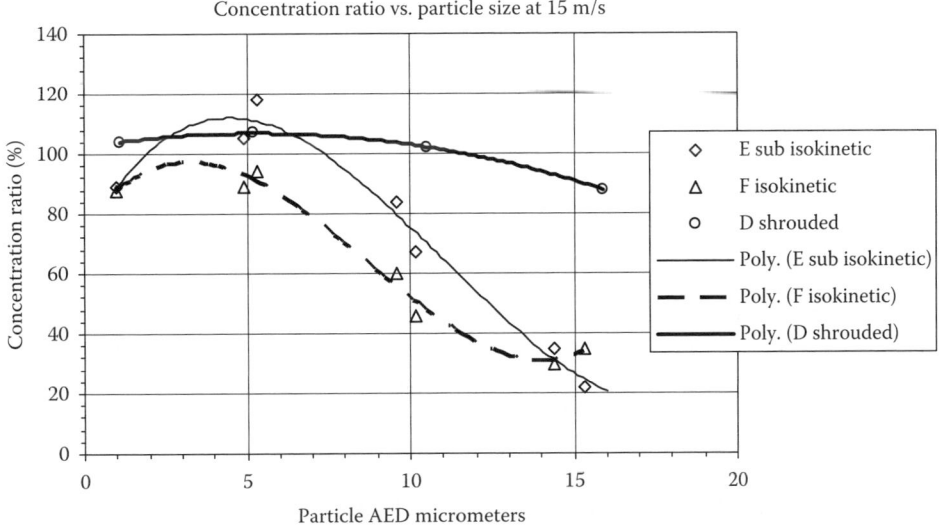

FIGURE 15.17 Comparison of data and polynomial fits of performance results for sampling nozzles at 15 m/s.

For sharp-edged nozzles, the model of Vincent (1989) provides a means to compute aspiration ratio accounting for yaw angle, θ, between the nozzle axis and the mean velocity vector of the airstream approaching the inlet of the nozzle:

$$A_e = 1 + \left[\frac{1}{1 + 1.05\text{Stk}[\cos\theta + 4(R\sin\theta)^{0.5}]}\right](R\cos\theta - 1), \quad (15.5)$$

where R is the velocity ratio U/U_0 and U is the mean velocity in the nozzle inlet. The dimensionless Stokes number is defined as

$$\text{Stk} = \frac{\rho_p D_p^2 U_0 C_c}{9\mu_{air} d}, \quad (15.6)$$

where ρ_p is the particle density, D_p is the particle diameter, U_0 is the particle velocity in the free stream, C_c is the Cunningham slip correction factor, d is the nozzle or tube diameter, and μ_{air} is the air viscosity.

TRANSPORT IN THE SAMPLING SYSTEM

As the sample stream passes through the elements that make up the sampling system, it is common that fractions of the airborne substances are lost to the internal walls. The fraction that passes through an element is said to be the penetrating fraction. The penetration, P_j, through the jth element is defined as

$$P_j = \frac{c_{e,j}}{c_{i,j}}, \quad (15.7)$$

where $c_{e,j}$ is the airborne substance concentration at the exit plane of the jth element and $c_{i,j}$ is the airborne substance concentration at the inlet plane of the jth element.

It is assumed that penetration through each element is independent of the penetration of the other elements, and that the overall penetration, P, is calculated as the product of all n individual element penetrations as

$$P = \sum_{j=1}^{n} P_j. \quad (15.8)$$

This is conceptually simple; however, of the possible airborne substances of interest to sample, only the penetration of particles through tubes has been empirically studied sufficiently to allow estimates of penetration. Particle penetration through tubes can be estimated through the use of hand calculations (Brockman, 1993) and software (McFarland et al., 2001) made available by the U.S. Nuclear Regulatory Commission.

A sample system that loses a significant fraction of the airborne substances of interest may provide samples of little value. Consequently, the system designer should conservatively select a minimum value for overall sample delivery, including the nozzles and transport system. An overall penetration of 50%, or greater, is recommended.

Usually, sample losses are assumed to be permanent. However, certain contaminants (iodine, large dry particles, ruthenium, and tritiated compounds) may deposit and later resuspend (become airborne again) at a later time and with some transformation of physical or chemical characteristics.

In the case of particles, other flow parameter factors being equal, the penetration is a strong function of particle size and declines with increasing AD as was illustrated in Figures 15.16 and 15.17.

Principles of Sampling Airborne Radioactivity from Stacks

The AD of particles in the effluent is usually unknown and subject to change depending on the status of abatement equipment and facility processes. Therefore, the recommended overall penetration of 50%, or greater, should be based on a large enough particle AD so that the actual sample penetration will vary within a smaller range as illustrated in Figures 15.16 and 15.17. The default design particle size, barring a measurement to the contrary, is recommended to be 10 µm AD.

Methods for estimating penetration of particles through straight tubes and bends are given in the following subsections. Contractions and expansions in the sampling line should be avoided. Particle loss in contraction fittings was discussed in Chapter 11.

STRAIGHT TUBES

The penetration of particles through a straight tube is calculated as

$$P = \exp\left(\frac{-\pi d v_e L}{Q}\right) \quad (15.9)$$

where v_e is the effective depositional velocity, Q is the flowrate through the tube, and L is the length of the straight section of the tubing.

The effective depositional velocity is the vector sum of gravitational settling (which is always downward), turbulent inertial deposition, and Brownian diffusion velocities (which are directed radially outward in a tube). It is assumed that the aerosol is well mixed across the cross section of the tube.

$$v_e = \frac{1}{2\pi}\int_0^{2\pi}(v_{bd} + v_{td} - v_{ge}\sin\alpha)\,d\alpha. \quad (15.10)$$

The effective deposition velocity v_e for an inclined tube was modeled by Anand et al. (1992) as the vector sum of gravity settling, turbulent diffusion and Brownian diffusion:
where α is the angular coordinate in the tube cross section relative to the horizontal line through the tube diameter, v_{bd} is the depositional velocity due Brownian diffusion, v_{td} is the deposition velocity due to turbulent inertial deposition, v_{ge} is the cross-stream component of gravitational settling velocity, and the integral is taken over the entire circumference of the tube (2π).

For a tube with its long axis inclined at an angle of ϕ relative to the vertical direction

$$v_{ge} = v_g \sin\phi, \quad (15.11)$$

where v_g is the gravitational sedimentation velocity. Equation 15.11 is subject to the constraint

$$(v_d - v_{ge}\sin\alpha) > 0, \quad (15.12)$$

where v_d is the depositional velocity due to the combined effects of thermal (Brownian) diffusion and turbulent inertial deposition.

If the constraint is not satisfied, then

$$v_e = 0 \quad (15.13)$$

The constraint is necessary because the model would otherwise predict that aerosol particles would enter the tube through its top surface (relative to the earth's surface).

The gravitational settling velocity used in Equation 15.11 is calculated using the gravitational acceleration term g as the simplified terminal settling velocity:

$$v_g = \frac{gC_c\rho_p U D_p}{18\mu_{air}}. \quad (15.14)$$

The model of Beal (1970) can be used to calculate the Brownian diffusion and turbulent deposition velocities. The Brownian diffusion velocity employed in Equation 15.10 is given by

$$v_{bd} = \sqrt{\frac{kT}{2\pi m}}, \quad (15.15)$$

where T is the temperature in Kelvin (K), k is the Boltzman constant, 1.38×10^{-16} (g cm)/(molecule °K s²), and m is the particle mass, g.

The turbulent deposition velocity is calculated as the friction velocity, v_*, times the average of the dimensionless velocity, v^+, of a particle toward the tube wall starting at its dimensionless stopping distance, S^+ and ending at half of the dimensionless diameter, d^+, where the particle would contact the wall.

$$v_{td} = \frac{v_*\left[v^+(d^+/2) + v^+(s^+)\right]}{4}. \quad (15.16)$$

The friction velocity v_* is calculated as a function of the mean stream velocity, U, and the Fanning friction factor, f:

$$v_* = \frac{U}{4}\sqrt{\frac{f}{2}}. \quad (15.17)$$

The Blasius estimate for the Fanning friction factor may be used:

$$f = \frac{0.316}{4\,\mathrm{Re}^{1/4}}. \quad (15.18)$$

The function for dimensionless velocity has two different forms depending on the dimensionless distance, y^+, from the tube wall:

$$v^+ = 0.05 y^+, \quad \text{where } 0 \leq y^+ \leq 10. \quad (15.19)$$

$$v^+ = 0.5 + 0.0125(y^+ - 10), \quad \text{where } 10 \leq y^+ \leq 30. \quad (15.20)$$

The dimensionless particle diameter is

$$d^+ = \frac{D_p v_*}{\nu}, \quad (15.21)$$

where ν is the kinematic viscosity of air (μ_{air}/ρ_{air}), that is, the viscosity of air divided by the density of air and D_p is as mentioned above, the particle diameter.

The stopping distance, S, is the distance a particle with an initial velocity will coast in a stagnant or slow-moving fluid while influenced by drag forces only:

$$S = \frac{0.05 U D_p^2 \rho_p \sqrt{f/2}}{\mu_{air}} + \frac{D_p}{2}, \quad (15.22)$$

Principles of Sampling Airborne Radioactivity from Stacks

where the particle initial radial velocity is assumed to be U and ρ_p is the particle density. The dimensionless stopping distance is defined as

$$S^+ = \frac{SU}{\nu}\sqrt{\frac{f}{2}}. \qquad (15.23)$$

where ν is the kinematic viscosity of air.

The dimensionless velocity terms are calculated by substituting both d^+ and S^+ in Equation 15.19 or 15.20 in turn. The turbulent deposition velocity is then calculated from Equation 15.16. Units used in equations of this section must be consistent.

BENDS

For 90° bends, Pui et al. (1987) found that the particle penetration could be correlated with the Stokes number as follows (when the ratio of bend radius to tube diameter is between 2.5 and 15 and for Reynolds number between 6000 and 10,000):

$$P = 10^{-0.963 STK} \qquad (15.24)$$

For 45° bends, the constant in the exponent is −0.482.

Most well-designed air sampling systems should be expected to deliver at least 50% of 10 μm AD particles. Because particle penetration decreases as particle size increases, 50% penetration of 10 μm AD should ensure even better delivery of typical aerosol in filtered exhaust systems and in the working environment. The equations for the transport of particles through the sampling system tubing are generally based on experiments using liquid aerosols for conservatism. Once a particle is deposited on a surface, it is assumed to remain on the surface. McFarland et al. (1991, 1997) provides a comparison of experimental measurements to the mathematical models.

SAMPLE COLLECTION

Particle samples are usually collected on filter paper in special leak tight holders or in continuous air monitors. It is a common misconception that filters work like sieves in that filters with a certain pore size will only collect particles larger than that pore size. In reality, filters with advertised mean pore size of 5 μm can provide excellent collection for particles of all sizes larger and smaller while providing comparatively high flowrate per unit area. This is because filters are porous structures that collect particles with several physical processes occurring simultaneously—direct interception, Brownian diffusion, electrical attraction, and gravitational sedimentation.

Filters can be made of several materials, for example, glass, quartz, mineral, plastic, and cellulose fibers. They are also made of silver membrane, sintered metal particles and polycarbonate sheets with microscopic holes. Selection of filters should take account of the planned analytical method, durability, composition, particle collection efficiency, and flowrate per area. If filters are going to be analyzed by direct counting, then a filter material with low self-absorption would be optimal for alpha spectroscopy. Detailed information on filter media is available in ANSI/HPS-N13.1-1999, Liu et al. (1983), and Lippmann (2001).

Vapor and gas samples are collected with a variety of special media, evacuated containers, and flow-through chambers. Vapor or gas sample collection media and devices are specialized by radionuclide or chemical form. Some of the common ones are listed in Table 15.2.

More information on gas and vapor sample collection is available in Brown and Monteith (2001). While that information is not specific for radionuclide collection, the collection of radionuclides as gases and vapors is typically based on the described technology.

TABLE 15.2
Vapor and Gas Sample Collectors

Vapor Species	Collection Medium
Organic radioiodine	Potassium iodide or triethylene-diamine-treated activated carbon
Elemental radioiodine	Plain or cadmium iodide treated activated carbon
Hypoiodous acid (hypothesized)	4-Iodophenol-treated alumina
Tritium oxide	Silica gel, molecular sieve, ethylene glycol bubbler, and condensers
Elemental tritium	Palladium or other catalysts to transform to oxide for collection as the oxide
Tritium in organic compounds	Platinum or aluminum oxide catalyst in the combustion chamber for oxidizing to oxide
Noble gas	Silver zeolite flowing or evacuated chambers

The desirable qualities of filter holders are discussed in Chapter 11. In-leakage and bypass leakage can cause serious problems in sample accuracy and analysis. Filter holders should be assessed for these types of leakage. Gaskets and sealing surfaces should be inspected regularly for deterioration. Proper installation of the filter may need to be checked with each filter change by visual inspection or an in-place leak test. These same cautions should be observed for the vapor and gas sample collectors.

SAMPLE CONTROLS

Estimating emissions from a stack requires knowing the fraction of the stack flow in the sample stream. This further requires knowledge of the stack flowrate and the sample flowrate, the ratio of these two being the sample fraction. For the ratio to be meaningful, both flowrates must be measured using the same units of gas density. That is, the volume units are in terms of consistent pressure, temperature and molecular weight. For stacks with potential for high-risk emissions, the sample ratio should be controlled. The accuracy of the stack and sample flowrates will determine the quality of the emissions estimate.

STACK FLOW MEASUREMENT

The stack flow can be estimated or measured by a variety of means:

- Fan capacity
- Periodic manual measurements using standard methods, for example, those from the U.S. EPA in 40 CFR 60, Appendix A, Methods 1, 1A, 2 and 2C, (40CFR60, Appendix A, Method 1A, 2000 and Method 2C, 2000)
- Continuous measurements

For stacks with significant flow variation during a year or with a risk of significant emissions, the stack flow should be monitored continuously or manually at intervals less than the expected frequency of the flow variation.

Continuous flow monitoring can be performed with sensors mounted in the effluent air flow stream. Typical sensors are based on the principles of thermal anemometry, differential pressure, or acoustic signal transmission, and sensors are available from several manufacturers. A review of the operating principles of these sensor types is found in ANSI/HPS N13.1-1999. Continuous flow sensors should be regularly checked for accuracy and annually audited.

For less variable stack flows, traditional measurements using Pitot tubes or other flow sensing devices are conducted annually or semiannually. These measurements are taken by inserting the sensor into the duct at a point relatively free of turbulence (e.g., several duct diameters downstream from the fan) and measuring the velocity pressure or air velocity at distance intervals measured

Principles of Sampling Airborne Radioactivity from Stacks 337

from the inner surface of the duct. The cross section of round ducts is subdivided into equal annular areas and the measurements are obtained at the centroid of each area. Rectangular cross-section ducts are divided using a grid and the measurement points are the centers of each rectangular grid area. The individual velocity measurements are averaged and the average flowrate is calculated by multiplying the average velocity times the total cross-sectional area. Corrections are often needed to account for actual gas density relative to standard conditions. The details for determining the measurement points, equipment, and methods are described in standard methods.*

SAMPLE FLOW MEASUREMENT

The sample flow sensor should be located downstream of the sample collector so as not to interfere with the transport of the sample. The sample flowrate is chosen to provide a sufficient amount of collected sample over the desired time interval for detection by a continuous air monitor or for laboratory analysis. The factors determining the sample size, and hence collection interval and flowrate, include the following:

- Stack flowrate
- Analytical sensitivity
- Desired minimum detectable concentration
- Flow limitations of the sample collector

The gas density units for sample flow measurement should be the same as those used for stack flow. For stacks where there is a risk of significant emissions, the sample flowrate should be recorded at least every 10 min. The sample volume is then based on an integration of the recorded readings. As a minimum, the flow reading should be recorded at the start and end of the sample collection period and the readings should be averaged. Sample flow control may be needed if the sample flowrate would otherwise vary more than ±15%. Where the stack flowrate can vary significantly and the sample flowrate/stack flowrate ratio must be controlled, the sample flowrate may be controlled with a feedback control loop based on readings from the stack flow sensor.

A variety of measurement technologies are typically available for sample flow meters:

- Thermal anemometry-based mass flow meters
- Differential pressure measurement
- Sonic nozzles
- Constant differential pressure (rotameters)

The topics of sample flow measurement with rotameters and mass flow meters, pumps, and pulsation control are discussed in Chapter 11.

LEAK TIGHTNESS

Leakage in sample collectors was discussed in the section "Sample Collection." Leakage in the sample transport lines and flow controls can bias the collected sample and the measured sample volume. The sample transport, collection, and flow control systems should be inspected for leaks when assembled, following significant maintenance, and on a regular basis thereafter.

Leak checks can be performed in different ways. Large leaks can be found by visual inspection of joints or by observing the presence of foreign matter on the air sample filters. Quick leak checks

---

* 40 CFR 60, Appendix A, Methods 1, 1A, 2, and 2C, or standards such as ISO 10780:1994E. *Stationary source emissions—Measurement of velocity and volume flowrate of gas streams in ducts.* International Standards Organization, Geneva, Switzerland.

are facilitated if the system is outfitted with one or more full-bore ball valves to isolate parts of the system.

- With a single valve closed, or the probe nozzle blocked, observe if the sample flow drops to zero. This will indicate that an acceptable vacuum inside the sampling train has been achieved due to the tightness of connections and integrity of the sampling lines.
- With two valves, isolate as much of the system as practical. Briefly apply the amount of vacuum observed in normal operation, and then monitor the vacuum decay. Coupled with an estimate of the contained volume, the rate of vacuum decay can be used to estimate a leak rate. It is left to the operator to determine if the leak rate is acceptable or must be reduced.
- Pressurize as much of the system as practical, equal to normal vacuum, and inspect using traditional leak checking methods such as soap bubbles, gas sniffing, or ultrasonic detectors.

Individual components or small assemblies such as filter holders and continuous air monitors can be tested in an operational mode using a tracer gas and an enclosing chamber as described by Karthik and McFarland (2004).

MATERIALS

The construction material used for the sampling nozzles and transport lines should not be reactive with the vapors, gases, or particles in the sample stream. In most instances, the material should be conductive to eliminate the buildup of static charge as the sample stream flows through it. For particle sampling, certain plastic tubing has been shown to build electrostatic charge that enhances particle deposition in the transport system (Charuau, 1982; Liu et al., 1985). Stainless steel (300 series) is useful in most applications. If the sample stream contains a variety of reactive chemical species, consideration may be needed for separate sampling systems.

Except where it is intentionally induced to collect certain vapor species, condensation in the sample transport should be avoided. This is usually accomplished with insulation and heating to maintain the sample stream temperature above the dewpoint.* If the resulting temperature is too high for the sample collector or monitor, dilution with dry air may be considered.

To facilitate decontamination and to minimize particle deposition, the internal surfaces should be as smooth as practical. Surface roughness is a measure of the magnitude of fine irregularities of surface texture. The technical definition is the arithmetic average of the absolute values of the measured profile height deviations taken over a length of material and measured from the graphical center line (see "roughness average" in ANSI/ASME B46.1-1985.) A value of 5×10^{-5} is satisfactory and typical of drawn tubing.

The entire system should be periodically inspected for buildup of deposits. Joints should be provided in the sample transport system to facilitate disassembly for visual inspection or insertion of fiber optic scopes. This is of particular concern in installations where the effluent stream is unfiltered or where background aerosol is present.

REFERENCES

40 CFR 60, Appendix A, Method 1, U.S. Environmental Protection Agency, *Method 1—Sample and Velocity Traverses for Stationary Sources*. U.S. Code of Federal Regulations, 2000.

40 CFR 60, Appendix A, Method 1A, U.S. Environmental Protection Agency, *Method 1A—Sample and Velocity Traverses for Stationary Sources with Small Stacks or Ducts*. U.S. Code of Federal Regulations, October 2000.

---

* The dewpoint is the temperature where 100% relative humidity is achieved. Relative humidity is the ratio of the amount of water vapor the air is holding to the amount of vapor it could hold at a particular temperature times 100%, for example, air at 86°F is at its dewpoint when it holds 30.4 g/m^3 of water vapor. The dewpoint for 68°F air is reached when there is 17.3 g/m^3 of water vapor present (warmer air can hold more water vapor).

40 CFR 60, Appendix A, Method 2, U.S. Environmental Protection Agency, *Method 2—Determination of Stack Gas Velocity and Volumetric Flow Rate*. U.S. Code of Federal Regulations, October, 2000.

40 CFR 60, Appendix A, Method 2C, U.S. Environmental Protection Agency, *Method 2C—Determination of Stack Gas Velocity and Volumetric Flow Rate in Small Stacks or Ducts (Standard Pitot Tube)*. U.S. Code of Federal Regulations, October, 2000.

ANSI/ASME B46.1-1985. *Surface Texture (Surface Roughness, Waviness, and Lay)*. American National Standards Institute/American Society of Mechanical Engineers, New York, 1985.

ANSI/HPS N13.1-1999. *Sampling and Monitoring Releases of Airborne Radioactive Substances from the Stacks and Ducts of Nuclear Facilities*. Health Physics Society McLean, VA.

Anand, N.K., McFarland, A.R., Kihm, N.K., and Wong, F.S. Optimization of aerosol penetration through transport lines. *Aerosol Sci. Technol.* 16:105–112, 1992.

Ballinger, M.Y., Barnett, J.M., Glissmeyer, J.A., and Edwards, D.L. Evaluation of sampling locations for two radionuclide air-sampling systems based on the requirements of ANSI/HPS N13.1-1999. *Health Phys.* 86(4):406–415, 2004.

Barnett, J.M., Ballinger, M.Y., and Recknagle, K.P. *Comparison of a Computational Fluid Dynamics Model with Exhaust Flow Data From a Scale Model Stack*. PNNL-SA-38036, Pacific Northwest National Laboratory, Richland, WA, 2003.

Beal, S.K. Deposition of particles in turbulent flow on channel or pipe walls. *Nucl. Sci. Eng.* 40:1–11, 1970.

Brockman, J.E. Sampling and transport of aerosols. In: K. Willeke and P. Baron (eds), *Aerosol Measurement*. van Nostrand Reinhold, New York, 1993.

Brown, R.H. and Monteith, L.E. Gas and vapor sample collectors. In: B.S. Cohen and C.S. McCamon (eds), *Air Sampling Instruments: For Evaluation of Atmospheric Contaminants*, 9th edition, American Conference of Governmental Industrial Hygienists, Inc., Cincinnati, Ohio, pp. 415–455. 2001.

Charuau, J. Etude du Depot des Particules dans les Conduits; Optimisation des Tubes de Prelevement des Aerosols Radioactifs. Report No. CEA-R-5118, Institute do Protection et de Surete Nuclearie, Departement de Protection, Centre d'Etudes Nucleaires de Fontenay-aux-Roses, Saclay, France, 1982.

Glissmeyer, J.A. and Ligotke, M.W. *Generic Air Sampler Probe Tests*. PNL-100816, Pacific Northwest National Laboratory, Richland, WA, 1995.

Glissmeyer, J.A. and Maughan, A.D. *Project W420 Air Sampler Probe Placement Qualification Tests for Four 6-inch Diameter Stacks: 296-A-25, 296-B-28, 296-S-22, and 296-T-18*. PNNL-12016, Pacific Northwest National Laboratory, Richland, WA, 1998a.

Glissmeyer, J.A. and Maughan, A.D. *Airborne Effluent Monitoring System Certification for New B-Plant Ventilation Exhaust Stack*. PNNL-12017, Pacific Northwest National Laboratory, Richland, WA, 1998b.

Glissmeyer, J.A. and Maughan, A.D. *Airborne Effluent Monitoring System Certification for New Canister Storage Building Ventilation Exhaust Stack*. PNNL-12166, Pacific Northwest National Laboratory, Richland, WA, 1999.

Glissmeyer, J.A. *Cold Vacuum Drying Facility Stack Air Sampling System Qualification Tests*. PNNL-13401, Pacific Northwest National Laboratory, Richland, WA, 2001a.

Glissmeyer, J.A. and Maughan, A.D. *Qualification Tests for the Air Sampling System at the 296-Z-7 Stack*. PNNL-13687, Pacific Northwest National Laboratory, Richland, WA, 2001b.

Glissmeyer, J.A., Maughan, A.D., and Jarvis, T.T. *Qualification Tests for the New Air Sampling System at the 296-Z-1 Stack*. PNNL-14057, Pacific Northwest National Laboratory, Richland, WA, 2002.

Glissmeyer, J.A. *Assessment of the 296-S-21 Stack Sampling Probe Location*. PNNL-16014, Pacific Northwest National Laboratory, Richland, WA, 2006.

Gong, H., Chandra, S., McFarland, A.R., and Anand, N.K. A predictive model for aerosol transmission through a shrouded robe. Aerosol Technology Laboratory Report 8838/12/95/HG, Department of Mechanical Engineering, Texas A&M University, College Station, Texas, 1995.

Han, T., O'Neal, D.L., and Ortiz, C.A. A generic-tee-plenum mixing system for application to single point aerosol sampling in stacks and ducts. *Health Phys.* 92(1):40–49, 2007.

ISO 10780:1994E. *Stationary Source Emissions—Measurement of Velocity and Volume Flowrate of Gas Streams in Ducts*. International Standards Organization, Geneva, Switzerland, 1994.

Karthik, V.V. and Mcfarland, A.R. A leak quantification method using sulfur hexafluoride as a tracer gas. *Health Phys.* 86(6):613–618, 2004.

Lippmann, M. Filters and filter holders. In: B.S. Cohen and C.S. McCamon (eds), *Air Sampling Instruments: For Evaluation of Atmospheric Contaminants*, 9th edition, pp. 281–314. American Conference of Governmental Industrial Hygienists, Inc., Cincinnati, OH, 2001.

Liu, B.Y.H., Pui, D.Y.H., and Rubow, K.L. Characteristics of air sampling filter media. In: *Aerosols in the Mining and Industrial Work Environments*. Ann Arbor Science, Ann Arbor, MI, pp. 989–1038, 1983.

Liu, B.Y.H., Pui, D.Y.H., Rubow, K.L., and Szymanski, W.W. Electrostatic effects in aerosol sampling and filtration. *Ann. Occup. Hyg.* 29:251–261, 1985.

McFarland, A.R., Gong, H., Muyshondt, A., Wente, W.B., and Anand, N.K. Aerosol deposition in bends with turbulent flow. *Environ. Sci. Technol.* 31(12):3371–3377, 1997.

McFarland, A.R., Gupta, R., and Anand, N.K. Suitability of air sampling locations downstream of bends and static mixing elements. *Health Phys.* 77(6):703–712, 1999a.

McFarland, A.R., Anand, N.K., Ortiz, C.A., Gupta, R., and Chandra, S. A generic mixing system for achieving conditions suitable for single point representative effluent air sampling. *Health Phys.* 76(1):17–26, 1999b.

McFarland, A.R., Mohan, A., Ramakrishna, N.H., Rea, J.L., and Thompson, R. *Deposition 2001a: An Illustrated User's Guide*. ATL Report 6422/03/01/ARM; Texas A&M University, College Station, TX, 2001.

McFarland, A.R., Ortiz, C.A., Moore, M.E., DeOtte, R.E., Jr., and Somasundaram, S. A shrouded probe aerosol sampler. *Environ. Sci. Technol.* 23:1847–1492, 1989.

McFarland, A.R., Wong, F.S., Anand, N.K., and Ortiz, C.A. Aerosol penetration through a model transport system: comparison of theory and experiment. *Environ. Sci. Technol.* 25(9):1573–1577, 1991.

Pui, D.Y.H., Romay-Novas, F., and Liu, B.Y.H. Experimental study of particle deposition in bends of circular cross section. *Aerosol Sci. Technol.* 7:301–315, 1987.

Rodgers, J.C., Fairchild, C.I., Wood, G.O., Ortiz, C.A., Muyshondt, A., and McFarland, A.R. Single point aerosol sampling: Evaluation of mixing and probe performance in a nuclear stack. *Health Phys.* 70:25–35, 1996.

Seo, Y., McFarland, A.R., Ortiz, C.A., and O'Neal, D.L. Mixing in a square and a rectangular duct regarding selection of locations for extractive sampling of gaseous contaminants. *Health Phys.* 91(1):47–57, 2006.

Vincent, J.H. *Aerosol Sampling Science and Practice*. John Wiley and Sons, New York, NY, pp. 105, 1989.

16 Methods for Comprehensive Characterization of Radioactive Aerosols
A Graded Approach

Mark D. Hoover

CONTENTS

Introduction 341
Hierarchy of a Graded Approach 342
    Initial Screening and Detection 342
    Comprehensive Characterization and Assessment 343
    Routine Monitoring and Control 343
Methods for Comprehensive Aerosol Characterization 344
    Optical Particle Counting 344
    Particle Collection for Microscopy 347
    Filtration 347
    Inertial Sampling 347
    Measurement of Electrical Properties 347
    Volumetric Grab Samples, Impingers, Cold Traps, and Adsorbers 348
    Analytical Chemical Techniques 348
Special Techniques for Radioactive Aerosols 349
    Detection of Individual Particles by Autoradiography 349
    Particle Solubility Measurement to Infer Biological Behavior 349
    Density Measurement by Isopycnic Gradient Ultracentrifugation 350
    Surface Area Measurement by ^{85}Kr Adsorption 351
Conclusions 351
References 351

INTRODUCTION

Previous chapters in this book have described important methods for detecting and determining the quantity and radioisotopic composition of airborne radioactivity. This chapter describes a variety of additional methods that can be used in a graded approach to judiciously characterize a comprehensive array of physical, chemical, and biological properties of interest. Although nearly all standard aerosol sampling and characterization techniques can be applied to the measurement of radioactive gases and particles, there are practical limitations on the quantities or qualities of material that may be significant or available. For example, the airborne particle number and associated mass of many radionuclides of concern may be below the limits of detection of optical monitoring devices that count the number of airborne particles per unit volume; piezoelectric monitoring systems that measure airborne mass on cascade impaction substrates in real time; gas pycnometry methods that determine particle density by volume displacement; or particle surface area determination methods that employ gas adsorption techniques. Additional limitations may relate to regulatory and administrative restrictions on when or in what quantity a radioactive material can be introduced into a facility or process. For example, commercial electron microscopy facilities and equipments that are readily available for evaluation of nonradioactive materials may not be accessible for evaluation of radioactive samples. Dedicated facilities with specially designed and controlled equipment may be needed. Cost, time, and efficiency factors associated with characterizing radioactive particles and gases make it especially important that any measurements of radioactive materials be soundly justified.

HIERARCHY OF A GRADED APPROACH

As illustrated in Figure 16.1 and detailed below, a graded approach to aerosol characterization can logically involve questions at three levels:

1. What sampling and analytical methods are needed for *initial screening and detection* of a radioactive aerosol of potential concern?
2. What methods are needed for *comprehensive characterization and assessment* of the aerosol so that its properties and control requirements are fully understood?
3. What subset of the screening and characterization methods are necessary and sufficient for *routine monitoring and control* of the material and situation of interest?

INITIAL SCREENING AND DETECTION

Sampling for initial screening and detection is performed to determine whether conditions of concern exist in a workplace or environmental situation. Typical components of a "Level 1" characterization include process knowledge as an indication of whether material of concern might be present and available for dispersion, gross mass, or radioactivity counting of air sampling results, and (when concentrations of inert background materials are low relative to concentrations of the aerosol of concern) optical particle counting or condensation particle counting to compare air concentrations before, during, and after process activities. Samples collected by filtration or other

A graded approach to aerosol characterization

Level 1	Level 2	Level 3
Initial screening and detection	Comprehensive characterization and assessment	Routine monitoring and control

• Process knowledge
• Gross mass or activity counting
• Optical particle counting
• Condensation particle counting
• Microscopy
	• Elemental composition
	• Chemical composition
	• Particle diameter
	  – Physical
	  – Aerodynamic
	  – Thermodynamic
	  – Electrical mobility
	• Morphology
	• Surface area
	• Biological solubility
	• Other relevant properties
		• A necessary and sufficient subset of Level 1 and 2 methods for the material and situation of interest

FIGURE 16.1 Hierarchy of a graded approach to aerosol characterization.

methods are typically examined by electron microscopy or other microscopic techniques to assess the presence of materials of concern.

COMPREHENSIVE CHARACTERIZATION AND ASSESSMENT

Comprehensive characterization and assessment is conducted when there is an indication that airborne radioactivity is present at a level of concern which requires additional information to evaluate the adequacy of controls or the potential consequences of exposure. As illustrated in Figure 16.1, the techniques used for "Level 2" characterization and assessment can be targeted to reveal information about the aerosol elemental composition; chemical composition; particle diameter (physical, aerodynamic, thermodynamic, or electrical mobility); particle morphology, surface area, and biological solubility; and other relevant properties that would influence decisions about management of control or consequences. The guiding principle is: Would information from application of a possible characterization method influence decisions regarding the need to control a material or steps that should be taken in the event that control is lost?

Examples of studies that have involved a suite of characterization methods include assessment of exposures of workers to aerosols during fabrication of nuclear fuel (Hoover et al., 1983); assessment of exposures of uranium milling workers to small amounts of aerosol during the extraction of uranium from ore (Eidson, 1984); assessment of potential inhalation exposures to radioactive aerosols from metal cutting techniques typically used in decommissioning nuclear facilities (Newton et al., 1987); characterization of enriched uranium dioxide particles from a uranium handling facility (Hoover et al., 1998); analysis of aerosol samples collected during an accidental release to provide particle size, solubility, and composition information to estimate and provide appropriate medical response for exposed individuals (Cheng et al., 2004); and assessment of aerosol generation and particle characterization for risk assessment of depleted uranium impact with armored vehicles (Parkhurst et al., 2005; Parkhurst and Guilmette, 2009). Each of these studies included process knowledge, filtration, and microscopy for initial screening and detection, and then augmented those results with a number of the more comprehensive characterization methods described below.

ROUTINE MONITORING AND CONTROL

The objective and challenge for routine monitoring and control is to select a necessary and sufficient subset of Level 1 and 2 methods for characterizing the material and situation of interest. If routine assessment of a parameter is not important for confirming or altering work practices or workplace controls, then methods to assess that parameter are not needed as part of the "Level 3" effort. Emphasis should be on selection of an economical and efficient set of measurements that will simultaneously confirm important aerosol properties and reveal any important changes in aerosol concentrations or characteristics that might require investigation or action. Investigation and responses can include reintroduction of Level 2 characterization methods. For example, microscopic examination of particle morphology can reveal if something in the chemistry, composition, or temperature history has changed. The appearance of irregular, rather than spherical particles might indicate that process temperatures are lower than normal, or that the residence time of an aerosol at elevated temperature is shorter than normal.

The graded approach to selection of individual sampling methods builds on the message of Chapter 1 regarding the importance of understanding the following fundamental objectives for sampling: *basic aerosol characterization* to understand the physicochemical and biological properties of aerosols that may be encountered; *worker health protection* to ensure that worker exposures are within allowed limits and as low as reasonably achievable (ALARA); *environmental monitoring* to ensure that environmental releases of aerosols are within allowed limits and ALARA for environmental and public health concerns; *process quality assurance and control* to ensure that processes and process controls are working properly; *emergency preparedness and response* to provide a basis for appropriate actions when things go wrong; *demonstration of compliance* to document that administrative and regulatory requirements are met; and, finally, as a fundamental underpinning, *research* to advance a comprehensive understanding of the behavior, measurement, and control of aerosols.

METHODS FOR COMPREHENSIVE AEROSOL CHARACTERIZATION

The scope of comprehensive characterization methods presented below is illustrative of currently available methods and includes information on limitations of each of the described methods.

OPTICAL PARTICLE COUNTING

Optical particle counters have not been widely used for radioactive aerosols, but there are circumstances under which they might provide useful information to help protect workers from exposures to radioactive aerosols, especially to warn of unusual particle releases. Optical particle counters have two main advantages: they sample air streams continuously, and they provide real-time information. Their main disadvantages are that radioactive particles cannot be differentiated from nonradioactive particles on the basis of light scattering behavior, and that light scattering generally provides an estimate of the physical size distribution of the observed particles, rather than an estimate of the aerodynamic size distribution, which is more relevant for assessing inhalation risks. A number of commercially available instruments are being used to detect nonradioactive aerosols in clean rooms, to monitor work area dust levels in industries such as mining and textiles, and to provide quality control monitoring for processes such as paint pigment preparation that fabricate or use fine particles. Optical monitoring has also been used in systems for generating and characterizing inhalation exposure atmospheres of radioactive aerosols (e.g., Hoover et al., 1988). Issues related to their application for radioactive aerosols include: (1) level of detection compared to allowed air concentrations for radioactive aerosols; (2) level of detection compared to background levels of nonradioactive aerosols in the workplace; (3) aerosol characterization requirements to determine relationships between radioactive and nonradioactive aerosols in the workplace; (4) calibration requirements to quantify instrument response to specific aerosols; and (5) health protection management strategies for using optical monitoring information in a total program for workplace control and worker protection.

Figure 16.2 illustrates a graphical scheme developed by Hoover and Newton (1991) to evaluate whether a given optical particle counter can meet useful requirements for a level of detection. The abscissa of the two-dimensional field describes the level of an airborne radionuclide concentration in relevant units such as Bq/m^3. This permits placement of a vertical line for any concentration level of interest, such as one derived air concentration (DAC) for a specific radionuclide. For convenience we will refer to this level-of-interest line as the "alarm limit." This line divides the field in half. The ordinate of the two-dimensional field describes the particle number concentration (particles/m^3) or mass concentration (mg/m^3) that corresponds to a given radionuclide concentration. The ordinate uses the units in which the optical counter measures or reports information, and it permits placement of two horizontal lines corresponding to the lower limit of detection and the upper limit of detection for the instrument. The lower limit of detection depends intrinsically on the sensitivity of the instrument (a function of flow rate, sensing volume dimensions, and internal signal-to-noise ratio), and it also depends in practice on the concentration of background dust (external signal-to-noise ratio). Except in very clean environments, the lower limit of detection is determined by the background aerosol concentration.

The upper limit of detection is caused by saturation of the counter (more than one particle in the sensing volume at once). In very dusty environments, the concentration of background aerosols may actually be *above* the upper limit of detection due to coincidence (the monitor is saturated by background aerosols).

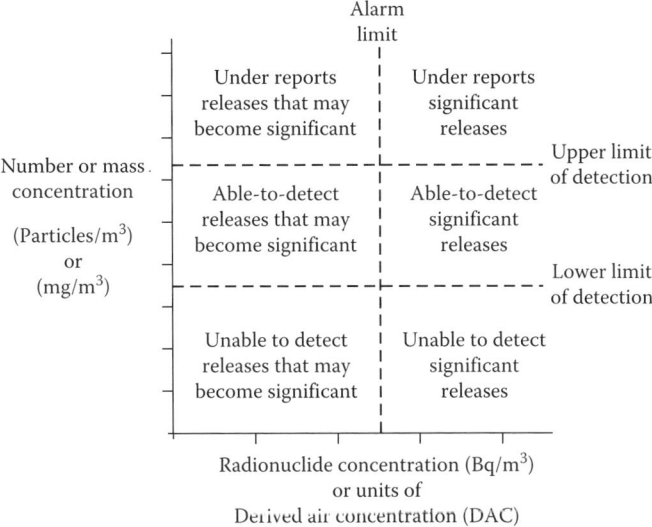

FIGURE 16.2 Format of a graphical scheme for evaluating whether an optical monitor could be useful for detecting radioactive or other toxic aerosols. (Adapted from Hoover, M.D. and Newton, G.J., *Preliminary Evaluation of Optical Monitoring for Real-Time Detection of Radioactive Aerosol Releases*, Inhalation Toxicology Research Institute, Albuquerque, NM, 1991. With permission.)

The combination of the horizontal limit-of-detection lines and the vertical alarm-limit line divides the field into six sectors. When a curve of mass versus activity or particle number versus activity (assuming a particle size) is added to this plot, it reveals the usefulness of the instrument for detecting a given radionuclide.

The ideal situation would involve a monitor whose effectiveness spans the *able-to-detect* sectors of Figure 16.2. This would involve a robust instrument that tracks routine airborne levels at concentrations well below the alarm limit, allows detection of increased airborne particle levels (thus, providing opportunity for preventive action or progressive responses), provides a reliable alarm when airborne concentrations exceed an action level, and continues to track airborne concentrations well above the alarm limit (thus providing a reliable information for estimating human exposures or environmental releases if accidents have occurred).

Table 16.1 illustrates the number of particles per cubic meter as a function of monodisperse particle size for selected toxic materials at their derived air concentration. Figure 16.3 illustrates a practical application of the evaluation scheme for applying an optical particle monitor to the three radioactive materials ($^{238}PuO_2$, $^{239}PuO_2$, and enriched uranium) and one nonradioactive material (beryllium). For convenience, the abscissa units of Figure 16.3 are DAC. This equalizes the scale for all radioactive materials and also allows treatment of nonradioactive materials. The particle size in the Figure 16.3 example is assumed to be monodisperse 3-µm-diameter spheres. Optical particle counters typically give information as "number greater than a given size." The saturation limit is assumed to be $10^8/m^3$. The lower limit of detection depends on the flow rate and sampling interval. For a flow rate of 0.003 m^3/min (nominal 0.1 cfm, typical of small, portable units) and a sampling interval of 1 min, detection of a single particle during the sample interval would correspond to a concentration of 347 particles/m^3. Figure 16.3 indicates that optical particle counting will not be effective for $^{238}PuO_2$ and $^{239}PuO_2$ because number concentrations of concern are below normal limits of detection. However, optical particle counting may be useful for enriched uranium and beryllium metal (assuming that background levels of other dusts are not excessive).

Parameters that still need to be evaluated include the unique response of light scattering devices to each material. This is of special concern because the particles are unlikely to be spherical.

TABLE 16.1
Number of Particles per Cubic Meter as a Function of Monodisperse Particle Size for Selected Toxic Materials at their DAC

Particle Diameter (μm)	$^{238}PuO_2$	$^{239}PuO_2$	Enriched Uranium	Beryllium Metal
10	0.0001	0.02	54	2065
5	0.0008	0.15	433	16,518
3	0.004	0.7	2007	76,471
1	0.1	19	54,180	2,064,715
0.5	0.8	150	433,443	16,517,716

Source: Adapted from Hoover, M.D. and Newton, G.J., *Aerosol Measurement: Principles, Techniques, and Applications*, 2nd ed., Baron, P.A. and Willeke, K., Eds., John Wiley & Sons, New York, 2001.

Notes: Insoluble ^{238}Pu has a specific activity of 6.44×10^{11} Bq/g and a DAC of 0.3 Bq/m^3. Insoluble ^{239}Pu has a specific activity of 2.26×10^9 Bq/g and a DAC of 0.2 Bq/m^3. For enriched uranium, the specific activity is 2.35×10^6 Bq/g (dominated by the contribution from ^{234}U, which is present at 1% by mass), and the DAC is 0.6 Bq/m^3. The effective density of beryllium metal aerosol particles with a slight oxide coating is 2 g/cm^3 (Hoover et al., 1989) and the occupational exposure limit for beryllium is 2 μg/m^3.

Correlations of optical diameter or real diameter with aerodynamic diameter are also important because movement of the aerosol through the workplace and inhalation of the aerosol by workers will depend on aerodynamic, not optical, diameter. Strategies also need to be developed on precisely how optical monitors would be integrated into a total monitoring program (placement, worker response to alarm, record keeping, and other considerations). The use of optical monitors as early warning devices to detect radioactive releases appears promising, especially for special applications, such as during maintenance operations to detect unusual leaks and to guide in the

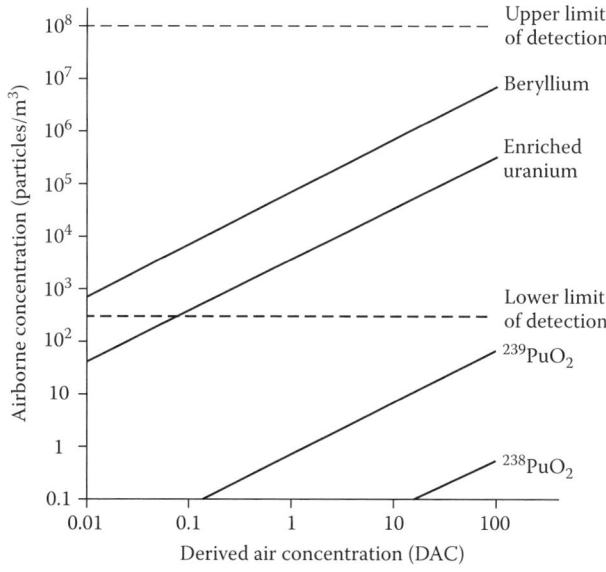

FIGURE 16.3 Illustration of the evaluation scheme for usefulness of optical particle monitors. Particles are assumed to be monodisperse with diameters of 3 μm. Sampling flow rate is 5×10^{-5} m^3/s (3 L/min). (Adapted from Hoover, M.D. and Newton, G.J., *Preliminary Evaluation of Optical Monitoring for Real-Time Detection of Radioactive Aerosol Releases*, Inhalation Toxicology Research Institute, Albuquerque, NM, 1991. With permission.)

Particle Collection for Microscopy

Collection of radioactive particles for morphological examination by transmission or scanning electron microscopy can easily be done using standard instruments such as the point-to-plane electrostatic precipitator (Morrow and Mercer, 1964). A small hood is normally sufficient for sample handling. The airflow rate through the hood opening should be maintained at the lowest effective level (approximately 75 linear feet per minute) to avoid problems in handling the small, fragile electron microscope grids. Standard Formvar®-coated, copper grids can be used for transmission electron microscopy. Degradation of the Formvar by radiation damage is usually only a problem for high-specific-activity radionuclides such as ^{238}Pu (half-life = 87.7 years).

Care should always be taken to use contamination control features, such as liquid nitrogen cold fingers, to minimize the spread of contamination within the microscope. High-specific-activity, α-emitting radionuclides such as ^{238}Pu are prone to migration from the collection grid by radioactive decay-induced recoil and fragmentation (spallation) of particles. The need for respiratory protection and radiation monitoring should be considered before servicing any microscope used for radioactive materials.

Filtration

Filtration is the most widely used method for collecting samples of radioactive aerosols. It is noted here for completeness and described in detail in Chapter 8 of this book. Filtration will typically be part of any Level 1, Level 2, or Level 3 characterization strategy.

Inertial Sampling

Inertial sampling using cascade impactors, spiral duct centrifuges, and cyclones has been the major approach for characterizing the aerodynamic particle size distribution of radioactive aerosols. A number of specialized versions of these instruments have been developed specifically for use with radioactive aerosols (e.g., Mercer et al., 1970; Kotrappa and Light, 1972). Special requirements for instruments used in handling radionuclides generally include being compact and easy to assemble and disassemble while wearing protective gloves in a confined space such as a glovebox enclosure. They also need to be easy to clean. Low sample collection rates are usually adequate for collection of small sample masses. Analytical methods for collected samples are straightforward by radioactive counting. The spiral duct centrifuge has been used to estimate the density or shape factor of individual particles. This works equally well for radioactive and nonradioactive particles. Because the aerodynamic diameter associated with each particle deposition location is known, electron microscopy can be used to determine the physical size and shape of particles found at those locations, and density or shape factor can be calculated (Stöber and Flachsbart, 1969).

Real-time inertial techniques, such as time-of-flight measurements of particles accelerated through a nozzle, are also useful for radioactive aerosols. This assumes a willingness to purchase dedicated instruments for use with radioactive aerosols, because decontamination of equipment for return to unrestricted use is not always easy.

Measurement of Electrical Properties

Measurement of aerosol electrostatic charge distribution can be done using a standard aerosol charge spectrometer (Yeh et al., 1976). The theory of Yeh et al. (1976, 1978) predicts that self-charging due to α- or β-emission in radioactive aerosols will occur in addition to friction charging due to comminution. Even when aerosols are created by highly charging processes like grinding, radioactive decay processes such as α- or β-decay may quickly result in a charge distribution that is near Boltzmann equilibrium. Measurements of particle charge distribution on a plutonium–uranium aerosol obtained by Yeh et al. (1978) with and without the use of a krypton-85 discharge unit were identical. At high α-radioactivity concentrations (>25 nCi/L), it is likely that sufficient ion pairs are present to reduce the charge on the aerosols to near Boltzmann equilibrium.

At lower radioactivity concentrations, this equilibrium condition may not be reached. Raabe et al. (1978) reported the anomalous results of two cascade impactor samples taken without the use of a ^{85}Kr discharge unit in the blending step of a mixed plutonium–uranium oxide fuel preparation process. The α-radioactivity concentration at the time those samples were taken was only 1–2 nCi/L. Because the activity median aerodynamic diameter of these samples was larger than observed in samples taken with a discharger, it is likely that anomalous deposition on the upper stages of the impactors occurred as a result of electrostatic charge effects. In the absence of advance information on the radioactivity concentration being sampled, inclusion of an in-line ^{85}Kr, ^{210}Po, or other type of discharge unit is therefore recommended as a standard procedure, even though it may not be needed to reduce the charge distribution to Boltzmann equilibrium. Fabrication and use of ^{85}Kr discharge units are described by Teague et al. (1978). Krypton-85 discharge units are commercially available (e.g., TSI, Shoreview, MN).

Users of ^{85}Kr discharge units should be aware that krypton is a gas which can leak from the metal tube that retains the gas within the discharge unit, and that ^{85}Kr has a radioactive half-life of 10.756 years, which results in a steadily decreasing source strength. Thus, it is prudent to monitor the source strength over time. This can be done by using a Geiger counter to periodically measure and maintain a control chart of the radiation dose rate at the surface of the discharge unit. The measured dose rate will be the result of bremsstrahlung from interaction of the pure β-emissions from ^{85}Kr with the structure of the discharge unit.

VOLUMETRIC GRAB SAMPLES, IMPINGERS, COLD TRAPS, AND ADSORBERS

Sampling techniques such as evacuated volumes, liquid bath impingers, cold traps, and activated charcoal adsorbers work equally well for capturing radioactive and nonradioactive vapors and particles. Standard radioactivity counting techniques can be applied, depending on sample geometry. The Lucas cell (Lucas, 1977; NCRP, 1988) is a grab-sampling approach in which radioactive particles or gases are drawn into a chamber whose interior walls are coated with a layer of crystalline ZnS(Ag) or other scintillator. Scintillators (also known as phosphors) are materials that absorb energy during the ionization process and reemit a fraction of it as light flashes (scintillations). Light emission occurs when electrons elevated to a higher energy state make the transition back to the ground state. Scintillation bursts can be detected with a photodiode or photomultiplier device and counted to determine the number of charged particles or gamma rays that passed through the material. The number, intensity, and duration of the scintillation can also be analyzed to determine the concentration and energy of the radiation being detected. Nearly 100% of the α-particles reaching the scintillator will result in a flash of light. Any flashes of light occurring within the chamber are observed by a detector. This method can be applied to radon gas, radon decay products, or other α-emitting radionuclides that can be drawn into the chamber. The half-life of the radionuclides being sampled influences the delay time or cleaning requirements before the cells can be reused.

ANALYTICAL CHEMICAL TECHNIQUES

Traditional analytical chemistry methods such as infrared spectrometry, flame or furnace atomic absorption spectrometry, energy-dispersive x-ray analysis, electron or neutron diffraction, and inductively coupled plasma (emission or mass) spectroscopy have sensitivities that are compatible with the small sample sizes usually associated with radioactive aerosols. Sensitivities of various techniques have been compiled and reported (EPA, 2006). Dedicated equipment is usually required for handling radionuclides, and appropriate controls must be applied. Techniques requiring tens or hundreds of milligrams, such as x-ray diffraction, have much more limited application for radioactive aerosols.

SPECIAL TECHNIQUES FOR RADIOACTIVE AEROSOLS

DETECTION OF INDIVIDUAL PARTICLES BY AUTORADIOGRAPHY

The interaction of ionizing radiation with photographic film or solid-state nuclear track detectors such as CR-39 (a polycarbonate plastic) creates tracks or material defects that can be photographically developed or chemically etched for examination by scanning electron microscopy or light microscopy. The location, number, and length of the tracks can be used to determine the position and estimate the radioactivity of individual radioactive particles. In one example, Voigts et al. (1986) identified single aerosol particles as the α-sources from industrial plume samples with particle number concentrations of 2000 particles/mm^2. Cohen et al. (1980) used cellulose nitrate track-etch film to measure the α-radioactivity on human autopsy specimens of the bronchial epithelium. In combination with microscopy, autoradiography can be used to determine how cells and organelles are irradiated by particles that are inhaled and deposited in the body. Figure 16.4 shows an example of autoradiography used with a histological slide of lung tissue to determine the microdosimetry of an inhaled uranium–plutonium oxide aerosol.

PARTICLE SOLUBILITY MEASUREMENT TO INFER BIOLOGICAL BEHAVIOR

Particle solubility and biokinetic studies can be done on all classes of aerosol particles, but measurement of dissolved material is especially straightforward for radioactive aerosols (Kanapilly et al., 1973). Samples can be sandwiched between filters and subjected to continuous solvent flow

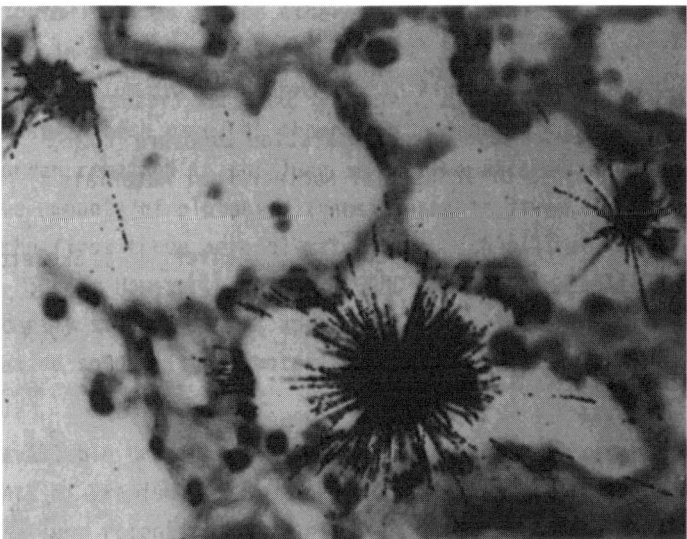

FIGURE 16.4 Autoradiograph of α-emitting particles of varied specific activity in the lungs of a rat following inhalation exposure to mixed uranium–plutonium oxide aerosols. (Adapted from Mewhinney, J.A., *Radiation Exposure and Risk Estimates for Inhaled Airborne Radioactive Pollutants Including Hot Particles*, Annual Progress Report, July 1, 1976–June 30, 1977. NUREG/CR-0010, Inhalation Toxicology Research Institute, Albuquerque, NM, 1978.)

(dynamic systems) or placed sequentially in fresh containers of solvent (static systems). Particles can also be placed in a tube with the solvent and periodically centrifuged to concentrate the particles in the bottom of the tube and to allow sampling of dissolved material from the supernatant. Radioactivity counting provides the very low limits of detection needed for accurate determination of particle dissolution, especially for highly insoluble materials, and radioactive aerosols have played a unique role in many unusual discoveries about aerosol behavior. The high degree of measurement sensitivity was a major factor in the studies of Mewhinney et al. (1987a) that showed a rapid initial release of material whenever particles were reintroduced into a solvent. That work provided an insight into the possible environmental effects of wet and dry weathering on particles released to the biosphere. Differences in bioavailability due to chemical form such as metals or oxides have been revealed by *in vitro* dissolution studies (*cf.*, Finch et al., 1988). Current practice involves conducting *in vitro* dissolution studies under conditions that simulate the important biological compartments of the extracellular fluid of the lung, which has near neutral pH (7.2–7.4), and the intracellular phagolysosomal environment, where phagocytized materials are subjected to a more acidic pH (4.5–5) (Ansoborlo et al., 1999). Stefaniak et al. (2005) have developed a simulated phagosomal stimulant fluid which complements the simulated lung fluid of Kanapilly et al. (1973). In addition, Stefaniak et al. (2007), building on the historical model of Mercer (1967) for the dependence of dissolution on particle size and surface area, provide an improved theoretical framework for the use of measured particle properties to evaluate analytical digestion methods for poorly soluble particulate materials.

DENSITY MEASUREMENT BY ISOPYCNIC GRADIENT ULTRACENTRIFUGATION

Isopycnic density-gradient ultracentrifugation has been shown to be a useful technique for measuring the density of small quantities (0.1–5 mg) of a variety of particles (Allen and Raabe, 1985; Finch et al., 1989). (Normal density measurement techniques, such as air or gas pycnometry or liquid displacement, are not suitable for the small sample volumes normally associated with radioactive aerosols.) Thallium formate has been the usual heavy-metal solution for this technique, but

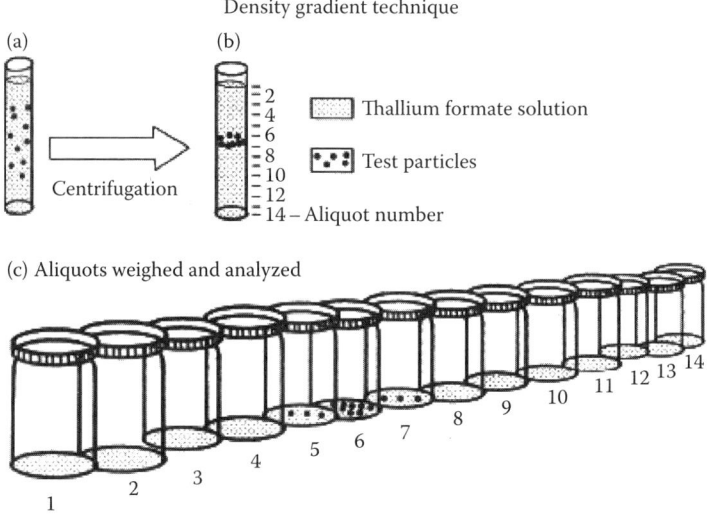

FIGURE 16.5 Illustration of the density gradient ultracentrifugation technique for (a) preparing a suspension of test particles in a thallium formate solution, (b) subjecting the suspension to centrifugation to create a density gradient, and (c) weighing and analyzing the solution aliquots to determine the density of individual particles. (Adapted from Finch, G.L. et al., *J Aerosol Sci* 20: 29–36, 1989.)

sodium metatungstate ($3Na_2WO_4 \cdot 9WO_3 \cdot xH_2O$) has been shown to be an economical, nontoxic alternative (Hoover et al., 1991). Figure 16.5 illustrates the technique. Particles are added to a centrifuge tube containing the heavy liquid. The tube is subjected to centrifugation to form a gradient of density from top to bottom. Density near the top of the tube normally approaches that of water, and higher density near the bottom of the tube can be higher than 3.0 g/cm^3. Particles move to the location within the tube where their density equals that of the surrounding liquid. Successive samples of known volume are then removed, weighed to confirm the density of the liquid in the sample, and then analyzed by radioactivity counting or other suitable methods to determine the fraction of the particle material in the sample. Refractive index measurements of the liquid samples can also be used to determine the density of the liquid samples, but that requires a dedicated refractometer and is usually more time-consuming than simple weighing. The density gradient ultracentrifugation technique provides information about the density distribution of particles within an aerosol sample.

SURFACE AREA MEASUREMENT BY ^{85}Kr ADSORPTION

Particle specific surface area (m^2/g) influences the rate of surface phenomena such as dissolution (*cf.*, Mercer, 1967). When adequate sample masses are available (10–50 mg, or more), measurement of specific surface area is reliable and straightforward. The most widely used approach involves the "BET" method (Brunauer et al., 1938) of calculating nitrogen adsorption onto the surface of a sample of known mass. Other gases such as oxygen, helium, carbon dioxide, krypton, and methane have also been used. A number of commercial instruments are available. Rothenberg et al. (1982, 1987) have focused attention on the special problems of surface area measurement when sample size is less than 10 mg, which is a typical restriction when characterizing radioactive particles. They have described and evaluated a method for adsorbing ^{85}Kr gas onto the sample surface (Rothenberg et al., 1987). Radioactive decay of ^{85}Kr emits a 0.514 MeV gamma ray that can be readily detected by a standard scintillation method such as NaI(Tl). They note that a 1-cm^2 monolayer of ^{85}Kr gas having a specific activity of 10 Ci/g will give approximately 10,000 disintegrations per min that can easily be measured with an uncertainty of less than 1%. The major statistical uncertainty is associated with blank correction for the sample holder. The ^{85}Kr-adsorption technique has been successfully applied to the characterization of small samples of mixed uranium and plutonium dioxide particles (Mewhinney et al., 1987b), and the technique can be used for samples as small as 1 mg, with specific surface areas greater than 1 m^2/g. However, the disadvantage of the method rests in the lack of a commercially available instrument, and the significant effort required for a user to set up the method.

CONCLUSIONS

The selection of methods for measuring radioactive aerosols and gases should be determined by the underlying objectives of the sampling effort. "What information is really needed?" This concept seems trivial, but it must be emphasized so that needed data will not be missed, and time and fiscal resources will not be wasted. A copy of the AEROSAMP spreadsheet for making the aerosol calculations described in this chapter can be obtained from the author.

REFERENCES

Allen, M.D. and Raabe, O.G., Slip correction measurements of spherical solid particles in an improved Millikan Apparatus. *Aerosol Sci Technol* 4: 269–286, 1985.

Ansoborlo, E., Henge-Napoli, M.H., Chazel, V., Gibert, R., and Guilmette, R.A., Review and critical analysis of available *in vitro* dissolution tests. *Health Phys* 77(6): 638–645, 1999.

Brunauer, S., Emmett, P.H., and Teller, E., Adsorption of gases in multimolecular layers. *J Am Chem Soc* 60(2): 309–319, 1938.
Cheng, Y.S., Guilmette, R.A., Zhou, Y., Gao, J., LaBone, T., Whicker, J.J., and Hoover, M.D., Characterization of Plutonium aerosol collected during an accident. *Health Phys* 87(6): 596–605, 2004.
Cohen, B.S., Eisenbud, M., and Harley, N.H., Measurement of the α-radioactivity on the mucosal surface of the human bronchial tree. *Health Phys* 39: 619–632, 1980.
Eidson, A.F., *Biological Characterization of Radiation Exposure and Dose Estimates for Inhaled Uranium Milling Effluents*. U.S. Nuclear Regulatory Commission Report, NUREG/CR-2539, LMF-108, National Technical Information Center, Springfield, VA, 1984.
EPA, *Inventory of Radiological Methodologies for Sites Contaminated with Radioactive Materials*, EPA-402-R-06-007. U.S. Environmental Protection Agency, Washington, DC, October 2006.
Finch, G.L., Mewhinney, J.A., Eidson, A.F., Hoover, M.D., and Rothenberg, S.J., *In Vitro* dissolution characteristics of beryllium oxide and beryllium metal aerosols. *J Aerosol Sci* 19: 333–342, 1988.
Finch, G.L., Hoover, M.D., Mewhinney, J.A., and Eidson, A.F., Respirable particle density measurements using isopycnic density gradient ultracentrifugation. *J Aerosol Sci* 20: 29–36, 1989.
Hoover, M.D., Newton, G.J, Yeh, H.C., and Eidson, A.F., Characterization of aerosols from industrial fabrication of mixed-oxide nuclear reactor fuels, in *Aerosols in the Mining and Industrial Work Environments*, Marple, V.A. and Liu, B.Y.H., Eds., Ann Arbor Science, Ann Arbor, 1983.
Hoover, M.D., Eidson, A. F., Mewhinney, J.A., Finch, G.L., Greenspan, B.J., and Cornell, C.A., Generation and characterization of respirable beryllium oxide aerosols for toxicity studies. *Aerosol Sci Tech* 9: 83–92, 1988.
Hoover, M.D., Finch, G.L., and Castorina, B.T., Sodium metatungstate as a medium for measuring particle density using isopycnic density gradient ultracentrifugation. *J Aerosol Sci* 22: 215–221, 1991.
Hoover, M.D. and Newton, G.J., *Preliminary Evaluation of Optical Monitoring for Real-Time Detection of Radioactive Aerosol Releases*, Inhalation Toxicology Research Institute, Albuquerque, NM, 1991.
Hoover, M.D. and Newton, G.J., Radioactive aerosols, in: *Aerosol Measurement: Principles, Techniques, and Applications*, 2nd ed., Baron, P.A. and Willeke, K., Eds., John Wiley & Sons, New York, 2001.
Hoover, M.D., Newton, G.J., Guilmette, R.A., Howard, R.J., Ortiz, R.N., Thomas, J.M., Trotter, S.M., and Ansoborlo, E., Characterization of enriched uranium dioxide particles from a uranium handling facility. *Radiat Prot Dosim* 79(1–4): 57–62, 1998.
Kanapilly, G.M., Raabe, O.G., Goh, C.H.T., and Chimenti, R.A., Measurement of the *in vitro* dissolution of aerosol particles for comparison to *in vivo* dissolution in the respiratory tract after inhalation. *Health Phys* 24: 497–507, 1973.
Kotrappa, P. and Light, M.E., Design and performance of the Lovelace aerosol particle separator. *Rev Sci Instrum* 43: 1106–1112, 1972.
Lucas, H.F., Alpha scintillation counting, in *Atomic Industrial Forum Workshop on Methods for Measuring Radiation in and around Uranium Mills*, Vol. 3, No. 9, Harwood, E.D., Ed., Atomic Industrial Forum, Washington, DC, 1977.
Mercer, T.T., On the role of particle size in the dissolution of lung burdens. *Health Phys* 13: 1211–1223, 1967.
Mercer, T.T., Tillery, M.I., and Newton, G.J., A multi-stage low flow rate cascade impactor. *J Aerosol Sci* 1: 9–15, 1970.
Mewhinney, J.A., *Radiation Exposure and Risk Estimates for Inhaled Airborne Radioactive Pollutants Including Hot Particles*, Annual Progress Report, July 1, 1976–June 30, 1977. NUREG/CR-0010, Inhalation Toxicology Research Institute, Albuquerque, NM, 1978.
Mewhinney, J.A., Eidson, A.F., and Wong, V.A., Effect of wet and dry cycles on dissolution of relatively insoluble particles containing Pu. *Health Phys* 53: 337–384, 1987a.
Mewhinney, J.A., Rothenberg, S.J., Eidson, A.F., Newton, G.J., and Scripsick, R., Specific surface area determination of U and Pu particles. *J Colloid Interface Sci* 116: 555–562, 1987b.
Morrow, P.E. and Mercer, T.T., A point-to-plane electrostatic precipitator for particle size sampling. *Am Ind Hyg Assoc J* 25: 8–14, 1964.
NCRP, *Measurement of Radon and Radon Daughters in Air*, NCRP Report No. 97. National Council on Radiation Protection and Measurements, Bethesda, MD, 1988.
Newton, G.J., Hoover, M.D, Barr, E.B., Wong, B.A., and Ritter, P.D., Collection and characterization of aerosols from metal cutting techniques typically used in decommissioning nuclear facilities. *Am Ind Hyg Assoc J* 48: 922–932, 1987.

Parkhurst, M.A., Daxon, E.G., Lodde, G.M., Szrom, F., Guilmette, R.A., Roszell, L.E., Fallo, G.A., and McKee, C.B., *Depleted Uranium Aerosol Doses and Risks: Summary of U.S. Assessments*, Battelle Press, Richland, WA, 2005.

Parkhurst, M.A. and Guilmette, R.A., Overview of the capstone depleted uranium study of aerosols from impact with armored vehicles: Test setup and aerosol generation, characterization, and application in assessing risk. *Health Phys* 96(3): 207–220, 2009.

Raabe, O.G., Newton, G.J., Wilkenson, C.J., and Teague, S.V., Plutonium aerosol characterization inside safety enclosures at a demonstration mixed-oxide fuel fabrication facility. *Health Phys* 35: 649–661, 1978.

Rothenberg, S.J., Denee, P.B., Cheng, Y.S., Hanson, R.L., Yeh, H.C. and Eidson, A.F., Methods for the measurement of surface areas of aerosols by adsorption. *Adv Colloid Interface Sci* 15: 223–249, 1982.

Rothenberg, S.J., Flynn, D.K., Eidson, A.F., Mewhinney, J.A., and Newton, G.J., Determination of specific surface area by krypton adsorption, comparison of three different methods of determining surface area, and evaluation of different specific surface area standards. *J Colloid Interface Sci* 116: 541–554, 1987.

Stefaniak, A.B., Brink, C.A., Dickerson, R.M., Day, G.A., Brisson, M.J., Hoover, M.D., and Scripsick, R.C., A theoretical framework for evaluating analytical digestion methods for poorly soluble particulate Beryllium. *Anal Bioanal Chem* 87(7): 2411–2417, 2007.

Stefaniak, A.B., Guilmette, R.A., Day, G.A., Hoover, M.D, Breysse, P.N., and Scripsick, R.C. Characterization of phagosomal simulant fluid for study of beryllium aerosol dissolution. *Toxicol In Vitro* 19(1): 123–134, 2005.

Stöber, W. and Flachsbart, H., Size-separating precipitation of aerosols in a spinning spiral duct. *Environ Sci Techol* 3: 1280–1296, 1969.

Teague, S.V., Yeh, H.C., and Newton, G.J., Fabrication and use of krypton-85 aerosol discharge devices. *Health Phys* 35: 392–395, 1978.

Voigts, Chr., Siegmon, G., Berndt, M., and Enge, W., Single alpha-emitting aerosol particles, in *AEROSOLS, Formation and Reactivity*, 2nd International Aerosol Conference Berlin, Pergamon Press, Oxford, 1986.

Yeh, H.C., Newton, G.J., Raabe, O.G., and Boor, D.R., Self-charging of ^{198}Au-labeled monodisperse gold aerosols studied with a miniature electrical mobility spectrometer. *J Aerosol Sci* 7: 245–253, 1976.

Yeh, H.C., Newton, G.J., and Teague, S.V., Charge distribution on plutonium-containing aerosols produced in mixed-oxide reactor fuel fabrication and the aboratory. *Health Phys* 35: 500–503, 1978.

Part IV

Nonroutine Radioactive Air Sampling

17 Emergency Situation Air Sampling

Robert B. Hayes

CONTENTS

Background and Introduction 357
Plume Measurement..358
    Measurement at the Source..358
    Plume Measurements...358
    Resuspension Measurements 359
    Current Guidance on Emergency Response Air Monitoring Procedures 360
    Anthropogenic and Natural Radioactivity Discrimination 360
References...360
Appendix: First Responder Radiological Monitoring..362
    Thomas F. O'Connell and Stephen P. Clendenin

    Initial Incident Response 362
    First Responder Radiation Hazard Assessment..363
    Stay Time 365
    Electronic Dosimeter Alarm Set Points 365
    References..367

BACKGROUND AND INTRODUCTION

There are two basic categories of air monitoring activities that could occur during and or after an airborne release of radioactive materials: (1) monitoring of the radioactive plume during its transport or release to the atmosphere, and (2) monitoring to determine resuspension of the released material after it has been deposited on the ground. An example of plume measurement would be the case for most accidental or incidental releases from nuclear facilities. These facilities would have real-time air monitoring equipment in place for monitoring planned releases and/or for verifying any lack of releases. A resuspension measurement would consist of taking an air sample downwind of the plumes footprint after the plume had settled out. Resuspension from surface radioactivity is an issue not only with known releases but also any unexpected, malicious use of radioactivity (because of the expected delay in recognizing that radioactivity was involved until after the plume has dispersed). Resuspension is more of a long-term air monitoring effort because ground deposition may prove to be a source term for indeterminate lengths of time after the event itself (depending on precipitation and other factors). These two categories (plume and resuspension) will be addressed independently.

    Real-time air monitoring versus grab sampling of air is also independently addressed insofar as the potential for rapid evaluation of those isotopes of interest can be discriminated from natural or background radioactivity already present in the air. Different technologies are available and employed in these cases depending on the levels and isotopes being measured. Examples include the use of the environmental continuous air monitor (ECAM) described by Rodgers et al. (1998)

For grab sampling, either fixed environmental air samplers or portable air samplers can be used for α-, β-, and γ-emitting isotopes (see Miller and Larsen, 2002 for further examples).

One of the first activities to be initiated when an atmospheric release of radioactivity is known to have taken place is that of predicting where the plume will be (or has been) transported. This can be done through various desktop simulation software programs such as the RASCAL or HOTSPOT (Napier, 2001) codes, each of which requires meteorology (wind speed and direction) be known along with the source term. Most federal exercises employ the capabilities offered by the National Atmospheric Release and Advisory Center (NARAC) (Maiello and Groves, 2006; Remick et al., 2005). By predicting where the plume will go or has gone, measurement activities can be optimized by placing technically qualified responders ("assets") in those positions most likely to safely verify and refine initial modeling calculations and assumptions. Much of the subsequent measurement efforts focus on replacing NARAC models with actual measurement results. The NARAC models can then be used as interpolation tools to approximate the missing data on the plume and/or its deposition footprint until a site can be considered fully characterized.

PLUME MEASUREMENT

MEASUREMENT AT THE SOURCE

Many if not most nuclear facilities have a credible release mechanism or even legal planned releases of radioactive waste (such as hospitals or universities) to either the environment (such as via the sewer or atmospheric releases through fume hoods or other evaporation pathways) or other legal waste disposal routes. If the source term for the air release was a known inventory item from one of these facilities, the source term would already be known *a priori*, not only in chemical form and isotopic distribution but also total activity. To an extent, this even applies to radon emissions from abandoned uranium mines insofar as site characterization has been carried out (NCRP, 1991). In these instances, the source term for the airborne radioactivity can be assumed to be very accurately, if not precisely known, especially if measurement of any residual source location activity is made so as to know just how much was dispersed from the source term.

Nuclear power plants continually monitor environmental radioactivity in and around their facility as do almost all other nuclear facilities (e.g., universities, hospitals with nuclear medicine departments, national laboratories, radioactive waste disposal sites, etc.). If any of these facilities were to be the source of a release, their air monitoring capability could be used to provide source-term-related information on the plume itself, while it was still in transport.

PLUME MEASUREMENTS

If the plume (the primary release pathway) were from a radioactive gas independent of a nuclear facility, then the source term may need to be estimated using solid-state retrospective dosimetry techniques on either elements exposed by the initial container or those of the plume passage itself. One example of this approach is to utilize Thermoluminescence or optically stimulated luminescence of ceramic materials which were exposed as these are effectively very sensitive dosimeters (as in the case of quartz from bricks in residential homes, Haskell et al., 1994).

If a release is only a temporary future possibility, prestaged real-time environmental continuous air monitors can be utilized. An example of these kinds of scenarios would be a rocket launch for a space mission employing plutonium for the spacecraft thermoelectric power source (such as the recent launch to Pluto; Goldman, 2006). In these instances (where a rocket payload contains radioactive material which would be a healthy and safety concern if the rocket were to fail and release its contents), emergency response capabilities are established in advance. Here, environmental continuous air monitors and grab samplers can be deployed prior to the potential event in calculated positions to optimize the available assets. In principle, this is very similar to any nuclear facility with a credible release path, although fixed nuclear facilities tend to overwhelmingly use fixed/permanent radiation monitoring equipment. In the case of a rocket launch, however, the source is already known with extreme precision and accuracy and only the plume would be measured postaccident to characterize terrestrial dispersion. Grab sampling methods or responses from continuous air monitors have a large variation in responses due to the heterogenous nature of the sampling media (the activity typically is in discrete dust-like particles with low volumetric concentration). Experiments show that, under laboratory conditions, the correlation between breathing zone air monitors from multiple workers, fixed air samplers, and installed continuous air monitors can have disagreements within a multiplicative factor of 5, even though all measurements correlated (within an order of magnitude) fairly well when evaluated over many orders of magnitude (Munyon and Lee, 2002).

In the case of an ongoing release, measurement of the plume can be a fairly straightforward effort. As occurred during the Los Alamos and Hanford fires (Farmer et al., 2003), the potential for continuous release was suspected to have resulted in uptakes to fauna from legacy radioactive contamination. In these cases, grab sampling was carried out in a coordinated and calculated fashion so as to optimize any potential need for implementing protective actions for populations or the environment. The location of the samples, their frequency, and type are all determined on a case-by-case basis by personnel trained for these tasks.

There is also the possibility that opportunistic measurements may be available for plume characterization by employing existing monitoring in place to support nearby nuclear facilities or by using other environmental measurement programs already established in the area. This of course assumes that the instrumentation is not only close enough but also is able to detect the type of radiation being emitted by the plume, that is, can acquire a sample that can be analyzed appropriately to generate useful information. For example, if the instrumentation is capable of only measuring airborne α-contamination but the release was that of a β-/γ-source (such as Cobalt-60), then the measurements would not be very useful.

Resuspension Measurements

If an appreciable plume of respirable radioactive contamination were to occur, this material would often (depending on meteorological conditions) deposit the majority of its particulate within the first 4 or 5 hours. This means that it is quite credible that measurement systems would not be deployed to the plume location until after the plume had been already plated out. This of course depends on the location and source of the plume as malicious incidents (e.g., radiological dispersion devices) could be expected to be identified more slowly as compared to accidental or inadvertent releases, which would be more likely to be reported right away. Once the plume has already passed, efforts focus on defining the deposition footprint including its spatial distribution and activity characterization (for chemical form, isotopics, and geometrical distribution). Grab sampling can then be used to estimate the amount of resuspended radioactivity so that decision makers can adequately address any possible public or environmental concerns that may require protection or prevention efforts.

The physics of resuspension measurements is related to that of plume dispersion measurements. As with air monitoring in controlled indoor activities with airborne contamination (Alvarez et al., 1994; Munyon and Lee, 2002), outdoor measurements involve sampling particles that are low in number per unit volume with variations in their individual activities (Arimoto et al., 2002). The low number per unit volume and the range in activities per particle result in a large spread in the expected sample activity when sampled air volumes are small. Resuspension measurements at the Nevada Test Site (Anspaugh et al., 1975) and after the Chernobyl release (Nair et al., 1997) show comparable orders of magnitude of the dispersion in the measurements.

One of the more powerful tools used in determining where resuspension measurements should be made is the U.S. Department of Energy Aerial Measurement System (AMS) (Maiello and Groves, 2006; Remick et al., 2005). With AMS, fixed wing aircraft and helicopters are able to deploy with sensitive airborne radiation detectors using sufficiently sophisticated telemetry and software to quantify the location, spatial distribution, migration, deposition, and dominant isotopes of the radioactive air mass. This system deployed on a helicopter can detect dispersed Americium-241 (a 60 keV γ-emitter) at about 0.2 µCi/m^2 (Remick et al., 2005). When coupled with model interpolation from NARAC and ground truth from surface measurements, this system can provide definitive source-term data for resuspension source terms and subsequent air sample measurements. Ground teams may come from a variety of sources, including federal, facility-specific, state, county, and various local responders. It is from these types of resources that resuspension measurements can be made in a truly optimized fashion to evaluate radiation doses to the public and the need for protective actions such as evacuation or shelter in place.

Current Guidance on Emergency Response Air Monitoring Procedures

Federal guidance for emergency response air monitoring is well-developed (U.S. DOE, 2005a). This source provides instruction on maintaining proper equipment set up, documentation, sample handling, and chain of custody protocols. Practical guidance (such as placing the exhaust from a portable generator downwind from the sampler) is also offered in each of these areas. The actual placement, duration, and use of the air sampling assets are driven by the priorities which are determined by the appropriate decision makers based on the incident specifics and the assets available (U.S. DOE, 2005b). The decision makers are recognized representatives of the affected states, Indian Tribes, counties, cities, and usually, relevant facilities, for example, a nuclear facility (a nuclear facility could be either federal or commercial), as appropriate. Typically, this results in air monitoring assets being deployed where either rapid knowledge of unsuspected contamination transport is a high priority (hospitals, schools, watersheds etc.) or where contamination transport is expected (down wind) to insure proper controls are in place. Optimization of asset use is a collaborative process based on expert input (when it arrives) regarding available instrumentation capabilities and limitations relative to the established priorities set by the decision makers.

Anthropogenic and Natural Radioactivity Discrimination

Typically, there may be difficulty measuring the activity of concern in an air measurement under emergency or other conditions especially when the radioactivity is dominantly an α- or β-emitter. Natural radon and its progeny will be incorporated with the activity to be measured. If the isotopes of concern have nominal γ-emissions, these isotopes can be measured using γ-spectroscopy which discriminates from the natural isotopes. Even if only x-rays are present in appreciable amounts (such as with Plutonium-238), photon spectroscopy is a ready method for assay of the anthropogenic activity on an air filter (Buchheit and Marianno, 2005).

A suite of technology has been developed for discriminating α- and β-air contamination from natural radon progeny for use in emergency monitoring. These include combined chemical and mass spectrometry (Farmer et al., 2003) and simple α-spectrometry of air filters (Hayes et al., 2005) to curve fitting the resultant decay curves of the same filters (Papp and Uray, 2002; Hayes and Chiou, 2003). Other options include portable liquid scintillation systems (Metzger et al., 1995; Martinez et al., 2003) and Frisch grid detectors (Scarpitta et al., 2003). A simple technique that may offer the required selectivity and sensitivity within a time-frame useful for emergency air sampling is to recount samples to allow for the decay of short-lived activity attributed to radon daughters. Samples may be recounted as little as 4 hours after initial collection.

REFERENCES

Alvarez, J.L., W.S. Bennett, and T.L. Davidson. 1994. Design of an airborne plutonium survey program for personnel protection. *Health Physics*. 66(6):634–642.

Anspaugh, L.R., J.H. Shinn, P.L. Phelps, and N.C. Kennedy. 1975. Resuspension and redistribution of plutonium in soils. *Health Physics.* 29(4):571–582.

Arimoto, R., T. Kirchner, J. Webb, M. Conley, B. Stewart, D. Shoep, and M. Wathall. 2002. ^{239,240}Pu and inorganic substances in aerosols from the vicinity of a waste isolation pilot plant: The importance of resuspension. *Health Physics.* 83(4):456–470.

Buchheit, R. and C. Marianno. 2005. Update of technology for use with FIDLER detectors. *Radiation Protection Management.* 22(4):19–22.

Farmer, D.E., A.C. Steed, J. Sobus, K. Stetzenbach, K. Lindley, and V.F. Hodge. 2003. Rapid identification and analysis of airborne plutonium using a combination of alpha spectroscopy and inductively coupled plasma mass spectrometry. *Health Physics.* 85(4):457–465.

Goldman, M. 2006. Pluto and plutonium. *Health Physics News.* 34:16.

Haskell, E.H., I.K. Bailiff, G.H. Kenner, P.L. Kaipa, and M.E Wrenn. 1994. Thermoluminescence measurements of gamma-ray doses attributable to fallout from the Nevada Test Site using building bricks as natural dosimeters. *Health Physics.* 66(4):380–391.

Hayes, R.B. and H.C. Chiou. 2003. Curve fitting air sample filter decay curves to estimate transuranic content. *Health Physics.* 86:80–91.

Hayes, R.B., A.M. Pena, and T.E. Goff. 2005. Use of alpha spectroscopy for conducting rapid surveys of transuranic activity on air sample filters and smears. *Health Physics.* 89:172–180.

Maiello, M.L. and K.L. Groves. 2006. Resources for nuclear and radiation disaster response. *Nuclear News* 49:29–34.

Martinez, B.A., D.E. Dry, S.D. Ware, R.C. Roback, M.M. Fowler, and G.H. Brooks. 2003. Liquid–liquid solvent extraction/alpha spectrometry procedure for determination of multiple radionuclides in terrorist bomb debris. LA-UR-03-4489. *The 2003 Radiochemistry Conference*, July 13–16, Carlsbad, NM.

Metzger, R.L., B.H. Jessop, and B.L. McDowell. 1995. Bioanalysis of uranium, plutonium, and curium on breathing zone air samples by solvent extraction and PERALS spectroscopy. *Radioactivity and Radiochemistry.* 6(3):46–55.

Miller, K.M. and R.J. Larsen. 2002. The development of field-based measurement methods for radioactive fallout assessment. *Health Physics.* 82(5):609–625.

Munyon, W.J. and M.B. Lee. 2002. Summary of stationary and personal air sampling measurement made during a plutonium glovebox decommissioning project. *Health Physics.* 82(2):244–253.

Nair, S.K., C.W. Miller, K.M. Thiessen, E.K. Garger, and F.O. Hoffman. 1997. Modeling the resuspension of radionuclides in Ukrainian regions impacted by Chernobyl fallout. *Health Physics.* 72(1):77–85.

Napier, B.A. 2001. Computer modeling codes for radiological and chemical airborne releases. *Health Physics.* 81(2S):S15-S17.

National Council on Radiation Protection and Measurements. 1991. *Exposures from the Uranium Series with Emphasis on Radon and its Daughters.* NCRP 77. Bethesda, MD.

Papp, A. and I. Uray. 2002. Sensitive method for determination of F18 attached to aerosol particles in a PET centre. *Nuclear Instruments and Methods A.* 480:788–796.

Remick, A.L., J.L. Crapo, and C.R. Woodruff. 2005. U.S. national response assets for radiological incidents. *Health Physics.* 89(5):471–484.

Rodgers, J.C., P.T., Wasiolek, S.D., Schery, and R.E., Alcantara. 1998. LA-UR-98–1684. High resolution real time optical studies of radiological air sample processes in an environmental continuous air monitor. *SPIA Symposium on Industrial and Environmental Monitors and Biosensors, November 1–6, 1998*, Boston, MA.

Scarpitta, S.C., R.P., Miltenberger, Rt., Gaschott, and N. Carte. 2003. Rapid analytical techniques to identify alpha emitting isotopes in water, air-filters, urine, and solid matrices using a Frisch grid detector. *Health Physics.* 84(4):492–501.

U.S. DOE. 2005a. *Federal Radiological Monitoring and Assessment Center Monitoring Manual Volume 2 Radiation Monitoring and Sampling.* DOE/NV/11718—181-Rev. 1. Available on-line at www.nv.doe.gov/library/publications/frmac. Accessed 4/3/07.

U.S. DOE. 2005b. *Federal Radiological Monitoring and Assessment Center, FRMAC Operations Manual.* DOE/NV/11718—080-Rev.2. Available on-line at www.nv.doe.gov/library/publications/frmac. Accessed 4/3/07.

APPENDIX: FIRST RESPONDER RADIOLOGICAL MONITORING

Thomas F. O'Connell and Stephen P. Clendenin

INITIAL INCIDENT RESPONSE

This appendix outlines the capabilities, equipment, and types of radiological measurements that can be performed by trained first responders such as law enforcement, fire services, and emergency medical services (HSPD, 2003).

When emergency responders arrive at the scene of an incident they are entering a world of potential unknown hazards. Some information may be known as a result of an initial phone call or incoming alarm but experience has taught responders that they cannot fully evaluate the situation until they arrive and perform what is known in responder jargon as a "size-up" (NFPA, 2004). Size-up is an attempt to identify potential threats to the life-safety of both the responders and the public. In a commonly taught approach to prioritizing response activities, it is explained that there are three incident priorities (FEMA, 2005):

1. Life safety–anything that can injure the responders and the victims
2. Incident stabilization–the process of mitigating the severity of the incident
3. Property conservation–the limitation of property damage caused by the event or the incident stabilization operations

Size-up continues throughout the initial response. It is the responsibility of every trained emergency responder to gather information from the emergency scene and feed this information up through the command structure to the Incident Commander (Figure A1) (Firescope California, 2004). This continuous information-gathering process allows the Incident Commander to develop and adjust the strategy of the response. All of the hazards that are present at a response scene need to be identified. Among the many observations that may be made as part of size-up is radiological monitoring made with portable instrumentation. Identification of such hazards gives the Incident Commander the information needed to notify and request additional response resources such as radiological subject matter experts as required by preestablished response planning.

As the incident progresses beyond the initial response phase or if the incident is of a terrorist or unknown origin, environmental monitoring for chemical, biological, radiological, and explosive hazards becomes critical information for consideration by the Incident Commander in the development of the response strategy. For example, environmental conditions are factored into the continued refinement of protective equipment used by response personnel and the initiation of protective actions for the general population. The measurement, collection, and the interpretation of this information must be on a real-time basis. It is realized that incidents are dynamic situations requiring that measurement, data collection, and analysis be reiterative.

Measurements made by the initial first-response organizations are a single point-in-time measurement. These measurements support the tactical decisions that must be made during the initial phases of a response. Unfortunately, first responders are usually not mandated, equipped nor have the time to perform on-scene air measurements for airborne radioactivity (some large municipalities may have this capability). They will depend on other "assets" such as subject matter experts, for example, health physicists or county, state, federal, and perhaps military responders to provide such expertise. It is recognized that airborne or "resuspension" measurements will, in the event of a radiological dispersion device incident, most likely be performed hours or days after initiation of the event. But, first responders can perform other measurements to assess a radiological hazard as outlined below.

Note: Although most chemical (gas) detection systems in use today meet the criteria to be operated in a potentially explosive environment, radiation detection equipment currently in use by

Emergency Situation Air Sampling

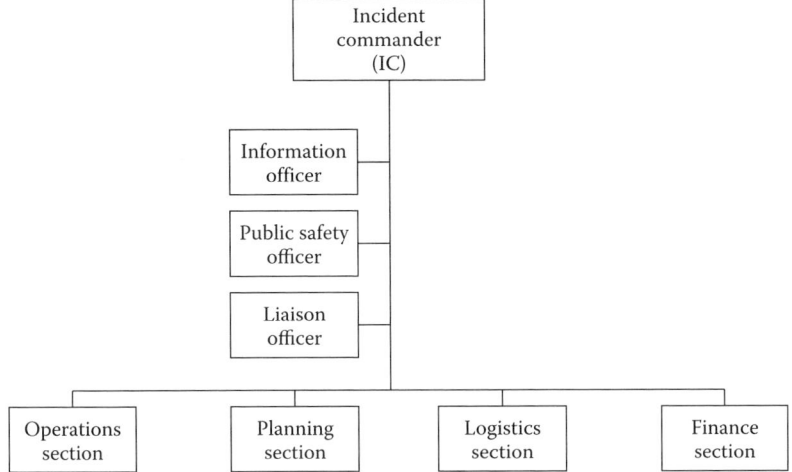

Incident Commander (IC): The individual with overall responsibility for the response. The IC is responsible for performing all of the duties of all of the Incident Command System positions until the IC delegates the responsibility to an individual that is qualified to fill the assigned position.

Public Information Officer: Serves as the press interface for incident information. The PIO prepares press releases for the IC for internal and external stakeholders, including the media or other organizations seeking information regarding the incident.

Safety Officer: Responsible for the monitoring and assessment of the overall work environmental and safety conditions and develops measures for assuring the safety of all personnel that has been dispatched to and assigned to positions within the incident.

Liaison Officer: Serves as the primary contact for supporting agencies assisting at an incident. All of the reporting agencies should report directly to the Liaison Officer, not the IC.

Operations Section: Conducts tactical operations to carry out the incident action plan. It is comprised of the groups that will perform the tactical operations that have been developed in the incident action plan.

Planning Section: Responsible for the preparation and the development of the Incident Action Plan (IAP). It is also responsible for collecting and evaluating information, documentation of the incident, and maintaining resource status. The Planning Section develops the IAP for the next operational period.

Logistics Section: Provides the materials to support the ability to perform the tactical operations and to support the response infrastructure. It also provides support, resources, and all other services that are needed to meet response objectives.

Finance Section: Monitors costs related to the incident. It will order the resources that are requested to enable the response operation to carry out the incident action plan objectives and to support the operation. The records maintained by this section allow for documentation to support cost recovery through accounting, procurement, time recording, and cost analyses.

FIGURE A1 Incident command structure.

the first-responder community has not. Requirements for multiple-hazards monitoring have been developed and are codified in the United States Department of Labor Occupation Safety and Health Administration regulations on hazardous waste operations and emergency response, 29 CFR 1910.120 (CFR, 1990).

FIRST RESPONDER RADIATION HAZARD ASSESSMENT

First responders cannot be fully protected by personnel protective equipment against ionizing radiation. Therefore, an assessment is required to determine if a radiation hazard is present. The objective of the initial radiological assessment is to determine if radiological agents are present in the area of responder operational activities (cold zone) and in the incident scene where response operations are being conducted (warm and hot zones) (see Figure A2; NFPA, 2008). The resulting

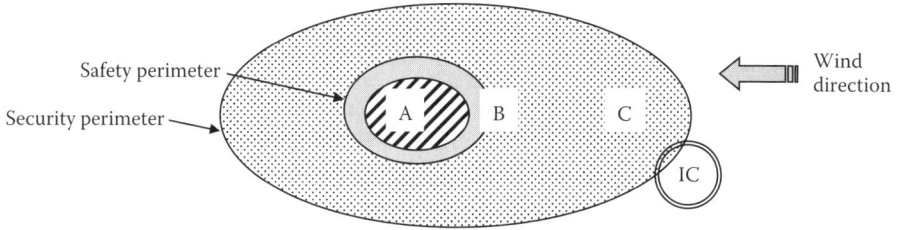

	Designation	Description	External Exposure/Contamination Limits and Recommended Actions
A	Hot Zone (IAEA: "Inner Cordoned Area")	Area impacted by incident	NCRP—10,000 mR/h. Only time sensitive, mission critical operations e.g., life-saving IAEA—>10,000 mR/h. Life saving only. Stay time <30 min Boundary of A and B <10 mR/h
B	Warm Zone (IAEA: "Safety Perimeter"—includes access and contamination control area)	Entry/egress to hot zone and contamination control/reduction area. Inner perimeter provides accountability of emergency personnel and control to impacted area	NCRP—10 mR/h, Evacuate public; minimize stay times IAEA—>10 mR/h Boundary of B and C <0.03 mR/h
C	Cold Zone (IAEA: "Outer Cordoned Area")	Non-impacted area. May be the location of Incident Command (IC) leadership. Outer perimeter provides security and reduces interference from non-emergency personnel/operations	NCRP—decontamination, equipment staging and support personnel areas here (areas should be near ambient background level) IAEA—dose rate should be near ambient background level; set up temporary morgue, public processing area, and waste storage area here

NCRP: National Council on Radiation Protection and Measurements—Key Elements of Preparing Emergency Responders for Nuclear and Radiological Terrorism, Commentary No. 19, December 2005.
IAEA: International Atomic Energy Agency—Manual for First Responders to a Radiological Emergency, October 2006.

FIGURE A2 General arrangement of emergency radiation control zones.

data are used by the Incident Commander to provide information to the local authorities that can be used for tactical and public safety decisions.

If it is determined during entry that radiation is present, a decision must be made about the continuation of operations in the radiation area pending additional input from subject matter experts. The local authorities must make notifications per established emergency response plans to the appropriate local, state, and federal radiation authorities for guidance and assistance. The determination is a fine balance between the hazards that are present, the risk to the responders from the hazards present, and the objectives of the responders to mitigate the incident.

Radiation measurements are needed to develop responder hot zone stay-times, for establishing the perimeters of incident zones, for establishing the locations for various incident facilities (such as the command post and staging areas) and for performing on-scene safety surveys. Further measurements are required to determine the presence of radiological contamination on personal, contamination on equipment, and on the outer surfaces of anything leaving the hot zone.

Pancake style and tube-style β–γ Geiger–Mueller probes are deployed for contamination detection and γ-dose rate measurements. γ-Scintillator probes (μR/h or mR/h) are used for low-level dose rate measurements. As a general rule of thumb, they are used in a radiation field of less than

Emergency Situation Air Sampling

10 mR/h. γ-Identifier units (spectrometers) are also used in radiation fields of less than 10 mR/h. In addition, electronic personal dosimeters are issued to hot zone entry team members.

As indicated, these measurements are primarily used to determine if γ-radiation is present because it is the one form of ionizing radiation that cannot be shielded using personal protective equipment.

There are three different areas to deploy radiation measurement and detection equipment:

1. Area impacted by the incident
2. Contamination reduction corridor/sample screening area
3. Areas outside the impacted area where safety surveys need to be conducted initially and periodically

Before entering the incident area, radiation detection equipment is checked to determine if it can be placed in service. One of the critical checkout parameters is a confident establishment of baseline (background) radiation levels present in nonimpacted areas of the incident. The initial entry team can advance into the incident scene using tube-type Geiger probes so as to assess and report the dose rate in mR/h to command. The information obtained will be used to establish the operating zones, identify high dose areas to avoid, and to develop stay-time tables for responders that are required to enter into the hot zone. Scans of potentially contaminated surfaces can be conducted with the Geiger–Mueller pancake probe. If wipe-tests are collected, they too can be measured with the pancake probe, though in many instances, a tube-type probe will suffice. When the radiation levels in the area interfere with the wipe test measurement, the sample screening will need to be conducted after the sample has been collected and moved to a lower radiation area.

STAY TIME

To determine down-range or "stay times," the baseline radiation level in the nonimpacted area is measured. The baseline radiation level (mrem/h) is subtracted from the radiation data from the impacted areas to obtain the net radiation level above the baseline.

Table A1 provides dosimeter alarm point guidance. Table A2 is a generic example of a down-range stay-time table based on the values listed in the *Reading* column of Table A1 (Ladd et al., 2005). Based on the radiation levels encountered in the field, incident-specific stay time determinations will be made on-scene. The following operational conditions must also be taken into consideration when considering the stay time that an entry team can spend in the hot zone.

1. Type of respiratory protection required (self-contained breathing apparatus, powered air purifying respirators or air purifying respirators) for entry into impacted area
2. Time to get from the entry location to the work location
3. Amount of time that has been projected to complete a mission
4. Time required for the entry team to leave the work area and arrive at the decontamination zone
5. Time needed to be surveyed and possibly decontaminated

ELECTRONIC DOSIMETER ALARM SET POINTS

An emergency responder wearing an electronic radiation dosimeter with an appropriately set alarm will be alerted when the dose from the radiation field reaches an established standard operating guideline level. The radiation dose values and dose rate tables contained in this section of the chapter, in conjunction with standard emergency response concepts, were used for the suggested initial set of preset alarm values (Table A1). These recommended alarm set points are based on generic incident response scenarios. Generic set points will allow for responder safety and protection and make it possible to monitor radiological conditions across a wide variety of incidents.

TABLE A1
Electronic Dosimeter Alarm Point Guidance

Alarm Set Point Type	Reading[a]	Comments[b,c]	Responder Actions
1st dose rate	2 mrem/h	This is a generally accepted value to be used to establish the hot zone (exclusion area) for a response to a transportation accident involving radiation	Continue rescue and investigation activities
2nd dose rate	10,000 mrem/h (10 Rem/h)	Recommended value listed in NCRP Report No. 138	Establish exclusion zone Leave the area unless rescue of known victims can be accomplished efficiently and within guidance values for accumulated dose alarms to responders
1st accumulated dose	2500 mrem (2.5 Rem)	This is one-half of the 5000 mrem annual regulatory exposure limit for occupationally exposed radiation workers	Accumulated doses greater than 10 Rem must be carefully considered
2nd accumulated dose	10,000 mrem (10 Rem)	This is less than one-half the 25,000 mrem dose value listed in EPA 400-R-92-001 for lifesaving or protection of large populations. Recommended value listed in NCRP Report No. 138	Seek expert advice

[a] At the listed values, no immediate health effects from the radiation exposure would be observed in the responder.
[b] Reading in Rem or mrem refers to all exposure pathways. If proper respiratory protection is being used then the internal pathway contribution to dose is minimal.
[c] Basis of the value used is taken from the document indicated: NCRP Report No. 138, Management of Terrorist Events Involving Radioactive Material, National Council on Radiation Protection and Measurements, Bethesda, MD, 2001; EPA Report EPA 400-R-92-001, Manual of Protective Action Guides and Protective Actions for Nuclear Incidents, U.S. Environmental Protection Agency, Office of Radiation Programs, Washington, DC, May 1992, www.epa.gov/radiation/rert/pags.htm.

TABLE A2
Down-Range Stay Time Table

Dose Rate	Accumulated Dose Target	Down Range Time
10,000 mrem/h (10 Rem/h)	25,000 mrem (25 Rem)	150 min (2.5 h)
10,000 mrem/h (10 Rem/h)	10,000 mrem (10 Rem)	60 min (1 h)
10,000 mrem/h (10 Rem/h)	5000 mrem (5 Rem)	30 min
10,000 mrem/h (10 Rem/h)	2500 mrem (2.5 Rem)	15 min
10,000 mrem/h (10 Rem/h)	1000 mrem (1 Rem)	6 min
2500 mrem/h (2.5 Rem/h)	25,000 mrem (25 Rem)	600 min (10 h)
2500 mrem/h (2.5 Rem/h)	10,000 mrem (10 Rem)	240 min (4 h)
2500 mrem/h (2.5 Rem/h)	5000 mrem (5 Rem)	120 min (2 h)
2500 mrem/h (2.5 Rem/h)	2500 mrem (2.5 Rem)	60 min (1 h)
2500 mrem/h (2.5 Rem/h)	1000 mrem (1 Rem)	24 min
1000 mrem/h (1 Rem/h)	25,000 mrem (25 Rem)	1500 min (25 h)
1000 mrem/h (1 Rem/h)	10,000 mrem (10 Rem)	600 min (10 h)

TABLE A2 (continued)
Down-Range Stay Time Table

Dose Rate	Accumulated Dose Target	Down Range Time
1000 mrem/h (1 Rem/h)	5000 mrem (5 Rem)	300 min (5 h)
1000 mrem/h (1 Rem/h)	2500 mrem (2.5 Rem)	150 min (2.5 h)
1000 mrem/h (1 Rem/h)	1000 mrem (1 Rem)	60 min (1 h)
200 mrem/h	25,000 mrem (25 Rem)	7500 min (125 h)
200 mrem/h	10,000 mrem (10 Rem)	3000 min (50 h)
200 mrem/h	5000 mrem (5 Rem)	1500 min (25 h)
200 mrem/h	2500 mrem (2.5 Rem)	750 min (12.5 h)
200 mrem/h	1000 mrem (1 Rem)	300 min (5 h)
2 mrem/h	25,000 mrem (25 Rem)	750,000 min (1.4 years)
2 mrem/h	10,000 mrem (10 Rem)	300,000 min (208 days)
2 mrem/h	5000 mrem (5 Rem)	150,000 min (104 days)
2 mrem/h	2500 mrem (2.5 Rem)	75,000 min (53 days)
2 mrem/h	1000 mrem (1 Rem)	30,000 min (21 days)

To calculate stay time in minutes divide the accumulated dose target (mrem) by the dose rate (mrem/h). The result equals the stay-time (down-range time) in hours. To calculate the number of minutes of down range time, multiply the down-range time by 60 min/h.

The main purpose of the personal dosimeter is to assess the responder's cumulative dose. The majority of electronic radiation dosimeters used in the first-responder community measure only γ-radiation. With most electronic alarming radiation dosimeters, alarm set points can be established for the cumulative dose mode and in the dose rate mode. Each mode usually has two alarm set points that can be programmed into the device. For the two accumulated dose alarm set points, refer to the values listed in Table A1. Table A1 values can be adjusted to apply further safety factors and to generate a set of cumulative dose alarm set points. The guidance takes into consideration the safety of the responder, the ability to safely accomplish the mission, and the flexibility the incident commander needs to safely manage the incident.

In an incident of long-time duration, additional radiological expertise from state, federal, and professional organizations should be available to assist in determining mission-specific dose limits. These dose limits would be based on the incident-specific radioactive materials, dose rates present in the operational area of the incident, the level of personal protective clothing required, the levels of contamination present, and other hazards in the incident's operational area.

REFERENCES

Code of Federal Regulations (CFR), *Federal Register, United States Department of Labor Occupation Safety and Health Administration Hazardous Waste Operations and Emergency Response*, 29 CFR 1910.120, March 6, 1990.

Federal Emergency Management Agency (FEMA), ICS-100 Introduction to ICS, September 2005.

Firescope California, Field Operations Guide ICS 420–1, Incident Command System Publication (ICSP), June 2004.

Homeland Security Presidential Directive 8 (HSPD), National Preparedness, United States of America, December 17, 2003.

Ladd, D., Clendenin, S., and O'Connell, T., Radiological capabilities, strategy and response, The Massachusetts Approach, *Fire Engineering Magazine*, August 1, 2005.

National Fire Protection Association (NFPA), Fundamentals of Fire Fighter Skills, Chapter 10, Jones and Bartlett Publishers, 2004.

National Fire Protection Association (NFPA), NFPA 472, Standard for Competence of Responders to Hazardous Materials/Weapons of Mass Destruction Incidents, 2008.

18 Monitoring Nuclear Fallout

Harold L. Beck

CONTENTS

Introduction 369
What is Fallout? 369
Fallout Production and Transport Mechanisms 371
Radioactive Composition of Fallout 374
Effects of Meteorology on Fallout 376
Detecting Fallout in the Atmosphere by Sampling Air or by Measuring the Amount
    of Fallout Actually Deposited on the Ground 378
Summary of Historical Fallout Monitoring Programs and Current Techniques 379
Sampling and Analyzing Air for Fallout 385
Summary 387
References 387

INTRODUCTION

An important historical application of environmental air sampling has been to detect radioactive debris injected into the atmosphere by weapons tests or monitoring for public health and safety following nuclear accidents. An important component of this effort is to determine the particle sizes of the debris. Determining concentrations of radionuclides at sampled locations is complicated due to rapidly changing conditions. Releases may be instantaneous, short-lived, or continuous for several days or even weeks. Released species may be undergoing rapid radioactive decay, nucleation, condensation, precipitation, nonuniform settling and plate-out, and transport due to transient meteorological conditions. This chapter discusses the sources and characteristics of radioactive "fallout" and the impact of meteorology on fallout transport and deposition. Techniques used to measure fallout in ground-level air, in the upper atmosphere, and other media are reviewed. In addition to historical methods, newer techniques used to detect and monitor fallout in the atmosphere are discussed.

WHAT IS FALLOUT?

Most radioactive debris (particulates) injected into atmosphere will eventually deposit onto the earth's surface. The generic term "fallout" has been generally applied to the debris itself, although technically, the material does not only fall out of the atmosphere, but is rather removed via a number of processes including gravitational settling, washout via precipitation scavenging, and eddy diffusion. A small component is gaseous that is removed by processes that require a great deal of time before a decline in concentration is discernable.

The term "fallout" has generally been associated with debris from atmospheric nuclear weapons tests. However, radioactive debris has also been injected into the atmosphere as a result of nuclear accidents. A well-known example is the Chernobyl nuclear reactor accident that occurred in Ukraine in 1986. The Chernobyl accident released significant quantities of radioactive debris into the lower troposphere, resulting in heavy fallout over much of Europe and detectable concentrations of radioactivity in air samples as far away as the continental USA (UNSCEAR, 2000a).

Another accident that resulted in significant radioactive fallout occurred in 1964. A U.S. satellite launch carrying a Pu-238-fueled thermoelectric generator failed to achieve orbit. The resultant burn-up of the generator upon reentry released 0.63 PBq (17,000 Ci) of Pu-238 into the atmosphere, resulting in fallout of Pu-238 over much of the earth's surface (Krey, 1967). A number of other accidents over the years also released radionuclides into the atmosphere, but in much smaller amounts, resulting generally only in local as opposed to widespread fallout.

Although accidents such as at Chernobyl can be a source of fallout, most of the radioactive debris injected into the atmosphere has been the result of nuclear weapons testing by the United States and the USSR, during the 1950s and early 1960s, as well as, to a lesser extent, from tests carried out by the United Kingdom, France, and China. Nuclear testing in the atmosphere by the United States, the United Kingdom, and the USSR ended in 1962 with enactment of the Nuclear Test Ban Treaty. Testing in the atmosphere by China and France ended in 1980. According to the United Nations Scientific Committee on the Effects of Nuclear Radiation (UNSCEAR, 2000b), nuclear testing from 1945 through 1980 resulted in the injection of fission and activation products corresponding to a total yield about 440 mega-tons (MT) TNT equivalent. About 190 MT TNT equivalent was from fission and 250 MT TNT equivalent was from fusion. Over 1500 PBq (4 MCi) of the long-lived radionuclides Sr-90 and Cs-137 were injected into the earth's atmosphere. Most of this activity was eventually deposited as "fallout" on the Earth's surface. Although large amounts of hundreds of other radionuclides are also produced in a nuclear weapons test, many of the shorter-lived nuclides decay to stable nuclides before they eventually fall out. The first nuclear weapons test, the TRINITY test conducted by the United States in the New Mexico desert, had a yield of about 20 kT and released about 0.12 PBq (3200 Ci) of Cs-137 into the atmosphere (along with other fission products, of course) (UNSCEAR, 2000b). For comparison, the Chernobyl reactor accident released about 85 PBq (0.23 MCi) of Cs-137 (UNSCEAR, 2000a).

The U.S. and USSR weapons tests were carried out at a number of sites around the world. Figure 18.1 shows the locations of major US, UK, USSR and French test sites as well as the site of the Chernobyl reactor accident. Table 18.1 summarizes the number of atmospheric weapons tests carried out by each country and the total fission yields. The UNSCEAR (2000b) and Beck and Bennett (2002) discuss the weapons debris injected into the atmosphere by each testing entity as a function of time and location of testing.

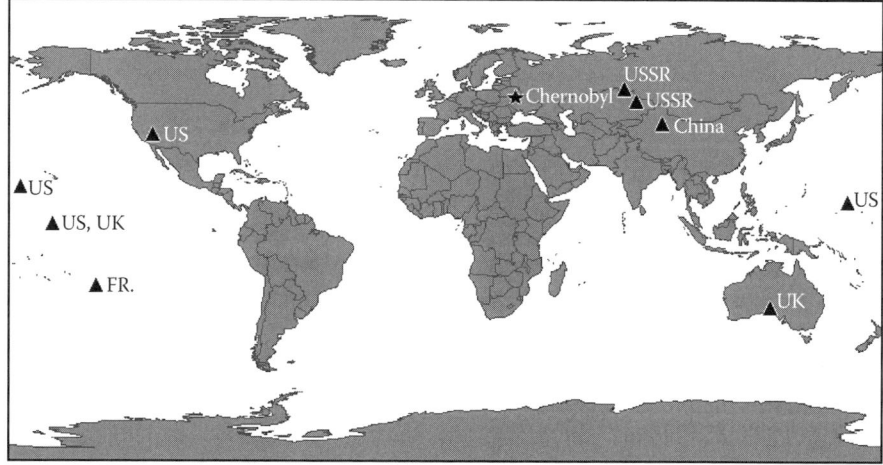

FIGURE 18.1 Locations of major nuclear weapons test sites and the site of the Chernobyl accident.

TABLE 18.1
Number and Yield of Nuclear Tests by Country, Test Site

Country	Site	Number	Total Yield, MT
USA	NTS	86	1.1
	RMI	65	108
	Johnston Is.	12	21
	Christmas Is.	24	23
	Other	10	<0.2
USSR	Semipalatinsk	116	6.6
	Novaya Zemlya	91	240
	Other	12	1
UK	Pacific	9	8
	Australia	12	0.2
France	Polynesia	41	10
	Algeria	4	0.07
China	Lop Nor	22	21
All		504	440 (~200 fission)

Source: Adapted from UNSCEAR. 2000a. *Sources and Effects of Ionizing Radiation. United Nations Scientific Committee on the Effects of Atomic Radiation.* Report to the General Assembly with Scientific Annexes. New York: United Nations Publication, E.00.IX.3; 2000. Annex J.

FALLOUT PRODUCTION AND TRANSPORT MECHANISMS

When radioactive debris is injected into the atmosphere, the locations where it will eventually be deposited depend on a number of complex factors. However, in general, the higher the debris injected into the atmosphere, and the smaller the size of the particles, the further the debris will travel before depositing on the earth's surface.

Often fallout from nuclear weapons tests has been characterized either as "local" fallout, that is, the fraction of the debris deposited in the immediate vicinity of the test site, "regional" or "tropospheric" fallout, that is, fallout that is deposited at distances of up to several thousand kilometers downwind of the test site but generally within days to weeks after the test, and "global" fallout. "Global" fallout generally refers to debris from a very high-yield test that was injected into the stratosphere as opposed to the troposphere. Debris injected into the stratosphere tends to remain there for periods of years. During that long residence time in the stratosphere, most of the shorter-lived radionuclides decay to stable nuclides. The typical half-time for stratospheric debris to deposit to the ground is on the order of 1 year. In contrast, fallout injected into the lower atmosphere (troposphere), that is not deposited locally, falls out with a half-life of about 30 days (Bennett, 2002).

The amount of "locally deposited fallout" depends on the amount of radioactive debris present or adsorbed to large particles. Large particles fall out rapidly as a result of gravitational settling. The fraction of debris on large particles in turn depends on the height above the ground of the explosion and the characteristics of the ground surface. Early U.S. nuclear tests were conducted at the Nevada Test site (NTS). These were generally of fairly low yield compared to later thermonuclear weapons tests carried out in the Pacific (a few kilotons to a maximum of 70 kilotons versus several megatons). Although many tests were conducted on 100–200 m steel towers, the resultant fireball contacted the ground surface, drawing large quantities of soil into the fireball. Owing to the extreme temperatures created by the nuclear detonation, all the fission and activation products are initially in a gaseous form. As the fireball cools, the various radionuclides condense onto the available suspended

FIGURE 18.2 Picture taken early after a weapons test at the NTS shows the development of the toroidal-shaped fireball and the stem resulting from entrainment of soil into the fireball by the resulting updraft.

particles. Primarily, the radionuclides condense onto particles of local terrestrial origin. Oceanic tests generated debris derived from coral and other shallow ocean deposits. The more the available particles, the more the heavy debris that is available to fall out locally. Figure 18.2 is a picture of one of these earlier NTS tests showing the mushroom cloud that that forms as the fireball cools and the stem that is formed by soil being entrained into the cloud. Generally, the stem was found to contain about 10% of the total radioactivity produced in the explosion, although this ratio probably varied from shot to shot.

As the fireball cools, it expands and rises to great altitudes (Glasstone and Dolan, 1977). Even relatively low-yield tests conducted near ground level at the NTS injected debris high into the troposphere. Typical cloud tops reached 10–12 km, while the bottoms of the mushroom clouds stabilized at altitudes of 7–8 km (DNA, 1979), even though most had burst heights within a few hundred meters of the ground surface. Clouds from tests conducted at airplane altitudes rose to even greater altitudes. For later thermonuclear tests in the Pacific, cloud tops were often over 30 km (DNA, 1979). After the cloud reaches its maximum altitude and stabilizes, generally in about 5–10 min, the diameter can range from 5–10 km for low-yield devices to tens of kilometers for high-yield thermonuclear tests.

Thus the pattern of local fallout depends on the size distribution of the particles containing radioactive debris, the height within the cloud or stem from which the particles originate, and the wind speed and direction. Figure 18.3 graphically depicts some of the mechanisms contributing to the pattern of local fallout.

Particle sizes immediately downwind from the NTS (local fallout) had activity median diameters (AMD)* up to several 100 μm, although the particle size decreased fairly rapidly with distance from

---

* The activity median diameter is the diameter for which half the total activity produced is on larger particles and half on smaller particles. The AMAD is the diameter of an equivalent particle of unit density at which half the activity is associated with smaller particles. For spherical particles, the AMAD is the AMD times the square root of the particle density. Generally, the distribution of activity was found to be close to lognormal.

Monitoring Nuclear Fallout

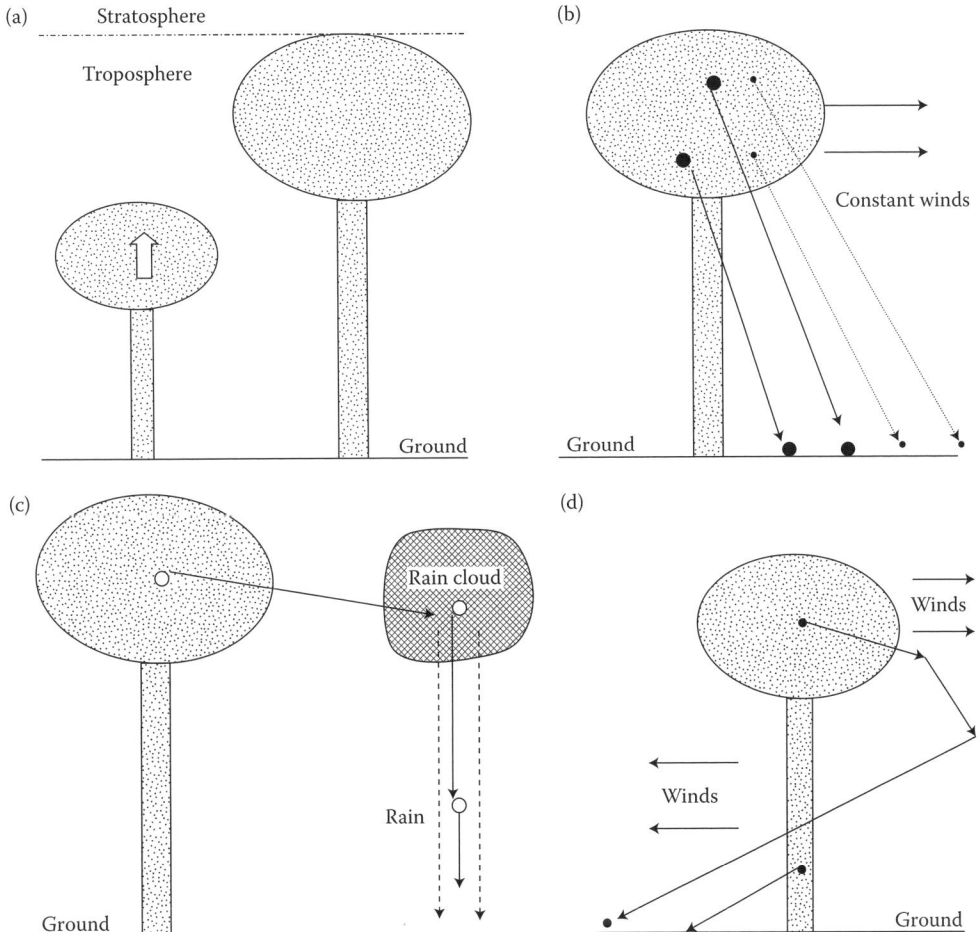

FIGURE 18.3 Graphical depictions of major mechanisms governing fallout from a weapons test: (a) As the cloud of debris cools, it rises and expands. For a typical 30–40 Kt test, the cloud top stabilized at about the level of the tropopause (~11 km) with a diameter of about 5–10 km. For higher yields, the cloud top extended well into the stratosphere. (b) For constant winds in the same direction at all altitudes, a large particle falls out sooner (closer) than a smaller particle; particles falling out from lower parts of the cloud land closer than the same-sized particle originating at a higher altitude. The distance traveled before falling out is proportional to the wind velocity. (c) Small particles can travel long distances before depositing, primarily as a result of precipitation scavenging. (d) Variations in wind speed and direction at different heights determine where a particle will eventually be deposited. The path of the particle is determined by the vector sum of its vertical velocity (due to gravity) and its horizontal velocity (wind speed and direction). The horizontal velocity can vary in speed and direction as the particle falls from one level in the atmosphere to a lower level.

the test site (Baurmash et al., 1958). Cederwall et al. (1990) measured particle sizes at St. George Utah, about 150 km from the NTS, from tests ANNIE in 1953 and found that the first few hours of fallout had an activity mean aerodynamic diameter (AMAD)[*] of about 76 μm, although the range of particle sizes was quite broad (Geometric Standard Deviation ~ 9). This was typical of NTS tests sampled in 1953. While the particle sizes in tropospheric fallout even much further downwind were

---

[*] ibid footnote as * in p. 370.

still on the order of several microns, the particle sizes (AMD) of debris in global fallout from the stratospheric reservoir were very small, fractions of a micron (Klement, 1965).

The smaller particles in the mushroom cloud will be effected less by gravity and the time for them to fall out will be effected more by wind and intermittent, precipitation scavenging. The very smallest particles can stay in the atmosphere for long periods, some even circumventing the globe prior to depositing. The influence of meteorological factors on fallout is discussed in more detail later in this chapter.

RADIOACTIVE COMPOSITION OF FALLOUT

The radionuclide composition of the fallout depends on a number of factors and differs considerably for local, regional, and global fallout. First, the distribution of fission products formed in the detonation varies somewhat with the fissile material used and with the energy of the neutrons that are generated during the fission process. The first nuclear weapons test (TRINITY) was fueled with Pu-239, while the device exploded over Hiroshima was fueled with U-235. Many of the early tests conducted at the NTS during the 1950s contained mixtures of both Pu-239 and U-235 (DHHS, 2006). Besides the fission products themselves, activation products are formed by neutrons interacting with various isotopes in the bomb casing, surrounding material, and soil (for tests conducted near the earth's surface). Activation products include radionuclides such as Co-60, Fe-55, Na-24, Mn-54, and Np-239. For the later high-yield thermonuclear tests conducted in the Pacific and the USSR, a considerable amount of the yield was from fission of U-238 by the higher-energy neutrons released in the fusion reactions. In these tests, Pu-239 or U-235 was generally used only in small amounts in a fission trigger to initiate the fusion process, which in turn initiated fission in a secondary blanket of U-238. In all nuclear tests, only a fraction of the fissile material was actually fissioned; hence large quantities of the fuel itself become part of the potential fallout, isotopes of plutonium being among the most significant.

The composition of the fallout at any location, however, will differ considerably from the debris in the fireball due to decay of short-lived radionuclides prior to deposition, as well as due to the variation in particle size distribution of the fallout versus distance. The latter phenomenon is generally referred to as "fractionation" (Hicks, 1982). As the fireball cools, the more refractory (high boiling point) atomic elements such as zirconium and cerium condense first, along with soil constituents and heavy elements such as iron from the bomb casing and tower. The refractory radionuclides thus tend to be distributed throughout these particles. The large particles that fall out locally are enriched in these refractory materials in comparison with the volatile elements such as iodine and cesium, each having a significant yield of isotopes due to fission. The latter have lower boiling points, and condense later. These tend to deposit on the surface of the already condensed soil and device-related particles. Because the surface to volume ratio of a spherical particle is inversely proportional to its radius, the activity of volatile nuclides relative to refractory nuclides tends to be enhanced on the smaller particles that travel further before depositing. It has been demonstrated that for a near-surface test, such as occurred at the NTS, as one moves further away from the test site, the ratio of refractory to volatile nuclides in the fallout tends to approach an asymptotic value of about 1/2 (Hicks, 1982), that is, about half of the refractory nuclides tends to fall out locally as a consequence of their enhancement on the larger particle sizes. Thus, the fallout occurring very close to a site of a surface or near-surface test is highly enriched in the refractory nuclides. Because Cs-137 and Sr-90 are not formed directly in the fission process but are decay products of gaseous precursors (Xe-137 and Kr-90), they tend to deposit onto particles at an even later time than other volatile elements. Cs-137 and Sr-90 are even more depleted relative to refractory nuclides at distances close to ground zero (GZ) than other volatile nuclides such as I-131 (Freiling et al., 1965). The exact degree of fractionation for a given test will depend on the amount and characteristics of the soil entrained into the fireball, which, as mentioned earlier, in turn depends on the height of the burst and the explosive yield.

Short-lived radionuclides contribute most of the fallout radioactivity in the first few hours after a test. Over 150 fission products have been shown to contribute to the radioactivity in the cloud in the first few minutes after detonation. Even after several days, several dozen radionuclides are still important contributors to fallout activity (Hicks, 1981). A number of fission and activation products created in a nuclear test have half-lives on the order of years and thus even debris injected into the stratosphere can still produce measurable radioactive fallout years after the test. Table 18.2 lists the radionuclides contributing the major fraction of the total fallout activity in unfractionated fallout at various times after the detonation (Hicks, 1981). Because the exact fractions will vary with the type of test and the amount of fractionation, the percentages given in the table are only order of magnitude values provided to illustrate the general importance of various nuclides as a function of time. Note that the nuclides that contribute most of the activity vary markedly with fallout age.

The composition and characteristics of fallout from accidental releases will of course differ from that of fallout. For example, the quantities of radionuclides released into the atmosphere from the Chernobyl accident, shown in Table 18.3, consisted of mainly volatile radionuclides (UNSCEAR, 2000a). Because of the buildup of longer-lived radionuclides in reactor fuel, the accident released larger quantities of Cs-137 relative to short-lived fission products than produced in a nuclear weapons test. Although the height of injection was much lower than for a typical atmospheric weapons test, much of the debris was apparently attached to smaller particles which allowed it to traverse relatively large distances before depositing (UNSCEAR, 2000a).

TABLE 18.2
Major Radionuclides in Fallout versus Time

	Approximate % of Total Activity			
Nuclide	12 h	5 days	30 days	1 year
Np-239	10	25	*	*
Sr-90	*	*	*	<2
Sr-91 + Y-91m	10	*	5	*
Y-92	10	*	*	*
Y-93	5	*	*	*
Zr-95 + Nb-95	*	*	10	20
Zr-97 + Nb-97m	20	*	*	*
Mo-99 + Tc-99m	*	20	*	*
Ru-103	*	*	10	*
Ru-106 + Rh-106	*	*	*	15
I-131	*	5	*	*
Te-132 + I-132	*	15	*	*
I-133	5	*	*	*
I-135	5	*	*	*
Cs-137	*	*	*	<2
Ba-140 + La-140	*	10	20	*
Ce-141	*	*	10	*
Ce-143 + Pr-143	*	10	10	*
Ce-144 + Pr-144	*	*	*	45
Others (all <5%)	35	15	35	20

Source: Adapted from Hicks, H.G. 1981. *Results of Calculations of External Radiation Exposure Rates from Fallout and the Related Radionuclide Composition.* Livermore, CA: Lawrence Livermore National Laboratory, Report UCRL-53152, parts 1–8. Available at http://www.osti.gov/opennet/

TABLE 18.3
Chernobyl Debris Activity

	Activity Released	
Radionuclide	PBq[a]	MCi
Xe-133	6500	176
Te-132	~1150	~31
I-131	~1760	~48
Sr-89	~115	~3
Ru-103	>168	>4
Ba-140	240	6
Zr-95	196	5
Mo-99	>168	>4
Ce-141	~116	~3
Np-239	945	26
Cs-137	~85	~2
Cs-134	~54	~1
Pu-239 + 240	0.07	0.002

Source: Adapted from UNSCEAR. 2000a. *Sources and Effects of Ionizing Radiation. United Nations Scientific Committee on the Effects of Atomic Radiation.* Report to the General Assembly with Scientific Annexes. New York: United Nations Publication E.00.IX.3, 2000. Annex J.

[a] Peta $(P) = 10^{15}$.

EFFECTS OF METEOROLOGY ON FALLOUT

Meteorology has a major impact on the amount of fallout deposited as a function of distance, as well as when the fallout is deposited. One major meteorological factor is of course the speed and direction of the winds at the levels where the radioactive debris cloud stabilizes. The general philosophy during both the U.S. and the USSR testing programs was to try to detonate devices when the winds at the test site were in directions away from major population centers. However, although information on wind speed and direction were available at the test sites, sufficient information was not always available on conditions downwind. Thus debris from a given altitude might leave the test site traveling in a particular direction, but change the direction as it proceeded downwind due to changes in the wind direction downwind.

A common phenomenon in the atmosphere is that the winds frequently differ in velocity and direction as a function of height above the ground. Figure 18.4 shows an example of calculated trajectories at various levels of the atmosphere for the NTS SIMON shot of April 15, 1953 (List, 1959), illustrating the different velocities and directions at various altitudes as a function of time after the shot. Thus, as debris falls from one altitude to a lower altitude due to gravitational or vertical diffusion, its trajectory will change (see Figure 18.3d). For example, for many tests carried out in the Marshall Islands at Bikini and Enewetak, the prevailing winds at lower altitudes, the trade winds, were generally from east to west. Thus the local fallout from the stem and lower parts of the mushroom cloud was generally transported toward the west in the general direction of Guam and the Philippines. Conversely, the debris in the upper levels of the mushroom clouds was generally transported toward the east toward Hawaii and the west coast of the United States. However, as the heavier particles from the upper levels fell to lower altitudes, they changed the direction, sometimes moving back toward the test site itself. This often resulted in light fallout occurring in the Marshall Islands several days after a test, sometimes even over GZ itself.

Monitoring Nuclear Fallout 377

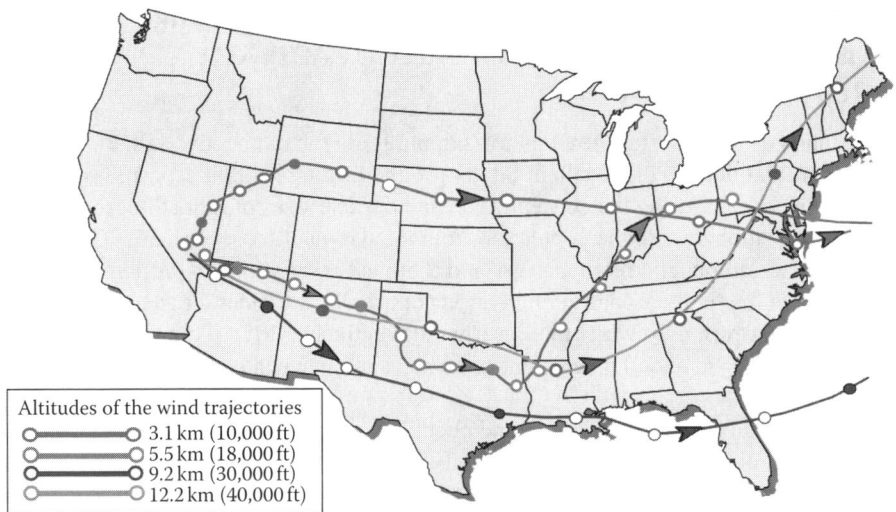

FIGURE 18.4 Calculated trajectories of debris at various altitudes from an NTS test demonstrating wind shear, that is, differing directions and velocities of air at different altitudes and distances from the NTS. The points indicate the position of the cloud at 6-h intervals. (Adapted from List, R.J. 1954. *The Transport of Atomic Debris from Operation Upshot-Knothole*. Washington, DC. Joint US Atomic Energy Commission US Weather Bureau Report NYO-4602 (del); 1954 (declassified with deletions April 22, 1959. Available at http://www.osti.gov/opennet/.)

One of the more important effects of meteorology on fallout is precipitation scavenging. Most of the very small fallout particles tend to remain in the atmosphere until they are removed primarily by precipitation scavenging. Studies of the deposition of global fallout, that is, the debris reentering the atmosphere from the stratosphere, were able to demonstrate that on the order of 90% of the fallout occurred during precipitation and only about 10% was in the form of dry fallout (Klement, 1965). The exact wet/dry ratio of course varied depending on the average annual precipitation. The more arid western states generally experienced a greater fraction of dry fallout. Most of the NTS fallout deposited downwind in the states immediately from the NTS occurred as dry fallout, while most of the fallout occurring in the eastern United States was associated with precipitation, again reflecting the smaller particles remaining in the atmosphere as the fallout moved east and higher average precipitation rates (Beck et al., 1990).

Because of the influences of meteorology on fallout transport and deposition, a number of unanticipated instances of heavy fallout occurred, some very far from a test site. Perhaps the two most publicized were the fallout encountered by the Japanese fishing boat, *Lucky Dragon* in March, 1954 (Eisenbud, 1997) and the fallout that occurred over the Albany NY area in April, 1953 (Hoecker and Machta, 1990). The *Lucky Dragon* was about 180 km east of the Bikini test site near Rongerik in the Marshall Islands shortly after the detonation of test BRAVO. With an estimated 15 Mt explosive yield (UNSCEAR, 2000b), BRAVO was the largest thermonuclear test detonated by the United States. Instead of the heaviest fallout occurring to the west of the test site, the lower-level winds shifted toward the east and both the fishing boat, as well as some of the Northern Marshall Island atolls directly downwind, experienced very heavy fallout. In fact, several of these atolls had to be evacuated. The other notable incident was the rainout that occurred in the Albany NY area a few days after the 1953 SIMON test at the NTS. As can be seen from Figure 18.4, the upper-level winds passed directly over this area. It turns out that the time the debris cloud passed over the area happened to coincide with heavy local thunderstorm activity resulting in significant precipitation scavenging and fallout, although not hazardous as was the case for the BRAVO fallout, was an order of magnitude greater than levels generally occurring at these great distances from the NTS (List, 1954; Hoecker and Machta, 1990).

DETECTING FALLOUT IN THE ATMOSPHERE BY SAMPLING AIR OR BY MEASURING THE AMOUNT OF FALLOUT ACTUALLY DEPOSITED ON THE GROUND

There are a number of important reasons for sampling air for radioactive fallout, as well as for measuring the actual deposition of fallout on the ground or the resultant exposure rate in air. One important reason for measuring the concentration and particle sizes of fallout in ground-level air is to estimate the radiation dose to the population from inhalation of those particles. This dose is very sensitive to the size, radionuclide composition, and chemical composition of the particles (solubility). Thus, a measure of the total concentration in air is not usually sufficient for an accurate estimate of inhalation dose. However, used with other information on likely AMD of the particles, and age and composition of the fallout, a total air concentration measurement can be used to estimate inhalation doses.

Furthermore, based on the particle size and chemical composition, it is possible to estimate the deposition velocity of the particles and from that the amount of the fallout that will deposit as a function of time. Deposition velocity depends on surface conditions, local meteorology, and particle size. Thus, depositional velocity is highly variable from site to site. It is not an easy parameter to quantify. An actual measurement of fallout deposition is preferable for estimating the external exposure rates from fallout on the ground surface or in the soil. An actual measurement is also useful for estimating the dose from nuclides that might be incorporated into food products and ingested by humans or animals.

Even though air sampling may not be sufficient for good quantitative estimates of deposition rates, air sampling is valuable because of the great sensitivity for documenting the presence and activity of debris at a specific site and for determining the approximate time of arrival of fallout. Often debris could be easily detected in a high-volume air sample when there was no detectable increase in external exposure rates at a site.

Another important reason for sampling air is to study the transport of debris through the atmosphere. Fallout provides a tracer of opportunity of atmospheric processes and has proved highly useful in the past for developing a better understanding of movement of particles and gases through the atmosphere, both over short distances and over very long distances. Because of the influence of meteorology on fallout transport and deposition, fallout debris has provided a tracer of opportunity for investigators studying atmospheric processes such as precipitation scavenging, stratospheric and tropospheric residence times, stratospheric–troposphere interchange rates, dry deposition velocities on various surfaces, and so on. (Klement, 1965; Machta, 2002).

Knowledge of the effects of meteorology on fallout transport may also be used to detect the origin of a clandestine source of radioactive debris in the atmosphere. Back-trajectory reconstruction enabled the identification of the location of the first Soviet nuclear test using meteorological data from fallout deposition and air activity (Machta, 2002). Similarly Larsen et al. (1994) used back-trajectory analysis to demonstrate that radioactive debris measured in high-volume air samples at Barrow Alaska originated from an accident at a nuclear fuel reprocessing facility at Tomsk, in the Soviet Union (see Figure 18.5). This has obvious application to nonproliferation enforcement as well as to tracking potential terrorist activities involving use of nuclear materials. Fallout has also served as an important tracer of opportunity for studying ocean currents and mixing (Livingston and Povince, 2002).

A major and currently very important application of sampling air for radioactive fallout is to detect the occurrence of clandestine weapons tests. Even underground nuclear tests may emit noble gases such as xenon and krypton that can be detected by collecting large volumes of air for laboratory analysis. Clandestine reprocessing of reactor fuel to produce Plutonium for nuclear weapons can also be potentially detected via the use of very sensitive air sampling and analysis techniques. Very low levels of radioactivity in air samples may be detected compared to that from deposition samples. Deposition sample sensitivity suffers from low total activity within the sample as well as

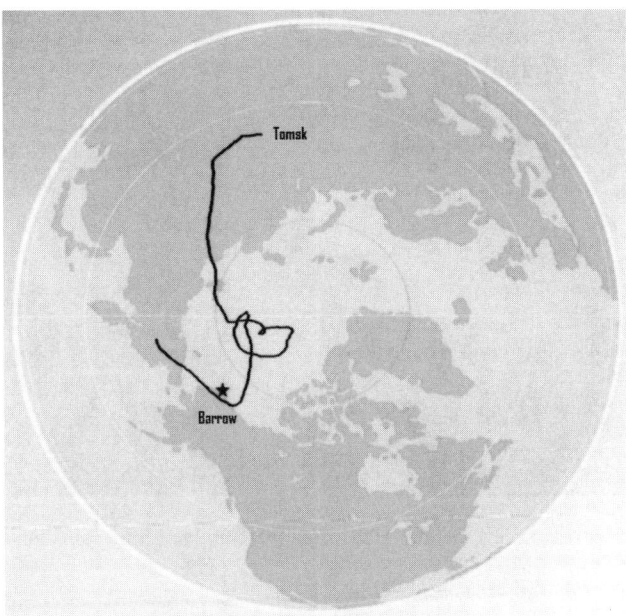

FIGURE 18.5 Calculated back trajectory of fallout detected at Barrow Alaska (Larsen et al., 1994). (Picture courtesy of USDHS EML.)

from the deposition of nonradioactive dust and pollen in the sample. Currently a number of nations maintain networks of continuous exposure rate monitoring to detect possible clandestine nuclear tests. The Comprehensive Nuclear Test Ban (CTBT) organization operates a worldwide network of air samplers (Matthews and Schultze, 2001).

SUMMARY OF HISTORICAL FALLOUT MONITORING PROGRAMS AND CURRENT TECHNIQUES

Fallout from nuclear weapons tests was extensively monitored during the 1950s through the 1980s throughout the world. Greatest attention was paid on areas immediately downwind from test sites (Beck and Bennett, 2002). The techniques used to monitor fallout evolved from fairly crude and relatively inaccurate methods in the 1950s to fairly sophisticated techniques in the 1980s, when the major powers were no longer testing nuclear weapons in the atmosphere. This evolution coincided with improvements in nuclear detection and analysis technology as well as improvements in air sampling technology, including the development of higher capacity pumps, improved filter media, and more sophisticated particle-sizing techniques.

Deposition of specific radionuclides on the ground surface typically involves sampling soil and precipitation, then performing isotopic analysis in a laboratory. However, at the inception of monitoring, most samples were analyzed only for total β-activity. Another strategy early in the testing era was the HASL* investigation of the geographical variations in the deposition of fission products using gummed-film measurements of deposited β-activity (Bouville and Beck, 2000). Daily measurements of β-activity deposited on the 1 ft^2 monitors at several hundred sites around the globe enabled HASL to document, at least semiquantitatively, the deposition of fallout from each

---

* The Health and Safety Laboratory (HASL) of the U.S. Atomic Energy Commission (AEC). The AEC was later incorporated into the Energy Research and Development Administration (ERDA) which in 1977 became part of the Department of Energy. HASL's name was changed to the Environmental Measurements Laboratory (EML) in 1977. EML was transferred to the Department of Homeland Security in 2003.

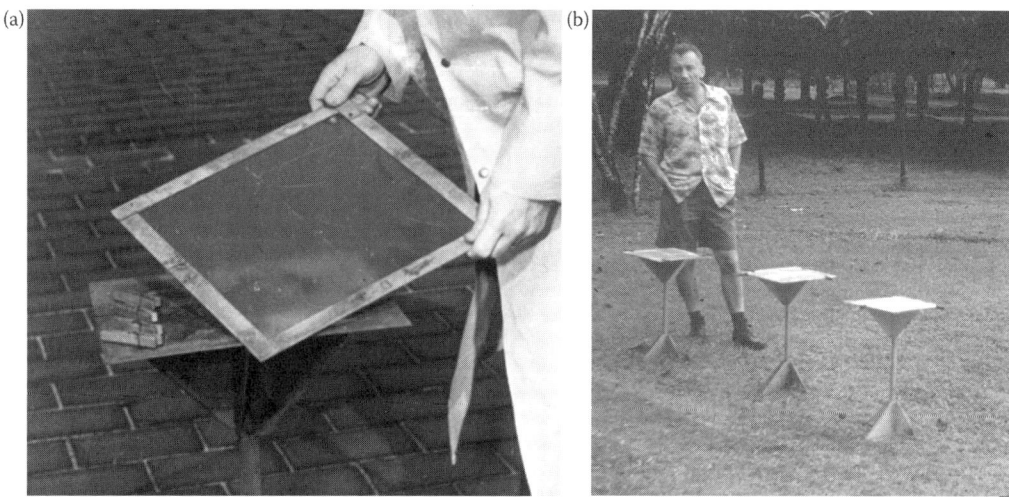

FIGURE 18.6 (a) Close-up picture of gummed film; (b) Gummed-film collectors deployed at Ponope. (Pictures courtesy of USDHS EML.)

test as the debris traveled around the world from the various test sites. Figures 18.6a and 18.6b show the gummed film and three gummed-film monitors deployed at Ponope, several hundred km southwest of the Bikini and Enewetak tests sites in the Marshall Islands.

For most of the 1950s, these gummed-film monitors were the only large-scale, nationwide, and international network of fallout monitors. However, many measurements were also made for regional-scale monitoring in areas close to the test sites. In the late 1950s, HASL began to replace the gummed-film monitors with precipitation collectors. These were generally large open-faced pots or ion exchange columns as shown in Figure 18.7 (EML, 1997). Collecting precipitation directly along with any dry fallout deposited in the pots allowed the sample to be taken back to the laboratory for more precise isotopic analysis, thus providing a more accurate estimate of the fallout deposition. Some sites were later equipped with wet/dry samplers (Figure 18.8) where a sensor was used to cover one of the pots during precipitation in order to collect dry fallout separately from wet fallout (EML, 1997; Miller and Larsen, 2002). Although some of the time resolution offered by the daily gummed-film readings was lost (the precipitation sampling was generally weekly or monthly), lack of time resolution was of less importance in the 1960s and 1970s because much of the fallout was from the stratospheric reservoir.

The HASL also periodically sampled soil worldwide beginning in the early 1950s to document the cumulative deposition of various long-lived radionuclides, in particular, Cs-137 and Sr-90. Beck and Bennett (2002) discuss these data and show the locations of these sampling sites.

Besides the global HASL gummed-film and precipitation sampling networks, the U.S. Public Health Service (PHS) began sampling precipitation and air at a number of sites throughout the United States in the late 1950s and continued through the early 1970s. The PHS also sampled milk from about 60 sites across the United States during that period (Beck and Bennett, 2002).

HASL conducted a number of surveys in the early 1960s, during the peak years of weapons fallout, using the then new technique called *in situ* γ-ray spectrometry (Miller and Larsen, 2002). This technique measured the flux of γ-rays from the soil surface *in situ* using large NaI(Tl) scintillation detectors in order to estimate the deposition density of specific radionuclides in the soil. From that data, the fraction of the fallout exposure rate due to various specific radionuclides could be estimated (EML, 1997). Zr-Nb-95 was the major contributor to the exposure rates from fallout during this period. Most of the short-lived radionuclides had decayed away because the majority of this fallout was debris originally injected into the stratosphere, reducing prompt deposition. In the 1970s, Germanium diodes (see Figure 18.9) began to replace NaI for *in situ* γ-ray spectrometry. The much

FIGURE 18.7 HASL/EML pots and columns used for collecting precipitation. (From Miller, K.M. and Larsen, R.J. 2002. *Health Phys.*, 82(5), 609–625. With permission.)

higher spectral resolution of these detectors enabled detection and quantification of not only the major fallout components but also minor contributing components. In recent years, *in situ* γ-ray spectrometry using large-volume Germanium diodes is routine not only for detecting fallout but also for assessing the concentration of radionuclides in soils in support of other contamination scenarios including environmental restoration efforts at weapons production facilities. *In situ* spectrometry was used extensively to study the deposition of debris from the Chernobyl accident.

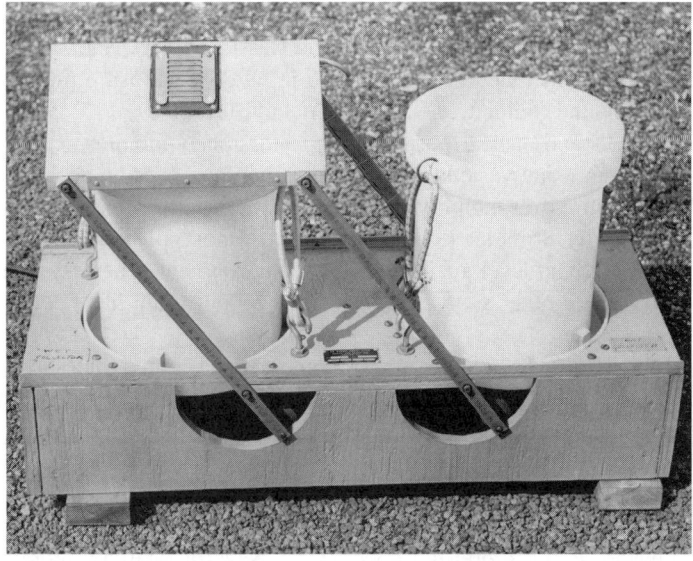

FIGURE 18.8 HASL/EML wet/dry collector. (From Miller, K.M. and Larsen, R.J. 2002. *Health Phys.*, 82(5), 609–625. With permission.)

FIGURE 18.9 *In situ* γ-ray spectrometer. (Adapted from EML. 1997. *Environmental Measurements Laboratory. Procedures Manual. HASL*-300. Available at National Technical Information Service (NTIS), Springfield, VA. Also available at http://www.eml.st.dhs.gov/)

As the global fallout levels increased as a result of the testing of high-yield devices in the Pacific and in the Soviet Union in the late 1950s and early 1960s, other nations began their own fallout monitoring programs. These included sampling of soil, precipitation and biota, surface water, foodstuffs, and so on, as well as air and milk (Beck and Bennett, 2002).

Besides monitoring soil, agricultural products, milk, and vegetation, HASL and other organizations made direct measurements of γ-ray exposure rates or integral exposure in air using survey meters, high-pressure ionization chambers, film badges, and more recently thermoluminescent detectors (TLDs) (Beck and Bennett, 2002). Much of the information on local fallout in the vicinity of weapons test sites is based on land and/or aircraft surveys of exposure rates using various types of survey meters. (Thompson et al., 1994; Breslin and Cassidy, 1955). Nationwide surveys carried out by HASL in the 1960s documented the increase in outdoor exposure rates across the United States from fallout (Beck and Bennett, 2002; Lowder et al., 1964).

Many nuclear facilities and nuclear power plants continuously monitor exposure rates near the site boundary and beyond, using various types of detectors in case of an accidental release. Some facilities integrated exposure over monthly or quarterly time intervals, using passive dosimetry, for example, TLD or Optically Stimulated Luminescence. Almost all nuclear facilities around the world routinely sample ground-level air in their local environment. These systems of course will also detect any fallout from other sources, and in fact, fallout often has been detected by these monitoring systems.

The EPA has also maintained a network of air samplers in the United States to monitor for fallout from weapons tests as well as debris from nuclear accidents such as Chernobyl (EPA, 2005). For example, Figure 18.10 shows β-activity measured by the EPA in routine air samples in the 1970s from Chinese weapons tests.

Air sampling has always been a major component of fallout assessment. Although early measurement protocols focused on total β-activity, some crude measurements of particle size using cascade impactors have also been reported. In 1957, the Naval Research Laboratory established a network of high-volume filtered air samplers along the 80th meridian, designed to detect the movement of fallout from Nevada tests across the United States (Lockhart et al., 1965). The filter

Monitoring Nuclear Fallout 383

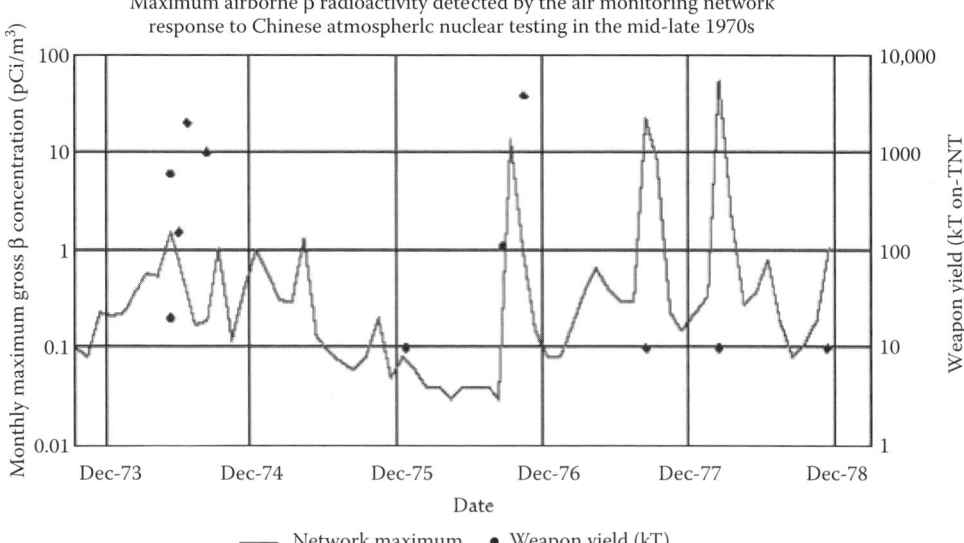

FIGURE 18.10 Activity measured in air by the EPA ERAMS network from fallout from Chinese weapons tests in the later 1970s. (Adapted from EPA. 2005. Real-Time Monitoring of Radiation in Air in the United States-Updating and Expanding the Environmental Radiation Ambient Monitoring System (ERAMS), Background Information for the Radiation Advisory Committee of the EPA Science Advisory Board, Office of Radiation and Indoor Air, United States Environmental Protection Agency, January 24, 2005.)

samples were generally analyzed for total β-activity as well as selected radionuclides. In 1963, the HASL assumed control of this network and eventually expanded it to additional sites around the world (Beck and Bennett, 2002; Miller and Larsen, 2002). Figure 18.11 shows a high-volume surface air sampling system used by HASL/EML. The HASL surface air sampling program (SASP) filters were analyzed not only for γ-ray emitters, using high-resolution low-background γ-ray spectrometry, but also occasionally for Plutonium isotopes. The data collected by these air sampling networks are presently available online at www.eml.st.dhs.gov/.

The EML SASP network was further enhanced by the addition of automatic sampling systems at remote sites (EML, 1997; Miller and Larsen, 2002). These Remote Air Monitoring Program (RAMP) systems were unique in that after the air is sampled, the filter is counted *in situ* with a high-resolution γ-ray spectrometer and the data transmitted via satellite back to the laboratory. Thus, detection of radioactive debris at very low levels within hours of its arrival at the monitoring site was possible.

Fallout from nuclear weapons tests and from the Chernobyl accident was detected at SASP sites in Alaska; fallout was detected from an accident at a nuclear fuel reprocessing facility in Tomsk in the former Soviet Union. Figure 18.12 shows the γ-ray spectrum of the nuclides detected (Larsen et al., 1994). As discussed earlier, meteorological back trajectories verified that the detected debris did in fact originate at the Tomsk site.

In addition to the NRL/HASL/EML worldwide sampling network, air sampling for fallout was also routinely carried out at a number of other sites around the world, including in the United Kingdom and at various U.S. national laboratories. For example, the Pacific Northwest National Laboratory (PNNL) sampled air for fallout in the 1960s using ultrahigh-volume samplers (~4 m^3 min^{-1}) and were thus able to detect extremely low levels of radioactive debris from tests in the Pacific and the USSR (Perkins et al., 1965).

Air sampling was also frequently used to detect fallout immediately downwind from test sites. High-volume and cascade impactor samples obtained downwind from the NTS in the 1950s provided estimates of the particle size distributions of the debris (Cederwall et al., 1990). Although these

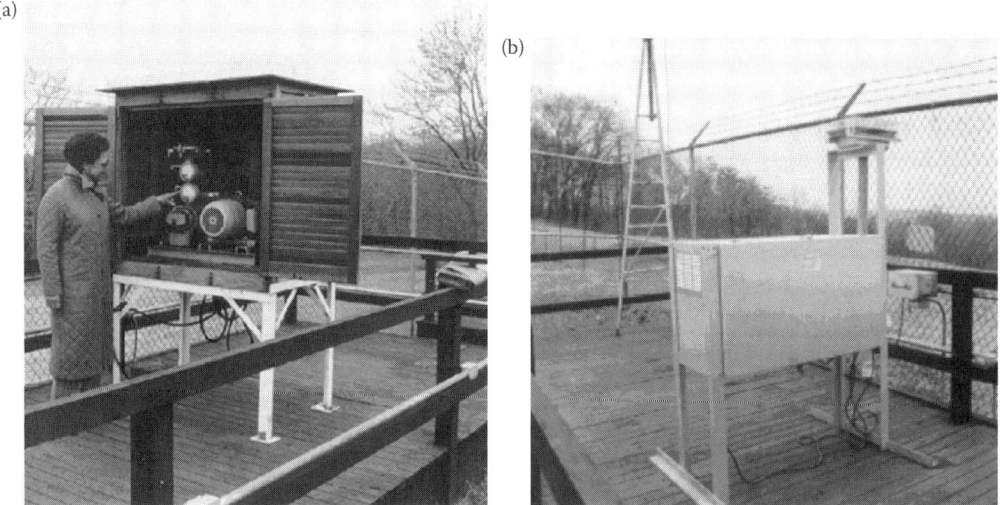

FIGURE 18.11 HASL/EML surface air sampling systems. (Adapted from EML. 1997. *Environmental Measurements Laboratory. Procedures Manual. HASL-300.* Available at National Technical Information Service (NTIS), Springfield, VA. Also available at http://www.eml.st.dhs.gov/)

early particle size distribution measurements were fairly crude, and as mentioned earlier in this chapter the particle size spanned a large range, they did allow estimates of inhalation doses to the local population. Recent developments in particle size measurements of air samples allow much more accurate estimates of particle sizes over a wide size range.

Another major application of air sampling during/after nuclear tests was the determination of the radionuclide composition of the debris at altitude. During the U.S. weapons atmospheric testing program, Air Force cloud sampling planes traversed the debris clouds shortly after the detonation,

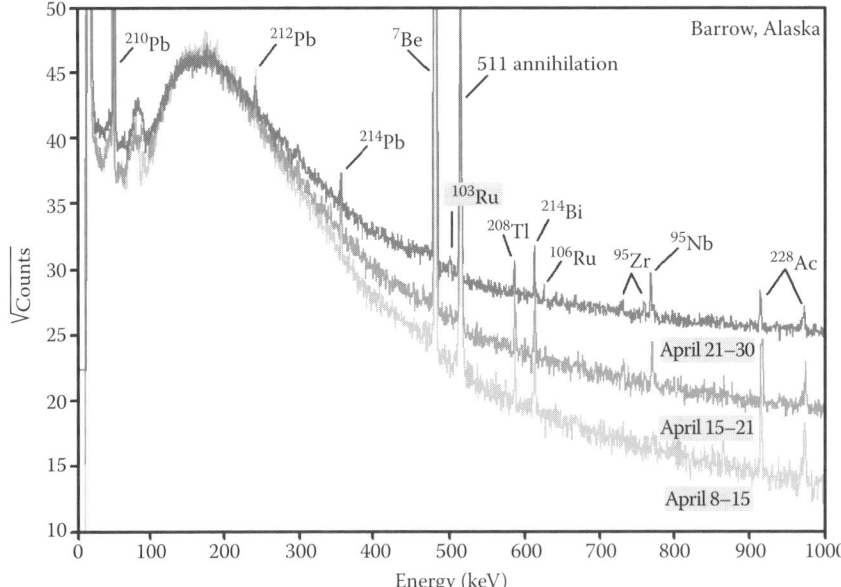

FIGURE 18.12 Spectrum of γ-rays in high-volume sample at Barrow Alaska, Picture courtesy USDHS EML. (From Larsen, R.J. et al., 1994. *J. Environ. Radioact.*, 23, 205–209. With permission.)

collecting air samples for laboratory analysis of the isotopic composition of mixed fission products and unfissioned fuel. These flights provided information on the yield of the explosion as well as the composition of the debris. This information has been used to estimate the composition of the deposition of fallout as a function of time from various types of tests (Hicks, 1982).

Aircraft and balloons were also used to sample stratospheric air to help determine the inventory of long-lived radionuclides in the stratosphere from the high-yield tests as well as to help study stratospheric residence times. Tracers such as W-185 and Rh-102m were added to several U.S. thermonuclear tests in the 1950s to aid in tracing the movement of the debris through the atmosphere through analyses of the collected air samples (Feely and Spar, 1960).

SAMPLING AND ANALYZING AIR FOR FALLOUT

Sampling air for fallout radionuclides often involves issues that are somewhat different from other air sampling applications. For the measurement of particles, the most common sampling methods are based on filtration of large volumes of air (collecting particulates on or near the surface of a filter). The large volumes required a major constraint upon the time resolution of measurements. One important issue involves the selection of filter media. The criteria for filter selection (EML, 1997) are good collection efficiency for submicron particles at the range of face velocities used, high particle and mass loading capacity, low-flow resistance, low cost, high mechanical strength, low-background activity, compressibility, low-ash content, solubility in organic solvents, nonhygroscopic properties, temperature stability, and availability in a variety of useful sizes. In the selection of a filter material, a compromise must often be made among the above criteria that best satisfies the particular sampling requirements. In order to sample over relatively long periods and filter large volumes of air at a high rate,* and to detect low levels of fallout in a reasonable time interval, a filter medium with relatively low resistance but good collection efficiency is required, particularly for the very small particles sizes associated with global fallout. Although membrane media with minimal penetration are useful for applications where filters are to be counted directly for α-particle emissions, these tend also to restrict flow and thus limit the volume of air which can be sampled. The filter media must also be suitable for counting in γ-ray spectrometers, that is, easily folded or crushed into a fixed counting geometry. Furthermore, if the radioactive material is to be chemically removed from the filter matrix for subsequent chemical analysis, a filter medium which easily dissolves, has low ash content, and whose ash does not interfere with the chemical separation is required. The media used most recently by EML in its high-volume systems was a synthetic fiber, Dynaweb, which meets most of these requirements. EML has carried out extensive studies of collection efficiency versus particle size for this filter media (EML, 1997).

Another issue for high-volume sampling is mass loading. To calculate air concentrations of radionuclides, it is necessary to accurately determine the total volume of the air that has been sampled. Thus, continuous high-volume samplers that used to collect samples over weeks or longer must be equipped with flow monitors or governors that either maintain a constant flow or measure the flow rate versus time.

Because the air movers must perform reliably over long periods of time under sometimes difficult environmental conditions, reliability is very important. Early air sampling systems used for fallout monitoring were often very unreliable and suffered frequent breakdowns. Details of the HASL/EML high-volume fallout monitoring system designs, air movers, and so on are given in the EML Procedures model (EML, 1997). These designs are similar to those used by other organizations for monitoring fallout in the environment.

Filters cannot be generally used for collecting gaseous radionuclides. Although some chemically impregnated filter materials are available, gaseous radionuclides such as radioactive xenon and krypton were generally sampled by either collecting whole air samples or by removing the gaseous

---

* HASL SASP samplers collected air at rates on the order of 1 m^3/min.

nuclides from an air stream using some type of absorption process. Besides collecting air on filters, aircraft and balloon sampling systems often included whole air samplers (Figure 18.13) (EML, 1997; Miller and Larsen, 2002). Large volumes could be collected by pressurizing the samples.

The most widely used absorbent has been activated charcoal. Activated charcoal is almost a universal absorbent and effectively removes nuclides such as radon, krypton, xenon, and gaseous iodine from an air stream with an acceptable efficiency. The efficiency depends on flow rate and can be increased by cooling the charcoal using dry ice or liquid nitrogen. Cold traps have also been used to measure tritium by condensing tritiated water vapor from the atmosphere. Tritium (H-3) is a long-lived ($t_{1/2}$ = 12.3 y) activation product produced in large amounts in thermonuclear weapons tests (UNSCEAR, 2000b). Although it is a weak β-ray emitter, its long half-life makes it a significant contributor to the total population radiation dose. Similarly, cryogenic separations have been used to remove noble gases such as Xe-133 from whole air streams. C-14 is another important contributor to the population dose from fallout (UNSCEAR, 2000b); it has also been sampled by drawing high volumes of air through a dry molecular sieve. $^{14}CO_2$ has often been sampled by bubbling air through a caustic solution of NaOH where the carbon is chemically trapped. Other detectors (e.g., silver zeolite) have been used to collect specific chemical species of iodines such as organics and elemental iodine, which may not be efficiently retained on filters or charcoal.

Sampling and analysis for Xe-133 have long been used for detecting the presence of fallout from recent events, and thus the occurrence of a nuclear test, or other nuclear activity because it is a noble

FIGURE 18.13 HASL Balloon aerosol and whole air sampling systems. (Adapted from EML. 1997. *Environmental Measurements Laboratory. Procedures Manual.* HASL-300. Available at National Technical Information Service (NTIS), Springfield, VA. Also available at http://www.eml.st.dhs.gov/)

gas and has a fairly long half-life (~5 days), and thus a relatively long residence time in the atmosphere. Routine monitoring of Xe-133 is an important component of the CTBT nonproliferation monitoring program (Matthews and Schultze, 2001).

Improvements in chemical techniques have resulted in very sensitive analyses of α- and β-emitters such as Sr-90 and Pu-239 + Pu-240. Ratios of plutonium isotopes in collected air and soil have been used to identify specific sources of fallout, for example, to distinguish fallout in soil due to NTS tests from global fallout (Beck and Krey, 1983). The development of more sensitive mass spectrometric techniques to measure the ratio in chemically separated plutonium samples has allowed this technique to gain widespread use in characterizing not only fallout samples, but also for distinguishing plutonium contamination downwind from the Rocky Flats nuclear fuel fabrication facility from global fallout (Krey, 1976). Sensitive accelerator mass spectrometry (AMS) techniques have also been used to investigate signatures of other radioactive debris that can be used to infer the origin of the activity (Beasley et al., 1998).

SUMMARY

Air sampling has been, and will continue to be, an important tool for detecting and characterizing radioactive fallout. This chapter has discussed the origins and characteristics of radioactive fallout and some of the techniques used in the past and present to measure fallout. Although not always sufficient by itself to fully characterize the deposition of fallout, air sampling is an invaluable technique for determining particle size distributions and isotopic composition and for tracing the movement of radioactive debris through the environment.

REFERENCES

Baurmash, L., Neel, J.W., Vance, W.K.III, Mork, H.M., and Larson, K.H. 1958. Distribution and characterization of fallout and airborne activity from 10 to 160 miles from Ground Zero, Spring 1955; Project 37.2 (CETG), Operation Teapot, WT-1178, November 1958; Atomic Energy Project, School of Medicine, University of California, Los Angeles, Unclassified. Available at http://www.osti.gov/opennet/.

Beasley, T.M., Kelley, J.M., Orlandini, K.A., Bond, L.A., Aarkrog, A., Trapeznikov, A.P., and Pozolotina, V.N. 1998. Isotopic Pu, U, and Np signatures in soils from Semipalatinsk-21, Kazakh Republic and the Southern Urals, Russia, *J. Environ. Radioact.*, 39, 215–230.

Beck, H.L., Helfer, I.K., Bouville, A., and Dreicer, M. 1990. Estimates of fallout in the western U.S. from Nevada weapons testing based on gummed-film monitoring data. *Health Phys.*, 59, 565–570.

Beck, H.L. and Krey, P.W. 1983. Radiation exposure in Utah from Nevada nuclear tests. *Science*, 220, 18–24.

Beck, H.L. and Bennett, B.G. 2002. Historical overview of atmospheric nuclear weapons testing and estimates of fallout in the continental United States. *Health Phys.*, 82, 591–608.

Bennett, B. G. 2002. Worldwide dispersion and deposition of radionuclides produced in atmospheric tests. *Health Phys.*, 82, 644–655.

Bouville, A. and Beck, H. L. 2000. The HASL gummed-film network and its use in the reconstruction of doses resulting from nuclear weapons tests. *Technology*, 7, 355–379.

Breslin, A. and Cassidy, M. 1955. *Radioactive Debris from Operation Castle*, Islands of the Mid-Pacific, NYO-4623.

Cederwall, R.T., Ricker, Y.E., Cederwall, P.L., Homan, D.N., and Anspaugh, L.R. 1990. Ground-based air-sampling measurements near the Nevada test site after atmospheric nuclear tests, *Health Phys.*, 59, 533–540.

DHHS. 2006. Appendix D of *A feasibility Study of the Health Consequences to the American Population from Nuclear Weapons Tests Conducted by the United States and Other Nations*. Vols. I and II. Department of Health and Human Services, Centers for Disease Control and Prevention and the National Cancer Institute. Washington, DC. Accessible at http://www.cdc.gov/nceh/radiation/fallout/default.htm.

DNA. 1979. *Defense Nuclear Agency: Compilation of Local Fallout Data from Test Detonations 1946–1962.* Extracted from DASA 1251, Volume I-Continental U.S. Tests; Volume II-Oceanic U.S. Tests, DNA 1251-EX, 1 May 1979.

Eisenbud, M. 1997. Monitoring distant fallout: The role of the Atomic Energy Commission Health and Safety Laboratory during the Pacific tests, with special attention to the events following BRAVO, *Health Phys.*, 73, 21–27.

EML. 1997. *Environmental Measurements Laboratory. Procedures Manual. HASL-300*. Available at National Technical Information Service (NTIS), Springfield, VA. Also available at http://www.eml.st.dhs.gov/.

EPA. 2005. Real-Time Monitoring of Radiation in Air in the United States-Updating and Expanding the Environmental Radiation Ambient Monitoring System (ERAMS), Background Information for the Radiation Advisory Committee of the EPA Science Advisory Board, Office of Radiation and Indoor Air, United States Environmental Protection Agency, January 24, 2005.

Feely, H.W. and Spar, J. 1960. Tungsten-185 from nuclear bomb tests as a tracer for stratospheric meteorology. *Nature*, 188, 1062–1064.

Freiling, E.C., Crocker, G.R., and Adams, C.E. 1965. Nuclear debris formation, in A.W. Klement (ed.), *Radioactive Fallout from Nuclear Weapons Tests—Proceedings of the 2nd Conference, USAEC-5 Conference Proceedings*, Conf. 765, Germantown, MD, pp. 1–41.

Glasstone, S. and Dolan, P.J. 1977. *The Effects of Nuclear Weapons*, 3rd ed. U.S. Department of Defense and U.S. Energy Research and Development Administration. Washington, DC. Available at http://www.dtra.mil/.

Hicks, H.G. 1981. *Results of Calculations of External Radiation Exposure Rates from Fallout and the Related Radionuclide Composition*. Livermore, CA: Lawrence Livermore National Laboratory; Report UCRL-53152, parts 1–8. Available at http://www.osti.gov/opennet/.

Hicks, H.G. 1982. Calculation of the concentration of any radionuclide deposited on the ground by off-site fallout from a nuclear detonation. *Health Phys.*, 42, 585–600.

Hoecker, W.H. and Machta, L. 1990. Meteorological modeling of radioiodine transport and deposition within the continental United States. *Health Phys.*, 59, 603–617.

Klement, A.W. (ed). 1965. Radioactive fallout from nuclear weapons tests. *Proceedings of the Second Conference. USAEC 5 Symposium Series*. Conf-765. USAEC, Germantown, MD. Available from NTIS.

Krey, P. 1967. Atmospheric burn-up of a plutonium-238 Generator, *Science*, 158(3802), 769–771.

Krey, P.W. 1976. Remote plutonium contamination and total inventories from Rocky Flats. *Health Phys.*, 30, 209–214.

Larsen, R.J., Sanderson, C.G., Lee, H.N, Decker, K.M., and Beck, H.L 1994. Fission products detected in Alaska following the Tomsk-7 accident, *J. Environ. Radioact.*, 23, 205–209.

List, R.J. 1954. *The Transport of Atomic Debris from Operation Upshot-Knothole*. Washington, DC. Joint US Atomic Energy Commission US Weather Bureau Report NYO-4602 (del); 1954 (declassified with deletions April 22, 1959. Available at http://www.osti.gov/opennet/.

Livingston, H.G. and Povinec, P.P. 2002. A millennium perspective on the contribution of global fallout to ocean science, *Health Phys.*, 82, 656–668.

Lockhart, L.B., Jr., Patterson, R.L., Jr., Saunders, A.W., Jr., and Black, R.W. 1965. Atmospheric radioactivity along the 80th Meridian west. In: Klement, A. (ed.), *Radioactive Fallout from Nuclear Weapons Tests. Symposium Proceedings of the Second Conference*, Washington, DC: U.S. Atomic Energy Commission; AEC Symposium Series Conf.-765, Germantown, MD, pp. 602–615.

Lowder, W.M., Beck, H.L., and Condon, W.J. 1964. The spectrometric determination of dose rates from natural and fallout gamma radiation in the United States, 1962–63, *Nature*, 202, 745–749.

Machta, L. 2002. Meteorological benefits from atmospheric nuclear tests. *Health Phys.*, 82, 635–643.

Matthews, M. and Schultze, J. 2001. The radionuclide monitoring system of the Comprehensive Nuclear-Test-Ban Treaty Organization: From sample to product, *Kerntechnik*, 66(3):96–101.

Miller, K.M. and Larsen, R.J. 2002. The development of field-based measurement methods for radioactive fallout assessment. *Health Phys.*, 82(5), 609–625.

Perkins, R.W., Thomas, C.W., and Nielson, J.M. 1965. Measurements of airborne radionuclides and determination of their physical characteristics, in Klement, A. (ed.), *Radioactive Fallout from Nuclear Weapons Tests. Symposium Proceedings of the Second Conference*, Washington, DC: U.S. Atomic Energy Commission; AEC Symposium Series Conf.-765, Germantown, MD, Nov., 1964, pp. 198–221.

Thompson, C.B., McArthur, R.D., and Hutchinson, S.W. 1994. *Development of the Town Data Base: Estimates of Exposure Rates and Times of Fallout Arrival near the Nevada Test Site*. Springfield, VA: National Technical Information Service; DOE/NVO-374 (1994). Available at http://www.osti.gov/opennet/.

UNSCEAR. 2000a. *Sources and Effects of Ionizing Radiation. United Nations Scientific Committee on the Effects of Atomic Radiation*. Report to the General Assembly with Scientific Annexes. New York: United Nations Publication, E.00.IX.3; 2000. Annex J.

UNSCEAR. 2000b. *Sources and Effects of Ionizing Radiation. United Nations Scientific Committee on the Effects of Atomic Radiation*. Report to the General Assembly with Scientific Annexes. New York: United Nations Publication, E.00.IX.3; 2000. Annex C.

Part V

Example Air Sampling Methods for Airborne Radioactivity

INTRODUCTION TO THE METHODS

Mark L. Maiello and Mark D. Hoover

Some of the following methods were originally presented in the book, *Methods of Air Sampling and Analysis* edited by the late J. P. Lodge, Jr. When that book was not revised for a 4th edition, the decision was made to present them here. All have been updated and a few new methods have been introduced.

The radionuclides presented here are some of those most often requiring air sampling. They also reflect the varied techniques and challenges associated with radioactive air sampling. Some of these radionuclides are associated with nuclear power production such as the radioiodines and the radioactive noble gases krypton and argon. Others, like tritium and carbon-14, are used in biomedical research. Still others such as radon and its progeny are naturally occurring and may be measured for their own sake or to determine possible interference with measurements for other nuclides. The emphasis has been placed on integrated measurements using physico-chemical means rather than electronic instrument detection. The exceptions are the methods associated with continuous air sampling for plutonium and personal air sampling that of course, rely heavily on commercially available equipment.

The format in this section of the book is derived from that used in *Methods of Air Sampling and Analysis*. The intent was to provide a didactic description rather than a narrative so that practical measurements could be realized. Those with an interest or a compelling need to measure the radionuclides in this section should find enough detail to accomplish the task.

Method 1

Determination of the Gross Alpha-Radioactivity Content of the Atmosphere*

CONTENTS

1 Principle of the Method 391
2 Range and Sensitivity 391
3 Interferences 392
4 Advantages/Disadvantages 393
5 Precision and Accuracy 393
6 Apparatus 394
7 Reagents 395
8 Procedure 395
9 Calibration and Standards 395
10 Effects of Storage 396
11 Calculation of Concentration 396
12 Cautions 398
References 398

1 PRINCIPLE OF THE METHOD

1.1 Air particulate matter collected on a filter of high surface retention is counted with an α-sensitive detector to establish the gross concentration of α-particle-emitting radionuclides present in the sample air. A specific α-emitter cannot be present in a greater concentration than the gross α-concentration of a mixture of unidentified radionuclides. Consequently, a gross analysis may eliminate the need for a more time consuming and expensive analysis for specific radionuclides. Gross α-analysis does not yield detailed information about the radionuclide composition of the air sample. However, it is a simple, rapid, inexpensive screening technique.

2 RANGE AND SENSITIVITY

2.1 Overloading the filter medium with airborne particles can define the upper measurement limit of α-emitting radionuclide concentrations as can the need to minimize the potential for detector contamination due to α-recoil (NCRP, 1985). The sensitivity of the method is adjustable by specification of the sampling flow rate, sampling time, and/or α-counter sensitivity, that is, the

---

* Method authors: Appreciation is extended to the writers of the Radioactivity section of the book *Methods of Air Sampling and Analysis* where this method originally appeared. Method reviewers and editors: Hung-Cheng Chiou, Paul Linsalata, Mark L. Maiello, Robert P. Miltenberger, and Theodore E. Rahon.

counting efficiency, the counter background, and the duration of the counting time. Present U.S. regulatory guidelines limit exposure of the general public to annualized concentrations of α-emitters in air as low as 0.4–0.004 mBq m^{-3} (10^{-14}–10^{-16} μCi mL^{-1})* for some transuranic nuclides (Federal Register, 1991). The investigator should be mindful of such limits even when conducting gross α-measurements, especially for nonoccupational environments.

2.2 The required volume of sample air is dictated by the lower limit of detection (LLD) of the α-counter and by extension, the minimum detectable concentration (MDA) (refer to the Chapter 5 for a brief discussion and sample calculation of LLD). For example, a 0.1 m^3 min^{-1} pump operating for 24 h (144 m^3 air) sampling 0.4 mBq m^{-3} (10^{-14} μCi mL^{-1}) would collect about 3.5 dpm of α-activity (assumed to be net activity above the background). If this activity is below the detection limit of the counter, the flow rate or sampling time should be increased so that sufficient activity is collected. Alternatively, a more sensitive α-analysis counting system could be used. The required volume of air should be calculated with consideration paid to filter loading and air flow-rate control.

3 INTERFERENCES

3.1 The principal interference is from the progenies of naturally occurring radon and thoron (see Method 7) that are usually present in the outdoor atmosphere in average concentrations of about 8 Bq m^{-3} (2×10^{-10} μCi mL^{-1}) and 15 Bq m^{-3} (4×10^{-10} μCi mL^{-1}),† respectively (NCRP, 1988; Schery, 1990). They may build up to more than 10 times these levels in closed indoor environments. Thus, α-emitting progeny of radon (^{222}Rn) and thoron (^{220}Rn) are present in the atmosphere in concentrations hundreds of times greater than the guidelines for many α-emitters of interest. Fortunately, the radioactive half-lives of the radon progeny are relatively short, the longest with relevance to air sampling being 26.8 min (^{214}Pb) and 19.7 min (^{214}Bi), which are intermediates in the decay chain to the two α-emitters ^{218}Po and ^{214}Po with 3.05 min and 1.6×10^{-4} second half-lives, respectively. The 10.6 h half-life of ^{212}Pb (thoron progeny) is short enough as is that of the radon progeny that their interference may be nullified by waiting until this radioactivity has decayed to negligible levels before counting the filter. This would require about 3 d for the thoron progeny and about 4 h for the radon progeny although 2 h is usually adequate for the latter (Harley, 1953). An evaluation of long-lived gross α-concentrations can be performed by making two counts at least 8 h apart (see calculation in Section **11.2**).

Another way to differentiate between these natural progenies and other α-emitters is to employ an α-spectrometer using a solid-state semiconductor detector (NCRP, 1985; Knoll, 2000) or perhaps a liquid scintillation counter (McDowell and McDowell, 1994; Kessler, 1989). Technically, this is no longer a gross α-measurement. It also presupposes that the energies of the natural emitters do not interfere with other α-emitters of interest to the investigator. The signals from semiconductor detectors are proportional to the energy of the α-particles. Most long-lived α-emitters have energies below 6 MeV, whereas the daughters of radon and thoron decaying by α-emission do so with equal or greater energies. It is therefore possible to gate the associated electronics or specify energy regions of interest to count only the pulses from low-energy α-emissions. This method works better when detecting ^{235}U but may not work well with measurements of ^{241}Am and ^{239}Pu due to the higher-energy α-emissions of these nuclides and the greater energy overlap with radon or thoron daughters. The reader should be aware of the limitations of the instruments being used especially when using energy discrimination techniques to avoid interference from radon and/or thoron.

3.2 Use of improper filter media is a potential source of error. α-Particles have a limited range and are stopped by very thin and lightweight materials. When airborne particles containing α-emitting radionuclides penetrate deep into the filter matrix, the α-radiation may be shielded from detection by the intervening filter mass. Therefore, the use of membrane-type filters that collect

---

* As of this writing, effluent regulatory limits are commonly expressed in μCi mL^{-1} in the United States.
† Mean thoron concentration at 1 m above the ground.

airborne material at the filter surface is recommended. Surface collection of dust particles allows the emitted α-particles to escape the filter and be counted by the detector rather than being "self-absorbed" at deeper levels. An additional advantage of membrane filters (with a pore diameter of 0.8 μm or less) is that collection efficiencies are over 99% (Hinds, 1982), thus precluding the need to include a collection efficiency correction factor in the radionuclide concentration calculations (Section **11**).

3.3 Dust loading also contributes to self-absorption and α-energy degradation. The decrease in α-count rate with the increase in surface density up to 700 g cm^{-2} of a radioactive uranium source is illustrated in NCRP Report No. 58 (1985). The total volume of air that can be sampled in a dust-laden atmosphere is thereby limited. Light dust loading (light gray air filter sample) can be counted in a liquid scintillation counter by adding the filter to aqueous-accepting scintillation fluid (McDowell and McDowell, 1994). Dust loading can also introduce unwanted pressure drops in sampling lines without automatic flow rate correction. This will result in errors in the concentration results (Craig, 1971).

4 ADVANTAGES/DISADVANTAGES

4.1 This method is applicable only for certain screening situations where radionuclide identification is not necessary. The accuracy of the measurement depends on the α-emitting radionuclide used to standardize the α-counter. This is so primarily because self-absorption is energy dependent. Most α-energies are between 4 and 8 MeV. If a 5 MeV α-standard is used to determine the counting efficiency, the field sample results (dpm) should agree within a factor of two with the true decay rate (NCRP, 1976). A dust-free (thin) sample analyzed in a counter calibrated with a standard geometrically and energetically similar, is recommended.

4.2 Meaningful predictions about the fate of an α-emitting radionuclide in the environment cannot be made unless an identification is made. It follows that without specific identification, the human dosimetric implications of a release of an (unknown) α-emitting radionuclide cannot be accurately achieved. Gross measurements should not be used for calculating radiation dose nor should they be used to show compliance with nuclide-specific regulations.

4.3 α-Spectroscopic analyses and/or radiochemical techniques are available from many commercial laboratories for the quantification and identification of individual α-emitters from an air-filter sample.

4.4 Refer to National Council on Radiation Protection and Measurements Report No. 50 (NCRP, 1976) for more discussion about the implications of the use of this method. The problems associated with air-filter mass loading, choice of counting standards, and interpretation of the results are reviewed.

5 PRECISION AND ACCURACY

5.1 The precision of the method is essentially a function of the volume of air sampled, the sample counting time, the background of the counter, and the counting efficiency of the detector. The uncertainty can be decreased (precision increased) by increasing the air volume sampled, increasing the counting time, decreasing the counter background, and/or increasing the counting efficiency. See Section **11.4** for an elementary standard deviation equation commonly used in radioactivity measurements.

Consider an α-counter operating with 35% efficiency (0.35 cpm resulting from one radioactive dpm, or 0.35 cpm/dpm), a background of 0.1 cpm (determined by counting a blank filter), and a counting time of 1 h for the sample and background. If a result of 2.0 dpm (including background) was obtained, the resulting uncertainty (1.96σ) would be 0.65 dpm, or 37.7 of the net results at the 95% confidence level (1.96σ). The same result obtained after a 10 min count in a different type of α-counter with 90% efficiency and a background of 0.5 cpm would have an uncertainty of 1.04 dpm, or 72.3%. Thus, the first counting method with the longer counting time and lower background will provide a more precise result.

5.2 The accuracy of a gross α-analysis is dependent primarily on the calibration of the α-counting system (the uncertainty in the counting efficiency due to the energy difference between the standard and the unknown in the sample), the accuracy of measurement of the volume of air sampled, and the appropriate consideration of an α-particle self-absorption correction.

6 APPARATUS

6.1 The basic requirements for the method are an air mover, a calibrated device for measuring airflow rate or volume (rotameter or gas meter), a particle filter, and an α-radiation counter. The size (diameter) of the filter should not be any larger than the sensitive area of the α-detector. Due to the limited range of α-radiation, a high surface-retention filter, such as a membrane-type filter, should be used and the dust buildup on the filter minimized so that the problem of α-particle self-absorption is avoided. A constant flow rate can be obtained using a critical flow venturi. A specialized filter head incorporating a venturi is described by Parulian et al. (1996).

6.2 If filters of appreciably less than 100% collection efficiency are employed, it is necessary to know or determine the collection efficiency as a function of filter face velocity and the physical properties of the material being sampled. The collection efficiency of polystyrene fiber, glass fiber, and cellulose ester membrane filter media is >99% at a particle diameter of 0.3 μm and a face velocity of 27 cm s^{-1} (Hinds, 1982). Such high values preclude the need for a collection efficiency factor in calculating airborne concentrations (see Section **11**). If in doubt about the collection efficiency, consult the filter manufacturer* or, if the expertise and equipment are on hand, perform a collection efficiency determination.

6.3 Because of the low penetration ability of α-radiation, a windowless or very thin window α-sensitive detector should be used to count the filter sample. Descriptions of the general types of counters follow:

- Gas proportional counter: Thin window or windowless operation, automatic sample changers available, requires that the counting chamber be purged with P10 gas (10% methane + 90% argon), typical background: <0.1 cpm, and nominal efficiency 50% (windowless) or 40% with window (Oxford Systems, 1998).
- ZnS scintillator with photomultiplier tube: Normally uses a light-tight Mylar window or light-tight box for windowless operation; usually a portable, low-cost system, typical background: 0.3 cpm, and nominal (4π) efficiency 37% (windowless) or 30% (window) (Ludlum Catalog, 1998).
- Solid-state (charged particle) detector: Uses a thin metallic window or vacuum chamber for windowless operation; can be portable, low-cost system; can be used for α-energy analysis to identify radionuclides in the sample; typical background: is 0.001–0.03 cpm; efficiency is >25% with ^{241}Am (Oxford Systems, 1998; Canberra, 1998).
- Liquid scintillation counter: The sample is submerged in an organic scintillation fluid in a 7–20 mL vial which is counted by electronically gated photomultiplier tubes to discriminate against β-radiation; automatic sample changers are available; modern units are equipped for spectroscopic energy analysis; typical background: is <20 cpm; nominal (4π) α-particle efficiency is 90% (Packard Catalog material, 1998); and dissolution of the filter or use of a compatible scintillation fluid may be necessary (Passo and Cook, 1996).

Note: All of the above devices require specialized knowledge to operate. The novice is advised to seek the assistance of individuals familiar with the operation of these units.

---

* For example, Gelman Sciences, Inc., USA, or suppliers such as HI-Q Environmental Products Co., USA, F & J Specialty Products, Inc., USA.

Method 1: Determination of the Gross Alpha-Radioactivity Content of the Atmosphere

7 REAGENTS

7.1 No reagents are needed for dry filter counting. If liquid scintillation analysis is to be performed, liquid scintillation fluid and vials are obtained commercially.

8 PROCEDURE

8.1 Calculate the volume of air necessary to collect measurable α-activity, considering the LLD of the counting system and the regulatory concentration values.

8.2 Record the time and the initial flow rate or the initial value of the flow totalizer when sampler is started.

8.3 Operate the sampler in a location representative of the atmosphere for which the concentration is to be established with the sample point preferably 2 m off the ground. If not using a totalizer, carefully time the air sampling period.

8.4 At the end of sampling, record the final airflow rate and time or the final totalizer value. Place the filter in a protective cover or container upon removal from its holder to minimize the possibility of damaging the filter or dislodging the collected activity.

8.5 If necessary, store the sample for 2 h to 3 days to allow radon and thoron progeny to decay away.

8.6 Determine the counter background, perform preuse performance check, and calibrate the instrument if required.

8.7 Prior to counting, place the filter sample in a planchet or holder for reproducible positioning under the detector, or if liquid scintillation analysis is to be used, place it in a scintillation vial and add an appropriate volume of scintillation fluid (15 mL in a 20 mL vial is sufficient).

8.8 Count the sample for an interval (or total count) sufficient for a statistically significant result.

8.9 If needed, determine the sample self-absorption correction factor and the filter collection efficiency.

8.10 Use the initial and final airflow rates to calculate a mean rate. Calculate the gross α-airborne concentration and its uncertainty (Section **11**).

9 CALIBRATION AND STANDARDS

9.1 The efficiency of an α-counter should be determined with an α-standard as close as possible to the sample in size and composition. It should be counted in the same geometry as the sample as well. Standards containing α-emitters are commercially available from various suppliers.* Depending on the counting system used, the counting efficiency may vary significantly with α-particle energy. The standard should be similar in energy to the suspected airborne radionuclide. In effect, the air concentration determined from a gross measurement is what would be measured if the radionuclide comprising the standard was being sampled. To make this apparent, the reported results should specify the standard radionuclide used.

9.2 Counter background should be determined over an interval similar to or longer than the sample counting time. Gross α-counters are usually quite stable and it is satisfactory to make only daily determinations of background, mainly to ensure that malfunction, electronic drift (detector voltage and/or amplifier gain), or radioactive contamination have not occurred. The background is determined by counting a clean filter or "blank." The blank filter should be of the same size and composition as the filter used for field measurements.

9.3 If a rotameter is used as the air flow measurement device, the reader is urged to consult the reference by Craig (1971) concerning the problems associated with the calibration of these devices.

---

* Suppliers include: GE Healthcare Life Sciences, USA and UK, Isotope Products Laboratories, USA and North American Scientific, USA. Generally, traceability to the U.S. National Institute of Standards and Technology is available.

For example, if the rotameter is not calibrated using exactly the same sampling line situation (pressure drop across the sampler) and sampling conditions (ambient temperature and pressure), significant errors may be introduced into the air concentration results. Craig advises use of flow rates that cause the lowest pressure drops possible, and meter readings at the upper end of the rotameter scale. It is also advisable to place a pressure gauge on the same side of the pump regulator valve as the rotameter float for the purposes of determining a correction for any pressure drop between the filter and the rotameter and for the detection of line breaks. The novice should also consult an experienced individual such as an industrial hygienist concerning these matters. The ASTM standard for calibration of a rotameter to standard conditions may be used (ASTM, 1996) although ASTM does not provide a discussion of calibration problems. Air flow meters should be calibrated annually and after repairs or modifications (NRC, 1992).

10 EFFECTS OF STORAGE

10.1 A properly protected air-filter sample should experience no effects other than radioactive decay during storage. Immediately after sample collection, the filter should be carefully placed in an acetate, plastic, or paper envelope. It must be handled carefully to avoid tearing and loss of mass.

α-Emitting samples, such as those of ^{210}Po, can lose activity to the immediate surroundings through the recoil action of α-particle decay (NCRP, 1985). This phenomenon can cause recoil contamination of the α-particle detector or any of the sample surroundings. The contamination can be greater with higher-activity sources. To mitigate the effect, the air filter can be protected with metallized Mylar or other protective films.

10.2 Scintillation fluid samples will evaporate over time. It may help to store the tightly sealed vials in a cool or cold location. Long-term storage is not recommended. Glass scintillation vials are subject to breakage.

11 CALCULATION OF CONCENTRATION

The gross α airborne concentration can be calculated using the following equations.

11.1 *Counting Efficiency:*

$$\text{EFF} = \frac{\text{CR}_{\text{STD}}}{\text{ACT}_{\text{STD}}}, \qquad (1)$$

where CR_{STD} is the net count rate* of standard [usually in units of counts per second (cps) or counts per minute (cpm)] or the net counts achieved in the counting interval, ACT_{STD} is the activity or disintegration rate of standard (dps or dpm) or the total disintegrations in the counting interval, EFF is the counting efficiency [net counts/disintegrations (c d^{-1}) or cps dps^{-1} or cpm dpm^{-1}].

Note: any units of time for the count or disintegration rates may be used as long as they cancel out in the equation.

11.2 *Sample Activity:*

$$\text{ACT}_{\text{SAM}} = \frac{(G/t_s) - (B/t_B)}{\text{EFF} \times \text{COR}_\alpha \times F_{\text{EFF}}}, \qquad (2)$$

where ACT_{SAM} is the sample activity (Bq) [1 dps = 1 Becquerel (Bq)], G is the gross counts, B is the background counts, t_S is the sample count time (s), t_B is the background count time (s), EFF is the

---

* The background count rate has been subtracted.

Method 1: Determination of the Gross Alpha-Radioactivity Content of the Atmosphere

counter efficiency [see above for units], COR_α is the α-particle self-absorption correction [if needed] and F_{EFF} is the filter collection efficiency [if needed].

If a long-lived α-emitter is to be determined in the presence of radon and thoron progenies, the following method (Cember, 1992; Allen, 1997) may be employed:

$$ACT_{MAT} = \frac{ACT_2 - ACT_1 e^{-\lambda \Delta t}}{1 - e^{-\lambda \Delta t}}, \quad (3)$$

where ACT_{MAT} is the activity of the long-lived material (Bq; see note below), ACT_1 is the activity on filter at time t_1 (Bq), ACT_2 is the activity on filter at time t_2 (Bq), Δt is the elapsed time between counts = $t_2 - t_1$ (h) and λ is the decay constant for $^{212}Pb = \ln 2/t_{1/2} = 0.693/10.6$ h $= 0.0653$ h^{-1}.

Allen recommends that the activity on the filter decay for 8 h to remove ^{222}Rn progeny and allow ^{220}Rn progeny to reach equilibrium with ^{212}Pb. Cember mentions waiting about 4 h for the decay of ^{214}Pb (t_1) and 20 h for the decay of ^{212}Pb (t_2).

Note: ACT_1 and ACT_2 must be corrected for background, counting efficiency, and filter efficiency as shown in Section **11.2** in order for the result to be meaningful. ACT_{MAT} may then be substituted for ACT_{SAM} in the following equations.

11.3 *Gross α air concentration:*

$$Conc_\alpha = \frac{ACT_{SAM}}{V_S},$$

where $Conc_\alpha$ is the α-particle air concentration (Bq m^{-3}) and V_S is the air volume sampled (m^3), [mean flow rate (m^3 min^{-1}) × air-sampling time (min)].

This result may be converted into µCi mL^{-1} by dividing ACT_{SAM} by $(1 \times 10^6$ mL m$^{-3} \times 2.22 \times 10^6$ dpm µCi$^{-1} \times 1/60$ min s$^{-1}) = 3.7 \times 10^{10}$.

11.4 *Uncertainty:* The statistical significance of this result should also be indicated. It can be determined from the standard deviation (SD) of the count. If the equation for ACT_{SAM} in Section **11.2** is used,

$$SD = \frac{\left[(G/t_S^2) + (B/t_B^2)\right]^{1/2}}{EFF \times COR_\alpha \times F_{EFF} \times V_S}, \quad (4)$$

where SD is the standard deviation of the air concentration (Bq m^{-3}), G is the gross counts, B is the background counts, t_S is the sample count time (s), t_B is the background count time (s), EFF is the counter efficiency [see above for units], COR_α is the α-particle self-absorption correction [if needed], F_{EFF} is the filter collection efficiency [if needed], and V_S is the mean air volume sampled (m^3).

The statistical significance of the final concentration result should be expressed as the uncertainty at a specified confidence level. For example, to express the uncertainty at the 95% confidence level (also referred to as the 1.96σ or "two sigma" level), multiply the SD result by 1.96. This may be converted into µCi mL^{-1} by dividing the uncertainty at 1.96σ (or the uncertainty calculated at any other confidence level) by $(1 \times 10^6$ mL m$^{-3} \times 2.22 \times 10^6$ dpm µCi$^{-1} \times 1/60$ min s$^{-1}) = 3.7 \times 10^{10}$.

If Equation 3 for ACT_{MAT} was applied, the SD is determined from a more complicated expression (Allen, 1997):

$$SD = \left[\left(\frac{1}{1 - e^{-\lambda \Delta t}}\right)^2 \times (SD_{ACT2})^2 + \left(\frac{e^{-\lambda \Delta t}}{1 - e^{-\lambda \Delta t}}\right)^2 \times (SD_{ACT1})^2\right]^{1/2}, \quad (5)$$

where SD_{ACT1} is the standard deviation of the first count activity and SD_{ACT2} is the standard deviation of the second count activity.

See Section **11.4** for Equation 4 to determine SD_{ACT1} and SD_{ACT2}.

12 CAUTIONS

12.1 This method may involve the use of hazardous chemicals and/or radioactivity. It is the user's responsibility to take appropriate safety precautions and to obtain necessary regulatory approvals from institutional and/or governmental agencies.

12.2 If the user is not trained or familiar with the calibration and operation of radiation analysis equipment, they should consult a qualified expert, such as a health physicist, for assistance in the proper and safe usage of the equipment and development of appropriate quality assurance protocols.

12.3 Radioactive waste and hazardous waste such as scintillation fluid require special disposal considerations that are regulated by governmental authorities. Consult a specialist if unfamiliar about the treatment of such wastes.

REFERENCES

Allen, D.E., Determination of MDA for a two-count method for stripping short-lived activity out of an air sample, *Health Phys.*, 73, 512–517, 1997.
ASTM, *Annual Book of ASTM Standards.* Section 11, *Water and Environmental Technology.* Designation D-3195–90, 1996, Standard Practice for Rotameter Calibration, ASTM, West Conshohocken, PA, 1996.
Canberra, Edition Ten Product Catalog, Canberra Industries, Inc., Meriden, CT, 1998.
Cember, H., *Introduction to Health Physics*, Pergamon Press, New York, NY, 1992.
Craig, D.K., The interpretation of rotameter air flow readings, *Health Phys.*, 21, 328–332, 1971.
Federal Register, Part VI, U.S. Nuclear Regulatory Commission, 10 CFR Part 20 et al., Code of Federal Regulations, Standards for Protection Against Radiation; Final Rule, May, 1991, Washington, DC, 1991.
Harley, J.H., Sampling and measurement of airborne daughter products of radon, *Nucleonics* 11, 12–17, 1953, Reprinted in *Health Phys.*, 38, 1068–1074, 1980.
Hinds, W.C., *Aerosol Technology*, John Wiley & Sons, New York, 1982.
Kessler, M.J., *Liquid Scintillation Analysis—Science and Technology*, Packard Instrument Co. (now Perkin-Elmer), Publication No. 169–3052, Meriden, CT, 1989.
Knoll, G.F., *Radiation Detection and Measurement*, 3rd ed., John Wiley & Sons, New York, 2000.
Ludlum Catalog, Ludlum Measurements, Inc., Sweetwater, TX, 1998.
McDowell, W.J. and McDowell, B.L., *Liquid Scintillation Alpha Spectrometry*, CRC Press, Boca Raton, FL, 1994.
NCRP, *Environmental Radiation Measurements*, National Council on Radiation Protection and Measurements, Report No. 50, NCRP, Bethesda, MD, 1976.
NCRP, *A Handbook of Radioactivity Measurements Procedures*, 2nd ed., National Council on Radiation Protection and Measurements Report No. 58, NCRP, Bethesda, MD, 1985.
NCRP, *Measurement of Radon and Radon Daughters in Air*, National Council on Radiation Protection and Measurements Report No. 97, NCRP, Bethesda, MD, 1988.
NRC, *Air Sampling in the Workplace, U.S. Nuclear Regulatory Commission, Regulatory Guide 8.25*, National Technical Information Service, Springfield, VA and U.S. Government Printing Office, Washington, DC, 1992.
Oxford Systems, *Instruments and Components for Radiation Detection and Measurement*, 5th ed., Oxford Instruments, Inc., Oak Ridge, TN, 1998.
Packard Catalog Material, Packard Instrument Company (now Perkin-Elmer), Meriden, CT, 1998.
Parulian, A., Rodgers, J.C., and McFarland, A.R., A constant flow filter air sampler for workplace environments, *Health Phys.*, 71, 870–878, 1996.
Passo, C.J. and Cook, G.T., *Handbook of Environmental Liquid Scintillation Spectrometry*, Packard Instrument Company (now Perkin-Elmer), Meriden, CT, 1996.
Schery, S.D., Thoron in the environment, *J. Air Waste Manage. Assoc.*, 40, 493–497, 1990.

Method 2
Determination of the Gross Beta-Radioactivity Content of the Atmosphere*

CONTENTS

1 Principle of Method 399
2 Range and Sensitivity 399
3 Interferences 400
4 Advantages/Disadvantages 400
5 Precision and Accuracy 401
6 Apparatus 401
7 Reagents 402
8 Procedure 402
9 Calibration and Standards 403
10 Effects of Storage 403
11 Calculation of Concentration 403
12 Cautions 405
References 405

1 PRINCIPLE OF METHOD

1.1 Air particulate matter collected on a filter of high retention is counted with a β-sensitive detector to establish the gross concentration of the β-emitting radionuclides present in the sampled ambient air. Since a given β-emitter cannot be present in a greater concentration than the gross β-concentration of a mixture of unidentified radionuclides, a gross analysis may eliminate the need for a more time-consuming and expensive analysis for specific radionuclides. Gross β analysis does not yield specific information about the radionuclide composition of the air sample. However, it is a simple, rapid, inexpensive screening technique.

2 RANGE AND SENSITIVITY

2.1 The method has the advantage of being suitable over the entire range of concentrations of airborne β-emitters ordinarily encountered in the environment (except for tritium—see Methods 3 and 4). The sensitivity of the method is adjustable by specification of sampling flow rate, sampling time, and/or β-counter sensitivity, that is, the counting efficiency, the counter background, and the duration of the counting. Present U.S. regulatory limits for exposure of the general public to annualized

---

* Method authors: Appreciation is extended to the writers of the Radioactivity section of the book *Methods of Air Sampling and Analysis* (Lodge, 1988) where this method originally appeared. Method reviewers and editors: Hung-Cheng Chiou, Paul Linsalata, Mark L. Maiello, Robert P. Miltenberger, and Theodore E. Rahon.

concentrations of most β-emitters in air are in the range of 3.7–3700 Bq m^{-3} (10^{-10}–10^{-7} µCi mL^{-1}) but may approach 0.037 Bq m^{-3} (10^{-12} µCi mL^{-1}) (Federal Register, 1991). The investigator should be mindful of such limits even when conducting gross β-measurements, especially in nonoccupational environments.

2.2 The required volume of sample air is dictated by the lower limit of detection (LLD) of the β-counter and by extension, the minimum detectable concentration (MDA) for the suspected β-activity. The reader may refer to Chapter 5 for a brief discussion and sample LLD calculation. The volume of required air must be estimated with consideration paid to filter loading and airflow-rate control.

3   INTERFERENCES

3.1 The principal interference is from the progenies of naturally occurring radon and thoron that are usually present in the outdoor atmosphere in concentrations of about 8 Bq m^{-3} (2×10^{-10} µCi mL^{-1}) and 15 Bq m^{-3} (4×10^{-10} µCi mL^{-1}),* respectively (NCRP, 1988; Schery, 1990). They may build up to more than 10 times these levels in closed indoor environments. Thus, β-emitting progeny of radon and thoron could be present in the atmosphere in concentrations equal to or greater than the guidelines for many β-emitters of interest. The β-particle-emitting progeny in the decay chains possesses relatively short half-lives, specifically 26.8 min (^{214}Pb) and 19.7 min (^{214}Bi) in the radon chain, and 10.6 h (^{212}Pb) in the thoron chain. Thus, the interferences from the β-emitting progeny may be nullified by waiting until these activities have decayed to negligible levels before counting the filter. This would require about 3 days for the thoron progeny and about 4 h for the radon progeny, although 2 h for the latter is usually adequate (Harley, 1953). An evaluation of long-lived gross β-concentrations can be performed by making two counts at least 8 h apart (see calculation in Method 1, Section **11.2**).

3.2 As explained in Method 1, filter samples with heavy dust loads may cause serious interference, particularly when measuring low-energy β-particles which, after radioactive emission, may be shielded from detection by the dust mass (NCRP, 1976). This "self-absorption" is difficult to correct for and, because β-particle energies vary widely, a maximum sample thickness cannot be stated. Dust loading can also introduce unwanted pressure drops in the sampling line, which cause errors in the airflow measurements and thus in the concentration results (Craig, 1971). The reader should note that the requirement for detection sensitivity (large volumes of sample air) conflicts with the need to obtain relatively dust-free filter samples.

4   ADVANTAGES/DISADVANTAGES

4.1 This method is applicable only for certain screening situations where radionuclide identification is not necessary. Because one standard is usually used to calibrate the β-counter, the gross β-results are reported as if the air concentration of the radionuclide comprising the standard was being measured. Using a single counting efficiency for all β-emitters is necessary because this is a gross measurement. To make this apparent, the reported results should specify the standard radionuclide used.

4.2 Meaningful predictions about the fate of the β-emitting nuclide in the environment cannot be made unless an identification is made. It follows that without specific identification, the human dosimetric implications of a release of an (unknown) β-emitting radionuclide cannot be accurately performed. Gross measurements should not be used for calculating radiation dose nor should they be used to show compliance with nuclide-specific regulations.

4.3 Techniques are available from many commercial laboratories for the measurement and identification of individual β-emitters from an air-filter sample (including γ-spectroscopy for β–γ-emitting radionuclides).

---

* Mean thoron concentration at 1 m above the ground.

4.4 Refer to NCRP Report No. 50 (NCRP, 1976) for a discussion concerning the implications of the use of this method. The problems associated with air-filter mass loading, choice of counting standards, and interpretation of the results are reviewed.

5 PRECISION AND ACCURACY

5.1 Precision of the method is essentially a function of the volume of air sampled, the sample counting time, the background of the counter, and the counting efficiency of the detector measuring a filter with a given dust loading. The uncertainty can be decreased (precision increased) by increasing the activity collected (air volume sampled), increasing the counting time, decreasing the background, and/or increasing the efficiency. Calculation of the uncertainty is shown in Section **11.4**.

For example, the uncertainty of the analysis of a sample with a total activity of 1000 dpm (sample + background), counted for 10 min in a Geiger–Mueller counter with 10% efficiency and a background of 40 cpm (also determined by a 10-min count), would be 73.3 dpm, or 12.2% of the net dpm at the 95% confidence level. The same 1000 dpm sample counted for 10 min in a gas proportional counter with 40% efficiency and a background of 1 cpm would have an uncertainty of 31.0 dpm, or 3.1%. Thus, the latter counting method will provide a more precise result. Increasing the counting time will also decrease the uncertainty of both measurements.

5.2 The accuracy of gross β-analysis is dependent on the accurate measurement of the air volume sampled, the appropriate use of any self-absorption correction, and the accurate determination of the β-particle counting efficiency. The counting efficiency (see Section **11.1**) of β-counters is energy dependent. In many cases, the decision to perform gross β-analysis precludes some knowledge of the β-emitters potentially present. β-Particle energies vary over a few orders of magnitude. Therefore, calibration of the counter may be difficult. One "solution" is to report gross β-air concentrations solely in terms of the β-emitting calibration standard employed (see Section **9.1**). Should the identity of the β-activity be ascertained, a β-energy versus counting efficiency calibration (energy correction) curve would be useful to determine dpm. Some manufacturers can provide these curves.

6 APPARATUS

6.1 The basic requirements for the method are an air mover, a calibrated device for measuring airflow rate or volume (rotameter or gas meter), a particle filter, and a β-particle radiation counter. The size (diameter) of the filter should not be larger than the sensitive area of the detector. Due to the limited range of low-energy β-radiation, a high surface-retention filter, such as a membrane filter, should be used to avoid self-absorption, so that air particulate collection is essentially on the surface of the filter. A constant flow rate can be obtained using a critical flow venturi. A specialized filter head incorporating a venturi is described by Parulian et al. (1996).

6.2 If filters of appreciably less than 100% collection efficiency are employed, it is necessary to know or determine the collection efficiency as a function of filter face velocity and the physical properties of the material being sampled. The collection efficiency of a polystyrene fiber, glass fiber, and cellulose ester membrane filter media is >99% at a particle diameter of 0.3 μm and a face velocity of 27 cm s^{-1} (Hinds, 1982). Such high values preclude the need for use of a collection efficiency factor in calculating concentrations (see Section **11**). If in doubt about the collection efficiency, consult the filter manufacturer* or, if the expertise and equipment are available, perform a filter efficiency determination.

6.3 β-Sensitive detectors such as end-window Geiger–Mueller tube, thin window gas proportional flow counter, plastic scintillator, solid state detector, or liquid scintillation counter can be used for counting the sample. Counting efficiencies for these detectors range from 5% to 90% depending on

---

* For example, Gelman Sciences, Inc., USA or suppliers such as HI-Q Environmental Products Co., USA, F & J Specialty Products, Inc., USA.

β-energy and counting geometry. Background count rates are also widely different, ranging from less than 1–50 cpm depending on the effectiveness of the detector's shielding from cosmic and terrestrial radiation.

- Geiger–Mueller detector: End-window type—performs better if enclosed in a lead shield to reduce ambient background count rate; typical background: 40 cpm; approximate β-particle efficiency is 5% for ^{14}C to 30% for high-energy β-emitters.
- Plastic scintillator: Scintillator is coupled to a photomultiplier tube; can be used in dual α-/β-counters for field measurements; similar background and efficiencies as the Geiger–Mueller detector.
- Gas proportional counter: Thin window or windowless operation; requires that the counting chamber be purged with P10 gas (10% methane + 90% argon); automatic sample changers available; typical β-particle background: <0.1 cpm; approximate β-particle efficiency 50% (windowless) or 40% (window) (Oxford, 1998).
- Liquid scintillation counter: Sample is submerged in an organic scintillation fluid in a 7–20 mL vial, which is counted by photomultiplier tubes; automatic sample changers available; computer driven, which provides significant analysis capability; typical β-particle background: <20 cpm; approximate β-particle efficiency 60–90% depending on the radionuclide (Packard, 1998). Dissolution of the filter and use of a compatible scintillation fluid may be necessary (McDowell and McDowell, 1994; Passo and Cook, 1996).

Note: All of the above devices require specialized knowledge to operate. The novice is advised to seek the assistance of individuals familiar with the operation of these units.

7 REAGENTS

7.1 No reagents are needed for dry filter counting. If liquid scintillation analysis is to be performed, liquid scintillation fluid and vials may be obtained commercially.

8 PROCEDURE

8.1 Calculate the volume of air necessary to collect measurable β-activity, considering the minimum detectable activity of the counting system and the regulatory concentration values.

8.2 Record the time and initial flow rate or initial value of the flow totalizer when the sampler is started.

8.3 Operate the sampler in a location representative of the atmosphere for which the concentration is to be established with the sample point preferably 2 m off the ground. Carefully time the air sampling period.

8.4 At the end of sampling, record the final airflow rate and time or the final totalizer value. Place the filter in a protective cover or container upon removal from its holder to minimize the possibility of damaging the filter or dislodging the collected activity.

8.5 If necessary, store the sample for 2 h to 3 days to allow radon and thoron progenies to decay away.

8.6 Determine the counter background, perform preuse performance check, and calibrate the instrument if required.

8.7 Prior to counting, place the filter sample in a planchet or holder for reproducible positioning under the detector. Care should be taken to avoid contaminating the detector. This can be done by covering the sample with a thin Mylar sheet. If liquid scintillation analysis is to be used, place the filter in a scintillation vial and add an appropriate volume of scintillation fluid.

8.6 Count the sample for an interval (or total count) sufficient for a statistically significant result, for example, long enough to make comparison with a regulatory limit meaningful.

Method 2: Determination of the Gross Beta-Radioactivity Content of the Atmosphere

8.7 If needed, determine the selfabsorption correction factor and the filter collection efficiency.

8.8 Use the initial and final airflow rates to calculate a mean rate or use the flow totalizer final value and time of sampling. Calculate the gross β-airborne concentration and its uncertainty (Section **11**).

9   CALIBRATION AND STANDARDS

9.1 The efficiency of a β-counter can be determined using one standard for reporting the results based on that standard or doing a nuclide-specific calibration. If counting is done through a thin Mylar cover, calibration must be done through Mylar of equal thickness or a similar absorber. Standards containing β-emitters are commercially available from various sources.* A set of radionuclide β-standards covering a range of energies may be obtained in order to determine the energy dependence of the β-counter.

9.2 Determine the background of the counter using blank filter media over an interval similar to or longer than the sample counting time. Most β-counters are relatively stable and it is sufficient to make one daily determination of background to ascertain that major shifts due to malfunction, electronic drift (small changes in detector voltage and/or amplifier gain), or radioactive contamination have not occurred.

9.3 If a rotameter is used as the air-flow measurement device, the reader is urged to consult Craig (1971) concerning the problems associated with the calibration of these devices. If the rotameter is not calibrated using exactly the same sampling line situation (pressure drop across the sampler) and sampling conditions (ambient temperature and pressure), significant errors may be introduced into the air concentration results. Craig advises use of flow rates that cause the lowest pressure drops possible and meter readings at the upper end of the rotameter scale. It is also advisable to place a pressure gauge on the same side of the pump regulator valve as the rotameter float for the purposes of determining a correction for any pressure drop between the filter and the rotameter and for the detection of line breaks. The novice should also consult an experienced individual such as an industrial hygienist concerning these matters. The ASTM standard for calibration of a rotameter to standard conditions may be used (ASTM, 1996), although ASTM does not provide a discussion of calibration problems. Airflow meters should be calibrated annually and after repairs or modifications (NRC, 1992).

10   EFFECTS OF STORAGE

10.1 Filter samples should be stored so that they are protected from damage. Immediately after sample collection, the filter should be carefully placed in an acetate, plastic, or paper envelope. It must be handled carefully to avoid tearing and loss of filter material.

10.2 Scintillation fluid samples will evaporate over time. It may help to store the tightly sealed vials in a cool or cold location. Long-term storage is not recommended. Glass scintillation vials are subject to breakage.

11   CALCULATION OF CONCENTRATION

Gross β-airborne concentration can be calculated using the following equations. If radon and/or thoron progenies are suspected or known to be in significant concentrations, the equations shown in Method 1, Section **11** can be used, or the sample must be held for progeny decay (see Method 1, Section **3**). Refer to the chapter in this book on radon characteristics for more information.

---

* GE Healthcare Life Sciences, USA and UK, Isotope Products Laboratories, USA and North American Scientific, USA. Generally, traceability to the U.S. National Institute of Standards and Technology is available.

11.1 Counter efficiency:

$$\mathrm{EFF} = \frac{\mathrm{CR_{STD}}}{\mathrm{ACT_{STD}}}, \tag{1}$$

where $\mathrm{CR_{STD}}$ is the net count rate* of standard [usually in units of counts per second (cps) or counts per minute (cpm)] or the total net counts achieved in the counting interval, $\mathrm{ACT_{STD}}$ is the activity or disintegration rate of standard (dps or dpm) or the total disintegrations in the counting interval, and EFF is the counting efficiency [net counts/total disintegrations (c d^{-1}) or cps dps^{-1}, or cpm dpm^{-1}]

Note: Any units of time for the count or disintegration rates may be used as long as they cancel out in the equation (total net counts can only be used with total disintegrations).

11.2 Sample activity:

$$\mathrm{ACT_{SAM}} = \frac{(G/t_S) - (B/t_B)}{\mathrm{EFF} \times F_{\mathrm{EFF}}}, \tag{2}$$

where $\mathrm{ACT_{SAM}}$ is the sample activity (Bq) [1 dps = 1 Becquerel (Bq)], G is the gross count, B is the background count, t_S is the sample count time (s), t_B is the background count time (s), EFF is the counter efficiency (with energy correction if needed; see Section **5.2**) [see above for units], and F_{EFF} is the filter collection efficiency correction [if needed]

11.3 Gross β-airborne concentration:

$$\mathrm{Conc}_\beta = \frac{\mathrm{ACT_{SAM}}}{V_S}, \tag{3}$$

where Conc_β is the gross β-airborne concentration (Bq m^{-3}) and V_S is the air volume sampled (m^3), [mean flow rate (m^3/min) × air sampling time (min)]

This result may be converted to μCi mL^{-1} by dividing $\mathrm{ACT_{SAM}}$ by (1 × 10^6 mL m^{-3} × 2.2 × 10^6 dpm μCi^{-1} × 1/60 min s^{-1}) = 3.7 × 10^{10}.

11.4 Uncertainty: The statistical significance of this result should also be indicated. It can be determined from the standard deviation (SD) of the count.

$$\mathrm{SD} = \frac{\left[(G/t_S^2) + (B/t_B^2)\right]^{1/2}}{\mathrm{EFF} \times F_{\mathrm{EFF}} \times V_S}, \tag{4}$$

where SD is the standard deviation of the air concentration (Bq m^{-3}), G is the gross count, B is the background count, t_S is the sample count time (s), t_B is the background count time (s), EFF is the counter efficiency (with energy correction if needed; see Section **5.2**) [see above for units], F_{EFF} is the filter collection efficiency correction [if needed] and V_S is the mean air volume sampled (m^3).

The statistical significance of the final concentration result should be expressed as the uncertainty at a specified confidence level. For example, to express the uncertainty at the 95% confidence level (also referred to as the 1.96σ or 2σ level), multiply the SD result by 1.96. This result may be converted to μCi mL^{-1} by dividing the uncertainty at 1.96σ (or the uncertainty calculated at any other confidence level) by (1 × 10^6 mL m^{-3} × 2.22 × 10^6 dpm μCi^{-1} × 1/60 min s^{-1}) = 3.7 × 10^{10}.

---

* The background count rate has been subtracted.

12 CAUTIONS

12.1 This method may involve the use of hazardous chemicals and/or radioactivity. It is the user's responsibility to take appropriate safety precautions and to obtain necessary regulatory approvals from institutional and/or governmental agencies.

12.2 If the user is not trained or familiar with the calibration and operation of radiation analysis equipment, they should consult a qualified expert, such as a health physicist, for assistance in the proper and safe usage of the equipment and development of appropriate quality assurance protocols.

12.3 Radioactive waste and hazardous waste such as scintillation fluid require special disposal considerations that are regulated by governmental authorities. Contact a specialist if unfamiliar about the treatment of such waste.

REFERENCES

ASTM. *Annual Book of ASTM Standards, Section 11, Water and Environmental Technology*. Designation D-3195—90, Standard Practice for Rotameter Calibration, ASTM, West Conshohocken, PA, 1996.

Craig, D.K., The interpretation of rotameter air flow readings, *Health Phys.*, 21, 328–332, 1971.

Federal Register, Part VI, U.S. Nuclear Regulatory Commission, 10 CFR Part 20 et al., Code of Federal Regulations, Standards for Protection Against Radiation; Final Rule, May, 1991,Washington, DC, 1991.

Harley, J.H., Sampling and measurement of airborne daughter products of radon, *Nucleonics*, 11, 12–17, 1953, reprinted in *Health Phys.*, 38, 1068–1069, 1980.

Hinds, W.C., *Aerosol Technology*, John Wiley & Sons, New York, 1982.

Lodge, Jr., J.P. (Ed.), *Methods of Air Sampling and Analysis*, 3rd ed., CRC Press, Boca Raton, FL, 1988.

McDowell, W.J. and McDowell, B.L., *Liquid Scintillation Alpha Spectrometry*, CRC Press, Boca Raton, FL, 1994.

NCRP, *Environmental Radiation Measurements*, National Council on Radiation Protection and Measurements Report No. 50, Bethesda, MD, 1976.

NCRP, *Measurement of Radon and Radon Daughters in Air*, National Council on Radiation Protection and Measurements Report No. 97, Bethesda, MD, 1988.

NRC, *Air Sampling in The Workplace, U.S. Nuclear Regulatory Commission, Regulatory Guide 8.25*, 1992, National Technical Information Service, Springfield, VA, USA and U.S. Government Printing Office, Washington, DC, 1992.

Oxford Systems, Instruments and Components for Radiation Detection and Measurement, 5th ed., Oxford Instruments, Inc., Oak Ridge, TN, 1998.

Packard, *Catalog Material*, Packard Instrument Company (now Perkin-Elmer), Meriden, CT, 1998.

Parulian, A., Rodgers, J.C., and McFarland, A.R., A constant flow filter air sampler for workplace environments. *Health Phys.*, 71, 870–878, 1996.

Passo, C.J. and Cook, G.T., *Handbook of Environmental Liquid Scintillation Spectrometry*, Packard Instrument Company (now Perkin-Elmer), Meriden, CT, 1996.

Schery, S.D., Thoron in the environment, *J. Air Waste Manage. Assoc.*, 40, 493–497, 1990.

Method 3
Determination of the Tritiated Water Vapor Content of the Atmosphere*

CONTENTS

1	Introduction	407
2	Principle of the Method	408
3	Range and Sensitivity	409
4	Interferences	410
5	Advantages/Disadvantages	410
6	Precision and Accuracy	411
7	Apparatus	411
8	Reagents	411
9	Procedure	412
10	Calibration	412
11	Calculations	414
12	Effects of Storage	415
13	Cautions	415
References		416

1 INTRODUCTION

1.1 There are a few general means of measuring tritium in air. The choice depends on several factors. One must decide whether only elemental tritium (HT,[†] T_2, or DT) or only tritiated water vapor (HTO, T_2O, or DTO) is to be measured, or whether a nondiscriminating (combined) measurement is to be made. Other considerations include the amount of time available for sampling, whether environmental or industrial-process sampling is to be conducted, whether other tritiated gases are present, and the desired tritium detection limit. A review by Wood et al. (1993) succinctly describes active tritium-in-air sampling using tritium oxide (HTO) collectors connected to air pumps and airflow measurement devices. The collectors can be traps (Figure 1) for concentration of the HTO onto a desiccant, traps cooled below the freezing point of water by liquid nitrogen for condensing the HTO out of the air stream and into the collector, or gas washing bottles (bubblers) filled with tritium-free water or other sorbents through which the HTO is passed and washed out of the sample air. A technique in which the tritiated gases in air are burned in an electrolytic cell in order to collect condensed tritiated water vapor has also been described (Brigoli et al., 1991). A passive method developed for both environmental (Harrison et al., 1989; Davis et al., 1992; Otlet et al., 1992; Wood

---

* Method authors: Appreciation is extended to the writers of the Radioactivity section of the book *Methods of Air Sampling and Analysis* (Lodge, 1988) where this method originally appeared. Method reviewers and editors: Hung-Cheng Chiou, Lawrence T. Dauer, Andrew Hull, Paul Linsalata, Mark L. Maiello, Donald M. Mayer, Robert P. Miltenberger, and Dennis M. Quinn.

[†] In Methods 3 and 4, HT is used as the abbreviation for elemental tritium and HTO is used for tritium oxide even though the other forms listed may be present.

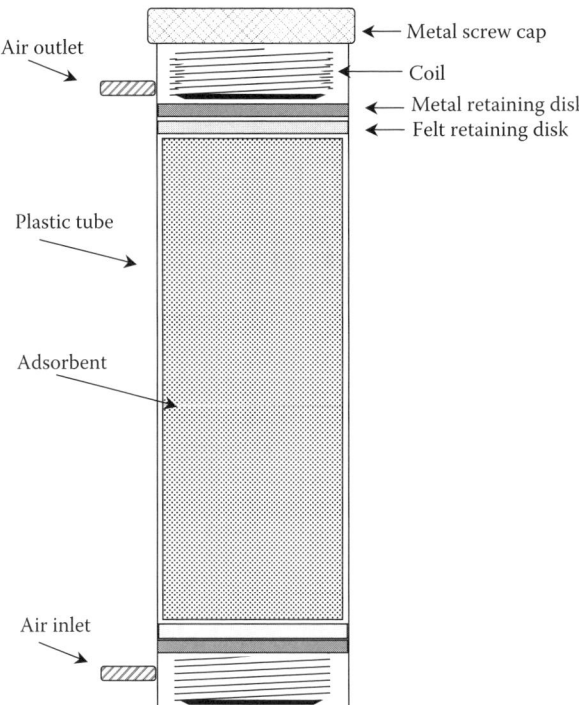

FIGURE 1 Cylindrical column for adsorption of elemental tritium or tritium oxide.

and Workman, 1992; Wood, 1996) and workplace sampling (Stephenson, 1990; Wood, 1991) consists of a liquid scintillation vial with a specially modified cap for diffusion of HTO into a sampling sink (water or silica gel) inside the vial. Real-time measurements of tritium in air (elemental and tritiated water vapor) can be made with commercially available tritium ion chamber monitors. These devices have usually been restricted to process monitoring due to a lack of sensitivity, but there have been recent improvements so that levels as low as 37 Bq m^{-3} (1×10^{-9} µCi mL^{-1}) can be measured. The most sensitive tritium monitors are expensive. All tritium monitors require careful maintenance and periodic calibration.

1.2 Method 3 describes an integrating method for HTO. Method 4 is an integrating method for the separate analysis of HTO and HT. Both methods require a liquid scintillation counter (LSC) for measuring the β-radioactivity of tritium. Improvements in LSC counting have led to the development of low-level liquid scintillation counters (LLLSC) the use of which enhances the sensitivity of the analyses. A chart modified from Wood et al. (1993) is shown in Figure 2 indicating the detection sensitivities of the various techniques.

1.3 For further information about tritium measurement methods, see the review by Budnitz (1974), NCRP Report No. 47 (NCRP, 1976), IAEA Technical Report No. 324 (IAEA, 1991), and Wood et al. (1993).

2 PRINCIPLE OF THE METHOD

2.1 Air is pumped at a sampling rate of approximately 1×10^{-4} to 2×10^{-4} m^3 min^{-1} through a trap containing 200 g of desiccant that collects moisture from the air. This method describes the use of silica gel as the desiccant. However, other common desiccants such as molecular sieves (alumino-silicates), anhydrous calcium sulfate, and activated alumina can be used. The stated flow rate is sufficient to allow continuous air sampling for 7–14 days at 50% relative humidity. Other methods of moisture collection such as cold traps, bubblers, and dehumidifiers may also be used

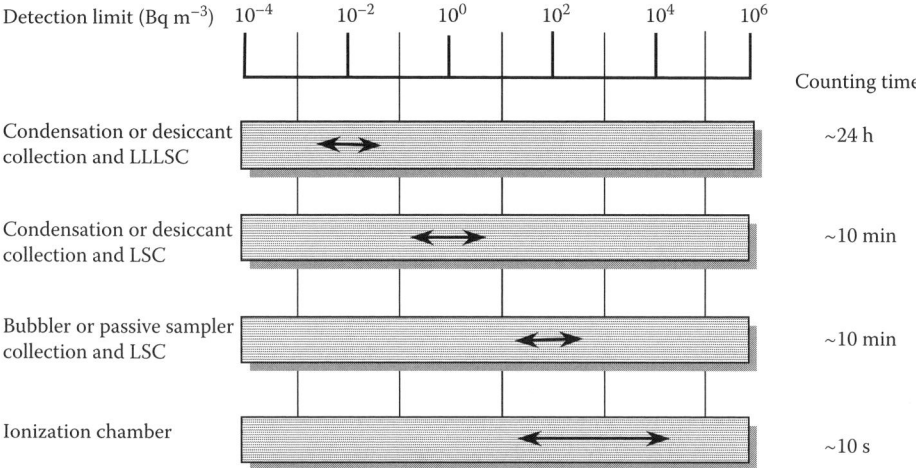

FIGURE 2 Detection limits of various tritium-in-air measurement techniques (water content of air = 5 gm^{-3}). (Reproduced from Wood, M.J. et al. *Health Phys.* 65, 610–627, 1993. With permission.)

with various effects on measurement accuracy, depending on completeness of collection (Budnitz et al., 1983).

2.2 The collected moisture is then desorbed from the trap by heating and condensation. Depending on the amount of moisture collected and the liquid scintillation cocktail used, some or all of the moisture is counted for tritium in an LSC. The airflow rate and total volume of moisture collected during the sampling period, or the absolute humidity, must be determined to convert the concentration of tritium in the collected moisture sample to its concentration in air. For long sampling periods, the determination of airflow rate and moisture collected may be more convenient than obtaining humidity measurements.

3 RANGE AND SENSITIVITY

3.1 The method is suitable for sampling periods of up to 7–14 days. The sampling period can be extended if low relative humidity conditions exist or if the trap size is increased. One reported variation uses a series of three columns, a sampling rate of 0.25×10^{-4} m^3 min^{-1}, and a sampling time of 4 weeks to obtain an air volume of 9.5 m^3, but with some suspected breakthrough problems (see below) (Patton et al., 1997). The sensitivity of the method is directly related to the volume of air sampled up to a limit. It has been shown that HTO passes through the column without collection (breakthrough) when the sample-air-volume to adsorbent-depth ratio in m^3 cm^{-1} reaches 0.36 at 20°C, 0.20 at 30°C, 0.15 at 40°C, and 0.077 at 50°C (all at 30% relative humidity) (Patton et al., 1997). The sensitivity is also an inverse function of temperature. For example, the collection efficiency of silica gel is dependent on air temperature as well as the air volume collected. At 20°C (30% relative humidity and 2.5×10^{-4} mL min^{-1} sampling rate), the collecting efficiency has been measured at 99.9% for sample volumes of 7.5 m^3 and 9.5 m^3. When the temperature reaches 40°C, the efficiency is 95% for 7.5 m^3, but just below 80% when a 9.5 m^3 volume is acquired (Patton et al., 1997). Relative humidity affects sensitivity as well. At 10% relative humidity, the equilibrium adsorption capacity of silica gel is 0.05 g of water per gram of adsorbent; at 50%, it is 0.26 g of water per gram of adsorbent (NCRP, 1976).

3.2 A sample of moisture from air at an average temperature of 30°C (86°F) and at 100% relative humidity containing a tritium in an air concentration of 740 Bq m^{-3} (2×10^{-8} µCi cm^{-3}), which is 0.1% of the derived air concentration (the DAC for the water vapor form of tritium = 7.4×10^5 Bq m^{-3}) (Federal Register, 1991), would have a liquid concentration of about 24×10^3 Bq L^{-1} (7×10^{-4} µCi cm^{-3}). This is sufficient to produce acceptable precision in less than 2 min of counting in most LSC systems.

3.3 In the workplace or other regulatory setting, the investigator should check that the lower limit of detection (LLD) or minimum detectable activity of the sampling/counting system is sufficient to detect the concentration guideline to which the measurement will be compared. The reader is urged to refer to Chapter 5 for a brief discussion and example calculation of LLD.

4 INTERFERENCES

4.1 None.

5 ADVANTAGES/DISADVANTAGES

5.1 Integrating adsorbent collection provides excellent sensitivity at less cost than real-time (ion chamber) monitoring, especially when considering low-level measurements of HTO in ambient air. The integrating adsorbent method is more specific for HTO than an ion chamber which may respond to any airborne activity (NCRP, 1976; Wood et al., 1997).

5.2 Solid adsorbents (desiccants) are more rugged than bubbler solutions and do not evaporate over time. Instead of delicate and difficult-to-clean glass impingers, plastic columns may be used to contain the solid adsorbent.

5.3 Use of an integrating adsorbent requires long sampling periods (often several weeks) in addition to time for laboratory sample analysis.

5.4 It is possible that some HTO will be collected under breakthrough conditions (the silica-gel adsorbent becomes saturated with water vapor). This may occur with long sampling periods, high relative humidity conditions, or a combination of both. Color-indicating silica gel is available that will show the migration of water vapor from the bottom to the top of the column. When the color change has occurred over the column length, breakthrough is possible.

5.5 If the sampling trap is made of plastic, be aware that if used with air streams of very high tritium concentrations, the tritium can exchange with plastic possibly causing cross-contamination of samples.

5.6 The traps must be maintained in a vertical position during sampling to avoid the problem of "air streaming." This occurs when a void space opens in the media along the length of a horizontally oriented trap. A void can open as the media is vibrated (tamped down) by pump action or by transportation of the sampler. Moisture cannot be collected if the sample air streams above the media rather than through it.

5.7 The method requires distillation at high temperatures. It is necessary to complete the distillation process to avoid enriching the remaining water in tritium and depleting the condensate. The lower vapor pressure of tritium oxide relative to ordinary water is responsible for this. Errors of 5–10% may occur if complete distillation is not achieved (NCRP, 1976).

5.8 Work at Los Alamos National Laboratory (Eberhart, 1999) has documented two environmental conditions when silica gel does not efficiently remove water vapor from the air. This chapter, along with EML, 1997, identifies low humidity as one condition where absorption efficiency can be as low as 5% instead of the assumed 100%. A second condition that reduces water removal efficiency is the high temperature of the silica gel. In the Eberhart paper, silica gel is shown to saturate in a few days. If the tritium in the atmosphere is at a steady state, then the underestimate of tritium concentration is on the order of a factor of 2. If releases are nonuniform, the underestimate of air concentration and dose can be more severe. If high temperatures and low humidity are encountered, the author recommends using the absolute humidity measurements to compensate for the lower collection efficiency.

5.9 In general, air sampling requires attention to measurement of airflow rates, correcting for in-line pressure drops, ambient air temperature, and pressure deviations from standard conditions, filter loading, and mechanical problems such as line leaks and pump failures. For occupational indoor air measurements, the airflow patterns of the workplace should be determined so that the air

Method 3: Determination of the Tritiated Water Vapor Content of the Atmosphere

6 PRECISION AND ACCURACY

6.1 Duplicate distillate samples of only 1 mL volume (as opposed to the 7 mL recommended here) with radioactivity in the range of tens to hundreds of pCi mL^{-1} counted for 10 min in an LSC should yield reproducible results to within 5% depending on such things as pipetting technique. This assumes a counting efficiency for ^{3}H of about 60% and background values in the ^{3}H counting window of <20 counts per minute (cpm) using standard scintillation fluids and LSCs. The third edition of this book stated that the 2σ uncertainty of the sample described in Section **3.2** would be less than $\sqrt{}2\%$, if a moisture aliquot of 7 mL is counted for 2 min at an efficiency of 25% using an LSC with a 7 cpm background in the tritium window.

6.2 The accuracy of the air concentration measurement is dependent on the calibration of the collection efficiency and the calibration of the LSC.

7 APPARATUS

7.1 Liquid scintillation counter.
7.2 Traps: 30.5×3.2 cm diameter (12 in. $\times$ 1.25 in.). The trap construction can be of aluminum, plastic, or glass.*
7.3 Low-volume air pump, flow rate indicator, critical or limiting orifice, and flow integration device (or temperature and relative humidity device). (Refer to Figure 3 for a diagram of the HTO sampling train.)
7.4 Distilling flask and condenser/heater.
7.5 Pipettes.
7.6 LSC vials, polyethylene or glass, 20 mL.
7.7 Rubber stoppers, tubing.

8 REAGENTS

8.1 Scintillation cocktails are available commercially.† Some counting cocktails contain organic constituents that may require special disposal procedures, whereas others may be free of such constituents

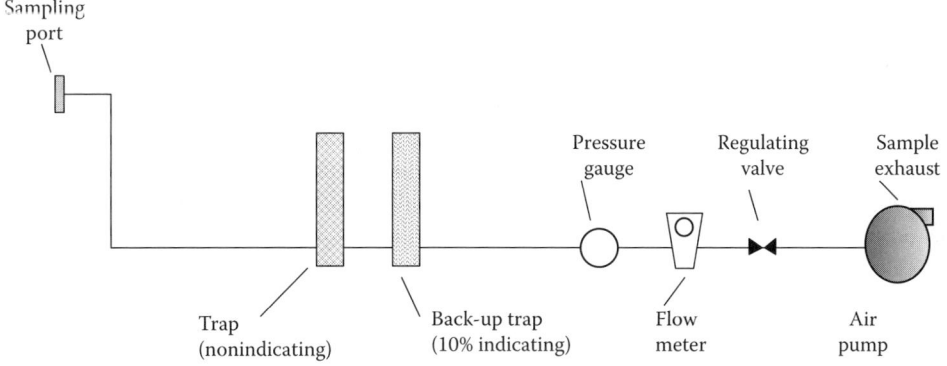

FIGURE 3 Sampling line for determining tritium oxide concentration in air.

---

* A manufacturer of suitable traps is W.A. Hammond Drierite Co., USA.
† Packard Instrument Company (now Perkin-Elmer), USA and Research Products International Corporation, USA.

(Klein and Gershey, 1990). Check with the manufacturer to verify that the cocktail will accept the volume of sample (distillate) recommended in Section **9.8**. An expert (health physicist or hazardous waste manager) with knowledge of governmental and institutional disposal regulations should also be consulted before choosing a cocktail.

8.2 Silica gel* 6–16 mesh, moisture indicating (cobalt chloride indicator).

8.3 Silica gel, 6–16 mesh, nonindicating.

9 PROCEDURE

9.1 Bake 6–16 mesh, nonindicating silica gel for 24 h at 250°C prior to use.

This step may be omitted if the user is willing to accept a 5% uncertainty in the total moisture content. Omission of this step may be highly desirable if the drying process cannot be performed in a low-airborne tritium environment. The silica gel can be contaminated by tritium in ambient air during the bake-out/cool-down step of the procedure.

9.2 Fill one trap (see Figure 1 for typical design) with nonindicating silica gel. Fill a second cylinder with a mixture of 10% indicating silica gel and 90% nonindicating silica gel. Since silica gel can adsorb about 20–30% of its own weight (NCRP, 1976; Budnitz et al., 1983; Dean, 1992), 200 g should not saturate during a sampling period as long as 7 days in extremely warm and humid conditions. Seal the inlet and the outlet until the initiation of sampling.

9.3 Unseal traps. Connect the outlet port of the sample trap to the inlet port of the back-up trap (containing the indicating silica gel). Refer to Figure 3.

9.4 Connect the outlet port of the back-up trap to the low-volume air pump. If necessary, connect the inlet port of the sample trap to a sampling line (this is used if the air stream is being sampled from a "remote" location). A glass fiber filter may be used at the inlet of the sampling line to trap particulate matter. Start the pump and record the date and starting time. Set the initial flow rate and record. An air flow meter such as a rotameter may be used to obtain readings at the initiation and termination of the sampling period with a simple average being taken as the characteristic flow rate (NRC, 1992). Over a typical sampling period of 7 days, it is advisable to observe the airflow rate periodically and to make adjustments as necessary. Constant flow air pumps may be employed for long periods of sampling.

9.5 At the end of the air sampling period, record the date and time sampling ends. Record the final flow rate. Remove the nonindicating silica gel trap and seal ports for transport to laboratory. Check the back-up trap for a change in the color of the indicating silica gel. If a change of color has occurred, remove back-up trap, seal ports, and transport to the laboratory for analysis.

9.6. Pour the silica gel into a 250–500-mL distilling flask or condenser.

9.7 Heat the silica gel until all moisture is removed (650°C). Weigh or measure total moisture collected. If the back-up trap was sent for analysis, heat its contents similarly, and add the condensed distillate from it to that from the first trap. Heating the silica gel to this temperature may make reuse of the adsorbent impossible. *Note:* During summer months, particularly in humid locations, it is possible to collect about 30–60 mL of moisture in a 1- or 2-week period.

9.8 Add 9 mL of scintillation solution to a 20-mL scintillation vial containing 7 mL of distillate.

9.9 Store vials in a dark location for at least 15 min prior to counting in an LSC.

9.10 Count for not less than 2 min in an LSC. (Deviation from ambient concentrations of tritium can be detected with count times of 100 min or more.)

10 CALIBRATION

10.1 Air sampling equipment: If a rotameter is used as the airflow measurement device, the reader is urged to consult Craig (1971) concerning the problems associated with the calibration of these

---

* J.T. Baker Co., Phillipsburg, New Jersey.

devices. For example, if the rotameter is not calibrated using exactly the same sampling line situation (pressure drop across the sampler) and sampling conditions (ambient temperature and pressure), significant errors may be introduced into the air concentration results. Craig advises use of flow rates that cause the lowest pressure drops possible and meter readings at the upper end of the rotameter scale. It is also advisable to place a pressure gauge on the same side of the pump regulator valve as the rotameter float for the purposes of determining a correction for any pressure drop between the filter and the rotameter and for the detection of line breaks. The novice should also consult an experienced individual such as an industrial hygienist concerning these matters. The ASTM standard for calibration of a rotameter to standard conditions may be used (ASTM, 1996), although ASTM does not provide a discussion of calibration problems. Airflow meters should be calibrated annually and after repairs or modifications (NRC, 1992).

10.2 COLLECTION EFFICIENCY

10.2.1 The efficiency of collection of tritium may be determined by delivering a known amount of the radionuclide as a gas into the trapping system. Small pressurized "lecture" bottles of tritiated methane can be purchased* to deliver a known activity (a "spike"). Oxidizing the tritiated methane by passing it through a tube furnace (900°C) before silica gel collection may produce more realistic results. An LSC measurement of the spiked distillate is performed in the same manner as for a field sample, and the total amount of activity collected is calculated for comparison to the delivered value. The ratio of collected activity to delivered activity is the collection efficiency.

10.3 LIQUID SCINTILLATION COUNTER

10.3.1 The manufacturers of LSCs provide procedures for the calibration of these devices. A full discussion of the various techniques that are employed in LSCs to refine the calibration is beyond the scope of this book. The reader is urged to refer to the operating manual of the LSC to be employed for this information. Here, we describe the determination of the counting efficiency, which constitutes a calibration of the LSC for specific counting conditions.

10.3.2 The calibration of LSCs for tritium is achieved by counting a tritium standard in the same size and type of scintillation vial as is used for field samples. Standards are available from manufacturers of LSCs, the U.S. National Institute of Standards and Technology, and from commercial vendors.[†]

10.3.3 When formulating a radioactive standard, consider that the chemical form must be soluble in scintillation fluid. Check with the standard and/or scintillation cocktail manufacturer if there is doubt. Immiscible chemical forms are unacceptable and will lead to inaccurate calibration of the LSC.

10.3.4 A 0.004 MBq (0.1 µCi) standard of ^3H is typically available from LSC manufacturers. A set of standards covering a realistic range of recovered volumes of distillate (say from 1 to 9 mL) may be necessary unless the LSC can correct for this. The type of scintillation fluid used in the standards must be the same as used for field samples.

10.3.5 Count the standards under the same LSC protocol (the β-energy window settings for ^3H) as are used for the field samples. Obtain the results in units of cpm and compare with the activity pipetted into the scintillation vial. The ratio of activity counted/activity added to the vial is the counting efficiency [cpm/disintegrations per minute (dpm), cps/dps, or cps/Bq] for the counting conditions (vial type and size, sample volume used, scintillation fluid type, and the LSC protocol) used. This result is stored in the LSC computer memory for application to field samples. If quenching

---

* Isotope Products Laboratories, Valencia, California.
† The National Institute of Standards and Technology, Gaithersburg, Maryland. Commercial suppliers include Analytics, Inc., USA and Isotope Products Laboratories, USA.

in field samples is a problem, it will be necessary to obtain a series of standards with varying degrees of known quench so that a calibration curve of efficiency versus quench indication can be stored in the LSC computer memory. LSC manufacturers can supply these standards.

10.3.6 Since a background sample will be needed for counting purposes, prepare one by adding 7 mL of low background, distilled (tritium-free) water to a scintillation vial with 9 mL of scintillation fluid. The volume of water used should match the volume of distillate used. A background sample is included at the start of every counting run in most LSC protocols.

11 CALCULATIONS

11.1 Many LSCs can be programmed to subtract background to obtain net cpm, to convert net cpm to dpm via a counting efficiency factor, and to change dpm to the concentration of radioactivity in the scintillation fluid (Bq mL^{-1}, µCi mL^{-1}). Converting cpm into concentration of radioactivity *in the air sample* may be out of the reach of many LSC units. Therefore, the user can be expected to perform some calculations. If the LSC cannot be programmed to do the conversions into net cpm and dpm, the user will have to do these manually as well. Since the half-life of ^{3}H is 12.35 a, calculations to correct for radioactive decay are unnecessary.

11.2 For manual calculations, the efficiency is determined as follows:

$$\mathrm{EFF} = \frac{\mathrm{CR}_{\mathrm{STD}} - R_{\mathrm{B}}}{\mathrm{ACT}_{\mathrm{STD}}}, \tag{1}$$

where EFF is the counting efficiency (cpm dpm^{-1}), $\mathrm{CR}_{\mathrm{STD}}$ is the gross count rate (cpm) of standard in scintillation solution, R_{B} is the background (cpm) of distilled water in scintillation solution, and $\mathrm{ACT}_{\mathrm{STD}}$ is the dpm of standard.

11.3 To determine the tritium concentration in the distilled moisture:

$$C_{\mathrm{M}} = \frac{R_{\mathrm{S}} - R_{\mathrm{B}}}{\mathrm{EFF} \times V \times K},$$

where C_{M} is the concentration of tritium in the collected moisture (Bq mL^{-1}), R_{S} is the gross sample count rate; that is, sample count plus background (cpm), V is the volume (mL) of sample counted, and K is the 60 s min^{-1} to convert dpm into dps (Bq).

11.4 The statistical significance of the result should be indicated. Many LSCs automatically calculate the standard deviation (SD) of the count rate and report count rate uncertainty (U) at the 95% confidence level (1.96σ or "two sigma," i.e., 1.96 × SD). Often, this value is reported as a percentage of the count rate result. This is convenient since further conversions of the count rate are needed to obtain a result in Bq m^{-3} of air. The two-sigma absolute value can then be easily found as a percentage of the final converted result in Bq m^{-3}.

The uncertainty (U) due to the counting error only is

$$U(\mathrm{cpm}) = \forall 1.96 \left[\left(\frac{R_{\mathrm{S}}}{t_{\mathrm{S}}} \right) + \left(\frac{R_{\mathrm{B}}}{t_{\mathrm{B}}} \right) \right]^{1/2}, \tag{2}$$

where t_{S} is the counting time of the standard (min) and t_{B} is the counting time of the distilled water background (min).

The uncertainty (U) must be converted from cpm to Bq mL^{-1} by dividing by (EFF × V × K). See Sections **11.2** and **11.3**.

11.5 The concentration of tritium oxide in air is calculated as follows using silica gel with 100% collection efficiency:

$$C_A = \frac{C_M \times W}{V_R \times T}, \tag{3}$$

where C_A is the concentration of tritium oxide in air in Bq cm^{-3}, W is the total volume of water collected (mL), V_R is the mean air sampling flow rate (cm^3 min^{-1}), and T is the air-sampling time (min).

To report the uncertainty of C_A in Bq cm^{-3}, multiply the uncertainty of C_M by $W/(V_R \times T)$.

11.6 If the collection efficiency or air sampling flow rate is unknown, the following equation can be used to determine the HTO air concentration:

$$C_A = C_M \times A \times 10^{-6} (\text{Bq cm}^{-3}), \tag{4}$$

where A is the absolute humidity during sample collection (g m^{-3}) (Table 1).*

To report the uncertainty of C_A, multiply the uncertainty of C_M by $A \times 10^{-6}$.

11.7 Appropriate factors for converting the tritium air concentrations into estimates of individual committed radiation dose to exposed personnel are provided by the U.S. Environmental Protection Agency (EPA) in Federal Guidance Report No. 11 (EPA, 1988) which has been adopted as part of the U.S. Code of Federal Regulations (Federal Register, 1991) and by many U.S. state radiation regulatory programs. The time of submergence in the tritium cloud (for elemental tritium exposure) or exposure to the tritium air concentration (for exposure to tritium oxide) must be known or accurately estimated to use these exposure-to-dose conversion factors. Tritium dosimetry must account for exposure due to inhalation and transference through the skin. Dose calculations are perhaps better performed by a qualified expert in health physics who can interpret the results or who could note the potential sources of error in measurement, calculation, and interpretation. Federal Guidance Report No. 11 also lists DACs and annual limits of intake (ALIs) that are also currently legally enforced. The DAC limit is particularly useful for comparing with measured tritium concentrations under occupational exposure circumstances.

For emissions from a stack, the air concentrations measured at the emission point may be transformed into individual and population doses by available software packages such as COMPLY (EPA, 1989a), AIRDOS-PC (EPA, 1989b), and CAP88-PC (EPA, 2006). These software packages vary in complexity and employ such user-supplied information as wind rose data, building dimensions, stack heights, and effluent velocities to calculate off-site doses.

12 EFFECTS OF STORAGE

12.1 Storage effects are minimized if the silica gel collection traps are *thoroughly* sealed and stored in a nontritiated atmosphere.

12.2 The liquid scintillation samples prepared for counting should be analyzed within a few days. Liquid scintillation fluid is volatile and slowly escapes from glass and plastic vials.

13 CAUTIONS

13.1 This method may involve the use of hazardous chemicals and/or radioactivity. It is the user's responsibility to take appropriate precautions and to obtain necessary regulatory approvals from institutional and/or governmental agencies.

---

* This is the mass of water vapor present in a unit volume of the atmosphere, usually expressed as grams per cubic meter. Representative values applicable at sea level are indicated in Table 1 (Marvin, 1941). More recent values may be calculated from steam tables (Mayer, 1979; Keenan et al., 1978).

TABLE 1
Weight (g m⁻³) of a Cubic Meter of Aqueous Vapor at Different Temperatures and Percentages of Saturation

Temperature (°F)	Percentage of Saturation									
	10	20	30	40	50	60	70	80	90	100
−20	0.039	0.076	0.114	0.151	0.190	0.229	0.266	0.304	0.341	0.380
−15	0.050	0.101	0.149	0.199	0.250	0.300	0.350	0.398	0.449	0.499
−10	0.064	0.130	0.197	0.261	0.325	0.392	0.458	0.522	0.586	0.653
−5	0.085	0.169	0.254	0.339	0.424	0.508	0.593	0.678	0.762	0.847
0	0.110	0.220	0.330	0.440	0.550	0.662	0.772	0.882	0.992	1.101
5	0.140	0.279	0.419	0.559	0.698	0.838	0.978	1.118	1.257	1.397
10	0.179	0.355	0.534	0.710	0.888	1.067	1.243	1.422	1.598	1.777
15	0.227	0.451	0.678	0.900	1.129	1.356	1.580	1.807	2.031	2.258
20	0.284	0.566	0.847	1.131	1.415	1.697	1.978	2.262	2.546	2.828
25	0.355	0.710	1.065	1.420	1.777	2.132	2.487	2.842	3.197	3.552
30	0.444	0.886	1.328	1.772	2.217	2.659	3.101	3.545	3.989	4.431
35	0.543	1.083	1.626	2.166	2.709	3.252	3.792	4.335	4.875	5.418
40	0.653	1.305	1.958	2.611	3.261	3.914	4.566	5.219	5.872	6.524
45	0.781	1.564	2.345	3.128	3.909	4.690	5.473	6.254	7.037	7.818
50	0.934	1.866	2.801	3.733	4.667	5.601	6.533	7.468	8.400	9.334
55	1.111	2.221	3.332	4.443	5.551	6.662	7.772	8.883	9.994	11.104
60	1.314	2.631	3.948	5.262	6.577	7.984	9.210	10.525	11.839	13.156
65	1.553	3.105	4.660	6.213	7.765	9.318	10.871	12.426	13.978	15.531
70	1.827	3.655	5.482	7.310	9.137	10.964	12.792	14.619	16.447	18.274
75	2.143	4.284	6.428	8.569	10.713	12.856	14.997	17.141	19.282	21.425
80	2.503	5.008	7.511	10.016	12.519	15.022	17.528	20.031	22.536	25.039
85	2.917	5.833	8.750	11.665	14.583	17.500	20.415	23.333	26.248	29.165
90	3.387	6.774	10.161	13.548	16.934	20.321	23.708	27.095	30.482	33.869
95	3.920	7.843	11.764	15.686	19.607	23.527	27.450	31.371	35.293	39.214
100	4.527	9.052	13.580	18.105	22.632	27.159	31.684	36.212	40.737	45.264

Source: Adapted from Marvin, C.F., *Psychrometric Tables*, Weather Bureau Publication #735, U.S. Government Printing Office, Washington, DC, 1941.

Note: To convert EF temperatures into EC, use EC = (EF − 32)/1.8.

13.2 If the user is not trained or familiar with the calibration and operation of radiation analysis equipment, they should consult a qualified expert, such as a health physicist, for assistance in the proper and safe usage of the equipment and development of appropriate quality assurance protocols.

REFERENCES

ASTM, *1996 Annual Book of ASTM Standards*. Section 11, *Water and Environmental Technology*. Designation D-3195-90, Standard Practice For Rotameter Calibration, ASTM, West Conshohocken, PA, 1996.

Brigoli, B., Campi, F., Foglio Para, A., and Tarani, S., Total tritium measurement in atmosphere, *Health Phys.*, 61, 105–110, 1991.

Budnitz, R.J., Tritium instrumentation for environmental and occupational monitoring, *Health Phys.*, 26, 165–178, 1974.

Budnitz, R.J., Nero, A.V., Murphy, D.S., and Graven, R., *Instrumentation for Environmental Monitoring*, Vol. 1, *Radiation*, Wiley Interscience, New York, 1983.

Craig, D.K., The interpretation of rotameter air flow readings, *Health Phys.*, 21, 328–332, 1971.

Davis, P.A., Cornett, R.J., Killey, R.W.D., Wood, M.J., and Workman, W.J.G., Environmental behaviour of tritium released to the atmosphere in winter, *Fusion Technol.*, 21(2, pt.2), 651–658, 1992.

Dean, J.A., *Lange's Handbook of Chemistry*, 14th ed., McGraw-Hill Book Company, New York, NY, 1992.

Eberhart, C., *Using Absolute Humidity and Radiochemical Analyses of Water Vapor Samples to Correct underestimated Atmospheric Tritium Concentrations*, Los Alamos National Laboratory, Los Alamos New Mexico, Report LA-UR-99-1107, 1999.

EML, *EML Procedures Manual*, HASL-300, Vol. II, 28th ed., U.S. Department of Energy-Environmental Measurements Laboratory, New York, NY, 1997.

EPA, *Limiting Values of Radionuclide Intake and Air Concentration and Dose Conversion Factors for Inhalation*, Submersion and Ingestion, U.S. Environmental Protection Agency, EPA-520/1-88-020, Federal Guidance Report No. 11, September, 1988.

EPA, *User's Guide for the COMPLY Code (Revision 2)*, U.S. Environmental Protection Agency, EPA-520/1-89-003, October, 1989a.

EPA, *User's Guide for AIRDOS-PC Version 3.0*, U.S. Environmental Protection Agency, EPA-520/6-89-035, December, 1989b.

EPA, *User's Guide for CAP88-PC Version 3.0*, U.S. Environmental Protection Agency, March, 2006. www.epa.gov/rpdweb00/docs/cap88/userguide_120907.pdf; http://www.epa.gov/rpdweb00/assessment/CAP88/index.html

Federal Register, Part VI, U.S. Nuclear Regulatory Commission, 10 CFR Part 20 et al., Code of Federal Regulations, Standards for Protection against Radiation; Final Rule, May 1991, Washington, DC, 1991.

Harrison, K.G., Marshall, M., Ryder, D.J., Myatt, J., Otlet, R.L., Punter, D.B., Walker, A.J., Forde-Johnston, J., and Reid, H.J., Recent developments in monitoring for tritium in air, in *4th International Symposium on Radiation Protection-Theory and Practice*, pp. 219–222, 1989.

IAEA, *Safe Handling of Tritium: Review of Data and Experience*, International Atomic Energy Agency. Technical Report Series No. 324. Vienna, Austria, 1991.

Keenan, J.H., Keys, F.G., Hill, P.G., and Moore, J.G., *Steam Tables, SI Units*, John Wiley and Sons, Inc., New York, NY, 1978.

Klein, R.C. and Gershey, E.L., Biodegradable liquid scintillation counting cocktails, *Health Phys.*, 59, 461–470, 1990.

Lodge, Jr., J.P. Ed., *Methods of Air Sampling and Analysis*, 3rd ed., CRC Press, Boca Raton, FL, 1988.

Marvin, C.F., *Psychrometric Tables*, Weather Bureau Publication #735, U.S. Government printing office, Washington, DC, 1941.

Mayer, C.A., *ASME Steam Tables*, 4th ed., American Society of Mechanical Engineers, New York, NY, 1979.

NCRP, *Tritium Measurement Techniques*, National Council on Radiation Protection and Measurements, Report No. 47, National Council on Radiation Protection and Measurements, Bethesda, MD, 1976.

NRC, *Air Sampling in the Workplace, U.S. Nuclear Regulatory Commission*, Regulatory Guide 8.25, National Technical Information Service, Springfield, VA and U.S. Government Printing Office, Washington, DC, 1992.

Otlet, R.J., Walker, A.J., and Caldwell-Nichols, C.J., Practical environmental, working area and stack discharge samplers, passive and dynamic, for measurement of tritium as HTO and HT, *Fusion Technol.*, 21(2, pt. 2), 550–555, 1992.

Patton, G.W., Cooper, A.T., and Tinker, M.R., Ambient air sampling for tritium-determination of breakthrough volumes and collection efficiencies for silica-gel adsorbent, *Health Phys.* 72, 397–407, 1997.

Stephenson, J., *Re-evaluation of the Diffusion Sampler for Tritiated Vapour*, Ontario Hydro, Health and Safety Division, Restricted Report HSD-SD-90-20, 1990.

Wood, M.J., *Field Evaluation of Passive HTO-in-Air Samplers at Chalk River Laboratories*. Mississauga, Ontario: Canadian Fusion Fuels Technology Project/AECL Report CFFTP-G-9117/AECL-10358, 1991.

Wood, M.J., Outdoor field evaluation of passive tritiated water vapor samplers at Canadian power reactor sites, *Health Phys.*, 70, 258–267, 1996.

Wood, M.J. and Workman, W.J.G., Environmental monitoring of tritium in air with passive diffusion samplers, *Fusion Technol.*, 21(2, pt. 2), 529–535, 1992.

Wood, M.J., Hong, A., Cross, W.G., Nunes, J.C., and Leon, J.W., Calibration of portable tritium-in-air monitor for various radioactive gases, *Health Phys.*, 72, 423–430, 1997.

Wood, M.J., McElroy, R.G.C., Surette, R.A., and Brown, R.M., Tritium sampling and measurement, *Health Phys.* 65, 610–627, 1993.

Method 4

Determination of the Elemental Tritium Content of the Atmosphere*

CONTENTS

1	Principle of the Method	419
2	Range and Sensitivity	420
3	Interferences	420
4	Advantages/Disadvantages	420
5	Precision and Accuracy	420
6	Apparatus	420
7	Reagents	421
8	Procedure	421
9	Calibration	422
10	Calculations	422
11	Effects of Storage	423
12	Cautions	423
References		424

1 PRINCIPLE OF THE METHOD

1.1 See Section **1** of Method 3 for a brief discussion of tritium-in-air sampling. Method 3 is a technique for measuring tritium oxide (HTO).

1.2 The method is based on a technique presented by Östlund (1974) and is used by others (Griesbach and Stencel, 1988). Air is pumped at a sampling rate of approximately 1×10^{-4} to 2×10^{-4} m³ min⁻¹ through a trap containing 200 g of desiccant that collects moisture from the air. The air is then mixed with enough additional hydrogen to sustain catalytic oxidation, and proceeds into a catalytic converter that oxidizes free hydrogen and tritium. The resulting water (^2H_2O and ^3H_2O) is then collected in a desiccant trap. This method describes the use of a palladium sponge catalyst, and silica gel is used as the desiccant; however, other catalysts and desiccants are available (see Method 3 for alternative desiccants and associated references). The stated flow rate is sufficient to allow continuous air sampling for 7–14 days at 50% relative humidity. The collected moisture is then removed from the trap and analyzed for tritium by mixing a suitable volume of it with a scintillation fluid and counting the mixture in a liquid scintillation counter (LSC). The air sampling rate and total volume of moisture collected during the sampling period must be determined to convert the concentration of tritium in the collected moisture sample to its concentration in air.

---

* Method authors: Appreciation is extended to the writers of the Radioactivity section of the book *Methods of Air Sampling and Analysis* (Lodge, 1988) where this method originally appeared. Method reviewers and editors: Hung-Cheng Chiou, Lawrence T. Dauer, Andrew Hull, Paul Linsalata, Mark L. Maiello, Donald M. Mayer, Robert P. Miltenberger, and Dennis M. Quinn.

2 RANGE AND SENSITIVITY

2.1 The method is suitable for sampling periods of up to 7–14 days over the entire significant range of elemental and tritiated water vapor concentrations in the air. The sampling period can be extended if low relative humidity conditions exist or if the trap size is increased. Sensitivity is a direct function of the volume of air sampled and an inverse function of the prevailing temperature and amount of carrier gas used. Typically, one would expect to collect 7–8 mL of water produced from the catalytic reaction between palladium and elemental hydrogen and tritium. An air sample containing elemental tritium in a concentration of 740 Bq m^{-3} (2×10^{-8} µCi cm^{-3}), or about a factor of 10^{-6} of the derived air concentration (DAC) for elemental tritium (Federal Register, 1991), would have a concentration in the water aliquot of about 1.5×10^5 Bq L^{-1} (4×10^{-3} µCi cm^{-3}). This is sufficient to produce acceptable precision in less than 2 min of counting in most LSC systems.

2.2 In the workplace or other regulatory settings, the investigator should check that the lower limit of detection (LLD) or minimum detectable activity (MDA) of the sampling/counting system is sufficient to detect the concentration guideline to which the measurement will be compared.

3 INTERFERENCES

3.1 None.

4 ADVANTAGES/DISADVANTAGES

4.1 The removal of atmospheric water (including HTO) is key to this method. For other comments, see the Advantages/Disadvantages of Method 3.

4.2 The trap must be maintained in a vertical position during sampling to avoid the problem of "air streaming." This occurs when a void space opens in the media along the length of a horizontally oriented trap. A void can open as the media is vibrated (tamped down) by pump action or by transportation of the sampler. Moisture cannot be collected if the sample airstreams above the media rather than through it.

5 PRECISION AND ACCURACY

5.1 Duplicate moisture samples of only 1 mL volume (as opposed to the 7 mL recommended here) with radioactivity in the range of tens to hundreds of pCi mL^{-1} counted for 10 min in an LSC should yield reproducible results within 5% depending on things such as the pipetting technique. This assumes a counting efficiency for ^{3}H of about 60% and background values in the ^{3}H counting window <20 cpm using modern scintillation fluids and LSCs. The third edition of this book stated that the 2σ uncertainty of the sample described in Section **2.1** would be less than $\forall 2\%$, if a moisture aliquot of 7 mL is counted for 2 min at an efficiency of 25% using an LSC with a 7 cpm background in the tritium window (Lodge, 1988).

5.2 The accuracy of the air concentration measurement is dependent on the calibration of the collection efficiency and the calibration of the LSC.

6 APPARATUS

6.1 Liquid scintillation counter.

6.2 Traps, 30.5×3.2 cm diameter (12 in. × 1.25 in.). The trap construction can be aluminum, plastic, or glass* (see Method 3, Figure 1 for a typical design). If the sampling device is to be used

---

* A manufacturer of suitable traps is W.A. Hammond Drierite Co., 138 Dayton Avenue, Xenia, OH 45385, USA.

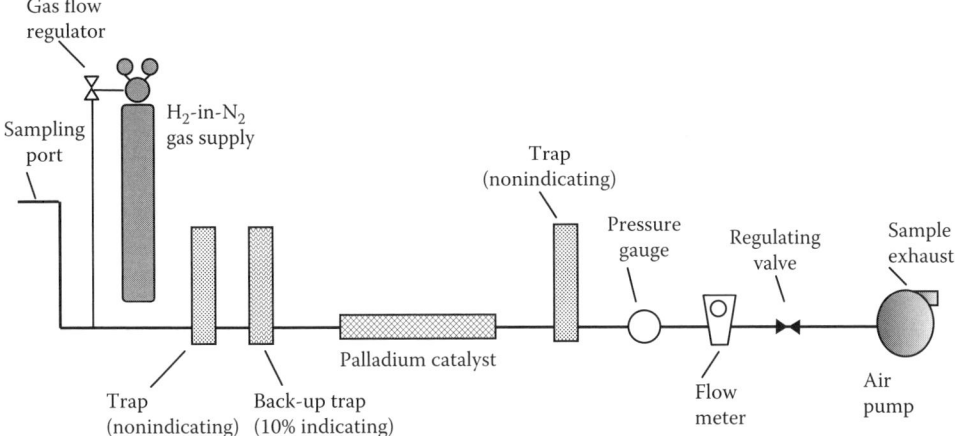

FIGURE 1 Sampling line for determining elemental tritium concentration in air.

to sample airstreams with very high tritium concentrations, tritium can exchange with plastic, thus causing cross-contamination of samples.

6.3 Low-volume air pump, airflow rate indicator, critical or limiting orifice, and airflow integration device (or temperature and relative humidity device).

6.4 Distilling flask and condenser.

6.5 Pipettes.

6.6 LSC vials, polyethylene, 20 mL.

6.7 Rubber Stoppers and Tubing.

6.8 Carrier gas of 3% hydrogen and 97% nitrogen (a size-H cylinder is adequate).

6.9 Palladium sponge catalyst (2 g).

6.10 Glass wool plugs.

7 REAGENTS

7.1 Scintillation fluids are available commercially.* Some fluids may require special disposal procedures; others may not (Klein and Gershey, 1990). An individual (health physicist or hazardous waste specialist) with knowledge of the governmental and institutional disposal regulations should be consulted before use of the fluid.

7.2 Silica gel,† 6–16 mesh, moisture indicating.

7.3 Silica gel, 6–16 mesh, nonindicating.

8 PROCEDURE

8.1 Bake 6–16 mesh, nonindicating silica gel for 24 h at 250°C prior to use. This step may be omitted if the user is willing to accept a 5% uncertainty in the total moisture content. Omission of this step may be highly desirable if the drying process cannot be performed in a low-background environment. The silica gel can be contaminated by ambient air in the bake-out/cool-down step of the procedure.

8.2 Fill two 30.5 cm traps (see Method 3, Figure 1 for a typical design) with nonindicating silica gel. Fill a third trap with a mixture of 10% indicating silica gel and 90% nonindicating silica gel. Since silica gel can adsorb about 20–30% of its own weight (Dean, 1992; Budnitz et al., 1983;

---

* For example, from Packard Instrument Company, Inc., USA, or Research Products International Corporation, USA.
† J.T. Baker, USA.

NCRP, 1976), 200 g should not saturate during a sampling period as long as 7 days in extremely warm and humid conditions. Seal the inlet and the outlet until the initiation of sampling.

8.3 Introduce the palladium sponge into a length of 6 mm Pyrex tubing, holding it in place with small portions of glass wool.

8.4 Unseal the traps. Assemble the sampling line (Figure 1) as follows: a T-connector, one arm of which is attached to the tank of H_2-in-N_2, the other to the sample inlet; a trap containing nonindicating silica gel; a back-up trap containing 10% indicating silica gel; the tube containing the palladium catalyst; the second trap containing nonindicating silica gel; and the sampling pump. Appropriate devices must be added to regulate and measure the air sample and H_2-in-N_2 flows.

8.5 Set the H_2-in-N_2 flow at 3×10^{-5} m^3 min^{-1}, then turn on the sampling pump. Begin timing the air-sample collection process. Adjust the air-sample flow rate to 1×10^{-4} m^3 min^{-1}. Record the initial total and air-sample flow rates.

8.6 At the end of air-sample collection, record the final total and ai-sample flow rates. Record the duration of the air sample collection period. Remove the nonindicating silica-gel traps, label them appropriately, and seal the ends. Check the back-up trap for a change in the color of the indicating silica gel. If a change of color has occurred, remove the back-up trap, seal the ends, and transport it to the laboratory for analysis.

8.7 If it is desired to determine the tritiated water content of the atmosphere, process the first trap and the back-up trap, if necessary, according to Method 3, Sections **9.6** through **9.10**.

To determine the elemental tritium content of the air, process the final trap according to Method 3, Sections **9.6** through **9.10**.

9 CALIBRATION

Refer to Section **10** of Method 3 and follow instructions.

10 CALCULATIONS

10.1 Efficiency:

$$\text{EFF} = \frac{\text{CR}_{\text{STD}} - R_{\text{B}}}{\text{ACT}_{\text{STD}}}, \tag{1}$$

where EFF is the counting efficiency (cpm dpm^{-1}), CR_{STD} is the gross count rate (counts per minute or cpm) of standard in scintillation solution, R_{B} is the background count rate (cpm) of distilled water in scintillation solution, and ACT_{STD} is the disintegrations per minute (dpm) of standard.

10.2 Elemental tritium concentration of moisture:

$$C_{\text{M}} = \frac{R_{\text{S}} - R_{\text{B}}}{\text{EFF} \times V \times K}, \tag{2}$$

where C_{M} is the concentration of elemental tritium in collected moisture (Bq mL^{-1}), R_{S} is the gross sample count rate (cpm), that is, sample count plus background, V is the volume (mL) of moisture-sample counted, and K is 60 s min^{-1} to convert dpm into dps.

10.3 The uncertainty at the 95% confidence level ($\forall 1.96\sigma$) due to counting statistics only is calculated as follows:

$$U(\text{cpm}) = \forall 1.96 \left[\left(\frac{R_{\text{S}}}{t_{\text{S}}} \right) + \left(\frac{R_{\text{B}}}{t_{\text{B}}} \right) \right]^{1/2}, \tag{3}$$

where t_{S} is the counting time of the standard and t_{B} is the counting time of the distilled water background.

Determination of the Elemental Tritium Content of the Atmosphere 423

The uncertainty (U) must be converted from cpm into Bq mL^{-1} by dividing by (EFF $\times V \times K$). See Sections **11.2** and **11.3** of Method 3 for definitions of these variables.

10.4 The concentration of elemental tritium in air is calculated as follows, assuming that silica gel with 100% collection efficiency is used.

$$C_A = \frac{C_M \times W}{V_R \times T}, \qquad (4)$$

where C_A is the concentration of elemental tritium in air (Bq cm^{-3}), W is the total volume of water collected (mL) from the final trap,* V_R is the mean air-sampling flow rate (cm^3 min^{-1}), and T is the air-sampling time (min).

To report the 2σ uncertainty of C_A in Bq cm^{-3}, multiply the 2σ uncertainty of C_M by $W/V_R \times T$.

10.5 Using the volume of water of the first two silica-gel traps in the same set of equations from above yields the tritium oxide content of the atmosphere.

10.6 Appropriate factors for converting the tritium air concentrations to estimates of individual committed radiation dose to exposed personnel is provided by the U.S. Environmental Protection Agency (EPA) in Federal Guidance Report No. 11 (EPA, 1988), which has been adopted as part of the U.S. Code of Federal Regulations (Federal Register, 1991) and by many U.S. state radiation regulatory programs. The time of submergence in the tritium cloud (for exposure to elemental tritium) or exposure to the tritium air concentration (for tritium oxide exposure) must be known or accurately estimated to use these exposure-to-dose conversion factors. Tritium dosimetry must account for exposure due to inhalation and transference through the skin. Dose calculations are perhaps better performed by a qualified expert in health physics who can interpret the results or who could note the potential sources of error in measurement, calculation, and interpretation. Federal Guidance Report No. 11 also lists DACs and annual limits of intake (ALIs) that are also currently legally enforced. The DAC limit is particularly useful for comparing with measured tritium concentrations. In addition, the U.S. DOE has officially adopted the newer ICRP methodologies in calculating its DACs, which resulted in a more conservative DAC for elemental tritium (Federal Register, 2007). The reader should determine which set of regulations to follow in choosing the proper DAC value for elemental tritium.

For emissions from a stack, the air concentrations measured at the emission point may be transformed into individual and population doses by available software packages such as COMPLY (EPA, 1989a), AIRDOS-PC (EPA, 1989b), and CAP88-PC (EPA, 2006; Federal Register, 2006). These software packages vary in complexity and employ such user-supplied information as wind rose data, building dimensions, stack heights, and effluent velocities to calculate off-site doses.

11 EFFECTS OF STORAGE

11.1 Storage effects are minimized if the silica gel collection traps are *thoroughly* sealed and stored in a nontritiated atmosphere.

11.2 The liquid scintillation samples prepared for counting should be analyzed within a few days. Liquid scintillation fluid is volatile and slowly escapes from glass and plastic vials.

12 CAUTIONS

12.1 This method may involve the use of hazardous chemicals and/or radioactivity. It is the user's responsibility to take appropriate precautions and to obtain necessary regulatory approvals from institutional and/or governmental agencies.

---

* The water content of this trap derives principally from H_2-in-N_2 that helps to support the catalytic reaction.

12.2 If the user is not trained or familiar with the calibration and operation of radiation analysis equipment, he/she should consult a qualified expert, such as a health physicist, for assistance in the proper and safe usage of the equipment and development of appropriate quality assurance protocols.

REFERENCES

Budnitz, R.J., Nero, A.V., Murphy, D.S., and Graven, R., *Instrumentation for Environmental Monitoring*, Vol. 1, *Radiation*, Wiley Interscience, New York, 1983.

Dean, J.A., *Lange's Handbook of Chemistry*, 14th ed. McGraw-Hill Book Company, New York, NY, 1992.

EPA, Limiting Values of Radionuclide Intake and Air Concentration and Dose Conversion Factors for Inhalation, Submersion and Ingestion, U.S. Environmental Protection Agency, EPA-520/1–88–020, Federal Guidance Report No. 11, September, 1988.

EPA, *User's Guide for the COMPLY Code* (Revision 2), U.S. Environmental Protection Agency, EPA-520/1-89-003, 1989a.

EPA, *User's Guide for AIRDOS-PC Version 3.0*, U.S. Environmental Protection Agency, EPA-520/6-89-035, 1989b.

EPA, *User's Guide for CAP88-PC, Version 3.0*, U.S. Environmental Protection Agency, March, 2006 (www.epa.gov/radiation/assessment/CAP88/index.html, last accessed 9/28/07).

Federal Register, Part VI, U.S. Nuclear Regulatory Commission, 10 CFR Part 20 et al., *Code of Federal Regulations, Standards for Protection Against Radiation*; Final Rule, May 1991, Washington, DC, 1991.

Federal Register, U.S. Environmental Protection Agency, National Emission Standards for Hazardous Air Pollutants (Radionuclides), Availability of Updated Compliance Model; Notices, February 2006. Washington, DC, 2006.

Federal Register, Part II, U.S. Department of Energy, 10 CFR Parts 820 and 835, Procedural Rules for DOE Nuclear Activities and Occupational Radiation Protection; Final Rule, June 2007. Washington, DC, 2007.

Griesbach, O.A. and Stencel, J.R., The PPPL Differential Atmospheric Tritium Sampler (DATS), in *Proceedings of the 22nd Midyear Topical Meeting of the Health Physics Society*, pp. 374–380, December 4–8, San Antonio, TX, Health Physics Society, McLean, VA, 1988.

Klein, R.C. and Gershey, E.L., Biodegradable liquid scintillation counting cocktails, *Health Phys.*, 59, 461–470, 1990.

Lodge, Jr., J.P. Ed., *Methods of Air Sampling and Analysis*, 3rd ed., CRC Press, Boca Raton, FL, 1988.

NCRP, *Tritium Measurement Techniques*, National Council on Radiation Protection and Measurements Report No. 47, National Council on Radiation Protection and Measurements, Bethesda, MD, 1976.

Östlund, H.G., A sampling system for atmospheric HT and HTO, *IEEE Trans. Nucl. Sci.*, 21, 510–512, 1974.

Method 5
Determination of Carbon-14 in Air*

CONTENTS

1	Introduction	425
2	Principle of the Method	425
3	Sensitivity and Range	426
4	Interference	426
5	Advantages/Disadvantages	427
6	Precision and Accuracy	427
7	Apparatus	428
8	Reagents	428
9	Procedure	428
10	Calibration	429
11	Calculation	430
12	Effect of Storage	432
13	Cautions	432
References		432

1 INTRODUCTION

1.1 This method can be employed for sampling room-air and stack-air emissions of Carbon-14 (^{14}C). Industrial and research uses of ^{14}C may result in many kinds of molecules tagged with ^{14}C. Without extensive experimental data it is not possible to state unequivocally the collection efficiency of this method for all molecular forms bearing the ^{14}C atom. The method employs a chemical trap described by Moghissi et al. (1971), Cantelow et al. (1972), and Pfeiffer et al. (1981). Also see Kennally (1969). Other methods using different trapping agents have been described (Qureshi et al., 1985; Kunz, 1985; Burchell and Judkins, 1996). Method 4 has been used within a research setting (Maiello and Linsalata, 2000).

1.2 This method is also applicable for determining the combined elemental (^{3}H) and tritium oxide (^{3}H$_2$O) concentration of the air if the elemental tritium is efficiently oxidized by heat (see below).

2 PRINCIPLE OF THE METHOD

2.1 To measure ^{14}C in air, the ^{14}C-bearing molecule is oxidized by heating it to a high temperature. The resulting ^{14}CO$_2$ is then bubbled through a known volume of 0.2 N solution of sodium hydroxide (NaOH). The reaction causes the ^{14}C to be trapped in the Na$_2$CO$_3$ molecule. The other product of the reaction is water. The reaction is summarized as

$$2\text{NaOH} + {}^{14}\text{CO}_2 \rightarrow \text{Na}_2{}^{14}\text{CO}_3 + \text{H}_2\text{O}$$

---

* Method authors: Mark L. Maiello and Paul Linsalata. Method reviewers and editors: Hung-Cheng Chiou and Robert P. Miltenberger.

The product solution is recovered and an aliquot is counted for the beta emissions of ^{14}C in a liquid scintillation counter (LSC). The concentration in the solution is converted by calculation using known factors such as the amount of air sampled to obtain the ^{14}C air concentration and the total activity discharged. The collection efficiency is estimated to be greater than 90%.

2.2 The conditions required to produce $^{14}CO_2$ are high temperature and sufficient residence time in the high-temperature environment. Practical application has indicated that 900°C at a sampling flow rate of about 2×10^{-4} m^3 min^{-1} is sufficient to convert and trap about 90% or more of the $^{14}CO_2$. However, the user is advised to verify this for the sampled molecular species. For routine surveillance, for example, for fume-hood emissions monitoring, sampling may be continuous at a 2×10^{-4} m^3 min^{-1} flow rate for the course of a work day. The maximum period measurable would be dependent on the evaporation rate of the solution and the amount of reactive contaminants captured from the air stream, for example, carbon, sulfur, and so on. The carbon species could theoretically consume the NaOH in the chemical reaction described above if a long sampling period was employed using a relatively small volume and normality or molarity of NaOH.

2.3 Gaseous tritium oxide produced in the tube furnace will become part of the water component of the 0.2 N NaOH solution. The collection efficiency for tritium is estimated to be in the 80–90% range.

3 SENSITIVITY AND RANGE

3.1 At a sampling flow rate of about 2×10^{-4} m^3 min^{-1} over a sampling period of 10 h using 150 mL of 0.2 N NaOH as the sampling medium, the minimum detectable concentration* of nonparticulate ^{14}C by liquid scintillation counting for 10 min is about 80 Bq m^{-3} of air (2×10^{-9} Ci m^{-3}). This result can be obtained using 1 mL aliquot of sample and 15 mL of an alkaline tolerant scintillation cocktail in a 20 mL glass scintillation vial. Other researchers using similar trapping systems (2 and 0.2 M NaOH) report detectable concentrations of between 2 and 7 Bq m^{-3} (Caron and Benz, 2002; Cantelow et al., 1972).

3.2 Under the same sampling and counting conditions stated in Section **3.1**, the minimum detectable concentration of tritium in air would also be about 80 Bq m^{-3} (2×10^{-9} Ci m^{-3}).

3.3 The upper range is dependent on the reactive capacity of the NaOH. For about 10 h of sampling at air concentrations of ^{14}C on the order of 37,000 Bq m^{-3} (1×10^{-6} μCi mL^{-1}), 150 mL of 0.2 N NaOH is sufficient. This is so because 0.015 moles of carbon atoms are required to saturate the trapping reaction. Calculations indicate that atmospheric CO_2 (0.034% by volume of air) can accomplish this in about 65 h of sampling at 280 mL min^{-1} (at 24°C) with 150 mL of 0.2 N NaOH. Even ^{14}C releases of 74×10^6 Bq m^{-3} (2×10^{-3} μCi mL^{-1}) are a small fraction of the moles needed to saturate the reaction. The upper limit can be extended by increasing the volume or normality of the NaOH.

3.4 In the workplace or other regulatory settings, the investigator should check that the minimum detectable activity of the sampling/counting system is sufficient to detect the concentration guideline to which the measurement will be compared.

4 INTERFERENCE

4.1 It is possible that other radioisotopes could be collected in the 0.2 N NaOH by simply being trapped in the fluid by physical or chemical means. The presence of a prefilter should prevent most particulate matter from entering the bubbler. Fortunately for the analyst, LSCs permit energy windows to be set so that only the counts from the desired radionuclide may be recorded. This is true if there is no overlap of the signals from the desired and the unwanted radionuclide in the window. If overlap occurs, it may be possible to allow the unwanted radionuclide to decay away if

---

* See Chapter 5 for a brief discussion of lower limit of detection and minimum detectable concentration.

its half-life is short enough. If tritium and ^{14}C are present simultaneously in the sampled air, a typical LSC could separate the counts from both to produce nuclide-specific air concentration results.

4.2 Alkaline solutions added to standard scintillation fluid will cause chemiluminescence, that is, scintillations due to chemical reactions of the sample with the scintillation fluid rather than due to radioactive decay. These false scintillations may be minimized by using a scintillation fluid specially made for ionic (alkaline) solutions like Na_2CO_3 (Packard, 1998; L'Annunziata, 1998).

4.3 Photoluminescence occurs when the scintillation fluid and sample combination are exposed to ultraviolet light. Prepared samples should be put in the dark for 15 min (dark adapted) to eliminate this interference (L'Annunziata, 1998).

4.4 Other interference or inaccuracies may occur when the location for sampling the air is chosen poorly. For occupational indoor measurements, the airflow patterns of the workplace should be determined so that the air sampler is appropriately located and representative of inhaled air (NRC, 1992). Outdoor sampling, especially during an accidental release, will require timely wind direction and other meteorological data. For stack monitoring, the flow rate through the effluent system must be measured by qualified individuals using calibrated instruments so that an accurate volume of the air discharged may be used to calculate the total activity of ^{14}C released to the environment. Isokinetic sampling conditions are not critical if it is known that the contaminant is a gas or vapor and is not associated with entrained particles.

5 ADVANTAGES/DISADVANTAGES

5.1 Disadvantages

The method is limited to sampling flow rates in the 200–500 mL min^{-1} range because more vigorous bubbling may introduce unwanted vibration of the sampling equipment and perhaps increase the chances for line leaks or insufficient mixing of the sample air. Since this is not a real-time method, the results will not be immediately available. Set-up time is required to replace the prefilter, to replenish the NaOH and the silica gel, and to count the ^{14}C in the aliquot. An air sampling system such as this will require attention to line leaks, filter loading, mechanical problems associated with the pump or tube furnace, and to proper measurement and calibration of the airflow rate. In addition, radiological counting with an LSC requires the implementation of quality control procedures.

5.2 Advantages

This air sampling system is relatively simple to construct and operate. It is also reliable.

6 PRECISION AND ACCURACY

6.1 An airflow measurement accuracy of about ±2–5% is obtainable with most commonly employed flow meters (rotameters) depending on the model and proper calibration (Dwyer, 2001).

6.2 LSCs are usually calibrated using appropriate ^{14}C and tritium standards. Quenched standards are also available for some counters to compile an efficiency curve that accounts for varying levels of quench. Precision and accuracy will depend on the quality of this calibration and the ability of the LSC to compensate for signal interference. Other factors include the type of scintillation fluid used, the radiation background level of the LSC/sample (fluid, sample, and vial) system, the quality of sample preparation and the counting time.

6.3 Samples should always be counted in duplicate and an average count value used as the final result. LSCs can also determine the counting uncertainties [standard deviations (SDs)] of each result.

6.4 The use of an accurate, calibrated, trigger-type pipette is recommended for preparing the Na_2CO_3 sample solution for liquid scintillation counting. Depending on brand, the accuracy of sample volumes of 1 mL or more is usually <± 0.7% with precision <0.3%. This is stated on supplied quality control certificates.

6.5 Considering calibration uncertainties, radioactive counting statistics, the need for sample airflow measurement, and sample preparation (pipetting), a final accuracy of ±10–20% is predicted.

7 APPARATUS

7.1 The sampling apparatus consists of a filter (0.1 μm pore size), a bench-top tube furnace, a bubbler, a pump, and an airflow measurement device (Figure 1). The bubbler should be able to hold at least 150 mL of NaOH.* The tube furnace is available from manufacturers[†] and laboratory supply vendors. A filter holder and a silica-gel column[‡] are also needed. An air pump and an airflow measurement device, for example, a rotameter, are required as well. These devices are also available from laboratory supply houses.

7.2 The filter is present to collect particles that may be contaminated with ^{14}C. Even if the user is not interested in collecting particulate matter, such a filter is useful for trapping radioactive dusts before they can enter and contaminate the sample line. If significant contamination is suspected to be in particulate form, the filter should also be counted in an LSC. Accurate air sampling of particulate matter from a stack will necessitate isokinetic sampling conditions.

7.3 The silica-gel column acts to protect the pump and the rotameter from moisture or vapor that may emanate from the bubbler and travel downline. Such moisture may contain small amounts of radioactivity.

7.4 A supply of 20 mL liquid scintillation vials (glass or plastic) and a reliable LSC are required to complete the analysis.

8 REAGENTS

8.1 The required chemicals are color-indicating silica gel,[§] a supply of 0.2 N NaOH,[§] and a chemiluminescent-resistant liquid scintillation fluid that can accept alkaline samples.** If a filter is to be counted, it may be analyzed using standard liquid scintillation cocktail.

9 PROCEDURE

9.1 PREPARATION

Calibrate the LSC (see Section **10.2**). Set up the counting protocol (energy window) of the radionuclide to be counted. Determine the air sampling collection efficiency of the radionuclide to be

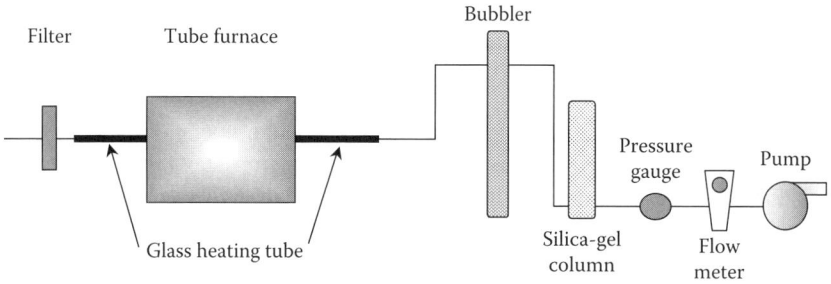

FIGURE 1 Sampling train for C-14 or tritium using NaOH as the trapping solution in the bubbler.

---

* One manufacturer of bubblers (also known as wash bottles, gas scrubbers, and absorption towers) is Kimbel/Kontes Glass, Inc., USA.
† For example, Lindberg/Blue M—A General Signal Co., USA.
‡ A manufacturer is W.A. Hammond Drierite Co., USA.
§ J.T. Baker, USA.
** Packard Instrument Company (now part of Perkin-Elmer), USA.

monitored. This entails running the apparatus under field conditions and introducing a known amount of radioactivity into the sampler (see Section **10**). The sampling fluid is counted and a comparison of the results with the introduced amount is made. The ratio of these values is the collection efficiency. For ^{14}C introduced as $^{14}CO_2$, greater than 90% collection efficiency is achievable. However, for other compounds, experiments may need to be conducted to verify that they are oxidized in the high-temperature furnace and collected as efficiently.

9.2 Sampling

Add 150 mL of NaOH to the bubbler. The tube furnace is activated and allowed to reach 900°C. Air is then drawn by the sampling system through the prefilter, then sequentially through the tube furnace, the bubbler, and the silica-gel column before being exhausted through the pump.

Sampling should be performed in a location free of major restrictions to the sampled airflow. Permanent outdoor samplers should be located in protective structures. Samplers located on building roofs or in mechanical rooms near the emission point may need security measures to assure that the equipment will not be disturbed. An airflow meter such as a rotameter may be used to obtain readings at the initiation and termination of the sampling period with a simple average being taken as the characteristic flow rate (NRC, 1992). An air volume totalizer may be used as an alternative.

For daily sampling, an air sampler that will provide a flow rate of about 2×10^{-4} m^3 min^{-1} should be employed. Along with other long-term sampling problems (Section **3.3**) is the buildup of dust on the filter. If this is a problem, a constant-flow pump may be used.*

Sampling may be terminated manually or by an electronic timer. Collect the sample fluid from the bubbler using a leak-proof container. At this time the filter can be changed before commencing another sampling run. If isokinetic sampling is being done, the filter may be collected for analysis in an LSC. The volume of trapping solution collected should be measured. The actual volume recovered should be used in the calculations of ^{14}C air concentration (over 10 h, about 4–5 mL is expected to be lost by evaporation depending on ambient temperature and humidity conditions).

9.3 Counting

9.3.1 As stated earlier, LSCs can distinguish ^{14}C counts from tritium counts. To obtain results for both radionuclides, samples can be run under "dual-count" protocols for simultaneous counting of each scintillation vial sample or the vials can be counted under separate nuclide-specific protocols. The protocols are usually preset by the manufacturer with energy windows for each radionuclide and usually make provision for the automatic subtraction of background counts and conversion to various units such as dpm or Bq mL^{-1} of sampling solution. Because many LSCs are computer programmable, further customization of protocols, for example, count time, number of times counted, reported units, and a variety of quality control data such as long-term tracking of the counting efficiency, is usually obtainable.

9.3.2 The field samples and the background sample should be counted using scintillation vials of the same size and composition and using the same volumes of scintillation fluid as used for the LSC-calibration standards.

9.3.3 After sampling, the bubbler must be cleaned with glassware detergent and rinsed with deionized water. Excess water may be removed from the interior surfaces by rinsing with acetone.

10 CALIBRATION

10.1 Air Sampling Equipment

If a rotameter is used as the airflow measurement device, the reader is recommended to consult Craig (1971) concerning the problems associated with the calibration of these devices. For example,

---

* Available from HI-Q Environmental Products Co., USA.

if the rotameter is not calibrated using exactly the same sampling line situation (pressure drop across the sampler) and sampling conditions (ambient temperature and pressure), significant errors may be introduced into the air concentration results. Craig advises use of flow rates that cause the lowest pressure drops possible and meter readings at the upper end of the rotameter scale. It is also advisable to place a pressure gauge on the same side of the pump regulator valve as the rotameter float for the purposes of determining a correction for any pressure drop between the filter and the rotameter and for the detection of line breaks. The novice should also consult an experienced individual such as an industrial hygienist concerning these matters. The ASTM standard for calibration of a rotameter to standard conditions may be used (ASTM, 1996) although ASTM does not provide a discussion of calibration problems. Airflow meters should be calibrated annually and after repairs or modifications (NRC, 1992).

10.2 Trapping Efficiency

10.2.1 The efficiency of collection of tritium and ^{14}C in the NaOH solution may be determined by delivering a known amount of the radionuclide as a gas into the measurement system. Small pressurized "lecture bottles" can be purchased* to deliver a known activity to the bubbler. ^{14}C is available this way as $^{14}CO_2$. An LSC measurement of the bubbler solution is performed in the same manner as a field sample and the total amount of activity sampled is calculated for comparison with the delivered value. The ratio of measured activity to delivered activity is the collection efficiency.

10.2.2 The collection efficiency calculated for the known activity of the delivered gas may not be the same for the many different gaseous compounds of ^{14}C encountered in the field or in stack emissions. However, the conversion of many compounds to $^{14}CO_2$ with subsequent trapping is estimated to be greater than 90% at the recommended temperature of 900°C.

10.3 LSC Efficiency

10.3.1 The manufacturers of LSCs provide procedures for the calibration of these devices. A full discussion of the various techniques that are employed in LSCs to refine the calibration is beyond the scope of this book. The user is urged to refer to the operating manual of the LSC for this information. The calibration is achieved by counting a ^{14}C or tritium standard available from the LSC manufacturer or from other vendors† in the same size and type scintillation vial as the field samples are to be counted. Typically, ^{14}C or tritium standards contain about 0.004 MBq (0.1 µCi) of radioactivity. Several standards each containing increasing amounts of the quenching agent are usually acquired to make an efficiency curve that is maintained in the LSC computer. The LSC can account for chemical quench in the samples by applying an appropriate efficiency factor for the quenching encountered.

10.3.2 Count the standards according to the procedure recommended by the LSC manufacturer so that the measured efficiencies are recorded in computer memory and applied appropriately in the sample analysis. The ratio of activity counted to activity in the standard is the counting efficiency (cpm/dpm, cps/dps, or cps/Bq) for the counting conditions (vial type and size, sample volume used, scintillation fluid type, and the LSC protocol used). Some LSCs track efficiency and other operating parameters, for example, background count rates, for quality assurance.

11 CALCULATION

11.1 Many LSCs can be programmed to subtract background to obtain net cpm, to convert net cpm to dpm via the counting efficiency factor (Section **10.3**), and to change dpm to the concentration of

---

* For example, Isotope Products Laboratories, USA.
† The National Institute of Standards and Technology, USA. Commercial suppliers include Analytics, Inc., USA.

radioactivity in the scintillation fluid (Bq mL^{-1}, ΦCi mL^{-1}). Converting cpm into concentration of radioactivity *in the air sample* may not be possible with all LSC units. Therefore, the user can be expected to perform some calculations. If the LSC cannot be programmed to do the conversions to net cpm and dpm, the user will have to do these manually as well. Since the half-lives of ^{14}C and ^{3}H are 5730 years and 12.35 years, respectively, calculations to correct for radioactive decay are unnecessary.

11.2 For manual calculations, the activity per unit volume (ACT$_{VOL}$) of air is determined as follows:

$$\text{ACT}_{VOL} = \frac{R_S \times V_{NaOH}}{\text{EFF} \times V_{scint} \times T \times V_{SR}}, \quad (1)$$

where ACT$_{VOL}$ is the activity per unit volume of air (Bq m^{-3}), R_S (cps) is the net counting rate of sample (total sample counting rate minus background counting rate) from the ^{14}C energy window or from the tritium energy window of the LSC, V_{NaOH} is the volume of 0.2 N NaOH recovered from bubbler after sampling (m^3), EFF is the counting efficiency of the LSC (cps Bq^{-1}), V_{scint} is the volume of NaOH pipetted to scintillation vial for counting (m^3), T is the total air sampling time (min), and V_{SR} is the mean air sampling rate (m^3 min^{-1}).

11.3 The uncertainty of the result should be indicated. Many LSCs automatically calculate the SD of the count rate and report count rate uncertainty at the 95% confidence level (1.96σ or "two sigma," i.e., 1.96 × SD). Often, this value is reported as a percentage of the count rate result. This is convenient since further conversions of the count rate are needed to obtain a result in Bq m^{-3}, as was done in Section **11.2**. The two-sigma value can then be easily found as a percentage of the final, converted result in Bq m^{-3}.

11.4 A manual determination of the SD of the count rate follows:

$$\text{SD} = \frac{\left((R_G/t_S) + (R_B/t_B)\right)^{1/2} \times V_{NaOH}}{\text{EFF} \times V_{scint} \times T \times V_{SR}}, \quad (2)$$

where SD is the standard deviation of the count (Bq m^{-3}), R_G is the gross count rate (cps), R_B is the background count rate (cps), V_{NaOH} is the volume of 0.2 N NaOH recovered from bubbler after sampling (m^3), t_S is the sample count time (s), t_B is the background count time (s), EFF is the counter efficiency (cps Bq^{-1}), V_{scint} is the volume of NaOH pipetted to scintillation vial for counting (m^3), T is the total sampling time (min), and V_{SR} is the mean air sampling rate (m^3 min^{-1}).

Note that only the counting error is considered here. This manually calculated uncertainty of the result should be explicitly reported at a specified confidence level, such as 95% (1.96σ or "two sigma"). Therefore, the uncertainty of the calculated concentration at 95% = 1.96 × SD.

11.5 Appropriate factors for converting the ^{14}C (or tritium) air concentrations to estimates of individual committed radiation dose to exposed personnel are provided by the U.S. EPA in Federal Guidance Report No. 11 (EPA, 1988), which has been adopted as part of the U.S. Code of Federal Regulations (Federal Register, 1991) and by many U.S. state radiation regulatory programs. The time of exposure to the ^{14}C (or tritium) air concentration must be known or accurately estimated to use these exposure-to-dose conversion factors with the EPA specified breathing rate for the working individual of 0.020 m^3 min^{-1}. Tritium dosimetry must account for exposure due to inhalation and transference through the skin. Dose calculations are perhaps better performed by a qualified expert in health physics who can interpret the results or who could note the potential sources of error in the measurement and calculations. Federal Guidance Report No. 11 also lists derived air concentrations (DACs) and annual limits of intake (ALIs) that are also currently legally enforced. The DAC limit is particularly useful for comparison with measured ^{14}C or tritium air concentrations.

For emissions from a stack, the air concentrations measured at the emission point may be transformed into individual and population doses by available software such as COMPLY (EPA, 1989a),

AIRDOS-PC (EPA, 1989b), and CAP88-PC (EPA, 2006; Federal Register, 2006). These software packages vary in complexity and employ such user-supplied information as wind rose data, building dimensions, stack heights, and effluent velocities to calculate off-site doses.

12 EFFECT OF STORAGE

12.1 The sampling solution will have to be stored in a leak-proof container tightly sealed to prevent evaporation and spillage.

12.2 The liquid scintillation samples prepared for counting should be analyzed within a few days. Liquid scintillation fluid is volatile and slowly escapes from glass and plastic vials.

13 CAUTIONS

13.1 This method involves the use of hazardous chemicals. It is the user's responsibility to take appropriate precautions and to obtain the necessary regulatory approvals from the local institutional and/or governmental agencies.

13.2 Liquid scintillation fluid containing <1.9 kBq g^{-1} (<0.05 µCi g^{-1}) of tritium and ^{14}C is currently regarded by the U.S. EPA as a hazardous waste, as opposed to a radioactive waste, and may be incinerated by licensed hazardous waste facilities. Arrangements must be made to have the liquid scintillation vials properly packaged in an approved shipping container and transported to the waste facility by a licensed hazardous waste broker.

13.3 If the user is not trained or familiar with the calibration and operation of radiation analysis equipment, a qualified expert, such as a health physicist, should be consulted for assistance in the proper and safe usage of the equipment and development of appropriate quality assurance protocols.

REFERENCES

ASTM, *1996 Annual Book of ASTM Standards*, Section 11, *Water and Environmental Technology*, Designation D-3195—90, Standard Practice for Rotameter Calibration, ASTM, West Conshohocken, PA, 1996.

Burchell, T.D. and Judkins, R.R., Passive CO_2 removal using a carbon filter composite molecular sieve, *Energy Convers. Manage.*, 37, 947–954, 1996.

Cantelow, H.P., Bolton, R.L., Peck, J.S., and Aune, R.G., Sampling system for tritium oxide and carbon-14 for environmental air, *Health Phys.*, 23, 384–385, 1972.

Caron, F. and Benz, M.L., Analysis and processing of samples for a carbon-14 monitoring program at a radioactive waste storage site, *Analyst*, 127, 1121–1128, 2002.

Craig, D.K., The interpretation of rotameter air flow readings, *Health Phys.*, 21, 328–332, 1971.

Dwyer Inc., *Catalog 2002 Edition*, Dwyer Instruments, Inc., Michigan City, IN, 2001.

EPA, *Limiting Values of Radionuclide Intake and Air Concentration and Dose Conversion factors for Inhalation, Submersion and Ingestion*, U.S. Environmental Protection Agency, EPA-520/1-88-020, Federal Guidance Report No. 11, September, 1988.

EPA, *User's Guide for the COMPLY Code (Revision 2)*, U.S. Environmental Protection Agency. EPA-520/1-89-003, 1989a.

EPA, *User's Guide for AIRDOS-PC Version 3.0*, U.S. Environmental Protection Agency. EPA-520/6-89-035, 1989b.

EPA, *User's Guide for CAP88-PC Version 3.0*, U.S. Environmental Protection Agency, 2006 (www.epa.gov/radiation/assessment/CAP88/index.html, last accessed 9/28/07).

Federal Register, Part VI, U.S. Nuclear Regulatory Commission, 10 CFR Part 20 et al. *Code of Federal Regulations. Standards for Protection Against Radiation*, Final Rule, May 1991.

Federal Register, U.S. Environmental Protection Agency, *National Emission Standards for Hazardous Air Pollutants (Radionuclides), Availability of Updated Compliance Model*; *Notices*, February 2006, Washington, DC, 2006.

Kennally, J.R., Letter to the editor, *Health Phys.*, 16, 813, 1969.

Kunz, C., Carbon-14 discharge at three light-water reactors, *Health Phys.*, 49, 25–35, 1985.

L'Annunziata, M.F., *Handbook of Radioactivity Analysis*, Academic Press New York, 1998.

Maiello, M.L. and Linsalata, P., An air sampler for ^{3}H and ^{14}C from a radiopharmaceutical laboratory, Poster presented at the *33rd Midyear Meeting of the Health Physics Society*, Virginia Beach, VA, 2000.

Moghissi, A.A., McNellis, D.N., Plott, W.F., and Carter, M.W., A rapid liquid scintillation technique for monitoring ^{14}C in air, in *Rapid Methods for Monitoring Radioactivity in the Environment*, pp. 391–394, IAEA/STI/PUB 289, International Atomic Energy Agency, Vienna, Austria (Notice Added to text 25 May 1973), 1971.

NRC, *Air Sampling in the Workplace, U.S. Nuclear Regulatory Commission, Regulatory Guide 8.25*, 1992, National Technical Information Service, Springfield, VA, USA and U.S. Government Printing Office, Washington, DC, 1992.

Packard, *A Catalog of Chemicals and Supplies for Life Science Research*, Packard Instrument Company (now part of Perkin-Elmer), Meriden, CT, 1998.

Pfeiffer, K., Rank, D., and Tschurlovits, M., A method for counting ^{14}C as $CaCO_3$ in a liquid scintillator with improved precision, *Health Phys.*, 32, 665–667, 1981.

Qureshi, P., Fritz, P., and Dsimmie, R.J., The use of CO_2 absorbers for the determination of specific ^{14}C activities, *Int. J. Appl. Isot.*, 36, 165–170, 1985.

Method 6
Determination of the Iodine-131 Content of the Atmosphere*

CONTENTS

1. Principle of the Method 435
2. Sensitivity and Range 436
3. Interferences 436
4. Advantages/Disadvantages 437
5. Precision and Accuracy 438
6. Apparatus 438
7. Reagents (Sampling Media) 438
8. Procedure 439
9. Calibration 440
10. Calculation 442
11. Effect of Storage 444
12. Cautions 444
References 444

1 PRINCIPLE OF THE METHOD

1.1 Radioactive iodine, principally ^{131}I and ^{133}I, may be released to the atmosphere in the inorganic elemental state as I_2, in an inorganic combined state as HOI or CsI, or in an organically combined state as CH_3I. All except CsI may be present in the vapor form or may be in part attached to particles. CsI would be expected to be present in a dissolved aqueous state following core damage at a water-cooled nuclear reactor and could be discharged to the atmosphere entrained in released steam, which then would condense into a particulate form. Thus, sampling for radio-iodines in the atmosphere presents a unique problem.

This method involves sampling for radio-iodine in its solid and gaseous states with a particulate filter in series with a chemically activated charcoal or silver-zeolite cartridge. Iodine attached to solid particles and possibly some vaporous iodine will be deposited on the particulate filter, while the remaining vaporous iodine will be adsorbed on the charcoal or silver-zeolite cartridge. The use of a silver-zeolite cartridge is the preferred methodology in order to minimize the difficulty analyzing for iodine due to the presence of noble gases in the sample as would be the case following a reactor accident.

The collected radio-iodine activity is then quantitatively determined in the laboratory by γ-spectroscopy. Alternatively, it may be determined in the field using a portable γ-spectrometer, although an unshielded detector is subject to a higher background signal and perhaps less sensitivity.

---

* Method authors: Appreciation is extended to the writers of the Radioactivity section of the book *Methods of Air Sampling and Analysis* (Lodge, 1988) where this method originally appeared. Method reviewers and editors: Hung-Cheng Chiou, Lawrence T. Dauer, Andrew Hull, Paul Linsalata, Mark L. Maiello, Donald M. Mayer, Robert P. Miltenberger, Dennis M. Quinn, and Theodore Rahon.

Although other fission product radio-iodines may be also present in an atmospheric release from an operating nuclear reactor or during the first few days following a reactor accident, ^{131}I is dosimetrically the most significant form due to its relatively longer half-life (8.04 days). Therefore, this method is principally directed toward its evaluation.

1.2 For routine environmental surveillance, samples may be collected continuously at low flow rates (perhaps 0.01–0.03 m^3 min^{-1}) for periods of up to 1 or 2 weeks. In order to make a rapid evaluation of suspected or known incidents or accidents in which radio-iodines may have been released to the atmosphere, samples may be collected at higher flow rates for periods as short as 5 min.

2 SENSITIVITY AND RANGE

2.1 At a sampling rate of 0.140 m^3 min^{-1}, the minimum detectable concentration (MDC) of nonparticulate ^{131}I by laboratory γ-spectroscopy of a 5-min sample is about 7 Bq m^{-3} (2×10^{-10} Ci mL^{-1}). For a 1-week sample collected at 0.028 m^3 min^{-1}, the MDC of a γ-spectroscopy system is about 2×10^{-3} Bq m^{-3} (5×10^{-14} Ci mL^{-1}).

2.2 The upper range is dependent on several simultaneous factors including:

1. Adsorption capacity of the charcoal or silver-zeolite cartridge as determined by the age of the sampling media and the interferences caused by the presence of other airborne constituents which may also be adsorbed; for example, water vapor
2. Airflow used to sample the radioiodine (the higher the airflow the less efficient the capture)
3. Chemical and physical state of the radioiodine
4. Amount of radioactivity collected (the higher the activity, the more counting dead-time the sample is subject to)
5. Pore volume between the charcoal and silver-zeolite granules (minimizing the pore volume will reduce the trapping of unwanted gases)

2.3 In the workplace or other regulatory setting, the investigator should check that the MDC (derived from the lower limit of detection or LLD) of the sampling/counting system is sufficient to detect the concentration guideline to which the measurement will be compared. The reader is urged to refer to Chapter 5 for a brief discussion of LLD.

3 INTERFERENCES

3.1 In nonaccident situations such as environmental compliance monitoring, the major interference may be the presence of other collected γ-emitting radionuclides (radon and thoron progeny), especially on the particulate filters. This effect can be diminished or eliminated by permitting a sufficiently long decay time (2 h to 4 days) prior to counting depending on whether radon or thoron is the source of the interference (see Harley, 1953 for a discussion of radon progeny decay). If a waiting period is implemented, the ^{131}I activity, if present, will have to be corrected for decay due to the short 8.04 day half-life. *Note:* Holding this type of sample for decay is not an option if other radio-iodines are present and require measurement.

Most γ-spectroscopy is currently performed with high-resolution Ge detectors and computer-assisted interpretation. When provided with an adequate library of potentially interfering nuclides, such a system is essentially interference-free.

3.2 In nuclear power plant accident situations, the major interference is due to adsorption of noble radioactive gases or vapors on the charcoal cartridge, particularly ^{133}Xe. Noble gases can also remain trapped in the pore volume between the sample media. This may be minimized by the use of silver-zeolite collecting medium (Cline, 1981). If radioactive gases are present in much larger concentrations than radioactive iodines, they may *still* interfere with analysis.

3.3 Sample loss is a possibility due to the varying affinity of the chemical forms of gaseous radioiodine for certain materials commonly used in sampling lines. At least one study (Kabat, 1983) evaluated the deposition of various metals and plastics and considered high-humidity conditions that would prevail in a nuclear power plant emergency. It was determined that aluminum, lined with Teflon, may be used with minimal losses when sampling for elemental radioiodine and HOI. Minimization of the length and diameter of the sampling line will decrease sampling losses. Polyvinyl chloride components of the sampling apparatus, for example, gaskets, and stainless-steel sampling lines, must be avoided. Sample losses are expected to be maximal under nuclear power plant accident conditions when high humidity is present and I_2 and perhaps HOI are the major forms of radioiodine.

4   ADVANTAGES/DISADVANTAGES

4.1 In general, air sampling requires attention to measurement of airflow rates, sample loss in the line, correcting for in-line pressure drops, ambient air temperature and pressure deviations from standard conditions, filter loading and mechanical problems such as line leaks and pump failures. Radiological counting also demands quality control procedures. Radiological counting in the field is always subject to scrutiny because it is more difficult to control the conditions of the measurement. For example, the background radiation rate is variable depending upon location (Maiello, 1997) and is undoubtedly different from that observed in the laboratory. Shielding to reduce background radiation and increase measurement sensitivities cannot be easily provided in the field. Other problems related to air sampling encountered in the field include accessibility to timely wind direction and supplementary meteorological data. For occupational indoor air measurements, the airflow patterns of the workplace should be determined so that the air sampler is appropriately located and representative of inhaled air (NRC, 1992).

4.2 Organic iodides can only be collected (chemisorbed) on charcoal impregnated with potassium iodide (KI), triethylenediamine (TEDA), or hexamethylene tetramine (HMTA).

4.3 Silver-zeolite traps vaporous iodine without collecting ^{133}Xe. ^{133}Xe will cause the following problems if it is collected on charcoal under nuclear power-plant emergency conditions (Cline, 1981): (1) Continuous iodine monitors exposed to high count rates from trapped ^{133}Xe can malfunction (a condition referred to as "paralysis" with consequent loss of real-time data. (2) Charcoal cartridges containing significant amounts of ^{133}Xe may not be counted quickly on Ge spectroscopy systems unless the cartridge is placed at a distance from the detector due to significant counting dead time. The counting system usually has to be recalibrated for the increased distance. Lower sensitivity for ^{131}I and longer counting times are the consequences.

4.4 Short sampling times may result in a distribution of ^{131}I activity on the upstream face of the cartridge. With longer sampling times, some migration of the iodine into the charcoal bed may occur. At least one study (using ^{125}I) indicates that an exponential distribution of activity occurs and that an underestimation of the activity may result if a uniform distribution is assumed (Button, 1982). A laboratory that measured ^{131}I in the United States from the Chernobyl nuclear accident used a technique that lessened the effect of the nonuniform distribution. At a point midway through the counting period, the charcoal cartridge was inverted. This laboratory reported that measured activity at the front of the cartridge was 40% higher than that measured from the rear (EML, 1986). When calibrating with a standard, for example, a filter/canister assembly containing a nonuniformally mixed iodine standard, inverting the filter/(charcoal or zeolite) canister can be done to determine a "front load calibration" and a "mid-point calibration" for breakthrough conditions (saturation of the collecting media). The iodine concentration is accurately determined with a front load calibration when the sample iodine resides mostly on the air-inlet side of the canister. When breakthrough occurs, a more uniform distribution of the iodine is present. If a front-loaded standard is used to estimate the iodine concentration with a breakthrough situation, the result will be underestimated.

Counting the filter on the front and back and comparing the response to the calibration standard can provide the analyst with the data to determine if breakthrough occurred.

4.5 It is probably necessary for the user to collect empirical data to determine the activity distribution in experimental cartridges exposed to radio-iodine. From this information, a correction can be applied to all field samples with nonuniform activity distributions assuming that the calibration of the counting system was obtained with a uniformly distributed standard of ^{131}I.

5 PRECISION AND ACCURACY

5.1 An airflow measurement accuracy of about 3% is obtainable with most commonly employed devices when properly calibrated.

5.2 With the use of charcoal treated with KI or TEDA, the collection of all forms of nonparticulate iodine will approach 100%.

5.3 A measurement accuracy of $\forall 5$–25% is achievable. This is dependent on the quality of the calibration of the counting device which in turn depends on how well the distribution of ^{131}I in the calibration standard conforms to the actual distribution of ^{131}I on the air-sample cartridge.

6 APPARATUS

6.1 The sampling apparatus consists of a particulate filter, a charcoal cartridge if ^{133}Xe is not present (or a silver-zeolite cartridge if it is) and an air sampling pump in series with associated holders and connections, and a flow measuring device (Figure 1).

6.2 The sampling medium consists of prefabricated plastic or metal-encased cartridges containing charcoal treated with either KI or TEDA or containing silver-zeolite. The charcoal cartridges and the entire air sampling system are available from several vendors.*

Most of the commercial vendors of these cartridges can also supply suitably sized particulate filters and filter holders. Since there are several standard sizes, all about 5 cm diameter × 2.5 cm thick (2″ × 1″), the user should be careful to purchase a match of particulate filter, cartridge and filter holder.

6.3 Owing to their superior resolution of photopeaks, Ge detectors are preferable over NaI. Most laboratories use computer-assisted spectroscopy systems that incorporate libraries of nuclide photopeaks to facilitate the interpretation of complex spectra.

For field evaluations, portable γ-spectroscopy systems are available that employ Ge detectors. These devices are coupled to equally portable multichannel analyzers that provide near laboratory-equivalent photopeak analysis capability. However, variations of background from place to place and less sensitivity for ^{131}I due to a higher background if the detector is unshielded are to be expected.

7 REAGENTS (SAMPLING MEDIA)

7.1 Glass fiber filter (0.8-μm pore size or smaller).

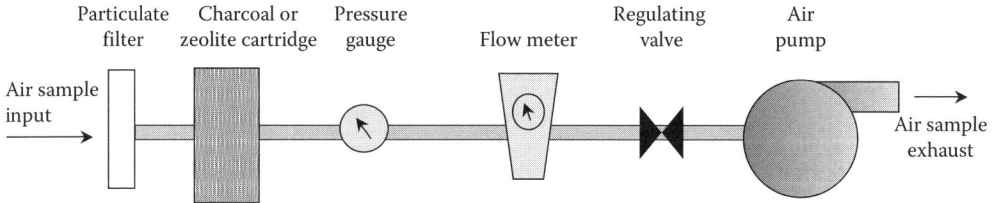

FIGURE 1 Sampling apparatus for the measurement of iodine in air.

---

* For example, HI-Q Environmental Products Co., USA, and F & J Specialty Products, Inc. USA.

Method 6: Determination of the Iodine-131 Content of the Atmosphere

7.2 Charcoal impregnated with TEDA or KI in ready-made cartridges.

7.3 Silver-zeolite in ready-made cartridges.

7.4 SAMPLING MEDIA CHARACTERISTICS

7.4.1 Manufacturers of charcoal-based sampling equipment may be contacted for technical information about the sampling characteristics of carbon (CSC, 1998). Although not scientific literature, technical catalogs from manufacturers provide some useful facts which the reader may attempt to verify by experiment or by researching published journals.

7.4.2 Analytical grade coconut charcoal (carbon) should be used for sampling. Charcoal ready-packed in sampling cartridges is available in several mesh sizes designated in order of smaller granule size as follows: 8×16, 20×40, and 30×50. The pressure drop through the cartridge increases with decreasing granule size. The 30×50 mesh size charcoal is too fine to use even with portable high-volume air movers. Silver-zeolite is available in 16×40, 30×50, and 50×80 mesh sizes.

7.4.3 Both charcoal and silver-zeolite are rated according to collection efficiency of CH_3I for various sample flow rates and mesh size. As the air sampling flow rate is increased, the retention or collection efficiency of the charcoal decreases. The user may consult such information in vendor catalogs before choosing a mesh size for the air sampling flow rates.

7.4.4 The recognized mechanisms of adsorption of radio-iodines onto charcoal are kinetic adsorption, isotopic exchange, and chemisorption (CSC, 1998).

7.4.4.1 Kinetic adsorption is due to the electrostatic attraction of the iodine molecule to the charcoal granule. Elemental radio-iodine is captured using this mechanism. High temperature or humidity may cause desorption of the iodine.

7.4.4.2 Isotopic exchange occurs when stable (nonradioactive) iodine, impregnated into the charcoal as KI_3, is replaced by the radioactive isotope in the form of CH_3I as the latter passes through the charcoal bed. This exchange may repeat, temporarily freeing the radioactive form, but ultimately the repetitive exchange slows the escape of the radioiodine until it undergoes radioactive decay. Methyl iodide can be adsorbed in this manner.

7.4.4.3 Chemisorption is the chemical attachment of the radio-iodine to a nonradioactive molecule such as TEDA or another tertiary amine impregnated into the charcoal. Methyl iodide can be adsorbed by this mechanism also. Desorption results in the loss of the impregnate and the radioiodine.

7.4.5 Charcoal canisters must be protected from the outside environment until required for sampling. Carbon adsorbs many unwanted substances (poisoning). Those substances intended for sampling may not be captured due to lack of adsorption sites caused by poisoning. Canisters must be tightly sealed using a vapor barrier of some kind especially if they are to be stored for months or years before use. Canisters should not be used if they are more than 5 years old (CSC, 1998).

7.4.6 Water is a commonly encountered charcoal poison. The collection efficiency of carbon will decrease with increasing atmospheric water vapor (humidity).

8 PROCEDURE

8.1 SAMPLING

8.1.1 Outdoor sampling should be performed in a location free of major airflow restrictions such as trees and structures. Permanent outdoor samplers should be located in protective structures. An airflow meter such as a rotameter may be used to obtain readings at the initiation and termination of the sampling period with a simple average being taken as the characteristic flow rate (NRC, 1992). An alternate method employs a dry gas meter inserted in the line.

For long-term sampling (1 day up to 1 week), an air sampler that will provide a flow rate of about $0.028\ m^3\ min^{-1}$ ($28\ L\ min^{-1} = 1$ CFM) should be employed. Owing to the possibility of the appreciable buildup of atmospheric dust on the particulate filter, a positive-displacement air sampler should be employed. Constant-flow samplers are available from several vendors (see footnote on p. 442).

For short-term sampling (5–10 min to 1–2 h), a greater flow rate of at least 0.085 m³ min⁻¹ (85 L min⁻¹ = 3 CFM) is desirable.

8.1.2 In a nonemergency situation, there should be time to calculate the volume of air necessary to collect measurable radioiodine activity, considering the minimum detectable activity of the counting system and any regulatory concentration values (refer to Chapter 5).

8.1.3 Carefully time the air sampling period. Record the initial airflow rate. At the end of sampling record the final airflow rate. If different, use the average value for subsequent calculations.

8.1.4 After sampling is complete, place the filter in a protective cover or container upon removal from its holder to minimize the possibility of damage or the dislodging of the collected activity. Collect the charcoal or silver-zeolite cartridge and store it in manner so it too will be protected from damage and loss of collection media.

8.1.5 Prior to counting, place the filter sample in a planchet or holder for reproducible positioning under the radiation detector. Care should be taken to avoid contaminating the detector. This can be done by covering the sample with a thin Mylar sheet. This sample yields results for particulate radioiodine.

8.1.6 Prior to counting, place the charcoal or silver-zeolite cartridge on the detector in a preselected geometry to provide a stable and reproducible position relative to the radiation detector. This sample yields results for vapor-phase radioiodine.

8.2 COUNTING

8.2.1 Counting of the cartridge must be performed in a counting geometry (sample-to-detector distance and alignment) identical to that used for calibrating the detector.

8.2.2 Count the cartridge over a time interval sufficient to detect activity, should it be present, above an MDC that meets regulatory guidelines, or other criteria. If using the technique mentioned above to lessen the effects of nonuniform activity distribution in the cartridge (Section **4.4**), stop the count midway through the measurement period to invert the canister on its counting platform. Then resume counting.

8.2.3 The filter should also be counted in a reproducible geometry identical to that used to calibrate the detector. Since a nonuniform activity distribution is not possible, the counting period need not be interrupted to invert the filter.

8.2.4 If only ^{131}I ($t_{1/2}$ = 8.04 d) is known to be present, the use of a NaI detector will yield sufficiently accurate results. However, the other short-lived radio-iodines may also be present in the air effluents from an operating reactor or from one that has recently been shut down. If present in appreciable quantities or if the radioactive noble gases (^{85}Kr and ^{135}Xe) have been adsorbed in the cartridge, they will produce interfering counts in the 0.364 MeV region of the principal ^{131}I photopeak (see Table 1). In such cases, a Ge detector must be used.

8.2.5 While the predominant ^{131}I photopeak at 0.364 MeV may be employed for ^{131}I quantitation, most computer-assisted systems include comprehensive libraries of nuclides which enable them to identify other radio-iodines (^{132}I-^{135}I) as well as other nuclides that may be present on the particulate sample or the cartridge.

9 CALIBRATION

9.1 If a rotameter is used as the air flow measurement device, the reader is urged to consult the reference by Craig (1971) concerning the problems associated with the calibration of these devices. For example, if the rotameter is not calibrated using exactly the same sampling line situation (pressure drop across sampler) and sampling conditions (ambient temperature and pressure), significant errors may be introduced into the air concentration results. Craig advises the use of

TABLE 1
Half-Lives and Primary γ-Ray Energies of Some Radioisotopes of Iodine and Two Important Noble Gases

Isotope	Half-Life	Photon Energy of Interest[a] (MeV)
^{131}I	8.04 days	0.364
^{132}I	2.30 h	0.668
^{133}I	20.8 h	0.530
^{134}I	52.6 min	0.847
^{135}I	6.61 h	1.26
^{85m}Kr	4.48 h	0.305
^{85}Kr	10.72 years	0.514
^{87}Kr	76.3 min	0.403
^{133m}Xe	2.19 days	0.233
^{135m}Xe	15.36 min	0.527
^{135}Xe	9.11 h	0.250
^{138}Xe	14.13 min	0.258, 0.434

[a] The energies of the listed noble gas emissions have the potential to interfere with some iodine isotope analyses.

flow rates that cause the lowest possible pressure drops and meter readings at the upper end of the rotameter scale. It is also advisable to place a pressure gauge on the same side of the pump regulator valve as the rotameter float for the purposes of determining a correction for any pressure drop between the filter and the rotameter and for the detection of line breaks. The novice should also consult an experienced individual such as an industrial hygienist concerning these matters. The ASTM standard for calibration of a rotameter to standard conditions may be used (ASTM, 1996) although ASTM does not provide a discussion of calibration problems. Airflow meters should be calibrated annually and after repairs or modifications (NRC, 1992).

9.2 Radiological calibration for ^{131}I is the determination of a counting efficiency (cpm dpm^{-1}) for the preselected sample-to-detector distance and alignment. It must account for the sample matrix and its container. It is achieved by counting a ^{131}I standard in the same counting geometry and container in which the filter or the cartridge is to be counted. The ratio of the count rate to the known disintegration rate of the standard is then computed. Standardized samples of ^{133}Ba, which has a prominent photopeak at 0.356 MeV and a half-life of 10.66 years, are sometimes employed as a substitute for ^{131}I. For the computer-assisted interpretation of complex spectra, a number of radionuclide standards or single standards containing multiple radionuclides of various energies are employed for calibration purposes. Counting efficiencies for nuclides not included in the standard are determined by interpolation and are done automatically by the computer.

9.3 Calibration of the detector for counting the particulate filter may be accomplished by using a thin, disc-shaped source of ^{133}Ba purchased from a manufacturer.* Finding a source with a diameter similar to that of the filter should be possible because manufactured sources are available in many diameters from about 0.6 in. (1.6 cm) to 5 in. (12.4 cm). These sources are incorporated into the metal surface of the backing. Depending on the half-life of the nuclide, these sources can last up to 15 years with reasonable care. Since the manufactured source is counted with a metal backing, the reader should use a metal planchet for the filter when counting it. An older method of calibration, which requires some care to perform, is to spike a blank filter with a standardized liquid of ^{131}I. In order to provide a known amount of activity to the filter, a volumetric or a mass determination of the

---

* Isotope Products Laboratories, USA.

amount of ^{131}I will have to be made. This requires some expertise to accomplish. The liquid will also have to be dispensed as uniformly as possible on the filter, another technique requiring careful manipulation. The amount of activity needed to calibrate a γ-spectroscopy system is low. No more than 0.1 µCi is required. *Note:* elemental iodine is volatile and represents an inhalation hazard. Consult a health physicist before attempting to make or use such a standard. Simulated charcoal or zeolite cartridge standards are available commercially (see footnote on p. 445).

9.4 Calibration of the charcoal and/or zeolite cartridges will necessarily be less accurate due to the problem of nonuniformly distributed activity in the sample cartridges. It is simply not possible to easily prepare a charcoal cartridge with a distribution that mimics the field samples. One method to attempt is to implant a disc source or a prepared filter standard (Section **9.3**) midway within a blank cartridge using charcoal above and below to support the source. Short sampling times might result in the activity being trapped at the charcoal surface. This could be mimicked by placing the standard source at the cartridge surface or at a very shallow depth in the cartridge. Of course, a graph showing the variation of counting efficiency with source depth could be made, but use of a manufactured disc or prepared filter source in this manner is still only a rough approximation of the nonuniform sample distribution. See Section **4.4** for an alternate method.

10 CALCULATION

10.1 If the collected sample contains a mixture of radio-iodines including ^{131}I or other radionuclides requiring immediate analysis, then it should be counted in a laboratory equipped with a multichannel analyzer and associated computer interpretation capability that incorporates a library of the major nuclides expected to be present in the sample.

The associated analysis programs usually provide for the input of variables such as flow rate or total flow, sampling time, detector efficiency, and time of analysis. They also provide outputs that include the necessary corrections for radioactive decay between the end of the sampling period and time of analysis and the measurement uncertainties of the calculated activities or air concentrations.

As indicated in Section **9.2**, the detector and multichannel analyzer combination must be calibrated so that the photopeaks and their corresponding energies are properly associated and so that counting efficiencies across a broad energy range are determined. From these data, photopeaks may be identified and photopeak heights may be transformed into activity values (air concentrations) via computer analysis.

10.2 For manual calculations, the counting efficiency of the detector may be found using:

$$\text{EFF} = \frac{\text{CR}_{\text{STD}}}{\text{ACT}_{\text{STD}}} \tag{1}$$

where CR_{STD} = net count rate of standard [usually in units of counts per second (cps) or per minute (cpm)] or the total counts achieved in the counting interval. To simulate the presence of ^{131}I, the principal emissions of the standard should be similar in energy to those measured for the analysis of ^{131}I ACT_{STD} = activity, or disintegration rate of standard (dps or dpm) or the total disintegrations in the counting interval EFF = counting efficiency [total counts/total disintegrations (c d^{-1}) or cps dps^{-1} or cpm dpm^{-1}].

If the standard has more than one photopeak emission but only one peak is used to determine the efficiency (because it is similar in energy to the ^{131}I photopeak of interest), the equation must be modified to include the γ-ray abundance (γ_{AB}) of the detected photon.

$$\text{EFF} = \frac{\text{CR}_{\text{STD}}}{(\text{ACT}_{\text{STD}} \times \gamma_{\text{AB}})} \tag{2}$$

Method 6: Determination of the Iodine-131 Content of the Atmosphere

Note: Any units of time for the count or disintegration rates may be used as long as they cancel out in the equation.

10.3 For manual calculations, the activity per unit volume of air at the time of counting (ACT_{VOL}) is determined as follows:

$$ACT_{VOL} = \frac{(G/t_S)-(B/t_B)}{EFF \times T_S \times V_{SR}} \qquad (3)$$

where ACT_{VOL} = activity per unit volume of air (Bq m^{-3}) [1 dps = 1 Becquerel (Bq)], G = gross counts of sample, B = background counts, t_S = sample count time (s), t_B = background count time (s), EFF = counter efficiency [see above for units], T_S = air sampling time (min), and V_{SR} = mean volume air sampling rate (m^3 min^{-1}).

10.4 If the sample cartridges are not counted immediately due to long-distance transport or other delays, account for the decay of ^{131}I awaiting counting by correcting the activity (ACT_{VOL}) to the moment sampling ends (ACT_{COR1}).

$$ACT_{COR1} = \frac{ACT_{VOL}}{e^{-\lambda t_1}} \qquad (4)$$

where t_1 = time between end of sampling and the start of counting (days) and λ = 0.0862 days^{-1} (^{131}I decay constant).

10.5 If the air sampling period is a significant portion of the half-life of ^{131}I, a correction should be made for the fraction of the sample activity that decays during sampling. The corrected activity is found as follows (Moe and Vallario, 1988):

$$ACT_{COR2} = ACT_{COR1} \left(\frac{\lambda T_S}{1 - e^{-\lambda T_S}} \right) \qquad (5)$$

where T_S = sampling time (days).

10.6 If the counting period is a significant fraction of the decay half-life of the ^{131}I, the measured activity per unit volume of air (ACT_{VOL}) must be corrected for decay during counting (NCRP, 1985). It is done in the same manner described for the correction due to decay during air sampling (Section **10.5**) except that the sampling time (T_S) would be replaced with the counting time (t_S) and the correction is applied to ACT_{VOL} in place of ACT_{COR1}. The result is the activity at the start of counting. If necessary, the corrections described in Sections **10.4** and **10.5** would also be applied to yield the activity at the start of sampling. These corrections should be made with a computer associated with the detector and multichannel analyzer of the measurement system. An explanation of activity buildup and decay during air sampling may be found in Moe and Vallario (1988).

10.7 The uncertainty of the result should be indicated. Taking into consideration counting error only, it can be determined from the standard deviation of ACT_{VOL} as follows:

$$SD = \frac{\left[(R_S/t_S)+(R_B/t_B)\right]^{1/2}}{EFF \times T_S \times V_{SR}} \qquad (6)$$

where SD = standard deviation (Bq m^{-3}), R_S = gross sample count rate (cps), R_B = background count rate (cps), t_S = sample count time (s), t_B = background count time (s), EFF = counter efficiency (c/day), T_S = air sampling time (min), and V_{SR} = mean air sampling rate (m^3 min^{-1}).

The confidence level of the uncertainty should be specified; for example, 95% (the 1.96σ or "two sigma" level). In this case, the SD value would be multiplied by 1.96. The SD value must be corrected for decay in the same manner as was the ACT_{VOL} result.

10.8 Appropriate factors for converting the iodine air concentrations into estimates of individual committed radiation dose to exposed personnel are provided by the U.S. EPA in Federal Guidance Reports Nos. 11 and 13 (EPA, 1988, 1999) which has been adopted as part of the U.S. Code of Federal Regulations (Federal Register, 1991) and by many U.S. state radiation regulatory programs. The time of exposure to the iodine air concentration must be known or accurately estimated to use these exposure-to-dose conversion factors with the EPA specified breathing rate for the working individual of 0.020 m^3 min^{-1}. This calculation is perhaps better performed by a qualified expert in health physics who can interpret the results or who could note the potential sources of error in the measurement and calculations. Federal Guidance Report No. 11 and the U.S. Code of Federal Regulations, 10 CFR Part 835 (Federal Register, 2007) also lists derived air concentrations (DACs) and/or annual limits of intake (ALIs) that are also currently legally enforced in the U.S. The DAC limit is particularly useful for comparison with measured iodine concentrations.

For emissions of radio-iodines from a stack, the air concentrations measured at the emission point may be transformed into individual and population doses by available software such as COMPLY (EPA, 1989a), AIRDOS-PC (EPA, 1989b), and CAP88-PC (EPA, 2007). These software packages vary in complexity and employ such user-supplied information as wind rose data, building dimensions, stack heights and effluent velocities to calculate off-site doses. For accident situations, programs such as HOTSPOT (Homann, 1994) can be used to estimate dose and expected air concentration down-wind of the accident.

11 EFFECT OF STORAGE

11.1 A properly protected air-filter sample should experience no effects other than radioactive decay during storage. The pre-filter should be carefully placed in an acetate, plastic or paper envelope. It must be handled carefully to avoid tearing and loss of mass. The charcoal cartridges are more rugged but should be protected from cracking, denting, deformation, and loss of contents.

12 CAUTIONS

12.1 This method may involve the use of hazardous chemicals and/or radioactivity. It is the user's responsibility to take appropriate safety precautions and to obtain necessary regulatory approvals from institutional and/or governmental agencies.

12.2 If the user is not trained or familiar with the calibration and operation of radiation analysis equipment, they should consult a qualified expert, such as a health physicist, for assistance in the proper and safe usage of the equipment and development of appropriate quality assurance protocols.

REFERENCES

ASTM, 1996 *Annual Book of ASTM Standards*. Section 11, *Water and Environmental Technology*. Volume 11.03, *Atmospheric Analysis; Occupational Health and Safety; Protective Clothing*. Designation D-3195–90, Standard Practice for Rotameter Calibration, ASTM, West Conshohocken, PA, 1996.

Button, T.M., Activated charcoal filter counting for radioactive effluent determination in protein iodinations, *Health Phys.*, 43, 853–857, 1982.

Cline, J.E., Retention of noble gases by silver-zeolite cartridges, *Health Phys.*, 40, 71–73, 1981.

Craig, D.K., The interpretation of rotameter air flow readings, *Health Phys.*, 21, 328–332, 1971.

CSC, *High Efficiency Gas Adsorbers*, Bulletin No. 283A, Flanders/CSC Corporation, Bath, NC, 1998.

EML, *A Compendium of the Environmental Measurements Laboratory's Research Projects Related to the Chernobyl Nuclear Accident, EML-460*, H.L. Volchok and N. Chieco, eds, National Technical Information Service, Springfield, VA, 1986.

EPA, *Limiting Values of Radionuclide Intake and Air Concentration and Dose Conversion Factors for Inhalation, Submersion and Ingestion*, U.S. Environmental Protection Agency, EPA-520/1-88-020, Federal Guidance Report No. 11, September, 1988.

EPA, *Users Guide for the COMPLY Code (Revision 2)*, U.S. Environmental Protection Agency, EPA-520/1-89-003, October, 1989a.

EPA, *Users Guide for AIRDOS-PC Version 3.0*, U.S. Environmental Protection Agency, EPA-520/6-89-035, December, 1989b.

EPA, *Cancer Risk Coefficients for Environmental Exposure to Radionuclides*, U.S. Environmental Protection Agency, EPA-402-R-99-001, Federal Guidance Report No. 13, September, 1999.

EPA, *User's Guide for CAP88-PC, Version 3.0*, U.S. Environmental Protection Agency, March, 2006 (www.epa.gov/radiation/assessment/CAP88/index.html, last accessed 9/28/07).

Federal Register, Part VI, U.S. Nuclear Regulatory Commission, 10 CFR Part 20 et al., Code of Federal Regulations, Standards for Protection Against Radiation; Final Rule, May, 1991.

Federal Register, Part II, U.S. Department of Energy, 10 CFR Parts 820 and 835, Procedural Rules for DOE Nuclear Activities and Occupational Radiation Protection; Final Rule, June 2007. Washington, DC, 2007.

Harley, J.H., Sampling and measurement of airborne daughter products of radon, *Nucleonics*, 11, 12–17, 1953: reprinted in *Health Phys.*, 38, 1068–1069, 1980.

Homann, S.G., *Hotspot Health Physics Codes for PC*, Lawrence Livermore National Laboratory, UCRL-MS-106315, March, 1994.

Kabat, M.J., Deposition of airborne radioiodine species on surfaces of metals and plastics. *Proceedings of the 17th DOE Nuclear Air Cleaning Conference*, W.M. First, ed., Vol. I, pp. 285–300, CONF 820833, National Technical Information Service, Springfield, VA, 1983.

Lodge, Jr., J.P. Ed., *Methods of Air Sampling and Analysis*, 3rd ed., CRC Press, Boca Raton, FL, 1988.

Maiello, M.L., The variations in long term TLD measurements of environmental background radiation at locations in New York State and northern New Jersey, *Health Phys.*, 72, 915–922, 1997.

Moe, H.J. and Vallario, E.J., *Operational Health Physics Training*, ANL-88-26, Argonne National Laboratory, U.S. Department of Energy & National Technical Information Service, Springfield, VA, 1988.

NCRP, *A Handbook of Radioactivity Measurement Procedures*, NCRP Report No. 58, 2nd ed., National Council on Radiation Protection and Measurements, Bethesda, MD, 1985.

NRC, *Air Sampling in the Workplace*, U.S. Nuclear Regulatory Commission, Regulatory Guide 8.25, June, National Technical Information Service, Springfield, VA and U.S. Government Printing Office, Washington, DC, 1992.

Method 7

Sampling Air for Argon-41, Krypton-85, and Other Gamma-Emitting Radioactive Gases Using Gamma-Spectroscopy*

CONTENTS

1	Principle of the Method	447
2	Range and Sensitivity	448
3	Interferences	448
4	Advantages and Disadvantages	448
5	Precision and Accuracy	449
6	Apparatus	449
7	Reagents	451
8	Procedure	451
9	Calibration and Standards	452
10	Calculations	453
11	Effects of Storage	454
12	Cautions	455
References		455

1 PRINCIPLE OF THE METHOD

1.1 A small pump is used to compress an air sample into a commercially available compressed-gas sampling cylinder that has been equipped with either manual or automated valves, pressure monitoring, and flow control attachments (Newton et al., 1994). The flow rate of the pump determines the time required to fill the cylinder to the recommended pressure. This is a convenient and effective system to collect gases of γ-emitting radionuclides such as ^{41}Ar (half-life = 1.827 h) or ^{85m}Kr (half-life = 4.48 h) and count them by intrinsic germanium γ-spectrometry.

1.2 The cylinder of collected gas can be manually removed for γ-radioactivity counting at a remote location or the system can be operated in an automated, repetitive mode, by extracting the gas to a shielded location containing the gas collection cylinder and counting system.

1.3 Remotely controlled, normally closed, electrically actuated, solenoid valves can be used to reduce external radiation exposure to personnel taking the samples in high radiation areas.

1.4 The procedure can be used for compliance with regulations such as the U.S. Environmental Protection Agency National Emission Standards for Hazardous Air Pollutants (NESHAP) (40 CFR, Part 61, subpart H) (Federal Register, 1995).

---

* Method authors: George J. Newton and Mark D. Hoover. Method reviewers and editors: Hung-Cheng Chiou, Mark L. Maiello, and Robert P. Miltenberger.

2 RANGE AND SENSITIVITY

2.1 Compression of the collected gas provides a linear increase in the detection limit of the method compared to systems that detect a radioactive gas in a flask or grab-sample bag at normal atmospheric pressure.

2.2 Overall range and sensitivity of the method depend on the fundamental geometry and detection efficiency characteristics of the intrinsic germanium counting system.

2.3 The decision level or critical level L_C (95% confidence that the observed concentration of radioactivity is not due to background) is about 4.1×10^{-11} Bq cm^{-3} (1.1×10^{-15} µCi mL^{-1}) for ^{85}Kr in the system described in this method, when the collected gas is compressed to a pressure of 10 atm. This is based on the standard statistical approach (Currie, 1968) of $L_C = 1.96\sigma$, where σ is the standard deviation of the background count rate for the γ-energy of interest for ^{41}Ar (1.290 MeV). It also assumes that the radioactivity counting for ^{41}Ar (radioactive half-life 1.83 h) is done within 1 h of sample collection.

2.4 The minimum detectable concentration, MDC (95% confidence that the stated concentration will provide a radioactivity counting response greater than the decision level, and therefore not be recorded as a false negative) can be considered to be about 9.6×10^{-11} Bq cm^{-3} (2.6×10^{-15} µCi mL^{-1}), based on the standard statistical approach (Currie, 1968) of MDC = 4.6σ, where σ is the standard deviation of the background count.

3 INTERFERENCES

3.1 Noble gases are expected to remain uniformly distributed in the collection cylinder. If the method is used for reactive radioactive gases, such as iodine, care should be taken to ensure that radioactive sample does not plate out or otherwise accumulate in the cylinder in a manner that alters the detection efficiency. Method 6 describes an alternate iodine collection method.

3.2 Unintentional collection and accumulation of particles or condensation of water or other liquids in the cylinder can decrease the volume available for gas collection. This would result in collection of smaller volumes of gas, and a resulting underestimation of the radioactivity concentration in the sample. A filter on the inlet side of the sample collection cylinder can be used to prevent particles from entering the cylinder. Physical confirmation of the absence of condensation of water vapor may be needed for applications involving high humidity or transfer of the gas sample to a location with a temperature that is below the dew point.

3.3 Standard approaches for γ-spectroscopy should be used to select the energy detection windows and background correction algorithms for the radionuclides of interest. The γ-energy spectrum and detection efficiency for all radionuclides that could be collected should be understood and accounted for.

4 ADVANTAGES AND DISADVANTAGES

4.1 Advantages

4.1.1 The measurement relies on γ-ray spectroscopy which is an accurate method of analysis when performed with properly calibrated, well-maintained equipment.

4.1.2 The sampling method can be used in areas of high radiation exposure due to the use of remotely controlled valve actuators.

4.1.3 The sampling method can be modified for repetitive grab sampling.

4.2 Disadvantages

4.2.1 Continuous sampling and measurement of radio-gases is not possible.

Method 7: Sampling Air for Gamma-Emitting Radioactive Gases

4.2.2 Precautions must be taken to avoid potential errors from high humidity and high particulate concentration.

5 PRECISION AND ACCURACY

5.1 The physical attributes of the sample collection cylinders (such as wall thickness, shape, and volume) should be identical to the attributes of the cylinder used for calibration.

5.2 Failure to adequately flush the cylinder before collection of a new sample can result in an improper determination of the radioactive gas concentration.

5.3 Failure to accurately measure and properly account for the time delay between sample collection and radioactivity counting can introduce errors.

5.4 The pressure sensor and valve closure mechanisms should be chosen to provide a known and reproducible compression of gas in the cylinder. Improper assumptions about the compression will result in a linear degradation of accuracy. Variable compression from sample to sample will result in a commensurate degradation of accuracy. Installation of a pressure gauge directly on the cylinder can enable monitoring of the pressure and application of standard ideal gas law corrections to the true gas volume, if needed.

5.5 The ideal gas law describes the relationship between the reading of the cylinder pressure gauge and the number of "ambient volumes" compressed into the cylinder (see Section **10.2**). The relationship is dependant on the ambient pressure and temperature at the time of sample collection and the pressure and temperature in the cylinder when the gauge is read. Ambient pressure and temperature and cylinder pressure and temperature must be accurately measured to determine the actual volume of air collected.

5.6 Other contributions to precision and accuracy can arise from minor variations in the geometry of the counting system and in the inherent Poisson counting characteristics of γ-spectroscopy.

5.7 An estimate of the accuracy of any particular measurement using this method is about $\forall 5\%$ of the true air concentration of ^{41}Ar or ^{85}Kr. This is based upon estimated relative uncertainties of $\forall 3\%$ in the true activity of the calibration gas, $\forall 2\%$ in the determination of the counting efficiency for the γ-spectrometer, $\forall 2\%$ in the mean counting statistics for a sample of interest, and $\forall 2\%$ in the determination of the pressure (and therefore volume) of the sample. The total estimated accuracy is calculated as the root mean square of the individual uncertainties as follows:

$$\text{Accuracy} = \sqrt{(0.03)^2 + (0.02)^2 + (0.02)^2 + (0.02)^2} = 0.05.$$

5.8 The precision of a measurement is estimated to be approximately $\forall 3\%$ of the result for ^{41}Ar or ^{85}Kr. This is based on repetitive counting of samples from a stable source of these radioactive gases.

6 APPARATUS

6.1 THE COMPRESSED-GAS SAMPLING SYSTEM

This system consists of three parts: an inlet module (Figure 1), a gas cylinder module (Figure 2), and an exhaust module (Figure 3). The modules are assembled via ball joint unions. The electrically actuated solenoid valves are energized by a 110 V AC line. Wherever possible, components are type 304 or 316 stainless steel. Typical pressure ratings are 9000 pounds per square inch gauge, 62 MPa for the plumbing components, 12 MPa for the sample cylinder, and 1 MPa for the polyethylene tubing. Thus, operating the sampling system at a pressure of 690 kPa (100 psig) provides a margin of safety relative to the standard rated operating pressures of the components.

6.1.1 The inlet module (Figure 1) consists of a tubing adapter for connecting the inlet module to the radioactive gas source, usually a stack. Gas enters the compressor and is pumped through the

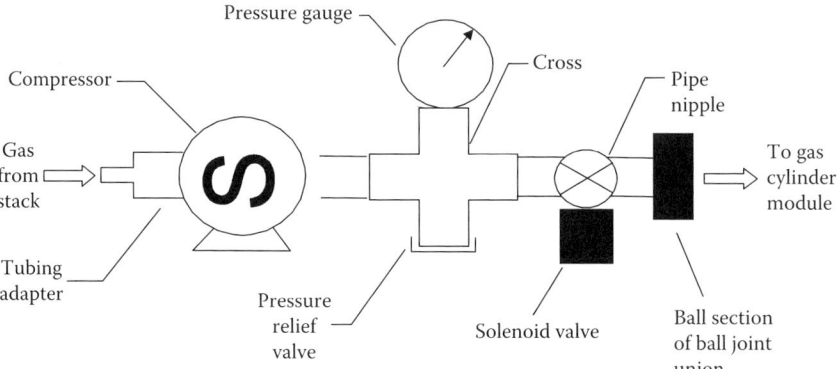

FIGURE 1 Schematic diagram of the inlet module of the compressed-gas-sampling system used to sample radioactive gases. Fittings are schedule 40 and type 304 stainless steel, 316 stainless steel, or brass.

entire gas sampling train to flush out clean air. The volume of the flush should be about 10 times the internal volume of the sampling train to achieve total washout of the clean air. The 1-L compressed gas cylinder accounts for most of the internal volume. A flush time of 2 min at the typical compressor flow rate of 5 L min^{-1}, that is, a total of 10 air changes, should be adequate.

6.1.2 The inlet module includes plumbing cross fitted with a pressure gauge and an adjustable pressure relief valve to ensure that the compressed-gas-sampling cylinder is not pressurized over 690 kPa. The pressure relief valve is typically set to open at 690 kPa. At 690 kPa, the 1-L sampling cylinder contains 6.8 L of ambient air when operated at sea level pressure (101.3 kPa or 14.7 psia). Operation at 690 kPa will involve more liters of ambient air when operated at elevations above sea level. For example, the cylinder will contain 9.3 L of ambient air when operated at Albuquerque, New Mexico, which has an elevation of 5300 ft. (1600 m) and an ambient atmospheric pressure of 83.3 kPa (625 mm Hg). The electrically actuated solenoid valve in the inlet module remains open during flushing and filling operations and is closed by the operator when the cylinder had been filled.

6.1.3 The inlet module is connected to the gas cylinder module by a ball joint union. A pair of manual ball valves allows the gas cylinder to be closed after filling and permits the gas cylinder module (Figure 2) to be removed for counting. Assurance of the true pressure in the cylinder at the time of radioactivity counting can be achieved by installing an additional pressure gauge between the manual ball valve on the inlet end and the exhaust end of the cylinder itself. If the actual pressure in the cylinder is not 690 kPa or the temperature of the cylinder is different from the temperature of the air at the point of collection, then a correction to the gas volume can be made using the ideal gas law (see Section **10.2**).

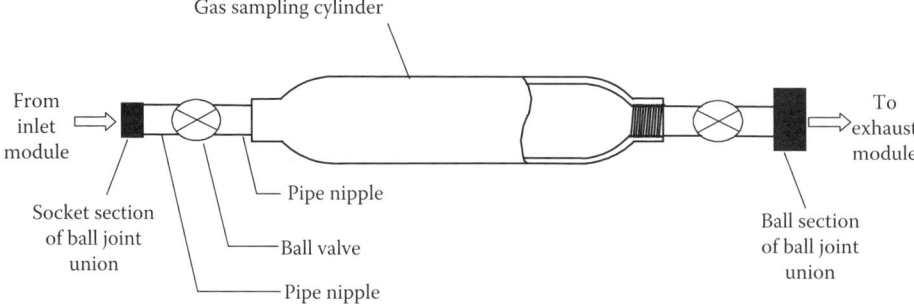

FIGURE 2 Schematic diagram of the gas cylinder module of the compressed-gas-sampling system used to sample radioactive gases. Fittings are schedule 40 and type 304 stainless steel, 316 stainless steel, or brass.

Method 7: Sampling Air for Gamma-Emitting Radioactive Gases

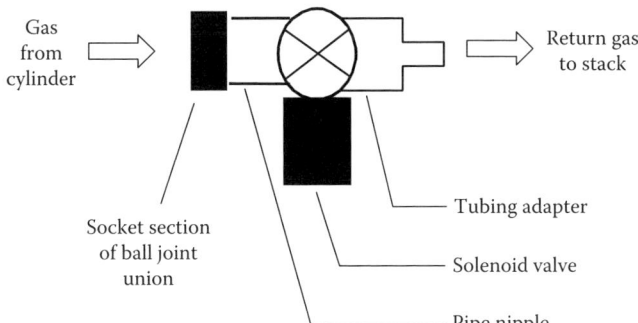

FIGURE 3 Schematic diagram of the exhaust module of the compressed-gas-sampling system used to sample radioactive gases. Fittings are schedule 40 and type 304 stainless steel, 316 stainless steel, or brass.

6.1.4 The gas cylinder module is connected to the exhaust module (Figure 3) via a ball joint union. In the example shown, the exhaust module contains an electrically actuated solenoid valve which remains open during flushing, and is closed for filling. A tubing adapter at the end of the exhaust module allows the flush gas to be returned to the facility off-gas stack or to be vented to a suitable location.

6.2 Description of the Counting System

The typical detector is a planar intrinsic germanium γ-detector. The detector is cooled for virtually noiseless operation using liquid nitrogen housed in a dewar. The entire arrangement must be shielded with lead or a graded-Z shield to decrease the interfering effects of terrestrial radiation, cosmic rays and the radiation originating in nearby building materials. Access through the shield must be provided for signal and power cables and to allow the gas cylinder to be placed in close proximity to the detector. The counting data are transmitted to the γ-spectrometer for display of the photopeak spectrum. A computer analyzes the photopeaks of interest and the surrounding spectrum in order to identify the radioactive gases, subtract the Compton scattered γ-rays and other background radiation to yield net results, and to calculate the radiogas air concentration based on the height of the photopeak(s).

7 REAGENTS

7.1 None.

8 PROCEDURE

8.1 Assemble the three-module sampling train via the ball-and-socket unions.
8.2 Open the manual ball valves.
8.3 Connect the solenoids to a 110 V line and energize the valves.
8.4 Connect the inlet module of the sampling train to the stack sampling port using 3/8-in. (0.95 cm) diameter polyethylene tubing.
8.5 Connect the exhaust module polyethylene tube to a return port on the stack or to an alternate, acceptable location for venting the gas.
8.6 Open the shut off valves at the stack ports and ensure that the gas flow path is open.
8.7 Energize the compressor and flush out the clean air in the sampling system with a minimum of 10 volumes.
8.8 Turn off the exhaust module solenoid, thereby closing off the exhaust portion of the sampling train.
8.9 Observe the pressure gauge until the pressure is 7.0 kg cm^{-2} (690 kPa or 100 psig).

8.10 Turn off the inlet module solenoid, thereby trapping the compressed-gas sample between the two solenoid valves.

8.11 At the earliest time possible, shut off the compressor and close the two manual ball valves on the gas sampling cylinder.

8.12 Disassemble the sampling train at the ball joint unions and cap the unions on the ends of the sampling cylinder module.

8.13 Record ambient temperature and pressure. Also record the time and date that the sample was obtained along with other identifying information.

8.14 Transport the sampling cylinder to the counting facility for quantification of the γ-emitting radioactive gas.

8.15 Measure the radioactivity in the cylinder as a function of energy.

8.16 Make appropriate corrections for time delays from sample collection to start of counting, decay during counting, and changes in detection efficiency as a function of γ-emission energy (see Section **10**).

9 CALIBRATION AND STANDARDS

9.1 Standard approaches can be used for the calibration of the germanium-based spectrometer. One method that can be used is to count actual samples of the radioactive gas of interest, such as ^{41}Ar obtained from a certified source. The radioactive concentration of the standard radiogas will probably be given by the manufacturer at standard temperature and pressure. Measurement of the local temperature and air pressure will be necessary to correct the volume of radioactive gas introduced into the sample cylinder. Manufacturers of standards often provide a worksheet to make this correction.

9.2 Another calibration technique is to count a simulated gas counting standard that was prepared using other radionuclides.

9.2.1 For example, Table 1 describes a multienergy standard sample that was prepared by coating nine different radionuclides onto 1/8-in diameter polyurethane beads* which were then placed into the standard gas sampling cylinder. The beads do not interfere with the γ-emissions. Most of the

TABLE 1
Example of a Mixed γ-Standard for Simulated Radioactive Gases in a Gas Cylinder[a]

Radionuclide	γ-Ray Energy (keV)	Half-Life	γ-Rays per Second	Total Uncertainty (Percent)
Cd-109	88	462.6 days	3000	4.8
Co-57	122	271.7 days	1682	4.8
Ce-139	166	137.64 days	1813	5.0
Hg-203	279	46.60 days	3355	4.8
Sn-113	392	115.08 days	2956	4.7
Sr-85	514	64.85 days	5614	5.0
Cs-137	662	30.0 years	1866	5.0
Y-88	898	106.61 days	7795	4.7
Co-60	1173	5.27 years	3060	4.7
Co-60	1332	5.27 years	3068	4.9
Y-88	1836	106.61 days	8205	4.7

[a] Emissions from ^{41}Ar (1294 keV) are simulated by the higher energy γ (1333 keV) from ^{60}Co. The prominent γ-ray emission of ^{85}Kr is at 1512 keV. An effective calibration date was provided with the standard to correct the emission rates of the individual radionuclides for radioactive decay. This reference standard was prepared by Analytics, Inc., Atlanta, GA, USA.

---

* For example, Analytics, Inc., Atlanta, Georgia.

attenuation comes from the cylinder walls. Table 1 lists the radionuclides in the γ-standard. Because the primary radionuclide of concern was ^{41}Ar which emits 1.290 MeV γ-rays, the higher energy γ-photons from ^{60}Co (1.332 MeV) are an adequate surrogate for ^{41}Ar.

9.3 The standard is placed on the germanium detector in a counting geometry duplicating that used for actual samples. Figure 4 is a schematic diagram of a convenient detector-to-source geometry. A count is commenced for a period of time that gives a preselected degree of precision. The ratio of the measured count rates to disintegration rates in the photopeak regions of the radioactive gas of interest are automatically calculated and input to the spectrometer computer. These counting efficiencies are used to calculate the actual activity of the radioactive gas. The computer produces a counting efficiency curve (stored in memory) to accommodate radioactive gas photon energies which fall between those used in the standard.

10 CALCULATIONS

10.1 Subtracting background radioactivity (the Compton background) from the photopeaks of interest is a necessary step to obtain the net activity due to the presence of the radioactive gas. This is usually performed using a background stripping algorithm which is part of the spectrometer analysis software. Alternatively, one may count a gas cylinder that contains all the radioactive gas constituents encountered in an actual sample except the radioactive gas of interest. The background count rate in the photopeak region for the radioactive gas of concern (which will include contributions from terrestrial radiation, cosmic-ray and shielding sources) can be noted and subtracted from that of an actual sample.

10.2 Owing to the short half-lives of ^{41}Ar and ^{85}Kr, two corrections must be made to the radioactivity counting results to account for decay of the radioactive gas during the time delays after collection of the gas. The first time delay, t, is the time between collection of the gas and the start of radioactivity counting. The second time delay is an additional Δt which occurs during the radioactivity counting process.

10.2.1 The correction that accounts for the decay of radioactivity during the radioactivity counting time uses the standard equation for radioactive decay during a counting interval Δt:

$$A(t) = \frac{\Delta N \lambda / (1 - \exp(-\lambda \Delta t))}{\varepsilon} \tag{1}$$

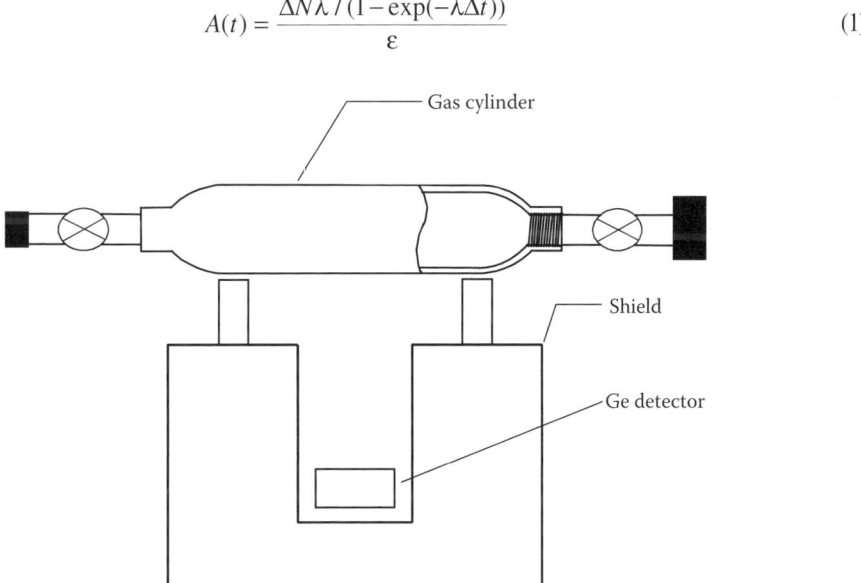

FIGURE 4 Schematic diagram of a detector-to-source geometry used to calibrate the spectrometer and measure the contents of the radiogas cylinder.

where ΔN is the number of counts during the counting time Δt, ε is the detection efficiency of the γ-counting system (counts per disintegration), λ is the decay constant for the radionuclide of interest, $A(t)$ is the radioactivity disintegrations per unit time at the start of the counting interval.

The values of Δt and λ can be in any consistent set of units (e.g., min and per min, or hours and per hour). The equation for the gas concentration in units of Bq m^{-3} requires that activity be expressed in disintegrations per second (see Section **10.4**).

10.2.2 The correction that accounts for the decay time (T) after collection of the sample and before the sample can be counted uses the standard radioactive decay equation:

$$A(0) = A(t) \exp(\lambda T) \tag{2}$$

where $A(0)$ is the activity collected in the cylinder, $A(t)$ is the radioactivity after a decay time of T at which radioactivity counting begins.

10.2.3 The combined equation is

$$A(0) = \frac{\Delta N \lambda \exp(\lambda T)}{(1 - \exp(-\lambda \Delta t))}, \tag{3}$$

10.2.4 It should be possible to program the spectrometer computer to make these corrections so that they are automatically accounted for in the final gas concentration result.

10.3 Using the ideal gas law, the *effective* volume of the cylinder V, is

$$V = V_c \frac{P_c T_a}{P_a T_c}, \tag{4}$$

where V_c is the actual volume of the cylinder, P_c and T_c are the pressure and temperature in the cylinder when the gauge is read and, P_a and T_a are the local ambient pressure and temperature where the sample was collected.

Thus, if the cylinder has a volume of 1 L, the cylinder pressure is 690 kPa (100 psi), the ambient pressure is 101.4 kPa, and the ambient temperature and the cylinder temperature are the same (i.e., they cancel in the equation above), then the effective (or pressurized) volume of the cylinder is

$$V = 1 \frac{690}{101.4} = 6.8 \, \text{L}.$$

If the ambient temperature (where the gas was collected) is 20°C (293 K) and the temperature in the cylinder (when the pressure gauge is being read) is 25°C (298 K), then the effective volume of the cylinder is

$$V = 1 \frac{(690)(293)}{(101.4)(298)} = 6.7 \, \text{L}.$$

10.4 After the net counts or the net count rate in a photopeak region are corrected for decay during sampling and counting, and then converted into total disintegrations by application of the counting efficiency for the radioactive gas of concern as noted above, then the air concentration is calculated using

$$C(\text{Bq m}^{-3}) = \frac{A(0)}{V(10^{-3})}. \tag{5}$$

11 EFFECTS OF STORAGE

11.1 If gas is collected in the cylinder and the cylinder is stored prior to radioactivity counting, then the proper corrections must be made to account for radioactive decay (see above).

11.2 If the sampling system itself is not used for an extended period of time, such as might occur if the system is used for periodic confirmatory measurements on an annual or an every-other-year basis, then:

11.2.1 The system should be inspected to ensure that all seals and valves are properly functioning.
11.2.2 The integrity of the polyethylene tubing sections should be inspected and parts replaced as needed because of normal aging, physical damage, or degradation from ultraviolet light.
11.2.3 The calibration of the pressure gauge(s) should be confirmed.
11.2.4 All portions of the 110 V electrical system for the pump and solenoid valve systems should be inspected to ensure adequate safety from electrical hazards.

12 CAUTIONS

12.1 This method involves the use of radioactivity. It is the user's responsibility to take appropriate safety precautions and to obtain necessary regulatory approvals from institutional and/or governmental agencies.

12.2 If the user is not trained or familiar with the calibration and operation of radiation analysis equipment, they should consult a qualified expert, such as a health physicist, for assistance in the proper and safe usage of the equipment and development of appropriate quality assurance protocols.

REFERENCES

Currie, L.A., Limits for qualitative detection and quantitative determination, applications to radiochemistry, *Anal. Chem.*, 40, 586–593, 1968.

Federal Register, *National Emission Standards for Hazardous Air Pollutants (NESHAP)*, National Emission Standards for Emissions of Radionuclides Other than Radon from Department of Energy Facilities, 40 CFR, Part 61, subpart H, U.S. Environmental Protection Agency & U.S. Government Printing Office, Washington, DC, 1995.

Newton, G.J., Hoover, M.D., Barr, E.B., McDonald, M.J., and Ghanbari, F., Design of a radioactive gas sampling system for NESHAP compliance measurements of ^{41}Ar, in *Inhalation Toxicology Research Institute Annual Report*, ITRI-144, Inhalation Toxicology Research Institute, Albuquerque, NM, 1994.

Method 8

Determination of the ^{222}Rn Content of the Atmosphere*

CONTENTS

1	Introduction	457
2	Method Summaries	459
3	Cautions	461
Part A:	Measurement of Airborne ^{222}Rn Decay Products by Filter Paper Collection and α-Activity Detection (Thomas Method or Modified Tsivoglou Method)	461
Part B:	Determination of Airborne ^{222}Rn by Its Passive Adsorption on Activated Charcoal	464
Part C:	Measurement of Airborne ^{222}Rn by Counting Damage Tracks Caused by ^{222}Rn and Progeny α-Particles on Special Plastic Film	467
Part D:	Determination of Airborne ^{222}Rn by α-Activity Measurement Using Ionization Chambers	468
Part E:	Measurement of Airborne ^{222}Rn by Grab-Sample Collection and α-Activity Measurement Using Scintillation Cells and Photomultiplier Tubes	471
Part F:	Measurement of ^{222}Rn by Electrostatic Precipitation and α-Energy Spectral Analysis Using a Solid State (Passive Ion-Implanted Planar Silicon) Detector	474
References		479

1 INTRODUCTION

1.1 These methods are useful in determining the potential hazard of radon† in certain occupational environments, in evaluating the radiation hazard of ambient radon to general population groups, and in atmospheric tracer studies involving radon.

1.2 Following the identification in 1984–1985 of unusually high levels of radon in homes located in the Reading Prong geological region of Eastern Pennsylvania, indoor radon has been recognized as the largest single contributor from natural background to the total radiation dose of the United States population (NCRP, 1987).

1.3 The U.S. Environmental Protection Agency (EPA; http://www.epa.gov/iaq/radon/index.html) has established a guide for indoor radon of ~150 Bq m^{-3} (4 pCi L^{-1}), above which remedial action is recommended (EPA, 1986). Surveys of residences show that about 10% of the homes in the United States may exceed EPA's remedial action guidelines (White et al., 1989; Cohen and Shah, 1991; Marcinowski et al., 1994). Housing with elevated radon concentrations has become a matter of

---

* Method authors: Appreciation is extended to the writers of the Radioactivity section of the book *Methods of Air Sampling and Analysis* (Lodge, 1988) where this method originally appeared. Method writers: Phillip Jenkins, Adam R. Hutter, Derek Lane-Smith, and Mark L. Maiello. Method reviewer: Hung-Cheng Chiou.

† Radon in these methods refers to the isotope ^{222}Rn and is not to be confused with other radon isotopes; for example, ^{220}Rn (thoron) or ^{219}Rn (actinon).

public health concern. As a consequence, studies of radon control and hazard evaluation have been conducted. In addition to studies related to health effects, radon may also be used as an atmospheric tracer in meteorological, geophysical and air pollution studies.

1.4 Radon is the immediate decay product of ^{226}Ra, which is an intermediate decay product of the 14-step natural ^{238}U decay chain. It is ubiquitous in the terrestrial environment. Gaseous radon, released to the atmosphere by emanation from soil, decays to atoms of particulate polonium, bismuth and lead. The decay series for radon is presented in Figure 1.

1.5 For the estimation of the effective dose equivalent due to the inhalation of ^{222}Rn and its progeny, it is necessary to measure their mean concentrations over an extended period of time. Temporal variations of concentration occur due to varying meteorological conditions, building ventilation rates, and the presence of certain building materials, among other factors. While in principle a mean concentration can be estimated from a series of grab samples or short-term measurements, methods that integrate over a long period of time are preferable. Short-term measurements are useful screening techniques for determining whether to investigate further with long-term methods.

1.6 In the United States, many reputable commercial radon-measurement device manufacturers, suppliers and related service laboratories participate in intercomparison test programs such as the National Environmental Health Association National Radon Proficiency Program in order to document the accuracy of their techniques. The prospective user is advised to obtain the commercial radon service's latest intercomparison results before engaging them, particularly for a large-scale radon measurement project.

1.7 Governmental agencies have access to a facility in which radon detectors can be exposed to known concentrations of radon over a range of relative humidities and/or temperatures. An exposure

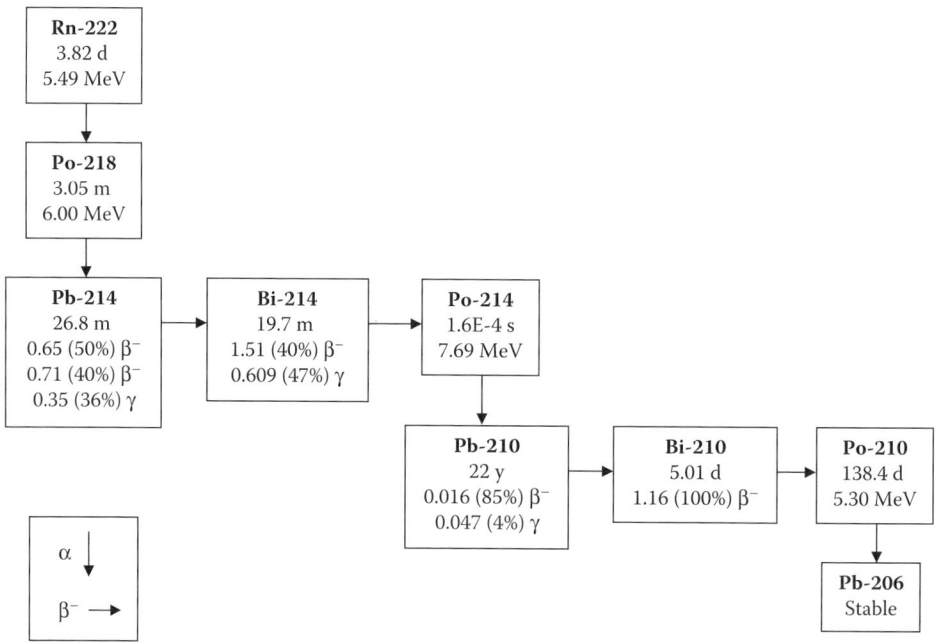

FIGURE 1 Decay scheme for ^{222}Rn and progeny. The half-lives are expressed in units of time: seconds (s), minutes (m), days (d), and years (y). The energies of the primary means of decay are indicated in MeV with associated abundances (abundances of α-dacays can be considered 100%). (Data from Lederer, C.M. and Shirley, V., *Table of Isotopes*, 7th ed., John Wiley & Sons, Inc., New York, 1978.)

room or chamber is available at the U.S. EPA Radiation and Indoor Environment National Laboratory in Las Vegas, NV. Nongovernmental entities may acquire radon exposure services from commercial establishments.* If the results of any of the devices or techniques described below are called into question during actual practice, the user should submit the devices to an available radon calibration facility to obtain confirmation of accuracy.

1.8 The U.S. EPA has published measurement protocols for certain devices, notably passive, low-cost devices used in residential surveys such as charcoal, α-track and electret radon detectors (EPA, 1992, 1993). These protocols should be reviewed before beginning any large-scale residential survey because they are often considered essential practices for validation of the results. An important protocol is the establishment of a quality assurance program. Depending on the measurement techniques, the determination of such items as predeployment background and charge stability and the running of field blanks, duplicate measurements and spikes exposed at a radon calibration facility are necessary components of a quality assurance program. Some localities in the United States, for example, the states of New Jersey, Pennsylvania, Ohio, Illinois, and others, license commercial radon measurement specialists through testing and continuing education programs. Such licensing is considered a prerequisite for residential survey purposes. In the United States, the user is urged to contact the local (state) radon measurement program in which the investigation is to take place in order to determine whether licensing and other quality assurance protocols are required.

1.9 Because this book is concerned with descriptions of sampling methods rather than those of sampling devices, a comprehensive survey of all radon and radon progeny measurement instruments is not appropriate. The methods described here were chosen primarily because they are the basis for many of the available commercial instruments (passive and active, continuous and grab sampling) and because most have some sort of historical or U.S. regulatory acknowledgment of credibility. Though no less important, instruments such as commercial working level monitors and the plethora of prototype or research radon measurement devices developed at universities and government laboratories are not mentioned.

1.10 Since the EPA's Radon Proficiency Program was terminated, two private organizations are currently offering proficiency listing, accreditation, and certifications in radon testing and mitigation, and are useful sources of general information on radon methods, equipment, and vendors. These two programs are: the National Environmental Health Association's (NEHA) National Radon Proficiency Program (http://www.radongas.org/) and the National Radon Safety Board (NRSB; http://www.nrsb.org). Also, the American Association of Radon Scientists and Technologists (AARST; http://www.aarst.org) maintains web pages that contain information and many links to sites associated with various aspects of radon measurement and mitigation.

1.11 There are several papers that compare methods or instruments (Ronca-Battista and Magno, 1988; Martz et al., 1991; George, 1996), which, in addition to this Method, may be helpful for deciding which is most appropriate for the desired application. A summary of advantages and disadvantages of the methods described below is shown in Table 1.

2 METHOD SUMMARIES

The investigator is provided below with a broad outline of methods employed in the determination of airborne radon, as well as their applications and limitations, so that there may be a basis for choosing one method for a particular application. Choice of the method depends on: (1) the level of radon concentration to be measured, (2) the accuracy required, (3) the available equipment, and (4) complexity of the technique.

---

* For example, Bowser-Morner, Inc., Dayton, Ohio www.bowser-morner.com

TABLE 1
Summary of Advantages and Disadvantages of the Detection Techniques Discussed in this Method

Method	Advantages	Disadvantages
Collection on membrane filter (Thomas Method or modified Tsivoglou method)	Yields concentrations of Rn progeny; results acquired in <1 day; precision above average (~10% at 1.96σ); operator performs analysis	For an estimate of radon concentration, must assume or know a Rn/progeny equilibrium factor; grab sample not reflective of natural Rn variations; requires specialized knowledge; calibrated pump, timer, counter, and careful attention to counting intervals needed; somewhat labor intensive; power source needed; Thoron progeny may interfere; MDC ~40 Bq m^{-3}
Charcoal Canister (passive collection)	Deployment does not require specialized knowledge; exposure requires 2–7 days; unobtrusive to homeowner; no power source needed; good for screening measurements of large numbers of homes; commercially available at low costs	Canister susceptible to humidity; requires calibration under varying conditions of humidity; requires laboratory for γ-analysis; duplicate measurements needed for verification of precision; if high readings are obtained, longer-term follow-up measurements are recommended; precision ~10% at 150 Bq m^{-3}; MDC ~5 Bq m^{-3} for 4-d measurement
α-track detectors	Deployment does not require specialized knowledge; integrated measurement (Rn variations averaged over 1 year+ possible) is more representative of human dose; environmental interferences are minimal; unobtrusive to homeowner; no power source needed; commercially available at low costs	Exposure of 1 month minimum required; requires laboratory for analysis; duplicate, simultaneous measurements recommended to verify precision; precision for typical 3 month measurement ~10%; MDC ~5–10 Bq m^{-3} (3 month measurement)
Pulse ionization chamber	MDC of commercial version very low: ~2 Bq m^{-3}; overall precision ~3% at 1.96σ; provides continuous monitoring; lab versions were used for extremely sensitive work; e.g., atmospheric tracer studies	Relatively expensive; lab models require specialized knowledge to operate; signal from thoron may interfere
Electret ion chamber	Deployment does not require specialized knowledge; no power source needed; analysis can be done by operator or by manufacturer; short- and long-term exposures possible; precision good: for 1 day exposure at 37 Bq m^{-3}~16%; MDC good: 7 Bq m^{-3} in 3 days of exposure	Operator voltage analyzer expensive; γ-radiation will affect measurement (should be corrected by a survey meter reading at measurement location); electret may discharge if contacted by dirt or if mishandled; calibration is altitude dependent
Scintillation cells	Simple methodology; operator performs all phases of analysis; results available in <1 day; screening method; precision ~10% at 1.96σ; MDC: ~20 Bq m^{-3}	Grab-sampling method; thoron progeny may interfere; requires laboratory grade PMT/counter and specialized knowledge; calibration is altitude dependent
Solid-state α-spectroscopy	MDC very low: ~2 Bq m^{-3} or less; background remains low even after long periods of use; spectroscopy allows only Po-218 to be measured if desired; thoron does not interfere and can be measured	Power source is required to operate; relatively expensive; use in high concentrations requires time to decay activity and purge sample air before further use at low concentrations; humidity of sample air must be kept <10%

Method 8: Determination of the ^{222}Rn Content of the Atmosphere

3 CAUTIONS

3.1 These methods may involve the use of hazardous chemicals and/or radioactivity. It is the user's responsibility to take appropriate precautions and to obtain the necessary regulatory approvals from institutional and/or governmental agencies.

3.2 If users are not trained or familiar with the calibration and operation of radiation analysis equipment, they should consult a qualified expert, such as a health physicist, for assistance in the proper and safe usage of the equipment and the development of appropriate quality assurance protocols.

3.3 If a rotameter is needed as the airflow measurement device, the reader is urged to consult the reference by Craig (1971) concerning the problems associated with the calibration of these devices. For example, if the rotameter is not calibrated using exactly the same sampling line situation (pressure drop across sampler) and sampling conditions (ambient temperature and pressure), significant errors may be introduced into the air concentration results. Craig advises use of flow rates that cause the lowest pressure drops possible and meter readings at the upper end of the rotameter scale. It is also advisable to place a pressure gauge on the same side of the pump regulator valve as the rotameter float for the purposes of determining a correction for any pressure drop between the filter and the rotameter and for the detection of line breaks. The novice should also consult an experienced individual such as an industrial hygienist concerning these matters. The ASTM standard for calibration of a rotameter to standard conditions may be used (ASTM, 1996) although ASTM does not provide a discussion of calibration problems. Airflow meters should be calibrated annually and after repairs or modifications (NRC, 1992).

PART A: MEASUREMENT OF AIRBORNE ^{222}Rn DECAY PRODUCTS BY FILTER PAPER COLLECTION AND α-ACTIVITY DETECTION (THOMAS METHOD OR MODIFIED TSIVOGLOU METHOD)

1 PRINCIPLE OF METHOD

1.1 This method is based on the collection of the short-lived radon progeny whose activity is then related to the parent gaseous radon concentration. Assuming either equilibrium conditions or a specific disequilibrium factor between radon and its progeny, the concentration of the ^{222}Rn can be estimated (Tsivoglou et al., 1953; Kusnetz, 1956; Raabe and Wrenn, 1969; Thomas and LeClare, 1970; Thomas, 1972; Rolle, 1972; Nazaroff, 1984). A disequilibrium factor of 0.4 is widely utilized for indoor radon; 0.7 for outdoor conditions (NCRP, 1987). Depending upon sensitivity requirements, collection times can be as short as 5 min. This method is simple and rapid. Most of the following sections describe the Thomas method, also referred to as the modified Tsivoglou method (Thomas, 1972).

2 RANGE AND SENSITIVITY

2.1 Lower Range

Assuming a 5-min sampling period, a sampling rate of 0.005 m^3 min^{-1}, a counter efficiency of 0.30 cpm dpm^{-1}, a counter background of 1 cpm, a counting time of 30 min and a delay of 2 min from sampling to start of counting, the lower limit of detection (LLD)* or minimum detectable concentration (MDC) for this method is ~40 Bq m^{-3} (1.0 pCi L^{-1}) (Thomas, 1972). The MDC can be significantly decreased if the sampling rate is significantly increased (10 L min^{-1} is recommended). For an example of the sensitivity that can be achieved using the Thomas method with some modifications,

---

* See Chapter 5 for a brief explanation of LLD. Also see NCRP Report No. 58 (NCRP, 1985), NCRP Report No. 97 (NCRP, 1988) and Martin 2000.

refer to the atmospheric tracer studies discussed in the recent literature (Collé et al., 1995; Hutter et al., 1995; Whittlestone and Zahorowski, 1998).

2.2 Upper Range
Not limited.

3 INTERFERENCES

3.1 The method assumes that only ^{218}Po, ^{214}Pb, and ^{214}Bi are present, that their concentrations do not change during the sampling period and that the counter has the same efficiency for the ^{218}Po and ^{214}Po α-particle emissions (small differences in efficiency will not introduce significant errors).

3.2 The presence of other long-lived α-emitting radionuclides, such as the progeny of ^{220}Rn (thoron), have to be corrected for by recounting after a delay of 4 h following the initial count. ^{210}Po, ^{239}U, ^{240}Pu, ^{235}U, ^{238}U, and ^{226}Ra are usually present in low enough concentrations that they offer minimal interference with the determination of ^{222}Rn.

4 PRECISION AND ACCURACY

4.1 Counting error is a function of the sampling and counting times, which affects precision but not accuracy. Accuracy is a function of how well one knows the counting efficiency and pump flow rate, and how well one can time the sampling and counting sequence. A precision of 10% at the 95% confidence level (1.96σ level) is obtainable. Flow measurements should be correct to ± 2%. Assuming a 4000 Bq m^{-3} (≈100 pCi L^{-1}) ^{222}Rn concentration, a sampling flow rate of 0.01 m^3 min^{-1}, a counter efficiency of 20% and a counting time of 30 min, the accuracy is <10% (95% confidence level) for each radon progeny (Thomas, 1972).

5 APPARATUS

5.1 Collection

A vacuum pump with a sampling rate of 0.005–0.015 m^3 min^{-1} (0.2–0.5 cfm), a 25 or 47 mm (–1″ or –2″) diameter membrane filter with a 0.8-mm pore size, a flow-rate meter and filter holder are required. Membrane filters are strongly recommended because of their >99% retention of submicron particles. There is also minimal particle penetration into the filter, thus preventing most "self-shielding" of the α-particle emissions. A dry-gas meter is the air-flow measurement device of choice for this method. If not available, a well calibrated and correctly installed rotameter may be used.

5.2 Alpha-Counting Equipment

The α-scintillation counter consists of a silver-activated zinc sulfide screen mounted on a 2.54-cm or 5.08-cm photomultiplier tube (PMT), preamplifier, high voltage power supply, scaler, timer and a light-tight counting enclosure. This type of counter is commercially available.* Typical counter backgrounds range from 0.05 to 0.5 counts min^{-1} depending on detector size and the presence or absence of α-contamination in the materials used to construct the counter. Counter efficiency should be determined with a suitable standard spread over a membrane filter or a plated source of the same geometry as the collected sample. Efficiencies of 40–50% are achievable. Alternatively, a PMT tube with a housing for a radon scintillation cell can be used in conjunction with ZnS disks to count the membrane filters. The advantage is that the filter can be placed in direct contact with the ZnS leaving no air space between the sample and the scintillator. Likewise the standard source can also be put in direct contact with the ZnS, eliminating any uncertainty about the spacing being consistent.

---

* For example, Ludlum Measurements, Inc., USA. www.ludlums.com

5.3 Alternative counting systems employ either a solid-state surface barrier detector (Negro and Watnick, 1978; Howard et al., 1990), or α-spectroscopy for precise determination of the individual progeny concentrations or the potential α-energy concentration (Martz et al., 1969; Jonassen and Hayes, 1974; Revzan and Nazaroff, 1983).

6 REAGENTS

6.1 None.

7 PROCEDURE

7.1 Sample for a collection period of 5 min (Thomas, 1972). Although it is in principle possible to sample for other time periods, the equations presented in Section **9** are valid for sampling times of 5 min only.

7.2 Remove sample from filter holder. Record sampling rate (m³ min⁻¹), duration of sampling, and the time of sample removal.

7.3 Count the sample for the time intervals specified in Section **9**.

7.4 If the presence of longer-lived α-emitters are of concern (Section **3**), recount the sample about 2–4 h after the end of sampling; that is, after the decay of the ^{222}Rn progeny (Harley, 1953). If ^{220}Rn progeny were present in the air sample, their concentrations may be calculated according to the procedures described by Raabe and Wrenn (1969) or by Knutson (1989).

8 CALIBRATION

8.1 The α-counter should be calibrated with an α-particle-emitting standard such as ^{244}Cm or ^{241}Am of known disintegration rate distributed over a geometric area similar to the area of the air filter sample. The counting efficiency is the ratio of the count rate of the standard to the known disintegration rate. Suitable standards are commercially available* and should be traceable to the National Institute of Standards and Technology (NIST).

9 CALCULATIONS

9.1 The total count method for determining the air concentrations of the ^{218}Po, ^{214}Pb, and ^{214}Bi requires counting in three different time intervals measured from the moment sampling ends. The number of net counts in the time intervals, 2–5 min, 6–20 min, and 21–30 min, counter efficiency, flow rate, and sampling time, are used for calculating the concentrations of the three progeny. For a sampling time of 5 min, the following equations apply (Thomas, 1972):

$$C_{218\text{Po}} = [10^{-3}/(V_R E)][6.24 N_1 - 3.034 N_2 + 2.868 N_3], \qquad (1)$$

$$C_{214\text{Pb}} = [10^{-3}/(V_R E)][0.044 N_1 - 0.762 N_2 + 1.817 N_3], \qquad (2)$$

$$C_{214\text{Bi}} = [10^{-3}/(V_R E)][-0.832 N_1 + 1.228 N_2 - 1.395 N_3], \qquad (3)$$

where E = counter efficiency (cpm/dpm); N_1 = net α counts during 2–5 min period; N_2 = net α counts during 6–20 min period; N_3 = net α counts during 21–30 min period; V_R = Sampling flow rate (m³ min⁻¹); and where all concentrations, C, are in Bq m⁻³. Equations 1 through 3 were modified from those published by Thomas (1972) to express the concentration and flow rate in SI units.

---

* Eckert and Ziegler, USA. www.ipl.isotopeproducts.com/new_ipl_site/

The equations for the error terms of C_{218Po}, C_{214Pb}, and C_{214Bi} are found in NCRP Report No. 97 (NCRP, 1988). From the individual radon progeny concentrations, the equilibrium equivalent concentration (EEC) of the ^{222}Rn may be calculated using Equation 9.17 from Chapter 9. The radon concentration may then be estimated by dividing the EEC by a known or assumed value of equilibrium.

10 Effect of Storage

10.1 Not applicable.

PART B: DETERMINATION OF AIRBORNE ^{222}Rn BY ITS PASSIVE ADSORPTION ON ACTIVATED CHARCOAL

1 Principle of the Method

1.1 The activated charcoal canister allows radon to diffuse into the charcoal where it is adsorbed. The method described here is passive, so no air pump or electrical power is required. At the end of the sampling period, the sealed canister is returned to a laboratory, where the sample is analyzed for γ-radiation emitted by radon progeny in the canister. Measured γ-activity is related to the amount of radon trapped in the canister and, in turn, is related to the length of the sampling period, as well as to the time elapsed since its end. The analysis uses known decay rates and calibration factors to determine the average radon concentration during the sampling period. Optimum sampling periods range from 2 to 7 days.

1.2 Gaseous radon, quantitatively adsorbed on activated carbon (usually coconut charcoal) is counted on a γ-spectrometer to establish the activity of the radon progeny ^{214}Bi and ^{214}Pb. The combined count rate generated by the 0.61 MeV ^{214}Bi γ-ray emission and that of the 0.295 and 0.35 MeV ^{214}Pb γ-rays can be determined. The abundances of these γ-rays and the typical counting efficiencies are sufficient to detect quantities of ^{214}Bi and ^{214}Pb that have reached equilibrium with the radon collected in the canister. This method is useful as a screening method to determine if long-period measurements are warranted. Screening a house for radon usually requires multiple canisters depending on the number of floors and the presence of a basement.

1.3 Radon collected in the canister may diffuse back out when a reduction of the ambient radon concentration occurs. This may lead to large errors when the device is used over extended sampling periods during which there has been a large change in radon concentrations. The back-diffusion rate increases and the effective sampling rate decreases as a function of temperature. Therefore, the effective sampling rates should be established at the anticipated temperature at which they will be exposed. Back-diffusion may be minimized at the cost of sensitivity by the employment of a diffusion barrier between the charcoal and the atmosphere being sampled (George and Weber, 1990).

2 Range and Sensitivity

2.1 Lower Range

The MDC for radon using the passive charcoal method is ~5 Bq m^{-3} (0.14 pCi L^{-1}) (Cohen and Cohen, 1983). Commercial laboratories may not be able to achieve this due to higher background levels.

2.2 Upper Range

One of the limitations of the passive charcoal method is the short time in which the measurement can be made, that is, <7 days (Cohen and Nason, 1986). Part of the reason for this is that the adsorption

Method 8: Determination of the ^{222}Rn Content of the Atmosphere

sites are finite in number and water vapor competes for collection at these sites. Tests have been made that show charcoal canisters effectively adsorb radon in atmospheres up to ~1500 kBq m^{-3} (40,000 pCi L^{-1}) over a 3-day period (Thomas, 1974). Concentrations higher than this may also be possible. In contrast with the α-track detector (Part C), the charcoal canister performance declines over extended sampling periods.

3 INTERFERENCES

3.1 Charcoal is an excellent adsorbent of many gases including water vapor. The radon adsorption efficiency is dependent on the ambient relative humidity and temperature, showing an inverse relationship with both. To minimize this effect, charcoal used for radon sampling should be dried before use and should be sealed from the atmosphere except during sampling (George, 1984).

4 PRECISION AND ACCURACY

4.1 Practical measurement uncertainty at 150 Bq m^{-3} (4 pCi L^{-1}) is generally 10% or less (George, 1984), provided that the sampling period does not exceed one week, and calibration factors are employed with corrections for humidity. Deployment of duplicate canisters or more, if feasible, will improve the precision if an average concentration is calculated.

5 APPARATUS

5.1 Investigators have fabricated passive activated-carbon monitors in different sizes and configurations. The passive radon charcoal canister formerly used by the U.S. Department of Energy Environmental Measurements Laboratory (HASL-300, 1992) consists of the following materials: (1) 236 mL (8-oz) metal can with a 10.2 × 2.9 cm lid, (2) 75 g of activated charcoal,* (3) 80-mesh metal screen with an openness of at least 30–50%, (4) removable, internally expanding retaining ring, (5) pad material attached to the inner surface of the lid, and (6) a 35-cm strip of pliant vinyl tape. Prefabricated charcoal canisters for passive radon sampling are available from many vendors, who may also supply γ-counting services.

5.2 A γ-ray spectrometer may be coupled to either a scintillation NaI(Tl) detector or a solid-state Ge detector. While better detection capability is normally obtained with a solid-state detector, this may be offset by its limited size, relative to the geometry of typical charcoal canisters.

The following components make up the counting system designed by the U.S. DOE (now U.S. Department of Homeland Security) Environmental Measurements Laboratory: (1) 8-cm × 8-cm sodium iodide detector inside a 9 cm thick lead counting shield, (2) high-voltage power supply, (3) preamplifier and amplifier, (4) multichannel pulse-height analyzer, (5) timer, (6) time-of-day clock, and (7) printer.

6 REAGENTS

6.1 None.

7 PROCEDURE

7.1 Charcoal canisters are exposed by removing the lid that covers the charcoal bed. After the exposure period is completed, the lid is replaced and sealed in place with vinyl tape. The canister is then

---

* For example, Calgon Carbon Corporation, USA. www.calgoncarbon.com

returned to the manufacturer or service laboratory for analysis. The charcoal canisters are counted with a NaI or Ge detector and a γ-spectrometer. Following sampling, at least 4 h should elapse prior to counting to allow for the radon progeny to reach radioactive equilibrium with the collected radon. A correction for the decay of radon from the midpoint of the sampling period to the time of counting must be applied. Because this method does not rigorously correct for decay during the counting period, the counting time should be kept small compared to the half-life of radon. An alternative method published by Jenkins (Jenkins, 1991) corrects for decay to the time when the canister is closed instead of the midpoint of the sampling period and also corrects for decay during the counting time allowing any counting time to be used.

8 Calibration and Standards

8.1 Effective Sampling Rate

An effective sampling rate (m³ min⁻¹) must be determined based on the exposures of charcoal canisters to a constant concentration of radon over different periods of time under controlled conditions of temperature and humidity in laboratory exposure chambers (see Introduction, Section **1.7**). The effects of humidity are established by making exposures for different sampling times at various relative humidities. This can be related to the amount of water adsorbed by weighing the canister before and after exposure. The basic equation for the effective sampling rate is given by George (1984). It is preferable that a constant value of the counting time be used for calibration and for analyzing field-sample canisters.

8.2 Gamma-Spectrometer Calibration

To calibrate the detector of the γ-spectrometer, a known quantity of NIST-certified ^{226}Ra solution may be added to the charcoal in a sampling canister, which should then be sealed. Following a sufficient time for radon and its immediate decay products to achieve equilibrium between themselves and the ^{226}Ra (~25 days), the canister may be placed on the detector to integrate the counts under the spectrometer peaks associated with the ^{214}Bi and the ^{214}Pb γ-rays. The counts from the γ-ray emissions at 0.242, 0.295, 0.352, and 0.609 MeV are summed. From this, the net count rate is determined and then divided by the amount of ^{222}Rn (equal to the ^{226}Ra) in the canister to establish the efficiency (cpm Bq⁻¹) for the specific counting geometry.

9 Calculations

9.1 The concentration of radon in air (C_{Rn} in Bq m⁻³), is calculated as follows (George, 1984):

$$C_{Rn} = \frac{N}{(E \times t_s \times S_{eff} \times D_f)} \qquad (4)$$

where S_{eff} = effective radon adsorption rate (m³ min⁻¹), N = net combined counting rate (cpm) of the γ-ray peaks at energies 0.242, 0.295, 0.352 MeV, and 0.609 MeV, t_s = exposure time of canister (min), E = γ-detector efficiency, (cpm Bq⁻¹), C_{Rn} = radon concentration (Bq m⁻³), D_f = decay correction, = exp[–λ(0.5t_s + t)], λ = decay constant for ^{222}Rn (min⁻¹) = 1.26 × 10⁻⁴ min⁻¹, and t = time from end of sampling to the beginning of counting (min).

Note: Results may be obtained using other means (Jenkins, 1991). The method described above assumes that the radon concentration remains constant during the sampling period (as does any method that collects radon over time), and so may suffer from fluctuations in the radon concentration during the sampling period.

Method 8: Determination of the ^{222}Rn Content of the Atmosphere

PART C: MEASUREMENT OF AIRBORNE ^{222}Rn BY COUNTING DAMAGE TRACKS CAUSED BY ^{222}Rn AND PROGENY α-PARTICLES ON SPECIAL PLASTIC FILM

1 Principle of Method

1.1 The passive α-track detector uses a small section of special film, either cellulose nitrate (LR-115*) or allyl diglycol carbonate (CR-39†) that is sensitive to α-particle radiation. The film is enclosed inside a small, lightweight, vented housing to allow ^{222}Rn to enter the detector. As the radon and its progeny decay, α-particles are emitted that create damage tracks on the film (Fleischer et al., 1965, 1972). Calculating the track density (number of damage tracks per unit area) gives the average radon concentration over the sampling period (Lovett, 1969). Since exposure is a function of both concentration and time, results are typically expressed in Bq m^{-3} • mo.‡ Screening a house for radon usually requires multiple detectors, depending on the number of floors and the presence of a basement.

2 Range and Sensitivity

2.1 Relatively speaking, α-track detectors are not sensitive so that long exposure intervals are required, often three months to one year. However, this is advantageous for the determination of the annual mean exposure of the general public to the continually varying concentrations of radon found in indoor environments.

2.2 The MDC for an exposure of three months is 5–10 Bq m^{-3} (0.1–0.3 pCi L^{-1}). Alpha-track detectors have a range of up to 1850 kBq m^{-3} (5.0 × 10^4 pCi L^{-1}) for a 2-month exposure (Alter and Fleischer, 1981).

3 Interferences

3.1 Determining the radon concentration from the track density is sometimes performed manually by scanning the film under a microscope. Using this method of analysis, the uncertainty introduced by human error is considerable and largely not quantifiable, since it is dependent on the skill of the analyst. Automated optical counters are available which produce an uncertainty of ~10%.

3.2 The build up of electrostatic charge on the plastic film during exposure can cause measurement anomalies. Carbon, impregnated in the plastic, has been used to help alleviate this.

3.3 A membrane at the vented opening of the housing is needed to allow ^{222}Rn to pass into the detector but not the ambient particulate progeny. This also prevents ambient ^{220}Rn progeny from contributing α tracks on the film. *Note:* There are devices that use a "thoron barrier." If a device with, and a device without the thoron barrier are simultaneously deployed, then the thoron concentration can be measured.

4 Precision and Accuracy

4.1 The precision of these instruments is dependent on the number of net tracks counted (Urban and Piesch, 1981). The 1.96σ error ranges from 100% if only four tracks are counted down to 20% if 100 are counted. Therefore, an important consideration in instances of low concentrations is the area of the film counted vs. the increasing cost for larger areas counted. The statistical uncertainty of these measurements is inversely proportional to the concentration. For concentrations at or above

---

* Produced by Dosirad Company, France. http://pagesperso-orange.fr/dosirad/soc_a.htm
† PPG Industries, Inc., USA, www.ppg.com
‡ •mo = multiplying by 1 month of exposure time. Similarly, •d = multiplying by 1 day.

150 Bq m^{-3} • months (4 pCi L^{-1} • months) that are monitored for at least three months, the uncertainty is <~10%, and it decreases as the concentration or the sampling period increases.

5 APPARATUS

5.1 Commercially supplied detector containing α-track plastic film inside a vented housing suitable for deployment in indoor environments.

6 REAGENTS

6.1 No reagents are required by the end user. The analysis is done by the manufacturer or a service laboratory and requires NaOH and considerable expertise to etch the plastic film.

7 PROCEDURE

7.1 As with charcoal canisters, these devices simply require unsealing and deployment for a predetermined exposure period, followed by resealing and returning to the manufacturer or service laboratory for analysis.

8 CALIBRATION

8.1 This is performed in a radon chamber where the radon level is accurately known (see Introduction, Section **1.7**). The individual concentrations of the radon decay products, the aerosol size and concentration, the relative humidity and the uniformity of radon concentration within the calibration chamber should also be well characterized. The radon concentration, time of exposure and the number of tracks counted per unit area are related by a calibration factor K (see Section **9.1**).

9 CALCULATIONS

9.1 Radon exposure is calculated using the following equation (Alter and Fleischer, 1981):

$$C = \frac{T - BA}{AK}, \qquad (5)$$

where C = exposure in Bq m^{-3} • months, T = total tracks counted, B = detector background (T mm^{-2}), A = area counted (mm^2), and K = calibration factor (T mm^{-2})/(Bq m^{-3} • months).

10 EFFECT OF STORAGE

10.1 The background of the plastic film, that is, the track density determined without exposure to radon, increases as storage time increases due to infiltration of radon into the detector storage container. Background track density variations should be scrutinized by the manufacturer as part of the company's quality assurance program. The user may investigate this by sending the manufacturer or service lab track detectors that have not been exposed and observing the results so obtained.

PART D: DETERMINATION OF AIRBORNE ^{222}Rn BY α-ACTIVITY MEASUREMENT USING IONIZATION CHAMBERS

1 PRINCIPLE OF METHOD

1.1 Pulse Ionization Chamber

1.1.1 Ionization chambers for the measurement of radon can be subdivided into two types, pulse or direct current. In both types, ambient radon progeny are prevented from entering the detection

Method 8: Determination of the ^{222}Rn Content of the Atmosphere

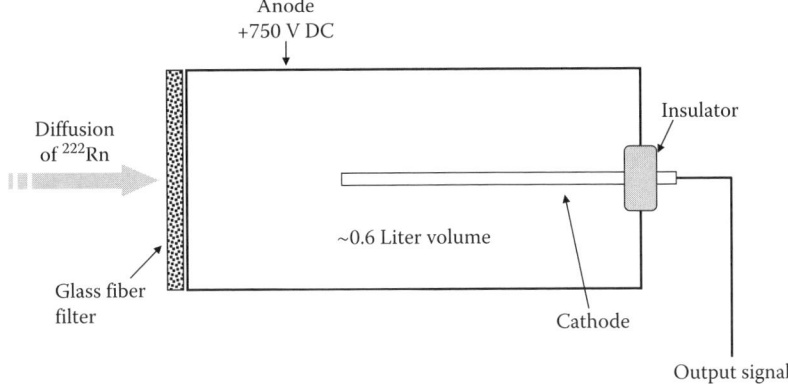

FIGURE 2 Simplified diagram of an ionization chamber used for Radon measurement.

chamber. In this volume of the pulse ionization chamber, electrons created through air ionization by the emission of α-particles are collected electrostatically and converted to electrical pulses in an electrometer that is calibrated to convert this signal into a radon concentration (Figure 2). Research laboratory ion chambers were designed for specialized work (Fisenne and Keller, 1985) in which the radon is introduced from grab-sampling devices. Many commercial models are available for continuous indoor monitoring using passive diffusion of the radon into the measurement volume. Some of these devices are designed to repel the radon progeny away from the sensitive volume so that the majority of the α-pulses originate from ^{222}Rn only; however, other monitors use pulse ion chambers that measure the α-signal from both radon and radon progeny.

1.2 Direct Current Ionization Chamber

1.2.1 In a direct current ionization chamber, or electret ionization chamber as it is more specifically described, the decay of radon also causes air ionization. The electrons thus formed are attracted to an electret that is a positively charged disk of Teflon®* located at the bottom of the measurement volume (Figure 3). The surface charge diminishes as the collection of electrons progresses. The

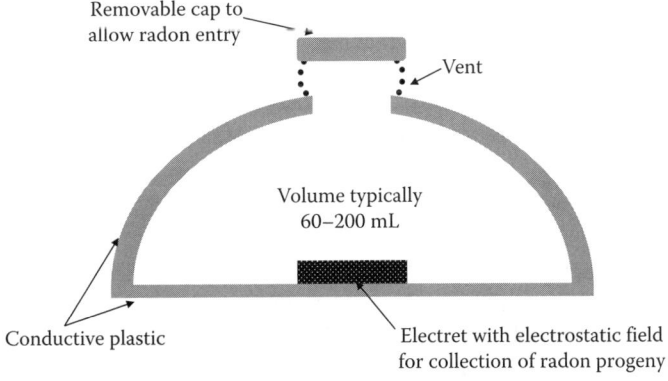

FIGURE 3 Simplified diagram of a passive electret-based Radon measurement device.

---

* Dupont Company, www2.Dupont.com

electret voltage decrease is proportional to the integrated radon concentration. The difference between the starting electret surface voltage and that at the end of the exposure period can then be converted to a radon concentration using a calibration factor. Short-term electret ionization chambers integrate exposures commonly from 1 to 7 days; although, long-term devices can be used for quarterly or annual measurements.

2 RANGE AND SENSITIVITY

2.1 Sensitivity

The pulse ionization chamber is perhaps the most sensitive of all radon detectors. A description of research-grade units is available (Fisenne and Keller, 1985). The sensitivity of the commercially available pulse and electret ion chambers is given below.

2.1.1 Lower Range

For a short-term electret ionization chamber, the MDC is ~7 Bq m^{-3} (~0.2 pCi L^{-1}) over three days of exposure (Kotrappa et al., 1988). For a commercial version of the pulse ionization chamber, the MDC is ~2 Bq m^{-3} (~0.05 pCi L^{-1}) (Genrich, 2008), 0.5 pCi L^{-1} (femto-TECH, 2008; Radalink, 2008).

2.1.2 Upper Range

Electret ionization chambers are specifically designed for different indoor exposure measurements, up to a maximum of ~1400 kBq m^{-3} • day (~4 × 10^4 pCi L^{-1} • day) (Kotrappa et al., 1988). For a commercial pulse ionization chamber, the upper limit of detection is claimed to be as high as ~2000 kBq m^{-3} (~5 × 10^4 pCi L^{-1} • day) (Genitron, 2008), although others assert such numbers as 92.5 kBq m^{-3} (Radalink, 2008) or 74 kBq m^{-3} (femto-TECH, 2008). It should be noted that measurements made at the upper range of the device could raise background levels so that the sensitivity for subsequent lower range measurements is compromised.

3 INTERFERENCES

3.1 In pulse ionization chambers, air ionization from thoron decay may cause an interfering signal. Organic and acidic vapors and extreme temperatures should be avoided.

3.2 Electret ionization chambers are sensitive to background γ-radiation. The background γ-exposure rate should be assessed so that an appropriate correction can be made to the measured radon concentration. This is best accomplished by measuring the background exposure rate at the measurement location using a calibrated NaI survey meter or portable ionization chamber. The electret must not be soiled or handled in order to avoid discharging it. The calibration of an electret ion chamber varies with altitude. A set of altitude correction factors is supplied by the manufacturer.

4 PRECISION AND ACCURACY

4.1 The pulse ionization chamber radon detectors offer the advantage of very low measurement uncertainties. For most commercially available units, under normal operating conditions for indoor radon measurements, the overall uncertainty is ~3% at the 95% confidence limit (Genitron, 1990).

4.2 For electret ionization chambers, measuring ~40 Bq m^{-3} (~1 pCi L^{-1}) for one day gives a total uncertainty of ~16%. Since the thickness of the Teflon is a critical factor in the radon detection, the uncertainty due to this factor alone is estimated to be ~7%. Calibration uncertainty is estimated to be ~5%, and a statistical "counting uncertainty" at ~40 Bq m^{-3} • 7 d (~1 pCi L^{-1} • 7 days) is only ~1% (Kotrappa et al., 1988).

Method 8: Determination of the ^{222}Rn Content of the Atmosphere

5 APPARATUS

5.1 Owing to the complexity of the pulse ionization instruments, they come from the manufacturer complete and ready to use.

5.2 The electret ionization chambers are small passive devices that are deployed similarly to charcoal canisters and α-track detectors. They also come complete and ready to use. An electret voltage reader can be purchased as an alternative to analysis by the manufacturer.

6 REAGENTS

6.1. None.

7 PROCEDURE

7.1 The procedure for measuring radon using a commercial pulse ionization chamber instrument is simply to turn it on (these complex devices are largely "black boxes" to the user). The manufacturer's instructions should be followed regarding overall operation. Research-lab pulse ionization chambers accept grab samples collected by another means, for example, evacuated flasks, and require specialized knowledge to operate.

7.2 Electret ionization chambers are passive detectors so that opening the protective housing is all that is required to initiate the measurement. The measurement is terminated by closing the housing. Analysis can be performed using a portable electret voltage reader supplied by the manufacturer or by shipping the detector to the manufacturer for read out.

8 CALIBRATION

8.1 Calibrations of pulse ionization chambers are performed using air with a known radon concentration, such as from a radon chamber (see Introduction, Section **1.7**). It may be necessary to apply dead-time corrections to pulse ion chamber results at the radon concentrations common in calibration chambers if the device does not do so automatically. Calibrations of electret ionization chambers are also performed in radon calibration chambers. Calibration data to convert the difference of the initial and final voltages of the electret to a radon concentration are supplied by the manufacturer to those users who have purchased the electret voltage reader.

9 CALCULATIONS

9.1 Not applicable.

10 EFFECT OF STORAGE

10.1 Not applicable.

PART E: MEASUREMENT OF AIRBORNE ^{222}Rn BY GRAB-SAMPLE COLLECTION AND α-ACTIVITY MEASUREMENT USING SCINTILLATION CELLS AND PHOTOMULTIPLIER TUBES

1 PRINCIPLE OF METHOD

1.1 The potential user is reminded that the diurnal concentration variations of radon will not be observable by this method unless several grab samples are collected over time. It is strongly recommended that a waiting time of 4 h be used to allow secular equilibrium to be established. Failure to do so can introduce a large uncertainty in the final result. A grab sample of air containing radon is collected in a cell that is lined with an Ag-activated ZnS phosphor (Lucas, 1957). The cell may be preevacuated or a pump may be used to pass sample air through the cell. The cell has a transparent window on the bottom that is placed on a photomultiplier tube (PMT) for counting. As the radon and its progeny decay, α-particles striking the phosphor cause the emission of low-energy photons that are detected by the PMT and converted to counts by attendant circuitry.

Electronic devices have been available for years that can be used to analyze grab samples; for example, the Pylon Model AB-5.* This device consists of a PMT, an air pump and a rechargeable battery. This device can also be used as a continuous radon monitor where air containing radon is either pumped continuously through the scintillation cell or passively diffuses into the cell. After equilibrium is established, hourly counts are then converted into measurements of radon concentration.

2 Range and Sensitivity

2.1 Lower Range

The range and sensitivity are largely determined by the volume of the scintillation cell and the background counts. Typical new commercially available scintillation cells have a volume of ~1.6×10^{-4} m^3, resulting in a MDC of ~20 Bq m^{-3} (0.5 pCi L^{-1}) (George, 1976). The background count rate increases with use.

2.2 Upper Range

This method is often used for studies involving soil gas, where the Rn concentrations have been found to exceed 1850 kBq m^{-3} (50,000 pCi L^{-1}) (Hutter, 1995). The upper range is limited only by the saturation of the PMT which typically occurs at greater concentrations than found in almost all indoor conditions. Use of the cells at the higher range can result in background count rates that may compromise use at the lower end of the range.

3 Interferences

3.1 The method assumes that only ^{222}Rn, ^{218}Po, ^{214}Pb, and ^{214}Bi are present and that their concentrations do not change during the sampling period. If a waiting time of 4 h is employed until equilibrium is achieved, and if the calibration is obtained under equilibrium, then differences in counting efficiencies among the α-particles from ^{222}Rn, ^{218}Po, and ^{214}Po have no effect, because it is the sum of the counting efficiencies that is used as a calibration factor. Problems with differences in the counting efficiencies arise only when trying to make measurements well before equilibrium is established. Thus, a waiting time of at least 4 h before doing any counting, for calibration and for measurements using this method is recommended.

3.2 The background of the scintillation cell increases with time as solid progeny adsorb onto the phosphor lining, which also causes an increase of the MDC. It is the *long-lived* progeny that buildup and the increase in background comes primarily from the alpha emitted by ^{210}Po.

3.3 If ^{220}Rn (thoron) is present in the air, it potentially could present a significant interference. The scintillation cell can be flushed with nitrogen or aged air and counted again after a 4-h delay to allow for the radon progeny to decay to insignificant levels. If the count rate is then significantly above background, this could be an indication of interference from thoron.

4 Precision and Accuracy

4.1 The overall uncertainty is ~10% at the 95% confidence limit (1.96σ), which includes sampling errors (Lucas, 1957). Calibration errors, a measure of the accuracy of the method, are typically ~5%, whereas counting errors are dependent upon the concentration measured. *Note*: The counting

---

* Pylon Electronics, Inc., Ottawa, Ontario, Canada. www.pylonelectronics.com

Method 8: Determination of the ^{222}Rn Content of the Atmosphere

efficiency is a function of altitude (barometric pressure). Care must be taken to calibrate at the same (or at least similar) barometric pressure as will be encountered in the field. Alternatively, a table of efficiency versus barometric pressure can be constructed for use with field barometric readings to determine the correct cpm/dpm ratio to use.

5 APPARATUS

5.1 Sample Collection
Scintillation cell(s) and a pump to either evacuate the scintillation cell for air sampling or to draw air through the cell are required.

5.2 Sample Analysis
A PMT/scaler combination to detect the photons resulting from α-particles interacting with the phosphor coating of the scintillation cell is needed to obtain a count rate. The PMT must be sealed in a light-tight enclosure.

6 REAGENTS

6.1 None.

7 PROCEDURE

7.1 The sample is drawn into the scintillation cell, either by opening a previously evacuated cell or by pumping air through a nonevacuated cell (George, 1976). If the pumping method is used, the flow rate and pumping time should ensure that the cell is flushed two or three times.

7.2 After a known amount of time, the scintillation cell is placed into a light-tight enclosure containing a PMT to obtain a count rate.

8 CALIBRATION

8.1 Scintillation cells and the PMT/scaler unit are calibrated using known radon concentrations such as can be found in a radon exposure chamber (see Introduction, Section **1.7**). Most commercially available PMT and cell combinations result in detection efficiencies on the order of 60%.

9 CALCULATIONS

9.1 Radon concentrations are calculated using decay equations for radon and its progeny. The ^{222}Rn concentration in Bq m^{-3} (C_{222}) can be determined using the following calculation:

$$C_{222} = \frac{N}{(E \cdot g)} \tag{6}$$

where N = net counts; E = calibration factor (cps/Bq/m^3); g = "effective counting time" (s) where $g = [\exp(-\lambda\ t_1) - \exp(-\lambda\ t_2)]/\lambda$, where λ = decay constant of ^{222}Rn (s^{-1}); t_1 = time from grab sample to beginning of count; and t_2 = time from grab sample to end of count.

This equation accounts for decay during the counting time, so any counting time can be used.
The calibration factor can be determined from filling cells at a reference laboratory.

10 EFFECT OF STORAGE

10.1 Scintillation cells should be flushed with low radon-concentration air (e.g., nitrogen or aged air) as soon as possible after sampling to limit the buildup of radon progeny plating out on the interior walls, thus minimizing the increase of the background counting rate.

PART F: MEASUREMENT OF ^{222}Rn BY ELECTROSTATIC PRECIPITATION AND α-ENERGY SPECTRAL ANALYSIS USING A SOLID-STATE (PASSIVE ION-IMPLANTED PLANAR SILICON) DETECTOR

1 Principle of Method

1.1 When ^{222}Rn (radon) atoms decay the polonium progeny are driven by an electric field onto the surface of a solid-state α-detector. Subsequent decays occur on this surface and the α-particles emitted are counted. A commercially available instrument (Lane-Smith, 2007) uses a solid-state high-resolution α-detector and precision multichannel analyzer. As a result, the instrument is able to distinguish the α-emissions of ^{218}Po (6.0 MeV), ^{216}Po (6.78 MeV), ^{214}Po (7.69 MeV), ^{212}Bi (6.05 MeV), ^{212}Po (8.78 MeV), and ^{210}Po (5.3 MeV). The count rate in energy window "A" (^{218}Po) is corrected for the presence of ^{212}Bi originating from ^{220}Rn (thoron) progeny on the detector by deducting half the count rate in window "D" (^{212}Po). Similarly, the count rate in window "B" (^{216}Po of the ^{220}Rn chain) is corrected for interference from window "C" (^{214}Po of the ^{222}Rn chain) by deducting a proportion (determined in the instrument calibration process) of the window "C" count rate. In practice, the radon concentration may be rapidly determined solely from the count rate in window "A" or over time from the sum of the count rates in windows "A" and "C."

2 Range and Sensitivity

2.1 For radon in air, (counting both ^{218}Po and ^{214}Po decays), the sensitivity historically has been 0.4 cpm/pCi L^{-1} (1.10 cpm per 100 Bq m^{-3}) (George, 1996). Modern instruments typically achieve 0.5 cpm/pCi L^{-1} or 1.25 cpm/100 Bq m^{-3}. Certain types of these devices are available with sensitivities in excess of 1.2 cpm/pCi L^{-1}. When counting only ^{218}Po decays, the sensitivity is 0.25 cpm/pCi L^{-1}.

2.2 By observing, and ignoring, ^{210}Po decays the instrument maintains an almost zero intrinsic background, equivalent to a radon concentration of less than 0.005 pCi L^{-1} (0.2 Bq m^{-3}), for the life of the instrument.

2.3 Lower Range

The minimum detectable concentration (MDC) is determined by the intrinsic background (0.005 pCi L^{-1}) of the instrument. An MDC for radon in air of 0.05 pCi L^{-1} (2 Bq m^{-3}) is typically achieved although with care, this can be decreased. This MDC will not change even with heavy use of the instrument at high radon concentrations because the consequent buildup of ^{210}Pb and ^{210}Po is observed by the instrument but not counted.

2.4 Upper Range

At very high radon concentrations this method starts to underreport the radon level (the response becomes nonlinear). To keep within a ±5% accuracy, an upper limit to the linear range is specified at 20,000 pCi L^{-1} (750,000 Bq m^{-3}). If a larger deviation from the linear is acceptable, the instrument can make measurements beyond 20,000 pCi L^{-1}.

Use at the high end of the range for an extended period produces a significant ^{210}Po count rate but has no impact on the background of the instrument because the 5.3-MeV α-emissions from ^{210}Po are excluded.

3 Interferences

None. The α-detector is unresponsive to γ- and β-radiation and only counts α-particles within narrowly defined energy ranges.

Method 8: Determination of the ^{222}Rn Content of the Atmosphere

4 ADVANTAGES/DISADVANTAGES

4.1 Advantages

The MDC is very low (typically 2 Bq m^{-3}) and remains low over the life of the instrument. A high-resolution α-energy spectrum can be obtained and observed. The gain of the electronics determines the position of the peaks in the spectrum. The electronics and software include active compensation for drift in position due to temperature changes. High accuracy (±5%) and sensitivity is attainable at relative humidities from 0% to 100%. The method can be adapted to real-time, continuous monitoring or grab sampling for radon in air. The results can be obtained quickly (<15 min response time). Thoron (^{220}Rn) does not interfere with the radon measurement and can be measured simultaneously. There is no background subtraction performed because there is no intrinsic background. The background count rate for instruments that do not employ spectral analysis originates with the buildup of ^{210}Pb and the α-decays of ^{210}Po supported by the ^{210}Pb. The α-spectral discrimination of the different polonium isotopes ignores the radioactive emissions of ^{210}Po effectively maintaining a zero background.

Alpha spectroscopy is also employed to reduce the time required between the measurements of high and low radon concentrations. The rate of decay of the signal from the previous sample has only a 3.05-min half-life when counting only the decay of ^{218}Po. In 30 min the count rate decreases by a factor of about 1000 allowing a radon concentration one thousandth of the high concentration to be detected (but at a lower precision than when the emissions of ^{214}Po are also counted).

4.2 Disadvantages

The instrument requires a battery power source for measurements away from AC power. Sample air is dried with desiccant that needs periodic regeneration or replacement.

If the unit is used to measure very high radon concentrations, time is required for activity on the detector to decay and the high concentration air must be purged from the device before low concentrations can be measured (see advantages above).

The relative humidity of the air admitted to the α-detection chamber must remain below 10% during the measurement (use of a dessicant is required).

Grab sampling is possible but not "instantaneous" (see section on grab sampling below) because a recovery period is required between grab samples. Modified grab sampling is possible by evacuating the measurement chamber and opening it at the sampling location or by using a sampling bag and a sampling pump to take a grab sample to be taken back to the instrument for analysis.

5 PRECISION AND ACCURACY

5.1 The accuracy of a radon result is determined by the accuracy of the instrument's calibration. The stated accuracy of the detector calibration is 5% (this is the stated accuracy of the "secondary standard" radon calibration chamber that is currently used). The precision of the instrument calibration is better than ±1% and repeatability better than ±2%. The stability of the instrument between calibrations also has an effect on accuracy. It is determined mainly by the stability of the physical dimensions of the measurement chamber and is only slightly dependent on the stability of the electronics. If the electronics gain varies it will move the spectrum relative to the counting windows. However, the spectra are so sharp and the low-energy tail so thin that even a significant drift in the position of the peaks has small impact on the sensitivity.

5.2 The precision of the radon concentration result is determined by the total number of counts included in a measurement and therefore is also a function of the duration of the measurement. This is a direct result of Poisson counting statistics. Statistical uncertainty is dominant until it becomes less than the calibration accuracy. With a *calibration* accuracy of ±5%, the measurement precision should have a two-sigma uncertainty of no more than ±2% to claim a measurement accuracy of ±5%

using the square root of the sum of the squares of both of these error terms. This would require the operator to achieve 10,000 or more counts in the reading. Thus, in order to attain the accuracy of which the instrument is capable it is necessary that the measurement precision be significantly narrower than the accuracy of the instrument. At 4 pCi L^{-1}, a 24-h sampling time would produce about 2900 counts. The standard deviation (σ), would be 53.8 counts and $2\sigma = 107.7$ counts or 3.7% (a value greater than 2%). This is less than the absolute accuracy of the instrument, so quoting the result as 4 pCi L^{-1} ± 3.7% is not correct. The result should be quoted as 4 pCi L^{-1} ± 5% or 4 ± 0.2 pCi L^{-1}. To be more rigorous, the combination of statistical and calibration uncertainties may be assessed as the square root of the sum of the squares, or ±6.2% in this case.

6 APPARATUS

6.1 Commercial devices incorporating this measurement technique consist of a case containing the measurement chamber (typically less than 1 L in volume), detector, air pump, batteries, and electronics. A drying unit containing Drierite is used to condition the sample air (see Figure 4).

7 REAGENTS

7.1 None.

8 PROCEDURE

8.1 General

Instruments such as these are computer controlled and user friendly. The reader follows the instructions in order to obtain accurate results. Certain precautions peculiar to instruments of this type are noted below.

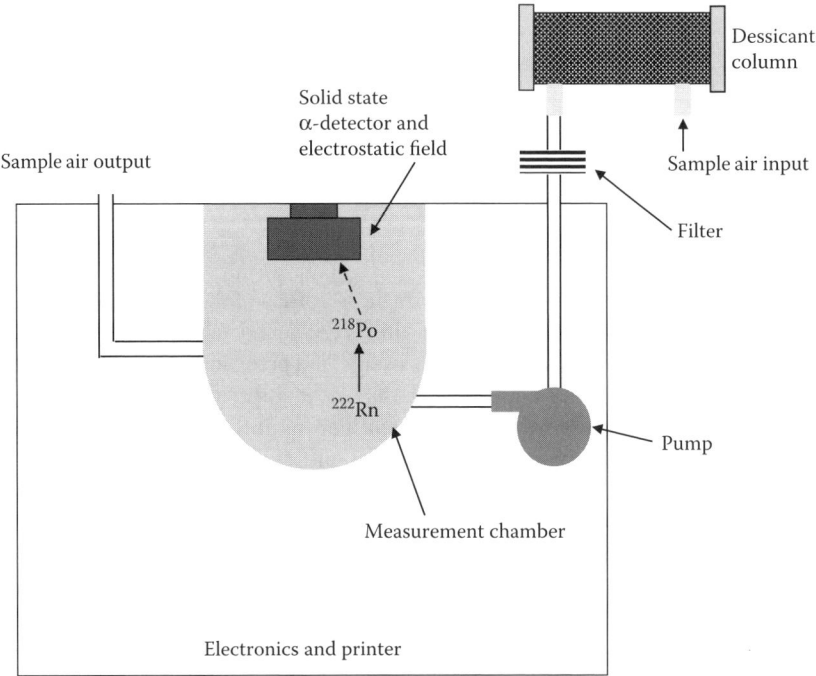

FIGURE 4 Simplified drawing of radon detector employing solid-state α-detector.

Method 8: Determination of the ^{222}Rn Content of the Atmosphere 477

8.2 Continuous Sampling

The air sample is drawn into the instrument by the internal pump. For a typical continuous air sample over a 2-day period, "the cycle time" of the instrument is set at one hour. A run consists of 48 cycles. At the end of each cycle data are stored in computer memory and printed out for the user. At the end of the run, a summary of the measurement is printed which includes the average and standard deviation of the radon readings, a bar chart of individual readings and a cumulative spectrum of all the counts in the run. The instrument prints out a summary of the sample (see Figure 5) including

- Date and time
- Machine serial number
- Average value for the test
- Bar chart of the individual readings
- Cumulative α-energy spectrum

The data collected, to the end of the last completed cycle, are automatically stored in the memory, and available for later display, printing, or downloading to a computer.

8.3 Grab Sampling

Grab sampling is conducted by pumping sample air for 5 min either from the sampling point or from a discreet sample such as an air-sample bag or by evacuating the chamber and opening it at the sampling point. After acquisition a 5-min delay is required to allow the ^{218}Po count rate to approach equilibrium with the radon concentration in the measurement chamber. Typically, four 5-min-cycle counts are made from which the radon concentration in the air sample is determined (a total of 30 min is required for the analysis of one grab sample). The sample must be purged from the chamber and the ^{218}Po count rate depleted before the next grab sample can be taken.

8.4 Precautions

Normal operation should be checked monthly. This is accomplished by examining the α-spectrum from a completed sampling protocol (no radon source is needed). The spectrum should look the same as previous spectra, with clean peaks in "normal" positions. The desiccant must be in condition to absorb moisture. The relative humidity of the air admitted to the α-detection chamber must remain below 10% during the measurement. Other instrument characteristics should be verified to be in the operable range before measurements commence such as battery voltage and pump current. In some devices, the user is alerted to malfunctions of other device components such as the electric field strength around the α-detector (this critical component of the device is continuously monitored by a microcontroller that maintains the voltage at the required level).

9 CALIBRATION

The U.S. EPA recommends that all continuous radon monitors be calibrated once per year. The device may be returned to the manufacturer to perform this procedure. In addition, the air pump that introduces sample air to the α-detection chamber requires a flow-rate calibration to determine that it is operating normally. This is typically done on a periodic basis of at least one year.

10 CALCULATIONS

The instrument performs all calculations internally. Count rates are converted into radon-in-air concentrations by multiplying sensitivity factors determined at calibration.

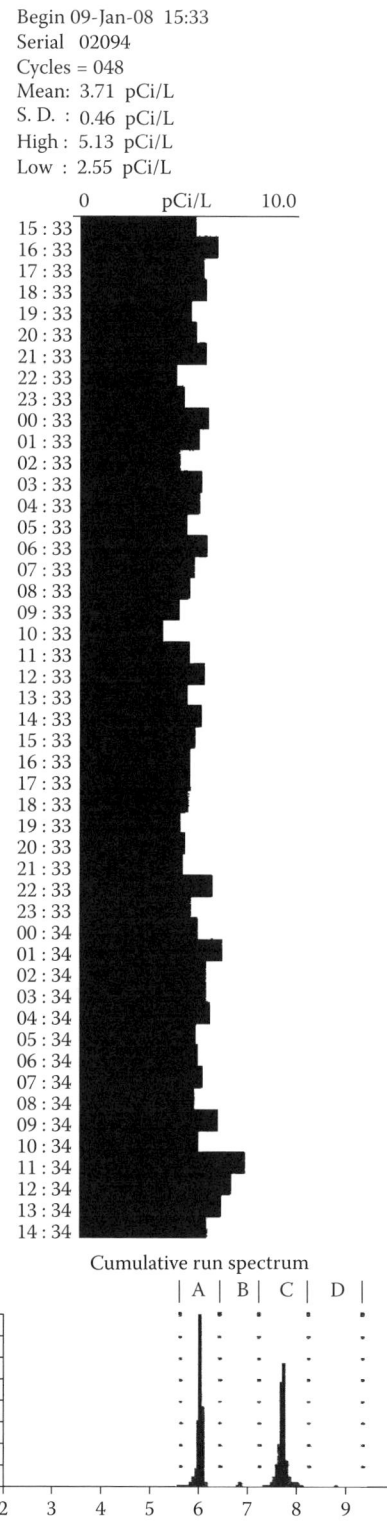

FIGURE 5 End-of-run printout for a typical solid-state radon detector. In the example shown it may be seen that the average level was 3.71 pCi/L, or 137 Bq m^{-3}.

11 EFFECT OF STORAGE

The instrument should be stored in a clean and dry environment. If left in storage for a long time, the batteries should be recharged every month.

REFERENCES

Alter, H.W. and Fleischer, R.L., Passive integrating radon monitor for environmental monitoring. *Health Phys.*, 40, 693–702, 1981.

ASTM, *1996 Annual Book of ASTM Standards*, Section 11, *Water and Environmental Technology*, Designation D-3195—90, *Standard Practice for Rotameter Calibration*, ASTM, West Conshohocken, PA, 1996.

Cohen, B.L. and Cohen, E.S., Theory and practice of radon monitoring with charcoal adsorption. *Health Phys.*, 45, 501–508, 1983.

Cohen, B.L. and Nason, R., A diffusion barrier charcoal adsorption collector for measuring Rn concentration in indoor air. *Health Phys.*, 50, 457–463, 1986.

Cohen, B.L. and Shah, R.S., Radon levels in United States homes by states and counties. *Health Phys.*, 60, 243–259, 1991.

Collé, R., Unterweger, M.P., Hodge, P.A., Hutchinson, J.M.R., Whittlestone, S., Polian, G., Ardouin, B., et al. An international intercomparison of marine atmospheric Rn-222 measurements in Bermuda. *JGR*, 100(D8), 16617–16638, 1995.

Craig, D.K., The interpretation of rotameter air flow readings. *Health Phys.*, 21, 328–332, 1971.

EPA, *A Citizens Guide to Radon: What It Is and What to Do About It*, U.S. Environmental Protection Agency Public Information Center, Mail Code PM-211B, 820 Quincy St., NW, Washington, DC 20011, 1986.

EPA, *Indoor Radon and Radon Decay Product Measurement Device Protocols*, U.S. Environmental Protection Agency, EPA Document Number 402-R-92-004, Washington, D.C. 20011, 1992 (available at http://www.epa.gov/iaq/radon/pubs/devprot1.html).

EPA, *Protocols for Radon and Radon Decay Product Measurements in Homes*, U.S. Environmental Protection Agency, EPA Document Number 402-R-93-003, Washington, DC 20011, 1993.

femto-TECH, Inc., P.O. Box 8257, 25 Eagle Court, Carlisle, OH 45005, www.femto-tech.com, 2008.

Fisenne, I.M. and Keller, H., *The EML Pulse Ionization Chamber Systems for ^{222}Rn Measurements*. U.S. Department of Energy Report EML-437, National Technical Information Service, Springfield, VA 22161, 1985.

Fleischer, R.L., Alter, H.W., Furman, S.C., Price, P.B., and Walker, R.M., Technological applications of science: The case of particle track etching. *Science*, 178, 255–263, 1972.

Fleischer, R.L., Price, P.B., and Walker, R.M., Solid state track detectors: Applications to nuclear science and geophysics. *Ann. Rev. Nucl. Sci.*, 15, 1, 1965.

Genitron (or Genrich V.), *Alpha Guard Multiparameter Radon Monitor*, Frankfurt, Germany: Genitron Instruments; File:EVALSYS2.Doc.Gmbh D-60488, 1990 and www.genitron.de, 2008.

George, A.C., Scintillation flasks for the determination of low level concentrations of radon, in *Proceedings of Ninth Midyear Health Physics Symposium*, Denver, CO, 1976.

George, A.C., Passive, integrated measurement of indoor radon using activated carbon. *Health Phys.*, 46, 867–872, 1984.

George, A.C., State-of-the-art instruments for measuring radon/thoron and their progeny in dwellings—a review. *Health Phys.*, 70, 451–463, 1996.

George, A.C. and Weber, T., An improved passive activated carbon collector for measuring environmental ^{222}Rn in indoor air. *Health Phys.*, 58, 583–589, 1990.

Harley, J.H., Sampling and measurement of airborne daughter products of radon. *Nucleonics*, 11, 12–17, 1953, reprinted in *Health Phys.*, 38, 1068–1074, 1980.

HASL-300, P.W. Krey and H.L. Beck, Eds. *EML Procedures Manual*. U.S. Department of Energy Report HASL-300, 27th ed., National Technical Information Service, Springfield, VA 22161, 1992.

Howard, A.J., Johnson, B.K., and Strange, W.P., A high-sensitivity detection system for radon in air. *Nucl. Instr. Meth. Phys. Res.*, A293, 589–595, 1990.

Hutter, A.R., Larsen, R.J., Maring, H., and Merrill, J.T., ^{222}Rn at Bermuda and Mauna Loa: Local and distant sources. *J. Radioanal. Nucl. Chem.*, 193(2), 309–318, 1995.

Hutter, A.R., A method for determining soil gas ^{220}Rn (thoron) concentrations. *Health Phys.*, 68, 835–839, 1995.

Jenkins, P.H., Equations for calculating radon concentration using charcoal canisters, *Health Phys.*, 61, 131–136, 1991.

Jonassen, N. and Hayes, E.I., The measurement of low-concentrations of the short lived ^{222}Rn in the air by alpha spectrometry. *Health Phys.*, 26, 104–110, 1974.

Knutson, E.O., *Personal Computer Programs for Use in Radon/Thoron Progeny Measurements*, U.S. Department of Energy Report EML-517, National Technical Information Service, Springfield, VA 22161, 1989.

Kotrappa, P., Dempsey, J.C., Hickey, J.R., and Stieff, L.R., An electret passive environmental ^{222}Rn monitor based on ionization measurement. *Health Phys.*, 54, 47–56, 1988.

Kusnetz, H.L., Radon daughters in mine atmospheres, a field method for determining concentrations. *Am. Ind. Hyg. Assoc. J.*, 17, 85–88, 1956.

Lane-Smith, D.R., Radon, Durridge Company document, File: RADON71001, 2007.

Lederer, C.M. and Shirley, V., *Table of Isotopes*, 7th ed., John Wiley & Sons, Inc., New York, 1978.

Lodge, Jr., J.P. Ed., *Methods of Air Sampling and Analysis*, 3rd ed., CRC Press, Boca Raton, FL, 1988.

Lovett, D.B.,Track etch detectors for alpha exposure estimation. *Health Phys.*, 16, 623–628, 1969.

Lucas, H.F., Improved low-level alpha scintillation counter for radon. *Rev. Sci. Instrum.*, 28, 680–683, 1957.

Marcinowski, F., Lucas, R.L., and Yeager, W.M., National and regional distributions of airborne radon concentrations in U. S. Homes. *Health Phys.*, 66, 699–706, 1994.

Martin, J.E., *Physics for Radiation Protection*, John Wiley & Sons, New York, 2000.

Martz, D.E., George, J.L., and Langner, G.H., Comparative performance of short-lived diffusion barrier charcoal canisters and long-term α-track monitors for indoor ^{222}Rn measurements. *Health Phys.*, 60, 497–505, 1991.

Martz, D.E., Hollerman, D.E., McCurdy, D.E., and Schiager, K.J., Analysis of atmospheric concentrations of RaA, RaB and RaC by alpha spectroscopy, *Health Phys.*, 17, 131–138, 1969.

Nazaroff, W.W., Optimizing the total-alpha three count technique for measuring concentrations of radon progeny in residences, *Health Phys.*, 46, 395–405, 1984.

NCRP, *A Handbook of Radioactivity Measurements Procedures*, 2nd ed., National Council on Radiation Protection and Measurements Report No. 58, National Council on Radiation Protection and Measurements, Bethesda, MD, 1985.

NCRP, *Exposure of the Population of the United States and Canada from Natural Background Radiation*, National Council on Radiation Protection and Measurements Report No. 94, National Council on Radiation Protection and Measurements, Bethesda, MD, 1987.

NCRP, *Measurement of Radon and Radon Daughters in Air*, National Council on Radiation Protection and Measurements Report No. 97, National Council on Radiation Protection and Measurements, Bethesda, MD, 1988.

Negro, V.C. and Watnick, S., "FUNGI"—A radon measuring instrument with fast response. *IEEE Trans. Nucl. Sci.*, NS-25, 757–761, 1978.

NRC, *Air Sampling in the Workplace*, U.S. Nuclear Regulatory Commission, Regulatory Guide 8.25, National Technical Information Service, Springfield, VA and U.S. Government Printing Office, Washington, DC, June, 1992.

Raabe, O.G. and Wrenn, M.E., Analysis of the activity of radon daughter samples by weighted least squares. *Health Phys.*, 17, 593–605, 1969.

Radalink. Inc., 5599 Peachtree Rd., Atlanta, GA 30341-2309, www.radalink.com, 2008.

Revzan, K.L. and Nazaroff, W.W., A rapid spectroscopic technique for determining the potential α-energy concentration of radon decay products. *Health Phys.*, 45, 509–523, 1983.

Rolle, R., Rapid working level monitoring. *Health Phys.*, 22, 233–238, 1972.

Ronca-Battista, M. and Magno, P., A comparison of the variability of different techniques and sampling periods for measuring ^{222}Rn and its decay products. *Health Phys.*, 55, 801–807, 1988.

Thomas, J.W., Measurement of radon daughters in air. *Health Phys.*, 23, 783–789, 1972.

Thomas, J.W., *Evaluation of Activated Carbon Canisters for Radon Protection in Uranium Mines*, U.S. Atomic Energy Commission Report HASL-280, 1974.

Thomas, J.W. and LeClare, P.C., A study of the two-filter method for radon-222. *Health Phys.*, 18,113–122, 1970.

Tsivoglou, E.C., Ayer, H.E., and Holladay, D.A., Occurrence of nonequilibrium atmospheric mixtures of radon and its daughters. *Nucleonics*, 11, 40–45, 1953.

Urban, M. and Piesch, E., Low level environmental radon dosimetry with a passive track etch device. *Radiat. Protect. Dosim.*, 1, 97–109, 1981.

White, S.B., Bergsten, J.W., and Alexander, B.V., Multi-state surveys of indoor ^{222}Rn. *Health Phys.*, 57, 891–896, 1989.

Whittlestone, S. and Zahorowski, W., Baseline radon detectors for shipboard use: Development and deployment in the First Aerosol Characterization Experiment (ACE 1). *J. Geophys. Res.*, 103(DD13), 16743–16752, 1998.

Method 9
A Procedure for Continuous Air Monitoring of Plutonium*

CONTENTS

1   Principle of the Method 481
2   Range and Sensitivity 482
3   Interferences 483
4   Precision and Accuracy 484
5   Apparatus 485
6   Reagents...485
7   Procedure 485
8   Calibration and Standards 486
9   Calculations 487
10 Effects of Storage..487
11 Cautions 487
References..488
Appendix: Determination of False Alarm Rate, Acceptable Alarm Set Point, and
   Alarm Response Time for α-Continuous Air Monitors 490

1 PRINCIPLE OF THE METHOD

1.1 This method uses guidance and recommendations supplied by the U.S. Department of Energy (DOE), the American National Standards Institute (ANSI), and the International Electrotechnical Commission (IEC), in addition to the findings and observations of individual researchers. Alpha-continuous air monitors (α-CAMs) are used for real-time air monitoring to primarily detect airborne plutonium and other α-emitting radionuclides such as uranium in nuclear facilities. Typical modern α-CAMs collect airborne particles on a filter that is located in front of a solid-state radiation detector and then use α-spectrometry to separate the radiation emissions of interest from the background interference of naturally occurring, α-emitting progeny of radon (^{222}Rn) and thoron (^{220}Rn).

1.2 Advances in microcomputers have enabled the use of an embedded multichannel spectrometer in the CAM, rather than single-channel analyzers. In a batch operation (collect and then count), the air gap between the filter and the detector can be evacuated to prevent attenuation of the energy of the α-particle emissions and to minimize any spectral overlap of radon progeny α-energy peaks onto the plutonium energy region (see, e.g., Prevo et al., 1987).

1.3 However, for real-time sampling of the aerosol, the region between the filter and the detector cannot be evacuated. Statutory requirements for real-time monitoring of α-emitting radionuclides in U.S. DOE facilities are included in the DOE occupational radiation protection standard (DOE, 2007),

---

* Method authors: Mark D. Hoover and George J. Newton. Method reviewers and editors: Hung-Cheng Chiou and Mark L. Maiello.

and guidance on conducting air monitoring in DOE facilities is contained in the associated implementation guide (DOE, 1998a). Type testing of α-CAMs is addressed in an international standard promulgated by the IEC (IEC, 1997). Test facilities meeting the requirements of that standard have been described by Hoover and Newton (1998) and Grivaud et al. (1998). Performance specifications for health physics instrumentation used for occupational airborne radioactivity monitoring are provided in ANSI N42.17B (1989). Specification and Performance of On-site Instrumentation for Continuously Monitoring Radioactivity in Effluents are proscribed in ANSI N48.18 (1980). Guidance for using α-CAMs and other air sampling and monitoring equipment for sampling of radioactive effluents from the stacks and ducts of nuclear facilities is given in ANSI N13.1 (1999). The U.S. Nuclear Regulatory Commission (NRC, 1992) has made general recommendations for sampling of radioactive aerosols in the workplace. A number of evaluations of the aerosol collection, detection, placement, and performance characteristics of α-CAMs have been published (see, e.g., Biermann and Valen, 1983; Unruh, 1986; McFarland et al., 1990, 1991, 1992; Moore et al., 1993; Hoover and Newton, 1991, 1992, 1993, 1994, 1995, 1998; Hoover et al., 1988, 1990, 1991, 1995; Crites, 1994; Whicker et al., 1997).

1.4 The generally accepted definition of "laboratory conditions" for the purpose of qualification testing of α-CAMs is a polonium-218 concentration of at least 0.1 pCi L^{-1} (3.7 Bq m^{-3}), and an airborne dust concentration of less than 10 μg m^{-3} (Hoover and Newton, 1994, 1998). This concentration of ^{218}Po would result from a ^{222}Rn concentration of about 0.2 pCi L^{-1} (7.4 Bq m^{-3}) under the typical condition of 50% equilibrium for ^{222}Rn in a ventilated, indoor environment.

2 RANGE AND SENSITIVITY

2.1 Instruments used for real-time air monitoring shall be appropriate for the type(s), levels, and energies of radiation encountered in the workplace and for the existing environmental conditions (DOE, 2007). Real-time air monitors should be capable of measuring 1 derived air concentration (1 DAC) when averaged over 8 h (8 DAC-h) under laboratory conditions (DOE, 1998a).

2.2 When monitoring for α-emitting radionuclides shows high radon and thoron concentrations, an alarm set point of up to 24 DAC-h may be acceptable (DOE, 1998b). In all cases, the actual alarm set point established for each CAM unit and the technical basis for the alarm set point should be documented (DOE, 1998a).

2.3 The Appendix of this procedure describes a method for determination of false alarm rates, acceptable alarm set points, and alarm response times for α- CAMs.

2.4 Continuous air monitoring for low airborne concentrations of high-specific-activity radionuclides, such as ^{238}Pu, will suffer from statistical limitations in achieving a representative sample (Birchall et al., 1991; Scott et al., 1997; DOE, 1998b; Scott and Fencl, 1999). This arises from the fact that a small number of particles per unit volume can account for a concentration of concern. For example, an average of 0.003 particles of ^{238}Pu per m^3 of air is equal to the ^{238}Pu-DAC of 0.3 Bq m^{-3}. This is based on the fact that a spherical particle of plutonium-238 dioxide (of nominal specific activity 6.44×10^{11} Bq g^{-1}, and nominal density of 10 g cm^{-3}) with a physical diameter of 3 μm and a corresponding aerodynamic equivalent diameter of about 10 μm (i.e., the aerodynamic diameter is approximately equal to the physical diameter times the square root of the density) has an activity of 105 Bq. Thus, there is a low probability that a CAM operating at 0.056 m^3 min^{-1} (2 cfm) will collect even a single particle in a reasonable amount of time. Such concerns do not exist for isotopes with lower specific activity (such as ^{239}Pu) where the aerosol cloud can be expected to have a larger number concentration and be more homogenous at concentrations of concern. In any case, the DOE Standard for Internal Dosimetry (DOE, 1998b) notes that air sampling data should not be used to assign radiation dose to workers except in cases (such as stable tritiated particulate) where no bioassay method is available.

Method 9: A Procedure for Continuous Air Monitoring of Plutonium

3 INTERFERENCES

3.1 As noted above, naturally occurring radon and thoron progeny radionuclides have α-emissions that can mask the presence of plutonium or uranium. Attempts to set alarms at unrealistically low levels can lead to false-positive reports. Conversely, exposures of concern may be missed if alarm set points are significantly greater than needed to compensate for background interferences. The α-emissions of naturally occurring radon progeny such as polonium-218 and thoron progeny such as bismuth-212 (with α-energies of 6.00 and 6.05 MeV, respectively) are similar enough in energy to the α-emissions of plutonium-239 (α-energy 5.16 MeV) and plutonium-238 (α- energy 5.50 MeV) to cause interference or false-positive reports of plutonium air concentrations. The two other naturally occurring isotopes of lesser concern for interference are the radon progeny polonium-214 (α-energy 7.68 MeV) and the thoron progeny polonium-212 (α-energy 8.78 MeV).

3.2 The presence of airborne dust can lead to burial and underestimation of plutonium or uranium deposited on the sample collection filter. The accumulation of ambient dust on the collection filter of an α-CAM leads to attenuation of α-energy, just as the air gap above the filter degrades the α-energy. Such burial of plutonium leads to underreporting of air concentrations, ranging from 10% to 100% when airborne dust concentrations are greater than 1 mg m^{-3} (Hoover et al., 1988, 1990).

3.2.1 α-Particles from plutonium that is buried by 2 mg cm^{-2} of salt on a filter are prevented from reaching the detector. This does not prevent the CAM from responding to large puff releases of radioactivity (which will be deposited on the surface of the collection substrate and detected before they can be attenuated by additional collection of dust), but it does raise the limit of detection for slow, continuous releases. Dust concerns are primarily associated with decommissioning activities where metal piping and structures are being cut (see Newton et al., 1987), with storage of transuranium wastes in underground salt mines (see Hoover and Newton, 1988, 1990, 1992), or with environmental restoration activities where soil is being disturbed.

3.2.2 Many CAMs use a nominal 4.3-cm (1.7 in.)-diameter detector with a nominal 4.3-cm-diameter filter. That results in an overall detection efficiency of 30–32%. The counting efficiency approaches 40% when measuring radioactivity from the center of the filter using the 4.3 cm filter/detector combination (Hoover et al., 1991).

3.2.3 Other CAMs use a 2.5-cm-diameter detector and a 2.5-cm-diameter filter. Use of a smaller detector and filter diameter leads to some improvement in spectral resolution (unless the detector is very far from the filter) because the maximum distance that α-emissions can travel (from one edge of the filter to the opposite edge of the detector) is shorter than in systems with larger filter and detector diameters. The smaller diameter detectors also provide a cost advantage compared to the cost of replacing the larger detectors, especially when they are used in harsh environments, such as those involving high concentrations of airborne salt dust. The 2.5-cm filter to 2.5-cm detector arrangement has an overall detection efficiency of approximately 20%, but may suffer from limitations in the amount of aerosol that can be collected per unit time on the smaller filter cross section.

3.2.4 A closer examination of detection efficiency as a function of filter diameter reveals that material collected at the filter edges contributes very little to overall efficiency. This is because of solid angle considerations that reduce both the efficiency at which α's are intercepted by the detector and the energy at which they are detected.

3.2.5 α's traveling from one edge of the filter to the opposite edge of the detector have the longest path through air, and thus suffer the greatest energy loss, perhaps enough to remove them from the plutonium energy region. With the combination of a large filter (4.3 cm) and small detector (2.5 cm), α's from the filter edge lose a significant amount of energy as they travel to the detector or are not intercepted by the detector at all. Plutonium collected at the center of the filter is detected in the plutonium region of interest (ROI) with little energy degradation at an efficiency of 30% in the case of a 2.5 cm detector and at an efficiency of 40% using a 4.3 cm detector. At the filter diameter of 2.5 cm, detection efficiency of the 2.5 cm detector has dropped to 15%, but only marginal energy degradation has occurred. For the 4.3 cm detection geometry, efficiency has dropped to about 35% with some energy degradation of those α-particles with very low-angle trajectories. Low-angle α-particles would not be detected using the 2.5 cm filter/detector geometry. At a filter diameter of 4.3 cm, only 0.04% of the emitted α-particles reach a 2.5 cm diameter detector and have energies in the plutonium ROI. Thus, the combination of a small diameter detector with a larger diameter filter provides little advantage in an α-CAM, especially when monitoring in the presence of dust (Hoover and Newton, 1993). The advantages of increased filter surface area over which dusts can deposit and the increase in detection efficiency can result in greatly improved overall performance despite some loss of spectral resolution.

3.3 Nonuniform deposition on the collection substrate of aerosol particles as a function of aerodynamic particle size may also interfere with the proper detection of α-emitting radionuclides in counting systems where the detection efficiency is not equal for all areas on the collection filter. This effect is minor when the detector and filter have similar diameters but can become significant if the collection filter is larger in diameter than the detector.

3.4 Radiation shielding and background correction for external sources or α-, β-, and γ-radiation are generally not required for α-detection systems involving solid-state detectors.

3.5 Radiofrequency interferences from portable radios, cellular phones, and other similar devices can cause spurious responses in CAMs that are poorly shielded or damaged, or from which the protective shields have been removed.

3.6 Electrical isolation of the CAM circuitry should be provided. Transients in the electrical power used to operate the CAM can also result in spurious responses.

3.7 Using a grounded front face on the detector is prudent to prevent the detector from acting as an antenna and being susceptible to radiofrequency interference. Grounding the detector surface also reduces electrostatic collection of dust and radon progeny on the detector surface.

3.8 Ambient light can cause false counts in the detection system if light is allowed to penetrate to the surface of a bare detector. Even if a detector has been manufactured with a light-tight coating (e.g., a thin layer of aluminum or gold), the detector can become light sensitive if the coating is abraded during cleaning, eroded by physical contact with moving airborne dust particles, or etched by acids or corrosion. Arc light flashes from welding operations may be registered by the detector if the high-intensity light can penetrate the detector housing or be transmitted to the detector through the aerosol inlet pathway.

3.9 Failure to select an appropriate number and placement of CAMs in the workplace can also result in an effective "interference" in the ability of the CAM to provide a proper estimate of air concentration.

4 PRECISION AND ACCURACY

4.1 The precision of an individual measurement can depend on normal variations in Poisson statistics.

4.1.1 Measurements become less precise at higher concentrations of radon and thoron background.

4.1.2 Longer integration periods reduce the uncertainties in precision, but may delay warning workers of airborne radioactivity.

4.2 Accuracy of the reported concentration for plutonium or other radionuclide of interest depends on the validity of the values that are calculated or assumed for the concentration calculation including the aerosol sampling flow rate, the efficiency of aerosol delivery to the collection substrate, the uniformity of the aerosol collection on the substrate, the radiation detection efficiency as a function of radionuclide position on the substrate, the presence of any dusts that might reduce the detection efficiency, and the ability of the background subtraction algorithm to make a proper correction for interference from radon and thoron decay products.

Method 9: A Procedure for Continuous Air Monitoring of Plutonium

4.3 Confirmatory radioactivity counting should be done on samples collected in the CAM to identify and correct any degradation of precision or accuracy. Refer to the Appendix of this method and Chapter 14 of this book for additional discussion of issues for precision and accuracy.

5 APPARATUS

5.1 See the chapter on CAM for a discussion of a typical commercially built model.

6 REAGENTS

6.1 None beyond appropriate sample collection substrates and cleaning solutions or solvents to remove accumulated dusts or radioactivity from the detector.

7 PROCEDURE

7.1 Select a CAM design that has been demonstrated to be appropriate for the type(s), levels, and energies of radiation(s) encountered in the workplace and suitable for operation under the existing environmental conditions, and follow the manufacturers' instructions for its use.

7.2 Determine the acceptable false alarm rates, acceptable alarm set points, and alarm response times for the system as it will be operated in the workplace (see the Appendix). For example, the false alarm rate might be less than 1 per year to provide confidence in the validity of the alarm and the alarm set point might be 40 DAC-h with a 5-min response time to limit the potential for individual worker exposures during any single event.

7.3 Place the CAMs in appropriate locations for collecting releases of radioactivity from potential sources. Select sampling locations based on a graded approach for placement including past experience, engineering judgement, qualitative airflow studies, quantitative aerosol dispersion studies, and computational fluid dynamics. Many users place CAMs near the room exhaust grates to ensure that all air exiting the room is sampled. This may result in dilution of releases from more central areas of the room. Therefore, low-level chronic and short duration puff releases could go undetected over the interval between CAM filter changes (Crites, 1994; Whicker et al., 1997). Other placement schemes may provide faster and more reliable alarm responses (Whicker et al., 1997). Identify and evaluate all pathways by which air can leave the room, including the inlets to glove box enclosure assemblies. Such inlets may draw unmonitored air from potential release locations and allow undetected exposures of workers. When performing continuous effluent monitoring from stacks or ducts, follow the recommendations of ANSI/HPS N13.1 (1999).

7.4 During placement, make use of design features such as a remote placement capability for the sampling head so that results can be read remotely and the CAM control unit can be located in a clean area (if possible). Select a reasonably small physical size for the sampling head to minimize interference with work and maximize the locations where the sampling head can be safely placed. Also, allow for potential mobility of the sampling head to adjust to changing work practices or equipment configurations. When possible, facility planning should support air sampling by providing for vacuum, power and data lines, so that many locations including the central locations of rooms can be readily sampled.

7.5 Perform functional checks of the air monitoring equipment to ensure that the equipment is operating properly prior to and during use.

7.5.1 Obtain guidance and procedures from the manufacturer regarding an appropriate method and frequency for verifying that the instrument is operating properly and able to provide the required alarms.

7.5.2 Use a frequency of functional checks that is commensurate with the overall reliability of the monitor, the degree to which the instrument monitors itself for failures, and the hazard potential of the material to be monitored. For example, monitors in locations where the airborne radioactivity concentrations may exceed 10 times the applicable DAC should be checked each working shift. Monitors in locations where the airborne radioactivity concentrations may exceed 25% of the applicable DAC but is unlikely to exceed 10 times the applicable DAC should be checked on a daily basis. Monitors in locations where airborne radioactivity concentrations are unlikely to exceed 25% of the applicable DAC should be functionally checked on at least a weekly basis.

7.5.3 Use a radioactive source to verify monitor operability. This may be as simple as confirming the presence of an appropriate response to ambient radon progeny.

7.5.4 Verify the operability of audible and visual alarms. This may be done electronically or by using a radioactive source to actuate the alarm. Verification should include all high, low, and failure alarms. The radiological response of the monitor during functional checks should be within ±20% of the response at the time of the calibration.

7.5.5 Include verification that the monitor is operating at the correct flow rate. This may be done by noting the indication of an internal flow meter (if the CAM includes such a device), by noting the appropriate pressure drop and reading for an external flow device such as a rotameter, or by attaching a calibrated flow meter to the CAM inlet. An adapter may be necessary for attaching a flow meter to inlets designed for area monitoring. Note that the pressure drop of the flow meter should not be large enough to reduce the flow rate of the CAM. Another viable approach is to install a critical flow venturi (CFV) between the CAM head and the vacuum line (Parulian et al., 1996). A CFV will assure a constant flow rate through the filter as long as the following two conditions apply: accumulation of dust on the filter does not significantly alter the pressure drop across the filter (which would change the critical flow conditions at the CFV), and the house vacuum system maintains sufficient negative pressure on the downstream side of the CFV to maintain a critical flow condition.

7.5.6 Compare the results against expected values determined during monitor calibration.

7.5.7 Record and review appropriate data from the functional checks. Quality control charts should be established as a minimum for source check data.

7.6 Perform periodic confirmatory measurements of the CAM collection filters (perhaps quarterly or semiannually) by radiochemical analyses or radioactivity counting to compare the actual amount of radioactivity on the filter with the radioactivity reported by the CAM.

8 CALIBRATION AND STANDARDS

8.1 Calibration devices for airflow rate should be traceable to the National Institute of Standards and Technology (NIST).

8.2 Airflow rate should be calibrated in (or converted to) ambient volumetric flow rate to allow a proper correlation to human breathing rates. Note that breathing rates for acclimated humans are similar for similar physical activities, regardless of barometric pressure (altitude) differences, and that compensation for air density is achieved by a natural biological change in the number of red blood cells. Use of a mass flow meter will require correction to local atmospheric conditions. Use of a bubble or piston flow meter will provide flow rates in ambient volume units.

8.3 Radioactivity counting efficiency should be determined with NIST-traceable electroplated sources.

8.4 The influence of source geometry on detection efficiency should be determined and the electroplated sources should be selected to provide a proper determination of system detection efficiency (Newton and Hoover, 1991). For example, if the collection filter has a diameter of 4.3 cm, then the detection efficiency of the CAM is likely to be overestimated if calibration is done with a 2.5 cm-diameter electroplated source.

8.5 Energy calibration of the CAM system for radiation detection is best done by collecting ambient radon progeny on the CAM substrate of interest. A correlation can then be established for the effective energy of the radionuclides of interest on electroplated sources. Note that the effective energy of α-emissions from electroplated sources are generally lower in energy than the expected "true" energy because of solid-state diffusion of the radionuclide into the surface of the electroplated

Method 9: A Procedure for Continuous Air Monitoring of Plutonium

source. The extent of this difference is generally associated with the amount and duration of heating during the preparation and annealing of the electroplated source. It is sometimes useful to compare the highest energies observed in the peaks from the filter and from the electroplated sources (where the peaks return to baseline on the high-energy side). These energies are associated with α-emissions nearest the top surface of the filter or electroplated source.

9 CALCULATIONS

9.1 The general calculation to convert a DAC in Bq m^{-3} into a radioactivity accumulation, A, in counts per minute on a CAM collection substrate is

$$A = \frac{\text{DAC} \cdot Q \cdot T}{E_R \cdot E_S}, \tag{1}$$

where Q is the air sampling flow rate, T is the duration of sampling, E_R is the radioactivity detection efficiency for the system, and E_S is the aerosol collection efficiency.

9.2 For example, if the integration period is 1 h, and the concentration of concern is 1 DAC, then A will be the activity associated with 1 DAC-h of integrated air concentration.

9.3 Each commercial CAM has instrument-specific calculations for radon progeny interference corrections (refer to Chapter 14 of this book for examples).

10 EFFECTS OF STORAGE

10.1 Note that the samples collected or observed by the CAM are analyzed in real time and may or may not be retained for later evaluation.

10.2 If the CAM is not used for an extended period of time, such as might occur for replacement or spare units, then

- **10.2.1** The proper functioning of the solid-state detector should be confirmed by testing with an electroplated source or ambient radon progeny.
- **10.2.2** The airflow and aerosol collection system should be leak tested to ensure that all seals are properly functioning and that there are no obstructions in the flow path.
- **10.2.3** Any flexible tubing sections should be inspected and tested for brittleness, aging, physical damage, or degradation from ultraviolet light and replaced as needed.
- **10.2.4** The calibration of any flow monitoring and control functions should be confirmed using NIST-traceable standards.
- **10.2.5** Any internal backup batteries for memory protection should be tested or replaced.
- **10.2.6** All portions of the 110-V electrical system should be inspected to ensure adequate safety from electrical hazards.

11 CAUTIONS

11.1 Modern real-time monitors provide a significant amount of automatic performance checking and verification. Nevertheless, operators should be aware of the underlying trends of performance and instrument response to avoid erroneous results.

11.2 This method may involve the use of hazardous chemicals and/or radioactivity. It is the user's responsibility to take appropriate precautions and to obtain necessary regulatory approvals from institutional and/or governmental agencies.

11.3 If the user is not trained or familiar with the calibration and operation of radiation analysis equipment, they should consult a qualified expert, such as a health physicist, for assistance in the proper and safe usage of the equipment and development of appropriate quality assurance protocols.

REFERENCES

ANSI, *Specification and Performance of On-Site Instrumentation for Continuously Monitoring Radioactivity in Effluent*, ANSI/IEEE N42.18, American National Standards Institute, New York, NY, 1980.

ANSI, *Performance Specification for Health Physics Instrumentation-Occupational Airborne Radioactivity Monitoring Instrumentation*, ANSI/IEEE N42.17B, American National Standards Institute, New York, NY, 1989.

ANSI, *Sampling and Monitoring for Releases of Airborne Radioactive Substances from the Stacks and Ducts of Nuclear Facilities*, ANSI/HPS N13.1, American National Standards Institute, New York, NY, 1999.

Biermann, A. and Valen, L., CAM particle deposition evaluation, in R. V. Griffin, Ed., *Hazards Control Department Technology Review*, Report UCRL-5007–83, Lawrence Livermore National Laboratory, Livermore, CA, pp. 79–82, 1983.

Birchall, A., James, A. C., and Muirhead, C. R., Adequacy of personal air samplers for monitoring plutonium intakes, *Radiat. Prot. Dosim.*, 39, 179–188, 1991.

Crites, T. R., Alpha air monitor alarm sensitivity: Operational experience, *Radiat. Prot. Dosim.*, 53, 65–68, 1994.

DOE, *Occupational Radiation Protection*, Title 10, *Code of Federal Regulation*, Part 835 (10 CFR 835), U.S. Department of Energy, Office of Federal Register, Washington, DC, 2007.

DOE, *Air Monitoring Implementation Guide for Use*, Title 10, *Code of Federal Regulations*, Part 835, Occupational Radiation Protection, DOE G 441.8 (formerly G-10 CFR 835/E2), U.S. Department of Energy, Washington, DC, 1998a.

DOE, *DOE Standard for Internal Dosimetry*, DOE-STD-1121-98, U.S. Department of Energy, Washington, DC, 1998b.

Grivaud, L., Fauvel, S., and Chemtob, M., Measurement of performances of aerosol type radioactive contamination monitors, *Radiat. Prot. Dosim.*, 79(1–4), 495–498, 1998.

Hoover, M. D. and Newton, G. J., Response of the Eberline Alpha 6 Continuous Air Monitor to low-level releases of plutonium, in *Inhalation Toxicology Research Institute Annual Report 1990–1991*, National Technical Information Service, Springfield, VA, pp. 20–22, 1991.

Hoover, M. D. and Newton, G. J., Influence of salt dust on alpha energy spectra in continuous air monitors for alpha-emitting radionuclides, in *Inhalation Toxicology Research Institute Annual Report 1991–1992*, National Technical Information Service, Springfield, VA, pp. 11–13, 1992.

Hoover, M. D. and Newton, G. J., Radioactive aerosols, in *Aerosol Measurement: Principles, Techniques, and Applications*, K. Willeke and P. A. Baron, Eds., pp. 768–798, Van Nostrand Reinhold, New York, NY, 1993.

Hoover, M. D. and Newton, G. J., Calibration and operation of continuous air monitors for alpha-emitting radionuclides, in *1993 Radiation Protection Workshop Proceedings*, pp. G27–G39, U. S. Department of Energy, Washington, DC, 1994.

Hoover, M. D., Newton, G. J., Fencl, A. F., and Marcinkovich, M. B., *Independent Evaluation of the Los Alamos National Laboratory Continuous Air Monitor Instrumentation at the Inhalation Toxicology Research Institute*, Report ITRI-951102, Inhalation Toxicology Research Institute, Albuquerque, NM, 1995.

Hoover, M. D., Newton, G. J., and Griffith, W. C., Influence of detector and collection filter geometry on counting efficiency for alpha continuous air monitors, in *Inhalation Toxicology Research Institute Annual Report 1990–1991*, National Technical Information Service, Springfield, VA, 1991.

Hoover, M. D., Newton, G. J., Yeh, H. C., Seiler, F. A., and Boecker, B. B., *Evaluation of the Eberline Alpha-6 Continuous Air Monitor for Use in the Waste Isolation Pilot Plant: Report for Phase I*, Report to the DOE Waste Isolation Pilot Plant Project Office from the Inhalation Toxicology Research Institute, Albuquerque, NM, 1988.

Hoover, M. D., Newton, G. J., Yeh, H. C., Seiler, F. A., and Boecker, B. B., *Evaluation of the Eberline Alpha-6 Continuous Air Monitor for Use in the Waste Isolation Pilot Plant: Report for Phase II*, Report to the DOE Waste Isolation Pilot Plant Project Office from the Inhalation Toxicology Research Institute, Albuquerque, NM, 1990.

Hoover, M. D. and Newton, G. J., Performance testing of continuous air monitors for alpha-emitting radionuclides, *Radiat. Prot. Dosim.*, 79(1–4), 499–504, 1998.

IEC, *Radiation Protection Instrumentation, Calibration and Verification of the Effectiveness of Radon Compensation for Alpha and/or Beta Measuring Instruments*, IEC Standard 61578, International Electrotechnical Commission, 1997.

McFarland, A. R., Bethel, E. L., Ortiz C. A., and Stanke, J. G., A CAM sampler for collection and assessment of alpha-emitting aerosol particles, *Health Phys.*, 61, 97–103, 1991.

McFarland, A. R., Ortiz, C. A., and Rodgers, J. C., Performance evaluation of continuous air monitor (cam) sampling heads, *Health Phys.*, 58, 275–281, 1990.

McFarland, A. R., Rodgers, J. C., Ortiz, C. A., and Moore, M. E., A continuous sampler with background suppression for monitoring alpha-emitting aerosol particles, *Health Phys.*, 62, 400–406, 1992.

Moore, M. E., McFarland, A. R., and Rodgers, J. C., Factors that affect alpha particle detection in continuous air monitors, *Health Phys.*, 65, 69–81, 1993.

Newton, G. J. and Hoover, M. D., *Spectral and Spatial Homogeneity Requirements for Electroplated Alpha-Emitting Calibration Sources for Continuous Air Monitors*, in Inhalation Toxicology Research Institute Annual Report 1991–1992, National Technical Information Service, Springfield, VA, pp. 5–8, 1991.

Newton, G. J., Hoover, M. D., Barr, E. B., Wong, B. A., and Ritter, P. D., Collection and characterization of aerosols from metal cutting techniques typically used in decommissioning nuclear facilities, *Am. Ind. Hyg. Assoc. J.*, 48, 922–932, 1987.

NRC, *Air Sampling in the Workplace, Regulatory Guide 8.25*, U.S. Nuclear Regulatory Commission, Washington, DC, 1992.

Parulian, A., McFarland, A., and Rodgers, J., A constant-flow air sampler for workplace environments, *Health Phys.*, 71, 870–878, 1996.

Prevo, C. T., Kaifer, R. C., Rueppel, D. W., Delvasto, R. M., Biermann, A. H., and Phelps, P. L., A transuranic aerosol measurement system: Preliminary results, *IEEE Trans. Nucl. Sci.*, NS34, 601–605, 1987.

Scott, B. R. and Fencl, A. F., Variability in PuO_2 inhalation: Implications for worker protection at the U.S. Department of Energy, *Radiat. Prot. Dosim.*, 83, 221–232, 1999.

Scott, B. R., Hoover, M. D., and Newton, G. J., On evaluating respiratory tract intake of high-specific activity emitting particles for brief occupational exposure, *Radiat. Prot. Dosim.*, 69, 43–50, 1997.

Unruh, W. P., *Development of a Prototype Plutonium CAM at Los Alamos*, LA-UR-90-2281, Los Alamos National Laboratory, Los Alamos, NM, December 15, 1986.

Whicker, J. J., Rodgers, J. C., Fairchild, C. I., Scripsick, R. C., and Lopez, R. C., Evaluation of continuous air monitor placement in a plutonium facility, *Health Phys.*, 72, 734–743, 1997.

APPENDIX: DETERMINATION OF FALSE ALARM RATE, ACCEPTABLE ALARM SET POINT, AND ALARM RESPONSE TIME FOR α-CONTINUOUS AIR MONITORS

Introduction

Poisson statistics have been shown to apply to the radioactivity counting of α-emissions from plutonium in modern, spectrometry-based α-continuous air monitors (CAMs) in the presence of an interfering background of radon progeny (Hoover and Newton, 1998). The standard deviation (SD) of the reported plutonium concentration can be determined by operating the CAM of interest in the presence of interference from radon progeny. The SD can also be used to calculate the statistical false alarm rate as a function of the alarm set point. Conversely, the SD of the reported plutonium concentration can be used to calculate the acceptable alarm set point (which can be viewed as a form of minimum detectable activity, MDA, or concentration) for a desired false alarm rate. The integration time of the counting system and the required number of positive reports for triggering an alarm can be used to determine the alarm response time for the CAM. Note that the *alarm response time* differs both from the *resolving time* of the system (normally defined as the smallest time interval which must elapse between the occurrence of two consecutive pulses or ionizing events and still be recognized as separate pulses or events) and from the *response time of a measuring system* (which is normally defined as the time required after a step variation in the measured quantity for the output signal variation to reach a given percentage, usually 90%, of its final value) for the first time (IEC, 1993).

Sensitivity

In general, for a modern, multichannel analyzer-based CAM, using 15-min averaging, in an effective radon concentration of 1 pCi L^{-1} (37 Bq m^{-3}), the SD of the background can be as low as 0.7 DAC-h (Hoover and Newton, 1998). This results in a traditional decision level, L_C (95% confidence that the observed plutonium concentration did not result from a statistical variation of the radon progeny background) of about 1.4 DAC-h (calculated as 1.96σ) (Currie, 1968). The validity of the false alarm rate and acceptable alarm set-point calculations have been evaluated at radon progeny concentrations up to about 10 pCi L^{-1} (370 Bq m^{-3}) (Hoover et al., 1995).

Precision and Accuracy

Individual differences between CAM units and their conditions of operation can influence the precision and accuracy of the false alarm rate, acceptable alarm set point, and alarm response time values. However, these differences can be detected and controlled by an effective receipt and inspection program and a well-controlled calibration and operating program. Factors such as electrical transients, radiofrequency noise, high light levels from arc welding or stroboscopic warning lights, and failure of CAM components, can be sources of false alarms or can alter the acceptable alarm set point.

Interferences

Interference with the proper reporting of the true concentration of ^{239}Pu (α-energy 5.15 MeV) and ^{238}Pu (α-energy 5.48 MeV) originates from the detection of radon progeny ^{218}Po (α-energy 6.0 MeV) and ^{214}Po (α-energy 6.78 MeV), and thoron progeny ^{212}Po (α-energy 8.78 MeV) and ^{212}Bi (α-energy 6.0 MeV). Other related concerns include the following:

- The ability of an α-CAM to provide a proper alarm for a true plutonium aerosol concentration in the presence of a background concentration of ambient radon progeny depends on

Method 9: A Procedure for Continuous Air Monitoring of Plutonium

the aerosol sampling flow rate, aerosol collection efficiency, smoothness of the sample collection filter, α-radioactivity detection efficiency, radioactivity counting time, amount of background dust, and fidelity of the background subtraction algorithm for radon progeny interference.
- The magnitude of the SD of the reported plutonium concentration increases with increasing radon progeny concentration and decreases with increasing counting or integration time.
- Statistical influences of Poisson variations in counting of plutonium will apply when the plutonium concentration is low.
- Statistical variations for the interference of radon progeny will dominate the calculation of the MDA and the associated false alarm rates.

The CAM configuration must be controlled and the environment in which it will be used must be known to ensure that the false alarm rate, acceptable alarm set point, and the alarm response time are relevant to the actual conditions needed for field use. The list below summarizes some of these issues.

- *Filter type:* Smoother filters provide better α-spectroscopy results as measured by the full-width-at-half-maximum (FWHM) of the energy peaks for the plutonium and radon progeny, and rougher filters will raise the false alarm rate and acceptable alarm set point (see Hoover and Newton, 1993).
- *Detector to filter geometry:* Increasing the detector to filter separation or increasing the diameter of the filter relative to the diameter of the detector will also decrease the detection efficiency and raise the FWHM for spectrometry. Decreasing the detection efficiency and increasing the FWHM will raise the false alarm rate and acceptable alarm set point.
- *Flow rate and Integration Time:* Decreasing the flow rate or decreasing the integration time will decrease the number of radioactive decays available for counting and this will raise the false alarm rate and acceptable alarm set point.
- *Presence of airborne dust:* Accumulation of dust on the filter can bury the radon progeny (most notably the ^{214}Po and ^{212}Bi, which have longer half-lives than the primary interfering radon progeny ^{218}Po). This will result in an increase in the FWHM of the interfering progeny, and thereby raise the false alarm rate and acceptable alarm set point. Accumulation of dust can also bury the collected plutonium and result in an underreporting of plutonium concentration.
- *Composition of the Radon Progeny Background:* Operating the CAM in an environment with a ratio of radon and thoron progeny significantly different from that assumed by the programmed background subtraction algorithm can alter the false alarm rate and acceptable alarm set point. (Note that the presence of the short-lived ^{216}Po progeny of ^{220}Rn can become detectable and alter the performance of the radon progeny correction algorithm if the concentration of ^{220}Rn is very high, as might occur in a workplace where thorium is present.)

APPARATUS

The test apparatus consists of the α-CAM of interest, along with its associated vacuum source and data collection and logging modules. Ambient radon progeny from laboratory air can be used for the basic determination of the false alarm rate and the acceptable alarm set point. Radon progeny at higher concentrations can be obtained from a radon calibration chamber. Facilities exist at locations worldwide such as Bowser-Morner, Inc. in Dayton, OH; the Laboratoire d'Essais Physiques des Instruments de mesure de la Contamination de l'Eau et de l'Air in Saclay, France (Grivaud et al., 1998); and at the Lovelace Respiratory Research Institute in Albuquerque, NM (Hoover and Newton, 1998).

Procedure

Part 1 of the procedure determines the SD of the reported plutonium concentration for the CAM of interest in the presence of radon progeny (for one or more concentrations of radon progeny) and uses that information to calculate the false alarm rate per year. Part 2 of the procedure inverts the equations to calculate the acceptable alarm set point for a given false alarm rate. Part 3 of the procedure determines the acceptable alarm response time based on information about the system integration time and number of positive reports required for an alarm to be triggered.

Throughout the process, it is assumed that the CAM of interest is operated with a specified flow rate, radon progeny concentration, filter type, and geometry.

Part 1: Determination of the False Alarm Rate per Year

For any counting time, T min, and effective radon progeny concentration, C in Bq m^{-3} or pCi L^{-1}, the SD of the reported plutonium concentration for the CAM, and $SD_{T,C}$ in Bq or DAC-h can be calculated from a knowledge of the SD of the reported plutonium concentration under a measured set of reference conditions (e.g., 15 min integration time and 1 pCi L^{-1} [37 Bq m^{-3}] effective radon concentration) as:

$$SD_{T,C} = SD_{15,1} * \text{SQRT}(C/1) * \text{SQRT}(15/T), \quad (A1)$$

where $SD_{15,1}$ is the SD of the background under the reference conditions of 15 min counting time in the presence of 1 pCi L^{-1} effective radon concentration (assuming 50% equilibrium between the radon parent and the progeny).

The false alarm rate per year, F, can be calculated as

$$F = MP^N, \quad (A2)$$

where M is the number of measurements per year, P is the probability of a false positive for a single measurement, and N is the number of sequential positive reports required for activation of the alarm.

The number of measurements per year, M, can be calculated as

$$M = 5.25 \times 10^5/T, \quad (A3)$$

where 5.25×10^5 is the number of minutes per year, and T is the number of minutes for integration of each measurement.

Note that the measurements are assumed as a first approximation to be independent. Thus, if T is 15 min, then $M = 35{,}000$ year^{-1}.

The probability, P, of a false-positive report for a single measurement can be determined as a two-step process:

Step 1: Divide the alarm set point, A, by the SD of the background to calculate the number of standard deviations (NSDs) that the alarm set point is above background:

$$\text{NSD} = \frac{A}{\text{SD}}. \quad (A4)$$

For a modern multichannel analyzer CAM when 15-min averaging is used and the effective radon concentration is 1 pCi L^{-1} (37 Bq m^{-3}), the SD of the background can be as low as 0.7 DAC-h (Hoover and Newton, 1998). Thus, if the alarm set point is 8 DAC-h then the set point is 11.4 SDs above the background.

If the effective radon concentration is 4 pCi L^{-1} (148 Bq m^{-3}), then the SD of the background may be 1.5 DAC-h, and the alarm set point is 5.3 SD above the background. *Note that if the CAM detector or the filter-retaining ring becomes contaminated with plutonium, then the CAM may erroneously report the presence of airborne plutonium. If the contamination cannot be removed, then the new "effective background" must be taken into account in setting the acceptable alarm set point.*

Step 2: Calculate the normal standard deviate or consult a handbook table for the normal standard deviate (see, e.g., CRC, 2002). The calculation can be done using the built-in statistical functions of many modern spreadsheet programs to obtain the fraction of occurrences greater than the stated number of SDs above the mean. Using Microsoft Excel spreadsheet as an example, the one-sided probability, $P = 1 - \text{NORMSDIST}(z)$, where z is the value of SD above the mean. One would simply type in any cell " $= 1 - \text{NORMSDIST}(z)$," and then strike the enter key to find the desired probability.

For example, if the set point is 1 SD above the mean, then the probability of a false positive is 16% [$P = 1 - \text{NORMSDIST}(1) = 0.16$]. If the set point is 5.3 SDs above the mean, then the probability of a false positive, $P = 5.8 \times 10^{-8}$ [$P = 1 - \text{NORMSDIST}(5.3) = 5.8 \times 10^{-8}$]. If the set point is 11.4 SDs above the mean, then the probability of a false positive, P is negligible.

The false alarm rate per year can now be calculated.

In the above example for a 4 pCi L^{-1} (148 Bq m^{-3}) radon concentration (not normally expected in the workplace), M was 35,000 measurements year^{-1}, P was a 5.8×10^{-8} probability of a false position on a single measurement, and $N = 1$ was the number of positive reports above the alarm set point that were needed to activate the alarm. Given these values, the false alarm probability per year is (Equation A2)

$$F = M*P^N = 35{,}000*(5.8 \times 10^{-8})^1 = 0.002 \text{ false alarms per year.}$$

Table A1 is an example of a false alarm rate summary for an α-CAM with an integration time of 15 min for alarm set points of 8 DAC-h and 24 DAC-h with assumed radon concentrations of 0.2, 1, and 4 pCi L^{-1}.

Part 2: Determination of the Acceptable Alarm Set Point

Now an acceptable alarm set point can be calculated for a range of radon progeny concentrations and a desired false alarm rate. The acceptable alarm set point, A, can be calculated by rearranging Equation A2 as follows to solve for the required probability of false alarm for a single measurement:

$$P = \left(\frac{F}{M}\right)^{1/N}. \tag{A5}$$

For example, if the desired false alarm rate is 1 per month, then F is 12 per year. Assuming that a 15-min averaging period will still be used, then M is still 35,000. Assuming that a single positive

TABLE A1
Example False Alarm Rate Summary for an α-CAM

Alarm Set point	False Alarm Rate per Year at the Assumed Radon Concentration		
	0.2 pCi L^{-1} (7.4 Bq m^{-3})	1 pCi L^{-1} (37 Bq m^{-3})	4 pCi L^{-1} (148 Bq m^{-3})
8 DAC-h	7 × 10^{-11} (Negligible)	16.8	3770
24 DAC-h	<1 × 10^{-12} (Negligible)	<1 × 10^{-12} (Negligible)	0.006

Note: Numbers less than 1×10^{-6} are considered negligible.

report will be used, then $N = 1$. Thus, the required probability of a false positive for a single measurement is

$$P = \left(\frac{12}{35,000}\right)^1 = 3.4 \times 10^{-4}.$$

We can now consult the table of false-positive fractions to find that this fraction corresponds to 3.4 SDs. Therefore, if the SD of the background is 1.5 DAC-h (associated with the relatively high concentration of 4 pCi L^{-1} [148 Bq m^{-3}]), then the alarm can be set at

$$A = 1.5*3.4 = 5.1 \text{ DAC-h}.$$

Note that one false alarm per instrument per month can correspond to hundreds per month per facility. Therefore, the calculation based on 1 year^{-1} per instrument is also useful. For the case of 1 false alarm rate per year, $F = 1$ year^{-1}, and assuming that a 15-min averaging period will still be used, then M is still 35,000. Assuming, again, that a single positive report will be used, then $N = 1$. Thus, the required probability of a false positive for a single measurement is:

$$P = \left(\frac{1}{35,000}\right)^1 = 2.9 \times 10^{-5}.$$

Again consulting the table of false-positive fractions, the number of SDs that corresponds to this fraction is approximately 4.0. Therefore, if the SD of the background is 1.5 DAC-h, then the alarm can be set at:

$$A = 1.5*4.0 = 6.0 \text{ DAC-h}.$$

Note that at higher concentrations of radon progeny, it is possible to increase the integration period or increase the number of positive reports (N) needed to trigger the alarm, in order to maintain a reasonable alarm set point and false alarm frequency.

Table A2 illustrates the results for acceptable alarm set points for a modern, multichannel analyzer CAM with an integration time of 15 min.

Part 3: Determination of the Alarm Response Time

Information about the alarming scheme can now be used to estimate the CAM alarm response time for a range of typical release conditions. Table A3 is an example of the format for a response time summary table. Estimates and supporting documentation should be provided for the time it will take the CAM of interest to actuate its alarm features using the standard instrument configuration and alarm settings.

One version of this table should be completed for each of the alarm types that are supported by the system. In completing the estimates, it should be assumed that the airborne concentration is distributed homogeneously around the CAM sample inlet(s). Also, for the purpose of these

TABLE A2
Example Acceptable Alarm Set-point Summary for an α-CAM

	Acceptable Alarm Set point (in DAC-h) at the Assumed Radon Concentration		
False Alarm Rate	0.2 pCi L^{-1} (7.4 Bq m^{-3})	1 pCi L^{-1} (37 Bq m^{-3})	4 pCi L^{-1} (148 Bq m^{-3})
1 per month	3.6	8.3	16.2
1 per year	4.3	9.6	19.4

TABLE A3
Format for an Alarm Response Time Summary for an α-CAM

Pu-239 Airborne Concentration (in DAC)	CAM Response Time at the Assumed Release Duration			
	1 min	10 min	1 h	10 h
1				
10				
100				
1000				
10,000				

Note: The information in this table is to be accompanied by the information on how the alarm works and how it can be adjusted by the user, or whether it is automatically adjusted by internal logic.

calculations, it should be assumed that the airborne concentration is constant throughout the entire release duration, and that the airborne radioactivity decreases immediately to zero at the end of the release (see Hoover and Newton, 1998).

EXAMPLE OF ALARM RESPONSE TIME

For illustration purposes, the following example assumes a very simple CAM that uses a fixed count cycle of 30 s and a single alarm mode with an 8 DAC-h set point. Thus, this CAM will alarm after any 30-s count cycle in which there was an accumulation on the filter of 8 DAC-h.

Sample calculation for 100 DAC release lasting 1 h:

Response Time = Time to Accumulate 8 DAC-h on the Filter	+ Count Time
= (8 DAC-h/100 DAC)	+ 30 s
= 4.8 min	+ 30 s
= 5.3 min	

TABLE A4
Example of an Alarm Response Time Summary for an α-CAM

Pu-239 Airborne Concentration (in DAC)	CAM Response Time at the Assumed Release Duration			
	1 min	10 min	1 h	10 h
1	No alarm	No alarm	No alarm	8 h
10	No alarm	No alarm	48.5 min	48.5 min
100	No alarm	5.3 min	5.3 min	5.3 min
1000	30 s	30 s	30 s	30 s
10,000	30 s	30 s	30 s	30 s

Note: The alarm in this example uses a fixed count cycle of 30 s and a single alarm mode with an 8 DAC-h set point. Thus, this CAM will alarm after any 30 s count cycle in which there was an accumulation on the filter of 8 DAC-h or more.

In this case, the CAM will require 5.3 min to respond to the 100 DAC-h release. Table A4 shows values entered for this simple example.

REFERENCES

CRC, *Standard Mathematical Tables and Formulae*, 31st ed., Daniel Zwillinger, Ed., Chapman & Hall/CRC Press, Boca Raton, FL, 2002.

Currie, L. A., Limits for qualitative detection and quantitative determination. Applications to radiochemistry, *Anal. Chem.*, 40, 586–593, 1968.

Grivaud, L., Fauvel, S., and Chemtob, M., Measurement of performances of aerosol type radioactive contamination monitors, *Radiat. Prot. Dosim.*, 79(1–4), 495–498, 1998.

Hoover, M. D. and Newton, G. J., Radioactive aerosols, in *Aerosol Measurement: Principles, Techniques, and Applications*, K. Willeke and P. A. Baron, Eds., pp. 768–798, Van Nostrand Reinhold, New York, NY, 1993.

Hoover, M. D., Newton, G. J., Fencl, A. F., and Marcinkovich, M. B., *Independent Evaluation of Los Alamos National Laboratory Continuous Air Monitor Instrumentation at the Inhalation Toxicology Research Institute*, ITRI-951102, Lovelace Inhalation Research Institute, Albuquerque, NM, November 30, 1995.

Hoover, M. D. and Newton, G. J., Performance testing of continuous air monitors for alpha-emitting radionuclides, *Radiat. Prot. Dosim.*, 79(1–4), 499–504, 1998.

IEC, *International Electrotechnical Vocabulary, Chapter 394, Nuclear Instrumentation, Instruments*, IEC 50–394–1993, International Electrotechnical Commission, Geneva, Switzerland, 1993.

Method 10
Personal Air Sampling for Particulate Radioactivity*

CONTENTS

1	Principle of the Method	497
2	Sensitivity and Range	498
3	Interferences	498
4	Precision and Accuracy	499
5	Apparatus	500
6	Reagents	500
7	Procedure	500
8	Calibration	501
9	Calculations	501
10	Effect of Storage	502
11	Cautions	502
References		502

1 PRINCIPLE OF THE METHOD

1.1 While it is most desirable to rely on bioassay measurements of a worker to determine the potential intake of airborne radioactive material, this is sometimes not practical. For certain radionuclides, their physical and metabolic properties preclude the use of *in vivo* and *in vitro* bioassay techniques. Because of this, it is sometimes necessary to rely on personal air sampling (PAS) as a surrogate and/or supplemental bioassay measurement.

Unlike general area air samples that are taken for the purpose of evaluating workplace controls through the measurement of air concentration, PAS is conducted to evaluate potential intakes of radioactive material. Therefore, special care is taken to ensure that the sample is representative of the breathing zone of the worker being monitored. This is accomplished by placing an air filter cassette as close to the nose and mouth of the worker as possible. The air is drawn through the filter cassette using a battery-powered pump that is typically attached to the worker's belt. The air in the vicinity of the worker's nose and mouth is then sampled during all work activities in which there is a potential for the generation of airborne radioactive particulate (Alvarez and Rich, 1984).

1.2 After the work is concluded and sufficient time has elapsed for the decay of naturally occurring ^{220}Rn and ^{222}Rn progeny, the filter is carefully removed from the cassette and quantitatively evaluated using an appropriate measurement instrument. For atmospheres containing a single radionuclide or that are well characterized, the filters can be measured for α- and/or β-activity using a gas flow proportional counter or another suitable radiation detection instrument (Knoll, 2000; NCRP, 1985).

---

* Method author: James W. Neton. Method reviewers and editors: Robert P. Miltenberger, Mark L. Maiello, and Hung-Cheng Chiou.

1.3 If liquid scintillation counting is employed for the measurement, sample preparation will be required. For α-particle emitters such as plutonium and uranium, various extractive agents are used to isolate the radioactivity into an organic, acidic or aqueous solution which is then analyzed using a compatible scintillation fluid (Passo and Cook, 1996). A photon–electron rejecting α-liquid scintillation spectrometer (PERALS) can be used for the analysis (Metzger et al., 1996). A liquid scintillation counter with α-/β-discrimination capability is suggested as an alternative (Passo and Cook, 1996).

1.4 Simple, relatively inexpensive counters using the solid, inorganic scintillator zinc-sulfide can be employed to provide a gross α-analysis of the filter. If the atmosphere contains complex mixtures of inadequately characterized radionuclides, α-, β-, or γ-spectroscopic methods should be employed.

1.5 The object of this method is the assessment of a worker's intake of inhaled radioactive particulates. Therefore, the measured activity on the air filter must be converted to an intake estimate. This estimate is typically expressed in derived air concentration-hours (DAC-h), which is the product of the fraction of the DAC measured on the filter and the worker's stay time in the atmosphere. The DAC is defined as that concentration of a radionuclide in air which, if breathed for a work year (2000 h), would result in an intake corresponding to the annual limit on intake (ALI) of the radionuclide. The ALI is the annual intake of a radionuclide that would result in a radiation dose to Reference Man (ICRP, 1975) equal to the most limiting value of committed organ dose or of committed effective dose equivalent.* DAC values are published in ICRP 30 (ICRP, 1979), in the U.S. Environmental Protection Agency (EPA) Federal Guidance Report No. 11 (EPA, 1988) and in the U.S. Department of Energy (DOE) Occupational Radiation Protection Standard, 10 CFR 835 (DOE, 2007). The reader should be aware that the DOE and EPA values can differ.

1.6 The significance of PAS results and appropriate follow-up measures should be evaluated against appropriate regulatory requirements and guidelines. Several guidance documents have been published to aid in the interpretation of positive bioassay results (ANSI, 1997; DOE, 1994; NRC, 1993).

2 SENSITIVITY AND RANGE

2.1 The sensitivity of the method is dependent on the air sampler flow rate and the sensitivity of the analytical measurement system used to perform the analysis. Current PAS pumps are capable of drawing up to 12 L min^{-1} of air.† A reasonably shielded proportional counter in a low-background area is capable of detecting approximately 26 mBq (7×10^{-7} μCi) of α-activity and 35 mBq (10×10^{-7} μCi) of β-activity for a filter measurement time of 50 min. A sample collected and measured under these conditions would correspond to an intake that is typically less than 2% of the established annual limits. Depending on the radionuclide of interest, measurement times of 10 min or less may be used to quantify intakes.

2.2 The upper range of a PAS sample is limited by the performance of the radioactivity detection system employed. Caution should be observed, however, in the collection and analysis of personal air samples that are taken in dusty environments. For radionuclides that decay by α-emission or low-energy β-decay, significant self-absorption can occur which may lead to negative bias in the data (Alvarez and Rich, 1984).

3 INTERFERENCES

3.1 When gross α- and/or β-counting is used to analyze the filters, interference from the progeny of naturally occurring ^{220}Rn and ^{222}Rn may lead to erroneously elevated results. To preclude this

---

* The ALI is the largest value of intake that results in a committed effective dose equivalent to the whole body of 0.05 Sieverts (5 rem) or a committed dose equivalent to organs or tissues of 0.5 Sieverts (50 rem) (ICRP 1979).
† A. P. Buck, Inc., USA.

possibility, it is sometimes necessary to delay measurement of the filter. For typical environmental radon levels, a delay time of 24 h is usually sufficient although waiting periods as short as 2 h may be adequate (Harley, 1953). In cases where ^{220}Rn progeny are significantly elevated and the radionuclide of interest has a relatively low DAC (i.e., ^{239}Pu) a delay time of 4 days or longer may be necessary.

3.2 In highly contaminated atmospheres, cross-contamination of the air filter by the worker is a concern. The filter may become contaminated through inadvertent physical contact or through resuspension of dust from a worker's clothing. In these cases, it is advisable to perform sampling using closed filter cassettes (Figure 1).

3.3 The collection of air samples in dusty environments may lead to a buildup of material on the filter that will reduce the detection efficiency of α-activity and low-energy β-activity. If α-spectrometry is employed as the method of analysis, dust loading will significantly degrade the resolution of the spectrum. The effect of dust loading on the detection of α-activity on air filters has been previously evaluated (Moore et al., 1993).

3.4 If γ-spectrometric analysis is employed, interference can result from the Compton scattering of high-energy photons into the photopeak region of the radionuclide of interest. Standard γ-ray analysis software can compensate for this problem. The effect can be minimized if the γ-measurements are performed using a high-resolution Ge detection system.

4 PRECISION AND ACCURACY

4.1 A flow measurement accuracy of about 5% should be obtainable for most commercially available flow-regulated PAS pumps.

4.2 A minimum detectable activity of about 26 mBq (7×10^{-7} µCi) for α-activity and 35 mBq (10×10^{-7} µCi) for β-activity should be achievable using a well-shielded low-background gas flow proportional counter. This assumes a 50 min sample counting time and detection efficiencies of 25% for α and 40% for β-activity. For radionuclides with relatively high DACs, measurement times of 10 min or less may be used.

4.3 A calibration accuracy of ±5% is achievable for most gas flow proportional counting systems. As with any radioactivity measurement, calibration sources should be traceable to the National Institute of Standards and Technology.

4.4 Accuracy may be affected due to self-absorption of activity on the filter. This is dependent on the type of filter material used and the dust-loading on the filter. For α-measurements, the use of membrane-type filters is recommended.

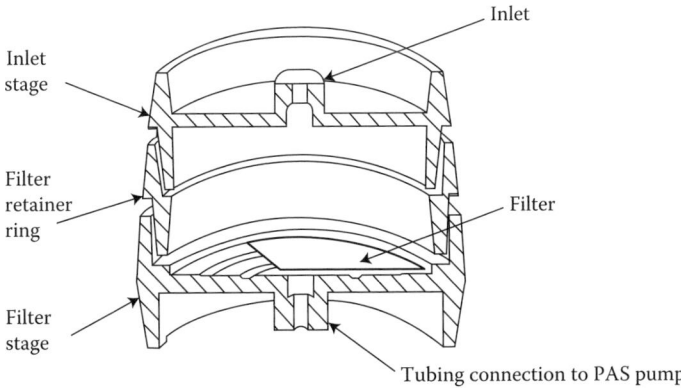

FIGURE 1 Closed-face filter cassette. (After Alvarez, J.L. and Rich, B.L. *Role of Personal Air Sampling in Radiation Safety Programs and Results of a Laboratory Evaluation of Personal Air-Sampling Equipment.* NUREG/CR-4033, National Technical Information Service, Springfield, VA, 1984.)

5 APPARATUS

5.1 A personal air sampler consists of a filter cassette connected via flexible plastic tubing to a small battery-operated air sampling pump that may or may not have a built-in rotameter. Personal air samplers are commercially available from a number of vendors.*

5.2 PAS pumps come with a variety of features. Pumps that provide flow regulation with a minimum flow rate of 3–4 L min^{-1} are recommended. When purchasing pumps, one should ensure that the batteries are capable of operating the pump for the expected duration of a worker's shift.

5.3 Figure 1 illustrates the typical configuration of a closed-face air filter cassette. Air filters used for PAS should have a high collection efficiency (>99%) for particles over a wide range of particle sizes. Glass fiber or cellulose ester membrane filters are recommended. When sampling for α or weak β-emitting radionuclides, membrane filters are superior to glass fiber filters.

6 REAGENTS

6.1 None.

7 PROCEDURE

7.1 Prior to issuing a PAS pump to an individual, verify that the battery is charged. Then, the flow rate should be established using a primary calibration device such as a bubble burette. Commercial bubble-type primary standard airflow calibrators are available that automatically calculate pump flow rates. Figure 2 provides a diagram of a simple setup for the flow rate calibration of a PAS pump. Although many PAS pumps have built-in rotameters, experience has shown that these devices may not be accurate after long-term use. Mass flow meters that detect the mass of air passing over a sensor may also be used to establish the pump flow rate; however, these devices are not considered primary calibration standards. Therefore, they should be periodically checked against a primary standard airflow calibrator such as a bubble burette.

7.2 At the beginning of work activity, the air filter cassette should be attached as close to the nose and mouth of the worker as possible. The filter cassette should be fixed to the lapel using a clip.

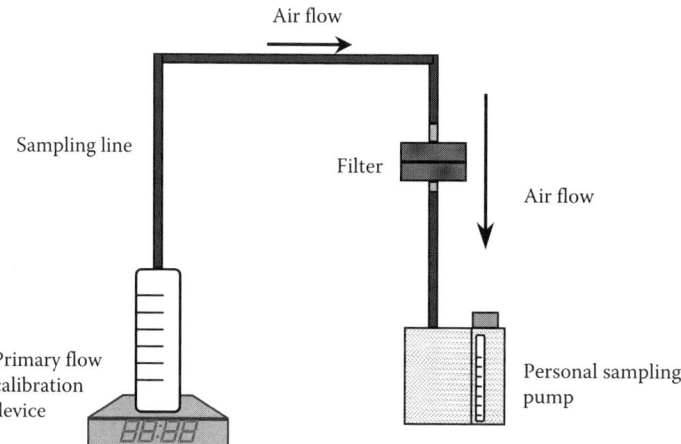

FIGURE 2 Setup for the calibration of a PAS pump. (After Alvarez, J.L. and Rich, B.L. *Role of Personal Air Sampling in Radiation Safety Programs and Results of a Laboratory Evaluation of Personal Air-Sampling Equipment.* NUREG/CR-4033, National Technical Information Service, Springfield, VA, 1984.)

---

* For example, Bladewerx, LLC, SKC, Inc., or Sensidyne Inc. (all US companies).

The flexible plastic tubing should be draped over the shoulder and connected to the pump which is typically attached to a worker's belt. The pump should be turned on and the starting time recorded. Workers should be instructed to use care to avoid disturbing the sample cassette.

7.3 After the work activity has been completed, the pump should be turned off and removed from the worker. The time that the PAS pump is shut off must be recorded, and the flow rate through the loaded filter is measured with the same device used in Section **7.1**. The average of the starting and ending flow rates should be calculated. The cassette should be removed from the tubing and sealed to avoid inadvertent contamination of the filter. Tweezers should be used when removing the filter from the cassette.

7.4 After an appropriate time has elapsed to allow for decay of interfering radon and/or thoron progeny, the filter is measured on an appropriately calibrated radiation detection instrument. The choice of the measurement system is dependent on the radionuclide(s) being measured, the desired sensitivity, and the number of samples being processed. Typical systems used are gas flow proportional counters, γ-spectroscopy systems, α-spectroscopy systems, and liquid scintillation counters.

8 CALIBRATION

8.1 The flow rate of the PAS pumps should be calibrated as indicated in Section **7**.

8.2 The efficiency of detection for the measurement should be established using a radioactive standard that is as close in composition and size to the PAS filter as possible. For the measurement of β-activity, care should be taken to ensure that the energy of the β-emitter used in the standard is close to that of the radionuclide being sampled.

8.3 The background of the detection system should be established prior to performing any sample measurements. Subsequent to this, the background should be periodically measured to ensure that the system has remained stable. To improve the detection ability, the background is typically measured for a longer period of time than that of the samples.

9 CALCULATIONS

9.1 The sample volume in liters (V) is calculated as

$$V = \frac{(F_i + F_f)}{2} \times t_S, \tag{1}$$

where F_i is the initial flow rate (L min^{-1}), F_f is the final flow rate (L min^{-1}), and t_S is the sampling time (min).

9.2 The air sample concentration in Bq m^{-3} (C) is calculated as

$$C = \frac{(R_{\text{Net}} / \text{Eff})}{(V / 1000 \, L \, m^{-3})}, \tag{2}$$

where R_{Net} is the net sample count rate (cps) and Eff is the detection efficiency (cps dps^{-1}).

9.3 The number of DAC-h is calculated as

$$\text{DAC-h} = \frac{C}{\text{Radionuclide Specific DAC (Bq m}^{-3}\text{)}} \times \frac{t_s}{60 \, \text{min} \, h^{-1}}. \tag{3}$$

9.4 The calculation above assumes that the worker is breathing a particle with an activity mean aerodynamic diameter (AMAD) of 1 μm. If the AMAD of the particulate matter in the sampled atmosphere varies considerably from this default value, the DAC should be adjusted according to the methodology described in Publication 30 of the International Commission on Radiation Protection (ICRP, 1979).

9.5 Decay corrections are necessary for radionuclides that have half-lives that are relatively short. This is seldom the case when sampling for radionuclides that have the potential to result in occupationally significant internal doses.

9.6 The statistical uncertainty of a result should be calculated as follows:

$$\text{SD} = \frac{\sqrt{R_S/t_S + R_B/t_B}}{(\text{Eff})(V/1000\,\text{L m}^{-3})}, \quad (4)$$

where SD is the standard deviation (Bq m^{-3}), R_S is the gross sample count rate (cps), R_B is the background count rate (cps), t_S is the sample count time (s), and t_B is the background count time (s).

The SD should be expressed as the uncertainty at a specified confidence level. For example, to express the uncertainty at the 95% confidence level (also referred to as the 1.96Φ or two sigma level), multiply the SD result by 1.96.

9.7 In advance of making measurements, the *a priori* detection capability [lower limit of detection (LLD)] of the system should be established. This will provide an indication if the system is sufficiently sensitive to detect the desired intake. For all measurements, the *a posteriori* decision level (L_c) should be calculated and used to determine the significance of a measured result. Methods for calculating these values have been previously described (Currie, 1968; Brodsky, 1992). A brief discussion of measurement sensitivity can be found in Chapter 5. Also see the text by Martin (2000).

10 EFFECT OF STORAGE

10.1 Under normal conditions, air filters should not be affected by long-term storage. Some radionuclides with short half-lives may decay appreciably if stored for long periods of time relative to their physical half-lives.

11 CAUTIONS

11.1 This method involves the use of radioactivity. It is the user's responsibility to take appropriate safety precautions and to obtain necessary regulatory approvals from institutional and/or governmental agencies.

11.2 If the user is not trained or familiar with the calibration and operation of radiation analysis equipment, they should consult a qualified expert, such as a health physicist, for assistance in the proper and safe usage of the equipment and development of appropriate quality assurance protocols.

REFERENCES

Alvarez, J.L. and Rich, B.L. *Role of Personal Air Sampling in Radiation Safety Programs and Results of a Laboratory Evaluation of Personal Air-Sampling Equipment.* NUREG/CR-4033, National Technical Information Service, Springfield, VA, 1984.

ANSI, *Internal Dosimetry for Mixed Fission and Activation Products*, HPS N13.42-1997, American National Standards Institute, Inc. and Health Physics Society, McLean, VA, 1997.

Brodsky, A., Exact calculations of probabilities of false positives and false negatives, *Health Phys.*, 63, 198–204, 1992.

Currie, L.A., Limits for quantitative detection and quantitative determination, applications to radiochemistry, *Anal. Chem.*, 40, 586–593, 1968.

DOE, *Implementation Guide: Internal Dosimetry Program, G-10 CFR 835/C1 Revision 1*, U.S. Department of Energy, Washington, DC, 1994.

DOE, *Occupational Radiation Protection*, Title 10, Code of Federal Regulation, Part 835 (10 CFR 835), U.S. Department of Energy, Office of Federal Register, Washington, DC, 2007.

EPA, *Limiting Values of Radionuclide Intake and Air Concentration and Dose Conversion Factors for Inhalation, Submersion and Ingestion*, U.S. Environmental Protection Agency, EPA-520/1-88-020, Federal Guidance Report No. 11, September, 1988.

Harley, J.H., Sampling and measurement of airborne daughter products of radon, *Nucleonics*, 11, 12–17, 1953, reprinted in *Health Phys.*, 38, 1068–1074, 1980.

ICRP, *Report of the Task Group on Reference Man*, Publication 23, International Commission on Radiological Protection, Pergamon Press, New York, 1975.

ICRP, *Limits for Intakes of Radionuclides by Workers*, Publication 30, International Commission on Radiological Protection, Pergamon Press, New York, 1979.

Knoll, G.F., *Radiation Detection and Measurement*, 3rd ed., John Wiley & Sons, New York, NY, 2000.

Martin, J.E., *Physics for Radiation Protection*, John Wiley & Sons, New York, USA, 2000.

Metzger, R.L., Jessop, B.H., and McDowell, B.L. *Solubility Testing of Actinides on Breathing-Zone and Area Air Samples*, NUREG/CR-6419, National Technical Information Service, Springfield, VA, 1996.

Moore, M.E., McFarland, A.R., and Rogers, J.C. Factors that affect alpha particle detection on continuous air monitor applications, *Health Phys.*, 65, 69–81, 1993.

NCRP, *A Handbook of Radioactivity Measurements Procedures*, National Council on Radiation Protection and Measurements Report No. 58, National Council on Radiation Protection and Measurements, Bethesda, MD, 1985.

NRC, *Acceptable Concepts, Models, Equations, and Assumptions for a Bioassay Program, Regulatory Guide 8.9, Revision 1*, U.S. Nuclear Regulatory Commission, Washington, DC, 1993.

Passo, C.J. and Cook, G.T. *Handbook of Environmental Liquid Scintillation Spectrometry*, Packard Instrument Company (now Perkin-Elmer), Meriden, CT, 1996.

Method 11

Real-Time Breathing Zone Monitoring for Personal Respiratory Protection*

CONTENTS

1	Principle of the Method	505
2	Sensitivity and Range	506
3	Interferences	506
4	Precision and Accuracy	507
5	Apparatus	507
6	Reagents	507
7	Procedure	507
8	Effect of Storage	508
9	Cautions	508
References		509

1 PRINCIPLE OF THE METHOD

1.1 In particularly hazardous environments where there is potential for the worker to rapidly accumulate significant dose, or in an effort to minimize the time in respiratory protection equipment to reduce the physical stress to the worker, the use of real-time breathing zone monitoring (BZM) is recommended. Personal air sampling (PAS), while providing good sensitivity and estimates of exposure, provides worker protection only in situations of accumulated chronic dose. Delays in the estimate of worker exposure can sometimes extend to 4 days if gross counting methods are used due to the requirement for radon product decay. Even in the case where spectroscopic analysis of the PAS filters are performed and radon subtraction algorithms are utilized allowing results within an hour, dose estimates are not known until after the intake has occurred. In contrast, the BZM samples at the worker's breathing zone and performs analysis of the accumulated filter activity in real time, alerting the worker to the estimated exposure as it occurs. Real-time dose information provides respiratory protection decision metrics such as stay time estimates in order to minimize worker dose. The BZM utilizes a battery-powered pump and sampling head like the PAS, but includes a detector, electronics, and analysis unit to perform spectroscopic analysis, radon subtraction, and dose estimation. The sampling head of a BZM is typically larger than the cassette used with PAS equipment due to the solid-state detector that is mounted above the collection surface of the filter. Detector electronics consist of a preamplifier and a multichannel analyzer (MCA) to quantify energy information for each α-count. Some BZM units are capable of audible reporting of sampling results. Another option, wireless reporting of results to a health physics data collection computer, is often utilized to augment real-time decision-making regarding worker safety.

---

* Method author: David Baltz. Method editor: Mark L. Maiello.

1.2 The analysis software must perform rapid spectroscopic analysis of the accumulating α-spectrum, identifying and accurately subtracting the counts due to radon progeny, while quantifying one or more radioisotopes of concern. Modern methods of spectroscopic analysis, such as alpha peak shape fitting (APSF), are typically used (Bortels and Collaers, 1987) to provide the sensitivity necessary to meet dose limits prescribed by regulatory agencies such as the U.S. Environmental Protection Agency (EPA), the U.S. Nuclear Regulatory Commission, and the U.S. Department of Energy (DOE).

1.3 α-spectroscopy requires peak energy resolution of 500 keV full-width-half-maximum or better to resolve and accurately subtract counts due to radon progeny. Surface barriers or ion-implanted solid-state detectors used in the BZM not only have excellent energy response, but are also desired for their small size and low power requirements.

1.4 The object of this method is the assessment of a worker's intake of inhaled radioactive particulates. Therefore, the measured activity on the air filter must be converted to an intake estimate. This estimate is typically expressed in derived air concentration-hours (DAC-h), which is the product of the fraction of the derived air concentration of the radionuclide in question as measured on the filter and the worker's stay time in the atmosphere. The DAC is defined as that concentration of a radionuclide in air that, if breathed for a work year (2000 h), would result in an intake corresponding to the annual limit on intake (ALI) of the radionuclide. The ALI is the annual intake of a radionuclide that would result in a radiation dose to reference man (International Commission on Radiological Protection, 1975) equal to the most limiting value of committed organ dose or of committed effective dose equivalent.* DAC values are published in ICRP 30 (ICRP, 1979), in EPA Federal Guidance Report No. 11 (EPA, 1988), and in DOE Occupational Radiation Protection Standard, 10 CFR 835 (DOE, 2007).

2 SENSITIVITY AND RANGE

2.1 The sensitivity of the method is dependent on the air sampler flow rate and the sensitivity of the measurement system used to perform the analysis. Current BZM pumps are capable of drawing up to 6 L min^{-1} of air. A 1-in. solid-state detector counting a 1-in. filter in a low radon-222 background is capable of detecting and alarming on approximately 1 DAC-h (3×10^{-6} µCi) of α-activity for a counting "sliding window time" of 4 h (see below). Since sampling is continuously performed, an alarm determination is made at least every minute. A sample collected and measured under these conditions would alarm on an intake approximately equal to the established daily limits of 8 DAC-h. Testing on a BZM at the Lovelace Respiratory Research Institute in 2006 showed an average time-to-alarm of 22 min when sampling approximately 20 times the DAC (about 89 dpm m^{-3}) of a ^{239}Pu aerosol at 2.9 L min^{-1} (Cheng and Holmes, 2006).

2.2 The upper range of a BZM is limited by the performance of the radioactivity detection system employed. The BZM readout may over-range if the activity exceeds more than 500,000 cpm.

3 INTERFERENCES

3.1 Real-time monitoring of α- and/or β-aerosols must utilize APSF algorithms to compensate for interference from the progeny of naturally occurring ^{220}Rn and ^{222}Rn to prevent erroneously elevated results. Since the activity of ^{220}Rn and ^{222}Rn progeny on the filter can be several orders of magnitude above the regulatory limits of the radionuclides of interest, very accurate algorithms for subtraction of the progeny signal are required. Commonly used with APSF algorithms are sliding-window averaging times to quantify minute radionuclide-of-interest activities amidst the predominant radon background. A sliding window algorithm performs spectroscopic analysis on

---

* The ALI is the largest value of intake that results in a committed effective dose equivalent to the whole body of 0.05 Sieverts (5 rem) or a committed dose equivalent to organs or tissues of 0.5 Sieverts (50 rem) (ICRP, 1979).

only the spectrum counts obtained in the last N minutes (or N seconds), rather than analyzing the gross spectrum since the last filter change. This has the effect of reducing the background and responding more quickly to elevated airborne concentrations. In cases where ^{220}Rn progeny are significantly elevated and the radionuclide of interest has a relatively low DAC, for example, ^{239}Pu, a sliding window time of 4 h or more may be necessary.

3.3 The collection of air samples in dusty environments may lead to a buildup of material on the filter that will reduce the detection efficiency of α-activity. Dust loading will significantly degrade the resolution of the spectrum and the sensitivity to radionuclides of interest.

3.4 Poor choice of filter type can also contribute to the interference of a BZM measurement. Surface-deposition-type filters such as PTFE membrane filters are the preferred collection media because of superior spectrum resolution properties-even on clean or lightly loaded filters. When used to monitor for ^{222}Rn and ^{220}Rn progeny in a Working-Level (WL) BZM, a glass fiber filter may be preferable in spite of the degraded spectral resolution. For a good WL measurement, the superior collection efficiency of glass fiber for the short-lived ^{218}Po nanoparticulates can take precedence over resolution.

4 PRECISION AND ACCURACY

4.1 A flow measurement accuracy of about 10% should be obtainable for most commercially available BZM pumps.

4.2 A minimum detectable activity of about 8 DAC-h (45 dpm) for ^{239}Pu should be achievable using a BZM. This assumes a 4-h sliding window time and detection efficiencies of 25% for α-particle radiation. Under these conditions, a BZM should alarm in 1 h or less.

5 APPARATUS

5.1 A breathing zone monitor consists of a sampling head (filter holder–detector–preamplifier) connected via a flexible umbilical to a small case consisting of a processor-Personal Digital Assistant (PDA), an MCA, and a battery-operated air sampling pump. Pumps must be capable of operating for the expected duration of a worker's shift. Breathing zone monitors are commercially available from only one vendor at the time of publication.* Figure 1 illustrates the configuration of a BZM.

5.2 Air filters used for BZM should have high collection efficiency (>99%) over a wide range of particle sizes. PTFE membrane filters are recommended. When sampling for α-transuranics in the presence of ^{220}Rn and ^{222}Rn progeny, membrane filters are superior to glass fiber filters.

6 REAGENTS

6.1 None.

7 PROCEDURE

7.1 Prior to issuing a BZM to an individual, verify that the battery is charged. The pump and the processor should be turned on and the spectrum analysis software should be activated. The user may perform a response check of the detector using an electroplated source that matches the selected radionuclide of interest. The response check verifies that the detector efficiency and energy peak location in the MCA is acceptable. Acceptable limits may be established by the manufacturer and the user to meet specific needs. After passing the response check, the BZM may be issued to the worker.

---

* Bladewerx LLC, USA.

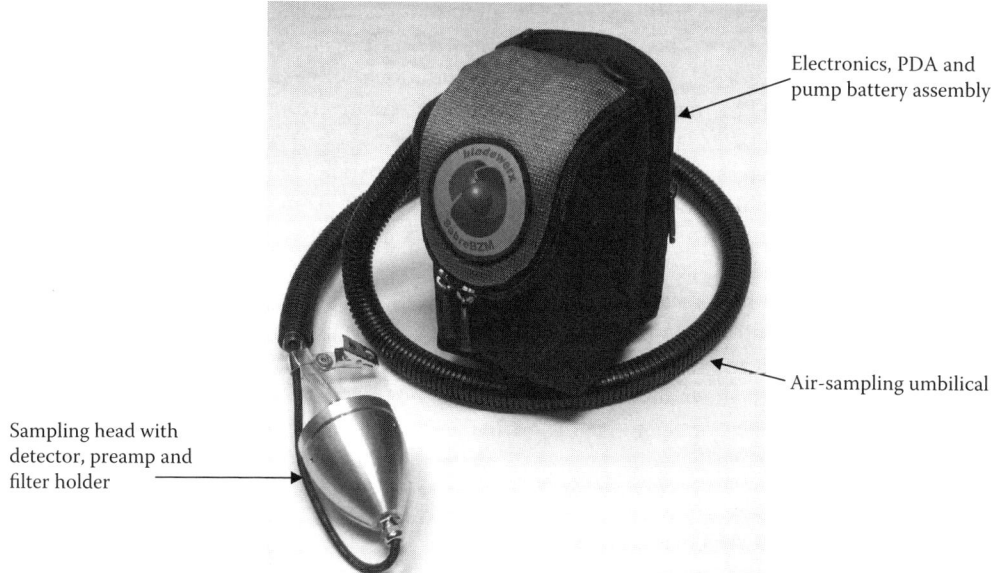

FIGURE 1   A Bladewerx SabreBZM breathing zone monitor.

7.2 At the beginning of work activity, the sampling head should be attached as close to the nose and mouth of the worker as possible. The sampling head should be fixed to the lapel using a clip. The BZM (electronics and pump) is typically attached to a worker's belt and the flexible air sampling umbilical routed behind the head and draped over the shoulder. Workers should be instructed to be careful to avoid disturbing the sampling head.

7.3 During the work period, a BZM may be reporting data wirelessly to radiation safety personnel present outside the work area. In this situation, the worker does not need to rely on the BZM voice output. If external monitoring of the workers by radiation safety personnel is not used, the worker may use a monaural earbud to hear local BZM voice reporting of alarms or dose status, even in noisy work conditions.

7.4 After the work activity has been completed, the BZM should be removed from the worker, the spectrum analysis software should be shut down, and the processor and pump should be turned off. The processor (usually a PDA) contains the data record for the worker and can at this point be uploaded to a data management system for analysis or archival purposes. Forceps should be used when removing the filter from the sampling head. The BZM filters may be discarded or saved for further analysis.

8   EFFECT OF STORAGE

8.1 Breathing zone monitors are battery powered and must be recharged before each use. Program storage is typically nonvolatile; so shelf life of a BZM is indefinite.

8.2 Under normal conditions, air filters should not be affected by long-term storage. Some radionuclides with short half-lives may decay appreciably if stored for long periods of time relative to their physical half-lives.

9   CAUTIONS

9.1 This method involves the use of radioactivity. It is the user's responsibility to take appropriate safety precautions and to obtain necessary regulatory approvals from institutional and/or governmental agencies.

9.2 If the user is not trained or familiar with the calibration and operation of radiation analysis equipment, they should consult a qualified expert, such as a health physicist, for assistance in the proper and safe usage of the equipment and development of appropriate quality assurance protocols.

REFERENCES

Bortels, G. and Collaers, P., Analytical function for fitting peaks in alpha-particle spectra from Si detectors, *International Journal of Radiation Applications and Instrumentation. Part A. Applied Radiation and Isotopes*, 38(10), 1987, 831–837.

Cheng, Y.S. and Holmes, T.D., *Evaluation of Two Alpha Continuous Air Monitors for the Los Alamos National Laboratory*, Lovelace Respiratory Research Institute, Albuquerque, NM, January 2006.

DOE, *Occupational Radiation Protection*, Title 10, *Code of Federal Regulation*, Part 835 (10 CFR 835), U.S. Department of Energy, Office of Federal Register, Washington, DC, 2007.

EPA, *Limiting Values of Radionuclide Intake and Air Concentration and Dose Conversion Factors for Inhalation, Submersion and Ingestion*, U.S. Environmental Protection Agency, EPA-520/1-88-020, Federal Guidance Report No. 11, September, 1988.

ICRP, *Limits for Intakes of Radionuclides by Workers*, Publication 30, International Commission on Radiological Protection, Pergamon Press, New York, 1979.

International Commission on Radiological Protection. Report on the Task Group on Reference Man. ICRP Publication 23, Pergamon Press, Oxford, 1975.

Appendix: Radionuclide Characteristics

Mark L. Maiello

CONTENTS

Selected Radionuclides 511
Abbreviations and Definitions 512
Dosimetry 512
Terminology 512
Example of ALI Calculation 513
References 513
Tritium ($^{3}_{1}$H) 515
Carbon-14 ($^{14}_{6}$C) 517
Argon-41 ($^{41}_{18}$Ar) 519
Krypton-85 ($^{85m}_{36}$Kr and $^{85}_{36}$Kr) 521
Iodine-131 ($^{131}_{53}$I) 523
Xenon-133 ($^{133m}_{54}$Xe and $^{133}_{54}$Xe) 525
Radon-222 ($^{222}_{86}$Rn) and Progeny 528
Polonium-218 ($^{218}_{84}$Po) 530
Lead-214 ($^{214}_{82}$Pb) 532
Bismuth-214 ($^{214}_{83}$Bi) 533
Polonium-214 ($^{214}_{84}$Po) 535
Lead-210 ($^{210}_{82}$Pb) 536
Plutonium ($^{238}_{94}$Pu) 538
Plutonium ($^{239}_{94}$Pu) 540

SELECTED RADIONUCLIDES

The following is a summary of the characteristics, with relevance to air sampling, of the radionuclides mentioned in this book. The summary includes:

Tritium (^{3}H)
Carbon-14 (^{14}C)
Krypton-85 (^{85}Kr)
Argon-41 (^{41}Ar)
Iodine-131 (^{131}I) and other radioiodines
Xenon-133 (^{133}Xe)
Radon-222 (^{222}Rn) and progeny (^{218}Po, ^{214}Pb, ^{214}Bi, ^{214}Po)
Plutonium-238, 239 (^{238}Pu, ^{239}Pu)

ABBREVIATIONS AND DEFINITIONS

The tables below use the following abbreviations and definitions:

- D—the lung clearance class referring to days.
- W—the lung clearance class referring to weeks.
- Y—the lung clearance class referring to years.
- F—a type of particulate aerosol with fast absorption to blood.
- M—a type of particulate aerosol with medium absorption to blood.
- S—a type of particulate aerosol with slow absorption to blood.
- f_1—the gastrointestinal uptake fraction (uptake from the small intestine to the blood).
- Mortality risk coefficient—(see Eckerman, 1999), an estimate of risk to an average member of the U.S. population per unit activity inhaled or per unit time-integrated activity concentration in air or soil for external exposures of *dying* from cancer due to intake or exposure to emitted radiations of the radionuclide in question.
- Morbidity risk coefficient—(see Eckerman, 1999), an estimate of risk to an average member of the U.S. population per unit activity inhaled or per unit time-integrated activity concentration in air or soil for external exposures of *experiencing* a cancer due to intake or exposure to emitted radiations of the radionuclide in question.

Please note that entry headings may differ from radionuclide to radionuclide due to the particular characteristics or chemical form of the radionuclide as inhaled by man.

DOSIMETRY

The Annual Limit on Intake (ALI) and Derived Air Concentration (DAC) values are set using either a stochastic dose limit of 5 rem (0.05 Sv) or a deterministic (organ or tissue) dose limit of 50 rem (0.5 Sv) per year, whichever is more limiting. Both are based on the dosimetry worked out by the International Commission of Radiological Protection (ICRP) in its Publications 26 and 30 (ICRP 1977, ICRP 1979). The exposure-to-dose conversion factor (DCF) values are the product of more recent dosimetry discussed in ICRP Publications 60, 68, and 72 (ICRP 1991, ICRP 1994, ICRP 1996). Thus, more modern values of ALI and DAC can be derived from these DCFs by calculation. However, the older ALI and DAC values are listed below because full conversion of the regulations into the newer ICRP dosimetry has not taken place in all radiation protection regulations, at least in the United States. For example, the U.S. Nuclear Regulatory Commission (NRC) and the agreement states of the United States (those with their own radiation protection regime established by agreement with the NRC), were using the ICRP 26-based ALIs and DACs at the time of publication of this book. ICRP 60 DACs are based on a particle size distribution with an AMAD of 5 µm.

TERMINOLOGY

Annual limits on intake (ALI)—The radioactivity which if internalized by reference man will deliver a dose equal to the most limiting of the following: 5 rem committed dose to the whole body or 50 rem committed dose equivalent to an organ. *Note*: There are separate ALI values for ingestion, immersion, and inhalation of the radioactivity.

Committed dose equivalent (CDE)—A dose equivalent (units of rem or Sieverts) accumulated by a tissue or organ over 50 years due to all sources of radioactive material that have been inhaled, ingested, or otherwise internalized to the body.

Committed effective dose equivalent (CEDE)—A dose equivalent (units of rem or Sieverts) accumulated by the whole body over 50 years due to all sources of radioactive material that have been inhaled, ingested, or otherwise internalized to the body. To calculate the CEDE, tissue-specific

Appendix: Radionuclide Characteristics

weighting factors which indicate the fraction of the overall health risk resulting from uniform whole body irradiation are multiplied with the corresponding individual tissue dose equivalent, then all are summed to determine the CEDE.

Dose conversion factor (DCF)—The dose per unit intake of a radionuclide. The dose may be the committed effective dose or the committed equivalent dose to a tissue.

Derived air concentration (DAC)—The air concentration of a single radionuclide, that if present for 40 h/week over 50 weeks and inhaled at 0.020 m^3 min^{-1} by reference man, will result in the uptake of one ALI.

Effective dose (E)—The sum of the products of equivalent tissue (or organ) doses (H_T) and the risk weighting factors (w_T) for those tissues (or organs). It is expressed mathematically as:

$$E = \sum_T w_T H_T$$

The unit of effective dose is the sievert or rem.

Equivalent dose (H) The product of absorbed dose (D) and the weighting factor for the type of radiation (w_R). The unit of equivalent dose is the sievert or rem.

EXAMPLE OF ALI CALCULATION

For ^{14}C in the form of CO_2, the effective DCF is 2.35×10^{-2} mrem μCi^{-1}. The ICRP 60 ALI can be calculated as:

5000 mrem/2.35×10^{-2} mrem μCi^{-1} = 212,766 μCi.
This is very similar to the ICRP 26 ALI of 2×10^5 μCi.

REFERENCES

Physical Data	Annual Limit on Intake (ALI)	Derived Air Concentration (DAC)	Dose Conversion Factor (DCF)	Mortality and Morbidity Cancer Risk Coefficients
Lederer, 1978	Eckerman, 1988 (FGR No. 11) based on ICRP 26 & 30)	Eckerman, 1988 (FGR No. 11) based on ICRP 26 & 30 and Eckerman, 1999 (FGR No. 13) based on ICRP 60, 68, & 72	Eckerman, 1999 (FGR No. 13) based on ICRP 60, 68, & 72 and Eckerman, 1993 (FGR No. 12) based on ICRP 26 & 30 for submersion data	Eckerman, 1999 (FGR No. 13) based on ICRP 60, 68, & 72

Note: FGR is the abbreviation for Federal Guidance Report (U.S. Environmental Protection Agency). The sources of the DAC values listed below are specified in the tables for each radionuclide as "ICRP 26" or "ICRP 60."

The sources for the table entries are as follows (refer to references below for details):

A Handbook of Radioactivity Measurements Procedures, NCRP No. 58, National Council on Radiation Protection and Measurements, Washington, DC, 1978.

Clarke, R.H., Dunster, J., Nenot, J., Smith, H., and Voeltz, G. The environmental safety and health implications of plutonium. *J. Radiol. Prot.* 16, 91–105, 1996.

Delacroix, D., Guerre, J.P., Leblanc, P., and Hickman, C. *Radionuclide and Radiation Protection Data Handbook 1998, Radiation Protection Dosimetry,* 76, 1–2, 1998.

DOE Handbook—Primer on Tritium Safe Handling Practices, DOE-HDBK-1079-94, U.S. Department of Energy, U.S. Dept. of Commerce, Technology Administration, National Technical Information Service, Springfield, VA 22161, December, 1994.

Eckerman, K.F., Leggett, R.W., Nelson, C.B., Puskin, J.D., and Richardson, A.C.B. *Cancer Risk Coefficients for Environmental Exposure to Radionuclides*, Federal Guidance Report No. 13. Oak Ridge National Laboratory and U.S. Environmental Protection Agency—Office of Radiation and Indoor Air, EPA 402-R-99-001, 1999.

Eckerman, K.F. and Ryman, J.C. *External Exposure to Radionuclides in Air Water and Soil*, Federal Guidance Report No. 12. Oak Ridge National Laboratory and U.S. Environmental Protection Agency—Office of Radiation and Indoor Air, EPA-402-R-93-081, 1993.

Eckerman, K.F., Wolbrast, A.B., and Richardson, A.C.B. *Limiting Values of Radionuclide Intake and Air Concentration and Dose Conversion Factors for Inhalation, Submersion, and Ingestion*, Federal Guidance Report No. 11. Oak Ridge National Laboratory and U.S. Environmental Protection Agency—Office of Radiation Programs, EPA-520/1-88-020, 1988.

Eisenbud, M. *Environmental Radioactivity*, 2nd ed., Academic Press, New York, 1973.

Exposure of the Population in the United States and Canada from Natural Background Radiation, NCRP No. 94, National Council on Radiation Protection and Measurements, Washington, DC, 1987.

Glasstone, S. and Jordan, W.H. *Nuclear Power and Its Environmental Effects*. American Nuclear Society, La Grange Park, IL, USA, 1980.

Harley, J. *Transuranic Data*. Health and Safety Laboratory, U.S. Energy and Research Administration, New York. HASL TM-75-9, November 4, 1975.

International Commission on Radiological Protection, *Recommendations of the International Commission on Radiological Protection*, ICRP Publication 26, *Annals of the ICRP*, Vol. 1, No. 3, Pergamon Press, NY, 1977.

International Commission on Radiological Protection, *Limits for Intake by Workers*, ICRP Publication 30, Part 1, Pergamon Press, Oxford, 1979.

International Commission on Radiological Protection, *Limits for Intake by Workers*, ICRP Publication 30, Part 2, Pergamon Press, Oxford, 1980.

International Commission on Radiological Protection, *Limits for Intake by Workers*, ICRP Publication 30, Part 3, Pergamon Press, Oxford, 1981.

International Commission on Radiological Protection, *Limits for Intake by Workers*, ICRP Publication 30, Part 4, Pergamon Press, Oxford, 1988.

International Commission on Radiological Protection, *1990 Recommendations of the International Commission on Radiological Protection*, ICRP Publication 60, Pergamon Press, Oxford, 1991.

International Commission on Radiological Protection, *Dose Coefficients for Intakes of Radionuclides by Workers*, ICRP Publication 68, Pergamon Press, Oxford, 1994.

International Commission on Radiological Protection, *Age-Dependent Doses to Members of the Public from Intake of Radionuclides, Part 5. Compilation of Ingestion and Inhalation Dose Coeffcients*, ICRP Publication 72, Pergamon Press, Oxford, 1996.

Iodine-129: Evaluation of Releases From Nuclear Power Generation, NCRP No. 75, National Council on Radiation Protection and Measurements, Washington, DC, 1983.

Krypton-85 in the Atmosphere—Accumulation, Biological Significance, and Control Technology, NCRP No. 44, National Council on Radiation Protection and Measurements, Washington, DC, 1975.

Lamarsh, J.R. *Introduction to Nuclear Engineering*. Addison-Wesley Publishing Co., Reading, MA, 1975.

Lederer, C. M. and Shirely, V.S. *Table of Isotopes*, 7th ed. John Wiley & Sons, New York, 1978.

Martin, M.J. and Blichert-Toft, P.H. *Nuclear Data Tables—Section A, Radioactive Atoms, Auger-Electrons, α-, β-, γ-, and X-Ray Data*, Academic Press, New York, 1970.

Measurement of Radon and Radon Daughters in Air, NCRP Report No. 97, National Council on Radiation Protection and Measurements, Bethesda, MD, 1988.

Sutlcliffe, W.G., Condit, R.H., Mansfield, W.G., Myers, D.S., Layton, D.W., and Murphy, P.W. *A Perspective on the Dangers of Plutonium*. Lawrence Livermore National Laboratory Report UCRL-JC-118825, April 1995.

Weast, R.C. and Astle, M.J. *Handbook of Chemistry and Physics*, 62nd ed., CRC Press, Boca Raton, FL, 1981.

Appendix: Radionuclide Characteristics

TRITIUM ($^3_1$H)

Simplified Decay Scheme

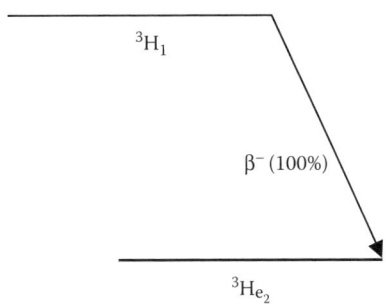

Physical Data

Physical Half-Life	Emission	Maximum β⁻ Emission Energy (MeV)	Average β⁻ Emission Energy (MeV)
12.33 years	100% β⁻	0.018610	0.00568

Biological Data for Inhalation

Form	Biological Half-Life	Annual Limit on Intake (ALI) (MBq), (µCi)		Derived Air Concentration (DAC) (ICRP 26)	
				(MBq m⁻³)	(µCi cm⁻³)
HTO (wv)	10 days	3000	8E+04	0.8	2E−05
HT (g)	–	–	–	2E+04	0.5

Note: wv = water vapor, g = gas. DAC for HT refers to submersion in HT cloud.

	Derived Air Concentration (DAC) (ICRP 60)					
	F		M		S	
Form	(MBq mL⁻¹)	(µCi mL⁻¹)	(MBq mL⁻¹)	(µCi mL⁻¹)	(MBq mL⁻¹)	(µCi mL⁻¹)
HTO (wv)	0.7E+00	2E−05	0.7E+00	2E−05	0.7E+00	2E−05
HT (g)	9E+03	2E−01	9E+03	2E−01	9E+03	2E−01

Exposure to Dose Conversion Factors for Inhalation: f_1 = 1.0, Form = Water Vapor Committed Dose Equivalent per Unit Uptake

Units	Gonad	Breast	Lung	Red Marrow	Bone Surface	Thyroid	Remainder	Effective
Sv Bq⁻¹	1.73E−11	1.73E−11	1.73E−11	1.73E−11	1.73E−11	1.73E−11	1.73E−11	1.73E−11
mrem µCi⁻¹	6.4E−02	6.4E−02	6.4E−02	6.4E−02	6.4E−02	6.4E−02	6.4E−02	6.4E−02

Exposure to Dose Conversion Factors for Submersion Dose Equivalent Rate per Unit Air Concentration

Units	Gonad	Breast	Lung	Red Marrow	Bone Surface	Thyroid	Remainder	Effective
Sv hr^{-1} per Bq m^{-3}	–	–	9.90E–15	–	–	–	–	1.19E–15
mrem h^{-1} per µCi cm^{-3}	–	–	3.66E+01	–	–	–	–	4.40E+00

Mortality and Morbidity Risk Coefficients for Inhalation via Environmental Exposure: $f_1 = 1.0$

Form	Mortality Bq^{-1}	µCi^{-1}	Morbidity Bq^{-1}	µCi^{-1}
HTO (wv)	1.04 E–12	3.85E–08	1.52E–12	5.62E–08
HT (g)	1.04E–16	3.85E–12	1.52E–16	5.62E–12

Sources

Occurs naturally in the environment via interaction of atmospheric gases with cosmic radiation. A typical reaction is

$$^{14}_{7}N + ^{1}_{0}n \rightarrow ^{3}_{1}H + ^{12}_{6}C$$

The natural production rate of tritium is 4×10^6 Ci year^{-1} and the equilibrium natural worldwide inventory is about 70×10^6 Ci. The NCRP reports a production rate of 0.2–0.5 atoms cm^{-2} s^{-1} (NCRP 1987). Tritium will convert into water and reach the surface of the Earth as rain. The residence time in the stratosphere is about 1 year before transfer to the troposphere where it may remain for approximately 30 days.

Man-made sources include nuclear weapons tests. The balance remaining from this source after decay (mostly in ocean water) is about 5×10^8 Ci. The nuclear power and defense industries release 1 to 2×10^6 Ci annually. Nuclear reactors produce 1 to 2×10^4 Ci for every 1000 MW(e) (megawatts of electricity) generated annually as a fission product. Tritium is also produced in reactor coolants. Light water reactors produce 500 to 1000 Ci year^{-1} per 1000 MW(e). Heavy water reactors produce 2×10^6 Ci year^{-1} per 1000 MW(e). Commercial producers of radioluminescent and neutron generator devices release about 1×10^6 Ci year^{-1} (DOE, 1994).

Aerosol Properties

Tritium (often abbreviated T) as the gas (HT) can convert into water vapor (tritium oxide, HTO). This can be accomplished by oxidation:

$$2HT + O_2 \rightarrow 2HTO$$
$$2T_2 + O_2 \rightarrow 2T_2O$$

or by exchange

$$HT + H_2O \rightarrow H_2 + HTO$$
$$T_2 + H_2O \rightarrow HT + HTO$$

Appendix: Radionuclide Characteristics

Special Considerations

The inhalation of tritium gas (HT) results in a small fraction dissolved in the bloodstream before exhalation with CO_2. Conversion to HTO is slight, perhaps only 0.003% of the total inhaled HT. Conversely, up to 99% of inhaled HTO is taken into the bloodstream and into body water within 1 to 2 hours. Immersion in HTO vapor results in about twice as much uptake through the lungs as through the skin (DOE 1994).

CARBON-14 ($^{14}_{6}C$)

Simplified Decay Scheme

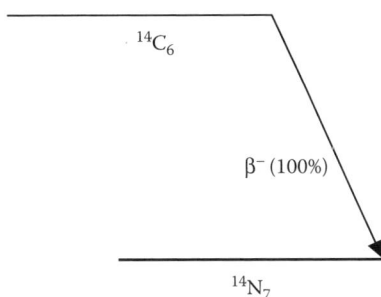

Physical Data

Physical Half-Life	Emission	Maximum β⁻ Emission Energy (MeV)	Average β⁻ Emission Energy (MeV)
5730 y	100% β⁻	0.155	0.0467

Biological Data For Inhalation

	Annual Limit on Intake (ALI)		Derived Air Concentration (DAC) (ICRP 26)	
Form	(MBq)	(μCi)	(MBq m⁻³)	(μCi cm⁻³)
CO	6E+04	2E+06	3E+01	7E–04
CO_2	8E+03	2E+05	3E+00	9E–05
Compounds	9E+01	2000	4E–02	1E–06

	Derived Air Concentration (DAC) (ICRP 60)					
	F		M		S	
Form	(MBq mL⁻¹)	(μCi mL⁻¹)	(MBq mL⁻¹)	(μCi mL⁻¹)	(MBq mL⁻¹)	(μCi mL⁻¹)
C (vapor)	–	–	3E–02	9E–07	–	–
CO	20E+00	7E–04	20E+00	7E–04	20E+00	7E–04
CO_2	3E+00	8E–05	3E+00	8E–05	3E+00	8E–05

Exposure to Dose Conversion Factors for Inhalation: $f_1 = 1.0$, Form = CO
Committed Dose Equivalent per Unit Uptake

Units	Gonad	Breast	Lung	Red Marrow	Bone Surface	Thyroid	Remainder	Effective
Sv Bq^{-1}	7.83E–13	7.83E–13	7.83E–13	7.83E–13	7.83E–13	7.83E–13	7.83E–13	7.83E–13
mrem µCi^{-1}	2.90E–01	2.90E–01	2.90E–01	2.90E–01	2.90E–01	2.90E–01	2.90E–01	2.90E–01

Exposure to Dose Conversion Factors for Inhalation: $f_1 = 1.0$, Form = CO_2
Committed Dose Equivalent per Unit Uptake

Units	Gonad	Breast	Lung	Red Marrow	Bone Surface	Thyroid	Remainder	Effective
Sv Bq^{-1}	6.36E–12	6.36E–12	6.36E–12	6.36E–12	6.36E–12	6.36E–12	6.36E–12	6.36E–12
mrem µCi^{-1}	2.35E–02	2.35E–02	2.35E–02	2.35E–02	2.35E–02	2.35E–02	2.35E–02	2.35E–02

Exposure to Dose Conversion Factors for Inhalation: $f_1 = 1.0$, Form = Organic Compounds
Committed Dose Equivalent per Unit Uptake

Units	Gonad	Breast	Lung	Red Marrow	Bone Surface	Thyroid	Remainder	Effective
Sv Bq^{-1}	5.64E–10	5.64E–10	5.64E–10	5.64E–10	5.64E–10	5.64E–10	5.64E–10	5.64E–10
mrem µCi^{-1}	2.09E+00	2.09E+00	2.09E+00	2.09E+00	2.09E+00	2.09E+00	2.09E+00	2.09E+00

Mortality and Morbidity Risk Coefficients for Inhalation via Environmental Exposure: $f_1 = 1.0$

Form	Mortality Bq^{-1}	Mortality µCi^{-1}	Morbidity Bq^{-1}	Morbidity µCi^{-1}
CO (g)	6.14E–14	2.27E–09	9.09E–14	3.36E–09
CO_2 (g)	3.68E–13	1.36E–08	5.39E–13	1.99E–08
Particulate	1.15E–11	4.26E–07	1.68E–11	6.22E–07

Note: AMAD for particulates = 1 µm.

Sources

Carbon-14 is a product of the interactions of cosmic rays (cosmic nucleons) with nitrogen-14 and oxygen-17 in the atmosphere. These reactions are:

$$^{14}_{7}N + {}^{1}_{0}n \rightarrow {}^{14}_{6}C + {}^{1}_{1}p,$$
$$^{17}_{8}O + {}^{1}_{0}n \rightarrow {}^{14}_{6}C + {}^{4}_{2}He.$$

NCRP No. 94 reports a total atmospheric production rate of 2.5 atoms cm^{-2} s^{-1} resulting in a total inventory of 6.8×10^4 kg. The cosmic ray background concentration is estimated as 200 Bq kg^{-1} (6 pCi g^{-1}) of total carbon.

It is quickly oxidized to CO_2. Since it is a gas, it is not significantly washed out of the atmosphere by precipitation. Instead, it is mixed into the atmosphere where it becomes available for incorporation into biological matter. The ultimate sink for C-14 is the deep ocean with approximately 90% of the global inventory residing there. Only 4% can be found on the land surface with the remaining 6% divided between the upper layer of the ocean, the ocean sediment, the troposphere, and the stratosphere.

Nuclear weapons testing has approximately doubled the atmospheric inventory of C-14. Prior to 1950, the global inventory was reported as 2.2×10^{30} atoms or 230 MCi. By 1962, 6×10^{28} atoms had been added to the atmosphere. This is being transferred to surface ocean water where it resides for up to 8 years before transferring to the thermocline. After a few thousand years it may reach the ocean floor.

A power plant nuclear reactor (light water type of 1000 MWe) produces about 8 Ci year^{-1}. The reaction is the same as that which produces natural C-14 in the atmosphere. Carbon-13, found in nature and also in graphite, captures neutrons to produce C-14 in the reaction

$$^{13}_{6}C + ^{1}_{0}n \rightarrow ^{14}_{6}C + \gamma.$$

Aerosol Properties

Carbon-14 may be oxidized to a gas such as CO or CO_2. It may also be incorporated into organic molecules (see below). At sufficiently high temperatures, these molecules may be in vapor form.

Special Considerations

Carbon-14 is used extensively in biomedical research where it is incorporated into organic compounds. In the course of synthetic chemistry involving such compounds or the precursors or wastes thereof, C-14 may be released to the occupational and environmental air. The organic compound will dictate the biokinetics of the inhaled C-14.

ARGON-41 ($^{41}_{18}$Ar)

Simplified Decay Scheme

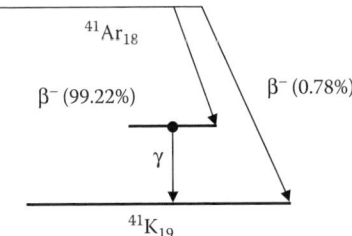

Physical Characteristics

Physical Half-Life	Emission	Maximum β⁻ Emission Energy (MeV)	γ Emission Energy (MeV)
1.827 h	β⁻ γ	0.155	1.29364

Biological Data For Inhalation

	Annual Limit on Intake (ALI)		Derived Air Concentration (DAC) (ICRP 26)	
Form	(MBq)	(μCi)	(MBq m⁻³)	(μCi cm⁻³)
Gas	–	–	0.1	3E–06

Note: DAC refers to submersion in Ar-41 cloud.

	Derived Air Concentration (DAC) (ICRP 60)					
	F		M		S	
Form	(MBq mL⁻¹)	(μCi mL⁻¹)	(MBq mL⁻¹)	(μCi mL⁻¹)	(MBq mL⁻¹)	(μCi mL⁻¹)
Gas	–	–	–	–	–	–

Exposure to Dose Conversion Factors for Submersion Dose Equivalent Rate per Unit Air Concentration

Units	Gonad	Breast	Lung	Red Marrow	Bone Surface	Thyroid	Remainder	Effective
Sv h⁻¹ per Bq m⁻³	1.90E–10	2.32E–10	2.20E–10	2.28E–10	2.47E–10	2.07E–10	2.24E–10	2.17E–10
mrem h⁻¹ per μCi cm⁻³	7.03E+05	8.58E+05	8.14E+05	8.43E+05	9.14E+05	7.66E+05	8.29E+05	8.03E+05

Mortality and Morbidity Risk Coefficients for Submersion in Contaminated Air

	Mortality		Morbidity	
Form	m³Bq⁻¹ s⁻¹	m³μCi⁻¹ s⁻¹	m³Bq⁻¹ s⁻¹	m³μCi⁻¹ s⁻¹
Gas	3.38E–15	1.25E–10	4.96E–15	1.84E–10

Sources

Argon-41 is a neutron activation product produced in nuclear reactors that use air as coolant, or in swimming pool teaching/research reactors by a reaction with Argon-40 (Lamarsh, 1975):

$$^{40}_{18}\text{Ar} + ^{1}_{0}n \rightarrow ^{41}_{18}\text{Ar} + \gamma.$$

Argon-40 is present to the extent of 1.3 weight percent in air, thus providing the neutron activation target for the production of Ar-41.

Properties

Argon is a very inert gas and is not known to form true chemical compounds. Like other noble gases, it can be trapped in organic compounds called clathrates. It is soluble in nonpolar solvents. The solubility increases with decreasing temperature (NCRP, 1975).

Appendix: Radionuclide Characteristics

Special Considerations

The short half-life of Ar-41 will require consideration during radioactive air sampling. Decay correction during sampling and analysis will be necessary.

KRYPTON-85 ($^{85m}_{36}$Kr AND $^{85}_{36}$Kr)

Simplified Decay Scheme

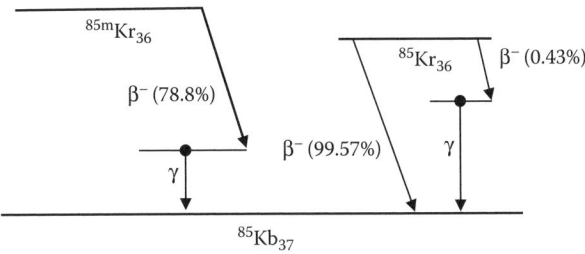

$^{85m}_{36}Kr$

Physical Half-Life	Maximum β⁻ Emission Energy (MeV)	γ Emission Energy (MeV)
4.48 h	0.840	0.151187

Biological Data for Inhalation

	Annual Limit on Intake (ALI)		Derived Air Concentration (DAC) (ICRP 26)	
Form	(MBq)	(μCi)	(MBq m⁻³)	(μCi cm⁻³)
Gas	–	–	0.8	2E–05

Note: DAC refers to submersion in Kr-85m cloud.

	Derived Air Concentration (DAC) (ICRP 60)					
	F		M		S	
Form	(MBq mL⁻¹)	(μCi mL⁻¹)	(MBq mL⁻¹)	(μCi mL⁻¹)	(MBq mL⁻¹)	(μCi mL⁻¹)
Gas	–	–	–	–	–	–

Exposure to Dose Conversion Factors for Submersion Dose Equivalent Rate per Unit Air Concentration

Units	Gonad	Breast	Lung	Red Marrow	Bone Surface	Thyroid	Remainder	Effective
Sv h⁻¹ per Bq m⁻³	3.35E–11	2.66E–11	2.57E–11	4.43E–11	4.72E–11	2.95E–11	2.25E–11	2.98E–11
mrem h⁻¹ per μCi cm⁻³	1.24E+05	9.84E+04	9.51E+04	1.64E+05	1.73E+05	1.09E+05	8.33E+04	1.10E+05

Mortality and Morbidity Risk Coefficients for Submersion in Contaminated Air

	Mortality		Morbidity	
Form	m³Bq⁻¹ s⁻¹	m³µCi⁻¹ s⁻¹	m³Bq⁻¹ s⁻¹	m³µCi⁻¹ s⁻¹
Gas	3.61E–16	1.33E–11	5.33E–16	1.97E–11

$^{85}_{36}Kr$

Physical Half-Life	Maximum β⁻ Emission Energy (MeV)	γ Emission Energy (MeV)
10.701 year	0.672	0.514

Biological Data for Inhalation

	Annual Limit on Intake (ALI)		Derived Air Concentration (DAC) (ICRP 26)	
Form	(MBq)	(µCi)	(MBq m⁻³)	(µCi cm⁻³)
Gas	–	–	5	1E–04

Note: DAC refers to submersion in Kr-85 cloud.

	Derived Air Concentration (DAC) (ICRP 60)					
	F		M		S	
Form	(MBq mL⁻¹)	(µCi mL⁻¹)	(MBq mL⁻¹)	(µCi mL⁻¹)	(MBq mL⁻¹)	(µCi mL⁻¹)
Gas	–	–	–	–	–	–

Exposure to Dose Conversion Factors for Submersion
Dose Equivalent Rate per Unit Air Concentration

Units	Gonad	Breast	Lung	Red Marrow	Bone Surface	Thyroid	Remainder	Effective
Sv hr⁻¹ per Bq m⁻³	5.18E–13	4.52E–13	4.31E–13	5.75E–13	6.15E–13	2.50E–13	4.20E–13 4.66E–11 →	4.70E–13 Skin
mrem h⁻¹ per µCi cm⁻³	1.92E+03	1.67E+03	1.59E+03	2.13E+03	2.28E+03	9.25E+02	1.55E+03 1.72E+05 →	1.74E+03 Skin

Mortality and Morbidity Risk Coefficients for Submersion in Contaminated Air

	Mortality		Morbidity	
Form	m³Bq⁻¹ s⁻¹	m³µCi⁻¹ s⁻¹	m³Bq⁻¹ s⁻¹	m³µCi⁻¹ s⁻¹
Gas	7.23E–18	2.67E–13	1.00E–17	3.70E–13

Sources

Kr-85 and Kr-85m are produced in nuclear reactors. In a reactor operated for 1000 MW for 1000 days, the reactor core will have 1.13×10^7 Ci of Kr-85m and 0.0390×10^7 Ci of Kr-85 (Lamarsh, 1975).

Kr-85 accumulation in the atmospheres was of more concern in the 1970s because a major manmade source was thought to be nuclear fuel reprocessing. Nuclear reactors were predicted to be only minor sources by comparison. Skin cancers, perhaps exacerbated by exposure to ultraviolet radiation, were of concern (NCRP, 1975).

The long half-life of Kr-85, its nonreactivity with other atmospheric constituents and its insolubility in water mean that the gas can accumulate in the atmosphere. It will mix with stable Kr and be dispersed from its point of origin. It is not found to any significant degree in plants or animals (Glasstone, 1980).

Two other isotopes of Krypton are also produced in nuclear reactors: Krypton-87 and Krypton-88. Both are of very short half-life.

Another source of Kr-85 is the fission process of nuclear weapons.

Kr-85 is produced naturally in the atmosphere by neutron interactions with natural, stable Kr-84. The total environmental inventory from this source is estimated at 10 Ci (NCRP 1975). It can also be produced by neutron induced and spontaneous fission of natural uranium. The equilibrium inventory in the first 3 m of soil and water is 2 Ci.

Properties

Krypton can be trapped in organic compounds called clathrates. It, like other noble gases, is soluble in nonpolar solvents. The solubility increases with decreasing temperature (NCRP, 1975).

Special Considerations

The dose to the body from inhaled Kr-85 and Kr-85m will be to the lungs. Some dissolution in body fluids and tissues is expected (NCRP, 1975). In the case of submersion in air containing radioactive krypton, the γ-ray emissions ensure that other organs and tissues can receive radiation doses. These are less important than the dose to the skin due to submersion.

IODINE-131 ($^{131}_{53}$I)

Simplified Decay Scheme

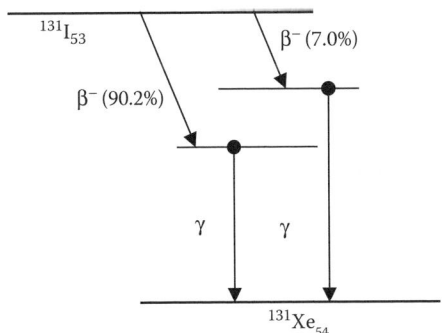

Physical Half-Life	Maximum β⁻ Emission Energy (MeV)	Principal γ Emission Energy (MeV)
8.040 d	0.6065	γ_1 0.364480
		γ_2 0.636973

Biological Data for Inhalation, Class D

	Annual Limit on Intake (ALI)		Derived Air Concentration (DAC) (ICRP 26)	
Form	(MBq)	(µCi)	(MBq m^{-3})	(µCi cm^{-3})
Particulate	2	50	7E–04	2E–08

	Derived Air Concentration (DAC) (ICRP 60)					
	F		M		S	
Form	(MBq mL^{-1})	(µCi mL^{-1})	(MBq mL^{-1})	(µCi mL^{-1})	(MBq mL^{-1})	(µCi mL^{-1})
Methyl	6E–04	1E–08	–	–	–	–
Vapor	–	–	6E–04	1E–08	–	–
Particulate	9E–04	2E–08	–	–	–	–

Exposure to Dose Conversion Factors for Inhalation: $f_1 = 1.0$, Committed Dose Equivalent per Unit Uptake

Units	Gonad	Breast	Lung	Red Marrow	Bone Surface	Thyroid	Remainder	Effective
Sv h^{-1} per Bq m^{-3}	2.53E–11	7.88E–11	6.57E–10	6.26E–11	5.73E–11	2.92E–07	8.03E–11	8.89E–09
mrem h^{-1} per µCi cm^{-3}	9.36E–02	2.92E–01	2.43E+00	2.32E–01	2.12E–01	1.08E+03	2.97E–01	3.29E+01

Mortality and Morbidity Risk Coefficients for Inhalation via Environmental Exposure

	Mortality		Morbidity	
Form	Bq^{-1}	µCi^{-1}	Bq^{-1}	µCi^{-1}
Vapor	1.48E–10	5.48E–06	1.36E–09	5.03E–05
Methyl iodide	1.10E–10	4.07E–06	1.06E–09	3.92E–05
Particulate $f_1 = 1$, F	5.55E–11	2.05E–06	5.27E–10	1.95E–05
Particulate $f_1 = 0.1$, M	1.29E–10	4.77E–06	2.20E–10	8.14E–06
Particulate $f_1 = 0.01$, S	1.40E–10	5.18E–06	1.69E–10	6.25E–06

Note: AMAD for particulates = 1 µm. F, M, and S refer to fast, medium, and slow absorption to the blood.

Mortality and Morbidity Risk Coefficients for Inhalation $f_1 = 1.0$

	Mortality		Morbidity	
Form	Bq^{-1}	µCi^{-1}	Bq^{-1}	µCi^{-1}
Particulate	5.55E–11	2.05E–06	5.27E–10	1.95E–05

Note: AMAD for particulates = 1 µm.

Appendix: Radionuclide Characteristics

Mortality and Morbidity Risk Coefficients for Submersion in Contaminated Air

Mortality		Morbidity	
m³Bq⁻¹ s⁻¹	m³μCi⁻¹ s⁻¹	m³Bq⁻¹ s⁻¹	m³μCi⁻¹ s⁻¹
9.14E−16	3.38E−11	1.35E−15	5.00E−11

Sources

There are at least 23 radioactive isotopes of iodine. Only one is stable: I-127. Only one radioactive isotope is naturally occurring: I-129 ($t_{1/2} = 1.57 \times 10^7$ years; this is the longest half-life of the radioiodines). Therefore, the sources of radioiodines such as I-131 (and including I-129) are nuclear reactors and nuclear weapons (NCRP, 1983).

Approximately 2.34×10^7 Ci of I-131 are produced in the core of a 1000 MW nuclear reactor after 1000 days of operation (Larmarsh, 1977).

Aerosol Properties

Radioiodines are released from reactors in elemental form and as organic compounds such as methyl iodide. The latter are formed from radiation influenced reactions of iodine with organic compounds in lubricating oils and greases, the methane in air or the carbon in steel (Glasstone, 1980). Iodine is a solid at room temperature, but it readily vaporizes.

Special Considerations

Elemental I-131 will deposit relatively rapidly, as compared to organic forms, on vegetation and ground surfaces. Therefore, the elemental form will most likely be ingested while the organic form, suspended for longer periods, will more probably be inhaled. Some inhalation is expected of both forms. In either case, the critical organ is expected to be the thyroid gland.

XENON-133 ($^{133m}_{54}$Xe AND $^{133}_{54}$Xe)

Simplified Decay Scheme:

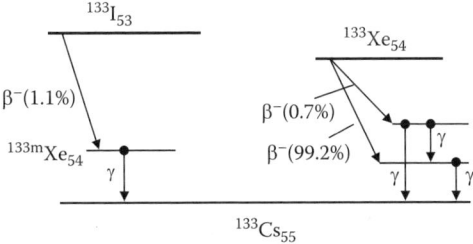

$^{133m}_{54}$Xe

Physical Half-Life	γ Emission Energy (MeV)
2.19 d	0.23324

Biological Data for Inhalation

	Annual Limit on Intake (ALI)		Derived Air Concentration (DAC)	
Form	(MBq)	(µCi)	(MBq m^{-3})	(µCi cm^{-3})
Gas	–	–	5	1E–04

Note: DAC refers to submersion in Xe-133m cloud.

	Derived Air Concentration (DAC) (ICRP 60)					
	F		M		S	
Form	(MBq mL^{-1})	(µCi mL^{-1})	(MBq mL^{-1})	(µCi mL^{-1})	(MBq mL^{-1})	(µCi mL^{-1})
Gas	–	–	–	–	–	–

Exposure to Dose Conversion Factors for Submersion Dose Equivalent Rate per Unit Air Concentration

Units	Gonad	Breast	Lung	Red Marrow	Bone Surface	Thyroid	Remainder	Effective
Sv h^{-1} per Bq m^{-3}	6.80E–12	4.88E–12	4.33E–13	7.37E–12	7.95E–12	4.89E–12	3.84E–12	5.38E–12
mrem h^{-1} per µCi cm^{-3}	2.52E+04	1.81E+04	1.60E+03	2.73E+04	2.94E+04	1.81E+04	1.42E+04	1.99E+04

Mortality and Morbidity Risk Coefficients for Submersion in Contaminated Air

	Mortality		Morbidity	
Form	m^{3}Bq^{-1} s^{-1}	m^{3}µCi^{-1} s^{-1}	m^{3}Bq^{-1} s^{-1}	m^{3}µCi^{-1} s^{-1}
Gas	6.30E–17	2.33E–12	9.34E–17	3.46E–12

$^{133}_{54}$Xe

Physical Half-Life	Principal γ Emission Energy (MeV)
5.245 d	γ$_1$ 0.07963
	γ$_2$ 0.08100
	γ$_3$ 0.16063

Appendix: Radionuclide Characteristics   527

Biological Data for Inhalation

	Annual Limit on Intake (ALI)		Derived Air Concentration (DAC) (ICRP 26)	
Form	(MBq)	(μCi)	(MBq m^{-3})	(μCi cm^{-3})
Gas	–	–	4	1E–04

Note: DAC refers to submersion in Xe-133 cloud.

	Derived Air Concentration (DAC) (ICRP 60)					
	F		M		S	
Form	(MBq mL^{-1})	(μCi mL^{-1})	(MBq mL^{-1})	(μCi mL^{-1})	(MBq mL^{-1})	(μCi mL^{-1})
Gas	–	–	–	–	–	–

Exposure to Dose Conversion Factors for Submersion Dose Equivalent Rate per Unit Air Concentration

Units	Gonad	Breast	Lung	Red Marrow	Bone Surface	Thyroid	Remainder	Effective
Sv h^{-1} per Bq m^{-3}	6.30E–12	5.62E–12	4.84E–12	1.08E–11	1.18E–11	7.12E–12	4.03E–12	6.07E–12
mrem h^{-1} per μCi cm^{-3}	2.33E+04	2.08E+04	1.79E+04	4.00E+04	4.37E+04	2.63E+04	1.49E+04	2.25E+04

Mortality and Morbidity Risk Coefficients for Submersion in Contaminated Air

	Mortality		Morbidity	
Form	m^{3}Bq^{-1} s^{-1}	m^{3}μCi^{-1} s^{-1}	m^{3}Bq^{-1} s^{-1}	m^{3}μCi^{-1} s^{-1}
Gas	6.59E–17	2.44E–12	9.86E–17	3.65E–12

Sources

Xenon isotopes are fission products; therefore, the sources are nuclear reactors and nuclear weapons.

Aerosol Properties

None. It is a noble gas capable, under certain conditions, of forming compounds such as a hydrate, deuterate and tetrafluoride form.

Special Considerations

Xenon is not toxic but the compounds it forms can be. They are strong oxidizers.

RADON-222 ($^{222}_{86}$Rn) AND PROGENY

Simplified Decay Scheme of Important Members of Chain

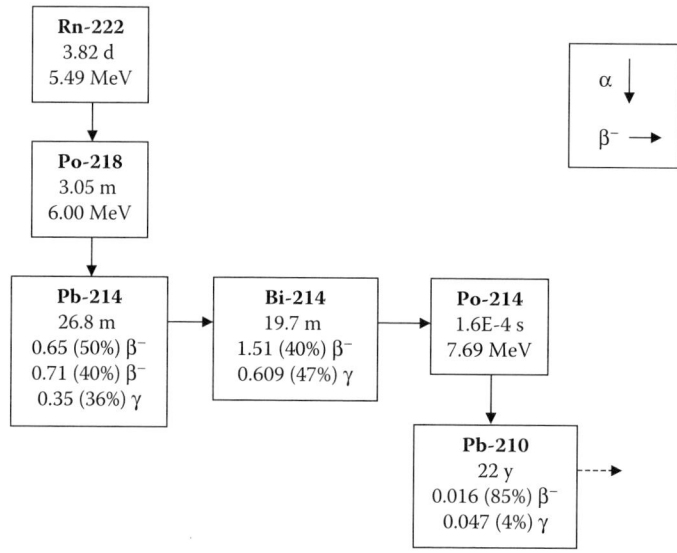

Radon-222 ($^{222}_{86}$Rn)

Physical Half-Life	α Emission Energy (MeV)
3.82351 days	5.48966

Biological Data for Inhalation

		Annual Limit on Intake (ALI)	Derived Air Concentration (DAC)	
Form			(MBq m^{-3})	(μCi cm^{-3})
Gas		4 WLM 4 WLM	–	–

Note: 4 WLM refers to the decay products of Rn-222.

	Derived Air Concentration (DAC) (ICRP 60)					
	F		M		S	
Form	(MBq mL^{-1})	(μCi mL^{-1})	(MBq mL^{-1})	(μCi mL^{-1})	(MBq mL^{-1})	(μCi mL^{-1})
Gas	3E–03	8E–08	–	–	–	–

Appendix: Radionuclide Characteristics

Exposure to Dose Conversion Factors for Submersion Dose Equivalent Rate per Unit Air Concentration

Units	Gonad	Breast	Lung	Red Marrow	Bone Surface	Thyroid	Remainder	Effective
Sv h^{-1} per Bq m^{-3}	6.73E–14	7.70E–14	6.70E–14	6.44E–14	1.19E–13	6.84E–14	6.37E–14	6.88E–14
mrem h^{-1} per µCi cm^{-3}	2.49E+02	2.85E+02	2.48E+02	2.38E+02	4.39E+02	2.53E+02	2.36E+02	2.55E+02

The WLM unit is used to for Rn-222 dosimetry (in the United States). NCRP No. 94 reports the following dose factors (dose to the bronchial epithelial cells) for radon and its decay products for a lifetime of exposure: 0.5 and 0.6 rad WLM^{-1} for males and females respectively. The estimated total effective dose equivalent from naturally occurring inhaled radionuclides (dominated by Rn-222 and progeny) for a person living in the United States or Canada is estimated at 200 mrem year^{-1} to the lung (NCRP, 1987).

Mortality and Morbidity Risk Coefficients for Submersion in Contaminated Air

Form	Mortality		Morbidity	
	m^{3}Bq^{-1} s^{-1}	m^{3}µCi^{-1} s^{-1}	m^{3}Bq^{-1} s^{-1}	m^{3}µCi^{-1} s^{-1}
Gas	9.67E–19	3.58E–14	1.42E–18	5.25E–14

Sources

Radon-222 is naturally produced from the direct decay of Radium-226. The ultimate source of Rn-222 is Uranium-238 which is the parent radionuclide of the "uranium series" of radioactive elements. About 0.1 MeV of recoil energy is available to the Rn-222 atom from Ra-226 decay. If the Rn-222 atom is located at the periphery of a soil grain, this energy may be sufficient to eject it into the pore space between grains. The presence of water in the pores actually enhances the emanation power of Rn-222 by stopping the recoiling atoms in the pore space. Transport through the soil is controlled by molecular/atomic diffusion and the effects of pressure induced flow. The mean flux of Rn-222 from soil is about 17 mBq m^{-2} s^{-1} (0.45 pCi m^{-2} s^{-1}). In the air, Rn-222 is transported by the turbulence of the atmosphere and the vertical and horizontal forces of the wind. Thus, there is a diurnal cycling of Rn-222 outdoor air concentrations with higher levels measured at night and early morning (stable atmosphere conditions) and lower levels during the day (more turbulent conditions due to solar heating). Average concentrations 3.3 ft above the ground (1 m) are about 4–15 Bq m^{-3} (0.1–0.4 pCi L^{-1}).

Other sources of Rn-222 are related to mining operations, for example, the production of phosphate residues and uranium mill tailings may concentrate sources of Rn-222 into specific areas. Rn-222 is also an occupational hazard inside uranium mines (NCRP 1988).

Aerosol Properties

None. Radon-222 is considered an inert noble gas. It is soluble in water.

Special Considerations

Rn-222 does not impart a significant radiation dose; the Rn-222 α-emitting progeny are much more potent. However, measurement of Rn-222 is an acceptable means of dose assessment since the Rn-222/progeny ratio can usually be estimated accurately for environmental situations. The use of Working Levels (WL) as a unit of exposure refers to radon progeny and is used in occupational assessments of dose.

The diurnal and seasonal variations of Rn-222 require that mean concentrations be measured (or predicted by models) so that meaningful dosimetry is obtained. Soil permeability also affects Rn-222 atmospheric concentrations because it influences emanation from the soil. Course soils are permeable; mud and clay tend to be impermeable (NCRP, 1987).

Entry into buildings is a factor of the parent concentration of Ra-226 in the ground, the rapidity of movement of Rn-222 through the building's subsurface walls and floors relative to its half-life, the presence of openings in the basement walls and floors, and the presence of a negative pressure inside the building relative to the soil air (NCRP, 1987).

POLONIUM-218 ($^{218}_{84}$Po)

Physical Half-Life	α Emission Energy (MeV)
3.05 m	6.00255

Biological Data for Inhalation

Form	Annual Limit on Intake (ALI)	Derived Air Concentration (DAC)	
		(MBq m^{-3})	(µCi cm^{-3})
Particulate	4 WLM 4 WLM	–	–

	Derived Air Concentration (DAC) (ICRP 60)					
	F		M		S	
Form	(MBq mL^{-1})	(µCi mL^{-1})	(MBq mL^{-1})	(µCi mL^{-1})	(MBq mL^{-1})	(µCi mL^{-1})
Particulate	–	–	–	–	–	–

Exposure to Dose Conversion Factors for Submersion Dose Equivalent Rate per Unit Air Concentration

Units	Gonad	Breast	Lung	Red Marrow	Bone Surface	Thyroid	Remainder	Effective
Sv h^{-1} per Bq m^{-3}	1.58E–15	1.79E–15	1.57E–15	1.54E–15	2.45E–15	1.62E–15	1.51E–15	1.61E–15
mrem h^{-1} per µCi cm^{-3}	5.85	6.62	5.83	5.69	9.07	6.01	5.58	5.98

Mortality and Morbidity Risk Coefficients for Inhalation

	Mortality		Morbidity	
Form	$m^3 Bq^{-1} s^{-1}$	$m^3 \mu Ci^{-1} s^{-1}$	$m^3 Bq^{-1} s^{-1}$	$m^3 \mu Ci^{-1} s^{-1}$
Particulate	–	–	–	–

Note: AMAD for particulates = 1 μm.

Sources

Po-218 arises from the decay of Rn-222. The ultimate source of Rn-222 is Uranium-238 which is the parent radionuclide of the "uranium series" of radioactive elements.

Aerosol Properties

Radioactive decay of Rn-222 results in the formation of a positively charged ion of Po-218. The recoil of the Po-218 caused by the α-emission will strip away the excess electrons left over after decay and one or two orbital electrons. At birth, over 90% of the Po-218 atoms are positively charged. The ion can react with oxygen to form PoO_2. It will neutralize and adsorb to atmospheric constituents such as water vapor. Attachment to atmospheric particles soon follows. Since there is a continual renewal of Po-218 atoms from decay of Rn-222, a minor fraction of the Po-218 atoms in the atmosphere will be unattached while the majority will be attached to aerosols.

The unattached fraction is of small diameter (2–20 nm), which supports rapid diffusion to surfaces (plateout) including the cells of the lung. This is the primary removal mechanism for the unattached fraction. Attached Po-218 is removed from the air primarily by radioactive decay (both are also removed by local ventilation). Before decay, the behavior of the attached fraction is controlled by the characteristics of the ambient aerosol. The characteristics of the ambient aerosol are the result of humidification, space heating, cooking, and other activities. Electrostatics also effect aerosol behavior. Some room surfaces can become charged and attract oppositely charged aerosols. Investigations indicate that attached Po-218 can regain a positive charge resulting in its loss from the atmosphere by plate out.

Interest in the unattached fraction of Po-218 and subsequent progeny in the chain, is due to the rapid diffusion that they exhibit. Since diffusion is an efficient mechanism of deposition in the lung, the unattached fraction can deliver more radiation dose to epithelial cells than the attached fraction, even though the air concentration of unattached atoms may be 1/10 that of attached atoms. Thus, measurements of the unattached fraction improve estimates of lung dosimetry. Measurements are difficult because the ion changes charge, chemical form and size immediately after formation (NCRP, 1988).

Special Considerations

The energy deposited by the α-particle emissions of Po-218 and Po-214 are of prime consideration to lung dosimetry. Therefore, special units have been devised that are based on the energy emitted by a particular air concentration of Rn-222 progeny (Po-218, Pb-214, Bi-214, Po-214). These units are the working level (WL), working level month (WLM), and the equilibrium equivalent concentration (EEC).

Working Level (WL): The WL is any combination of short-lived Rn-222 progeny that result in the ultimate emission of 1.3E + 05 MeV of potential α-energy.

$$WL = 1.05E{-}03(A) + 5.16E{-}03(B) + 3.79E{-}03(C)$$

where A, B, and C = the concentrations of Po-218, Pb-214, and Bi-214 in pCi L^{-1}.

$$WL = 2.8E{-}05(A) + 1.4E{-}04(B) + 1.0E{-}04(C)$$

where A, B, and C = the concentrations of Po-218, Pb-214, and Bi-214 in Bq m^{-3}.

A related unit is the Working Level Month (WLM) which is an exposure of 1 WL for a working month of 170 h. Thus, the formula for WLM is:

$$WLM = WL\left(\frac{\text{hours exposed}}{170}\right).$$

For radiation protection purposes, the WL and WLM are usually preferred over the EEC.

Equilibrium Equivalent Concentration (EEC): If a measurement of Rn-222 progeny is obtained, one can calculate Rn-222 concentration (in equilibrium with progeny nuclides) that would release the equivalent potential α-particle energy per unit volume of air as the measured mixture. This is the EEC.

$$EEC = 0.105(A) + 0.516(B) + 0.379(C)$$

where A, B, and C = the concentrations of Po-218, Pb-214 and Bi-214 in pCi L^{-1} or Bq m^{-3}.

$$EEC \text{ (in pCi L}^{-1}) = 100(WL)$$

$$EEC \text{ (in Bq m}^{-3}) = 3700(WL).$$

LEAD-214 ($^{214}_{82}$Pb)

Physical Half-Life	Principal β-Emission Energy (MeV)	γ-Emission Energy (MeV)
26.8 m	0.73	Multiple emissions 0.2–0.8

Biological Data for Inhalation, Class D

	Annual Limit on Intake (ALI)		Derived Air Concentration (DAC)	
Form	(MBq)	(µCi)	(MBq m^{-3})	(µCi cm^{-3})
Particulate	30	800	0.01	3E–07

	Derived Air Concentration (DAC) (ICRP 60)					
	F		M		S	
Form	(MBq mL^{-1})	(µCi mL^{-1})	(MBq mL^{-1})	(µCi mL^{-1})	(MBq mL^{-1})	(µCi mL^{-1})
Particulate	1E–03	4E–08	–	–	–	–

Appendix: Radionuclide Characteristics

*Exposure to Dose Conversion Factors for Inhalation: $f_1 = 0.2$,
Dose Equivalent Rate per Unit Air Concentration*

Units	Gonad	Breast	Lung	Red Marrow	Bone Surface	Thyroid	Remainder	Effective
Sv Bq^{-1}	1.63E–10	1.62E–10	1.49E–08	4.63E–10	3.88E–09	1.62E–10	2.62E–10	2.11E–09
mrem µCi^{-1}	6.03E–01	5.99E–01	5.51E+01	1.71E+00	1.44E+01	5.99E–01	9.69E–01	7.81E+00

Mortality and Morbidity Risk Coefficients for Inhalation

	Mortality		Morbidity	
Form, Type	Bq^{-1}	µCi^{-1}	Bq^{-1}	µCi^{-1}
Particulate, F, $f_1 = 0.2$	1.13E–10	4.18E–06	1.24E–10	4.59E–06
Particulate, M, $f_1 = 0.1$	9.31E–10	3.44E–05	9.81E–10	3.63E–05
Particulate, S, $f_1 = 0.01$	1.02E–09	3.77E–05	1.08E–09	4.00E–05

Note: AMAD for particulates = 1 µm.

Mortality and Morbidity Risk Coefficients for Submersion in Contaminated Air

	Mortality		Morbidity	
Form	m^{3}Bq^{-1} s^{-1}	m^{3}µCi^{-1} s^{-1}	m^{3}Bq^{-1} s^{-1}	m^{3}µCi^{-1} s^{-1}
Particulate	5.85E–16	2.16E–11	8.62E–16	3.19E–11

Sources

Pb-214 arises from the decay of Po-218 which is a decay product of Rn-222. The ultimate source of Rn-222 is Uranium-238 which is the parent radionuclide of the "uranium series" of radioactive elements.

Aerosol Properties

The parent Po-218, if attached to an aerosol, may cause the daughter Pb-214 to detach due to the recoil energy of α-particle decay. The Pb-214 may then follow the pattern of growth and attachment observed for Po-218.

Special Considerations
See entry for Po-218.

BISMUTH-214 ($^{214}_{83}$Bi)

Physical Half-Life	Principal β-Emission Energies (MeV)	γ-Emission Energy (MeV)
19.7 m	1.51 (40%)	Multiple emissions
	1.02 (23%)	0.5–2.1
	1.88 (9%)	

Biological Data for Inhalation, Class D and W

	Annual Limit on Intake (ALI)		Derived Air Concentration (DAC)	
Form	(MBq)	(μCi)	(MBq m^{-3})	(μCi cm^{-3})
Particulate	30	800	0.01	3E–07

	Derived Air Concentration (DAC) (ICRP 60)					
	F		M		S	
Form	(MBq mL^{-1})	(μCi mL^{-1})	(MBq mL^{-1})	(μCi mL^{-1})	(MBq mL^{-1})	(μCi mL^{-1})
Particulate	6E–04	1E–08	4E–04	1E–08	–	–

Exposure to Dose Conversion Factors for Inhalation: $f_1 = 0.05$,
Dose Equivalent Rate per Unit Air Concentration

Units	Class	Gonad	Breast	Lung	Red Marrow	Bone Surface	Thyroid	Remainder	Effective
Sv Bq^{-1}	D	5.08E–11	5.10E–11	1.22E–08	5.10E–11	5.08E–11	5.07E–11	9.43E–10	1.78E–09
mrem μCi^{-1}	D	1.88E–01	1.80E–01	4.51E+01	1.89E–01	1.88E–01	1.88E–01	3.49E+00	6.59E+00

Exposure to Dose Conversion Factors for Inhalation: $f_1 = 0.05$,
Dose Equivalent Rate per Unit Air Concentration

Units	Class	Gonad	Breast	Lung	Red Marrow	Bone Surface	Thyroid	Remainder	Effective
Sv Bq^{-1}	W	1.51E–11	1.55E–11	1.32E–08	1.55E–11	1.54E–11	1.54E–11	2.77E–10	1.68E–09
mrem μCi^{-1}	W	5.59E–02	5.74E–02	4.88E+01	5.74E–02	5.70E–02	5.70E–02	1.02E+00	6.22E+00

Mortality and Morbidity Risk Coefficients for Inhalation $f_1 = 0.05$

	Mortality		Morbidity	
Form, Type	Bq^{-1}	μCi^{-1}	Bq^{-1}	μCi^{-1}
Particulate, F	2.86E–10	1.06E–05	3.05E–10	1.13E–05
Particulate, M	7.45E–10	2.76E–05	7.84E–10	2.90E–05
Particulate, S	7.96E–10	2.95E–05	8.38E–10	3.10E–05

Note: AMAD for particulates = 1 μm.

Appendix: Radionuclide Characteristics

Mortality and Morbidity Risk Coefficients for Submersion in Contaminated Air

	Mortality		Morbidity	
Form	m³Bq⁻¹ s⁻¹	m³μCi⁻¹ s⁻¹	m³Bq⁻¹ s⁻¹	m³μCi⁻¹ s⁻¹
Particulate	3.98E–15	1.47E–10	5.85E–15	2.16E–10

Sources

Bi-214 arises from the β–γ decay of Pb-214. Po-218, a decay product of Rn-222, gives rise to Pb-214. The ultimate source of Rn-222 is Uranium-238 which is the parent radionuclide of the "uranium series" of radioactive elements.

Aerosol Properties

Unlike Pb-214 which is the product of α-emitting Po-218, the β–γ energies involved in the production of Bi-214 are too weak to cause detachment of Bi-214 from an aerosol.

Special Considerations

See entry for Po-218.

POLONIUM-214 ($^{214}_{84}$Po)

Physical Half-Life	α Emission Energy (MeV)
163.72 μs	7.6871

Biological Data for Inhalation

	Annual Limit on Intake (ALI)		Derived Air Concentration (DAC)	
Form	(MBq)	(μCi)	(MBq m⁻³)	(μCi cm⁻³)
Particulate	–	–	–	–

	Derived Air Concentration (DAC) (ICRP 60)					
	F		M		S	
Form	(MBq mL⁻¹)	(μCi mL⁻¹)	(MBq mL⁻¹)	(μCi mL⁻¹)	(MBq mL⁻¹)	(μCi mL⁻¹)
Particulate	–	–	–	–	–	–

Exposure to Dose Conversion Factors for Submersion Dose Equivalent Rate per Unit Air Concentration

Units	Gonad	Breast	Lung	Red Marrow	Bone Surface	Thyroid	Remainder	Effective
Sv h⁻¹ per Bq m⁻³	1.44E–14	1.63E–14	1.43E–14	1.40E–14	2.26E–14	1.47E–14	1.37E–14	1.47E–14
mrem h⁻¹ per μCi cm⁻³	5.33E+01	6.05E+01	5.29E+01	5.18E+01	8.35E+01	5.44E+01	5.07E+01	5.44E+01

Mortality and Morbidity Risk Coefficients for Inhalation

	Mortality		Morbidity	
Form	Bq⁻¹	µCi⁻¹	Bq⁻¹	µCi⁻¹
Particulate	–	–	–	–

Note: AMAD for particulates = 1 µm.

Mortality and Morbidity Risk Coefficients for Submersion in Contaminated Air

	Mortality		Morbidity	
Form	Bq⁻¹	µCi⁻¹	Bq⁻¹	µCi⁻¹
Particulate	2.09E–19	7.73E–15	3.07E–19	1.14E–14

Sources

Po-214 arises from the β-γ decay Bi-214. Bi-214 is the decay product of Pb-214 which is derived from the decay of Po-218. The latter follows from the decay of Rn-222. The ultimate source of Rn-222 is Uranium-238 which is the parent radionuclide of the "uranium series" of radioactive elements.

Aerosol Properties

Unlike Pb-214 which is the product of α-emitting Po-218, the β-γ energies involved in the production of Po-214 are too weak to cause detachment of Po-214 from an aerosol.

Special Considerations

See entry for Po-218.

LEAD-210 ($^{210}_{82}$Pb)

Physical Half-Life	β⁻Emission Energy (MeV)	γ Emission Energy (MeV)
22.26 y	0.015	0.0465

Biological Data for Inhalation, Class D

	Annual Limit on Intake (ALI)		Derived Air Concentration (DAC)	
Form	(MBq)	(µCi)	(MBq m⁻³)	(µCi cm⁻³)
Particulate	0.009	0.2	4E–06	1E–10

	Derived Air Concentration (DAC) (ICRP 60)					
	F		M		S	
Form	(MBq mL⁻¹)	(µCi mL⁻¹)	(MBq mL⁻¹)	(µCi mL⁻¹)	(MBq mL⁻¹)	(µCi mL⁻¹)
Particulate	5E–06	1E–10	–	–	–	–

Appendix: Radionuclide Characteristics

Exposure to Dose Conversion Factors for Inhalation: $f_1 = 0.2$, Dose Equivalent Rate per Unit Air Concentration

Units	Gonad	Breast	Lung	Red Marrow	Bone Surface	Thyroid	Remainder	Effective
Sv per Bq	3.18E−07	3.18E−07	3.18E−07	3.75E−06	5.47E−05	3.18E−07	4.69E−06	3.67E−06
mrem per µCi	1.18E+03	1.18E+03	1.18E+03	1.39E+04	2.02E+05	1.18E+03	1.74E+04	1.36E+04

Mortality and Morbidity Risk Coefficients for Inhalation

	Mortality		Morbidity	
Form, Type	Bq^{-1}	µCi^{-1}	Bq^{-1}	µCi^{-1}
Particulate, F, $f_1 = 0.2$	1.82E−08	6.73E−04	2.47E−08	9.14E−04
Particulate, M, $f_1 = 0.1$	6.84E−08	2.53E−03	7.48E−08	2.77E−03
Particulate, S, $f_1 = 0.01$	4.06E−07	1.50E−02	4.28E−07	1.58E−02

Note: AMAD for particulates = 1 µm.

Mortality and Morbidity Risk Coefficients for Submersion in Contaminated Air

	Mortality		Morbidity	
Form	m^{3}Bq^{-1} s^{-1}	m^{3}µCi^{-1} s^{-1}	m^{3}Bq^{-1} s^{-1}	m^{3}µCi^{-1} s^{-1}
Particulate	5.85E−16	2.16E−11	8.62E−16	3.19E−11

Sources

Pb-210 arises from the decay of Rn-222 which produces four intermediate progeny of short half-life in succession (Po-218, Pb-214, Bi-214, and Po-214), before yielding Pb-210. The ultimate source of Rn-222 is Uranium-238 which is the parent radionuclide of the "uranium series" of radioactive elements.

Aerosol Properties

The α-emission of Po-214 may free some Pb-210 from aerosol particles. Due to the long half-life of Pb-210, this is of minor dosimetric significance (NCRP, 1988).

Special Considerations

See entry for Po-218.

PLUTONIUM ($^{238}_{94}$Pu)

Simplified Decay Scheme

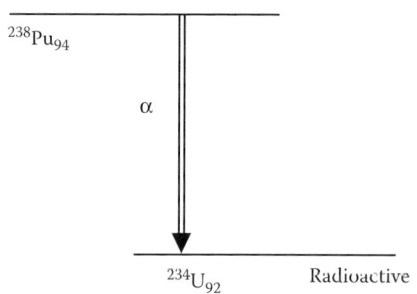

Physical Half-Life	α Emission Energies (MeV)
87.71 years	5.498 (71.1%)
	5.454 (28.7%)

Biological Data for Inhalation

			Annual Limit on Intake (ALI)		Derived Air Concentration (DAC)	
Form	Class	f_1	(MBq)	(μCi)	(MBq m^{-3})	(μCi cm^{-3})
Particulate	W	1E–03	3E–04	7E–03	1E–07	3E–12
Particulate	Y	1E–05	7E–04	2E–02	3E–07	8E–12

	Derived Air Concentration (DAC) (ICRP 60)					
	F		M		S	
Form	(MBq mL^{-1})	(μCi mL^{-1})	(MBq mL^{-1})	(μCi mL^{-1})	(MBq mL^{-1})	(μCi mL^{-1})
Particulate	–	–	2E–07	6E–12	1E–06	5E–11

Exposure to Dose Conversion Factors for Inhalation: f_1 as Indicated
Dose Equivalent Rate per Unit Air Concentration

Units	f_1	Gonad	Breast	Lung	Red Marrow	Bone Surface	Thyroid	Remainder	Effective
Sv Bq^{-1}	1E–03	2.80E–05	1.00E–09	1.84E–05	1.52E–04	1.90E–03	9.62E–10	7.02E–05	1.06E–04
mrem μCi^{-1}	1E–03	1.04E+05	3.70E+00	6.81E+04	5.62E+05	7.03E+06	3.56E+00	2.60E+05	3.92E+05
Sv Bq^{-1}	1E–05	1.04E–05	4.40E–10	3.20E–04	5.80E–05	7.25E–04	3.86E–10	2.74E–05	7.79E–05
mrem μCi^{-1}	1E–05	3.85E+04	1.63E+00	1.18E+06	2.15E+05	2.68E+06	1.43E+00	1.01E+05	2.88E+05

Appendix: Radionuclide Characteristics

Mortality and Morbidity Risk Coefficients for Inhalation

Form, Type	Mortality		Morbidity	
	Bq^{-1}	µCi^{-1}	Bq^{-1}	µCi^{-1}
Particulate, F, $f_1 = 0.0005$	1.19E–06	4.40E–02	1.41E–06	5.22E–02
Particulate, M, $f_1 = 0.0005$	8.04E–07	2.97E–02	9.07E–07	3.36E–02
Particulate, S, $f_1 = 0.00001$	9.06E–07	3.35E–02	9.60E–07	3.55E–02

Mortality and Morbidity Risk Coefficients for Submersion in Contaminated Air

Form	Mortality		Morbidity	
	m^{3}Bq^{-1} s^{-1}	m^{3}µCi^{-1} s^{-1}	m^{3}Bq^{-1} s^{-1}	m^{3}µCi^{-1} s^{-1}
Particulate	1.34E–19	4.96E–15	2.28E–19	8.44E–15

Sources

Plutonium isotope 238 was artificially produced in 1940 by deuteron bombardment of uranium inside the 60-inch cyclotron in Berkeley, California (Weast and Astle, 1981).

It is also produced by neutron irradiation of Neptunium-237, abbreviated symbolically as

$$^{237}\text{Np (n,}\beta\text{) }^{238}\text{Pu}$$

in which a β-particle is released in the process. In nuclear reactors, formation can occur as:

$$^{238}\text{U (n,2n) }^{237}\text{U} \xrightarrow{\beta} {}^{237}\text{Np (n,}\gamma\text{)} \rightarrow {}^{238}\text{Np} \xrightarrow{\beta} {}^{238}\text{Pu}.$$

Plutonium produced for nuclear weapons is typically 0.01% Pu-238 by weight (2.3% by Pu radioactivity) and 93.3% by weight Pu-239 (79.2% by activity). The activity ratio ^{238}Pu/^{239,240}Pu is 0.023. The ratio in nuclear weapons debris found in stratospheric air samples has been measured at 0.03 (Harley, 1975).

Isotope-238 is a minor constituent of nuclear weapons fallout, but plutonium was dispersed world wide from the testing of nuclear weapons in the 1950s and 1960s (Eisenbud, 1973). Pu-238 has been used as a power source in unmanned spacecraft. The failure of the SNAP-9A satellite in 1964 distributed 17 kCi of Pu-238, mostly in the Southern hemisphere (Harley, 1975).

Aerosol Properties

Plutonium is metallic and dense (specific gravity of isotope 239 is about 19.8). Pu-238 is a member of the actinide series of rare earth elements. It forms insoluble fluorides, hydroxides, and oxides. It can be found in the environment adsorbed to soils in the oxide form (PuO_2) (Eisenbud, 1973). It is usually stored in bulk as the oxide but for military purposes, may be stored as the metal (Clarke et al., 1995). The metal oxidizes in air.

Special Considerations

Plutonium-238 undergoes spontaneous fission. One gram possesses a fission rate of 1100 fissions s^{-1}. The specific activity of Pu-238 is 17.4 Ci g^{-1} (Harley, 1975).

Inhalation is the major pathway into human tissue (Clarke et al., 1995). Plutonium is hazardous, but not as acutely hazardous as some common chemicals. About 20 mg would need to be inhaled to produce fatal pulmonary edema or fibrosis in approximately a month's time. Inhaling such a high amount is considered to be unrealistic. To acquire just 0.1 mg, over 700,000 particles of the correct diameter (<3 μm) would have to be inhaled (Sutlcliffe et al. 1995).

Resuspension of a fraction of the Pu that has settled out of the air can be caused by wind or human activity.

Chronic exposure to small amounts of Pu should not be ignored. Once inhaled into the deep lung, Pu may migrate to the liver and bones where it can continue to deliver an α-particle radiation dose. About 33% of the Pu entering the bloodstream from the lungs may go to the liver and 50% to the bones. Retention half-times are 20 and 50 years respectively (Clarke et al. 1995). The likelihood of lung, bone, or liver cancer is increased by these exposures (see Sutlcliffe et al., 1995 for risk calculations).

PLUTONIUM ($^{239}_{94}$Pu)

Simplified Decay Scheme

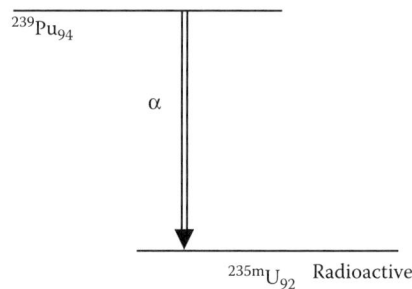

Physical Half-Life	α Emission Energies (MeV)
2.4131E+04 years	5.1554 (73.3%)
	5.1429 (15.1%)
	5.1046 (11.5%)

Biological Data For Inhalation

			Annual Limit on Intake (ALI)		Derived Air Concentration (DAC)	
Form	Class	f_1	(MBq)	(μCi)	(MBq m^{-3})	(μCi cm^{-3})
Particulate	W	1E–03	2E–04	6E–03	1E–07	3E–12
Particulate	Y	1E–05	6E–04	2E–02	3E–07	7E–12

	Derived Air Concentration (DAC) (ICRP 60)					
	F		M		S	
Form	(MBq mL^{-1})	(μCi mL^{-1})	(MBq mL^{-1})	(μCi mL^{-1})	(MBq mL^{-1})	(μCi mL^{-1})
Particulate	–	–	2E–07	5E–12	2E–06	6E–11

Appendix: Radionuclide Characteristics

Exposure to Dose Conversion Factors for Inhalation: f_1 as Indicated
Dose Equivalent Rate per Unit Air Concentration

Units	f_1	Gonad	Breast	Lung	Red Marrow	Bone Surface	Thyroid	Remainder	Effective
Sv Bq^{-1}	1E–03	3.18E–05	9.22E–10	1.73E–05	1.69E–04	2.11E–03	9.03E–10	7.56E–05	1.16E–04
mrem µCi^{-1}	1E–03	1.18E+05	3.41+00	6.40E+04	6.25E+05	7.81E+06	3.34E+00	2.80E+05	4.29E+05
Sv Bq^{-1}	1E–05	1.20E–05	3.99E–10	3.23E–04	6.57E–05	8.21E–04	3.75E–10	3.02E–05	8.33E–05
mrem µCi^{-1}	1E–05	4.44E+04	1.48E+00	1.20E+06	2.43E+05	3.04E+06	1.39E+00	1.12E+05	3.08E+05

Mortality and Morbidity Risk Coefficients for Inhalation

	Mortality		Morbidity	
Form, Type	Bq^{-1}	µCi^{-1}	Bq^{-1}	µCi^{-1}
Particulate, F, $f_1 = 0.0005$	1.26E–06	4.66E–02	1.49E–06	5.51E–02
Particulate, M, $f_1 = 0.0005$	7.94E–07	2.94E–02	8.99E–07	3.33E–02
Particulate, S, $f_1 = 0.00001$	8.45E–07	3.13E–02	8.96E–07	3.32E–02

Mortality and Morbidity Risk Coefficients for Submersion in Contaminated Air

	Mortality		Morbidity	
Form	m^{3}Bq^{-1} s^{-1}	m^{3}µCi^{-1} s^{-1}	m^{3}Bq^{-1} s^{-1}	m^{3}µCi^{-1} s^{-1}
Particulate	1.65E–19	6.11E–15	2.56E–19	9.47E–15

Sources

Plutonium exists in trace quantities in naturally occurring uranium ores. Isotope 239 can be formed from natural U-238 via neutron interaction:

$$^{238}U(n,\gamma)^{239}U \xrightarrow{\beta} {}^{239}Np \xrightarrow{\beta} {}^{239}Pu$$

in which U-239 is formed with the emission of γ-radiation followed by β-decay to Neptunium-239 which decays with β-particle emission to Pu-239 (Eisenbud, 1973).

Pu-239 is produced in nuclear reactors by neutron irradiation of Uranium-238. It is used in nuclear weapons. It can be employed as fuel in "fast" nuclear reactors (Eisenbud, 1973).

Plutonium produced for nuclear weapons is typically 0.01% Pu-238 by weight (2.3% by Pu radioactivity) and 93.3% by weight Pu-239 (79.2% by activity). The activity ratio ^{238}Pu/^{239,240}Pu is 0.023. The ratio in nuclear weapons debris found in stratospheric air samples has been measured at 0.03 (Harley, 1975).

Aerosol Properties

Plutonium is metallic and dense (specific gravity of isotope 239 is about 19.8). Plutonium is normally found in the oxide form (PuO$_2$). It is usually stored in bulk as the oxide but for military purposes, may be stored as the metal (Clarke et al., 1995). The metal oxidizes in air.

Special Considerations

Plutonium-239 undergoes spontaneous fission. One gram possesses a fission rate of 0.01 fissions s^{-1}. The specific activity of Pu-239 is 0.0613 Ci g^{-1} (Harley, 1975).

Inhalation is the major pathway into human tissue (Clarke et al., 1995). Plutonium is hazardous, but not as acutely hazardous as some common chemicals. About 20 mg would need to be inhaled to produce fatal pulmonary edema or fibrosis in approximately a month's time. Inhaling such a high amount is considered to be unrealistic. To acquire just 0.1 mg, over 700,000 particles of the correct diameter (<3 μm) would have to be inhaled (Sutlcliffe et al., 1995).

Resuspension of a fraction of the Pu that has settled out of the air can be caused by wind or human activity.

Chronic exposure to small amounts of Pu should not be ignored. Once inhaled into the deep lung, Pu may migrate to the liver and bones where it can continue to deliver an α-particle radiation dose. About 33% of the Pu entering the bloodstream from the lungs may go to the liver and 50% to the bones. Retention half-times are 20 and 50 years respectively (Clarke et al., 1995). The likelihood of lung, bone, or liver cancer is increased by these exposures (see Sutlcliffe et al., 1995 for risk calculations).

Glossary

Mark L. Maiello and Morgan Cox

A list of scientific terms, radiological terms, and those related to airborne radioactivity measurements follows. The sources for this glossary include the following:

Aerosol Technology by William C. Hinds, John Wiley & Sons, New York, 1982.
American National Standard for Portable Radiation Detection Instrumentation for Homeland Security, Institute of Electrical and Electronic Engineers, New York, 2004.
Concise Dictionary of Atomics by Alfred Del Vecchio, Philosophical Library, New York, 1964.
Exposure of the Population in the United States and Canada from Natural Background Radiation, National Council on Radiation Protection and Measurements, NCRP Report 94, 1987.
Fundamentals of Physics by David Halliday and Robert Resnick, John Wiley & Sons, Inc., New York, 1974.
Handbook of Chemistry and Physics, 62nd ed., CRC Press, Boca Raton, FL, 1981.
International Electrotechnical Vocabulary—Part 393: *Nuclear Instrumentation—Physical Phenomena and Basic Concepts*, IEC-60500-393, 2003.
International Electrotechnical Vocabulary—Part 394: *Nuclear Instrumentation—Instruments, Control Systems and Equipment*, IEC-60500-394, 2006.
Introduction to Health Physics by Herman Cember, 2nd ed., McGraw-Hill, Inc., New York, 1992.
Liquid Scintillation Analysis, Michael J. Kessler, Ed., Publication No. 169-3052, Rev. G Packard Instrument Co., Inc. 1989.
Measurement of Radon and Radon Daughters in Air, National Council on Radiation Protection and Measurements, NCRP Report 97, 1988.
Radiation Protection by William H. Hallenbeck, Lewis Publishers, Boca Raton, FL, 1994.
Sampling and Monitoring Releases of Airborne Radioactive Substances from the Stacks and Ducts of Nuclear Facilities, The Health Physics Society, ANSI N13.1-1999, McLean, VA.
The Effects of Nuclear Weapons, 3rd ed. Samuel Glasstone and Philip J. Dolan, U.S. Energy and Research Development Administration and Department of Defense, 1977.
The Occupational Environment—Its Evaluation and Control by Salvatore DiNardi, Ed., AIHA Press, Fairfax, VA, 1997.

A

Absolute pressure: The actual pressure above a perfect vacuum at a point in a fluid or gas.
Absorption: The interactions include the photoelectric effect, the Compton effect, and pair production.
Acceleration due to gravity (g): The time rate of change of velocity (of a particle) due to gravity: 9.806 m s^{-2} (31.174 ft s^{-2}).
Acceptance test: Evaluation or measurement of performance characteristics of a measurement system to verify that certain stated specifications and contractual requirements are met as agreed between the manufacturer of the system and the purchases/user. Notes: Acceptance testing generally involves performing a suite of tests on the completed system. Each individual test, known as a case, exercises a particular operating condition of the user's

environment or feature of the system, and will result in a pass or fail outcome. It is also known as functional testing, black-box testing, release acceptance, usability testing, QA testing, application testing, confidence testing, final testing, validation testing, or factory acceptance testing.

Accredited calibration laboratory: A calibration laboratory that has been accredited by an authoritative body (e.g., Health Physics Society, American Association of Physicists in Medicine, National Institute of Standards and Technology or NIST), with respect to its qualifications to perform instrument calibrations.

Accuracy: The magnitude of the difference between a measured value and the known or true value usually expressed as percent, that is, [(measured value – true value)/true value] × 100%. Notes: This formula is also known as the "percent error." A measurement is usually stated as being within a percent of the true value, for example, "±5%."

Activity median aerodynamic diameter (AMAD): The diameter of a unit density sphere with the same terminal settling velocity in air as that of an aerosol particle whose activity is the median for the entire aerosol.

Advective flow: In meteorology, the horizontal motion of atmospheric constituents as opposed to convective flow.

Aerodynamic equivalent diameter: The diameter of a unit density spherical particle having the same gravitational settling velocity as the particle of interest.

Aerosol: Solid and/or liquid particles suspended in a gaseous medium.

Air monitoring instrument: A radiation monitor for the continuous measurement of the radioactive emission rate of airborne materials. Note: These are also termed continuous air monitors or CAMs.

Air sampler: A device designed to collect airborne radioactive contamination contained in a known volume of air on a filter, on an impaction surface or into a chamber.

Airborne radioactivity area: A room, enclosure, or area in which airborne radioactive materials, composed wholly or partly of licensed material, exist in concentrations that (1) exceed the derived air concentration (DAC) limits or (2) would result in an individual present in the area without respiratory protection exceeding, during the hours the individual is present, 0.6% of the annual limits on intake (ALI) or 12 DAC-h.

ALARA: As Low as Reasonably Achievable—The objective of radiation protection programs concerning exposure and dose of personnel. Exposures should be ALARA with consideration made for the benefits achieved through the use of radiation, and by taking into account of any related societal and economic factors.

Alpha decay: Radioactive transformation by emission of an α-particle that decreases the number of protons of the radioactive element by two, thus transforming the radioactive element into another. Note: This transformation occurs with massive radionuclides possessing a relatively low ratio of neutrons to protons.

Alpha particle (α): A form of ionizing radiation consisting of two neutrons and two protons. Note: Its relatively large mass (mass = 4.00277 amu) and charge (2e+ or 3.2×10^{-19} Coulombs) causes it to have a short range in air and the least penetrability of the four types of ionizing radiation. The energies of α-particles are in the range of ~4 – ~8 MeV.

Ambient: Pertaining to the uncontrolled conditions of the environment as in ambient air pressure or temperature. Note: The term "ambient monitoring" may refer not only to the sampling of the environment but also to the immediate occupational workplace.

Amplitude: A term commonly associated with electromagnetic radiation; the magnitude or "height" of the wave; more generally the magnitude of an electrical signal such as current or voltage.

Analog-to-digital converter (ADC): An electronic device that can digitize an analog signal from such devices as photomultiplier tubes in preparation for multichannel analysis and/or computer analysis.

Anemometer: An air speed measurement device.

Annual limits on intake (ALI): The radioactivity which if internalized by reference man will deliver a dose equal to the most limiting of the following: 5 rem committed dose to the whole body or 50 rem committed dose to an organ. Note: There are separate ALI values for ingestion, immersion, and inhalation of the radioactivity.

Attached daughters: The short-lived decay products of radon-222 that are attached to the ambient aerosol. Note: These are also currently termed radon progeny.

Atomic mass (A): The integer nearest in magnitude to the atomic weight of an atom.

Atomic mass unit (amu): 1/12 of the mass of the carbon-12 atom = $1.6605402 \times 10^{-24}$; g = 931.478 MeV.

Atomic number (Z): The number of protons in the nucleus of an atom of an element.

Atomic weight (A): The weight of an atom relative to the carbon-12 atom, which is taken to be 12.000000 units. Note: The atomic weight includes the masses and binding energies of the extranuclear (orbiting) electrons.

Attenuation: The decrease in intensity of a signal, beam, or wave as a result of absorption and scattering out of the path of a detector. Notes: It does not include the reduction due to the geometric spreading of the radiation energy, that is, the inverse square law with distance effect. For γ- and x-rays, attenuation is the loss of photons and photon energy during the passage through material(s).

Attenuation, photon: The interactions include the photoelectric effect, the Compton effect, and pair production.

B

Background radiation: Ionizing radiation from natural and man-made sources, excluding the radiation to be measured.

Backpressure: The forces of resistance that a pump must overcome to impart motion to air. Notes: These forces include gravity, pump friction, and the resistance of tubing and media that the air is drawn through. Air density also influences backpressure.

Becquerel (Bq): The SI unit of radioactive decay equivalent to 1 disintegration per second.

Beta decay: Radioactive decay in which an intranuclear neutron is transformed into a proton and an electron. Notes: The emitted electron—the radiation—is called a β-particle. The addition of one proton transforms the radioactive element into another element. This type of decay occurs with radionuclides having a surplus of neutrons.

Beta particle (β): A form of ionizing radiation with mass = 1e (0.00055 amu). Notes: If it is negatively charged (–1e or 1.6×10^{-19} Coulombs) it is a negatron (β^-). If it is positively charged (+1e), it is a positron, (β^+). A radionuclide may decay either by negative β- or positron emission.

Bias: With regard to measurement results, a consistent error as compared to the true result (absolute bias). Note: Relative biases are the differences observed between different results without knowledge of the true result.

Bioassay: The measurement and quantification of radioactive material that has entered the body by sampling urine, fecal matter, and other specimens. Whole body counting for photon-emitting radionuclides is also considered a form of bioassay.

Biological half-life ($t_{1/2\ bio}$): The time required for a biological system, for example, the human body, to eliminate a substance. Note: In the case of a radioactive compound, the biological half-life and the radioactive half-life combine to produce an effective half-life of the substance.

Blank: A sampling medium, for example, a reagent, sampling solution, filter, swab, and so on, treated analytically as a sample to determine a background signal from the analytical system. Note: Subsequent true samples are corrected by the background signal so obtained.

Boyle's law: The relationship of gas pressure and volume for a fixed mass of gas at constant temperature whereby the absolute pressure varies inversely as the volume or that $P_1V_1 = P_2V_2$.

Breakthrough: Sample loss from the failure of a sampling medium, for example, a trap such as a filter or a bubbling solution to capture from the effluent stream or the environment, the hazard of interest.

Breathing zone: A sphere with radius of about 10 in centered at the nose. Note: This volume contains the air inhaled by workers during the course of occupational activities.

Bremsstrahlung: The continuous spectra of radiation (x-rays) produced by the change in trajectory of β-particles when they interact with matter. Note: This is the literal translation from German: braking radiation.

Brownian motion: The irregular motion of an aerosol particle in still air caused by random variations in the continuous impacts of gas molecules with the particle.

Bubble flow meter: A device that is used to calibrate the flow rate of air pumps by measurement of the time required for a "frictionless" soap bubble to traverse through a known volume of a burette while under the suction of the pump. Note: Thus the volume displacement per unit time of the pump at a particular flow rate setting is measured.

C

Calibration: A set of operations that establishes the relation between the values of the quantities indicated by a measuring instrument or system, and the conventionally true value of the quantity or variable measured. Notes: The relationship can be a ratio between the output of the measurement device and the known value of the parameter, for example, a radiation exposure, to which that the instrument was exposed. This ratio is used to translate the instrument response into an accurate reading of the measured parameter. Or, the act of calibration may mean adjustment of the instrument by some means (mechanical or electronic) so that the readings equate to the extent possible with the actual magnitude of the parameter measured. Calibrations are performed with calibration standards whose values can be confidently believed to be accurate.

Calibration standard: A radioactive source having an accepted or reference value for use in calibrating a measurement instrument or system. Also see standard.

Cascade impactor: A device that divides the particle size distribution of an aerosol sample into a series of contiguous groups according to aerodynamic diameter. Note: It accomplishes this by using a series of impaction stages with decreasing particle cut size.

Celsius temperature scale (°C): The Celsius degree has the same magnitude as the Kelvin degree. Notes: The triple point of water (the state where liquid water, ice, and water vapor are in equilibrium) corresponds to 0.01°C. The ice point of water is 0°C, and the steam point is 100°C. To convert to degrees Celsius from degrees Kelvin use °C = K − 273.15°. To convert from degrees Fahrenheit to degrees Celsius use °C = [(°F − 32) × 5]/9. This was formerly known as the Centigrade scale.

Channel: One of many multichannel analyzer memory locations for storage of a count corresponding to a specific level of radiation energy or division of time. Note: Counts are accumulated over time to quantitatively indicate the distribution of the radiation energies.

Channeling: The flow of effluent air or sampled air around a solid capture or drying medium such as silica gel because of loose or otherwise improper packing of the medium in the holding canister.

Charles' law: The relationship for a given mass of gas held at constant pressure whereby the volume is directly proportional to the absolute temperature or $V_1T_1 = V_2T_2$.

Cocktail: A synonym for liquid scintillation counting fluid which captures α- and β-particles emitted by radioactive material and converts the particle kinetic energy into visible light.

Coefficient of variation: The precision of a set of measurements calculated as the standard deviation of the data divided by the mean multiplied by 100%.

Collective dose (units of person-rem or person-Sieverts): The sum of products of a specific dose equivalent and the number of individuals receiving a specific dose equivalent in an exposed population.

Collision diameter: The distance between the centers of two molecules at the instant of their collision. Note: The collision diameter of air is 3.7×10^{-8} cm.

Committed dose equivalent (CDE): A dose equivalent (units of rem or Sieverts) accumulated by a tissue or organ over 50 years due to all sources of radioactive material that have been inhaled, ingested, or otherwise internalized to the body. Note: CDE is usually determined using the methods of internal dosimetry that include computer modeling of the biokinetics of radionuclides.

Committed effective dose equivalent (CEDE): A dose equivalent (units of rem or Sieverts) accumulated by the whole body over 50 years due to all sources of radioactive material that have been inhaled, ingested, or otherwise internalized to the body. Note: To calculate the CEDE, tissue-specific weighting factors that indicate the ratio of cancer risk/rem, are multiplied with the corresponding individual tissue dose equivalent; then all are summed to determine the CEDE.

Condensation nuclei: Any small particle or ion capable of serving as a site for the condensation of vapor.

Contamination: Loose, unconfined, residual radioactive materials (usually the result of sloppy handling practices) in locations where they are not wanted. Note: Radioactivity that can be removed by nondestructive means, for example, washing with a solvent or detergent, is referred to as removable contamination.

Continuous air monitor (CAM): An air sampling system capable of yielding near real-time concentrations of a radioactive material (or other hazardous substance) in the occupational environment, in an effluent stream, or in the environment. Notes: These systems are usually used when a serious hazard to human health exists, or perhaps in laboratories where small changes in air concentrations would significantly effect low-level measurements. They are usually equipped with an alarm and date-recording capability.

Continuous sampling: Uninterrupted sampling for a hazardous air constituent usually over a relatively long and meaningful time period, for example, a period of known release of the hazard or over a period that reveals changes in the concentration of the hazard that may be associated with man-made activities or natural phenomena. Note: As opposed to grab sampling.

Convection: The process of heat transfer by the circulation and movement of gaseous or liquid material from a region of relatively high temperature to a region of lower temperature. In meteorology, convective movement is vertical as opposed to the horizontal movement of advective motion.

Conventionally true value (CTV) of a quantity: The commonly accepted estimate of the value of that quantity. Note: This and the associated uncertainty will preferably be determined by a national or transfer standard, or by a reference instrument that has been calibrated against a national or transfer standard, or by a measurement quality assurance (MQA) interaction with the National Institute of Standards and Technology (NIST) or an accredited calibration laboratory.

Cosmic radiation: The component of naturally occurring radiation that originates in outer space. Notes: Often called "cosmic rays," this radiation is actually composed of protons, α-particles, and nuclei of atomic elements. The sources of cosmic radiation are divided into those of galactic origin and those of solar origin. The latter are associated with solar flares. "Primary" cosmic radiation interacts with the earth's atmosphere to form "secondary" cosmic rays consisting of electrons, γ-rays, neutrons, and mesons. Cosmic radiation is highly penetrating with energies as high as 10^{19} electron volts.

Coulomb: The unit of electric charge. The amount of charge flowing through a wire in 1 s when a current of 1 amp is present. Note: 1 Coulomb is the flow of 6.27×10^{18} electrons in 1 s.

Count: The response, for example, electrical discharge, pulse, or creation of a photon, in a radiation detection device caused by the interaction of ionizing radiation.

Criticality: The condition of a nuclear reactor when the rate of neutron production equals the rate of neutron loss, that is, the "nuclear chain reaction" of fission is sustained and controlled.

Critical mass: The minimum mass of fissile uranium or plutonium isotopes needed to maintain the chain reaction consisting of production of neutrons and the fission of uranium or plutonium atoms by the neutrons. Notes: The critical mass must be in a shape, concentration, and configuration that achieves the self-sustaining condition of criticality. Critical mass is required for nuclear reactor fuel to heat water into steam and is also a requirement for a nuclear weapon to detonate properly.

cpm: The counting rate recorded by a nuclear detector of a sample or standard in units of counts per minute. Note: Other units of time may be used, such as counts per second (cps).

Curie (Ci): A unit of radioactive decay equivalent to 3.7×10^{10} disintegrations per second (dps). Note: For practical use in air measurements, fractions of this unit are often used to express concentrations such as 10^{-6} (μ:Ci) and 10^{-12} (pCi) per unit volume of air.

Cunningham correction factor: A correction factor that accounts for the fact that particles with diameters approaching the mean free path of the gas settle faster than predicted by Stoke's law. Note: Important at standard conditions for particles <1 μm in diameter.

Cutpoint: A term related to the size analysis of an aerosol, or the aerodynamic particle diameter at which 50% of the aerosol particles are collected in a particle sizing device.

D

Dead time: The time that a radiation detection system is processing an input signal caused by a radiation interaction. Notes: During this time, the system is unable to process other signals that may be generated, therefore losing that information. Many radiation counting systems compensate for this by automatically counting for a longer period than was preset by the operator.

Decade: A range of values for which the upper value is a power of 10 above the lower value.

Decay: The process of radioactive transformation whereby a radioactive element transforms by emission of radiation into another element that may or may not be radioactive.

Decay products: Nuclides that are formed by the radioactive decay of some other radionuclide. Notes: In the case of radium-226, there are 10 successive daughters because these nuclides are also radioactive. The decay chain ends in the stable isotope, lead-206. Also referred to as radon progeny or daughter products.

Decontamination: The cleansing or removal of hazardous substances, for example, radioactivity from surfaces and areas where the radioactivity is unwanted.

Deep dose equivalent (DDE): The dose equivalent (in units of rem or Sieverts) obtained from a radiation dose originating external to the body and determined at a tissue depth of 1 cm.

Detection limits: The extremes of detection or quantification. Notes: minimum detection is usually described by two criteria: (1) the measured radioactivity is above the background radioactivity signal (the critical level) and (2) the specified minimum sample count rate of the measurement device that must be present to yield a net count large enough to imply the presence of radioactivity (the lower limit of detection). The upper detection limit is the maximum level at which the instrument meets the required accuracy.

Derived air concentration (DAC): The air concentration of a single radionuclide, which if present for 40 h per week over 50 weeks and inhaled at 0.020 m^3 min^{-1} by reference man, will result in the uptake of one ALI.

Desiccant: A material that can absorb water vapor from the air, such as silica gel.

Detector: A device that produces a measurable current or voltage pulse, or a change in one of these attributes after interaction with ionizing radiation.

Deuterium: The hydrogen isotope with atomic number 1, atomic weight 2, and atomic mass 2.0147(1/2H).

Diffusion: The mass transfer of one gas through another in the absence of fluid flow; the result of the motion of the gas molecules in a concentration gradient; or the random movement of an aerosol particle caused by collisions with individual gas molecules.

Diffusion coefficient: The constant of proportionality relating the mass transfer of aerosol particles or gas molecules through a gas such as air under the influence of a concentration gradient. It can be expressed in the units of cm^2 s^{-1}. It appears in Fick's first law of diffusion as D_{ba}:

$$J = D_{ba} \frac{dC}{dx},$$

where the flux of molecules or particles (mass transfer) is represented by J (molecules or particles/cm^2 s) and the concentration gradient is dC/dx.

Dose: The energy absorbed from any form of radiation in 1 g of any material. Notes: One rad equals 100 ergs of energy absorbed per unit gram of material. The units of dose are rads or the SI units of grays (100 rads = 1 gray). The dose equivalent, given in rem or Sieverts (100 rem = 1 Sievert), is a measure of the biological damage to living tissue from the radiation exposure. To obtain the dose equivalent, the dose is multiplied by factors that account for the effectiveness of the radiation to cause a specific form of damage.

Dose conversion factor (DCF): Committed dose equivalent (CDE) per µCi of intake for a specific radionuclide, tissue or organ, and route of exposure (units of rem µCi^{-1}or Sv Bq^{-1}).

Dose equivalent (H): The product of the absorbed dose in rad multiplied by a radiation weighting factor to account for the biological effectiveness of the radiation. Note: The weighting factor is a function of the type and energy of the radiation and the organ(s) affected.

dpm: The nuclear disintegration rate in disintegrations per minute of a radioactive source. Note: Other units of time may be used, such as disintegrations per second (dps).

Dust: A dry particle aerosol produced by such mechanical processes as grinding, pulverizing, or physical breaking.

Dyne: A unit of force. Note: One dyne imparts an acceleration of 1 cm s^{-2} to a mass of 1 g.

E

Effective center: For a given set of irradiation conditions, the point within a detector where the response is equivalent to that which would be produced if the entire detector were located at the point.

Effective dose (E): The sum of the products of equivalent tissue (or organ) doses and the risk weighting factors (w_T) for those tissues (or organs). It is expressed mathematically as

$$E = \sum_T W_T H_T.$$

The unit of effective dose is Sievert or rem.

Effective half-life ($t_{1/2\,\text{eff}}$): The time to remove 50% of a radionuclide from a living system taking into account the radioactive half-life and the biological half-life of the radioactive substance. Note: In health physics, it is expressed mathematically as

$$t_{1/2\text{eff}} = \frac{t_{1/2} \times t_{1/2\text{bio}}}{t_{1/2} \times t_{1/2\text{bio}}}.$$

Efficiency: The ratio of counts recorded by a detection system to the known number of nuclear disintegrations occurring in a source. Notes: Usually a counting standard of certifiable nuclear disintegration rate is used for determining efficiency. Counting efficiency depends on many factors, including the intrinsic efficiency of the detector, the type of radiation detected, and the geometrical arrangement of the sample in relation to the detector. It is essential to determine the efficiency using the same sample geometry and sample-to-detector geometry to routinely verify its value.

Effluent: Material, solid, liquid, or gas, discharged into the environment.

Electrical mobility : The ability of a particle to move in an electric field, or the velocity of a particle with an electrical charge moving in an electrical field of unit strength as given by $Z = V_{TE}/E$, where V_{TE} is the terminal electrostatic velocity and E is the electric field strength. Note: The units can be cm^2 V^{-1} s^{-1}.

Electron (e$^-$): The fundamental atomic particle with negative charge. Note: Electrons have a rest mass of about 9.1×10^{-28} g. The charge on the electron is 1.60×10^{-19} Coulomb.

Electron capture: Radioactive decay by the capture of an orbital electron by the nucleus. Notes: The electron unites with an intranuclear proton to form a neutron. The decrease in proton number by one transforms the radioactive element to another element with lower atomic number.

Electron volt (eV): The amount of kinetic energy obtained by an electron by passage through a potential difference of 1 V. Note: It is used as a measure of the kinetic energy of all types of radiation in units of kilo-eV (keV) and mega-eV (MeV).

Energy calibration: A procedure used in pulse-height analysis to correlate each channel of a multichannel analyzer to a specific energy.

Energy dependence: Variation in instrument response as a function of radiation energy for a constant radiation type and exposure rate referenced to air. Note: Also termed energy response.

Equilibrium equivalent concentration (EEC): The radon-222 concentration in equilibrium with its short-loved decay products, which has the same potential α-energy per unit volume as exists in a sample mixture. Note: Measured in units of activity per unit volume.

Equilibrium factor (F): The ratio of the potential α-energy of the short-lived progeny of radon-222 to the total potential α-energy of the progeny when in equilibrium with radon-222. Notes: Equilibrium of the progeny with radon-222 ($F = 1$) occurs when the activity concentrations of polonium-218 (Po-218), lead-214 (Pb-214), bismuth-214 (Bi-214), and radon-222 (Rn-222) are equal. If F is unknown, it is usually assigned a value of 0.5. The following formula may be used to compute F by entering the activity concentrations in pCi L^{-1} for the quantities in parentheses:

$$F = \frac{0.105(\text{Po}218) + 0.516(\text{Pb}214) + 0.379(\text{Bi}214)}{\text{Rn}222}.$$

Equivalent dose (H): The product of absorbed dose (D) and the weighting factor for the type of radiation (w_R). The unit of equivalent dose is Sievert or rem.

Erg: A unit of work or energy where 1 erg = 10^{-7}J. Note: An erg is the work done by a force on 1 dyne when its point or application is moved 1 cm in the direction of the applied force.

F

Fahrenheit scale (°F): The temperature scale that uses 32°F for the temperature at which water freezes and 212°F at which water boils. Note: To convert from degrees Celsius to degrees Fahrenheit, use °F = 32°F + 9/5°C.

Fissile: Atoms that can split (fission) spontaneously or when bombarded by neutrons of proper energy. Note: Common fissile materials are Uranium-235 and Plutonium-239.

Fission: The process whereby an atom splits into roughly two equal masses spontaneously or by absorption of a neutron, a photon, or a fast-charged particle of proper energy. Notes: Fission results in the emission of neutrons. If enough neutrons are released and not lost by other mechanisms, a fission chain reaction can be sustained.

Fission products: The result of the fission process includes energy, neutrons, and two atoms. Notes: The latter are termed fission products. These are rarely of equal mass or atomic number. Most range from about atomic number 34–58. Almost all fission products have high kinetic energy and undergo β-decay.

Fixed contamination: Radioactivity that cannot be removed by nondestructive means, for example, washing.

Flow rate: The rate at which a fluid (air) passes a cross-sectional area of a conduit (duct or stack). Note: The rate of mass of the fluid or the rate of the volume of the fluid may be specified for the measurement.

Fluid: A state of matter that flows under pressure, for example, gases and liquids.

Frequency: A term used to describe electromagnetic radiation as the number of completed cycles of the wave per second, that is, the number of times from minimum amplitude through maximum positive amplitude through maximum negative amplitude and back to minimum amplitude in 1 s.

Full-width–at-half-maximum (FWHM): The width of a spectroscopic peak, in number of channels, at one-half of its maximum amplitude. Note: This measurement is done with background radiation subtracted from the peak counts.

Fume: A solid aerosol of small diameter produced by vaporizing a solid.

G

Gain: The number of channels or voltage levels into which the analog to digital converter (ADC) input is divided.

Gamma radiation (γ-radiation): High-energy, short-wavelength electromagnetic radiation emitted from the nucleus of radioactive atoms. Notes: γ-Radiation (or rays) frequently accompanies α- and β- emissions. γ-Rays undergo interactions with matter which include the photoelectron effect, Compton scattering, and pair-production. γ-Rays are best shielded by dense materials such as lead.

Gas: A state of matter lacking cohesive forces, definite shape, or volume. It can be defined as a material that is neither solid nor liquid at 25°C and 760 mm Hg. Notes: If confined to a vessel, a gas will expand to fill the volume and will exert a force on the walls of the container. This force per unit area is measured as the gas pressure. Gases affect the aerosol particles suspended within them. Gases impart resistance to particles, share energy with them, and gas temperature gradients impart a force on particles.

Gauge pressure: The difference between the actual pressure and atmospheric pressure, that is, the pressure relative to atmospheric. "Gauge pressure positive" is the difference in pressure above atmospheric. "Gauge pressure negative" is the difference between the pressure achieved in an evacuated system and atmospheric.

Geiger–mueller counter: A radiation detection and measuring instrument consisting of a gas-filled probe and a meter. Notes: The probe is usually tube shaped or disc shaped (pancake). It contains an anode maintained at a high potential (typically, 900 V) relative to the cathode (the inner surface of the probe). When ionizing radiation passes through the probe, a pulse of current is generated by gas ionization and the subsequent collection of the electron avalanche. This current is measured by the meter. The number of pulses per second (the count rate) measures the intensity of the radiation field. It was named for Hans Geiger and W. Mueller, who invented it in the 1920s. It is sometimes referred to simply as a Geiger counter or a GM.

Geometric mean or, as used in aerosol science, the geometric mean diameter (d_g): The nth root of the product of N values, that is, $\sigma_g = (\sigma_1 \sigma_2 \sigma_3 \ldots \sigma_N)^{1/N}$. Note: The N values in aerosol science are particle diameters (d_i).

$$d_g = \exp\left[\frac{\sum n_i \ln d_i}{N}\right].$$

Geometric standard deviation: The measure of dispersion of a lognormal distribution. Note: In aerosol science, the dispersion of the natural logarithms of the diameters of aerosol particles in a particular sample.

$$\sigma_g = \exp\left[\frac{\sum n_i (\ln d_i - \ln d_g)^2}{N-1}\right]^{1/2}.$$

Grab sample: A single sample of a constituent in fluid flow such as air (or in the occupational or natural environment) that is obtained in short period of time and is only representative of the concentration of the constituent at the moment of collection and measurement. Note: It does not account for changes over time to the constituent or to the environment in which it is released.

Gram (g): A unit of mass. Kilogram is the base unit of mass in the international system (SI) of units and is equal to the mass of the international prototype of kilogram. A gram is 0.001 of a kilogram.

Gram atomic weight (gram-atom): The mass in grams numerically equal to the atomic mass of an element and containing Avogadro's number of atoms.

Gram molecular weight (gram-mole): The mass in grams numerically equal to the sum of the atomic masses of all the atoms in a molecule and containing Avogadro's number of molecules.

H

Half-life ($t_{1/2}$): The time observed for the quantity, or the population of atoms, and thereby the radioactivity, of a radionuclide to be reduced by 50%.

Health physics: The applied science concerned with the study, recognition, evaluation, and control of health hazards that may arise from exposure to naturally occurring or man-made ionizing radiation. Notes: Emphasis is placed on the safety concerns regarding the medical, industrial, or scientific use and application of ionizing radiation and the related environmental issues. It is sometimes referred to as radiation safety, radiological health, or radiological engineering.

HVAC: Heating, ventilating, and air conditioning; the air distribution system that treats incoming air to provide desired temperature, humidity, and cleanliness for building occupants. Note: HVAC systems also remove air to provide the required "air changes" per unit time and either negative or positive pressures relative to ambient conditions.

Glossary

Hydrostatic pressure: The pressure of a nonmoving (static) gas (or liquid) confined within a volume. The pressure is exerted in all directions.

I

I.D.: Inside diameter
Ideal gas: A gas that behaves according to a simple relationship between temperature (T), pressure (P), and volume (V), namely, $PV = nRT$, where n is the mass of the gas in moles and $R = 8.314$ J/mol °K.
Impaction: Removal of a particle from an airstream by striking an object in the airstream.
Impeller: A rotor for transmitting motion as in the rotating component of a pump, fan, compressor, or blower.
Indication (or indicated) value: A scale or decade reading or the displayed value of an instrument readout. Also see Reading.
Inhalable particles: Originally defined as dusts with aerodynamic diameters less than 15 μm. Notes: Now considered as dusts with median aerodynamic diameters of about 10 μm. This term was derived by the United States Environmental Protection Agency to distinguish this particle size group that is trapped in the nose, throat, and upper respiratory tract from "respirable particles" that penetrate deeper into the lung (see definition of respirable particles).
Instrument: A complete system consisting of one or more assemblies designed to quantify one or more characteristics of ionizing radiation or radioactive material.
Intensity: The number of photons or particles passing through a unit area per unit of time.
Interception: Removal of a particle from an airstream by a trajectory that brings it within one radius or less of the impacting surface while not deviating from the air streamline.
Intrinsic efficiency: The probability that a photon of a given energy impinging on the front of a detector will be completely absorbed by the detector element.
Inverse square law: When radiation from a point source is emitted uniformly in all directions, the amount received per unit area at any given distance from the source, assuming no absorption, is inversely proportional to the square of that distance.
Ion: An atom that has become electrically charged by losing one or more orbital electrons (positive ion) or by gaining one or more orbital electrons (negative electron).
Ionization: The process by which a neutral atom is transformed into an ion (see definition of ion above). Notes: It can occur when atoms absorb the energy imparted to them by photons or through collisions with particulate radiation. An example is the interaction of particulate or photonic radiation with air. The air molecules become positive ions when the interactions result in the emission of electrons from the air molecules. The positive ion and the emitted electron are known as an "ion pair."
Ionization energy (W): The average amount of energy required to produce an ion pair in a gas. Note: This is about 34 eV for air.
Isobar: One or more nuclides with the same number of nucleons (protons and neutrons) in their atomic nuclei.
Isokinetic: The condition in which the velocity of air in a sampler input port (nozzle) is equal to the velocity of air in the duct or stack being sampled, so that the inertia of any suspended particles remain undisturbed thus preventing their possible loss to the measurement. Note: a synonymous term is "isomean velocity."
Isomer: One or more isobars existing at a different quantum state and exhibiting different radioactive properties. Isomers at higher energy states frequently decay to a lower energy state by undergoing an "isomeric transition."
Isotope: A variety of the same element having the same number of nuclear protons but a different number of neutrons. Note: Isotopes of an element have the same chemical properties.
Isotropic: Having the same properties in all directions.

J

Joule: A unit of work or energy. Note: A Joule is the energy required to transfer a static charge of 1 Coulomb between two points that are held at a potential difference of 1 V.

K

K-capture: The radioactive transition whereby a nucleus of an atom reduces its charge by 1 via the capture of an electron from the innermost "K" shell of the extranuclear electrons. Note: The vacancy left in the K-shell is filled by an electron from another shell with the concomitant release of an x-ray photon ("K x-ray"), which brings the atom to a ground state. Light nuclei undergo positron emission, which also reduces the nuclear charge by 1. Heavier nuclei tend to undergo K-capture.

K-shell: The innermost shell of electrons orbiting the atom (extranuclear atoms). Note: The maximum number of electrons that can populate this shell is two.

Kelvin temperature scale: The fundamental temperature scale used in science. Notes: The Kelvin degree is the same magnitude as the Celsius degree (°C). One Fahrenheit (°F) degree is 5/9 as large as one Celsius degree. The triple point of water (the state where liquid water, ice, and water vapor are in equilibrium) is defined as 273.16°K. The temperature at which ice and air saturated water are in equilibrium (the ice point) is 0.01°C or 273.16°K or 32.0°F. The temperature at which liquid water and steam (at 1 atm of pressure) are in equilibrium (the steam point) is 100°C or −173.16°K or 212°F. Absolute zero is −273.15°C or 0°K or −459.67°F.

Kinetic energy: The energy of a subatomic particle, a photon, an object, or a mechanical system due to its motion. Note: Kinetic energy is numerically equal to 1/2 the product of the velocity squared and the mass of the moving body or system.

L

Laminar flow: Fluid flow in smooth layers with no fluid mixing between layers; associated with a Reynolds number <2300.

Liter (L): A unit of volume in the metric system although considered outside the international system (SI) of units. A milliliter (mL) is one-thousandth of a liter. $1 L = 0.001 m^3$ or $1000 cm^3$.

Lognormal distribution: The set of size values typically used to describe the particle diameter distribution of single source aerosols. Notes: It is determined by taking the logarithm of particle diameter and plotting it against the fraction of particles exhibiting that diameter. When done so, the skewed distribution of diameter observed for most aerosols (usually skewed to higher particle diameters and not centered around 0 diameter) becomes normally distributed allowing normal distribution parameters such as means and standard deviations to be calculated for it.

Luminescence: The emission of light or other radiations from a substance due to a cause other than the temperature of the substance. Note: For example, luminescence may be caused by the absorption of energy from another form of radiation such as ionizing radiation.

M

Manometer: An instrument used to make accurate readings of gas pressure.

Mass median aerodynamic diameter (MMAD): The aerodynamic particle diameter at which 50% of the aerosol mass lies above this diameter and 50% of the aerosol mass lies below this diameter.

Median: As used in aerosol science, the diameter at which one-half of the total number of particles in a collected sample are smaller and one-half are larger.

Megaton: A unit of explosive energy equivalent to the energy released in the detonation of 10^6 tons of TNT.

Meter (m): A unit of length in the international system of (SI) units. One meter is the length traveled by light in a vacuum during a time interval of 1/299, 792, and 458 of a second. It is equal to 3.281 ft. A centimeter (cm) is one-hundredth of a meter.

Micron (μm): A unit of length equal to 10^{-6} m.

Mist: A liquid aerosol in droplet form produced by mechanical processes such as spraying, bubbling, or splashing.

Mode: The most frequent diameter counted in a particular aerosol sample, that is, the diameter with the highest frequency as indicated in a frequency distribution curve.

Mole: The mass of a substance containing Avogadro's number of atoms or molecules (6.022×10^{23}). Note: One mole of a substance is equivalent to one molecular weight of the substance.

Molecular weight: The sum of the atomic weights of the atoms comprising a molecule.

Monitoring: The means provided to continuously indicate the state or condition of a system or assembly, or specifically the real-time measurement of radioactivity or radiation level.

Monodisperse: Descriptive of an aerosol only containing particles of a single size.

Morbidity: The age- and gender-specific or total incidence of a specified disease in the population.

Mortality: The age- and gender-specific or total rate at which people die from a specified disease in the population.

Multichannel analyzer (MCA): An instrument that collects, stores, and analyzes time-correlated and energy-correlated events such as the radiation interactions inside a detector.

N

Neutron (n): The fundamental nuclear particle with a mass of 1.675×10^{-24} g. Note: It possesses no electrical charge.

Noise: Signals caused by other processes other than by radiation interaction. Note: The processes may be the result of electronic equipment responding to the ambient environment such as temperature changes or lighting conditions.

Nonionizing radiation: Radiation with insufficient energy to remove orbital electrons from atoms or molecules. Notes: Photons with energy less than 12.4 eV (a wavelength of 100 nm that falls into the category of ultraviolet radiation or UV) are considered nonionizing. The other categories of nonionizing radiation include visible light, infrared radiation (IR), radio frequency radiation, and extremely low-frequency radiation. Laser radiation is also considered nonionizing because it may be emitted in the UV, IR, or visible wavelengths depending on its manufacture.

Nozzle: An opening that allows aerosol particles to be drawn from an effluent flow into a sampling line allowing them to be analyzed and collected. Note: Nozzles can affect sampling if wall losses to the internal surfaces of the nozzle are significant and the aerosol concentration at the nozzle inlet is much less than the aerosol concentration in the effluent flow stream.

Nucleons: The collective term for the protons and neutrons that comprise the nucleus of an atom. Note: The number of nucleons in an atom is equivalent to the atomic mass number (A) of the element or isotope of an element.

Nucleus: The core of an atom containing protons (positive charge) and neutrons (no charge). Notes: The nucleus is therefore positively charged. The extranuclear electrons that orbit the nucleus (negatively charged), if equal to the number of protons in the nucleus, will result in a neutral atom. The number of protons in a nucleus determines the chemical element. Relative to the entire atom, the nucleus is extremely small, but it possesses the majority of the atomic mass.

Nuclide: An atom that must be capable of existing for a measurable time.

O

OD: Outside diameter.

P

Pair production: The formation of an electron and a positron by a γ-ray passing in proximity to an atomic nucleus. Note: The electron and the positron are created by the conversion of the energy of the photon into mass via the Einstein expression of $E = mc^2$.

Parent element: A term, usually used in reference to a radionuclide, that decays to give rise to another radioactive element (the daughter) that may or may not be radioactive itself.

Partial pressure: The pressure of a gas in a mixture of gases that it would produce if it were isolated from the mixture and allowed to occupy the same volume and temperature as the mixture.

Particle: That portion of an aerosol that is suspended in the supporting gas. Notes: Particles may range from 0.001 µm to over 100 µm in diameter (fine particles < 2 µm < coarse particles). Particles may be liquid or solid or a combination thereof. The particles of an aerosol need not be chemically homogeneous.

Particle bounce: The rebound of particles from a collecting surface.

Passive detector: An instrument designed to estimate the concentration of radioactivity in air, for example, radon-222 or radon-222 decay products, without any moving parts, for example, a sampling pump.

Peak: The spectroscopic display of the accumulation of counts in a relatively narrow energy band of a multichannel analyzer system. Notes: When the measurement system is properly calibrated, the peak energy corresponds to the radiation energy interacting with the detector of the system. The height of the peak indicates the prevalence of the radiation and, when the system is properly calibrated, can yield the amount of radioactivity present (activity per unit volume in the case of air samples).

Photomultiplier tube (PMT): An electronic device that responds to the flashes of light produced by a scintillator by emitting electrons that are amplified in number to produce a measurable current.

Photon: A packet or "quantum" of electromagnetic energy possessing no electric charge. Notes: Photons may also be described mathematically as an electromagnetic wave. The energy of the photon may be converted into mass. This equivalent mass is described as $m = h\nu/c^2$, where ν is the frequency of the wave and c is the speed of light in a vacuum.

Pitot tube: A device used to determine the velocity of effluent air at a particular point within a duct or stack by measurement of the static air pressure and the impact air pressure. Velocity pressure (P_v) is the difference between these two air pressure parameters. Note: The linear effluent air velocity (V) is then $V = (2g_c P_v)^{1/2}$ where g_c is the gravitational constant in units consistent with P_v.

Polydisperse: Descriptive of an aerosol comprised of particles of many sizes.

Positron decay: Radioactive decay by emission of a positron (a positively charged β-particle). Note: In this decay, a proton is lost and a neutron formed transforming the radioactive element into another element. Positron decay is accompanied by two γ-ray photons when the positron interacts with an electron (almost instantaneously after emission).

Potential alpha energy concentration (PAEC): The air concentration of radon-222 decay products, in terms of the α-energy released during the complete decay through polonium-214, usually measured in units of working level. Note: The energy sum includes contributions from the polonium-214 α-particles that arise from decay of the β-emitting decay products in the sample air.

Precision: For repeated measurements of one parameter using one measurement system, the degree of agreement of the measurements; usually expressed using the relative standard deviation (RSD), also known as the coefficient of variation (CV) such that CV = (SD/mean)100%.

Pressure: The force exerted by a fluid or gas per unit surface area transmitted to a confining boundary. Note: Common units are lb in^{-2}, Newtons m^{-2}, dyn cm^{-2}, bars (1 bar = 10^6 dyn cm^{-2}, atmospheres (1 atm = 14.7 lb in^{-2}), and mm Hg (760 mm Hg = 1 atm).

Pressure drop: Within an air sampling system, the difference in air pressure across a component of the system, for example, the difference in pressure from one side of a filter to the other.

Prompt neutrons: Neutrons emitted immediately (<10^{-6} s) after the fission process or from the production of new fission products.

Proton (p): The fundamental nuclear particle possessing positive charge (1.6 × 10^{-19} Coulomb) and a rest mass of 1.672 × 10^{-24} g.

PsiA: Pounds per square inch absolute. The pressure measured above a perfect vacuum.

PsiG: Pounds per square inch gauge. The pressure indicated on a gauge using atmospheric pressure as the baseline. Gauge pressure can be negative (below atmospheric) or positive (above atmospheric).

Pulse-height analysis: The evaluation of counting data by a multichannel analyzer. Notes: Radiation interactions with the associated detector are separated into energy bins with upper and lower bounds defined by a difference in voltage. If the voltage pulse of a radiation interaction falls in the range defined by the bin, it is tallied and stored as a count in a memory channel corresponding to a particular energy level. The cumulative total of counts of all the energies registered by the detector is displayed as a spectrum.

Q

Quantum: A discrete packet of energy or other physical quantity.

Quantum jump: A change in the energy state of atom usually associated with the absorption of energy (a quantum change to a more "excited" state). Note: The release of energy, which occurs when electrons fill vacant orbitals, is also a quantum jump (to a stable state).

Quenching: In a liquid scintillation counter, the suppression of radiation energy transfer to the scintillant due to chemical interference or due to absorption of the light photons produced by the scintillation fluid by the semiopaque or colored nature of the sample. Note: Quenching also refers to the termination of discharge in the Geiger–Mueller tube.

R

Rad: One rad is equivalent to absorbing 100 ergs of energy per gram of tissue (or any material so long as the material is specified). Note: A milli-rad is 0.001 rad. The unit of absorbed ionizing radiation dose.

Radiation: Released and propagated electromagnetic energy. Notes: In reference to radioactivity, the energetic subatomic particles or photons (quanta) released in the process of radioactive decay. The particulate forms of radiation include α-particles, β-particles, and neutrons; the photonic forms comprise of x-rays and γ-rays. Radiation may be ionizing as in the case of the above forms of radiation, that is, energetic enough to remove electrons from atoms or nonionizing as in the case of radio waves.

Radiation safety: The applied science that is concerned with the protection of individuals and the public from the harmful effects of ionizing and nonionizing radiation. Notes: Radiation safety, also known as health physics, is interdisciplinary, drawing upon diverse sciences such as physics, biology, chemistry, meteorology, geology, the engineering disciplines and astrophysics to explain, control, and carefully harness natural and artificial sources of radioactivity. The "philosophy" of health physics is to maintain radiation doses to man and the environment as low as reasonably achievable while still accruing significant benefits from sources of radiation.

Radioactivity: The transformation of one element into another or into another isotope by the release of ionizing radiation. Notes: Radioactive transformations include positron emission, electron or "K" capture, α-particle emission, β-particle emission, γ-ray emission and combinations thereof. The form of radioactive transformation is dependent on the neutron to proton ratio of the atomic nucleus and the mass–energy relationship between the parent nucleus, the resultant daughter nucleus, and the emitted particle. The radioactive characteristics of an atom are the result of nuclear considerations only and cannot be changed by physical or chemical means.

Radioisotope: A radioactive isotope of an element produced artificially as in a nuclear reactor or cyclotron.

Radon: Naturally occurring radioactive noble gas with three isotopes, $^{222, 220, 219}$Rn.

Radon progeny: Decay products or daughters of radon gas.

Range: All values lying between the lower detection threshold and the upper measurement limit of a radiation detection system.

Reading: The indicated or displayed value of an instrument reading.

Readout: The portion of the instrument that provides a visual display of the instrument response with units displayed and/or recorded by the instrument as a result of the instrument's response to some influence quality.

Rem: The unit of ionizing radiation dose to man that produces the same biological effect as exposure to 1 Roentgen of x-rays or γ-rays. Note: Rem is approximately equivalent to rad for all forms of radiation, except α-particles and neutrons. A milli-rem is 0.001 rem.

Reproducibility: The precision of a series of measurements often expressed as the standard deviation.

Resistance: A property of conductors that is dependent on material composition, dimensions, and temperature. Note: Resistance determines the current produced in the conductor for a given difference of potential (voltage). The unit is ohm (Ω).

Resolution: The ability of a spectroscopy measurement system to differentiate the energies of two peaks that are similar in energy. Note: The full-width-at-half-maximum (FWHM) parameter is an indicator of a measurement system's ability to resolve emissions of like energy.

Respirable particles: Those particles that are capable of reaching the alveolar region of the lung. Notes: The aerodynamic diameter of respirable particles must be <10 μm and is usually <5 μm. The term "respirable particles" was originally derived by the United States Atomic Energy Commission to distinguish the portion of inhaled insoluble dust reaching the deep region of the lung devoid of ciliary action and the protective mucus layer found in the head and tracheobronchial regions.

Rest mass: The mass of a motionless particle. Note: According to relativistic physics, mass increases with velocity according to a factor of $[1 - v^2/c^2]^{-1/2}$, where v is the velocity of the particle and c is the velocity of light in a vacuum.

Roentgen (R): The unit of x-ray or γ-ray exposure that produces, through ionization of air molecules, 0.000258 Coulomb of positive or negative static charge in 1 kg of air at a temperature of 0 °C and an atmospheric pressure of 760 mm Hg. Note: This is equivalent to the absorption of 87.7 ergs g^{-1} of air or 0.877 rad (see the above definition of rad).

S

Sampling: The process of collecting a sample from a volume of air (or other media) for the purposes of detection and/or analysis.

Scattering: A process that can change the course of a radioactive emission such as a β-particle or a γ-photon. Notes: Scattering is caused by collisions with atoms and molecules or interactions with electromagnetic fields. When an emission is scattered into the sensitive volume of a detector when it had no motion in that direction, it has undergone backscattering, a potentially important effect for α-counting using a plastic scintillator. Scattering is dependent on many factors, including the sample mounting material, the detector shielding material (if any), and the detector/sample geometry.

Scintillation cell: A container lined with a phosphor that scintillates due to interaction with α-particles. Note: The radon contained in the container can be determined by using a photomultiplier tube to count the scintillations produced.

Scintillation counter: A term generally applied to portable survey meters that use an inorganic crystal such as sodium iodide as the radiation detector. Notes: The interaction of γ-ray photons or β-particles in the crystal produces photons of light that are intercepted by a photomultiplier tube, converted to electrons, and then recorded. The intensity of the radiation determines the amount of photon production in the crystal and thus the production of electrical signal is ultimately recorded. Scintillation counters are more sensitive to γ-rays than Geiger–Mueller counters.

SCFM: The volume flow rate of air in cubic feet per minute at standard conditions of temperature and pressure.

Secular radioactive equilibrium: The 1:1 ratio of radioactivity of a radionuclide (the "daughter" or product) produced from another radionuclide (the "parent") that is established over about seven daughter half-lives when the parent has a very long half-life relative to the daughters. Note: The formation of the product atoms equals the decay rate of the parent atoms under this condition.

Sedimentation: The settling of an aerosol particle through a gaseous medium due to the influence of gravity.

Sedimentation velocity: The maximum velocity attained by an aerosol particle settling under the force of gravity while in a quiescent fluid, for example, air.

Self absorption: The loss of radiation by absorption from the materials that constitute a radioactive source. Note: This phenomenon is most important for β-emitting sources mixed into a thick matrix.

Shell: Refers to the arrangement of extranuclear electrons around the nucleus of an atom. Notes: This arrangement can be explained as concentric shells or orbitals that can each "hold" a limited number of electrons. From inner to outer shells, they are designated as K, L, M, N, O, P, and Q.

Shielding: Usually refers to massive material surrounding a dangerous source of radioactivity in order to absorb hazardous radiation. Notes: Alternatively, the material may be erected around a sensitive radiation detector to prevent environmental or other sources of radiation from interfering with measurements. The massive material may be lead, concrete, or other dense substances. Other materials that absorb a particular form of radiation may be used, for example, water and other hydrogenous materials are good shields for certain neutron energies, while β-particles may be effectively stopped by certain thicknesses of Lexan. To protect sensitive instruments, shielding may be graded using a combination of lead, copper, and other metals to significantly reduce the environmental radiation field inside the shield.

Shroud: An aerodynamic decelerator surrounding a nozzle used to negate nozzle wall losses, off-angle flow direction, changes in sampling flow rate, or changes in the effluent stream velocity.

Single channel analyzer (SCA): An instrument that accepts voltage or current pulses originating from a detector and accumulates a total within a preset counting time. Note: SCAs may be equipped with adjustable lower limit and upper limit acceptance levels (a window) to reject electronic pulses outside these settings.

Smoke: The product of combustion of carbon-based material consisting of a mixture of solid and liquid aerosol particles, gases, and vapors.

Somatic cells: The cells of the body, except those involved in reproduction (the gametes).

Somatic effects: Biological effects of radiation concerned with the exposed individual as distinguished from genetic effects that may affect succeeding generations.
Specific activity: The amount of radioactivity in terms of disintegrations per unit time (or Curies or Becquerels) per unit of weight of the material in which the radioactivity resides, for example, mCi per gram.
Specific gravity: The ratio of the mass of a body to the mass of an equal volume of water at 4°C or another specified temperature. Note: The dimension of specific gravity is unity.
Specific ionization: The average number of ion pairs produced per unit distance of travel by a charged particle.
Spectrometer: An analysis device used to separate radiation energies into memory channels and to display them visually as peaks. Note: The height of the peaks correspond to the cumulative number of counts the associated radiation detector registers. See multichannel analyzer.
Spectrum: A distribution of peaks indicating the relative intensity of detected radiation (cumulative counts) usually displayed as a function of energy by a multichannel analyzer and associated software.
Stable isotope: The isotope of an element that does not undergo radioactive decay.
Standard air: As used in ventilation engineering, dry air at 70°F (21°C) and 29.92 in Hg (760 mm Hg). Note: Also defined as air at 68°F (20°C) and 29.92 in Hg (760 mm Hg) at a relative humidity of 50%.
Standard conditions: The reference temperature and pressure of air that is used to supply a basis for comparison of air sampling measurements taken at different temperatures and pressures. Notes: For example, measured volumes of gases are usually corrected to 0°C and 760 mm Hg. For industrial hygiene purposes, standard temperature and pressure (STP) may be defined as 25°C and 760 mm Hg although 20°C is sometimes used.
Standard (instrument or source): (1) National standard—a standard determined by a nationally recognized competent authority to serve as the basis for assigning values to other standards of the quantity concerned. In the United States this is an instrument, source, or other systems or devices maintained and promulgated by the National Institute of Standards and Technology (NIST). (2) Primary standard—a standard that is designated or widely acknowledged as having the highest metrological qualities and whose value is accepted without reference to other standards of the same quantity. (3) Secondary standard—a standard whose value is assigned by comparison with a primary standard of the same quantity. (4) Reference standard—a standard, generally having the highest metrological quality available at a given location or in a given organization, from which measurements made there are derived. (5) Transfer standard—A standard used as an intermediary to compare standards. Note: If the intermediary is not a standard, the term *transfer device* should be used. (6) Working standard—A standard that is used routinely to calibrate or check material measures, measuring instruments, or reference materials. A working standard is usually traceable to NIST.
Static pressure (SP): In general, the pressure on a stationary surface that is tangent to the mass-flow velocity vector. Note: For a fluid in motion such as air, it is measured as the pressure exerted perpendicular to the internal walls of a duct by the flowing air forced into motion by a blower or fan.
Stochastic effects: The biological effects of radiation that include cancer and hereditary effects. Note: Such effects do not require a dose threshold to be achieved before they manifest. The probability of occurrence increases with dose, and the magnitude of the effect is independent of dose.
Stokes diameter: The spherical diameter of a real aerosol particle (which may be any shape) that has the same density and settling velocity as the aerosol particle.
Stokes kaw: A solution to the general differential equations of fluid motion describing the behavior of an aerosol particle using the assumptions that it is a rigid sphere in an incompressible fluid under constant motion distant from walls and other particles and that inertial forces

are negligible compared to viscous forces. Note: The drag force on the particle is expressed as $F_D = 3\pi\eta Vd$ where η is the viscosity of the fluid, V is the particle velocity, and d is the particle diameter.

Streamline flow: The flow of a fluid whereby the velocity (direction and speed) in the fluid is everywhere constant.

Strong mixing: A high level of turbulence in the lower atmosphere causing rapid lateral and vertical mixing of a pollutant into the air; associated with light winds, clear days, and a negative vertical temperature gradient in the atmosphere.

T

Thermodynamic equivalent diameter: The diameter of a unit density sphere having the same diffusion coefficient as the particle of concern.

Thermoluminescent dosimeter (TLD): A crystalline (solid state) integrating radiation detector where energy is absorbed and can be determined later by thermal excitation of the detector.

Total effective dose equivalent (TEDE): The dose equivalent (units of rem or Sieverts) that is the sum of the committed effective dose equivalent (CEDE) and the deep dose equivalent (DDE). It is used to evaluate stochastic effects.

Traceability: The property of the result of a measurement or the value of a standard whereby it can be related to stated references, usually national or international standards, through an unbroken chain of comparisons all with stated uncertainties.

Transient radioactive equilibrium: The condition that exists between a product radionuclide and the parent radionuclide from which it is formed whereby formation of the product atoms equals the decay rate of the parent atoms. Notes: The product radioactivity will appear to decrease with the same rate as the parent radioactivity. This condition arises when the relative magnitudes of the half-lives between the parent and the product are of any value, except that defined by secular equilibrium.

Tritium ($_1H^3$): The radioisotope of hydrogen with mass number 3.

Turbulent flow: The flow of a fluid characterized by erratic velocity (direction and speed) of the fluid elements over time.

U

Unattached fraction: The fraction of airborne radon or thoron decay products or progeny that exists as free atoms, ions, or clusters that are not attached to particulates.

Uncertainty: The estimated bounds of the deviation from the conventionally true value, generally expressed as a percent of the mean, and ordinarily taken as the square root of the sum of the square of the following two components: (1) those components (random errors) that are evaluated by statistical means and (2) those components (systematic errors) that are evaluated by other means.

Upper measurement limit: The maximum level at which the instrument meets the required accuracy. Note: It may also be considered as the range of the instrument.

Upper respiratory tract: The mouth, nose, sinuses, and pharynx.

V

Vacuum: The condition where air pressure is greatly reduced from the normal atmospheric pressure of 760 mm of mercury (Hg). Notes: In reality, the term "vacuum" is used loosely to identify "partial vacuums" that may approach 10^{-5} or 10^{-6} mm Hg. Lower pressures can be achieved but an absolute vacuum (all matter, gas, and air removed) remains illusive.

Vapor pressure: The pressure exerted when a solid or liquid is in equilibrium with its own vapor. Note: Vapor pressure is the fixed quantity at a given temperature.

Variance: Square of the standard deviation.

Velocity pressure (VP): The pressure caused by moving air.

Viscosity: The resistance to change of form for fluids. Notes: As described by Newton's law of viscosity, if two plates are separated by a thin layer of fluid, a force is required to move one plate relative to the other stationary plate in order to maintain the constant velocity of the moving plate. This force is proportional to viscosity (η), area of the plates, and relative velocity of the plates and inversely proportional to the plate separation. Thus viscosity is associated with the transfer of momentum from a faster moving layer of fluid to a slower moving layer. The viscosity of air is at 20°C is 1.81×10^{-4} dyn s cm^{-2} (Poise).

Volt: The unit of electrical potential difference between two points on a wire carrying 1 amp of current and dissipating one 1 W of power between the two points.

Volume flow rate: The quantity of flowing air usually measured in cubic feet per minute (CFM) or cubic meters per second (m^{-3} s^{-1}).

W

Wavelength : A term used to describe electromagnetic radiation by the physical distance between two successive peaks or troughs of the wave. Note: The wavelength is related to the frequency of the electromagnetic wave by the formula $\lambda = c/\nu$, where λ is the wavelength, c is the velocity of light, and ν is the frequency.

Wall losses: Loss of an aerosol sample to the internal surfaces of a sampling system. Note: It is the ratio of the sum of nozzle loss (or inlet port), sampling line loss, and losses to other system components to the aerosol concentration at the inlet plane of the nozzle (or inlet port) of the sampling system.

Watt: The unit of rate of work performed (power dissipated). Note: The power corresponding to the production of energy at 1 J (10^7 ergs) s^{-1}.

Weak mixing: A low level of turbulence in the lower atmosphere inhibiting lateral and vertical mixing of a pollutant into the air; associated with nighttime conditions and a positive vertical temperature gradient in the atmosphere.

Wheatstone bridge: A four-armed electrical circuit with resistances in each arm. Notes: Two resistances (r) are known; the third is variable, but can be easily measured; and the fourth is unknown (r_x), but can be determined by balancing the current in the bridge and using the equation

$$\frac{r_1}{r_2} = \frac{r_3}{r_x}.$$

Whole body: The head and trunk of the body including the male gonads, arms above the elbow, and legs above the knee.

Window: The range of energies, between a lower level and an upper level, accepted for counting by a spectrometer or a single channel analyzer.

Working level (WL): Any combination of short-lived radon decay products in 1 L of air that will result in the ultimate emission of 1.3×10^5 MeV (2×10^{-5} J m^{-3}) of potential α-energy.

Working level month (WLM): The cumulative exposure equivalent to exposure at one working level for a working month of 170 h.

X

X-rays: High-energy, short-wavelength electromagnetic radiation (identical to γ-rays) emitted by electron transitions between orbitals of an atom (characteristic x-rays) or when energetic electrons are decelerated abruptly by a target as in an x-ray tube.

Index

Note: n denotes footnote.

A

AARST. *See* American Association of Radon Scientists and Technologists (AARST)
Absolute air pressure, 234. *See also* Barometer
    critical flow meter, 251
    measurement, 234
Absorbed fraction, 215
Absorption, 216. *See also* Clearance
    in bloodstream, 213
    classes, 214
    inhaled material, 216
    respiratory tract, 214
Accelerator mass spectrometry (AMS), 387
Acceptance test, 49
Accuracy, 298, 462. *See also* Precision
    air concentration measurement, 411, 420
    bias and precision impact, 299
    detection, 311
    gross α-analysis, 394
    gross β-analysis, 401
    mass flow meters, 232
    rotameters, 226, 227
    total estimated, 449
ACFM. *See* Actual CFM (ACFM)
ACGIH. *See* American Conference of Governmental Industrial Hygienists (ACGIH)
Actinon (^{219}Rn), 201, 182
    decay series, 201, 202
    deposition effect, 204
    f factors, 202–204
    h factors, 199–205
    ingrowth, 207, 208
    PAEC, 204
    progeny buildup, 205
    progeny in air, 206
    recurrence formula, 202
Activity mean aerodynamic diameter (AMAD), 373, 501
Activity median aerodynamic diameter (AMAD), 38, 372n
Activity median diameters (AMD), 372
Actual CFM (ACFM), 306
Acute alarm, 310. *See also* chronic alarm
    evaluation, 311
AD. *See* Aerodynamic diameter (AD)
ADC *See* Analog-to-digital converter (ADC)
AEC. *See* Atomic Energy Commission (AEC)
AED. *See* Aerodynamic equivalent diameter (AED)
Aerial Measurement System (AMS), 359
Aerodynamic diameter (AD), 96, 294, 317n. *See also* Stokes' diameter; Thermodynamic diameter
    comparison, 142
    dusts, for, 136

    equivalent, 139–141
    respirable size range, 147
    uranium oxide, 140
Aerodynamic equivalent diameter (AED), 140, 287
Aerodynamic factors, 135, 136, 138
    length-to-diameter ratio, 141
    particle density, 142
Aerodynamic processes, 212
Aerosol, 4, 77–78, 136
    characterization, 5, 79
    classifications, 136
    count mean diameter, 143
    count mode diameter, 143
    dispersion, 138, 272
    distribution and release, 147
    electron micrographs, 78
    features, 280
    fractional moments, 80
    gamma distribution, 82, 83
    Junge distribution, 82
    lognormal distribution, 81–82, 83, 143, 144, 145, 146
    log-probability plot, 145, 146
    molecular parameters, 87
    multimodal distributions, 83
    multiple modes, 145
    normal distribution, 145–146
    origin, 78
    phase space, 79, 106
    property, 103
    sample analysis, 7
    size distributions, 78
    transport, 88, 102, 272
    transport equation, 112
    trimodal size distribution, 84
Aerosol transport, 102, 272
    aerosol phase space, 106
    angular particle density, 107
    computation, 88
    deposition velocity, 110
    fixed spatial coordinate system, 106
    fundamental mechanisms, 272
    net flow rate, 110
    particle angular flux, 108
    particle deposition and GDE, 110
    particle size spectrum moments, 102–106
    scalar particle density, 107–108
    total particle density, 107, 109
AFNOR. *See* Association Française de Normalisation (AFNOR)
Agreement State, 12, 36
AIHA. *See* American Industrial Hygiene Association (AIHA)
Airborne ^{222}Rn decay products measurement. *See* Thomas method; Tsivoglou method, modified

Airborne radioactivity sampling, 3
  aerosol characterization, 5
  compliance demonstration, 7
  environmental monitoring, 6
  materials, 338
  measurement standards, 25–32
  objectives, 3
  probe placement, 316
  quality assurance and control, 6
  radiological emergencies, 6
  research, 8
  sample, 335, 336, 337
  sampling train, 327
  stack flow measurement, 336–337
  transport in sampling system, 332
  vapor and gas sample collectors, 336
  worker health protection, 5
Airborne release fraction (ARF), 34, 149
Air cleaning, 171, 173. *See also* Air filters
  considerations, 173
  fundamental mechanism, 139
  HVAC filters, 170–171
  system design, 171
Air filters, 158
  ANSI/ASHRAE standard 52.2 MERVs, 172
  cloth filters, 171
  HEPA filters, 171
  HVAC filters, 170–171
  ULPA filters, 171
Airflow studies, 273
  improved air quality, 274
  limitation, 280
  sampler placement, 273, 276
  tracers, 273, 280
  types, 273
  ventilation placement, 273
Airline respirators. *See* Supplied-air respirators (SARs)
Air monitoring
  anthropogenic discrimination, 360
  environmental, 6
  federal guidance, 360
  instrument, 7
  natural radioactivity discrimination, 360
  radioactive plume measurement, 357, 358
  resuspension, 357
  standards, 23
Air Monitoring Users Group (AMUG), 170
Air pressure, 234. *See also* Backpressure
  gauge pressure, 234
  rotameters, for, 227
Air-purifying respirators (APRs), 176
  efficiency, 177
  protection factors, 174
  resistance to degradation, 176
Air sample, 14
  alpha-CAM practice, 285
  compression, 447
  concentration calculation, 501
Air sampler. *See also* Sampling pumps
  calibration, 255
  daily sampling, for, 429
  FAS units, 287
  flow rate, 60
  long-term sampling, for, 439
  personal, 500
Air sampler placement, 276, 278
  airflow studies, 273–280
  continuous air monitors, 278
  optimization monitoring, 279
  retrospective air samplers, 278
  room airflow testing, 280
  ventilation-induced airflow, 272
Air sampling, 382–387
  activated charcoal, 386
  analysis for Xe-133, 386, 387
  β-activity measurement, 383
  chemical techniques, 387
  cloud sampling planes, 384, 385
  cold traps, 386
  fallout detection, 383
  field equipment, 14
  filter materials, 385
  γ-ray spectrum, 384
  HASL balloon aerosol, 386
  HASL/EML network, 383, 384
  high-volume sampling, 385
  mass loading, 385
  programs, 271
  RAMP systems, 383
  SASP filters, 383
  tritium (H-3), 386
Air sampling calibration. *See also* General calibration
  auditability, 255–256
  calibration source certificate, 253
  environmental conditions, 252, 254
  quality assurance program, 256–258
  rule, 228
  traceability, 252
  uncertainty, 254–255
ALARA. *See* As Low As Reasonably Achievable (ALARA)
Alarm response time, 490, 495. *See also* Alarm set point
  α-CAM, 495
  determination, 494
  sample calculation, 495–496
Alarm set point, 365, 494
  α-CAM, 494
  determination, 493
  dose mode, 367
  false alarm probability, 493, 494
  SD, 490
ALI. *See* Annual limits on intake (ALI)
Alkaline earth elements, 209
Alpha-CAM. *See* Alpha continuous air monitoring
Alpha continuous air monitoring, 285, 286
  airborne dust, 491
  alarm capability, 287
  alarm levels, 310
  alarm response history, 307
  alarm response time, 490, 494, 495
  alarm set point, 490, 493, 494
  α-continuous air monitors, 490
  α-energy peak spectrum, 291
  apparatus, 305
  application, 288–290
  background correction algorithm, 294
  calibration and gain control, 308

Index

components, 285
deposition on filter, 293
design, 287
detector to filter geometry, 491
exposure in DAC-h calculation, 309
false alarm rate determination, 492–493
filter type, 491
flow rate and integration time, 491
function, 287
interference, 291–293, 297, 490–491
placement, 307
practice overview, 285
precision and accuracy, 298, 490
procedure, 492
radon progeny background composition, 491
resolving time, 490
response time, 490
sampling rate and placement, 290–291
sensitivity, 290, 490
surveillance practice, 287
test apparatus, 491
Alpha counters, 18
α-emitters, 13, 287. *See also* β-emitters
Alpha particles (α), 55–56. *See also* β-particle
  backscatter, 68
  decay, 56
  emitting standard, 463
  measurement, 58
  peak shape function, 302
Alpha peak shape fitting (APSF), 506
Alpha-radiation counters, 68
Alpha-radioactivity
  accuracy, 394
  activity loss, 396
  advantages, 393
  air concentration SD, 397
α-scintillation counter, 462
α-spectrum, 287
  apparatus, 394
  calibration, 395–396
  cautions, 398
  continuous air monitoring, 286
  counting efficiency, 396
  disadvantages, 393
  dispersion in energies, 291
  dust loading interference, 393
  filter media interference, 392–393
  gas proportional counter, 394
  gross α-airborne concentration, 397
  inert dusts effects, 305
  liquid scintillation counter, 394
  long-lived material activity, 397
  peak-fitting algorithm, 303
  precision, 393
  principle, 391
  procedure, 395
  radionuclides interference, 392
  radon progeny, 291, 292
  range and sensitivity, 391–392
  reagents, 395
  ROI method, 300
  sample activity, 396–397
  sample air collection, 392
  solid-state detector, 394
  standards, 395
  storage effects, 396
  uncertainty, 397–398
  ZnS scintillator, 394
Alveolar–interstitial (AI) region, 212–215
AMAD. *See* Activity mean aerodynamic diameter (AMAD); Activity median aerodynamic diameter (AMAD)
Ambient temperature and pressure (ATP), 265
AMD. *See* Activity median diameters (AMD)
American Association of Radon Scientists and Technologists (AARST), 459
American Conference of Governmental Industrial Hygienists (ACGIH), 38
American Industrial Hygiene Association (AIHA), 34
American National Standards Institute (ANSI), 33, 288, 481
  accredited organizations, 34
  applicable standards, 25–27
  national standards organization, 33
  standard 52.2 MERVs, 172
American Nuclear Society (ANS), 34
American Society of Heating, Refrigerating and Air-Conditioning Engineers (ASHRAE), 34, 172
American Society of Mechanical Engineers (ASME), 170
American Society for Testing and Materials. *See* ASTM international
AMS. *See* Accelerator mass spectrometry (AMS); Aerial Measurement System (AMS)
AMUG. *See* Air Monitoring—Users Group (AMUG)
Analog-to-digital converter (ADC), 67
Anemometer, 14
  sonic, 281
  thermal, 275
  types, 273
Angular
  current, 109
  flow, 317
  flow rate, 109
  particle angular flux, 108
  particle density, 107
Annual limits on intake (ALI), 217, 415, 431, 444, 506
  See also Derived air concentration (DAC)
  calculation, 218, 288
  dose limit, 532
ANS. *See* American Nuclear Society (ANS)
ANSI. *See* American National Standards Institute (ANSI)
APF. *See* Assigned protection factor (APF)
a posteriori decision limit. *See* Critical measurement level (L_c)
APRs. *See* Air-purifying respirators (APRs)
APSF. *See* Alpha peak shape fitting (APSF)
ARF. *See* Airborne release fraction
Argon (Ar), 539
  exposure to DCF, 540
  inhalation biological data, 540
  properties, 540
  risk coefficients, 540
  simplified decay scheme, 539
  sources, 539, 540
  special considerations, 541

ASHRAE. *See* American Society of Heating, Refrigerating and Air-Conditioning Engineers (ASHRAE)
As Low As Reasonably Achievable (ALARA), 4, 11, 150, 288, 343
ASME. *See* American Society of Mechanical Engineers (ASME)
Aspiration efficiency, 329–330
Assigned protection factor (APF), 173–174
    respirators, for, 174–175
    respiratory protection, 176
Association Française de Normalisation (AFNOR), 34
    standards, 30
ASTM international, 34
    standards, 26–27
Atomic Energy Commission (AEC), 379n
Atomic number (Z), 56
ATP. *See* Ambient temperature and pressure (ATP)
Attached daughters, 292, 294, 553. *See also* Radon progeny; Unattached fraction
Attenuation, 453
Avogadro's number, 87

B

Background radiation rate, 437
Backpressure, 240. *See* Pressure drop
    pump and filter, 241
Barometer, 234
Bateman equations, 184
Becquerel (Bq), 61, 396
Bernoulli Principle, 265
β-emitters, 15
β-particle, 56–57
    β-decay, 184
    β-emitting progeny, 400
Beta-radioactivity. *See also* Alpha-radioactivity
    advantages, 400
    air concentration SD, 404
    apparatus, 401
    β-emitters concentrations, 399
    β-emitting progeny interference, 400
    β-sensitive detectors, 401–402
    calibration and standards, 403
    cautions, 405
    concentration, 403–405
    counter efficiency, 404
    disadvantages, 400–401
    filter samples interference, 400
    gas proportional counter, 402
    Geiger–Mueller detector, 402
    gross β-airborne concentration, 404
    gross β-analysis accuracy, 401
    liquid scintillation counter, 402
    precision, 401
    principle, 399
    procedure, 402–403
    range and sensitivity, 399–400
    reagents, 402
    sample activity, 404
    storage effects, 403
    uncertainty, 404
Bias, 299, 304
Bismuth (Bi), 483, 553
    aerosol properties, 555
    exposure to DCF, 554
    inhalation biological data, 554
    risk coefficients, 554–555
    sources, 555
    special considerations, 555
Bladewerx SabreBZM, 508
Blanc's law, 89
Blank, 15, 295, 301, 395
Blowers. *See* Dynamic pumps
BN. *See* Bureaux de Normalisation (BN)
BNEN. *See* Bureau de Normalisation des Equipements Nucléaires (BNEN)
Boltzmann. *See also* Maxwell–Boltzmann distribution
    constant, 84
    distribution function, 84
    probability distribution, 85
    transport equation, 90
Boyle's gas law, 228
BR. *See* Breathing rate (BR)
Breakthrough, 173, 437
Breathing rate (BR), 150
    acclimated humans, for, 486
    EPA specified, 431
    normal, 288
Breathing zone (BZ), 5, 271, 306
    aerosol removal, 275
    exposure to aerosols, 138
    monitor, 507
Breathing zone monitoring (BZM), 505
    ALI, 506
    α-spectroscopy, 506
    apparatus, 507
    APSF algorithms utilization, 506–507
    cautions, 508–509
    DAC-h, 506
    detectable activity, 507
    detector electronics, 505
    flow measurement accuracy, 507
    membrane filters interference, 507
    PAS, 505
    procedure, 507, 508
    range, 506
    sampling head, 505
    sensitivity, 506
    spectroscopic analysis, 506
    storage effect, 508
Bremsstrahlung radiation, internal, 57
British Standards Institution (BSI), 32
Bronchial (BB) region, 212, 215. *See also* Thoracic region
Brownian motion, 91, 141, 212
BSI. *See* British Standards Institution (BSI)
Bureau de Normalisation des Equipements Nucléaires (BNEN), 34
Bureaux de Normalisation (BN), 34
Bypass leakage, 223. *See also* in-leakage
BZ. *See* Breathing zone (BZ)
BZM. *See* Breathing zone monitoring (BZM)

C

Calibration laboratory, 252–254
    quality assurance program, 256
Calibration temperature and pressure (CTP), 265

Index

CAM. *See* Continuous Air Monitor (CAM)
Carbon (C), 137, 537
    activity per unit volume, 431
    advantages, 427
    aerosol properties, 539
    air sampling equipment, 429–430
    apparatus, 428
    calculation, 430–432
    cautions, 432
    counting, 429
    determination, 425
    disadvantages, 427
    EFF, 431
    exposure to DCF, 538
    filter, 428
    inhalation biological data, 537
    interferences, 426–427
    LSC efficiency, 430
    nuclear weapons testing, 539
    physical data, 537
    power plant nuclear reactor, 539
    precision and accuracy, 427–428
    preparation, 428, 429
    principle, 425–426
    reaction, 425
    reagents, 428
    risk coefficients, 538
    rotameter, 429–430
    sampling, 429
    sampling train, 428
    sensitivity and range, 426
    silica-gel column, 428
    simplified decay scheme, 537
    sources, 538–539
    special considerations, 539
    standard deviation, 431
    storage effect, 432
    trapping efficiency, 430
CDE. *See* Committed dose equivalent (CDE)
CEDE. *See* Committed effective dose equivalent (CEDE)
Cellulose membrane filters, 160, 163
CEN. *See* Comité Européen de Normalisation (CEN)
CENELEC. *See* Comité Européen de Normalisation Electrotechnique (CENELEC)
CFCs. *See* Chlorinated fluorocarbons (CFCs)
CFR. *See* Code of Federal Regulations (CFR)
CFV. *See* Critical flow venture (CFV)
Channel, 58, 63
Charles' law, 228
Chemiluminescence, 62, 427
Chernobyl accident, 138, 369
Chlorinated fluorocarbons (CFCs), 173
Chronic alarm, 310
    setting, 311
    threshold, 294
Cilia, 209. *See also* Bronchial (BB) region
    clearance compartments, 215
    function, 213
Clathrates, 540, 543
Clearance, 213
    compartments, 215
    half times, 216
    respiratory tract, 213
Cloth filters, industrial cleanable, 171

Coagulation, 112, 115
    Brownian, 124–127, 129
    coefficient, 116
    Fuchs' method, 127
    gravitational, 127, 129
    kernel, 116, 124
    method of moments, 120
    particle diameter sectionalization, 117
    sectional coagulation coefficients, 120
    sectional method, 117
    simultaneous coagulation mechanisms, 132
    Smoluchowski coagulation equation, 116
    sum kernel, 129–130
    turbulent diffusion, 130
Cocktail. *See* Scintillation fluid
Code of Federal Regulations (CFR), 12, 35
Coefficient of variance (COV), 301–302, 317
Collision diameter, 86–87
Collision frequency, 86
    diffusion coefficient, 89
    mean free path, 87
Colloids, 77. *See also* Aerosols
Comité Européen de Normalisation (CEN), 32
Comité Européen de Normalisation Electrotechnique (CENELEC), 32
Committed dose equivalent (CDE), 532
Committed effective dose, 210
    calculation, 218
Committed effective dose equivalent (CEDE), 288
Composite membrane filters, 164
Comprehensive Nuclear Test Ban (CTBT), 379
Compton scattering, 18, 64, 499
Conditional probability, 152, 153–154. *See also* Initiating-event probabilities
Contamination control, 13. *See also* Decontamination
    features, 347
    ventilation hoods, 13
Continuity equation, 110
Continuous air monitor (CAM), 160, 181, 260, 278, 481
    airborne dust, 483
    airflow rate calibration, 486
    α's travel interference, 483–484
    ambient light interference, 484
    apparatus, 485
    calibration and standards, 486–487
    cautions, 487
    detection efficiency, 483
    detector, 483
    DF, 278
    EF, 279
    electrical isolation, 484
    grounded front face, 484
    laboratory conditions, 482
    nonuniform deposition interference, 484
    potential locations, 279
    precision and accuracy, 484–485
    principle, 481–482
    procedure, 485–486
    radioactivity accumulation, 487
    radiofrequency interferences, 484
    radon and thoron progeny, 483
    range and sensitivity, 482
    real-time air monitors, 482
    storage effects, 487

Continuous sampling, 477. *See also* Continuous air monitor (CAM)
Continuum regime, 141. *See also* Mean free path drag force, 93
  Stokes' law, 91
Convection, 88
  deposition by, 111
  free space, 93
Convective fluid motion, 111
Convolution, 302
  Gaussian peak, 303
  normalized Gaussian function, 302
Counter, 15
  alpha-radiation, 68
  backgrounds, 462
  efficiency, 404
  Geiger, 15
  GPC, 16, 67–68
  LLLSC, 408
  optical particle, 344
  scintillation, 14
  types, 394
Counting damage tracks method
  airborne ^{222}Rn measurement, 467
  apparatus, 468
  calibration, 468
  interferences, 467
  passive α-track detector, 467
  precision and accuracy, 467–468
  procedure, 468
  radon exposure calculation, 468
  range and sensitivity, 467
  reagents, 468
  storage effect, 468
Counting efficiency, 61, 396, 442, 463
  estimation, 15
  reduction, 62, 166
Count mean diameter, 143
Count mode diameter, 143
Count rate (cpm), 72, 477
  background, 402
  CAM, 181
  errors, 71
  net, 72
  SD determination, 431
COV. *See* Coefficient of variance (COV)
Creeping flow. *See* Stokes' flow
Critical flow venturi (CFV), 486
  application, 251
Critical level. *See* Decision level (DL)
Critical measurement level (L_c). *See* Decision level (DL)
Critical orifice, 232–233
Cryostats, 65
CTBT. *See* Comprehensive Nuclear Test Ban (CTBT)
CTP. *See* Calibration temperature and pressure (CTP)
Cunningham slip correction factor, 128. *See also* Slip; Stokes' law
  diffusion regime, 125
  parameters, 95

D

DAC. *See* Derived air concentration (DAC)
DAC-h. *See* Derived air concentration hours (DAC-h)

Damage ratio (DR), 149
Darcy's law, 165
DCF. *See* Dose conversion factor
Decay. *See* Radioactive decay
Decision level (DL), 72, 294
  binary decision, 294
  blank count confidence interval, 295, 296
  determination, 295
Decontamination, 14, 336
Degree of deviation, 95
Denuder, 327n
Department of Energy (DOE), 35, 36, 73, 143, 272, 288, 481, 498, 506
Deposition velocity, 110
  determination, 378
  inclined tube, 333
  turbulent, 334–335
Depth filters. *See* Fibrous filters
Derived air concentration (DAC), 241, 344, 420, 431, 444, 498, 533
  calculation, 487
  determination, 217, 218
  occupational, 186, 196, 204
  radionuclide, 288
  sensitivity, 297
  tritium, 409
Derived air concentration-hours (DAC-h), 310, 500, 508
  number, 503
  plutonium DAC-h, 309
Desiccant, 14, 408
Detection limit, 69. *See also* Lower limit of detection (LLD)
  tritium-in-air measurement techniques, 409
Detector, 48, 58
  calibration, 261
  counting efficiency, 442
  filter geometry, 491
  germanium γ-detector, 451
  integrating, 66
  NaI (Tl) detectors, 63
  passive detector, 471
  passive α-track detector, 467
  photon detectors, 63
  radon detector, 476
  silicon surface barrier, 68
  solid-state detector, 68, 505
  TLD, 382
Determinants of exposure, 147–148
Dewpoint, 338n
DF. *See* Dilution factor (DF)
Diaphragm pumps, 237, 239
Diffusion, 88, 141. *See also* Brownian motion; Thermophoresis
  Blanc's law, 89
  diffusive deposition, 114
  equation, 91
  Fick's law, 88, 89, 111
  gaseous medium, 141
  isotropic, 124
  one dimension, 92
  process, 89
  Stefan–Maxwell equation, 89
  thermal, 138
  thermophoresis, 88

Diffusion coefficient, 89–90
    drag force, 97
    low pressures, 90
    particle, 91
    reciprocity condition, 89
    relative, 124
    spheroid particles, 97–98
    turbulent, 131
Digital volt meter and probe (DVM), 260
Dilution factor (DF), 278
Dimensionless stopping distance, 335
Dioctyl phthalate (DOP), 171
Direct current ionization chamber. *See* Electret ionization chamber
Directional cosine, 110
DL. *See* Decision level (DL)
DOE. *See* Department of Energy (DOE)
DOP. *See* Dioctyl phthalate (DOP)
Dose coefficient, 217, 218
Dose conversion factor (DCF), 150, 532
Dosimetry, 17, 532
DR. *See* Damage ratio (DR)
Drag coefficient, 94
Drag force, 93. *See also* Stokes' law
    drag coefficient, 94
    irregular particle, 98
    net, 97
    Stokes' formula, 101
    stopping distance, 334
Dufour effect, 88
Dust, 136
    accumulation, 165, 167
    airborne, 483, 491
    background, 344
    dust mask, 177
    loading, 393, 400
DVM. *See* Digital volt meter and probe (DVM)
Dynamic pumps, 236
Dynamic shape factor, 97, 101
    elongated bodies, 100
    spheroids particles, 99–100, 102

E

EC. *See* Electron capture (EC)
ECAM. *See* Environmental continuous air monitor (ECAM)
Eccentricity, 99
EEC. *See* Equilibrium equivalent concentration (EEC)
EF. *See* Exposure Fraction (EF)
Effective dose (E), 210, 218. *See also* Committed effective dose
    estimation, 458
    unit, 533
Efficiency, 177, 422
    alpha, 261
    aspiration, 329
    calibration, 261
    collection, 413
    collision, 128
    computational, 115
    counter, 404

counting, 422
detection, 483
detector counting, 442
filtration, 158, 212
geometric, 308
LSC, 430
modeling aspiration, 331
particle collection, 165
trapping, 430
Electret filters. *See* Electrostatic filters
Electret ionization chamber, 469
    background γ-radiation, 470
    calibration, 471
    MDC, 470
    radon measurement, 66
    uncertainty, 470
Electrets, 469
Electrometer, 469
Electron (e⁻), 56
    elementary electron charge, 66
    masses, 56
    orbital, 57
Electron capture (EC), 57
Electrostatic filters, 164
Electrostatic precipitation method
    accuracy, 475
    advantages, 475
    airborne ^{222}Rn measurement, 474
    alpha spectroscopy, 475
    apparatus, 476
    calculations, 477–478
    calibration, 477
    continuous sampling, 477
    disadvantages, 475
    end-of-run printout, 478
    general procedure, 476
    grab sampling, 477
    interferences, 474
    MDC, 474, 475
    precautions, 477
    precision, 475–476
    principle, 474
    radon detector, 476
    ranges, 474
    sensitivity, 474
    storage effect, 479
Elements. *See also* Particles
    actinides, 209
    alkaline earth elements, 209
    in bloodstream, 210
    fate, 210
    in generic mixing, 324
    iodine, 209
    molecular parameters, 87
    in organ and tissues, 209
    radiation dose, 210
    retention in body, 216
    uranium metabolic model, 217
EML. *See* Environmental Measurements Laboratory (EML)
Endcap, 65
Energy calibration, 486
Energy Research and Development Administration (ERDA), 379n

Environmental continuous air monitor (ECAM), 6, 357, 358
Environmental Measurements Laboratory (EML), 73n, 379n
Environmental Protection Agency (EPA), 35, 36, 328, 415, 423, 457, 498, 506
Environmental Radiation Ambient Monitoring System (ERAMS), 383
EPA. *See* Environmental Protection Agency (EPA)
Equilibrium equivalent concentration (EEC), 186, 464, 551, 552
Equilibrium factor (F), 186
Equivalent diameter, 79, 101, 139–141
    aerodynamic, 140
    mass, 96
    surface, 99
    thermodynamic, 141
    volume, 96, 98
Equivalent dose (H), 210, 533
ERAMS. *See* Environmental Radiation Ambient Monitoring System (ERAMS)
ERDA. *See* Energy Research and Development Administration (ERDA)
EU. *See* European Union (EU)
European Union (EU), 32, 35
Exponential tail, 301
Exposure
    characteristics, 147
    DAC-h, 309
    duration (T), 150
Exposure fraction (EF), 279
Exposure pathway model, 147–148
    exposure determinants, 147–148
    job-exposure matrix, 148
    key-parameter equation, 148
Extrathoracic region, 211–212

F

Fabrics, 178
Fallout, 369
    air sampling, 382, 385
    calculated back trajectory, 379
    debris, 378
    deposition of global, 377
    local, 371
    mechanisms governing, 373
    meteorological effects, 376–377
    monitoring, 379
    production, 371
    radionuclide composition, 374, 375
    regional, 371
    transport mechanism, 371
    tropospheric, 371
Fallout, radionuclide, 375
    accidental releases, 375
    activation products, 374
    Chernobyl debris activity, 376
    composition, 374
    fission products distribution, 374
    fractionation, 374
    radionuclides contribution, 375
    refractory nuclides, 374
    sampling air, 385

False alarm rate, 492
    α-CAM, 493
    in CAM, 297
    determination, 490
    false alarm probability, 493
    measurements per year, 492
    normal standard deviate calculation, 493
    NSD, 492
    plutonium concentration SD, 492
False-negative error, 294
Fanning friction factor, 332
Fans. *See* Dynamic pumps
FAS. *See* Filter air sampler (FAS)
FDA. *See* Food and Drug Administration (FDA)
Federal Guidance Recommendations (FGR), 39
FGR. *See* Federal Guidance Recommendations (FGR)
Fibrous filters, 164
Fick's law, 89. *See also* Blanc's law; Stefan–Maxwell equation
    diffusion, 111
    flow rate, 125
    mass flux, 88
    mean square distance, 91
    multicomponent gas mixture, 89
FIDLER. *See* Field instrument for the detection of low-energy radiation (FIDLER)
Field blank, 15
Field instrument for the detection of low-energy radiation (FIDLER), 18
Filter air sampler (FAS), 287, 312
Filter media
    α-spectroscopy, 166
    analytical and radiochemistry issues, 169
    avoiding unexpected changes, 170
    cellulose filters, 293
    fibrous, 305
    filter samples transportation, 169
    filter surface, 168
    filter type influence, 167
    improper, 392
    particle collection efficiency, 165
    physical characteristics, 160, 161–162
    pressure drop characteristics, 164
    radiation shielding, 166
    scanning electron photomicrographs, 163
    selection, 160, 169–170, 385
Filters, 335. *See also* Filter media
    air, 158
    gaseous radionuclides, 385
    with low porosity, 241
Filtration, 157. *See also* Filter media
    air filters, 158
    building filtration, 171
    considerations, 172
    fundamentals, 158
    materials, 157
    mechanisms, 158, 159
    MPPS, 158
    nuclear air and gas treatment, 170
    pore size, 158–160
    respiratory protection, 173
    strength, support, and sealing, 160

Index

Fissile, 374
Fission products, 374, 375, 547
Flow rate
   adequate, 233
   air sampler, 60, 429
   air sampling, 249
   ambient sampling, 251
   angular, 109
   net, 110
   particle, 110
   pump, 447
   rotameters, 227, 268
   standardized sampling, 251
   volumetric, 227, 233
Flow straighteners, 326n
Fluence, 109
Fluence rate. *See* Flux
Fluoropore, 166. *See also* Filter media—selection
   efficiency, 168
   filter, 168
   Teflon™ membrane filter, 163, 165
Flux, 109
   convective, 113
   mass, 88–89
   particle angular, 108
   scalar, 108
   total, 108
Fogs, 137. *See also* Mists
Food and Drug Administration (FDA), 178
Form drag, 101
Fractionation, 374
Freemolecular regime, 141
Full-width-at-half-maximum (FWHM), 64, 293, 491, 526
Fumes, 136
FWHM. *See* Full-width-at-half-maximum (FWHM)

G

Gain, 111, 475
Gamma distribution, 82. *See also* Lognormal distribution
   arbitrary units, 83
Gamma radiation. *See* Gamma rays
Gamma rays, 57
Gamma-spectroscopy
   advantages, 448
   air concentration, 454
   background radioactivity subtraction, 453
   calibration, 452
   cautions, 455
   compressed-gas sampling system, 449–451
   counting system, 451
   cylinder volume, 454
   detector-to-source geometry, 453
   disadvantages, 448–449
   γ-emitting radionuclides, 447
   germanium γ-detector, 451
   ideal gas law, 449
   interferences, 448
   MDC, 448
   mixed γ-standard, 452, 453
   precision and accuracy, 449
   principle, 451
   procedure, 53–54
   radioactive decay equation, 454
   radioactivity counting, 453
   range and sensitivity, 448
   sample collection cylinders, 449
   storage effects, 454–455
Gas, 137
   aerodynamic factors, 135, 136, 138
   concentration, 454
   dispersal factors, 138
   equilibrium factor, 186
   multiplication, 67
   particle collection mechanism, 138–139
   pycnometry methods, 342
   thermodynamic behavior, 139
Gas proportional counter (GPC), 16, 67–68, 394, 402
Gas sampling system, 449–451
   exhaust module, 451
   gas cylinder module, 450
   inlet module, 450
   parts, 449
Gastrointestinal tract (GI tract), 209, 215–216. *See also* Particles
Gauge pressure, 234. *See also* Manometer
Gay-Lussac law, 228
GDE. *See* General Dynamic Equation (GDE)
Gear pumps, 237
   external, 238
   internal, 237
Geiger Mueller (GM) counter, 16, 364, 402
Gel layer. *See* Mucous
General calibration procedures, 246, 260
   air sampling flow rates, 249
   documentation, 262
   efficiency calibration, 261
   electrical testing, 261
   equipment, 260
   example calibration record, 262–263
   flow calibration, 262
   hazard analysis, 260
   identified hazards and mitigation, 260
   inspection and repair, 260
   performance check, 261
   procedure, 260
   pulser calibrations, 261
   purpose, 260
   radiation detector systems, 246, 247
   references, 260
   scope, 260
General Dynamic Equation (GDE), 110, 117. *See also* Coagulation
   full form, 112
   simplified form, 111
Geometric mean, 81
Geometric shape factor, 99
Geometric standard deviation, 147
Germanium detectors, 65–66, 451, 453. *See also* Photon detectors
GI tract. *See* Gastrointestinal tract (GI tract)
GM counter. *See* Geiger Mueller (GM) counter
GPC. *See* Gas proportional counter (GPC)

Grab-sample collection method
  airborne ^{222}Rn measurement, 471–472
  calibrations, 473
  interferences, 472
  PMT, 472
  PMT/scaler combination, 473
  precision and accuracy, 472–473
  procedure, 473
  Pylon Model AB-5, 472
  radon concentration calculation, 473
  ranges and sensitivity, 472
  scintillation cell, 472, 473
  storage effect, 473
Grab sampling, 234, 477
  CAM, 359
  disadvantage, 475
  pump for, 240
  volumetric grab samples, 348
Gravity, 78
Gross sample count rate, 414, 422
Gross α-air concentration, 397
Ground zero (GZ), 374
GZ. *See* Ground zero (GZ)

H

Half-life, 58. *See also* Radioactive decay
  actinium, 201
  iodine isotopes, 441
  long half-life effect, 61
  noble gases, 441
  radiation sources, 250
  radionuclide, 59
  short half-life effect, 61
  thorium, 182, 193
  uranium, 182, 183
Half-value time, 121
Harmonic mean diameter, 105
Harmonization, 51
  benefits, 51
  CEN, 32
  instrument types or classes, 51
  life-cycle process, 51, 52
HASL. *See* Health and Safety Laboratory (HASL)
Haze, 137. *See also* Mists
HD. *See* Hydraulic diameter (HD)
Health and Safety Laboratory (HASL), 379n. *See also* Environmental Measurements Laboratory (EML)
  balloon aerosol, 386
  β-activity measurements, 379
  gummed-film collectors, 380
  in situ γ-ray spectrometry, 380, 381–382
  precipitation collectors, 380, 381
  SASP filters, 383
  surface air sampling systems, 384
  wet/dry collector, 381
Health Physics Society (HPS), 33. *See also* American National Standards Institute (ANSI)
Heating, ventilation, and air conditioning (HVAC) filters, 170–171
HEPA filters. *See* High-efficiency particulate air (HEPA) filters
Hexamethylene tetramine (HMTA), 437

High-efficiency particulate air (HEPA) filters, 171
High voltage (HV), 247
HMTA. *See* Hexamethylene tetramine (HMTA)
HPS. *See* Health Physics Society (HPS)
HT. *See* Tritiated hydrogen gas (HT)
HTO. *See* Tritium oxide (HTO)
Human respiratory tract, 211
  absorbed fraction, 215
  absorption, 213, 214
  air monitoring, 217
  amount of air breathed, 218
  clearance, 213, 215
  dose estimation, 214–215
  filtration efficiency, 212
  functional residual capacity, 212
  lower, 211–212
  models, 210
  morphology, 211
  particle collection mechanisms, 139
  particle deposition, 141
  radionuclides, 215
  regional deposition, 212
  sequestered material, 214
  tidal volume, 212
  upper, 211
HV. *See* High voltage (HV)
HVAC filters. *See* Heating, ventilation, and air conditioning (HVAC) filters
Hydraulic diameter (HD), 318
Hydrosol. *See* Liquisol

I

IAEA. *See* International Atomic Energy Agency (IAEA)
IC. *See* Incident Commander (IC)
ICRP. *See* International Commission on Radiological Protection (ICRP)
ICRU. *See* International Commission on Radiation Units and Measurements (ICRU)
Ideal gas law, 87, 266, 449
  air, 228
  effective volume, 454
IEC. *See* International Electrotechnical Commission (IEC)
IEEE. *See* Institute for Electrical and Electronic Engineers (IEEE)
Incident Commander (IC), 362, 363
Industrial hygiene standard conditions, 228
Inertials, 136. *See also* Dust
Inhalable particles, 281
Initiating-event probabilities, 152, 153
In-leakage, 251, 336
In situ γ-ray spectrometry, 380
Institute for Electrical and Electronic Engineers (IEEE), 33
Instrument, 33
Interception, 138
Internal bremsstrahlung radiation, 57
Internal void fraction, 96
International Atomic Energy Agency (IAEA), 38
International Commission on Radiation Units and Measurements (ICRU), 38
International Commission on Radiological Protection (ICRP), 36, 37, 38, 210

Index

International Electrotechnical Commission (IEC), 23, 481
  national standards organizations, 24–25
International Organization for Standardization (ISO), 23, 32, 49, 252
  standards, 29–30
International standards, radioactivity, 23
  air monitoring standards, 23
  ANSI, 25–26
  ASTM, 26–27
  European standards, 30
  French national standards, 30–32
  IEC, 24, 27–29
  ISO, 32, 29–30
  regional international standards, 32
Iodine (I), 543
  aerosol properties, 545
  concentration, 437
  exposure to DCF, 544
  inhalation biological data, 544
  measurement in air, 438
  radioactive, 435
  radioisotopes, 441
  risk coefficients, 544–545
  simplified decay scheme, 544
  sources, 545
  special considerations, 545
  thyroid affinity, 209, 216
  volatility, 138
Ionization chamber, 66–67, 468
  airborne ^{222}Rn measurement, 468
  apparatus, 471
  calibrations, 471
  commercial, 67
  direct current, 66, 469
  electret, 469–470
  humidity, 67
  interferences, 470
  passive electret-based device, 469
  precision and accuracy, 470
  procedure, 471
  pulse, 66, 468–469
  ranges, 470
  sensitivity, 470
ISO. *See* International Organization for Standardization (ISO)
Isokinetic, 329
  probe (F), 330
  sampling, 328
Isotropic
  eddies energy dissipation rate, 131
  particles, 124

K

Kelvin temperature scale, 230
Kernel approximation, 123. *See also* Coagulation
Kinetic energy, 55, 85
Krypton, 137, 541
  discharge units, 348
  exposure to DCF, 541, 542
  inhalation biological data, 541, 542
  properties, 543
  risk coefficients, 542

  simplified decay scheme, 541
  sources, 543
  special considerations, 543

L

Laminar flow, 326n
LAN. *See* Local Area Network (LAN)
Lead, 552, 556
  aerosol properties, 553, 557
  background radiation reduction, 63, 65
  dose conversion factors, 553, 557
  inhalation biological data, 552, 556
  risk coefficients, 553, 557
  sources, 553, 557
  special considerations, 553, 557
Leak path factor (LPF), 150
Length/hydraulic diameter (L/HD), 324
Levenberg–Marquardt method. *See* Nonlinear least squares method
L/HD. *See* Length/hydraulic diameter (L/HD)
Life-cycle steps, 43
  acceptance testing, 49
  development and application, 45, 46–47
  functional checks, 50
  initial calibration, 49
  instrumentation, for, 45
  maintenance and recalibration, 50
  mission evaluation, 44
  operational experience, 50
  periodic performance testing, 51
  production control testing, 49
  prototype testing, 48
  research and development, 45
  training, 49
  type testing, 48
Linear pumps, 239
Liquid scintillation counter (LSC), 13, 58, 419, 426
  apparatus, 411, 420
  calibration, 413
  cocktail scintillates, 62
  computers in, 62
  counting, 429
  cpm to dpm conversion, 15
  dust loading counting, 393
  efficiency, 430
  interfering effects, 62
  LLLSC, 408
  methodology, 394, 402
  radiation detection, 61
  sample preparation, 62
Liquisol, 77–78. *See also* Aerosols
LLD. *See* Lower limit of detection (LLD)
LLLSC. *See* Low-level liquid scintillation counters (LLLSC)
Lobe pumps, 237, 238
Local Area Network (LAN), 305
Lognormal distribution, 81
  aerosol number distribution, 81
  arbitrary units, 83
  feature, 143
  normal distribution and, 146
  statistics, 82
  volume distribution, 83

Lost opportunity for alarm, 295
Lower limit of detection (LLD), 71–73, 392, 400, 410, 420, 436, 461, 502
   calculation, 289
   DAC, 241
   factors influencing, 344, 345
Low-level liquid scintillation counters (LLLSC), 408
LPF. See Leak path factor (LPF)
LSC. See Liquid scintillation counter (LSC)

M

Manometer, 232, 234, 241
MAR. See Material-at-risk (MAR)
Mass diameter, 80, 104, 105, 144
Mass-equivalent diameter, 96
Mass flow meters, 230, 231
   application, 500
   calibration, 251
   deviations in accuracy, 232
   Wheatstone bridge circuit, 231
Mass transfer. See Aerosol—transport
Material-at-risk (MAR), 149
Maximum permissible concentration (MPC), 185
Maxwell–Boltzmann distribution, 85, 88, 126
Maxwellian. See Maxwell–Boltzmann distribution
MCA. See Multichannel analyzer (MCA)
MDA. See Minimum detectable activity (MDA)
MDC. See Minimum detectable concentration (MDC)
Mean, 254
Mean free path, 86, 141
   determination, 87
Mean particle diameter, 104
Media blank, 15
Median diameter, 80, 144
Mega-tons (MT), 370
MERV. See Minimum Efficiency Reporting Value (MERV)
Minimum detectable activity (MDA), 71, 294, 420, 490
Minimum detectable concentration (MDC), 392, 400, 436, 448, 461
   advantages, 475
   determination, 474
Minimum Efficiency Reporting Value (MERV), 172–173
Mists, 137
Mole fraction, 89
Moments transformation, 121
Momentum balance equation, 93
Monitoring, 3
Monodisperse aerosols. See Particles, monodisperse
Monodisperse system, 79
   binary coagulation, 121
Monte Carlo simulation
   conditional probability, 152, 153
   initiative-event probabilities, 152, 153
   input and output parameters, 153
   statistical distributions, 151, 154
   triangular distribution, 151
   uncertainty modeling, 151
Morbidity risk coefficient, 532
Mortality risk coefficient, 532
Most penetrating particle size (MPPS), 158
MPC. See Maximum permissible concentration (MPC)
MPPS. See Most penetrating particle size (MPPS)

MT. See Mega-tons (MT)
Mucous, 215
Multichannel analyzer (MCA), 58, 505
   applications, 45, 287, 292
   computer-based, 58
Multiple-frame-of-reference method, 264
   example calculations, 269
   factors influencing reliability, 270
   frames of reference, 265
   ideal gas law, 266
   objectives, 264
   operation, 265
   recommendations, 270
   rotameter equation, 266
   rotameter setup for calibration, 264
   scale factor equation, 267

N

NAFA. See National Air Filtration Association (NAFA)
NaI (Tl) detectors, 63. See also Photon detectors
   counting system, 64
   detection system calibration, 64
   PMT, 63
   radiation detection, 63
   spectral peaks resolution, 63–64
NARAC. See National Atmospheric Release and Advisory Center (NARAC)
National Air Filtration Association (NAFA), 172
National Atmospheric Release and Advisory Center (NARAC), 358
National Council on Radiation Protection and Measurements (NCRP), 38, 211
National Emission Standards for Hazardous Air Pollutants (NESHAP), 36, 447
National Environmental Health Association's (NEHA), 459
National Nuclear Security Administration (NNSA), 170
National Personal Protection Technology Laboratory (NPPTL), 177
National Radon Safety Board (NRSB), 461
National Institute of Health (NIH), 274
National Institute for Occupational Safety and Health (NIOSH), 171
National Institute of Standards and Technology (NIST), 34, 251, 463, 486
National standards, radioactivity, 32. See also International Standards, radioactivity
   ANSI-accredited organizations, 34
   France, 30, 34
   organizations, 24–25
   US, 33. See American National Standards Institute (ANSI)
National Standards System Network (NSSN), 34
National Technology Transfer and Advancement Act (NTTAA), 35
Navier–Stokes equation. See Momentum balance equation
NCRP. See National Council on Radiation Protection and Measurements (NCRP)
Negatron, 56. See also Beta particles
NEHA. See National Environmental Health Association's (NEHA)

Index

NESHAP. *See* National Emission Standards for Hazardous Air Pollutants (NESHAP)
Net activity, 60
Net count rate, 72, 396
Nevada Test site (NTS), 369
    conducted tests, 372
Newtonian fluids, 86
NIH. *See* National Institute of Health (NIH)
NIOSH. *See* National Institute for Occupational Safety and Health (NIOSH)
NIST. *See* National Institute of Standards and Technology (NIST)
NNSA. *See* National Nuclear Security Administration (NNSA)
Noble gases, 137, 376
    argon, 540
    half-life, 441
    krypton, 543
    radioactive, 440
    radon, 182
    transportation
Noise, 65, 240
    electronic, 297
Nonlinear least squares method, 302–303
Nonspherical particles, 96–97, 99
Normal stress. *See* Form drag
Normal temperature and pressure (NTP), 266, 269–270
No-slip condition, 94, 113
Nozzle, 327–328
    air velocity, 329
    angular flow, 317
    aspiration ratio computation, 332
    axis, 328
    performance results, 331
    placement, 316
    sampling, 330
NPPTL. *See* National Personal Protection Technology Laboratory (NPPTL)
NRC. *See* Nuclear Regulatory Commission (NRC)
NRSB. *See* National Radon Safety Board (NRSB)
NSDs. *See* Number of Standard Deviations (NSDs)
NSSN. *See* National Standards System Network (NSSN)
NTP. *See* Normal temperature and pressure (NTP)
NTS. *See* Nevada Test site (NTS)
NTTAA. *See* National Technology Transfer and Advancement Act (NTTAA)
Nuclear Regulatory Commission (NRC), 12, 35, 36, 272, 482
    acceptable surface contamination levels, 15
    Agreement State, 12
    ALI, 288, 532
    respiratory protection, 174, 176
Nuclear weapons tests, 370, 379. *See also* Fallout
    C-14 atmospheric inventory, 539
    Chernobyl accident, 369–370
    fallout transport mechanisms, 373–374
    NTS test effects, 371–372
    number and yield, 371
    test sites, 370
    TRINITY test, 370
Nucleons, 56
Nucleus, 55
Number density, 78, 79
    dilute suspensions, 84
Number of standard deviations (NSDs), 492

O

Occupational inhalation limit. *See* Derived Air Concentration (DAC)
Occupational safety, 16
Occupational Safety and Health Administration (OSHA), 35, 37
    respiratory protection, 176, 177
Office of Management and Budget (OMB), 35
OMB. *See* Office of Management and Budget (OMB)
Operating temperature and pressure (OTP), 265, 269
Optical particle counters, 344–346
OSHA. *See* Occupational Safety and Health Administration (OSHA)
OTP. *See* Operating temperature and pressure (OTP)
Oversight, 12

P

Pacific Area Standards Congress (PASC), 32
Pacific Northwest National Laboratory (PNNL), 383
PAEC. *See* Potential alpha energy concentration (PAEC)
PAPRs. *See* Powered air-purifying respirators (PAPRs)
Paralysis, 437
Particle (s), 209
    angular flux, 108
    coating, 137
    gastrointestinal tract, 209, 215–216
    impact on body, 209
    polydisperse, 129
    respiratory tract, 209
    resuspension, 142–143
    systemic metabolism, 216–217
Particle collection
    efficiency, 165
    mechanism, 138–139
    morphological examination, for, 347
Particle density, 96, 116, 121
    on air sampling devices, 140
    angular, 107
    average, 113
    distribution function, 105–106
    gradient, 114
    momentum-dependent, 108
    scalar, 107, 108
    total, 107, 109
Particle deposition, 110
    average particle density, 113
    boundary conditions, 111
    coagulation, 112
    continuity equation, 110
    diffusive boundary layer thickness, 114
    minimization, 338
    surface deposition, 114
Particle motion
    Boltzmann distribution function, 84
    Brownian, 132
    convection, 88
    Cunningham's slip correction, 95

Particle motion (*Continued*)
   diffusion, 88
   drag force, 93
   fluid properties, 84
   in gas, 84
   Maxwellian, 85
   mean free path, 86
   nonspherical particles, 96
   particles and molecules diffusion, 88
   terminal settling velocity, 140
   vertical displacement vs. diffusion mechanisms, 139
Particle size, 141
   aerosol property, 103
   deposition mechanisms, 212
   distribution, 212
   harmonic mean diameter, 105
   integral property, 103
   lognormal distribution, 143
   mean particle diameter, 104
   mean particle surface, 104
   mean square diameter, 104
   particles mean volume, 104
   spectrum moments, 102–106
Particle size distribution, 38, 118, 129
   characteristics, 212
   continuous, 117
   lognormal aerosol, 144
   moments, 102
Particles, monodisperse, 123, 212. *See also* Monodisperse system
   coagulation coefficient, 116
   particle density, 121
PAS. *See* Personal air sampling (PAS)
PASC. *See* Pacific Area Standards Congress (PASC)
Passive charcoal method, 464
   airborne ^{222}Rn determination
   back-diffusion, 464
   charcoal canister, 464, 465
   components, 465
   effective sampling rate, 466
   γ-ray spectrometer, 465, 466
   gaseous radon adsorption, 464
   interference, 465
   limitations, 464
   passive activated-carbon monitors, 465
   precision and accuracy, 465
   procedure, 465–466
   radon concentration calculation, 466
   ranges, 464–465
Passive detector, 473. *See also* Ionization chamber
PDA. *See* Personal Digital Assistant (PDA)
PDCA. *See* Plan-Do-Check-Act (PDCA)
Peak-fitting method, 302, 303
Penetration, 225, 332. *See also* Stokes number (Stk)
   bends, 335
   particles, 166, 332
   penetrating fraction, 332
   straight tube, 333
PERALS. *See* Photon–electron rejecting α-liquid scintillation spectrometer (PERALS)
Personal air sampling (PAS), 306, 497, 505
   air sample, 499, 501
   apparatus, 500
   calibration, 500, 501
   cautions, 502
   closed-face filter cassette, 499
   cross-contamination, 498
   DAC, 498, 501
   γ-spectrometric analysis, 499
   interference, 498–499
   liquid scintillation counting, 498
   precision and accuracy, 499
   principle, 497–498
   procedure, 500–501
   range, 498
   sample volume calculation, 501
   sensitivity, 498
   standard deviation (SD), 502
   statistical uncertainty, 502
   storage effect, 502
Personal Digital Assistant (PDA), 507
Personal protective equipment (PPE), 297
Petri-slides, 169
PHA. *See* Pulse-height analysis (PHA)
Phosphors. *See* Scintillators
Phoswich, 18. *See also* Scintillation probes
Photomultiplier tube (PMT), 61, 63, 462, 472
   alpha-radiation emissions detection, 18
   radon measurement, 471
   scintillations detection, 61
Photon detectors, 63
   germanium detectors, 65
   NaI (Tl) detectors, 63
Photon–electron rejecting α-liquid scintillation spectrometer (PERALS), 498
PHS. *See* Public Health Service (PHS)
Piezoelectric monitoring systems, 341–342
Piston pumps, 236, 239, 240
Plan-Do-Check-Act (PDCA), 49
Plume measurement, 357, 358
   AMS, 359–360
   environmental CAM, 358–359
   grab sampling methods, 359
   resuspension measurements, 359
   at source, 358
Plutonium (Pu), 17, 558, 560
   aerosol properties, 559, 561
   α-CAMs, 481
   background SD, 300
   contamination detection, 18
   dioxide aerosol, 166
   exposure to DCF, 558, 561
   inhalation biological data, 558, 560
   isotopes, 17
   risk coefficients, 559, 561
   ROI, 300
   simplified decay scheme, 558, 560
   sources, 559, 561
   special considerations, 559–560, 562
PMT. *See* Photomultiplier tube (PMT)
PNNL. *See* Pacific Northwest National Laboratory (PNNL)
Polar axis, 96–97
Polonium (Po), 166
   aerosol properties, 551, 556
   exposure to DCF, 550, 555
   inhalation biological data, 550, 555
   risk coefficients, 551, 556

Index 577

sources, 551, 556
special considerations, 551–552, 556
Positron, 56. *See also* Beta particles
Potential alpha energy concentration (PAEC), 185–186
  actinon, 204
  radon, 185
  thoron, 196
Powered air-purifying respirators (PAPRs), 176
PPE. *See* Personal protective equipment (PPE)
Precision, 70, 255, 298
Press-fit, 223
Pressure drop, 240. *See* Backpressure
  across orifice, 232
  characteristics, 164
  dust loading, 400
  venturi meters, 233
Probe placement, 316
  configurations, 318, 319–321, 326
  downdraft fans, 322
  general, 316
  generic mixing tests, 324
  horizontal stack, 323
  idealized sampling system, 327
  mixing, 316, 317
  mixing demonstration method, 318
  scale models, previously tested, 324
  static mixer, 325
  tested mixing elements, 325
Probit, 145
Production control testing, 49
Public Health Service (PHS), 380
Public Law (P.L.) 104–113. *See* National Technology Transfer and Advancement Act (NTTAA)
Pulse-height analysis (PHA), 63
Pylon Model AB-5, 472

Q

Quenching, 62–63, 413–414

R

Radiation, 55
  detection systems review, 61
  dose (REM), 222
  energy, 58
  safety, 11
Radiation detection, 61
  alpha-radiation counters, 68
  gas proportional counters, 67–68
  ionization chambers, 66
  LSC, 61
  photon detectors, 63
Radiation detector systems, 247
  calibration, 250
  daily operational test, 248, 249
  precalibration, 246, 247
  radioactive reference sources, with, 247
Radiation Safety Officer (RSO), 12
Radiation safety program, 11, 12
  basic components, 12
  contamination control, 13
  dosimetry, 17
  fundamental elements, 11

occupational safety, 16
personal protection, 16
regulatory concerns, 12
sampling and handling plutonium, 17
training, 12
Radioactive fallout, 369, 371. *See also* Fallout
  air sampling, 378, 382–387
  AMD, 372n
  back-trajectory analysis, 378, 379
  Chernobyl accident, 369–370
  concentration measurement, 378
  deposition velocity, 378
  detection in atmosphere, 378
  EPA involvement, 382
  exposure rates, 382
  global fallout, 371
  HASL investigation, 379–382
  local fallout, 371, 372
  meteorological effects, 376–377
  nuclear test effects, 370
  Pu-238 generator accident, 370
  radiation dose, 378
  radionuclide, 374–376, 379
  removing processes, 369
  transport mechanisms, 373–374
  tropospheric fallout, 371
  weapon test sites, 370–371
Radioactive aerosols, 38, 271
  α-emitting particles autoradiograph, 349
  analytical chemistry methods, 348–349
  BET method, 351
  characterization and assessment, 343
  detection, 342–343, 349
  electrical properties measurement, 347–348
  filtration, 157, 347
  graded approach, 342, 343
  guiding principle, 343
  isopycnic density-gradient ultracentrifugation, 350–351
  krypton, 348, 351
  Lucas cell, 348
  optical particle counters, 344–346
  particle solubility measurement, 349, 350
  radioactive particles collection, 347
  real-time inertial techniques, 347
  routine monitoring and control, 343
  sampling, 343–344, 347, 348
  scintillators, 348
  spiral duct centrifuge, 347
  surface area measurement, 351
Radioactive debris. *See* Radioactive fallout
Radioactive decay, 55, 57, 70, 454. *See also* Half-life
  alpha particles, 55
  beta particles, 56
  counting interval, 453
  decay products, 182
  equation, 454
  f factors, 184, 195, 202
  gamma rays, 57
  law, 59
  positron emission, 56
  rate of collection, 59–60
  recurrence formula, 184
Radioactive gases, 137, 437, 448

578 Index

Radioactive iodine, 435. *See also* Iodine (I)
   activity, 443
   air concentrations, 444
   apparatus, 438
   calibration, 437, 440, 441, 442
   cartridge counting, 440
   cautions, 444
   charcoal canisters protection, 439
   chemisorption, 439
   detector counting efficiency, 442
   EPA software packages, 444
   exposure time calculation, 444
   flow measuring device, 438
   grade coconut charcoal use, 439
   interference, 436–437
   isotopic exchange, 439
   kinetic adsorption, 439
   MDC, 436
   organic iodides collection, 437
   precision and accuracy, 438
   problems ^{133}Xe, 437
   radioisotopes half-lives, 441
   radiological counting, 437
   sample loss, 437
   sampling, 435–436, 438, 439–440
   silver-zeolite cartridge, 435
   standard deviation, 443–444
   storage effect, 444
   upper range, 436
   water, 439
Radioactive materials, airborne, 12, 143, 288, 345
   aerodynamic factors, 135, 136, 138, 139–141
   dusts, 136
   fumes, 136
   gases, 137
   mists, 137
   mixed physical forms, 137
   smokes, 136
   vapors, 137
Radioactivity, 15, 166, 532
   alpha particles, 55–56
   beta particles, 56–57
   counting statistics, 69
   gamma rays, 57
   half-life, 58, 61
   radiation detection, 61
   radiation energy, 58
   radioactive decay, 55, 59
Radioactivity, airborne. *See* Monte Carlo simulation
Radioactivity counting statistics, 69
   detection limits, 71–73
   error terms, 69
   quality assurance, 73
   significant figures, 69
Radioactivity, internalized, 210
   dose coefficient, 218
   effective dose, 210
   radionuclide, 217
Radio frequency (RF), 149, 297, 305
Radioisotope physical forms, 137
Radiological emergencies, 6
Radiological monitoring, first responder, 362
   alarm point guidance, 366
   dosimeter alarm set points, 365
   dosimeter purpose, 367
   down-range stay time table, 366–367
   emergency radiation control zones, 364
   environmental monitoring, 362
   incident command structure, 362, 363
   incident priorities, 362
   initial incident response, 362–363
   operational conditions, 365
   radiation hazard assessment, 363–365
   radiation measurement areas, 365
   radiation measurements techniques, 364–365
   single point-in-time measurement, 362
   size-up, 362
   stay times, 365
Radionuclide, 57–58, 217, 374, 391
Rad Net, 305
Radon (^{222}Rn), 182
   Bateman equations, 184
   biological data for inhalation, 548
   decay scheme, 458
   decay series, 183
   deposition effect, 187
   detection techniques merits/demerits, 460
   determination method, 459
   EEC, 186
   exposure to DCF, 549
   f factors, 184–185
   h factors, 190
   ingrowth, 187, 191–192
   intercomparison test programs, 458
   measurement protocols, 459
   method cautions, 461
   PAEC, 185
   progeny, 181, 182, 188–189, 190
   radon exposure services, 458, 459
   Radon Proficiency Programs, 459
   recurrence formula, 184
   risk coefficients, 549
   rotameter, 461
   simplified decay scheme, 548
   sources, 549
   special considerations, 550
   temporal variations occurrence, 458
Radon gas
   energy measurement, 348
   formation, 13
   specifications for measuring, 26
Radon progeny, 181, 182
   α-emitters, 292
   concentration, 186
RAMP. *See* Remote Air Monitoring Program (RAMP)
Reciprocity condition, 89. *See also* Diffusion coefficient
Recurrence formula
   actinon, 205
   Bateman equations, 184
   decay equation, 202
   radon, 190
   thoron, 194–196
Regional fallout. *See* Radioactive fallout—tropospheric fallout
Region of interest (ROI), 167, 287, 483
   method, 300
Regulations, radioactivity

Index

CFR, 12, 35
DOE, 35, 73, 143, 272
government, 35
NTTAA, 35
perspectives, 39
recommendations and guidance, 37
regulatory authorities, 23
regulatory concerns, 12
relationships among, 22
US radiation protection regulations, 35
Relaxation time, 93, 94, 122
REM. *See* Radiation—dose (REM)
Remote Air Monitoring Program (RAMP), 383
Resolution, 64
Respirable size range, 147
Respirable fraction (RF), 149
Respirable particles, 280
    dispersal, 272
Respirator filters, 178
Respiratory protection
    APF, 173, 176
    electrostatic formulations, 164
    filter media classes, 176
    filtration, 173–174
    guidance, 34, 36, 37, 143, 147
    legal requirements, 174
    masks, 177
    physical stress reduction, 505
Response time of a measuring system, 490
Resuspension, 357
Retrospective modeling, 148, 153
Reynolds number (Re), 225, 228, 324n
RF. *See* Radio frequency (RF); respirable fraction (RF)
Risk of Health Effect equation, 149
Rocking piston pumps, 239
ROI. *See* Region of interest (ROI)
Rotameters, 265, 226–227
    Bernoulli Principle, 265
    calibration, 264
    equation, 266
    equation of air flow, 227, 229
    factors influencing reliability, 270
    flow rates, 268
    materials used, 227
    overcome pressure drop, 230
    scale readings, 266, 267
Rotary vane pumps, 236–237
RSO. *See* Radiation Safety Officer (RSO)

S

Sampling, 3
Sampling pumps, 233
    backpressure, 240
    characteristics, 234
    choice, 240
    diaphragm pumps, 237, 239
    free air capacities, 235
    gear pumps, 237, 238
    linear pumps, 239
    lobe pumps, 237, 238
    miniature, 240
    oil versus oil-less pumps, 239
    piston pumps, 239
    pulsation control, 242
    rocking piston pumps, 239
    rotary vane pumps, 236–237
    terminology, 234
    types, 235
Sampling train, 14, 327
    alignment, 328
    aspiration efficiency, 329
    components, 222
    contraction fittings, 225
    critical orifices and venturi meters, 232
    filter holders, 223–224
    flow measurement, 226
    inlet port, 222
    isokinetic concept, 328–329
    mass flow meters, 230
    modeling aspiration efficiency, 331
    nozzles, 327, 330–331
    pumps, 233–240
    rotameters, 226
    sampling illustration, 328
    subisokinetic concept, 329
    superisokinetic concept, 329
    transmission ratio, 329, 330
    tritium, for, 428
    tubing, 224–226
    wall losses, 329
SARs. *See* Supplied-air respirators (SARs)
SASP. *See* Surface air sampling program (SASP)
Saturation vapor pressure, 137
Scalar particle density, 107–108
Scale factor equation, 267
SCBA. *See* Self-contained breathing apparatus (SCBA)
Scintillation cell, 204
    in detection techniques, 460
    interferences, 472
    nitrogen flushing, 193, 201
    radon, 186, 462
    significance of volume, 472
Scintillation counter, 14
Scintillation fluid, 61, 62, 411, 423. *See also* Liquid scintillation counter (LSC)
    photoluminescence, 427
    tritium in, 432
Scintillation probes, 16
Scintillators, 348
    β-detectors, 18
SCs. *See* Subcommittees (SCs)
SD. *See* Standard deviation (SD)
Secular radioactive equilibrium
Self-contained breathing apparatus (SCBA), 176
Sequestered material, 214
Settling velocity, 79, 97, 128
    cross-stream component, 333
    gravitational, 333
    irregular particle, 98
    terminal, 94, 140
    uranium oxide, 140
Shielding
    background reduction, 437
    radiation, 402, 484
    radio frequency noise, 297
    self shielding, 462
SI. *See* System of Units (SI)

Sieverts per becquerel (Sv/Bq), 217
Significant figures, 69
Single channel analyzer (SCA)
Size-up, 362
Slip, 95. *See also* No-slip condition
Smog, 137. *See also* Mists
Smoke, 136
 generators, 280–282
Smoluchowski coagulation equation, 116
Sol layer. *See* Mucous
Solid adsorbents. *See* Desiccant
Solid-state detector, 68, 505
 front face polarity, 48
 silicon and germanium, 58
Source tissue, 215
Specific effective energy, 215
Sphericity, 99. *See also* Geometric shape factor
Spheroid particles correction factor, 98
Standard deviation (SD), 70, 254, 404, 414, 490
Standard deviation unit. *See* Probit
Standard error, 70
Standard temperature and pressure (STP), 252, 266
Static pressure (SP)
Stefan–Maxwell equation, 89
Stokes' diameter, 96
Stokes–Einstein equation, 91
Stokes' flow, 96
Stokes' law, 93
 modified, 95
Stokes number (Stk), 225
 dimensionless, 332
Stopping distance, 334. *See also* Dimensionless stopping distance
STP. *See* Standard temperature and pressure (STP)
Subcommittees (SCs), 25
Subisokinetic, 329
Sum kernel, 129
Superisokinetic, 329
Supersaturated vapors, 137
Supplied-air respirators (SARs), 176
Surface air sampling program (SASP), 383
Surface area diameter, 144
Surgical masks, 177–178
Sv/Bq. *See* Sieverts per becquerel (Sv/Bq)
System of Units (SI), 252

T

T. *See* Exposure—duration (T)
TA. *See* Time to alarm (TA)
Tails, 147. *See also* aerosol—distribution and release
TC. *See* Technical Committee (TC)
Technical Committee (TC), 25
TEDA. *See* Triethylenediamine (TEDA)
Teflon membrane filters, 164, 168
Terminal settling velocity, 140
 uranium oxide, 140
Thermal diffusion. *See* Thermophoresis
Thermodynamic diameter, 141, 142
 particle density, 141
 particle size, 141
Thermodynamic processes, 212
Thermoluminescent detectors (TLD), 382
Thermophoresis, 88

Thomas method, 461
 alpha-counting equipment, 462
 calibration, 463
 counting systems, 463–464
 dry-gas meter, 462
 interference, 462
 membrane filters, 462
 precision and accuracy, 462
 principle, 461
 procedures, 463
 range, 461–462
Thoracic region, 211
 alveolar–interstitial (AI), 212
 bronchial (BB) region, 212
 systemic, 210
Thorium, 182. *See also* Thoron (^{220}Rn)
Thoron (^{220}Rn), 182, 193
 decay series, 193, 194
 deposition, 198, 204
 f factors, 194–196
 h factors, 198–199
 ingrowth, 197, 200
 as interference, 392, 400
 PAEC, 196
 progeny, 198, 200
 recurrence formula, 194
Three Mile Island accident, 138
Time to alarm (TA), 279
TLD. *See* Thermoluminescent detectors (TLD)
Total deposition, 141
Total effective dose equivalent (TEDE)
Traceability chain, 252
Track-etch filters, 164
Transition regime, 141
Transmission ratio, 329
Triangular distribution, 151
Triethylenediamine (TEDA), 437
TRINITY test, 370, 374. *See also* Nuclear weapons tests
Tritiated hydrogen gas (HT), 137
Tritiated water (HTO), 137
Tritiated water vapor content determination
 absolute humidity, 415n
 apparatus, 411
 aqueous vapor cubic meter weight, 416
 cautions, 415, 416
 collection efficiency, 413
 color-indicating silica gel, 410
 counting efficiency, 414
 cylindrical column, 408
 detection limits, 409
 distillation, 410
 EPA software packages, 415
 liquid scintillation counter, 413–414
 LSC calibration, 413
 merits/demerits, 410–411
 precision and accuracy, 411
 principle, 408, 409
 procedure, 412
 range and sensitivity, 409–410
 reagents, 411, 412
 rotameter calibration, 412–413
 sampling line, 411
 sampling period, 409
 scintillation cocktails, 411

Index

solid adsorbents, 410
storage effects, 415
tritium concentration, 414
tritium oxide concentration, 415
tritium-in-air sampling, 407
uncertainty, 414
workplace sampling, 408
Tritium ($^3H_1$), 535
   aerosol properties, 536
   exposure to DCF, 535–536
   inhalation data, 535
   inhalation results, 537
   man-made sources, 536
   natural production, 536
   physical data, 535
   risk coefficients, 536
   simplified decay scheme, 535
Tritium
   apparatus, 420–421
   calculations, 422–423
   cautions, 423–424
   concentration, 414, 422, 423
   counting efficiency, 422
   DAC, 409
   HTO concentration, 415
   merits/demerits, 420
   precision and accuracy, 420
   principle, 419
   procedure, 421–422
   range and sensitivity, 420
   reagents, 421
   sampling line, 421, 422, 428
   scintillation fluid, 432
   storage effects, 423
   tritium oxide (HTO), 407, 419
   tritium-in-air sampling, 407
   uncertainty, 422–423
Tritium oxide (HTO), 407, 419
Tsivoglou method, modified, 461. *See* Thomas method
   alpha-counting equipment, 462
   calibration, 463
   dry-gas meter, 462
   interference, 462
   membrane filters, 462
   precision and accuracy, 462
   principle, 461
   procedures, 463
   range, 461–462
   total count method, 463–464
Two sigma, 414, 431

U

ULPA filters. *See* Ultralow penetration aerosol (ULPA) filters
Ultralow penetration aerosol (ULPA) filters, 171
Unattached fraction, 189, 292, 551
United Nations Scientific Committee on the Effects of Nuclear Radiation (UNSCEAR), 370
UNSCEAR. *See* United Nations Scientific Committee on the Effects of Nuclear Radiation (UNSCEAR)
Uranium, 182, 183. *See also* Radon (^{222}Rn)
U.S. National Committee (USNC), 33
USNC. *See* U.S. National Committee (USNC)

V

Vacuum pumps. *See* Sampling pumps
Vacuum rating, 234. *See also* Sampling pumps
Van der Waals forces, 169
vapor pressure, 137
Vapors, 137
Variable-area meters. *See* Rotameters
Velocity differentials, 115, 124
Velocity pressure (VP)
Ventilation-induced airflow, 272
   air exchange rate, 275, 277
   evaluation, 274–275
   factors influencing, 274
   goals, 274
   time-dependent concentration model, 275, 276
Venturi meter, 232–233
Versapor 3000T, 170
Virtual impaction, 293
Viscosity, 86
Volume-equivalent diameter, 96
Volume fraction, 80, 104
Volumetric flow rate, 233

W

Wall losses, 329
Well-mixed hypothesis, 112
WG. *See* Working Group (WG)
Wheatstone bridge, 231
Windowless counting, 68
Windscale facility accident, 137
Wipe testing, 15
   field blank, 15
   media blank, 15
WL. *See* Working Level (WL)
WLM. *See* Working Level Month (WLM)
Worker health protection, 4, 5, 7, 343
Working Group (WG), 25
Working Level (WL), 185, 507, 550
Working Level Month (WLM), 186, 551

X

Xenon (Xe), 545
   decay scheme, 545
   dose conversion factors, 546, 547
   inhalation data, 546, 547
   risk coefficients, 546, 547
   sources, 547
   special considerations, 547
X-rays, 57
   air molecule ionization, 66
   anthropogenic activity determination, 350
   NaI scintillation detector, 16, 63

Z

z-score, 145, 147